OCEANOGRAPHY AND MARINE BIOLOGY

AN ANNUAL REVIEW

Volume 24

OCEANOGRAPHY AND MARINE BIOLOGY

AN ANNUAL REVIEW

Volume 24

HAROLD BARNES, *Founder Editor*

MARGARET BARNES, *Editor*
The Dunstaffnage Marine Research Laboratory
Oban, Argyll, Scotland

ABERDEEN UNIVERSITY PRESS

FIRST PUBLISHED IN 1986

British Library Cataloguing in Publication Data

Oceanography and marine biology: an annual
review.—Vol. 24
1. Oceanography—Periodicals
2. Marine biology—Periodicals
551.46′005 GC1

ISBN 0 08 032458-4
ISSN 0078-3218

PRINTED IN GREAT BRITAIN
in 10 point Times Roman type
BY ABERDEEN UNIVERSITY PRESS
ABERDEEN

PREFACE

This series of Annual Reviews has again benefited from contributions sent by scientists from many parts of the world. Working with them has been a pleasure and they have all willingly acceded to editorial requests. Few editors can be so well served by her colleagues and I am grateful to them all; Drs A. D. Ansell, R. N. Gibson, and T. H. Pearson deserve special thanks. The publishers continue to produce this series with meticulous care which reduces editorial worries.

CONTENTS

Oceanogr. Mar. Biol. Ann. Rev., 1986, **24**, 11–64
Margaret Barnes, Ed.
Aberdeen University Press

MARINE MINERALS IN THE PACIFIC

G. P. GLASBY
New Zealand Oceanographic Institute, Department of Scientific and Industrial Research, P.O. Box 12–346, Wellington, New Zealand

INTRODUCTION

The Pacific Basin, with an area of 177 million km^2, constitutes 35% of the earth's surface making it easily the largest feature on the globe (and 21% larger than the earth's total land surface). It has also become a region of increasing economic importance. It has been stated, for example, that "the rapid industrial growth in the countries of eastern and southeastern Asia has transformed the Pacific Basin into the world's most dynamic region in terms of economic and social development" (Benjamin & Hewett, 1982). Data presented by Crawford (1981) for 1976 showed that the nations of the Pacific Basin had a population of 668·1 millions (or 16.5% of the world total) with a total gross national product of $U.S. $2{\cdot}69 \times 10^{12}$. The population is expected to rise to 26% of the world total by the year 2000 (Philpott, 1982). In terms of marine resources, the Pacific region has about 54% of the estimated world fish reserves and contributes about 54% of the world catch (Philpott, 1982). Of course, hydrocarbons remain the principal offshore resource in this region (Hedberg, Moody & Hedberg, 1979; U.N. Ocean Economics and Technology Branch, 1981; Curlin, 1984; Ives, 1984; Siddayo, 1984) but this is beyond the scope of this article. Wijkman (1982) has estimated that the oceans now produce products with 5% of world income. The importance of the Pacific Basin in global terms cannot, therefore, be doubted. In this respect, it is worth recalling the comment of J. M. Hay (U.S. Secretary of State 1898–1905) that "the Mediterranean is the ocean of the past, the Atlantic is the ocean of the present, but the Pacific is the ocean of the future".

Yet, in spite of the fact that the 1970s was the International Decade of Ocean Exploration, in which extensive exploration of the ocean floor by both academic and industrial groups took place, offshore mining of hard minerals in the Pacific remains minimal (Thiede, 1983). Japan is a significant producer of offshore aggregate and phosphorite nodules have attracted commercial attention off California and New Zealand but the exploitation of deep-sea minerals, particularly manganese nodules, seems to be at least a decade away, if not substantially longer. It is significant that one of the pioneering firms in marine mineral exploration, Global Marine, is now involved in kelp farming (Ashkenazy, 1981). To give some idea of the importance of marine mining, Archer (1974) estimated that, excluding oil and gas, marine mining contributed 0·3% of the total value of minerals produced in 1970 (*cf.* Glasby 1979). (The figure quoted by Archer is actually 0·03% but this is an error.) If sub-sea bed coal and iron ore are included, the figure is closer to 1%. Listings of the value of offshore minerals have been given by Archer (1973a, 1974) and Wang & McKelvey (1976). Cruickshank (1973) has also listed all offshore exploration and dredging activities taking place in 1973. In one specific case where the non-fuel mineral resources of a country have been assessed (namely Colombia with an Exclusive Economic Zone (E.E.Z.) in the Pacific of 320 000 km^2), only marine salt was considered likely to be of significant economic value (Broadus, 1984a). It is seen, therefore, that the ocean floor yields disproportionately few economic minerals compared with the land. This low level of production is reflected in the fact that less than 2 pages of a 650-page volume on the management of the North Pacific is devoted to ocean mining (Miles *et al.*, 1982). The earlier optimism of workers such as Mero (1965), which inspired so much confidence, therefore now seems to have been misplaced (*cf.* Ware, 1966; Moore, 1972, 1983; Cruickshank, 1973). Craven (1982) has commented that the mineral resources of the Pacific region are at present derived almost exclusively from land-based sources (*cf.* Henrie, 1976; Philpott, 1982). This viewpoint is confirmed in Table I (*cf.* Anonymous, 1979). In spite of these negative comments, Goeller & Zucker (1984) point to marine sources, particularly sea water and manganese nodules, to extend the reserves of a number of elements substantially in the long-term (*cf.* Schott, 1976).

Five factors are involved in determining the viability of a mineral resource: geological, technical, economic, legal, and environmental. In fact, two of these have dominated the search for deep-sea minerals in the 1970s which has concentrated largely in the Pacific—the geological factor (or the prospecting for suitable ore deposits) and the legal factor (or the

TABLE I

Pacific mineral resources expressed as a percentage of total world resources: data from Crawford (1981)

	Mine output	Refined metal output	Reserves
Copper	38	38	32
Lead	37	35	56
Zinc	38	34	46
Bauxite	32	—	19
Aluminium	—	46	—
Silver	38	—	37
Nickel	60	—	41
Tin	69	75	47
Iron Ore	27	—	27
Raw Steel	—	33	—
Coal-A	13	—	—
Coal-B	20	—	34
Cobalt	19	—	37

negotiations for an internationally acceptable Law of the Sea which would permit mining in the open ocean). The development of systems for deep-sea mining or for processing deep-sea minerals has played a minor rôle and has not been taken beyond pilot plant stage. None the less, it was the technical advances of the post-war period which enabled the exploration effort of the 1970s to proceed (*cf*. Bascom, 1969; Bäcker, 1985).

One of the achievements of the 1970s was to build on the theory of plate tectonics, first formulated by Morgan (1968) and Le Pichon (1968), so that the principal tectonic features of the Pacific are now well known and mapped (*e.g*. Doutch, 1981). The Pacific is made up of eight plates or part plates (Pacific, Indian, Philippine, Eurasian, American, Cocos, Nazca, Antarctic), of which the Pacific Plate is the largest, making up 22% of the earth's surface. Kennett (1982) has provided an excellent summary of the geological evolution of the Pacific and it is not intended to repeat this here. In addition to explaining the geology of the Pacific, plate tectonics has become an important conceptual model in the search for marine minerals, not only for metallic deposits (Pereira & Dixon, 1971; Blissenbach, 1972; Sawkins, 1972, 1984; Sillitoe, 1972a,b,c; Blissenback & Fellerer, 1973; Katili, 1973; Mitchell & Bell, 1973; Rona, 1973a,b, 1976, 1977, 1983a, 1984; Bonatti, 1975, 1978, 1981; Hammond, 1975a,b; Mitchell & Garson, 1976, 1981; Rona & Neuman, 1976a,b; Emery & Skinner, 1977; Garson & Mitchell, 1977; Wollard, 1981; Robbins, 1983) but also hydrocarbons (McDowell, 1971; Thompson, 1976; Emery, 1980; Bois, Bouche & Pelet, 1982) and phosphorite (Cook & McElhinny, 1979).

The Third United Nations Conference on the Law of the Sea (UNCLOS) which was first convened in 1973 represents one of the most protracted negotiations in the history of diplomacy. Because of the impasse in these negotiations with the U.S.A. and several other industrialized nations declining to sign the final treaty in 1982, no internationally accepted treaty exists to cover deep-sea mining (Lucas, 1982; Mann Borgese, 1983a); a mini-treaty exists between U.S.A., Belgium, France, Germany, Italy, Japan, Holland, and U.K. (Comptroller General, 1983; Mann Borgese, 1983b; van Dyke, 1985), (*cf*. Broadus & Hoagland, 1984; Stavridis, 1985). It has been stated that the Law of the Sea treaty greatly discourages the high-risk capital investment necessary for the development of ocean resources in the international deep sea bed (Takeuchi, 1979; Halbach & Fellerer, 1980; Knecht & Bowen, 1982; Welling, 1982a; National Advisory Committee on Oceans and Atmosphere, 1983). Craven (1982) has outlined six possible scenarios that could be adopted by mining companies to this impasse. Although other factors (economic and technical) are involved, it is certainly true that, since about 1982, interest in manganese nodules as a potential economic resource has sharply declined and the decade of intensive prospecting for nodules by the industrialized nations appears to be over for the present (with the exception of Japan). One possibility, which has been evident from about 1982, is a redirection of effort to the mineral resources of the 200-mile Exclusive Economic Zones (E.E.Z.s) which come under national jurisdiction and together make up 40% of the total area of the Pacific (scenario 4 of Craven) (*cf*. McKelvey, 1980; Cronan, 1985). The importance of these E.E.Z.s to individual Pacific nations can readily be seen by inspection of their areas (Table II; see Mann Borgese, 1983a for

TABLE II

Areas of Pacific Exclusive Economic Zones (E.E.Z.s) (10^6 km^2): data from Gocht & Wolf (1982); see also Couper (1983); data for the New Zealand E.E.Z. is taken from Blezard (1980)—the area given by Gocht & Wolf ($6 \cdot 96 \times 10^6$ km^2) is incorrect.

Chile	3·17	Taiwan	0·16
Peru	1·07	Vietnam	0·72
Ecuador-Guatemala	2·45	Philippines	1·81
Mexico	2·40	Malaysia	0·51
U.S.A. (total)	15·48	Indonesia	5·25
U.S.A. (coast)	0·73	Papua New Guinea	2·16
Alaska	3·07	Australia	3·03
Hawaiian Islands	2·58	New Zealand	4·05
Islands	9·09	French Islands	7·30
Canada	0·50	U.K. Islands	5·00
U.S.S.R.	3·20	Solomon Islands	1·56
Japan	3·89	Nauru	0·31
Korea (South)	0·28	Fiji	1·06
Korea (North)	0·14	Western Samoa	0·17
Peoples Republic of China	1·15	Vanuatu	0·67
		Tonga	0·61

Pacific Ocean	177×10^6 km^2	(100%)
Economic Zones (total)	71×10^6 km^2	(40%)
Economic Zones of Coastal States	20×10^6 km^2	
Economic Zones of Islands and Archipelagos	51×10^6 km^2	

diagrammatic representation). The largest beneficiary in terms of absolute areal increase is the U.S.A. but, in terms of the relative areal increase (area of E.E.Z./land area), the largest beneficiaries are the island nations and archipelagos. For instance, the Cook Islands with a land area of 240 km^2 have an E.E.Z. of 1·2 million km^2. Whether these increases in areas translate into mineral wealth is another matter (*cf.* Glasby, 1986; Odunton, in press). Certainly, the impasse in the Law of the Sea has made the resources of these zones more attractive relative to those of the deep sea and the United States government, in particular, has pursued this line. The proclamation of the E.E.Z. in March 1983 for the United States and its territories (including the Pacific Trust Territories) gave the United States control of the resources of an offshore area of 15·7 million km^2 compared with a total onshore areas of 9·3 million km^2 (Edgar, 1983; McGregor & Offield, 1983; Rowland, Goud & McGregor, 1983; Anonymous, 1983a, 1984a; Champ, 1984; Champ, Dillon & Howell, 1984; McGregor & Lockwood, 1985). The importance of the resources of this area including an estimated 35% of the economically recoverable oil and gas yet to be found in the United States, major sources of strategic metals such as cobalt, manganese and nickel in seafloor crusts, pavements and nodules, massive sulphide deposits actively forming today and major concentrations of heavy minerals in nearshore sand bodies has been stressed by Edgar (1983). Production of oil and gas in the U.S. E.E.Z. for example, was valued at $U.S. 26 000 million in 1984 (and came mainly from the Gulf of Mexico). This compares with fish landings which were valued at $U.S. 2500 million

(Champ *et al.*, 1984) (*cf.* Curlin, 1984). The above six publications regard the U.S. E.E.Z. as a "vast new frontier". A further recent report on U.S. objectives on marine minerals has been published by the National Advisory Committee on Oceans and Atmosphere (1983). In addition, concern in the United States about strategic minerals (Zablocki, 1978; Dames & Moore, 1980; Holden, 1981; Anonymous, 1982a; Cruickshank, 1982a,b; Pendley, 1982; National Advisory Committee on Oceans and Atmosphere, 1983; Moore, 1984; Office of Technology Assessment, 1985) (see, however, Raymond, 1976; Clark, 1982) supports offshore mineral exploration (*cf.* Blissenbach, 1977; Cameron, 1977; Pryor, 1979; Takeuchi, 1979; Marjoram & Ford, 1980). Whilst no hard minerals leasing in U.S. Federal waters has occurred since 1968, preparations are being made for eventual lease offerings of massive sulphide deposits in the Gorda Ridge area (Rowland *et al.*, 1983), sand and gravel off Alaska (National Advisory Committee on Oceans and Atmosphere, 1983) and cobalt-rich crusts off Hawaii and Johnston Island (*cf.* Charlier, 1983; Anonymous, 1984a). The Minerals Management Service will be responsible for these lease sales (Anonymous, 1985a) (for a discussion of earlier leasing procedures on the U.S. continental shelf, see Adams *et al.*, 1975; U.S. Department of the Interior, 1979).

In spite of these optimistic views, it should be remembered that only about 15% of the money so far spent on the search for marine minerals (manganese nodules, manganese crusts, and sulphides) has been spent on the crust and sulphides (Johnson & Clark, 1985; Johnson, Clark & Otto, in press) and that this money has been spent almost exclusively by government agencies. This contrasts with the search for manganese nodules where a large input of private capital was involved.

The last decade has also seen a substantial increase in mineral exploration activity in the southwestern Pacific under the auspices of CCOP/SOPAC (Committee for Co-ordination of Joint Prospecting for Mineral Resources in South Pacific Offshore Areas). This was well illustrated at the 1983 Workshop in Suva, Fiji, which emphasized the increased level of activity (Intergovernmental Oceanographic Commission, 1983). The driving force for this is the desire of the island nations of the southwestern Pacific to ascertain and develop their mineral resources. In this regard, it has been stated that "the South Pacific is one of the exciting new frontiers of science" (Clark, 1980).

None the less, it is worth reconsidering the perceptive (and overlooked) views of Medford who, in 1969, wrote a paper very much against the conventional wisdom of the time (Medford, 1969a,b). Medford stated that, whilst the arguments in favour of exploitation of offshore minerals were tempting, the fact was that the grades of ore mined on land had been falling by about 2·5% per annum (*cf.* Bäcker & Schoell, 1974) and that it might be cheaper to mine lower grade, but extensive, open pit ores on land than richer ores elsewhere. The analysis suggested that oil, gas and aggregate could be economically recovered from the sea in competition with terrestrial deposits but pointed to caution for other minerals. Fifteen years later, the validity of Medford's conclusions is plain for all to see.

Archer (1973a,b) has concluded that the production costs of marine minerals are generally greater than those of terrestrial minerals and that this disadvantage is only overcome in certain cases. Furthermore, few, if any, minerals are likely to approach exhaustion by the end of this century (Archer, 1973b). In this regard, it is worth bearing in mind Coene's (1968) comments that minerals have a value independent of origin and that marine mining will proceed only if it is profitable. A similar conservative view on the potential value of marine minerals was also expressed by Ensign (1966). The most detailed analysis of the reasons for the delay in deep-sea mining over the past decade has, however, been presented by Crockett, Chapman & Jones (1984). These authors showed that it has been the softness of demand for nodule metals and rising capital costs rather than legal problems or the worries of land-based producers that have set back manganese nodule mining (*cf.* Johnson & Clark, 1985; Johnson *et al.*, in press).

Whilst knowledge of the distribution and genesis of Pacific marine minerals has increased substantially over the last two decades (as is apparent from re-reading the work of Mero, 1965), availability of this information is patchy. Much of the effort over the last decade has been directed to deep-sea deposits such as manganese nodules and more recently to metalliferous sediments and cobalt-rich crusts and this information is readily accessible. Data on shelf minerals, on the other hand, are much less available. Much of the world's continental shelves have been surveyed only at the reconnaissance level (*cf.* Pepper, 1958; Dunham, 1969). In the Pacific, for instance, not much work has been carried out off the coasts of Latin America (Ericksen, 1976; Palacio, 1980). Even in the U.S. E.E.Z., it has been estimated that a comprehensive research programme to assess the mineral resources would cost $U.S. 250 million, with geophysical exploration costing $U.S. 1250 million and exploratory drilling $U.S. 5000 million (Anonymous, 1983a) (*cf.* National Advisory Committee on Oceans and Atmosphere, 1983). In 1984, a survey the Pacific margin of the U.S. E.E.Z. was carried out using GLORIA (EEZ-SCAN '84 Scientific Team, 1985). Cruickshank (1982b) considers that mineral exploration systems are generally in a primitive state considering the magnitude of unexplored seafloor areas.

The object of this report is to bring together some of the available information on marine non-fuel minerals of the Pacific in an attempt to give a more balanced picture of the mineral availability and prospects for exploitation of this region and to update the earlier definitive accounts of Mero (1965), McKelvey & Wang (1970) (*cf.* McKelvey *et al.*, 1969; Wenk, 1969; Wang & McKelvey, 1976; Schott, 1976; Charlier, 1978, 1983; Cronan, 1980a; Couper, 1983; Thiede, 1983; Gopalakrishnan, 1984). It is impossible to be comprehensive; the literature on manganese nodules alone is huge (Fellerer, 1980; Thiede, 1983). Fisk (1982) noted that, during much of the 1970s, over 250 publications on nodules were appearing per year. Rather, the aim is to highlight some of the areas of interest, the progress that has been made over the last decade and prospects for the future. The author has recently completed extensive reviews of the nearshore mineral and manganese nodule resources of the southwestern Pacific (Glasby, 1986; Glasby, Exon & Meylan, 1986) and it is not the intention to duplicate this work here.

Before commencing this review, some examples of extraction of minerals which do not form superficial deposits on the sea floor will be mentioned (*cf.* Cruickshank, 1973). Gold was mined in the Treadwell Mine off Alaska from 1885 until subsidence caused closure in 1917 (Cruickshank, 1971). Barite was also mined below sea level on Castle Island near Petersburg, Alaska, from the late 1950s to 1981. Beginning with conventional mining of a small island, this extended to drilling and blasting of the sea bed after exhaustion of material above sea level. In 1970 and 1971, production was 122 000 and 93 000 tonnes, respectively (Stevens, 1970; Thompson & Smith, 1970; Archer, 1973a; Magnuson, 1974; U.S. Dept of the Interior, 1979; Earney, 1980; Kent, 1980; Moore, 1984). The mine was closed because of the decreasing amounts of ore and the decreasing use of barite in drilling because of environmental concerns. A further deposit of barite in shallow waters near Petersburg has recently been reported (National Advisory Committee on Oceans and Atmosphere, 1983). Also off Alaska, the solution mining of a hypothetical offshore high-grade copper-nickel sulphide deposit has been described (National Academy of Sciences, 1975). Lode copper has been reported in Prince William Sound, Alaska, and there are probably shallow lodes off several islands in the Northern Marianas (Moore, 1983). Scheelite is recovered under Bass Strait, off King Island, Tasmania (Earney, 1980). Silica stone containing 99·5% SiO_2 was also dredged in Omura Bay, Kyushu, Japan, in water depths of up to 20 m between 1960 and 1969. Between 1964 and 1969, approximately 10 000 tonnes were recovered (Archer, 1973a). Exploitation of offshore coal deposits in Japan began in 1860 and, in 1969, 22 mines were being worked in offshore areas. In 1963, production was almost 10 million tonnes (or 20.6% of the total Japanese output of coal) (Tokunaga, 1967, 1969; Wenk, 1969; Wang & McKelvey, 1976; Earney, 1980; Couper, 1983). Total coal production in Japan has in fact declined from a maximum of 55 million tonnes in 1961 to 18·6 million tonnes in 1977. Of the 29 coal mines operating in Japan in 1977, four were offshore producing 9·7 million tonnes (or 52%) of the total 18.6 million tonnes produced in that year. Offshore mining is, therefore, playing an increasing rôle in coal production in Japan, although this production has difficult economic and technological problems because the galleries must extend further and further from land. Coal has also been mined from sub-sea underground mines off Chile (Toenges *et al.*, 1948; Pepper, 1958; Cruickshank, 1971; Wang & McKelvey, 1976) and off British Columbia (Pepper, 1958; Cruickshank, 1971). A map showing the locations of offshore coal fields in the Pacific is given in Tokunaga (1967). One land-based iron mine extending offshore of western Kyushu, Japan, has also been identified but is no longer operating (Earney, 1980).

SEA WATER

With a total volume of $1 \cdot 37 \times 10^9$ km^3, sea water is a huge resource of salt as well as magnesium metal and its compounds, bromine, water, and potassium. The world's oceans contain some 5×10^{16} tonnes of mineral matter, of which 85·2% is sodium chloride (Mero, 1965). The earliest recorded extraction of salt from the sea appears to have been in China around 2200 BC (Wenk, 1969; Schott, 1976; Mero, 1978) and salt is at present recovered in Indonesia by traditional methods (Eiseman, 1982).

Archaeological evidence has shown that salt was produced from sea water by Polynesians in Hawaii over a long period (Ellis, 1969). In 1978–1980, the annual production of minerals from sea water was 69·6 million tonnes of which salt made up 94% (McIlhenny, 1981). This production had a value of \$U.S. $1 \cdot 3 \times 10^9$. Nearly 37% of all salt produced and 99% of all bromine comes from the oceans. Detailed reviews of the extraction of minerals from sea water have been presented by Mero (1965), Shigley (1968a,b), Lefond (1969), McIlhenny (1975, 1981), Schott (1976), Estrup (1977), and Balas (1984) and these will not be repeated here.

The production centres of chemicals from sea water have been mapped by Lefond (1969) and McIlhenny (1981) and listed by Cruickshank (1973). In the Pacific region, there are quite large solar marine salt operations in Baja (California), Taiwan, North and South Korea, China and Mexico. Lesser salt operations are found around San Francisco Bay (Leslie Salt Co.) and San Diego Bay in the U.S.A., and in El Salvador, Guatamala, Honduras, Nicaragua, Panama, Colombia, Ecuador, Peru, Philippines, Indonesia, New Zealand, and Australia (*cf.* Cruickshank, 1971; Magnuson, 1974; Kostick, 1980, 1981; Anonymous, 1983b; Couper 1983; Broadus, 1984a; Driessen, 1984). Mexico is in fact the largest producer of solar salt with a production of 5·63 million tonnes in 1978 (Carlos Ruiz, 1982). Magnesium oxide has been produced from solar salt bitterns at Chula Vista, California, by FMC Corp. and in South San Francisco by Merck & Co. Kaiser operated a large sea-water magnesia plant at Moss Landing, California. There are several such magnesia plants in Japan (Asahi and Ube Companies) as well as a bromine plant. Gypsum and crude potassium minerals have been recovered on a small-scale from marine solar salt bitterns, particularly in Taiwan.

An example is the producton of chemicals from sea water in Japan. In Japan, the total amount of common salt for food is produced from sea water by the ion-exchange resin membrane method and the total production capacity is $1 \cdot 2 \times 10^6$ tonnes per year. Common salt for industrial use (7×10^6 tonnes/yr) is imported. Magnesium hydroxide and magnesia clinkers are also extracted from sea water with a production of $0 \cdot 6 \times 10^6$ tonnes/yr. 12 000 tonnes/yr of bromine were also produced from sea water; this just meets domestic demand.

In addition, Japan has scarce uranium deposits on land so that the demand for natural uranium for present and future nuclear power generation must be met by imports. The scale of nuclear power generation in 1977 was 8×10 kW and the demand for natural uranium was 2100 tonnes U_3O_8 (1600 tonnes U). Research and development projects on extraction techniques of uranium from sea water are now underway as one of the ways of securing alternative uranium resources in future (*cf.* Hill, 1977; Brin, 1982; Georghiou, Ford, Gibbons & Jones, 1983).

Uranium is dissolved in sea water in trace concentrations (0·003 mg/l) but the total amount in the world oceans is enormous (4×10^9 tonnes). The adsorption method of uranium extraction is now thought the most promising. Extraction experiments have already succeeded at laboratory level but there are many problems to be solved for commercial recovery. Present research projects are directed to the study of adsorbent materials and development of the technological system for uranium extraction from sea water including construction of a pilot plant. There has also been some interest in uranium in organic-rich diatomaceous muds (as off Namibia) which have a uranium content in the range 21 mg/kg (Schott, 1976) or in phosphorite nodules such as off the Chatham Rise which have a uranium content in the range 7–435 mg/kg (Cullen, 1978) but this is beyond the scope of this paper (*cf.* Bäcker & Schoell, 1974; Thiede, 1983).

Japan also has no lithium deposits and all demand is met by imports. This amounted to 780 tonnes as lithium oxide in 1978. Lithium is dissolved in sea water at a concentration of 0·17 mg/l and its extraction is also a target. Research on the extraction technique by adsorption and other methods is being carried out as in the case of uranium, although many problems remain to be solved.

In Japan, the consumption of water has increased as a result of rising living standards and industrial expansion and a shortage of fresh water has often occurred in many districts. The situation has been remedied mainly by the construction of new dams and waterways but, in some local areas where they have difficulty obtaining fresh water, desalination is used for producing fresh water for domestic and industrial use. Among the desalination methods such as evaporation, reverse osmosis, and electro-osmosis, the multistage flash type evaporation method is mainly used for large-scale processing.

Japanese industry has built many desalination plants in Japan and overseas. For example, two plants with capacities of 2650 tonnes and 1200 tonnes per day, respectively, have operated since 1967 and 1975 in western Japan. In Hong Kong, Japanese-built facilities with a total capacity of 30 000 tonnes per day were constructed in 1975 (*cf.* Magnuson, 1974). In general, the cost of desalination is several times higher than that of dam construction or other methods. To meet future needs, improvement and development of desalination technology to reduce costs are therefore expected.

SAND AND GRAVEL

Sand and gravel rank first in tonnage amongst the minerals (Alexandersson & Klevebring, 1978). These materials are quarried or extracted as close as possible to the point of consumption in view of their widespread availability and low value. About $8 \cdot 5 \times 10^9$ tonnes of sand and gravel are estimated to have been produced worldwide in 1980 (Holdgate, Kassas & White, 1982). Offshore mining of sand and gravel generally takes place where there are inadequate deposits on land (or where environmental problems drive up the cost of production) and to supply specific metropolitan areas or development projects. It has been estimated that offshore aggregate mining accounts for 60% of the value of world mineral production from marine sediments (Odunton, in press) (*cf.* Archer, 1973a; Wang & McKelvey, 1976).

In the Pacific, Japan is the major producer of offshore sand and gravel. In Japan, demand for fine aggregate as construction material has markedly increased, but there have been constraints and limitations in onshore resources. Offshore sources have, therefore, gradually become important.

This tendency is, however, not the same in all districts. Offshore production is most marked in southwestern Japan, where 70–80% of the total production is from offshore areas. Sands as fine aggregate resources are classified in commercial terms as river sand, hill sand (Pleistocene sand beds or weathered granitic rocks), land sand (alluvial plain sand beds) and marine sand (sea-bed and beach sands). For the marine sands, it is now virtually prohibited to mine beach sands because of the need for preserving coastline stability. The sea-bed sand is of two types, i.e. deposited by recent river supply and coastal erosion and ancient beach deposits submerged due to post-glacial sea level changes.

In 1981, the total production of sand as fine aggregates in Japan was $98{\cdot}0 \times 10^6$ m^3 of which $35{\cdot}3 \times 10^6$ m^3 (36%) was marine sand. This compares with a figure of 8.9×10^6 m^3 of marine sand in 1968 when it constituted about 20% of total production. The mining of marine sand is carried out by the sand pumping or bucket dredge methods. Because the operating capability of the pump is now around 40 m depth, the present offshore mining is restricted to areas shallower than 20–30 m and is carried out by many small private companies. For the future, investigation and evaluation of the potential resources in deeper areas and the development of the technology for mining in these areas is desirable, particularly to help prevent coastal erosion resulting from mining (Narumi, 1984). Since 1975, the Geological Survey of Japan has been evaluating offshore aggregate resources at depths greater than 20 m. Promising offshore deposits tend to be on submarine terraces at depths of 40–50 m in many of the areas surveyed. An account of these resources and their working has been given by Usami *et al.* (1983) and Arita & Kinoshita (1984) (*cf.* Cruickshank & Hess, 1975; Nat. Acad. Sci., 1975; Cruickshank, 1979; Padan, 1983).

A major potential use of offshore aggregate is in the Beaufort Sea, off Alaska, where it is proposed to build a number (of the order of 15–20) of artificial islands to facilitate oil drilling operations. Assuming 20 artificial islands of 8000 m^2 are constructed in water depths of 20 m and projecting 10 m above sea level, it can be calculated that approximately $4{\cdot}8 \times 10^6$ m^3 of gravel will be required. Nearly 10×10^6 m^3 of gravel have already been mined onshore in the development of the Prudhoe Bay oilfield. One possible source of gravel is a sheet of Pleistocene gravel that lies buried beneath the coastal plain and the continental shelf eastwards from Colville River Valley and central Harrison Basin. This gravel deposit is several tens of metres thick but lies beneath a 3–10 m thick overburden of Holocene mud, sand and peat. The aggregate resources of the Alaskan shelf are only poorly surveyed but the results have been summarized by Molnia (1979) (*cf.* Gerwick, 1983).

In Hawaii, beach sand is vital for the tourist industry. For this purpose, calcareous sand formed by erosion of coral reefs is preferable to that formed by the erosion of basalt; the latter type of beaches are common on the island of Hawaii. In the island of Oahu which is the main tourist destination in Hawaii, there are only 106 km of sandy beaches totalling $1{\cdot}79$ km^2, of which only $0{\cdot}65$ km^2 are in public parks. This is considered insufficient for the tourist numbers visiting Hawaii and there is, therefore, a critical need for sand for recreational purposes. Many of the major tourist beaches in Hawaii have been significantly modified from their natural state.

Waikiki Beach, for instance, is now almost entirely artificial with imported sand, groynes, and seawalls along most of the beach. Many of these beaches have histories of net erosion and sand loss, especially during storms, and must be periodically replenished. In the period 1952–1978, $0{\cdot}42$–$0{\cdot}50 \times 10^6$ m^3 of beach sands were replenished and this nourishment must be carried out on a continuing basis.

Offshore beach sands are essential to any programme of beach replenishment. On Oahu, back beach sands are in the final stages of depletion, beach sand has been protected by legislation since 1975, crushed basalt is unsuitable for beach use and offshore sand is protected by legislation if the deposit is within 1000 ft (305 m) of the shoreline or in water depths of less than 30 ft (9 m) and by a system of permits if further offshore. In 1978, no offshore sand recovery was taking place in Hawaii. Recently, however, a large volume of coral rubble was dredged during the construction of the deep draft harbour at Barber's Point, Oahu. $7{\cdot}3 \times 10^6$ m^3 were recovered within the harbour basin and $1{\cdot}2 \times 10^6$ m^3 were dredged from the entrance channel. This material has not yet been released by the U.S. Army Corps of engineers but most will eventually be sold as aggregate (J. Wiltshire, pers. comm). A potential source of suitable sand is Penguin Bank, a sunken former reef 20 km × 50 km in area 40 km from Honolulu at a depth of 50–60 m. In excess of 350×10^6 m^3 of sand is available. Environmental considerations and conflicts of interest affect its exploitation as the bank is an important fishing area and provides 90% of Hawaii's kona crab catch. Detailed reports on Hawaii beach deposits have been presented by Moberly, Campbell & Coulbourn (1975) and Dollar (1979) which outline the nature of the problems very well.

In California, marine sand and gravel have been produced in San Francisco and San Pedro Bay (McIlhenny, 1981; Evans, Dabai & Levine, 1982). In the Los Angeles area, the expected depletion of onshore aggregate deposits has prompted surveys for offshore deposits. Estimates show approximately 20–120×10^6 m^3 of gravel, 64–320×10^6 m^3 of coarse sand, and 880—4400×10^6 m^3 of fine sand on the shelf between Point Conception and the U.S.-Mexican border. These offshore deposits may, however, be uneconomic to mine (*cf.* Mokhtari-Saghafi & Osborne, 1980). High-grade silica sands which have a potential market for specialist applications are also found in Monterey Bay.

Summaries of U.S. marine aggregate production have been given by Mero (1965), Archer (1973a, 1974), Yeend (1973), Cruickshank (1974), Magnuson (1974), Cruickshank & Hess (1975), National Academy of Sciences (1975), Duane (1976), U.S. Department of the Interior (1979), Manheim & Hess (1981), Tepordei (1980, 1981), Anonymous (1984a, 1985), Champ, Dillon & Howell (1984), Clague, Bischoff & Howell (1984), and Cruickshank & Olson (1984). Estimates given for offshore sand and gravel resources for Alaska, Hawaii, and the Pacific coast of the U.S.A. were 80×10^{10}, 37×10^{10} and $2{\cdot}4 \times 10^{10}$ tonnes, respectively (Cruickshank & Hess, 1975) (*cf.* Champ *et al.*, 1984). At present, the U.S. uses just under 10^9 tonnes of aggregate per year for construction alone (Anonymous 1982b; Champ *et al.*, 1984). In 1973, marine aggregate made up 7% of total consumption in the three Pacific states (Magnuson, 1974). Herbich (1975) has presented a detailed review of offshore dredging techniques. A review of the regulatory requirements of dredging in the U.S.A. is given by Herbich (1985).

In Canada, possible aggregate deposits exist off Victoria, British Columbia. Well-sorted sands that could supply markets as far away as Alaska (Shelf Working Group, 1980; Hale & McLaren, 1984) may be found off the western coast of Vancouver Island.

In Hong Kong, about 1·5 million tonnes of aggregate were dredged in 1972 (Archer, 1974). In Taiwan, onshore deposits of sand and gravel met the demand for aggregate in 1975. A huge volume of offshore sand and lesser amounts of gravel are, however, available to meet future demands, if required (Boggs, 1975). In Korea, surveys of silica sand have been reported by the Delegation of the Republic of Korea (1981).

Most aggregate in Australia is produced onshore (Glasby, 1986). In 1979, Consolidated Gold Fields Australia Ltd and ARC Marine Ltd were awarded exploration licences by the New South Wales Government for an area of 84 km^2 off Catherine Hill Bay (between Newcastle and Sydney) for coarse aggregate (gravel) and of 52 km^2 off Broken Bay (Sydney) for fine aggregate (sand) (Brown, 1981). A major survey of the Broken Bay deposit delineated an area of 22·8 km^2 in suitable water depths containing 48×10^6 tonnes of aggregate per vertical metre of sediment (Anonymous, 1980a; Brown, 1980a,b, 1981). In 1981, this project was determined by various government departments to be environmentally "not proven" and a final decision on it will not be reached until 1986 or 1987 (D. J. Debney, pers. comm.).

In the southwestern Pacific region (including New Zealand), the production of offshore aggregate has been summarized by Glasby (1986); in many of the more populated islands there is a shortage of aggregate due to the small land area available for mining. Beach mining is taking place on several islands and this is resulting in erosional problems which are unacceptable in such small islands. There has been some search for offshore aggregate which would mitigate these problems, for example off Tonga (Gauss, Eade & Lewis, 1983). The low capital availability in these islands has, however, meant that optimum solutions are not always sought. This problem has been discussed at length by Glasby (1986) (*cf.* Eade, 1984).

The above examples describe instances in which calcium carbonate is used as aggregate (as in Hawaii). It should be emphasized that, in many offshore areas *e.g.* off Taiwan (Boggs, 1975) or New Zealand (Summerhayes, 1967), calcareous ooze is abundant. Because of its low value, however, commercial extraction is generally not worthwhile where suitable land-based deposits exist. In Indonesia, coral limestone from reefs is used extensively and this has led to serious shore erosion (Polunin, 1983). Similar effects have been documented in the Philippines (Wells, 1982; Gomez, 1983; Meith & Helmer, 1983). Calcium carbonate can also be used in cement manufacture. For example, oyster shells are used in the manufacture of cement in the San Francisco Bay area (Mero, 1965; Cruicksank, 1971) and it has been common practice in tropical areas (*e.g.* in Hawaii, Fiji, and China) to collect modern coral for lime burning and the manufacture of mortar and cement, although this is becoming more open to environmental objections (Archer, 1973a; Kent, 1980).

In addition to the above, Cruickshank (1979) lists offshore and coastal aggregate mining taking place in China, Indonesia, Papua New Guinea,

Solomon Islands, Trust Territories, Vietnam, and Western Samoa but gives no indication of the quantities recovered.

PLACER DEPOSITS

Placer deposits are economic concentrations of heavy minerals separated from the lighter gangue components during weathering, transport, and sorting. For placer deposits to form, a specific source rock is required. The importance of placers can be seen from the fact that 75% of the world tin production, 11% of the gold and 13% of the platinum comes from placers (Tagg, 1979): most of this is from streams.

Emery & Noakes (1968) divided detrital minerals into three categories; heavy heavy minerals (gold, tin, platinum, chromite), light heavy minerals (rutile, zircon, ilmenite, magnetite, monazite), and gems (diamonds, sapphires). The heavy heavy minerals with specific gravities greater than 6·8 occur principally in stream beds less than 15 km from source whereas light heavy minerals with specific gravities between 4·2 and 5·3 travel long distances and economic deposits occur in high-energy accreting beaches. The value of light heavy minerals produced from placers is much less than that of heavy heavy minerals, only 7% in 1960–1965 (*cf*. Tagg, 1979).

From this, Emery & Noakes (1968) proposed a number of guidelines for offshore prospecting. Economic deposits offshore are likely to be associated with stream or beach deposits on land such that there is a source for the offshore deposits. The heavy heavy minerals are likely to be close to source; the seaward continuation of the deposit is unlikely to exceed 8 km for tin or 15 km in the case of gold. Fossil strand lines, formed as a result of the post-glacial sea-level rise, are likely to have been reworked by bottom currents obscuring the original distribution. Elsewhere, the submerged beach may receive a protective cover of marine sediments or glacial till which would increase the cost of economic recovery. Placers are also not likely to be found on reef-protected shorelines (as in the equatorial Pacific), even where provenance is favourable. In general, offshore deposits are likely to be of significantly lower grade than onshore deposits (although in some cases they may be richer, *e.g.* Indonesia and Namibia) and, by definition, exploitation of offshore deposits is likely in high-energy environments. This is well illustrated off eastern Australia where constant swell and wind conditions are encountered (Brown, 1981) or off Taiwan which lies in the typhoon belt (Macdonald, 1971) (*cf*. Archer, 1973a). Summaries of offshore deposits are given by Mero (1965), Duane (1976), Hails (1976), Schott (1976), Burns (1979), Moore (1972, 1979), Cronan (1980a), Hale & McLaren (1984), Sutherland (1985), and Odunton (1984, in press) and the distribution of placer deposits in the Pacific has been listed by Cruickshank (1973, 1979) and mapped by Burns (1979). It is significant that, of the 21 prospective areas for placers listed by Moore (1979), 10 are in the Pacific (*cf*. Moore, 1972). Moore (1972) has emphasized the need to be selective in the areas chosen for exploration based on a firm understanding of how the placer deposit is formed.

Placer deposits on the western coast of the U.S.A. as far as Alaska have been mapped by Holser, Rowland & Goud (1981) and Rowland, Goud &

McGregor (1983) and their occurrence and distribution summarized by Pepper (1958), Anonymous (1968, 1984a, 1985a), McKelvey (1968), Cruickshank (1971), Phillips (1979), Tagg (1979) U.S. Department of the Interior (1979), Earney (1980), Manheim & Hess (1981), Beauchamp & Cruickshank (1983), Moore (1983), Champ *et al.* (1984), Clague *et al.* (1984), and Cruickshank & Olson (1984). In general, the offshore deposits are not well surveyed (*cf.* Phillips, 1979). The potential of the Alaskan shelf stems in part from its huge size (75% of the total shelf area of the U.S.A., Sharma, 1977; Molnia, 1979; Tagg, 1979). Gold and heavy-mineral sand deposits occur rather extensively in relict beaches, buried river channels and reworked Pleistocene gravels bordering northern California, Oregon, Washington, and Alaska. The greatest potential for gold, platinum and tin is on the Alaska shelf area where 17 target areas have been defined (Reimnitz, von Huene & Wright, 1970; Cobb, 1973; Reimnitz & Plafker, 1976; Tagg, 1979; U.S. Department of the Interior, 1979); these include Nome (gold), Goodnews Bay (platinum and gold) and western Seward Peninsula (tin and gold) (Moore, 1979). The Nome deposit has been described by Mero (1965), Emery & Noakes (1968), Nelson & Hopkins (1972), Archer (1973a), National Academy of Sciences (1975), Burns (1979), and Odunton (in press). The first attempts at marine mining off Nome appear to have been made soon after 1900 (Tagg, 1979). The Gulf of Alaska has been estimated to contain tens, if not hundreds, of tonnes of recoverable gold. Platinum placers have been investigated in Chagvan Bay by Moore & Welkie (1976), Owen (1978, 1980), and Moore (1984) (*cf.* Mertie, 1976). The placer gold deposits of Alaska have been described by Brooks and others (1913). The potential of the Alaskan offshore placers is, however, limited by the fact that the sea is ice-covered for several months of the year. Some of the factors which influence placer mining, such as grade and volume of the deposit, water depth, length of working season, presence of ice, distance from sources of supply, weather and thickness of overburden have been summarized by Tagg (1979). Tagg is pessimistic for the prospect of offshore mining of Alaskan placers.

Chromite concentrations in heavy-mineral sands ranging from 7–50% have been reported in beach sands and terraces off Oregon and California (Pardee, 1934; Griggs, 1945; Wilson & Mero, 1966; Bowman, 1972; Burns, 1979; Komar & Wang 1984). Identified sources are 30 million tonnes of chromite, 170 tonnes of gold, and 10 tonnes of platinum. During World War II, 51 000 tonnes of chromite concentrate, containing 37–39% Cr_2O_3, were recovered from Oregon beaches. Along the Oregon coast, gold and platinum were discovered in the beach sands in 1852. No records were kept of the early (and richest) mining but, in 1903–1929, 75·1 kg of gold and 55·2 kg of platinum were recovered (Pardee, 1934). Titanium and zircon sand concentrates are widespread on the Pacific coast, in places reaching high grades. A summary of the distribution of these placers off Washington, Oregon, and California is given by Phillips (1979) (*cf.* Gray & Kulm, 1985). Kent (1980) has related these western coast placers to the onshore geology (*i.e.* the Cordillera belt; *cf.* Anonymous, 1966). It has been estimated that the U.S. Pacific continental shelf contains $2{\cdot}06 \times 10^9$ m^3 of heavy mineral sand. Wilcox, Mead & Sorensen (1972), however, concluded, on the basis of an economic analysis, that several of the U.S. offshore placer deposits would be mined at a loss, although there are probably insufficient data to justify such a conclusion.

In Micronesia, J. R. Moore (University of Texas) has been investigating the occurrence of placer deposits (gold, rare earths and platinium group metals) since 1976 (Moore, 1979, 1983).

In Canada, some possibility of gold deposits off western Vancouver Island and in some western coast fjords is suggested on the basis of suitable coastal source rocks but the prospects for commercial development are unknown (Shelf Working Group, 1980 Hale & McLaren, 1984; Samson, 1984).

In the U.S.S.R., the distribution of magnetite, titanomagnetite, and ilmenite in the sediments of the Bering Sea has been discussed briefly by Lisitsyn (1969) and of gold in the Bering Sea by Nelson & Hopkins (1972). Archer (1973a) mentions the discovery of gold and tin deposits within territorial waters in the Sea of Japan, off Kamchatka and Sakhalin Island (*cf.* Anonymous, 1970; Igrevskiy, Budnikov & Levchenko, 1972; Kogan, Naprasnikova & Ryabtseva, 1974; Mann Borgese, 1978; Earney, 1980). A map showing the distribution of these deposits is given by Bakke (1971).

In eastern Asia, much of the prospecting for offshore placers have taken place under the auspices of, or has been reported by, the Committee for Coordination of Joint Prospecting for Mineral Resources in Asian Offshore Areas (CCOP) (Emery & Noakes, 1968; Kim, 1970; Macdonald, 1971; Noakes, 1972, 1977; Johannas, 1982). In addition, the United Nations Economic and Social Commission for Asia and the Pacific (ESCAP) has compiled a mineral distribution map of Asia which superimposes the mineral distribution on the geology (Anonymous, 1979). This serves as useful background information (*cf.* Hutchison & Taylor, 1978; Ishiara, 1978). In this section, it is not proposed to discuss the placers of the southeastern Asian tin belt (Thailand, Malaysia, and Indonesia) (Hosking, 1971; Archer, 1973a; Hails, 1976; Schott, 1976; Noakes, 1977; Sujitno, 1977; Arman, 1978; Burns, 1979; Anonymous, 1981) which lie outside the area of interest.

In 1968, Emery & Noakes reported that titaniferous magnetite is mined from beach deposits in Japan, the Philippines, and Taiwan and from the sea floor off Japan but annual production was less than 2 million tonnes per year. In Japan, several million tonnes of ironsands were recovered during the 1960s at Ariake Bay and Kagoshima Bay off the southern part of Kyusha Island (Mero, 1965; Okano, Shimazaki & Maruyama, 1968; Archer, 1973a; Duane, 1976; Anonymous, 1979; Burns, 1979; Kent, 1980; McIlhenny, 1981). At Ariake Bay, the ironsands average 3–5% titaniferous magnetite and are about 600 m from the coast at depths of 50 m. At Kagoshima Bay, the ironsands are at depths of 15–20 m. In the Philippines, titaniferous magnetite was mined at Lingayen Gulf, northwestern coast of Luzon Island, in water depths of 3–10 m. Reserves were estimated to be 7 million tonnes of contained metal (Anonymous, 1979; Burns, 1979). In 1970, the Philippines were contracted to supply about 600 000 tonnes of beach magnetite to Japan, although this has now ceased (Glasby, 1982a). A survey of Philippine beach deposits has been reported by The Philippine Delegation (1981; see Hale & McLaren, 1984). Beaches of northern Taiwan have reserves of 500 000 tonnes of heavy minerals, of which the main constituent is titaniferous magnetite. Until 1970, these beaches had been mined manually but mechanization was becoming necessary as the higher-grade deposits were worked out (Macdonald, 1971). Boggs (1975) has delineated a number of areas off Taiwan which are prospective for heavy-mineral deposits (*cf.* Soong, 1978). In the Republic of Korea, beach placers are predominately magnetite and ilmenite in the north and monazite and zircon in the central western coast (Overstreet, 1967; Kim, 1970; Macdonald, 1971; Koo, 1980; Neary & Highley, 1984). In Vietnam, only limited information is available but the coast of southern Vietnam from Hue to the Mekong Delta is a mineral-sand province supplied with ilmenite, zircon, rutile, and monazite (Macdonald, 1971; Noakes, 1972). In China, a report on heavy minerals in the southern Yellow Sea has been given by Manyun *et al.* (1983) (*cf.* Delegation of China, 1981 Liu, 1985; Tan *et al.*, 1985).

In Australia, important deposits of rutile, zircon, and ilmenite are mined from the eastern coast beaches between Sydney and Frazer Island (Gardner, 1955; Hails, 1976; Pinter, 1976; Alexandersson & Klevebring, 1978; Burns, 1979; Morley, 1981; von Stackelberg, 1982; Towner, 1984a,b,c). In 1979, rutile produced from these beaches accounted for about 40% of world production. There has been considerable interest in establishing whether the offshore areas also contain economic concentrations of these minerals (Brown & MacCulloch, 1970; Brown, 1971, 1981; Noakes & Jones, 1975; Jones & Davies, 1979; Marshall, 1980).

In 1980, an extensive survey of four prospective areas in this region was undertaken by R.V. SONNE as part of a co-operative project between the Federal Geological Survey (Germany) and the Bureau of Mineral Resources (Australia). The results confirm that the total content of heavy minerals is lower, and the proportion of rutile and zircon in the heavy fraction much lower, than in the onshore deposits. Prospects for finding economic-grade deposits offshore are, therefore, poor. The onshore migration of Holocene beach ridges which started when sea level reached present water depths of 60–80 m coupled with the density sorting of the heavy-mineral suites has led to the concentration of the economically-interesting minerals in the onshore beach deposits rather than in the offshore areas (von Stackelberg, 1982). In addition, offshore exploration for gold off southern New South Wales failed to locate economic deposits and only low-grade deposits of tin were indicated in the submerged river channels off northeastern Tasmania (Brown, 1971; Jones & Davies, 1979). Tin deposits off Cape York, Queensland, have also been reported (Anonymous, 1969). Gold had been mined in the black beach sands of northern New South Wales following its discovery in 1870 but these deposits had been largely worked out by 1896 (Morley, 1981). Monazite is also an economically-recoverable by-product in these beaches (Overstreet, 1967; Neary & Highley, 1984).

In New Zealand, titanomagnetite is mined from dune- and beach-sands at Waipipi and Taharoa on the western coast of the North Island for export to Japan as well as for domestic consumption. Offshore deposits are again of lower grade than the associated onshore deposits and are considered to be sub-economic (McDougall, 1961; McDougall & Brodie, 1967; Summerhayes, 1967; Lewis, 1979; Carter, 1980; N.Z. Geological Survey, 1981). There has also been some interest in placer gold deposits off the western coast of the South Island, Fouveaux Strait, Otago, and the Coromandel Peninsula, and in ilmenite from the western coast of the South Island.

Within the southwestern Pacific, a number of beach deposits have attracted interest. These include beaches near Singatoka, Fiji, (magnetite), San Jorge Island, Solomon Islands (chromite-bearing sediments), and southeastern Papua New Guinea (titaniferous magnetite). The placer deposits to eastern Australia, New Zealand, and the southwestern Pacific have been discussed more fully elsewhere (Glasby, 1986).

From the above discussion, it is seen that the exploitation of offshore placer deposits in the Pacific region has been minimal with such deposits being recovered only in two countries, Japan and the Philippines. Elsewhere, with the possible exception of Alaska, offshore deposits are of lower grade and lesser potential than associated onshore deposits. Moore's (1972) prediction of large-scale placer mining in high-latitude coastal waters by 1980, has not been fulfilled, although this stems, in part, from the State of Alaska placing a moratorium on coastal permits in the mid-1970s (Tagg, 1979).

PHOSPHORITE

Phosphate rock is one of the most important non-metallic mineral commodities. In 1978, about 125 million tonnes of phosphate rock and concentrates were produced, of which 85% was used by the fertilizer industry (Anonymous, 1980b, 1982b, 1983b). 51·5 million tonnes of this phosphate were exported. The U.S.A., U.S.S.R., and Morocco are the major suppliers of phosphate rock, producing 76% of the world total in 1978. Nine other countries produced over one million tonnes in that year (Anonymous, 1980b) (*cf.* Alexandersson & Klevebring, 1978, Lawver, McClintok & Snow, 1978; Sheldon, 1982; Cook & Shergold, 1983; Northolt & Hartley, 1983). Canada, Japan, Australia, Korea, Mexico, and New Zealand are the main importing nations in the Pacific (Sheldon, 1982; Anonymous 1983b). The phosphate produced is dominantly of sedimentary origin. Only two Pacific nations produce significant quantities of phosphate rock, Nauru Island and Ocean Island (White, 1964). The Nauru Island deposits have been mined since 1906. In 1978, Nauru produced almost 2 million tonnes of phosphate rock which was exported to its traditional markets of Australia and New Zealand as well as to Japan and the Republic of Korea (Anonymous, 1980b). Phosphate was also mined on Enderbury Island in the Phoenix Islands (Burnett, 1980a). The mining of this deposit peaked in 1870 and ceased soon afterwards. These deposits are all phosphatic guano. In the Pacific, such deposits occur in the equatorial region, principally south of the equator (Cook, 1975a; Burnett, 1980a; Baturin, 1982; Roe & Burnett, 1985). A combination of high oceanic productivity and low rainfall appears to be necessary for the formation of these deposits, although fossil deposits do not necessarily meet these optimum environmental conditions at present. One promising deposit of this type has been found at Matahiva lagoon in the western part of the Tuamotu Islands (Bender, 1982; Scolari, 1982). About 14 million tonnes of phosphate containing 33% P_2O_5 were established. This deposit was probably formed above sea level, drowned by rising sea level and protected from erosion by a cover of sand. The results of a substantial survey for phosphate on the islands of the southwestern Pacific are reported by White & Warin (1964); the only exploitable resources discovered during this search were located on Bellona Island in the Solomon Islands.

Onland deposits are of economic importance in the Pacific with China and Vietnam major producers of phosphate rock (Anonymous, 1980b). In Australia, the Duchess Deposit in Queensland is known to contain in excess of 500 million tonnes of phosphate and was mined between 1976–1978 (Anonymous, 1980b; Cook & Nicholas, 1980; Rogers & Crase, 1980). After production of $1 \cdot 07 \times 10^6$ tonnes of ore, the operation closed in June 1978 due to large financial losses. At present, Australia imports its phosphate fertilizer but Cook & Nicholas (1980) suggest that, with the cessation of phosphate mining at Ocean and Christmas Islands and with the relatively small reserves at Nauru Island (sufficient for approximately 20 years at the present rate of extraction), mining of these deposits is likely to recommence in the not-too-distant future.

In the marine environment, there are two types of phosphorite deposits: continental margin and seamount deposits. Continental margin deposits are generally in low- to mid-latitudes in a depth range of 100–500 m. They can be divided into youthful (less than a few thousand years) and much older (mid-Tertiary) deposits. In the Pacific, the Peru and east Australian nodules appear to be youthful whereas the Californian and Chatham Rise deposits appear to be older (Burnett, 1980a). Seamount deposits (often phosphatized limestone) appear to be fairly common and a number of occurrences have been reported (*cf.* Baturin, 1982). None the less, deposits from this type of environment have not been well surveyed. Marine phosphorites have been subject to numerous reviews (Cook, 1976; Burnett & Sheldon, 1979; Cook & McElhinny, 1979; Manheim & Gulbrandsen, 1979; Bentor, 1980; Northolt, 1980; Sheldon & Burnett, 1980; Kolodny, 1981; Sheldon, 1981a,b, 1982; Baturin, 1982; Burnett, Roe & Piper, 1983), of which that by Baturin (1982) is particularly useful because of its description of several of the deposits discussed here. An index to the occurrence of onland, insular, and submarine phosphorite deposits in the Pacific is given by Lee (1980, pp. 124–143). Sheldon (1981a) has estimated that the circum-Pacific region contains over 26×10^9 tonnes of phosphate rock of which less than 20% occurs as sea-floor deposits. Mero (1965) estimated that the continental shelves of the world contain 3×10^{11} tonnes of phosphorite.

The southern California deposits are considered to be amongst the most significant offshore deposits in the world, especially the San Juanito sands and the Roger Bank nodules (Dietz, Emery & Shepard, 1942; Emery, 1960; Mero, 1965; Wilson & Mero, 1966; Wenk, 1969; Inderbitzen, Carsola & Everhart, 1970; Cruickshank, 1971; Archer, 1973a; Pasho, 1973; Magnuson, 1974; Burnett, 1980a; Manheim & Hess, 1981; Cruickshank & Rowland, 1983; Rowland & Cruickshank, 1983; Anonymous, 1984a, 1985a). The deposits occur off southern California between 32°30′ N and 38° N from the inner edge of the continental slope to within a few km of the coast. 95% of the deposits lie in the depth range 30–330 m. They are generally found on the tops and sides of banks, on deep escarpments, on the walls of submarine canyons and at the break of the continental shelf, and have a P_2O_5 content in the range 22–33%. They are believed to represent the replacement of Miocene limestone. Based on 330 samples of phosphorite,

Emery (1960) estimated this region to contain about 1×10^9 tonnes of phosphorite whereas Mero estimated about $3 \cdot 4 \times 10^9$ tonnes. This compares with a more recent economic evaluation of the resources as containing 65 million tonnes of phosphate nodules and 52 million tonnes of phosphate sand (U.S. Department of the Interior, 1979). Phosphate deposits are also found off Baja California, Mexico, between 24 and 26° N (D'Anglejan, 1967) where it is estimated that the deposits total $1 \cdot 5 - 4 \times 10^9$ tonnes.

There are phosphorite deposits off Peru and Chile mainly as nodules associated with laminated, organic-rich diatomaceous ooze (Manheim, Rowe & Jipa, 1975; Burnett, 1977, 1980b). They are between 5 and 22° S on the continental shelf and upper part of the slope in a depth range of about 100–400 m. Radiometric dating indicates that they are Holocene deposits (Veeh, Burnett & Soutar, 1973; Burnett, Veeh & Soutar, 1980; Burnett, Beers & Roe, 1982; Kim & Burnett, 1985). The nodules are generally irregularly shaped, flattened (about 1–2 cm thick) and average 22.5% P_2O_5 (Burnett, 1977).

The Chatham Rise is a structure 130 km wide and 800 km long situated east of New Zealand. Phosphorites from the Rise are eroded nodular remnants of Miocene and Oligocene limestone and are found in patches along some 400 km of the Rise crest in water depths of 370–400 m. In parts of the Rise, phosphorite concentrations of 75–80 kg/m^2 have been measured. It has been estimated that the Rise has a total reserve of phosphorite of about 100 million tonnes with an average P_2O_5 content of 21–22% although recent investigations suggest lesser amounts (Bender, 1982; Bäcker, 1985). These deposits have been surveyed during U.S., New Zealand, and German cruises (Buckenham, Rogers & Rouse, 1971; Cullen, 1975, 1979, 1980, 1984; Karns, 1976; Pasho, 1973, 1976; Kudrass, 1980, 1982; Kudrass & Cullen, 1982; von Rad & Kudrass, 1984) and the agronomic potential of the material assessed (Mackay, Gregg & Syers, 1984; Mackay, Syers & Gregg, 1984).

Phosphatic-glauconitic-goethitic nodules are found off northern New South Wales, Australia, between 29 and 32° S in water depths of 200–400 m (Kress & Veeh, 1980; Marshall, 1980; Marshall & Cook, 1980; O'Brien & Veeh, 1980; Cook & Marshall, 1981). The ages of the nodules fall into two groups (about 50 000 yr and 200 000 yr) and P_2O_5 contents of the nodules are low (in the range of 7–12%) reflecting dilution of the phosphatic component by allogenic components and iron oxides.

Phosphorite nodules have also recently been reported from off central California (Mullins, Green & Martin; 1981; Mullins & Rasch, 1985) and from the Sea of Japan (Bersenev, Shkol'nik & Gusev, 1985).

From these descriptions, continental margin (and in particular the southern California and Chatham Rise) phosphorites appear to be the most promising for economic exploitation (*cf.* Mullins & Rasch, 1985). The relatively low cost of phosphate rock (the export price from Morocco and the United States in 1978 was in the range of $U.S. 31–45 per tonne, Anonymous, 1980b), the relatively abundant terrestrial reserves and the high capital costs of offshore mining (even at depths less than 400 m) have meant that it has not yet proved economic to mine offshore phosphorite deposits. A particularly useful overview of the problems involved has been given by Overall (1968a,b,c).

Perhaps the most promising is the Chatham Rise deposit off New Zealand because it could supply a specific local market which at present imports over one million tonnes of phosphate fertilizer per year with a consequent saving in transport costs and overseas exchange (*cf.* Bender, 1982). Preliminary designs of hydraulic-lift systems to mine these deposits have been made (*cf.* Von Rad & Kudrass, 1984) and the economic potential of these deposits has been assessed by Poynter (1983). In spite of substantial exploration effort, exploitation of this deposit remains speculative and interest in the deposit is at present in abeyance. It should be emphasized, however, that this deposit is small compared with major onland deposits which are generally measured in thousands of millions of tonnes.

The southern California deposits have been the subject of a number of preliminary economic assessments (Mero, 1965; Sorensen & Mead, 1969; National Academy of Science, 1975; Trondsen & Mead 1977; U.S. Department of the Interior, 1979), and the Baja California deposit has been evaluated by Overall (1968c) but there are no plans to mine either of these deposits. According to Sorensen & Mead (1969), the quality and quantity of phosphorite off southern California are too low to warrant commercial exploitation in the foreseeable future. The United States is the world's principal phosphate producer, recovering about 50 million tonnes (or about 40% of the total world output) in 1978 (Anonymous, 1980b, 1982b, 1984a; Champ, Dillon & Howell, 1984). The offshore deposits would be in direct competition with these abundant and readily available terrestrial deposits (*cf.* Cathcart & Gulbrandsen, 1973; Stowasser, 1980, 1981), although environmental factors are tending to discourage onshore production.

In spite of the interest in offshore phosphorites as an economic resource stretching back over 20 years, the fact remains that commercial exploitation of these deposits is still a matter for the future.

PRECIOUS CORAL

Precious coral harvesting has been a moderately-sized industry in parts of the Pacific since Japanese exploitation began in 1848. The industry follows a pattern of discovery, exploitation, and depletion such that deposits of precious coral off Japan, Okinawa, and Taiwan are now largely depleted. In the last two decades, peak years of production have been in 1968 and 1969 when 150 000 kg of pink coral were recovered from depths of 400 m on Milwaukee Banks and the surrounding Emperor Seamounts northwest of Midway Island in the Hawaiian Archipelago and in the years following 1978 when a second discovery of precious coral was made at depths of 1000–1500 m in the Milwaukee Banks area. In 1980, approximately two-thirds of the Pacific harvest was landed in Taiwan and the rest in Japan. Approximately 206 250 kg of Midway coral with a value of $U.S. 20·6 million and 19 750 kg of far western Pacific coral with a value of $U.S. 19·7 million were recovered. The total value of the industry in the Pacific (after processing and export) was about $U.S. 50 million. The harvest of Midway coral in 1981 was about 281 000 kg or about 94% of total world production. This represented a glut in production and led to a price reduction for this coral from $U.S. 70 per kg in 1980 to $U.S. 25 per kg in 1981. The development of the precious coral industry in the Pacific has been recorded in considerable detail by Dr R. W. Grigg (*e.g.* Grigg, 1971, 1981, 1983; 1984a), and Wells (1981) has reported on the trade in such corals. Precious corals have also been described off Hawaii (Grigg, 1974, 1976, 1977a,b, 1979; Grigg & Pfund, 1980), off Micronesia (Grigg, 1975), and in the southwestern Pacific (Eade, 1978; Grigg & Eade, 1981; Glasby, 1986). A small domestic industry in Hawaii which utilized a submersible to mine pink and gold corals ceased operations in 1979 due to rising operational costs. Black coral is still harvested in Hawaii, although imports from Tonga and the Philippines have reduced the profit margins. In 1982, five divers produced about 3000 kg of coral. Stony corals, particularly *Pocillopora meandrina*, continue to be harvested for the tourist industry. In 1974 (most recent figure), about 8000 stony coral colonies were harvested in Hawaii netting about $U.S. 70 000 (J. Wiltshire, pers. comm). Grigg has been one of the principle proponents of the conservation of precious coral resources based on scientifically-controlled harvesting rather than in the uncontrolled use of tangle nets (*e.g.* Grigg, 1977a,b, 1983, 1984b). It must be stressed that the precious coral industry is characterized by large fluctuations in supply and price. It is not the intention to duplicate the detailed descriptions of the industry and the nature of the resource as referenced above.

MANGANESE NODULES

The potential importance of the deep-sea mineral resources is apparent from the fact that 53·4% of the surface of the Earth lies at water depths greater than 3000 m (Galtier, 1984) (*cf.* Menard & Smith, 1966). Of the deep-sea materials of possible economic interest, Dunham (1969) quotes the following figures for total abundances; manganese nodules 10^{12} tonnes, phosphorite 10^{10} tonnes, globigerina ooze 10^{14} tonnes, diatomaceous ooze 10^{13} tonnes, and red clay 10^{15} tonnes. Of these, manganese nodules have attracted the greatest interest over the last two decades.

Deep-sea manganese nodules are found in abundance principally in intra-plate regions (Blissenbach & Fellerer, 1973). The principal factor controlling the abundance of nodules on the sea floor is the sedimentation rate (*cf.* Frazer & Fisk, 1981; Cronan, 1982; Glasby *et al.*, 1982). Because of its huge size, the Pacific Ocean has the lowest average sedimentation rate of the oceans (Ewing, Houtz & Ewing, 1969; Chester & Aston, 1976). It is, therefore, the major area of occurrence of nodules. Mero (1965) estimated that the Pacific contained $1{\cdot}656 \times 10^{12}$ tonnes of nodules, although this is now considered to be an over-estimate (*cf.* Frazer, 1980; Glasby, Meylan, Margolis & Bäcker, 1980; Wedepohl, 1980; Heath, 1981).

Manganese nodules were first recovered in the Pacific in 1874 (Murray & Renard, 1891) and Agassiz (1906) was able to delineate the approximate geographic distribution of nodules in the eastern equatorial Pacific. None the less, it was Mero's (1965) classic book *The Mineral Resources of the Sea* devoting almost half the text to manganese nodules which first attracted serious attention to these deposits as an economic resource. Unfortunately, in his enthusiasm to promote manganese nodules, Mero made some statements which would today seem extraordinarily optimistic (*cf.* Archer, 1981). For example, Mero stated that "even assuming a world population of 20 billion people consuming metals at a rate equal to that in the United States at the present time, the reserves of most of the industrially important metals would still be measured in terms of thousands of years". Apart from Mn, Ni, Co, and Cu, Mero considered that Mo, Pb, Zn, Zr, R.E.E. and possibly Fe, Al, Ti, Mg, and V could be recovered as by-products from nodules. Such statements, well intentioned as they were, led to the idea of a cornucopia of metals on the deep-sea floor which has constituted such a stumbling block at UNCLOS. Mero himself has subsequently criticized the international politicians who over-valued nodules as a resource (Mero, 1978). Only few workers at that time, however, cast doubts on nodule mining as a potentially profitable economic venture (Sorensen & Mead, 1968; Medford, 1969b).

None the less, Mero's book stimulated a decade of intense exploration activity in the Pacific, principally in the equatorial North Pacific between the Clarion and Clipperton fracture Zones, throughout the 1970s by U.S. (Horn, 1972; Horn, Horn & Delach, 1973; Bischoff & Piper, 1979; Heath, 1981; Haynes *et al.*, 1982; Piper *et al.*, 1985), German (Fellerer 1979a; Halbach & Fellerer, 1980; Schott, 1980; Derkmann, Fellerer & Richter, 1981; Halbach, 1984; Glasby, 1984), French (Hein, 1977; Bastien-Thiry, Lenoble & Rogel, 1977; Lenoble, 1979, 1981a,b; Bléry-Oustrière, Beurrier & Alsac, 1980; Anonymous, 1984b; Le Gouellec, Herrouin & Charles, 1984; Pautot & Hoffert, 1984), Japanese, and Russian companies and academic groups. In the United States, four consortia applied for exploration licences under the Deep Seabed Hard Mineral Resources Act (1980) in 1982 (Anonymous, 1983c). These had a planned expenditure of in excess of $U.S.200 million. In addition, the Geological Survey of Japan has carried out work on the Central Pacific Basin (Moritani & Nakao, 1981; Nakao & Moritani, 1984) and later in the southwestern Pacific (Mizuno & Nakao, 1982) (*cf.* Georghiou *et al.*, 1983). The position of the U.S.S.R. on deep-sea mining appears unclear (Linebaugh, 1980), although the Russians have done substantial work on the Pacific nodules (Anonymous, 1976; Bezrukov & Skornyakova, 1976; Kazmin, 1984). In 1979, the DMITRIY MENDELEYEV surveyed a polygon centred at 9°50′ N: 146°26′ W to study the nodule distribution (Neprochnov, 1980; Skornyakova *et al.*, 1981, 1983; Skorniakova *et al.*, 1985). The U.K. has an interest in nodule mining through the Kennecott consortium (Cameron *et al.* 1980; Marjoram & Ford, 1980), although no British ship has prospected for nodules in the Pacific since H.M.S. CHALLENGER. South Korea and China appeared to be interested in obtaining "pioneer investor" status (Ford & Spagni, 1983; Spagni & Ford, 1984) although this now seems to have been abandoned. In 1983, the Chinese survey vessel HIANG YANG HONG 16 surveyed nodules in the area between 7–13° N and 167–178° W. The largest haul of nodules during this cruise was 185 kg (*cf.* Wang, 1982; Chuanzhu, 1983; The Sedimentation Laboratory, South China Sea Institute of Oceanology, 1983; Xu & Qinchen, 1983). A further cruise to the North Pacific is planned for 1985. South Korea plans to test sampling equipment in the Clarion-Clipperton Zone in 1984. Taiwan has also carried out some limited sampling in the western Pacific (Chen & Lin, 1981; Chen, 1984). Of the four nations seeking "pioneer investor" status, three (Soviet Union, Japan, and France) are likely to be most interested in Pacific nodules. Summaries of some of these national programmes are given by Comptroller General (1983) and Broadus & Hoagland (1984). Work in other parts of the Pacific such as the Peru Basin, equatorial South Pacific and southwestern Pacific tended to follow that in the Clarion-Clipperton Zone (Friedrich, Glasby, Plüger & Thijssen, 1981; Glasby, 1982a,b; Exon, 1983). McKelvey, Wright & Bowen (1983) have documented other areas in the Pacific containing nodules with high Ni + Cu contents and Exon (1983) and Cronan (1984) have pointed to prospective areas in the equatorial and southwestern Pacific.

This increased activity in nodule prospection in the 1970s is reflected in the number of nodule analyses in the public domain which increased from 522 and 2680 in the decade from 1972 (Fisk, 1982). The exploration for economic-grade nodules, however, tended to give undue emphasis to anomalous nodule types (high grade-high abundance) located in the Clarion-Clipperton Zone. McKelvey *et al.* (1983), for instance, showed that, of the 1700 Pacific nodules analysed for Ni + Cu, 404 are from the Clarion-Clipperton Zone, an area about $2 \cdot 5 \times 10^6$ km^2. This bias in nodule recovery is well illustrated by Frazer (1980, Fig. 2) and McKelvey *et al.* (1983, Fig. 1). According to United Nations Ocean Economics and Technology Branch (1982), the North Pacific has 45 reported nodule occurrences per 10^6 km^2 compared with 11 per 10^6 km^2 in the South Pacific (*cf.* Exon, 1983). In spite of this bias, the number of analyses in the public domain in the prime nodule area in the Clarion-Clipperton Zone is only 330 per 10^6 km^2, well below the number normally expected for mine site evaluation (the number of reported nodule occurrences in this region is 250 per 10^6 km^2; United Nations Ocean Economics and Technology Branch, 1982). Additional data are, of course, held by the private consortia but are proprietary information. Even so, in spite of the limited data available, the principal factors controlling nodule distribution and composition in the Pacific now seem to be moderately well understood (Cronan, 1980a, 1982; Frazer & Fisk, 1981, Glasby & Thijssen, 1982; Glasby *et al.*, 1982).

Considerable effort has been directed at the evaluation of nodules as an economic resource. This includes regional distribution and geochemistry of the nodules and mine site evaluation (Archer, 1975, 1981; Pasho & McIntosh, 1976; Frazer, 1977, 1980; McKelvey, Wright & Rowland, 1979; U.N. Ocean Economics and Technology Office, 1979; Frazer & Fisk, 1981; Lenoble, 1981a,b; Charles River Associates, 1982; U.N. Ocean Economics and Technology Branch, 1982), mining economics (Clauss, 1973; Drechsler, 1973; Meiser & Müller, 1973; Kildow, Dar, Bever & Capstaff, 1976; Mero, 1977a,b; Nyhart *et al.*, 1978; Smale-Adams & Jackson, 1978; Zablocki, 1978; Ocean Ass. Japan, 1979; Peterson, 1979; Antrim, 1980; Pasho & King, 1980; Boin, 1981; Cameron *et al.* 1981; Pasho, 1982; Hillman, 1983; McKelvey *et al.* 1983; Marsh, 1983; Crockett, Chapman & Jones, 1984; Donges, 1985; Johnson & Clark, 1985; Johnson, Clark & Otto, in press), technology (Yamakado, Handa & Usami, 1978; Brockett & Petters, 1980; Halkyard, 1980; Flipse, 1981; Amann, 1982; Handa, Miyashita, Oba & Yamakado, 1982; Noorany & Fuller, 1982; Usami, Saito & Oba, 1982; Welling, 1982b; Le Gouellec, Herrouin & Charles, 1984; Takahara, Handa,

Ishii & Kuboki, 1984; U.N. Ocean Economics and Technology Branch, 1984; Bardey & Guevel, 1985; Kaufman *et al.* 1985; Ligozat, 1985), extractive metallurgy (Agarwal *et al.* 1979; Hubred, 1980; Humphrey, 1982; Shinn, Sedwick & Zeitlin, 1984; Haynes *et al.*, 1985), pilot-scale mining operations (Anonymous, 1977; Clauss, 1978; Ocean Association of Japan, 1979; Kaufman, 1980; Boin, 1981; Ozturgut, Lavelle & Erickson, 1981; Cruickshank, 1982b; Humphrey, 1982), catalytic applications (Dodet *et al.*, 1984), and environmental impact studies (Amos *et al.*, 1977; Stephen-Hassard *et al.*, 1978; Bischoff & Piper, 1979; Lane, 1979; Ichiye & Carnes, 1981; Jenkins, Jugel, Keith & Meylan, 1981; Morgan, 1981; Ozturgut, Lavelle & Burns, 1981; Ozturgut, Lavelle & Erickson, 1981; Curtis, 1982; Humphrey, 1982; Gopalakrishnan, 1984; Wiltshire, 1984). The capital costs of nodule mining have been estimated to be in the range $U.S. 1500–1700 million (Shusterich, 1982; Flipse, 1983; Hillman, 1983). Fellerer (1979b) has estimated that, of the total costs, about 2% is for exploration, 58–60% for nodule processing and the rest for the development of mining systems. Thus, whilst mine sites have probably been evaluated in some detail in the equatorial Pacific by the various consortia, it should be emphasized that this represents only a relatively minor cost in the development of the total mining operation (*cf.* Johnson *et al.*, in press). Shusterich (1982) has indicated the U.S. companies have already spent approximately $U.S. 200 million in investigating various aspects of nodule mining. The total amount spent worldwide has been estimated to be almost $U.S. 500 million (U.N. Ocean Economics and Technology Branch, 1984) (*cf.* Broadus & Hoagland, 1984; Johnson & Clark, 1985; Johnson *et al.*, in press).

Archer (1981) has proposed that the requirement for "first generation" mine sites are Ni + Cu contents greater than 2·25% and abundances greater than 10kg/km^2. On this basis, he was able to calculate that the total quantity of nodules available for first-generation mining might be of the order of $15–25 \times 10^9$ tonnes (wet). Since a first-generation mine site is assumed to contain enough nodules to supply 3 million tonnes per year (on a dry wt basis) for 20–25 years, it can be seen that Archer's calculations indicate a maximum of about 150–250 possible first-generation mine sites and represent the same order of magnitude of reserves of copper, nickel, and cobalt as the land-based reserves (*cf.* Antrim, 1980). (These calculations are, however, as Archer stresses, subject to considerable uncertainty.) The industry would, however, commence on a more modest scale than this (Blissenbach, 1977). Estimates of the first-generation mine sites have been listed by Lenoble (1981b) (*cf.* Frazer, 1980). To give some idea of the relative scale of operations involved, world production of copper, nickel, and cobalt in 1981 was 8·17 million tonnes, 772 000 tonnes, and 69 000 tonnes, respectively, of which the United States produced 18·8%, 1·6%, and 0%, respectively (Anonymous, 1982b). The United States is, in fact, the world's leading copper producer mining 278 million tonnes of ore with an average yield of copper of 0·51%.

In nodule mining, a recovery rate of 3 million tonnes per year (on a dry wt basis) is generally assumed (Archer, 1973a, 1981; Blissenbach, 1977; White, 1982). This high volume is required to reduce unit costs and make the system economic (Meiser & Müller, 1973; Welling, 1977) but the overall feasibility of attaining this has been questioned (Glasby, 1983) (*cf.* Brockett

& Petters, 1980). Pilot tests have so far operated only at a rate of 30 tonnes/h (or roughly 5% of the required rate for full-scale mining). Only one author, Medford (1969a,b), argued against large-scale mining on the grounds that it would depress metal prices and therefore decrease the profitability of the operation (*cf.* Crockett *et al.,* 1984).

In spite of this intensive activity, it is now clear that manganese nodule mining will almost certainly not take place in the twentieth century and that earlier estimates that economic-scale mining might begin in the mid-1980s are incorrect (*cf.* Moore, 1972; Magnuson, 1974; Swan, 1974; Blissenbach, 1977; Mero, 1977b; McKelvey, 1980; White, 1982). Most of the nations with strong interest in nodule mining have terminated their programmes; the last German nodule cruise, for example, took place in 1980 (*cf.* Shusterich, 1982). The Japanese appear to be the only ones at present carrying out nodule exploration cruises in the Pacific. In fact, nodules may now be seen as a medium-grade ore comparable with the nickel laterites of New Caledonia or the porphyry copper deposits of the Americas (Charlier, 1983; Glasby, 1983). Crockett *et al.* (1984), for example, have calculated that one tonne of nodules containing 1·3% Ni, 1·1% Cu, and 0·25% Co would be worth $U.S. 134·7 at 1984 prices (compared with the value of $U.S. 192 per tonne calculated by Bischoff *et al.*, 1983a) (*cf.* Clark, Johnson & Chinn, 1984). This compares with an average value of $U.S. 13·70 for metallic ores in 1980 (Anonymous, 1982b). Interestingly, an even higher value for the metals in nodules was calculated in 1972 (Dorr, Crittenden & Worl, 1973). The large capital investment for the next phase of development, the formidable technological problems involved, the depressed metal prices at present and the uncertainties induced by the Law of the Sea regime will argue against an early resumption of activity (Anonymous, 1982b; Shusterich, 1982; Glasby, 1984).

One stumbling block to the setting up of a nodule mining industry is the recently negotiated Law of the Sea regime. Although grandiose claims have been made for nodule mining by the proponents of this regime (*cf.* Mann Borgese, 1978), the fact remains that nodule mining is unlikely to make more than a modest contribution to a new international economic order (Blissenbach, 1977; Mero, 1978; Johnston, 1979; LaQue, 1980; Cameron *et al.*, 1980; Shusterich, 1982; White, 1982; Langevad, 1983) and the requirements of the regime may lead to a situation where no nodule mining takes place (Craven, 1982; White, 1982) or where inefficient production results (Wijkman, 1982). Some of industry's objections to the regime and the strong bias against the involvement of private enterprise in nodule mining have been outlined by Blissenbach (1977), Greenwald & Nordquist (1979), Kollwentz (1979), Kaufman (1980), Boin (1981), Amann (1982), Comptroller General (1983), and Welling (1983). Whilst it is accepted that the United Nations should control nodule mining in international waters by licensing and taxation policies, the idea of a huge, politicized bureaucracy attempting to run a high capital, forefront technology industry such as nodule mining under the conditions of enforced technology transfer at a profit as envisaged under the Parallel System of exploitation strikes one as being somewhat Gilbertian. One cannot help but come to the conclusion that the premisses on which the international sea-bed regime are based are unrealistic (*cf.* Lucas, 1982). This is well illustrated in the papers of Lévy

(1979a,b) which state that the object of a deep sea bed resource policy should be:

(1) to maximize the revenues for the International Authority,
(2) to encourage the development of nodule resources,
(3) to ensure the participation of the developing countries in the nodule industry, and
(4) to minimize the impact of nodule mining on the mineral exports of developing countries.

As both Fukami (1979) and Takeuchi (1979) clearly state, this makes no provision for the essential precursor, namely to provide guarantees to private or national ventures for the reasonable profitability of nodule mining.

Similar unrealistic statements from other proponents of the regime are apparent. In an extraordinary display of logic, Mann Borgese in consecutive papers claims that "there is not going to be any commercial mining of manganese nodules for the rest of this century" (Mann Borgese, 1983b) and that "it is likely that, by the end of the century, deep seabed mining will come into its own" (Mann Borgese, 1983c). Mann Borgese (1983c) also claims that "in the long term, it is reasonable to assume that mining will be largely displaced from land to sea", an argument which she fails to support with any evidence whatsoever (*cf.* Archer, 1974, 1979).

In response to this regime, there has been some attempt to locate economic-grade nodules within E.E.Z.s. Mann Borgese (1983a) reports that Mexico and Chile have found rich fields of nodules in their E.E.Z.s (*cf.* Craven, 1982; McKelvey *et al.*, 1983; Cronan, 1985), France has nodules in Polynesian waters and the U.S. can assert claims offshore from the Hawaiian and other Pacific islands to hundreds of thousands of square miles of ocean floor likely to have commercially exploitable nodules. Other areas of nodule occurrence within E.E.Z.s have been given by Fisk (1982), Charlier (1983), Anonymous (1984a), and Cronan (1984). The statement by Mann Borgese is, however, a half truth. By virtue of the fact that nodules occur in highest abundance where sedimentation rates are lowest and that any land mass introduces terrigenous sediment to the adjacent deep-sea floor, it is unlikely that the highest nodule abundances, necessary for economic extraction, lie within E.E.Z.s.

In summary, it appears that, in spite of all the optimistic predictions made over the last 20 years and the decade of intense exploration and development activity through the 1970s, there will be a gap of at least a generation before manganese nodule mining takes place in the Pacific.

DEEP-SEA RADIOLARITE

Diatomite is mined in the western U.S.A. for use in filtration, as an abrasive and as a paint filler (Hoover, 1979). Arrhenius (1977) and Baak (1978) have presented an interesting economic analysis of the potential of deep-sea radiolarite recovered as a by-product of manganese nodule mining. In these calculations, it is assumed that an operation yielding 1·5 million tonnes of manganese nodules on a dry weight basis could yield seven times its weight of deep-sea sediment. In a radiolarian-rich area (where economic-grade nodule deposits are principally found), the sediment might contain 30% radiolarite and 90% of this radiolarite might be recoverable. It can then be estimated that this operation would yield 2.8 million tonnes of radiolarite, enough to satisfy the demand of the circum-Pacific market. Its main use would be in the building materials (87%), filter aids, and fillers in polymers. The value of the radiolarite from such an operation is then of the same order as that of manganese nodules. None the less, Cook (1975b) regards this as a resource for the distant future only. It should be pointed out, however, that, in the most efficient nodules mining systems (such as designed by C. G. Welling, Mann Borgese, 1983a) the object is to separate the nodules from the sediment in order to maximize the efficiency of the operation (*cf.* Ozturgut, Lavelle & Erickson, 1981; Amann, 1982).

COBALT-RICH MANGANESE CRUSTS

The fact that cobalt is enriched in manganese crusts from elevated volcanic areas such as the mid-Pacific mountains, the Hawaiian chain and the island groups of the South Pacific has been known for a considerable time (Menard, 1964; Mero, 1965, 1978; Cronan, 1972, 1977; Horn, Delach & Horn, 1973; Glasby, 1976; Arrhenius *et al.*, 1979; Craig, Andrews & Meylan, 1982, McKelvey, Wright & Bowen, 1983; Aplin & Cronan, 1985) but this observation was regarded as an academic curiosity. The presence of a huge amount of cobalt in Pacific Ocean manganese nodules ($5{\cdot}8 \times 10^9$ tonnes) has, of course, been known since the work of Mero (1965) (*cf.* Vhay, Brobst & Heyl, 1973).

In 1981, R.V. SONNE surveyed the manganese crusts on the Line Islands Ridge and mid-Pacific mountains, during the Midpac '81 cruise (Halbach, 1982, 1984; Halbach, Manheim & Otten, 1982; Halbach, Segl, Puteanus & Mangini, 1983; Halbach & Manheim, 1984; Halbach & Puteanus, 1984; Halbach, Puteanus & Manheim, 1984). The results of this cruise led to the U.S. Geological Survey to take considerable interest in these deposits as a possible mineral resource (Edgar, 1983; Rowland, Goud & McGregor, 1983; Anonymous, 1984a; Commeau *et al.*, 1984). Rowland *et al.* (1983) reported that these crusts average about 20 mm in thickness and are fairly uniformly distributed in the depth range 1000–2600 m (Halbach & Manheim, 1984). They are known to occur on a number of the 200 islands that lie within the U.S. E.E.Z. Abundances of up to 16 kg of ore per m^2 of crustal surface have been reported (Champ, Dillon & Howell, 1984; Halbach & Manheim, 1984). This abundance can, in favourable cases, be much higher than in deep-sea nodules (McKelvey *et al.*, 1983) and it has been suggested that a single seamount could yield enough ore for a commercial mining operation (up to 4 million tonnes of ore, Champ *et al.* 1984; Halbach & Manheim, 1984). This would be important as the United States has no domestic source of cobalt and cobalt is listed as one of the five most important strategic minerals to the U.S.A. (Holden, 1981). Furthermore, these deposits contain high concentrations of platinum which might be of economic interest (Halbach, Puteanus & Manheim, 1984). This finding is also significant in as much as the Pacific is estimated to contain in excess of 20 000 seamounts (Batiza, 1982; Jordan, Menard & Smith, 1983) and 200 in the U.S. E.E.Z. alone (Rowland *et al.* 1983). Bathymetrically favoured areas in the southwestern Pacific include the Line Islands, Phoenix Islands, Cook Islands, northern Tonga, and Samoa. Cronan (1984) has identified optimum areas for further search in the southwestern Pacific. To illustrate this interest in cobalt-rich crusts, the U.S. Department of the Interior Minerals Management Service has carried out a major survey of Pacific crusts in conjunction with the U.S. Geological Survey involving two cruises of S.P. LEE (Anonymous, 1984a,c). Results of the cruise of S.P. LEE to the Neckar Ridge, Horizon Guyot and S.P. Lee Guyot in 1983 have been presented by Hein *et al.* (1985), Hein, Manheim, Schwab & Davis (1985) and Manheim, Hein, Marchig & Puteanus (in press). A database for the crusts has also been prepared (Manheim, Ling & Lane, 1983) and surveys of the crusts as an economic resource have been given by Clark *et al.* (1984, 1985) and Johnson & Clark (1985). An overview of these deposits has been presented by Manheim (in press). At least one consortium, International Hard Minerals Co., has begun first phase feasibility investigations into crust mining (Manheim, in press).

In Hawaii, the Minerals Federal Management Service has set up a Federal/State Task Force to assess these deposits (Smith, Holt & Paul, 1985). Three cruises of R.V. KONA KEOKI were undertaken in 1984. Preliminary results have been presented by Johnson *et al.* (1985) and McMurtry *et al.* (1985). An Environmental Impact Statement to assess the impacts of exploration, mining and processing of these crusts was prepared in late 1985.

None the less, care should be taken to put the economic viability of this type of deposit in perspective. The idea that these crusts are uniformly distributed with water depth does not seem to accord with existing information and the factors controlling their distribution and geochemistry are likely to be complex. Furthermore, from the data of McKelvey *et al.* (1983), it appears that the mean cobalt content of manganese crusts and nodules from 170 stations taken at depths less than 2000 m is 0·53% with a standard deviation of ±0·37%. Although a plot of the cobalt content of Pacific nodules and crusts with water depth does indeed show a number of samples (15) in depth range 1000–2600 m containing in excess of 1% cobalt, these represent only a minority (<20%) of the total number of samples in this depth range. In the Line Islands Archipelago, seven samples collected between 1000 and 2000 m had a mean Co concentration of 1·09% and an average thickness of 14 mm (Aplin & Cronan, 1985). For the samples collected during the cruise of S.P. LEE, the corresponding average figures are Neckar Ridge (0·59% Co, 25 mm thick), Horizon Guyot (0·73% Co, 15 mm thick), S.P. Lee Guyot (1·29% Co, 8 mm thick), Colahan Seamount (0·49% Co, 6 mm thick), and Abbott Seamount (0·68% Co, 10 mm thick) (Hein *et al.*, 1985). For the Midpac 1981 cruise, the average cobalt content of the crusts was 0·79% (Halbach & Manheim, 1984). These figures show that the term "cobalt-rich crust" is something of a misnomer. From the S.P. Lee data, only samples from the Neckar Ridge have the 20 mm average crust thickness mentioned by Rowland *et al.* (1983).

In addition, mining of any economic-grade deposit is likely to prove extremely difficult in areas of rugged terrain at water depths greater than

1000 m on the scale required (0·75–1·5 million tonnes/yr, Clark *et al.*, 1984) (although the scale proposed by these authors is somewhat arbitrary). Their suggestion that mining might take place using the Continuous Line Bucket (CLB) system also seems unconvincing (although preliminary tests have been reported by Masuda, 1982) and mining technology for these deposits has not yet been properly evaluated (Anonymous, 1984a). Furthermore, many of these deposits are located in remote areas, have variable thicknesses on a local scale, contain an appreciable percentage of gangue substratum material and, in general, are poorly surveyed. It is not clear, therefore, that these deposits will ever turn out to be a mineable ore.

It is, however, instructive to consider some of the factors which have contributed to the recent assessment of cobalt-rich crusts as an economic resource (which are somewhat different from those listed by Clark *et al.*, 1984).

(1) Commercial exploration for manganese nodules ended in most industralized countries in 1982. Research vessels were, therefore, free to be redirected to exploration of this new resource.
(2) The importance of strategic minerals (of which cobalt is one of the most important) has been stressed by the Reagan administration.
(3) The declaration of the U.S. E.E.Z. in 1983 stimulated considerable exploration activity in this zone by the U.S. Geological Survey.
(4) The high price of cobalt ($U.S. 27 560 per tonne in 1984) compared with that of nickel ($U.S. 7056 per tonne) and copper ($U.S. 4433 per tonne) (Crockett, Chapman & Jones 1984). Cobalt-rich crusts, therefore, became more valuable than manganese nodules by the 1980s (Clark *et al.*, 1984; Halbach & Manheim, 1984). It is probable that such deposits, if mined, would be mined for cobalt alone and it seems misleading to suggest a value for the crusts which implies that manganese (which has large reserves on land) would also be recovered (*cf.* Halbach & Manheim, 1984); this more than trebles the apparent value of these deposits. The elevated cobalt prices in the period 1978–1982 following the invasion of Zaire in 1978 also served to favour cobalt-rich crusts (Manheim, in press).

The Midpac '81 cruise of R.V. SONNE was, therefore, most timely in stimulating interest in this new resource for other than strictly economic reasons. This conclusion is supported by the almost complete absence of private capital in the search for these deposits.

METALLIFEROUS SEDIMENTS AND MASSIVE SULPHIDE DEPOSITS

Metalliferous sediments are unconsolidated accumulations of variable proportions of hydrothermal, detrital, hydrogenous, and biogenous material in which the transition metal content is elevated above that of "normal" pelagic sediments (Meylan, Glasby, Knedler & Johnston, 1981). The nature of deep-sea metalliferous sediments was first recognized by Boström & Peterson (1966). In the Pacific, these sediments occur as both superficial and basal deposits and are formed principally at divergent plate margins (oceanic spreading centres) and also in island arcs and marginal basins. Over the last few years, these deposits have been subject to intensive investigation (Bäcker, 1973, 1982; Heath & Dymond, 1977; Corliss, Lyle, Dymond & Crane, 1978; Rona, 1978, 1983a,b, 1984; Cronan, 1980a; Rona & Lowell, 1980; Bonatti, 1981; Fyfe & Lonsdale, 1981; Goldie & Bottrill, 1981; Kulm, Dymond, Dasch & Hussong, 1981; Bougault, 1982; Cann & Strens, 1982; Edmond, Von Damm, McDuff & Measures, 1982; Humphrey, 1982; Löfgren & Boström, 1982; Macdonald, 1982; Strens & Cann, 1982; Bender, 1983; Hatem, 1983; Moorby & Cronan, 1983; Rona, Boström, Laubier & Smith, 1983; Thompson, 1983; Ballard, 1984; Halbach, 1984; Richards & Strens, 1985). It is now thought that the intensity of hydrothermal activity is related to the ocean spreading rate (Ballard & Francheteau, 1982a,b; Marchig & Gundlach, 1982; Ballard, 1984; Richards & Strens, 1985) and the sequence of hydrothermal precipitation has been described as sulphides, iron silicates, iron oxides, and manganese oxides (Cronan, 1980b) (*cf.* Bäcker, 1982). Examples of hydrothermal manganese deposits in the Pacific have been reported by Moore & Vogt (1976), Grill, Chase, Macdonald & Murray (1981), and Lalou, Brichet, Jehanno & Perez-Leclaire (1983). The GEOMETEP and GARIMAS cruises of R.V. SONNE have been an attempt to map metalliferous sediments at selected areas along the East Pacific Rise and Galapagos Rift (Marchig & Gundlach, 1982; Bäcker & Marchig, 1983; Gundlach, Marchig & Bäcker, 1983; Kunzendorf, Walter, Stoffers & Gwozdz, 1984; Marchig, Gundlach & Bäcker, 1984; Tufar, Gundlach & Marchig, 1984; Bäcker, Lange & Marchig, 1985; Marchig, Möller, Bäcker & Dulski, 1985; Walter & Stoffers, 1985). In addition to divergent plate margins, back-arc basins have become a focus for exploration for such deposits (Glasby, 1971; Cronan, 1976, 1983; Green, Hulston & Crick, 1978; Bonatti, Kolla, Moore & Stern, 1979; Leinen & Anderson, 1981; Cronan *et al.* 1982; Edmond, 1982; Yuasa & Yokota, 1982; Brocher, *et al.*, 1983; McMurtry *et al.*, 1983; Rona, 1984; Von Stackelberg *et al.*, 1985) and hydrothermal deposits associated with seamounts have recently been reported (Lonsdale & Batiza, 1980; Lonsdale, Burns & Fisk, 1980; Lonsdale, Batiza & Simkin, 1982; Malahoff, McMurtry, Wiltshire & Yeh, 1982; Batiza, 1983, 1985; De Carlo, McMurtry & Yeh, 1983; Exon & Cronan, 1983; McMurtry, Epp & Karl, 1983). Greenslate, Frazer & Arrhenius (1973) have suggested that metalliferous sediments from areas such as the Bauer Deep may have economic potential but this idea has been rejected (Glasby, 1976; McKelvey, 1980; Field, Wetherell & Dasch, 1981) (*cf.* McMurtry, 1981).

From an economic standpoint, the discoveries by submersible of hydrothermal vents on the crest of the Galapagos Rift in 1977 (Ballard, 1977; Corliss *et al.*, 1979; Edmond *et al.*, 1979a,b; Natland *et al.*, 1979) and of massive sulphide deposits at the crest of the East Pacific Rise in 1978 (Francheteau *et al.*, 1979) have been the most significant development. Sulphide deposits have now been discovered at the crest of the East Pacific Rise at 21° N (Francheteau *et al.*, 1979; Bischoff, 1980; Hekinian *et al.*, 1980; Lupton *et al.*, 1980; Macdonald, Becker, Spiess & Ballard, 1980; Mottl, 1980; Spiess *et al.*, 1980; Ballard *et al.*, 1981; Haymon & Kastner, 1981; Macdonald & Luyendyk, 1981; Styrt *et al.*, 1981; Francheteau & Ballard, 1983; Goldfarb, Converse, Holland & Edmond, 1983; Haymon,

1983; Kerridge, Haymon & Kastner, 1983; Lafitte, Maury & Perseil, 1983; Oudin, 1983; Converse, Holland & Edmond, 1984; Crawford, Hollingshead & Scott, 1984; Zierenberg, Shanks & Bischoff, 1984; Lafitte, Maury, Perseil & Boulegue, 1985), at 13° N (Francheteau & Ballard, 1983; Hekinian *et al.*, 1983a,b; Lafitte, Maury & Perseil, 1984; Ballard, Hekinian & Francheteau, 1984; Choukroune, Francheteau & Hekinian, 1984; Janecky & Seyfried, 1984; Michard *et al.*, 1984; Hekinian & Fouquet, 1985; Hekinian, Francheteau & Ballard, 1985; Lalou, Brichet & Hekinian, 1985) at 7° N (Boulegue *et al.*, 1984), at 9° S (Cronan & Varnavas, 1981), at 18·5 and 21·5° S (Tufar *et al.*, 1984; Bäcker *et al.*, 1985; Renard *et al.*, 1985) and at 20° S (Ballard, *et al.*, 1981b; Francheteau & Ballard, 1983; Bougault, Renard & Hekinian, 1984), in the Gulf of California (Lonsdale *et al.*, 1980; Campbell & Gieskes, 1984; Lonsdale, 1984; Koski *et al.*, 1985; Lonsdale & Becker, 1985), on the Juan de Fuca Ridge (Koski, Normark, Morton & Delaney, 1982; Normark *et al.*, 1982, 1983; Galerne, 1983; Hatem, 1983; Kingston, Delaney & Johnson, 1983; Koski, Clague & Oudin, 1984; Canadian American Seamount Expedition, 1985; Koski, Normark & Morton, 1985; Lupton, Delaney, Johnson & Tivey, 1985; Tivey & Delaney, 1985; Hannington, 1986), on the Explorer Ridge (Anonymous, 1985b) and on the Galapagos Rift (Lonsdale, 1977; Malahoff, 1981, 1982a; Ballard, van Andel & Holcomb, 1982; Malahoff, Embley, Cronan & Skirrow, 1983; Maris *et al.*, 1984) (*cf.* Edmond, 1982; Fisk, 1982). The Explorer Ridge deposits are the largest so far located (Anonymous, 1985b). Maps showing the location of these deposits are given in Edgar (1983), Bischoff *et al.* (1983a), and Ballard (1984). These deposits are extremely localized and difficult to find (Rona, 1983b). Models for the formation are presented by Bischoff (1980), East Pacific Rise Study Group (1981), Ballard & Francheteau (1982a,b), Ballard (1984), and Bischoff & Pitzer (1985) and for submarine exhalative deposits by Finlow-Bates (1980).

Because of their high contents of zinc and copper, these deposits have attracted considerable economic interest (Cann, 1980; Duane, 1982; Humphrey, 1982; Malahoff, 1982b; Malahoff, McLain & Ranson, 1982; Welling, 1982a; Bischoff *et al.*, 1983a,b; Anonymous, 1984a; Koski, Normark & Morton, 1985; Johnson & Clark, 1985), not only because of their intrinsic worth but also because they might serve as a guide to prospecting for similar deposits on land (Goldie & Bottrill, 1981; Oudin, Picot & Poult, 1981; Crerar *et al.* 1982; Malahoff, 1982a; Rona, 1982, 1983a; Strens & Cann, 1982; Hatem, 1983; Chyi, Crerar, Carlson & Stallard, 1984; Haymon, Koski & Sinclair, 1984; International Crustal Research Drilling Group, 1984; Oudin & Constantinou, 1984; Varnavas & Panagos, 1984; Alabaster & Pearce, 1985; Coombs *et al.*, 1985; Goodfellow, 1985; Richards & Strens, 1985; Scott, 1985). Preliminary metallurgical studies on sulphides have been carried out by Sawyer *et al.* (1983) and Ergunalp & Weber (1985). Malahoff (1982b) described the largest deposits then known which is on the Galapagos Rift. The deposit is said to be 1000 m long, 35 m high, and 200 m wide at a water depth of 2850 m, to have an average specific gravity of 5·8 and to contain an average 10% Cu, 35% Fe, and 1% Zn. This would correspond to a deposit of about 20 million tonnes (Bischoff *et al.*, 1983a). Subsequent work, however, suggests that these figures are highly exaggerated (H. Bäcker, pers. comm.).

Furthermore, the description of the size and character of deposits does not accord with the more detailed description given by Malahoff *et al.* (1983). Rona (1983b) has suggested that such a field might typically contain 10 to 100 mounds each weighing several thousand tonnes. Such fields may be well separated from each other (Cann, 1980; Strens & Cann, 1982; Welling, 1982b; Rona, 1983b; Sleep & Morton, 1983; Broadus, 1984b; Richards & Strens, 1985). A preliminary assessment of the value of these deposits was also given by Cruickshank (1982a).

One of the most detailed considerations of the economic worth of these deposits is that of Bischoff *et al.* (1983a,b) who analysed massive suplphide deposits from the 21° N, Juan de Fuca Ridge, and Galapagos Rift (*cf.* Morgan & Selk, 1984; Ergunalp & Weber, 1985). The first two deposits are primarily Zn, Fe, and sulphur with important minor amounts of Ag, As, Cd, and germanium (30% Zn, 0·5% Cu, and 0·02% Ag). The last deposit is primarily iron, copper, and sulphur with important minor contents of cobalt and molybdenum (5% Cu, 0·2% Zn, 0·02% Co). Bischoff *et al.* estimated the worth of the first two deposits to be $U.S. 348 per tonne and of the last to be $U.S. 85 per tonne; this compares with a value of $U.S. 192 per tonne for manganese nodules (not including the value of the manganese). These authors stress, however, that too little is known of these deposits to make a detailed economic evaluation. In particular, more detailed surveys of the size and grade of these deposits are required. The metal grade and mineral associations of these deposits vary considerably even within one mound. From a mining standpoint, the shallower depth of the deposits compared with manganese nodules may be countered by the fact that they occur in hard, rugged volcanic terrains and the deposits themselves tend to be hard and cemented (*cf.* Crawford *et al.* 1984). For this reason, Preussag AG has found it necessary to develop an electro-hydraulic grab with an underwater television camera and Bedford Institute of Oceanography an electric rock drill (Koski, Normark & Morton, 1985) specifically to sample this type of deposit. No mining technology for these deposits is, however, yet available (Anonymous, 1984a). None the less, Welling (1982a) makes the point that the sulphides are much more concentrated than nodules with approximately 1000 times the mass per unit area of sea floor. It should be remembered, however, that these deposits have an economic value only if they can be mined more cheaply than land-based zinc and copper deposits. In a detailed economic analysis, Broadus (1984b) came to the firm conclusion that these deposits are still largely a scientific phenomenon with little current economic interest (*cf.* Cann, 1980; Hatem, 1983; Broadus & Bowen, 1984; Crockett *et al.*, 1984; Johnson & Clark, 1985: Johnson, Clark & Otto, in press), although Mann Borgese (1983c) assumes a value of $U.S. 6700 million for the Galapagos Rift sulphides. Private industry has indicated that it has no interest in exploiting active hydrothermal vents but may be interested in older vents which have deposited all their sulphide (R. J. Greenwald, pers. comm.).

One feature of these deposits is that parts of the ridges on which they are found lie within national E.E.Z.s (Rona 1983a). For example, the Galapagos Rift may be claimed by Ecuador, 21° N by Mexico, Juan de Fuca Ridge by Canada and the U.S.A., and the Gorda Ridge by the U.S.A. (Bischoff *et al.*, 1983a; Mungall, 1983; Rowland, Goud & McGregor, 1983).

Koo (1980) has also pointed to the possible occurrence of exhalative-sedimentary deposits off Korea. The U.S. government has recently announced plans for leasing polymetallic sulphides on the Gorda Ridge (Rowland *et al.*, 1983; Cruickshank, 1984; Cruickshank & Zippin, 1984), and the Minerals Management service set up a Joint Federal/State Task Force to assess these deposits (Smith, Holt & Paul, 1985). This was done in spite of the fact that no sulphide deposits had been discovered there (Malahoff, 1981; Anonymous, 1984a, 1985a; Beauchamp, 1984). Sulphides were in fact first recovered from this area in late 1985 during a cruise of the S.P. LEE (Morton, 1985). None the less, marine sulphides are likely to attract continuing interest.

SUBMARINE GEOTHERMAL RESOURCES

In addition to the minerals associated with submarine hydrothermal activity, Williams (1976) has pointed to this activity as a possible geothermal resource and estimated that submarine hydrothermal discharge accounts for about 20% of the Earth's heat loss (*cf.* Cruickshank 1974; East Pacific Rise Study Group, 1981; Macdonald, 1982; Sleep & Morton, 1983; Rona, 1984; Richards & Strens, 1985). On the Galapagos spreading centre, for example, $2 \cdot 6 \times 10^7$ cal./s are released by hydrothermal water per km of ridge length. In the Gulf of California, it is calculated that sufficient heat is contained within the reservoir to generate approximately 180 000 MW of electricity for 30 years. In the Pacific, Williams (1976) has suggested that sub-seabed geothermal reservoirs near coastal areas such as in the Gulf of California, Sea of Japan, Sea of Okhotsk, Andaman Sea, and Juan de Fuca Ridge may be exploitable in the future, although this would seem to be a resource for the distant future. Koo (1980) has pointed to potentially economic geothermal zones off Korea. Rona (1983a) has, however, indicated that it is only where segments of the divergent plate form volcanic islands such as Iceland or intersect continents such as at the head of the Gulf of California that the geothermal energy is accessible for development. Cann (1980) has suggested that submarine geothermal systems could be used to produce metals rather than heat.

SUMMARY

From the previous considerations, it is clear that, whilst knowledge of the distribution of marine minerals resources in the Pacific has increased substantially over the last twenty years, actual production is restricted to aggregate, placers, and precious coral as well as salt and chemicals from sea water. The anticipated development of deep-sea minerals such as manganese nodules has not taken place and is unlikely to do so this century. In spite of the vastly increased level of knowledge of the distribution and mode of formation of Pacific marine minerals, sampling levels in most of the Pacific remain much below those considered necessary for the evaluation of land-based mineral resources. There still remains considerable scope for the systematic surveying of Pacific mineral resources and the development of technology for their eventual recovery.

ACKNOWLEDGEMENTS

I wish to acknowledge a number of colleagues who kindly sent material which could be incorporated in this paper. In particular, I thank T. Moritani (Geological Survey of Japan) who supplied much of the information on Japanese offshore resources, W. F. McIlhenny (Dow Chemical Co.) who supplied information on the production of chemicals from sea water, D. J. Debney (Renison Goldfields Consolidated Ltd) who supplied information on the marine aggregate project off Sydney and J. Wiltshire (State of Hawaii Department of Planning and Economic Development) who supplied data on Hawaiian offshore resources. I also thank H. Bäcker (Preussag AG), J. M. Broadus (Woods Hole Oceanographic Institution), M. J. Cruickshank (U.S. Geological Survey), M. Fisk (United Nations), and S. A. Moorby (Imperial College, London) who reviewed the manuscript.

REFERENCES

Adams, M. V., John, C. B., Kelly, R. F., LaPointe, A. E. & Meurer, R. W., 1975. *U.S. geol. Surv., Circ.*, No. 720, 32 pp.

Agarwal, J. C., Beecher, N., Davies, D. S., Hubred, G. L., Kakaria, V. K. & Moslen, J. H., 1979. *Mar. Mining*, **2**, 119–130.

Agassiz, A., 1906. *Mem. Mus comp. Zool. Harv.*, **33**, 1–75.

Alabaster, T. & Pearce, J. A., 1985. *Econ. Geol.*, **80**, 1–16.

Alexandersson, G. & Klevebring, B. -I., 1978. *World Resources Energy and Minerals.* de Gruyter, Berlin, 248 pp.

Amann, H., 1982. *Phil. Trans. R. Soc. Ser. A*, **307**, 377–403.

Amos, A. F., Roels, O.A., Garside, C., Malone, T. C. & Paul, A.Z., 1977. In, *Marine Manganese Deposits*, edited by G. P. Glasby, Elsevier, Amsterdam, pp. 391–437.

Anonymous, 1966. *A symposium on the tectonic history and mineral deposits of the Western Cordillera in British Columbia and neighbouring parts of the United States*. Can. Inst. Min. & Metall., Montreal, 353 pp.

Anonymous, 1968. *Ocean Ind.*, **3**(1), 39–42.

Anonymous, 1969. *Ocean Ind.*, **4**(8), 28 only.

Anonymous, 1970. *Ocean Ind.*, **5**(12), 25–26.

Anonymous, 1976. *Trans. P.P. Shirshov Inst. Oceanol.*, **109**, 288 pp.

Anonymous, 1977. *Ocean Ind.*, **12**(6), 79–82, 84.

Anonymous, 1979. *Mineral distribution map of Asia Second Edition (Revised).* United Nations Economic and Social Commission for Asia and the Pacific (ESCAP). 4 map sheets (scale 1:5 000 000) + explanatory brochure, 110 pp.

Anonymous, 1980a. *Environmental Impact Statement Marine Aggregate Project.* Consolidated Gold Fields Australia & ARC Marine Ltd, 3 volumes.

Anonymous, 1980b. *U.N. Mineral Resources Development Ser.*, No. 47, 73–81.

Anonymous, 1981. In, *Studies in East Asian Tectonics and Resources (SEATAR)*, CCOP Tech. Publ., 7a, 2nd edition, pp. 167–171.

Anonymous, 1982a. *Economic dependence on six imported strategic non-fuel minerals.* Report to the Subcommittee on Economic Stabilization of the Committee on Banking, Finance and Urban Affairs, House of Representatives 97th Congress, Second Session, Washington, D.C., 54 pp.

Anonymous, 1982b. *Minerals Yearbook. Vol. 1 Metals and Minerals.* U.S. Bureau of Mines, Washington, D.C., 968 pp.

Anonymous, 1983a. *Ocean Sci. News,* **25** (44), 1–6.
Anonymous, 1983b. *World mineral statistics 1977–1981.* Her Majesty's Stationery Office, London, 273 pp.
Anonymous, 1983c. *Ocean Ind.,* **18**(1), 44, 46, 49, 50.
Anonymous, 1984a. *U.S. geol. Surv. Circ.,* No. 929, 308 pp.
Anonymous, 1984b. *Les nodules polymétalliques Faut-ils exploiter les mines océaniques? Rapport de l'Academie des Sciences,* Gauthier-villars, Paris, 180 pp.
Anonymous, 1984c. *EOS Trans. Am. geophys. Un.,* **65**(4), 25 only.
Anonymous, 1985a. *Federal Register (U.S.),* Jan. 15, 2264–2266.
Anonymous, 1985b. *Mar. Pollut. Bull.,* **16,** 45 only.
Antrim, L., 1980. In, *Deepsea Mining,* edited by J. T. Kildow, MIT Press, Cambridge, Mass., pp 84–106.
Aplin, A. C. & Cronan, D. S., 1985. *Geochim. Cosmochim. Acta,* **49,** 427–436.
Archer, A. A., 1973a. *Ocean Management,* **1,** 5–40.
Archer, A. A., 1973b. *Mining Engng,* **25**(12), 31–32.
Archer, A. A., 1974. *Mining Mag.,* **130**(3), 150–163.
Archer, A. A., 1975. *CCOP/SOPAC Tech. Bull.,* **2,** 21–38.
Archer, A. A., 1979. In, *The Deep Seabed and its Mineral Resources,* Proc. 3rd Internat. Ocean Symp. 1978 Tokyo, pp. 54–61.
Archer, A. A., 1981. *Trans. Instn Min. Metall.,* **90A,** 1–6.
Arita, M. & Kinoshita, Y., 1984. In, *Proc. 19th Session CCOP,* 42–48.
Arman, M., 1978. *CCOP Tech. Bull.,* No. 12, 33–42.
Arrhenius, G., 1977. *Scripps Inst. Oceanogr.,* SIO Ref. No. 77–127, 7 pp.
Arrhenius, G., Cheung, K., Crane, S., Fisk, M., Frazer, J., Korkisch, J., Mellin, T., Nakao, S., Tsai, A. & Wolf, G., 1979. *Proc. Colloq. int. C.N.R.S.,* No. 289, 333–356.
Ashkenazy, I., 1981. *Oceans,* **14**(3), 44–49.
Baak, T., 1978. *Scripps Inst. Oceanogr.,* SIO Ref. No. 78–36, 77 pp.
Bäcker, H., 1973. *Erzmetall,* **26,** 544–555.
Bäcker, H., 1982. *Erzmetall,* **35,** 91–97.
Bäcker, H., Lange, J. & Marchig, V., 1985. *Earth Planet. Sci. Lett.,* **72,** 9–22.
Bäcker, H. & Marchig, V., 1983. *Meerestechnik,* **14,** 34–140.
Bäcker, H. & Schoell, M., 1974. *Chem. Z.,* **98,** 299–305.
Bakke, D. R., 1971. *Offshore,* **31**(12), 34–42.
Balas, D., 1984. *Proc. 2nd Internat. Seminar Offshore Mineral Resources (GERMINAL),* 341–357.
Ballard, R. D., 1977, *Oceanus,* **20**(3), 35–44.
Ballard, R. D., 1984. *Oceanus,* **27**(3), 7–14.
Ballard, R. D. & Francheteau, J., 1982a. In, *Marine Mineral Deposits—New Research Results and Economic Prospects,* edited by P. Halbach & P. Winter, Proceedings of the Clausthaler Workshop, Verlag Glückauf GmbH, Essen, pp. 137–176.
Ballard, R. D. & Francheteau, J., 1982b. *Mar. Technol. Soc. J.,* **16**(3), 8–22.
Ballard, R. D., Francheteau, J., Juteau, T., Rangan, C. & Normark, W., 1981. *Earth Planet. Sci. Lett.,* **55,** 1–10.
Ballard, R. D., Hekinian, R. & Francheteau, J., 1984, *Earth Planet. Sci. Lett.,* **69,** 176–186.
Ballard, R. D., Morton, J & Francheteau, J., 1981. *EOS Trans. Am. geophys. Un.,* **62**(45), 912–913 (Abstr.).
Ballard, R. D., van Andel, T. H. & Holcomb, R. T., 1982. *J. geophys. Res.,* **87,** 1149–1161.
Bardey, R. R. & Guevel, P., 1985. *Proc. Offshore Technol. Conf.,* OTC 4900, 19–32.
Bascom, W., 1969. *Scient. Am.,* **221**(3), 198–204, 206, 208, 210, 213, 214, 216, 217.
Bastien-Thiry, H., Lenoble, J. P. & Rogel, P., 1977. *Engng Min. J.,* **178**(7), 86, 87, 171.

Batiza, R., 1982. *Earth Planet. Sci. Lett.,* **60,** 195–206.
Batiza, R., 1983. *Oceans '83,* 797–800.
Batiza, R., 1985. *Mar Mining,* **5,** 181–190
Baturin, G. N., 1982. *Phosphorites on the Sea Floor, Origin, Composition and Distribution.* Elsevier, Amsterdam, 343 pp.
Beauchamp, R. G., 1984. *Proc. offshore Technol. Conf.,* OTC 4778, 17–24.
Beauchamp, R. G. & Cruickshank, M. J., 1983, *Oceans '83,* 698–702.
Bender, F., 1982. *The non-living resources potential of the South Pacific.* Unpubl. lecture delivered at Regional Conference, Suva, Fiji, June 1982.
Bender, M. L., 1983. *J. geophys. Res.,* **88b,** 1049–1056.
Benjamin, R. & Hewett, R., 1982. *The changing industrial structure of the Pacific Basin.* Hubert H. Humphrey Institute of Public Affairs, Minnesota, 24 pp.
Bentor, Y., 1980. Editor. *Spec. Publ. Soc. Econ. Paleontol. Mineral.,* **29,** 249 pp.
Bersenev, I. I., Shkol'nik, E. L. & Gusev, V. V., 1985. *Dokl. Earth Sci. Sect.,* **271,** 71–74.
Bezrukov, P. L. & Skornyakova, N. S., 1976. *Mem. Am. Assoc. Petrol. Geol.,* **25,** 376–381.
Bischoff, J. L., 1980. *Science,* **207,** 1465–1469.
Bischoff, J. L. & Piper, D. Z., 1979. Editors. *Marine Geology and Oceanography of the Pacific Manganese Nodule Province.* Plenum Press, New York, 842 pp.
Bischoff, J. L., Rosenbauer, R. J., Aruscavage, P. J., Baedecker, P. A. & Crock, J. G., 1983a. *U.S. geol. Surv. Open-File Rep.,* No. 83-324, 32 pp.
Bischoff, J. L., Rosenbauer, R. J. Aruscavage, P. J., Baedecker, P. A. & Crock, J. G., 1983b. *Econ. Geol.,* **78,** 1711–1720.
Bléry-Oustrière, P., Beurrier, M. & Alsac, C., 1980. *Documents du BGRM,* No. 28, 213 pp.
Blezard, R. H., 1980. *N. Z. Jl mar. freshw. Res.,* **14,** 137–138.
Blissenbach, E., 1972. *Oceanol. Int.,* **72,** 412–416.
Blissenbach, E., 1977. In, *Technology Assessment and the Oceans,* edited by P. D. Wilmot & A. Slingerland, IPC Science and Technology Press, Guildford, Surrey, pp. 82–87.
Blissenbach, E. & Fellerer, R., 1973. *Geol, Rdsch.,* **62,** 812–840.
Boggs, S., 1975. *Acta oceanogr. Taiwanica,* No. 5, 18 pp.
Boin, U., 1981. *Inter Ocean '81,* IO 87-303, 82–90.
Bois, C., Bouche, P. & Pelet, R., 1982. *Bull. Am. Assoc. Petrol. Geol.,* **66,** 1248–1270.
Bonatti, E., 1975. *Ann. Rev. Earth Planet. Sci.,* **3,** 401–431.
Bonatti, E., 1978. *Scient. Am.,* **238**(2), 54–61.
Bonatti, E., 1981. In, *The Sea, Vol. 7,* edited by C. Emiliani, John Wiley & Sons, New York, pp. 639–686.
Bonatti, E., Kolla, V., Moore, W. S. & Stern, C., 1979. *Mar. Geol.,* **32,** 21–37.
Boström, K. & Peterson, M. N. A., 1966. *Econ. Geol.,* **61,** 1258–1265.
Bougault, H., 1982. *Ann. Mines,* **189**(11–12), 91–122.
Bougault, H., Renard, V. & Hekinian, R., 1984. *Proc. 2nd Int. Seminar Offshore Mineral Resources* (GERMINAL) 499–505.
Boulegue, J., Perseil, E. A., Bernat, M., Dupré, B., Stouff, P. & Francheteau, J., 1984. *Earth Planet. Sci. Lett.,* **70,** 249–259.
Bowman, K. C., 1972. *Prepr. 8th Ann. Congr. Exposition Mar. Technol. Soc.,* 237–253.
Brin, A., 1982. *Ann. Mines,* **189**(11–12), 83–90.
Broadus, J. M., 1984a. *Woods Hole Oceanogr. Inst. Tech. Rep.,* WHOI-84-21, 49–82.
Broadus, J. M., 1984b. *Proc. 2nd Int. Seminar Offshore Mineral Resources (GERMINAL),* 559–576.
Broadus, J. M. & Bowen, R. E., 1984. *Oceanus,* **27**(3), 26–31.

Broadus, J. M. & Hoagland, P., 1984. *San Diego Law Rev.*, **21**, 541–576.
Brocher, T. M., Taylor, B. & Kroenke, L. W., 1983. *Hawaii Inst. Geophys.* Unpubl. Rep., 52 pp.
Brockett, F. H. & Petters, R. A. 1980, *Mar. Technol.*, 80, 308–313.
Brooks, A. H. & others, 1913. *Bull. U.S. geol. Surv.*, No. 542, 308 pp.
Brown, G. A., 1971. *Underwater J.*, **3**(4), 166–176.
Brown, G. A., 1980a. *Marine Aggregates for the Sydney Market.* Paper presented at Institute of Quarrying N.S.W. Branch Bowral Symposium, 5 July 1980, 27 pp.
Brown, G. A., 1980b. *Quarry Mine Pit*, **19**(10), 10–12, 14, 22.
Brown, G. A., 1981. *Exploration and Exploitation of Mineral Sands off the East Coast of Australia.* Paper presented at SEATEC III Symposium, Singapore, March 1981, 29 pp.
Brown, G. A., & MacCulloch, I. R. F., 1970. *Prepr. Marine Technology 1970*, **2**, 983–1001.
Buckenham, M. H., Rogers, J. & Rouse, J. E., 1971. *Proc. Australas. Instn. Min. Metall.*, 1971, 1–13, (Pap. A.I.M.M. Conf. 8).
Burnett, W. C., 1977. *Bull. geol. Soc. Am.*, **88**, 813–823.
Burnett, W. C., 1980a. In, *Fertilizer Mineral Potential in Asia and the Pacific*, edited by R. P. Sheldon & W. C. Burnett, East-West Center, Honolulu, pp. 119–144.
Burnett, W. C., 1980b. *J. geol. Soc. Lond.*, **137**, 757–764.
Burnett, W. C., Beers, M. J. & Roe, K. K., 1982. *Science*, **215**, 1616–1618.
Burnett, W. C., Roe, K. K. & Piper, D. Z., 1983. In, *Coastal Upwelling, Part A*, edited by E. Suess & J. Thiede, Plenum Press, New York, pp. 377–397.
Burnett, W. C. & Sheldon, R. P., 1979. Report on the Marine Phosphatic Sediments Workshop February 9–11, 1979, Honolulu, Hawaii U.S.A. East-West Center, Honolulu, 65 pp.
Burnett, W. C., Veeh, H. H. & Soutar, A., 1980. *Spec. Publ. Soc. Econ. Paleontol. Mineral.*, **29**, 61–71.
Burns, V. M., 1979. *Min. Soc. Am. Short Course Notes*, **6**, 347–380.
Cameron, E. N., 1977. *Mar. Mining.*, **1**, 73–84.
Cameron, H., Ford, G., Garner, A., Gibbons, M. & Marjoram, T., 1980. *Manganese Nodule Mining.* PREST Marine Resources Project (University of Manchester), 33 pp.
Cameron, H., Georghiou, L., Perry, J. G. & Wiley, P., 1981. *Engng Costs Production Econ.*, **5**, 279–287.
Campbell, A. C. & Gieskes, J. M., 1984. *Earth Planet. Sci. Lett.*, **68**, 57–72.
Canadian American Seamount Expedition, 1985. *Nature, Lond.*, **313**, 212–214.
Cann, J. R., 1980. *J. geol. Soc. Lond.*, **137**, 381–384.
Cann, J. R. & Strens, M. R., 1982. *Nature, Lond.*, **298**, 147–149.
Carlos Ruiz, F., 1982. *Prof. Pap. U.S. geol. Surv.*, No. 1193, 63–67.
Carter, L., 1980. *N.Z. Jl Geol. Geophys.*, **23**, 455–468.
Cathcart, J. B. & Gulbrandsen, R. A., 1973. *Prof. Pap. U.S. geol. Surv.*, No. 820, 515–525.
Champ, M. A., 1984. Chairman. Exclusive Economic Zone Papers presented at Oceans '84 (Reprinted by NOAA Ocean Assessments Division, Rockville, Md), 149 pp.
Champ, M. A., Dillon, W. P. & Howell, D. G., 1984. *Oceanus*, **27**(4), 28–34.
Charles River Associates, 1982. *Charles River Associates Rep.*, No. 383, 4–1–4–21.
Charlier, R. H., 1978. *Ocean Yb.*, **1**, 160–210.
Charlier, R. H., 1983. *Ocean Yb.*, **4**, 75–120.
Chen, J.-C., 1984. *Proc. Pacific Congr. on Marine Technology*, PACON 84, MRM3/15–21.
Chen, J.-C. & Lin, F.-J., 1981. *Acta oceanogr. Taiwanica*, No. 12, 15 pp.
Chester, R. & Aston, S. R., 1976. In, *Chemical Oceanography, Vol. 6*, edited by J. P. Riley & R. Chester, Academic Press, London, 2nd edition pp. 281–390.

Choukroune, P., Francheteau, J. & Hekinian, R., 1984. *Earth Planet. Sci. Lett.*, **68**, 115–127.

Chuanzhu, Q., 1983. *Tropic Oceanol.*, **2**(4), 269–277 (in Chinese, English Abstr.).

Chyi, M. S., Crerar, D. A., Carlson, R. W. & Stallard, R. F., 1984. *Earth Planet. Sci. Lett.*, **71**, 31–45.

Clague, D., Bischoff, J. & Howell, D., 1984. Exclusive Economic Zone Papers presented at Oceans '84 (Reprinted by NOAA Ocean Assessments Division, Rockville, Md), pp. 79–84.

Clark, A., Johnson, C. & Chinn, P., 1984. *Nat. Resourc. Forum*, **8**(2), 163–174.

Clark, J. P., 1982. *Oceanus*, **25**(3), 18–21.

Clark, W. J. 1980. *CCOP/SOPAC Tech. Bull.*, No. 3, 3–4.

Clauss, G., 1973. *Schiff u. Hafen*, **2**, 119–126, 179.

Clauss, G., 1978. *Mar. Mining*, **1**, 189–208.

Cobb, E. H., 1973. *U.S. geol. Surv. Bull.*, No. 1374, 213 pp.

Coene, G. T., 1968. *Ocean Ind.*, **3**(11), 53–58.

Commeau, R. F., Clark, A., Johnson, C., Manheim, F. T., Aruscavage, P. J. & Lane, C. M., 1984. Exclusive Economic Zone Papers presented at Oceans '84 (Reprinted by NOAA Ocean Assessments Division, Rockville, Md), pp. 62–71.

Comptroller General, 1983. *Uncertainties Surround Future of U.S. Ocean Mining.* Report to the Congress of the United States. U.S. General Accounting Office GAO/NSIAID-83-41, 55 pp.

Converse, D. R., Holland, H. D. & Edmond, J. M., 1984. *Earth Planet. Sci. Lett.*, **69**, 159–175.

Cook, P. J. 1975a. *Proc. 3rd Session CCOP/SOPAC*, 75–85.

Cook, P. J. 1975b. In, *Resources of the Sea*, edited by M. K. Banks & T. G. Dix, Royal Society of Tasmania, Hobart, pp. 39–64.

Cook, P. J. 1976. In, *Handbook of Strata-Bound and Stratiform Ore Deposits, Vol. 7*, edited by K. H. Wolf, Elsevier, Amsterdam, pp. 505–535.

Cook, P. J. & Marshall, J. F., 1981. *Mar. Geol.*, **41**, 205–221.

Cook, P. J. & McElhinny, M. W., 1979. *Econ. Geol.*, **74**, 315–330.

Cook, P. J. & Nicholas, E., 1980. In, *Fertilizer and Mineral Potential in Asia and the Pacific*, edited by R. P. Sheldon & W. C. Burnett, East-West Center, Honolulu, pp. 49–79.

Cook, P. J. & Shergold, J. H., 1983. *Geotimes*, **28**(9), 14–15.

Coombs, D. S., Dowse, M., Grapes, R., Kawachi, Y. & Roser, B., 1985. *Chem. Geol.*, **48**, 57–78.

Corliss, J. B., Dymond, J., Gordon, L. I., Edmond, J. M., von Herzen, R. P., Ballard, R. D., Green, K., Williams, D. & Bainbridge, A., 1979. *Science*, **203**, 1073–1083.

Corliss, J. B., Lyle, M., Dymond, J. & Crane, K., 1978. *Earth Planet. Sci. Lett.*, **40**, 12–24.

Couper, A., 1983. Editor. *The Times Atlas of the Oceans.* Times Books Ltd, London, 272 pp.

Craig, J. D., Andrews, J. E. & Meylan, M. A., 1982. *Mar. Geol.* **45**, 127–157.

Craven, J. P., 1982. *The Management of Pacific Marine Resources, Present Problems and Future Trends.* Westview Press, Boulder, Colorado, 105 pp.

Crawford, A. M., Hollingshead, S. C. & Scott, S. D., 1984. *Mar. Mining*, **4**, 337–354.

Crawford, J. 1981. Editor. *Pacific Econonomic Cooperation: Suggestions for Actions.* Heinemann, Singapore, 246 pp.

Crerar, D. A., Namson, J., Chyi, M. S., Williams, L. & Feigenson, M. D., 1982. *Econ. Geol.*, **77**, 519–540.

Crockett, R. N., Chapman, G. R. & Jones, A. L. P., 1984. *British Geological Survey: Mineral Intelligence, Statistics and Economics*, Rep. No. 2/52 (on open file).

Cronan, D. S., 1972. In, *Ferromanganese Deposits on the Ocean Floor,* edited by D. R. Horn, National Science Foundation, Washington, D.C., pp. 19–30.
Cronan, D. S., 1976. *Nature, Lond.,* **262,** 567–569.
Cronan, D. S., 1977. In, *Marine Manganese Deposits,* edited by G. P. Glasby, Elsevier, Amsterdam, pp. 11–44.
Cronan, D. S., 1980a. *Underwater Minerals.* Academic Press, London, 362 pp.
Cronan, D. S., 1980b. *J. geol. Soc. Lond.,* **137,** 369–371.
Cronan, D. S., 1982. In, *Marine Mineral Deposits—New Research Results and Economic Prospects,* edited by P. Halbach & P. Winter, Proceedings of the Clausthaler Workshop, Verlag Glückauf GmbH, Essen, pp. 86–101.
Cronan, D. S., 1983. *CCOP/SOPAC Tech. Bull.,* No. 4, 55 pp.
Cronan, D. S., 1984. *Sth Pacif. mar. Geol. Notes,* **3**(1), 1–17.
Cronan, D., 1985. *New Sci.,* **106** (1459), 34–38.
Cronan, D. S., Glasby, G. P., Moorby, S. A., Thomson, J., Knedler, K. E. & McDougall, J. C., 1982. *Nature, Lond.,* **298,** 456–458.
Cronan, D. S. & Varnavas, S. P., 1981. *Oceanol. Acta,* SP, 47–58.
Cruickshank, M. J., 1971. *J. of the West,* **10,** 23–34.
Cruickshank, M. J., 1973. In, *Undersea Technology Handbook Directory, 1973.* Compass Publications, Arlington, Virginia, A15–A28.
Cruickshank, M. J., 1974. In, *The Geology of Continental Margins,* edited by C. A. Park and C. L. Drake, Springer-Verlag, New York, pp. 965–1000.
Cruickshank, M. J., 1979. In, *Proceedings of the Workshop on Coastal Area Development and Management in Asia and the Pacific,* edited by M. J. Valencia, Manila, Philippines 3 to 12 December 1979, pp. 27–31.
Cruickshank, M. J., 1982a. *Ocean Ind.,* **17**(3), 28 only.
Cruickshank, M. J., 1982b. *Ocean Ind.,* **17**(1), 81, 84, 86, 88, 90.
Cruickshank, M. J., 1984. *Ocean Ind.* **19**(7), 99–101.
Cruickshank, M. J. & Hess, H. D., 1975. *Oceanus,* **19**(1), 32–44.
Cruickshank, M. J. & Olson, F. L, 1984. *Sea Technol.* **25**(8), 10–12, 16.
Cruickshank, M. J. & Rowland, T. J., 1983. *Spec. Publ. Soc. Econ. Paleontol. Mineral., No. 33, 429*–436.
Cruickshank, M. J. & Zippin, J. P., 1984. *Proc. Offshore Technol. Conf.,* OTC 4779, 25–29.
Cullen, D. J., 1975. *NZOI Oceanogr. Summ.,* No. 8, 6 pp.
Cullen, D. J., 1978. *Mar. Geol.,* **28,** M67-M76.
Cullen, D. J., 1979. *N.Z. agric. Sci.,* **13,** 85–91.
Cullen, D. J. 1980. *Spec. Publ. Soc. Econ. Paleontol. Mineral.,* **29,** 139–148.
Cullen, D. J., 1984. *Proc. 2nd Int. Seminar Offshore Mineral Resources (GERMINAL),* 287–299.
Curlin, J. W., 1984. *Exclusive Economic Zone Papers presented at Oceans '84* (Reprinted by NOAA Ocean Assessments Division, Rockville, Md), pp. 47–49.
Curtis, C., 1982. *Oceanus,* **25**(3), 31–36.
Dames, & Moore, 1980. *U.S. Natn Tech. Inf. Serv. Rept,* PB81–180119.
D'Anglejan, B. F., 1967. *Mar. Geol.,* **5,** 15–44.
De Carlo, E. H., McMurtry, G. M. & Yeh, H. -W., 1983. *Earth Planet. Sci. Lett.,* **66,** 438–449.
Derkmann, K. J., Fellerer, R. & Richter, H., 1981. *Ocean Mgmt,* **7,** 1–8.
Dietz, R. S., Emery, K. O. & Shepard, F. P., 1942. *Bull. geol. Soc. Am.,* **53,** 815–847.
Dodet, C., Noville, F., Crine, M., Marchot, P. & Pirard, J. P., 1984. *Colloids & Surfaces,* **11,** 187–197.
Dollar, S. J., 1979. University of Hawaii Sea Grant College Program TP–79–01, 106 pp.
Donges, J. B., 1985. Editor. *The Economics of Deep-Sea Mining.* Springer-Verlag, Berlin, 378 pp.

Dorr, J. V. N., Crittenden, M. D. & Worl, R. G., 1973. *Prof. Pap. U.S. geol. Surv.*, No. 820, 385–399.
Doutch, H. F., 1981. Plate Tectonic Map of the Circum-Pacific Region Southwest Quadrant. Circum-Pacific Council for Energy and Mineral Resources (Chart).
Drechsler, H. D., 1973. *Marit. Stud. Mgmt,* **1,** 53–66.
Driessen, A., 1984. In, *Australian Mineral Industry Annual Review of 1982,* Australian Government Publishing Service, Canberra, pp. 213–216.
Duane, D. B., 1976. In, *Marine Sediment Transport and Environmental Management,* edited by D. J. Stanley and D. J. P. Swift, John Wiley & Sons, New York, pp. 535–556.
Duane, D. B., 1982. *Mar. Technol. Sci. J.,* **16**(3), 87–91.
Dunham, K. C., 1969. *Q. Jl geol. Soc. Lond.,* **124,** 101–129.
Eade, J. V., 1978. *Proc. Seventh Session CCOP/SOPAC,* 51–53.
Earney, F.C.F., 1980. *Petroleum and Hard Minerals from the Sea.* Edward Arnold, London, 291 pp.
East Pacific Rise Study Group., 1981. *Science,* **213,** 31–40.
Edgar, N. T., 1983. *U.S. geol, Surv. Circ.,* No. 906, 23 pp.
Edmond, J. M., 1982. *Oceanus,* **25**(2), 22–27.
Edmond, J. M., Measures, C., McDuff, R. E., Chan, L. H., Collier, R., Grant, B., Gordon, L. I. & Corliss, J. B., 1979a. *Earth Planet. Sci. Lett.,* **46,** 1–18.
Edmond, J. M., Measures, C., Mangum, B., Grant, B., Sclater, F. R., Collier, R., Hudson, A., Gordon, L. I. & Corliss, J. B., 1979b. *Earth Planet. Sci. Lett.,* **46,** 19–30.
Edmond, J. M., Von Damm, K. L., McDuff, R. E. & Measures, C. I., 1982. *Nature, Lond.,* **297,** 187–191.
EEZ-SCAN '84 Scientific Team, 1985. *Geotimes,* **30**(1), 13–15.
Eiseman, F. B., 1982. *Oceans,* **15**(6), 36–37.
Ellis, W., 1969. *Polynesian Researchers Hawaii,* 1842 edition. Re-issued by Charles E. Tuttle, Rutland, Vermont, 471 pp.
Emery, K. O., 1960. *The Sea off Southern California.* John Wiley & Sons, Inc., New York, 366 pp.
Emery, K. O., 1980. *Bull. Am. Assoc. Petrol. Geol.,* **64,** 297–315.
Emery, K. O. & Noakes, L. C., 1968. *CCOP Tech. Bull.,* No. 1, 95–111.
Emery, K. O. & Skinner, B. J., 1977. *Mar. Mining,* **1,** 1–71.
Ensign, C. O., 1966. In, *Trans. 2nd Ann. Mar. Technol. Soc.* Suppl., pp. 65–82.
Ericksen, G. E., 1976. *Mem. Am. Assoc. petrol. Geol.,* **25,** 527–538.
Estrup, C., 1977. In, *Technology Assessment and the Oceans,* edited by P. D. Wilmot & A. Slingerland, IPC Science & Technology Press, Guildford, Surrey, pp. 88–92.
Evans, J. R., Dabai, G. S. & Levine, C. R., 1982. *Calif. Geol.,* **35**(12), 259–276.
Ewing, M., Houtz, R. & Ewing, J., 1969. *J. geophys. Res.,* **74,** 2477–2493.
Exon, N. F., 1983. *Mar. Mining,* **4,** 79–107.
Exon, N. F. & Cronan, D. S., 1983. *Mar. Geol.,* **52,** M43–M52.
Fellerer, R., 1979a. *Documents du BRGM,* No. 7, 427–436.
Fellerer, R., 1979b. *Documents du BRGM,* No. 7, 450–451.
Fellerer, R., 1980. *Geol. Jb.,* **D38,** 35–76.
Field, C. W., Wetherell, D. G. & Dasch, E. J., 1981. *Mem. Geol. Soc. Am.,* **154,** 315–320.
Finlow-Bates, T., 1980. *Geol. Jb.,* **D40,** 131–168.
Fisk, M., 1982. *Mineral resources in the South Pacific: Polymetallic Nodules, Crusts and Sulphides.* Paper presented at the Law of the Sea Seminar sponsored by the Pacific Islands Association, The Harley Hotel, New York, 15–16th November, 1982.
Flipse, J. E., 1981. *Mar. Mining,* **2,** 311–314.
Flipse, J. E., 1983. In, *The Law of the Sea and Ocean Development Issues in the*

Pacific Basin, edited by E. L. Miles & S. Allen, Proceedings Law of the Sea Institute Fifteenth Annual Conference 5–8th October, 1981, Honolulu, Hawaii, pp. 322–381.

Ford, G. & Spagni, D., 1983. *New Sci.,* **100,** 17 only.

Franchetau, J. & Ballard, R. D., 1983. *Earth Planet. Sci. Lett.,* **64,** 93–116.

Francheteau, J., Needham, H. D., Choukroune, P., Juteau, T., Séguret, M., Ballard, R. D., Fox, P. J., Normark, W., Carranza, A., Cordoba, D., Guerrero, J., Rangin, C., Bougault, H., Cambon, P. & Hekinian, R., 1979. *Nature, Lond.,* **277,** 523–528.

Frazer, J. Z., 1977. *Mar. Mining,* **1,** 103–123.

Frazer, J. Z., 1980. In, *Deepsea Mining,* edited by J. T. Kildow, MIT Press, Cambridge, Mass., pp. 41–83.

Frazer, J. Z. & Fisk, M. B., 1981. *Deep-Sea Res.,* **29A,** 1533–1551.

Friedrich, G., Glasby, G. P., Plüger, W. L. & Thijssen, T., 1981. *Inter Ocean '81,* IO 81–302, 72–87.

Fukami, H., 1979. In, *The Deep Seabed and its Mineral Resources,* Proc. 3rd Int. Ocean Symp. 1978 Tokyo, pp. 110–111.

Fyfe, W. S. & Lonsdale, P., 1981. In, *The Sea, Vol, 7,* edited by C. Emiliani, John Wiley & Sons, New York, pp. 589–638.

Galerne, E., 1983. *Sea Technol.,* **24**(11), 40, 42.

Galtier, L., 1984. *Proc. 2nd Int. Seminar Offshore Mineral Resources (GERMINAL),* 119–142.

Gardner, D. E., 1955. *Bull. Aust. Bur. Mineral Resources,* No. 28, 103 pp.

Garson, M. S. & Mitchell, A. H. G., 1977. *Spec. Publ. geol. Soc. Lond.,* **7,** 81–97.

Gauss, G. A., Eade, J. & Lewis, K., 1983. *Sth Pacif. Mar. Geol. Notes,* **2**(10), 155–184.

Georghiou, L., Ford, G., Gibbons, M. & Jones, G., 1983. *Mar. Policy,* **7,** 239–253.

Gerwick, B. C., 1983. *Oceans '83,* 657–660.

Glasby, G. P., 1971. *N.Z. Jl mar. freshw. Res.,* **5,** 483–496.

Glasby, G. P., 1976. *N.Z. Jl Geol. Geophys.,* **19,** 707–736.

Glasby, G. P., 1979. *Endeavour,* N.S., **3,** 82–85.

Glasby, G. P., 1982a. *Mar. Mining,* **3,** *379*–409.

Glasby, G. P., 1982b. *Mar. Mining,* **3,** 231–270.

Glasby, G. P., 1983. *Mar. Mining,* **4,** 73–77.

Glasby, G. P., 1984. *Mar. Mining,* **4,** 355–402.

Glasby, G. P., 1986. In, *Sedimentation and Mineral Deposits in the Southwestern Pacific Ocean,* edited by D. S. Cronan, Academic Press, London pp. 149–181.

Glasby, G. P., Exon, N. F. & Meylan, M. A., 1986. In, *Sedimentation and Mineral Deposits in the Southwestern Pacific Ocean,* edited by D. S. Cronan, Academic Press, London, pp. 237–262.

Glasby, G. P., Meylan, M. A., Margolis, S. V. & Bäcker, H., 1980. In, *Geology and Geochemistry of Manganese, Vol. 3,* edited by I. M. Varentsov & Gy. Grasselly, Hungarian Academy of Sciences, Budapest, pp. 137–183.

Glasby, G. P., Stoffers, P., Sioulas, A., Thijssen, T. & Friedrich, G., 1982. *Geo-Mar.. Letts,* **2,** 47–53.

Glasby, G. P. & Thijssen, T., 1982. *Neues Jb. Mineral.,* **145,** 291–307.

Gocht, W. & Wolf, A., 1982. Editors. *Report of the Training Programme for the Management and Conservation of Marine Resources. (Class A: Ocean Mining).* Research Institute for International Technical and Economic Cooperation of the Aachen Technical University (RWTH), 223 pp.

Goeller, H. E. & Zucker, H., 1984. *Science,* **223,** 456–462.

Goldfarb, M. S., Converse, D. R., Holland, H. D. & Edmond, J. M., 1983. *Econ. Geol. Monogr.,* **5,** 184–197.

Goldie, R. & Bottrill, T. J., 1981. *Geosci. Canada,* **8**(3), 93–104.

Gomez, E. D., 1983. *Ocean Mgmt,* **8,** 281–295.

Goodfellow, W., 1985. *GEOS,* **14**(1), 17–20.
Gopalakrishnan, C., 1984. Editor. *The Emerging Marine Economy of the Pacific.* Butterworth Publishers, Boston, 246 pp.
Gray, J. J. & Kulm, L. D., 1985. State of Oregon Department of Geology and Mineral Industry Chart GMS–37.
Green, D. C., Hulston, J. R. & Crick, I. H., 1978. *BMR Jl Aust. Geol. Geophys.,* **3,** 233–239.
Greenslate, J. L., Frazer, J. Z. & Arrhenius, G., 1973. In, *The Origin and Distribution of Manganese Nodules in the Pacific and Prospects for Exploration,* edited by M. Morgenstein, Honolulu, Hawaii, pp. 45–69.
Greenwald, R. J. & Nordquist, M. H., 1979. In, *Deep Ocean Mining,* edited by J. E. Flipse, The American Society of Mechanical Engineers, New York, pp. 23–31.
Grigg, R. W., 1971. University of Hawaii Sea Grant Program Rep., UNIHI–SEAGRANT–AR–71–022, 12 pp.
Grigg, R. W., 1974. *Proc. 2nd Int. Coral Reef Symp.,* **2,** 235–240.
Grigg, R. W., 1975. *Sea Grant Technical Report,* UNIHI–SEAGRANT–AR–75–03, 14 pp.
Grigg, R. W., 1976. *Sea Grant Technical Report,* UNIHI–SEAGRANT–TR–77–03, 48 pp.
Grigg, R. W., 1977a. *Hawaii's Precious Corals.* Island Heritage, Norfolk Is., Australia, 64 pp.
Grigg, R. W., 1977b. *Proc. 3rd Int. Coral Reef Symp.,* **2,** 609–616.
Grigg, R. W., 1979. *Nat. Geogr.,* **155**(5), 718–732.
Grigg, R. W., 1981. *Status of the precious coral industry in the Pacific: 1981.* Submitted to the Western Pacfic Regional Fishery Management Council in fulfilment of Contract No. WPC-00181, 15 pp.
Grigg, R. W., 1983. *Status of the Precious Coral Industry in the Pacific: 1982.* Prepared for "Consulta Technica sobre los vecursos del Coral rojo del Mediterraneo occidental y su explotacion racional", Palma de Mallorca, 13–16th December, 1983.
Grigg, R. W., 1984a. In, *The Emerging Marine Economy of the Pacific,* edited by C. Gopalakrishnan, Butterworth Publishers, Boston, pp. 77–85.
Grigg, R. W. 1984b. *Mar. Ecol. P.S.Z.N.I.,* **5**(1), 57–74.
Grigg, R. W. & Eade, J. V., 1981. In, *Report on the Inshore and Nearshore Resources Training Workshop Suva, Fiji, 13–17 July 1981,* CCOP/SOPAC Report, 13–18.
Grigg, R. W. & Pfund, R., 1980. UNIHI–SEAGRANT–MR–O8–04, 333 pp.
Griggs, A. B., 1945. *U.S. Geol. Surv. Bull.,* No. 945E, 113–150.
Grill, E. V., Chase, R. L., Macdonald, R. D. & Murray, J. W., 1981. *Earth Planet. Sci. Lett.,* **52,** 142–150.
Gundlach, H., Marchig. V. & Bäcker, H., 1983. *Erzmetall,* **36,** 495–500.
Hails, J. R., 1976. In, *Handbook of Strata-Bound and Stratiform Ore Deposits, Vol. 3,* edited by K. H. Wolf, Elsevier, Amsterdam, pp. 213–244.
Halbach, P., 1982. In, *Marine Mineral Deposits—New Research Results and Economic Prospects,* edited by P. Halbach & P. Winter, Proceedings of the Clausthaler Workshop, Verlag Glückauf GmbH, Essen, pp 60–85.
Halbach, P., 1984. *Ocean Mgmt,* **9,** 35–60.
Halbach, P. & Fellerer, R., 1980. *Geojournal,* 4.5, 407–422.
Halbach, P. & Manheim, F. T., 1984. *Mar. Mining,* **4,** 319–336.
Halbach, P., Manheim, F. T. & Otten, P., 1982. *Erzmetall,* **35,** 447–453.
Halbach, P. & Puteanus, D., 1984. *Earth Planet. Sci. Lett.,* **68,** 73–87.
Halbach, P., Puteanus, D. & Manheim, F. T., 1984. *Naturwissenschaften,* **71,** 1–6.
Halbach, P., Segl, M., Puteanus, D. & Mangini, A., 1983. *Nature, Lond.,* **304,** 716–719.
Halkyard, J. E., 1980. *Mar. Technol.* **80,** 303–307.

Hammond, A. L., 1975a. *Science,* **189,** 779–781.
Hammond, A. L., 1975b. *Science,* **189,** 868, 869, 915, 917.
Handa, K., Miyashita, Y., Oba, S. & Yamakado, N., 1982. In, *Proc. Offshore Techol. Conf., OTC* 4262, pp. 457–459.
Hannington, M. D., 1986. M.S. thesis, University of Toronto, Canada, 473 pp.
Hatem, M. B., 1983. Editor. *Marine Polymetallic Sulfides, A National Overview and Future Needs Workshop Proceedings January 19–20, 1983.* Maryland Sea Grant Publication Number UM-SG-TS-83-04, 160 pp.
Haymon, R. M., 1983. *Nature, Lond.,* **301,** 695–698.
Haymon, R. M. & Kastner, M., 1981. *Earth Planet. Sci. Lett.,* **53,** 363–38.
Haymon, R. M., Koski, R. A. & Sinclair, C., 1984. *Science,* **233,** 1407–1409.
Haynes, B. W., Barron, D. C., Kramer, G. W., Maeda, R. & Magyar, M. J., 1985. *U.S. Bureau of Mines Report of Investigations,* RI 8938, 15 pp.
Heath, G. R., 1981. *Econ. Geol.,* 75th Anniv. Vol., 736–765.
Heath, G. R. & Dymond, J., 1977. *Bull. geol. Soc. Am.,* **88,** 723–733.
Hedberg, H. D., Moody, J. D. & Hedberg, R. M., 1979. *Bull. Am. Assoc. Petrol. Geol.,* **63,** 286–300.
Hein, J. R., Manheim, F. T., Schwab, W. C. & Davis, A. S., 1985. *Mar. Geol.* **69,** 25–54.
Hein, J. R., Manheim, F. T., Schwab, W. C., Davis, A. S., Daniel, C. L., Bouse, R. M., Morgenson, L. A., Sliney, R. E., Clague, D., Tate, G. B., & Cacchione, D. A., 1985. *U.S. geol. Surv. Open File Rep.,* 85–292, 129 pp.
Hein, P., 1977. *Centre National Pour l'Exploitation des Oceans,* Rapp. No. 35, 74 pp.
Hekinian, R., Fevrier, M., Avedik, F., Cambon, P., Charlou, J. L., Needham, H. D., Raillard, J., Merlivat, L., Moinet, A., Manganini, S. & Lange, J., 1983a. *Science,* **219,** 1321–1324.
Hekinian, R., Fevrier, M., Bischoff, J. L., Picot, P. & Shanks, W. C., 1980. *Science,* **207,** 1433–1444.
Hekinian, R. & Fouquet, Y., 1985. *Econ. Geol.,* **84,** 221–249.
Hekinian, R., Francheteau, J. & Ballard, R. D., 1985. *Oceanol. Acta,* **8,** 147–155.
Hekinian, R., Francheteau, J., Renard, V., Ballard, R. D., Choukroune, P., Cheminee, J. L., Albarede, F., Minster, J. F., Charlou, J. L., Marty, J. C. & Boulegue, J., 1983b. *Mar. Geophys. Res.,* **6,** 1–14.
Henrie, T. A., 1976. *Mem. Am. Assoc. Petrol. Geol.,* **25,** 18–23.
Herbich, J. B., 1975. *Coastal and Deep Ocean Dredging.* Gulf Publishing Co., Houston, Texas, 622 pp.
Herbich, J. B., 1985. *Dock & Harbour Authority,* **66**(771), 55–57.
Hill, R. M., 1977. In, *Technology Assessment and the Oceans,* edited by P. D. Wilmot & A. Slingerland, IPC Science & Technology Press, Guildford, Surrey, pp 93–101.
Hillman, C. T., 1983. *U.S. Bur. Mines Inform. Circ.,* IC 8933, 60 pp.
Holden, C., 1981. *Science,* **213,** 305–307.
Holdgate, M. W., Kassas, M. & White, G. F., 1982. Editors. *The World Environment 1972–1982.* A Report by The United Nations Environment Programme. Tycooly International Publishing Ltd, Dublin, 637 pp.
Holser, A. F., Rowland, R. W. & Goud, M. R., 1981. *U.S. geol. Surv. Misc. Field Map Studies,* MF-1360.
Hoover, R. B., 1979. *Nat. Geogr.,* **155**(6), 870–878.
Horn, D. R., 1972. Editor. *Ferromanganese Deposits on the Ocean Floor,* National Science Foundation, Washington, D.C., 293 pp.
Horn, D. R., Delach, M. N. & Horn, B. M., 1973. *Int. Decade Ocean Explor. Tech. Rep.,* No. 3, 51 pp.
Horn, D. R., Horn, B. M. & Delach, M. N., 1973. *Int. Decade Ocean Explor. Tech. Rep.,* No. 8, 20 pp. + 20 profiles.
Hosking, K. F. G., 1971. *CCOP Tech. Bull.,* **5,** 112–129.

Hubred, G. L., 1980. *Mar. Mining,* **2,** 191–212.
Humphrey, P. B., 1982. Editor. *Marine Mining: A New beginning*. State of Hawaii Department of Planning and Economic Development, Honolulu, 319 pp.
Hutchison, C. S. & Taylor, D., 1978. *J. geol. Soc. Lond.,* **135,** 407–428.
Ichiye, T. & Carnes, M., 1981. In, *Marine Environmental Pollution, 2 Dumping and Mining,* edited by R. A. Geyer, Elsevier, Amsterdam, pp. 475–517.
Igrevskiy, V. I., Budnikov, N. P. & Levchenko, V. A., 1972. *Int. Geol. Rev.,* **15**(10), 1186–1196.
Inderbitzen, A. L., Carsola, A. J. & Everhart, D. L., 1970. *Prepr. Offshore Technol. Conf.,* **2,** 287–304.
Intergovernmental Oceanographic Commission, 1983. *Intergovernmental Oceanographic Commission Workshop Rep.,* No. 35, 6 pp. + annexes.
International Crustal Research Drilling Group, 1984. *Geotimes,* **29**(5), 12–14.
Ishiara, S., 1978. *J. geol. Soc. Lond.,* **135,** 389–406.
Ives, D., 1984. *Petrol. Gaz.,* **24**(4), 4–9.
Janecky, D. R. & Seyfried, W. E., 1984. *Geochim. cosmochim. Acta,* **48,** 2723–2738.
Jenkins, R. W., Jugel, M. K., Keith, K. M. & Meylan, M. A., 1981. *The Feasibility and Potential Impact of Manganese Nodule Processing in the Puna and Kohola Districts of Hawaii.* State of Hawaii Department of Planning and Economic Development, Honolulu, 271 pp.
Johannas, A., 1982. *Prof. Pap. U.S. geol. Surv.,* No. 1193, 192–200.
Johnson, C. J. & Clark, A. L., 1985. *Nat. Resour. Forum,* **9**(3), 179–186.
Johnson, C. J., Clark, A. L. & Otto, J. M., in press. *J. Bus. Admin.* (*Univ. of British Columbia*).
Johnson, C. J., Clark, A. L., Otto, J. M., Pak, D. K., Johnson, K. T. M. & Morgan, C. L., 1985. *Minerals Policy Program, Resource Systems Institute, East-West Center, Honolulu, Hawaii,* 135 pp. + 6 appendices.
Johnston, J. L., 1979. In, *Deep Ocean Mining,* edited by J. E. Flipse, The American Society of Mechanical Engineers, New York, pp. 47–60.
Jones, H. A. & Davies, P. J., 1979. *Mar. Geol.,* **30,** 243–268.
Jordan, T. H., Menard, H. W. & Smith, D. K., 1983. *J. geophys. Res.,* **88B,** 10508–10518.
Karns, A. W., 1976. *Mem. Am. Ass. petrol. Geol.,* **25,** 395–398.
Katili, J. A., 1973. *CCOP Tech. Bull.,* No. 7, 23–37.
Kaufman, R., 1980. *Oceans '80,* 49–54.
Kaufman, R., Latimer, J. P. & Tolefson, D.C., 1985. *Proc. Offshore Technol. Conf.,* OTC 4901, 33–51.
Kazmin, Yu. B. 1984. Editor. *Trans. All-Union Research Institute for Geology and Mineral Resources of the World Ocean,* No. 192, 175 pp. (in Russian).
Kennett, J. P., 1982. *Marine Geology.* Prentice-Hall, Englewood Cliffs, N.J., 1813 pp.
Kent, P., 1980. *Minerals from the Marine Environment.* Edward Arnold, London, 88 pp.
Kerridge, J. F., Haymon, R. M. & Kastner, M., 1983. *Earth Planet. Sci. Lett.,* **6,** 91–100.
Kildow, J. T., Dar, V. K., Bever, M. B. & Capstaff, A. E., 1976. MIT Report prepared for U.S. Ocean Mining Administration and Bureau of Mines. U.S. Bureau of Mines open-File Rep., 89–78, 485 pp.
Kim, K. H. & Burnett, W. C., 1985. *Geochim. Cosmochim. Acta,* **49,** 1073–1081.
Kim, W. J., 1970. *CCOP Tech. Bull.,* No. 3, 127–136.
Kingston, M. J., Delaney, J. R. & Johnson, H. P., 1983. *Oceans '83,* 811–815.
Knecht, R. W. & Bowen, R. E., 1982. *Mar. Technol. Soc. J.,* **16**(4), 31–40.
Kogan, B. G., Naprasnikova, L. A. & Ryabtseva, G. I., 1974. *Int. geol. Rev.,* **17**(8), 945–950.

Kollwentz, W. M., 1979. In, *The Deep Seabed and its Mineral Resources,* Proc. 3rd Int. Ocean Symp. 1978 Tokyo, 48 only.
Kolodny, Y., 1981. In, *The Sea, Vol. 7,* edited by C. Emilani, John Wiley & Sons, New York, pp. 981–1023.
Komar, P. D. & Wang, C., 1984. *J. Geol.,* **92,** 637–655.
Koo, J., 1980. *Acta oceanogr. Taiwanica,* No. 11, 19–31.
Koski, R. A., Clague, D. A. & Oudin, E., 1984. *Bull. geol. Soc. Am.,* **95,** 930–945.
Koski, R. A., Lonsdale, P. F., Shanks, W. C., Berndt, M. E. & Howe, S. S., 1985. *J. geophys. Res.,* **90,** 6695–6707.
Koski, R. A., Normark, W. R. & Morton, J. L., 1985. *Mar. Mining,* **5,** 147–164.
Koski, R. A., Normark, W. R., Morton, J. L. & Delaney, J. R., 1982. *Oceanus,* **25**(3), 42–48.
Kostick, D. S., 1980. *U.S. Bureau of Mines Bull.,* No. 671, 769–780.
Kostick, D. S., 1981. In, *Minerals Yearbook 1980, Vol. 1, Metals and Minerals,* U.S. Bureau of Mines, Washington, D.C., pp. 683–694.
Kress, A. G. & Veeh, H. H., 1980. *Mar. Geol.,* **36,** 143–157.
Kudrass, H.-R., 1980. *Geol. Jb.,* **D38,** 128–136.
Kudrass, H.-R., 1982. In, *Marine Mineral Deposits—New Research Results and Economic Prospects,* edited by P. Halbach & P. Winter, Proceedings of the Clausthaler Workshop, Verlag Glückauf GmbH, Essen, pp. 45–59.
Kudrass, H.-R. & Cullen, D. J., 1982. *Geol. Jb.,* **D51,** 3–41.
Kulm, L. D., Dymond, J. Dasch, E. J. & Hussong, D. M., 1981. Editors. *Mem. geol. Soc. Am.,* **154,** 824 pp.
Kunzendorf, H., Walter, P., Stoffers, P. & Gwozdz, R., 1984. *Chem. Geol.,* **47,** 113–133.
Lafitte, M., Maury, R. & Perseil, E. A., 1983. *Schweiz. mineral. petrogr. Mitt.,* **63,** 203–214.
Lafitte, M., Maury, R. & Perseil, E. A., 1984. *Mineral. Deposita,* **19,** 274–282.
Lafitte, M., Maury, R., Perseil, E. A. & Boulegue, J., 1985. *Earth Planet. Sci. Lett.,* **73,** 53–64.
Lalou, C., Brichet, E., Jehanno, C. & Perez-Leclaire, H., 1983. *Earth Planet. Sci. Lett.,* **63,** 63–75.
Lalou, C., Brichet, E. & Hekinian, R., 1985. *Earth Planet. Sci. Lett.,* **75,** 59–71.
Lane, A. L., 1979. In, *Deep Ocean Mining,* edited by J. E. Flipse, The American Society of Mechanical Engineers, New York, pp. 33–45.
Langevad, E. J., 1983. *Nat. Resour. Forum,* **7,** 227–238.
LaQue, F. L., 1980. *Mar. Technol. Soc. J.,* **14**(1), 16–19.
Lawver, J. E., McClintok, W. O. & Snow, R. E., 1978. *Miner. Sci. Engng,* **10,** 278–294.
Lee, A. I. N., 1980. Editor. *Fertilizer Mineral Occurrences in the Asia-Pacific-Region.* East-West Center, Honolulu, 156 pp.
Lefond, S. J., 1969. *Handbook of World Salt Resources.* Plenum Press, New York, 384 pp.
Le Gouellec, P., Herrouin, G. & Charles, G., 1984. *Proc. 2nd Internat. Seminar Offshore Mineral Resources (GERMINAL),* 587–594.
Leinen, M. & Anderson, R. N., 1981. *EOS Trans. Am. geophys. Un.,* **62**(45), 913 (Abstr.).
Lenoble, J.-P., 1979. *Documents BRGM,* No. 7, 403–425.
Lenoble, J.-P., 1981a. *Trans. Instn Min. Metall.,* **90A,** 6–12.
Lenoble, J.-P., 1981b. *Ocean Mgmt,* **7,** 9–24.
Le Pichon, X., 1968. *J. geophys. Res.,* **73,** 3661–3697.
Lévy, J.-P., 1979a. In, *The Deep Seabed and its Mineral Resources,* Proc. 3rd Internat. Ocean Symp. 1978 Tokyo, pp. 96–105.
Lévy, J.-P., 1979b. *Ocean Mgmt,* **5,** 49–78.
Lewis, K. B., 1979. *Mar. Geol.,* **31,** 31–43.

Linebaugh, R. M., 1980. *Mar. Technol. Soc. J.,* **14**(1), 20–24.
Lisitsyn, A. P., 1969. *Recent Sedimentation in the Bering Sea.* Israel Program for Scientific Translations, Jerusalem, 614 pp.
Löfgren, C. & Boström, K., 1982. In, *The Dynamic Environment of the Ocean Floor,* edited by K. A. Fanning & F. T. Manheim, D. C. Heath and Co., Lexington, Mass., pp. 369–380.
Lonsdale, P., 1977. *Earth Planet, Sci. Lett.,* **36,** 92–110.
Lonsdale, P., 1984. *Oceanus,* **27**(3), 21–24.
Lonsdale, P. & Batiza, R., 1980. *Bull. geol. Soc. Am,* **91,** 545–554.
Lonsdale, P., Batiza, R. & Simkin, T., 1982. *Mar. Technol. Soc. J.,* **16**(3), 54–61.
Lonsdale, P. & Becker, K., 1985. *Earth Planet. Sci. Lett.,* **73,** 211–225.
Lonsdale, P., Bischoff, J. L., Burns, V. M., Kastner, M. & Sweeney, R. E., 1980. *Earth Planet. Sci. Lett.,* **49,** 8–20.
Lonsdale, P., Burns, V. M. & Fisk, M., 1980. *J. Geol.,* **88,** 611–618.
Lucas, R., 1982. In, *Marine Mineral Deposits—New Research Results and Economic Prospects,* edited by P. Halbach & P. Winter, Proceedings of the Clausthaler Workshop, Verlag, Glückauf GmbH, Essen, pp. 202–242.
Lupton, J. E., Delaney, J. R., Johnson, H. P. & Tivey, M. K., 1985. *Nature, Lond.,* **316,** 621–623.
Lupton, J. E., Klinkhammer, G. P., Normark, W. R., Haymon, R., Macdonald, K. C., Weiss, R. F., & Craig, H., 1980. *Earth Planet. Sci. Lett.,* **50,** 115–127.
Macdonald, E. H., 1971. *CCOP. Tech. Bull.,* No. **5,** 13–111.
Macdonald, K. C., 1982. *Ann. Rev. Earth Planet. Sci.,* **10,** 155–190.
Macdonald, K. C., Becker, K., Spiess, F. N. & Ballard, R. D., 1980. *Earth Planet. Sci. Lett.,* **48,** 1–7.
Macdonald, K. C. & Luyendyk, B. P., 1981. *Scient. Am.,* **244**(5), 86–100.
Mackay, A. D., Gregg, P. E. H. & Syers, J. K., 1984. *N.Z. Jl agric. Res.,* **27,** 65–82.
Mackay, A. D., Syers, J. K. & Gregg, P. E. H., 1984. *N.Z. Jl expl Agric.,* **12,** 131–140.
Magnuson, W. G., 1974. Chairman. *The Economic Value of Ocean Resources to the United States.* U.S. 93rd Congress 2d Session, U.S. Senate Committee on Commerce, U.S. Government Printing Office, Washington, D.C., 109 pp.
Malahoff, A., 1981. *Proc. 13th Ann. Offshore Technol. Conf.,* No. 4129, 115–121.
Malahoff, A., 1982a. *Mar. Technol. Soc. J.,* **16**(3), 39–45.
Malahoff, A., 1982b. *Proc. 14th Ann. Offshore Technol. Conf.,* No. 4293, 725–730.
Malahoff, A., Embley, R. W., Cronan, D. S. & Skirrow, R., 1983. *Mar. Mining,* **4,** 123–137.
Malahoff, A., McLain, C. E. & Ranson, M. A., 1982. *Oceans '82,* 1222–1224.
Malahoff, A., McMurtry, G. M., Wiltshire, J. C. & Yeh, H.-W., 1982. *Nature, Lond.,* **298,** 234–239.
Manheim, F., Rowe, G. T. & Jipa, D. 1975. *J. sedim. Petrol.,* **45,** 243–251.
Manheim, F. T., in press. Science.
Manheim, F. T. & Gulbrandsen, R. A., 1979. *Min. Soc. Am. Short Course Notes,* **6,** 151–173.
Manheim, F. T., Hein, J. R., Marchig, V. & Puteanus, D., in press.
Manheim, F. T. & Hess, H. D. 1981. *Proc. Offshore Technol. Conf., OTC 4131,* 129–133.
Manheim, F. T., Ling, T. H. & Lane, C. M., 1983. *Oceans '83,* 828–831.
Mann Borgese, E., 1978. *Courier,* No. 49, 54–56.
Mann Borgese, E., 1983a. *Scient. Am.,* **248**(3), 28–35.
Mann Borgese, E., 1983b. *Ocean Yb.,* **4,** 1–14.
Mann Borgese, E., 1983c. *Ocean Yb.,* **4,** 15–44.
Manyun, L., Siqi, S. & Jinying, L., 1983. *Mar. Geol. Quatern. Geol.,* **3**(1), 55–65 (in Chinese, English Abstr.).

Marchig, V. & Gundlach, H., 1982. *Earth Planet. Sci. Lett.,* **58,** 361–382.
Marchig, V., Gundlach, H. & Bäcker, H., 1984. *Mar. Geol.,* **56,** 319–323.
Marchig, V., Möller, V. P., Bäcker, H. & Dulski, P., 1984. *Mar. Geol.,* **62,** 85–104.
Maris, C. R. P., Bender, M. L., Froelich, P. N., Barnes, R. & Luedtke, N.A., 1984. *Geochim. Cosmochim. Acta,* **48,** 2331–2346.
Marjoram, T. & Ford, G., 1980. *Mar. Technol.* **80,** 392–395.
Marsh, J. B., 1983. In, *The Law of the Sea in the 1980s,* edited by C.-h. Park, The Law of the Sea Institute, University of Hawaii, Honolulu, pp 210–239.
Marshall, J. F., 1980. *Bull. Aust. Bur. Mineral Resour.,* No. 207, 39 pp.
Marshall, J. F. & Cook, P. J., 1980. *J. geol. Soc. Lond.,* **137,** 765–771.
Masuda, Y., 1982. In, *Marine Mining: a New beginning,* edited by P. B. Humphrey, State of Hawaii Development of Planning & Economic Development, Honolulu, pp. 61–71.
McDougall, J. C. & Brodie, J. W., 1967. *N.Z. Oceanogr. Inst. Mem.,* No. 40, 56 pp.
McDougall, J. L., 1961. *N.Z. Jl Geol. Geophys.,* **4,** 283–300.
McDowell, A. N., 1971. *Oil Gas J.,* **69**(26), 114–116.
McGregor, B. A. & Lockwood, M., 1985. Mapping and Research in the Exclusive Economic Zone. U.S. Geological Survey/National Oceanic and Atmospheric Administration Booklet, 40 pp.
McGregor, B. A. & Offield, T. W., 1983. *The Exclusive Economic Zone: An exciting new frontier.* U.S. Department of the Interior/Geological Survey Booklet, 20 pp.
McIlhenny, W. F., 1975. In, *Chemical Oceanography, Vol. 4,* edited by J. P. Riley & G. Skirrow, Academic Press, London, 2nd edition, pp. 155–218.
McIlhenny, W. F., 1981. In, *Encyclopedia of Chemical Technology, Vol. 16,* edited by Kirk-Othmer, John Wiley & Sons, Inc., 3rd edition, pp. 227–296.
McKelvey, V. E., 1968. *Ocean Ind.,* **3**(9), 37–43.
McKelvey, V. E., 1980. *Science,* **209,** 464–472.
McKelvey, V. E., Tracey, J. I., Stoertz, G. E. & Vedder, J. G., 1969. *U.S. Geol. Surv. Circ.,* No. 619, 26 pp.
McKelvey, V. E. & Wang, F. F. H., 1970. *U.S. geol. Surv. Misc. Geol. Invest. Map* I–632 (4 maps).
McKelvey, V. E., Wright, N. A. & Bowen, R. W., 1983. *U.S. geol. Surv. Circ.,* No. 886, 55 pp.
McKelvey, V. E., Wright, N. A. & Rowland, R. W., 1979. In, *Marine Geology and Oceanography of the Pacific Manganese Nodule Province,* edited by J. L. Bischoff & D. Z. Piper, Plenum Press, New York, pp. 747–762.
McMurtry, G. M., 1981. In, *Energy Resources of the Pacific Region,* edited by M. T. Halbouty, American Association of Petroleum Geologists, Tulsa, Oklahoma, pp. 83–97.
McMurtry, G. M., De Carlo, E. H., Kim, K. H., Vonderhaar, D. & Malahoff, A., 1985. *EOS Trans. Am. geophys. Un.,* **66,** 1083 (abstract)
McMurtry, G. M., De Carlo, E. H. & Kroenke, L. W., 1983. *EOS Trans. Am. geophys. Un.,* **64,** 1018 (Abstr.).
McMurtry, G. M., Epp, D. & Karl, D. M., 1983. *Sea Grant Quarterly,* **5**(4), 1–18.
Medford, R. D., 1969a. *Mining Mag.,* **121**(5), 369, 371, 373, 375, 377, 379, 381.
Medford, R. D., 1969b. *Mining Mag.,* **121**(6), 474, 475, 477, 479, 480.
Meiser, H. J. & Müller, E., 1973. In, *The Origin and Distribution of Manganese Nodules in the Pacific and Prospects for Exploration,* edited by M. Morgensein, Honolulu, Hawaii, pp. 115–124.
Meith, N. & Helmer, R., 1983. *Ocean Yb.,* **4,** 260–294.
Menard, H. W., 1964. *Marine Geology of the Pacific.* McGraw-Hill, New York, 271 pp.

Menard, H. W. & Smith, S. M., 1966. *J. geophys. Res.*, **71**, 4305–4325.
Mero, J. L., 1965. *The Mineral Resources of the Sea*. Elsevier, Amsterdam, 312 pp.
Mero, J. L., 1977a. In, *Marine Manganese Deposits*, edited by G. P. Glasby, Elsevier, Amsterdam, pp. 327–355.
Mero, J. L., 1977b. In, *Technology Assessment and the Oceans*, edited by P. D. Wilmot & A. Slingerland, IPC Science and Technology Press, Guildford, Surrey, pp. 77–81.
Mero, J. L., 1978. *Mar. Mining*, **1**, 243–255.
Mertie, J. B., 1976. *Prof. Pap. U.S. geol. Surv.*, No. 938, 42 pp.
Meylan, M. A., Glasby, G. P., Knedler, K. E. & Johnston, J. H., 1981. In, *Handbook of Strata-Bound and Stratiform Ore Deposits, Vol. 9*, edited by K. H. Wolf, Elsevier, Amsterdam, pp. 77–178.
Michard, G., Albarède, F., Michard, A., Minster, J.-F., Charlou, J.-L. & Tan, N., 1984. *Earth Planet. Sci. Lett.*, **67**, 297–307.
Miles, E., Gibbs, S., Fluharty, D., Dawson, C. & Teeter, D., 1982. *The Management of Marine Regions: The North Pacific*. University of California Press, Berkeley, 656 pp.
Mitchell, A. H. & Bell, J. D., 1973. *J. Geol.*, **81**, 381–405.
Mitchell, A. H. G. & Garson, M. S., 1976. *Miner. Sci. Engng*, **8**, 129–169.
Mitchell, A. H. G. & Garson, M. S., 1981. *Mineral Deposits and Global Tectonic Settings*. Academic Press, London, 405 pp.
Mizuno, A. & Nakao, S., 1982. Editors. *Geol. Surv. Japan Cruise Rep.*, No. 18, 399 pp.
Moberly, R., Campbell, J. F. & Coulbourn, W. T., 1975. *Hawaii Inst. Geophys. Rep.*, HIG-75-10, 36 pp.
Mokhtari-Saghafi, M. & Osborne, R. H., 1980. *Oceans '80*, 55–59.
Molnia, B. R., 1979. *U.S. Department of Commerce*, NTIS No. PB81-192569, 70 pp.
Moorby, S.A. & Cronan, D. S., 1983. *Mineral. Mag.*, **47**, 291–300.
Moore, J. R., 1972. *Proc. R.Soc. Edinb.*, **72B**, 193–206.
Moore, J. R., 1979. *Documents BRGM*, No. 7, 131–163.
Moore, J. R., 1983. *Oceans '83*, 1145–1150.
Moore, J. R., 1984. In, *American Strategic Minerals*, edited by G. J. Mangone, Crane, Russack & Co, New York, pp. 85–108.
Moore, J. R. & Welkie, C. J., 1976. In, *Proc. Symp. on Sedimentation*, Alaska Geological Society, pp. K1–K17.
Moore, W. S. & Vogt, P. R., 1976. *Earth Planet. Sci. Lett.*, **29**, 349–356.
Morgan, C. L., 1981. In, *Marine Environmental Pollution, 2 Dumping and Mining*, edited by R. A. Geyer, Elsevier, Amsterdam, pp. 415–436.
Morgan, C. L. & Selk, B. W., 1984. *Proc. Offshore Technol. Conf.*, OTC 4777, 9–16.
Morgan, W. J., 1968. *J. geophys. Res.*, **73**, 1959–1982.
Moritani, T. & Nakao, S., 1981. Editors. *Geol. Surv. Japan Cruise Rep.*, No. 17, 281 pp.
Morley, I. W., 1981. *Black Sands A History of the Mineral Sand Mining Industry in Eastern Australia*. University of Queensland Press, St Lucia, 278 pp.
Mottl, M. J., 1980. *Oceanus*, **23**(2), 18–27.
Mullins, H. T., Green, H. G. & Martin, J. H., 1981. *Sea Grant College Program 1978–1980*, Biennial Report (Sea Grant Rep. No. R-CSGCP-004), 67–68.
Mullins, H. T. & Rasch, R. F., 1985. *Econ. Geol.*, **80**, 696–715.
Mungall, C., 1983. *GEOS*, **12**(3), 8–9.
Murray, J. & Renard, A. F., 1891. *Rep. Sci. Res. H.M.S. Challenger, Deep-Sea Deposits*, 525 pp.
Nakao, S. & Moritani, T., 1984. Editors. *Geol. Surv. Japan Cruise Rep.*, No. 20, 272 pp.

Narumi, Y., 1984. *Proc. 2nd Int. Seminar Offshore Mineral Resources* (GERMINAL), 119–142.
National Academy of Sciences, 1975. *Mining in the Outer Continental Shelf and in the Deep Ocean.* National Academy of Sciences, Washington, D.C., 119 pp.
National Advisory Committee on Oceans and Atmosphere, 1983. *Marine Minerals. An Alternative Mineral Supply. National Ocean Goals and Objectives for the 1980s.* U. S. Govt. Printing Office, Washington, D.C., 53 pp.
Natland, J. H., Rosendahl, B., Hekinian, R., Dmitriev, Y., Fodor, R. V., Goll, R. M., Hoffert, M., Humphris, S. E., Mattey, D. P., Petersen, N., Roggenthen, W., Schrader, E. L., Srivastava, R. K. & Warren, N., 1979. *Science,* **204,** 613–616.
Neary, C. R. & Highley, D. E., 1984. In, *Rare Earth Element Geochemistry,* edited by P. Henderson, Elsevier, Amsterdam, pp. 423–466.
Nelson, C. H. & Hopkins, D. M., 1972. *Prof. Pap. U.S. Geol. Surv.,* No. 689, 27 pp.
Neprochnov, Yu.P., 1980. *Oceanology,* **20,** 236–238.
Noakes, L. C., 1972. *CCOP Tech. Bull.,* No. 6, 161–173.
Noakes, L. C., 1977. *CCOP Tech. Bull,* No. 11, 157–168.
Noakes, L. C. & Jones, H. A., 1975. In, *Economic Geology of Australia and Papua New Guinea 1. Metals,* edited by C. L. Knight, The Australasian Institute of Mining and Metallurgy, Parkville, Victoria, pp. 1093–1104.
Noorany, I. & Fuller, J. T., 1982. *Proc. Offshore Technol. Conf.,* OTC 4261, 445–456.
Normark, W. R., Lupton, J. E., Murray, J. W., Delaney, J. R., Johnson, H. P., Koski, R. A., Clague, D. A. & Morton, J. L., 1982. *Mar. Technol. Soc. J.,* **16**(3), 46–53.
Normark, W. R., Morton, J. L., Koski, R. A., Clague, D. A. & Delaney, J. R., 1983. *Geology,* **11,** 158–163.
Northolt, A. J. G., 1980. *J. geol. Soc. Lond.,* **137,** 793–805.
Northolt, A. J. G. & Hartley, K., 1983. *Phosphate Rock: A Bibliography of World Resources.* Mining Journal Books, London, 147 pp.
Nyhart, J. D., Antrim, L., Capstaff, A., Kohler, A. D. & Leshaw, D., 1978. *MIT Sea Grant Program,* MITSG 78-4.
N.Z. Geological Survey, 1981. *Alpha* (N.Z. Dept. Sci. Industrial Res.), No. 22, 4 pp.
O'Brien, G. W. & Veeh, H. H., 1980. *Nature, Lond.,* **288,** 690–692.
Ocean Association of Japan, 1979. *Proc. 3rd Int. Ocean, Symp. 1978 Tokyo,* 128 pp.
Odunton, N., in press. U.N. Ocean Economics and Technology Branch, New York.
Office of Technology Assessment, 1985. *Strategic Minerals: Technologies to Reduce U.S. Import Vulnerability.* Congress of the United States, Washington, D.C., 55 pp.
Okano, T., Shimazaki, Y. & Maruyama, S., 1968. *Bull. geol. Surv. Japan,* **19**(6), 47–54.
Oudin, E., 1983. *Mar. Mining,* **4,** 39–72.
Oudin, E. & Constantinou, G., 1984. *Nature, Lond.* **308,** 349–353.
Oudin, E., Picot, P. & Poult, G., 1981. *Nature, Lond.,* **291,** 404–406.
Overall, M. P., 1968a. *Ocean Ind.,* **3**(9), 44–48.
Overall, M. P., 1968b. *Ocean Ind.,* **3**(10), 60–64.
Overall, M. P., 1968c. *Ocean Ind.,* **3**(11), 51–52.
Overstreet, W. C., 1967. *Prof. Pap. U.S. geol. Surv.,* No. 530, 327 pp.
Owen, R. M., 1978. *Mar. Mining,* **1,** 259–282.
Owen, R. M., 1980. *Mar. Mining,* **2,** 231–249.
Ozturgut, E., Lavelle, J. W. & Burns, R. E., 1981. In, *Marine Environmental Pollution, 2 Dumping and Mining,* edited by R. A. Geyer, Elsevier, Amsterdam, pp. 437–474.

Ozturgut, E., Lavelle, J. W. & Erickson, B. H., 1981. *Mar. Mining,* **3,** 1–17.
Padan, J. W., 1983. *Proc. Offshore Technol. Conf.,* OTC 4495.
Palacio, F. J., 1980. *Oceanus,* **23**(2), 39–49.
Pardee, J. T., 1934. *U.S. geol. Surv. Circ.,* No. 8, 41 pp.
Pasho, D. W., 1976. *N.Z. Oceanogr. Inst. Mem.,* No. 77, 28 pp.
Pasho, D. W., 1982. Internal Rep. 1982–1 (Canada Oil and Gas Lands Administration).
Pasho, D. W. & McIntosh, J. A., 1976. *Can. Inst. Min. Metall. Bull.,* **69**(773), 15–16.
Pautot, G. & Hoffert, M., 1984. *Les nodules du Pacifique Central dans leur environment géologique Campagnes COPANO-1979.* Publications du Centre National Pour l'Exploitation des Océans (C.N.E.X.O.), 202 pp.
Pendley, W. P., 1982. *Oceanus,* **25**(3), 12–17.
Pepper, J. F., 1958. *Bull. U.S. geol. Surv.,* No. 1067, 43–65.
Pereira, J. & Dixon, C. J., 1971. *Mineral. Deposita,* **6,** 404–405.
Peterson, G., 1979. In, *The Mineral Resources Potential of the Earth,* edited by F. Bender, E. Schweizerbart'sche Verlagsbuchhandlung, Stuttgart, pp. 69–78.
Phillips, R. L. 1979. *Program Feasibility Document—OCS Hard Minerals Leasing Heavy Minerals and Bedrock Minerals on the Continental Shelf off Washington, Oregon and California (Appendix 8).* NTIS No. PB81–192601, 56 pp.
Philpott, B., 1982. Victoria University of Wellington Project on Economic Planning Discussion Paper, No. 25, 54 pp.
Pinter, J., 1976. *Mem. Am. Assoc. Petrol. Geol.,* **25,** 418–421.
Piper, D. Z., Swint, T. R., Sullivan, L. G. & McCoy, F. W., 1985. *Manganese nodules, seafloor sediment, and sedimentation rates of the Circum-Pacific region.* Am. Assoc. Petrol. Geol. Circum-Pacific Map Series.
Polunin, N.V.C., 1983. *Oceanogr. Mar. Biol. Ann. Rev.,* **21,** 455–531.
Poynter, F. A., 1983. M.Sc thesis, Centre for Resource Management, University of Canterbury and Lincoln College, 179 pp.
Pryor, R. N., 1979. *Trans. Instn Min. Metall.,* **88A,** 43–46.
Raymond, R. C., 1976. *Mar. Technol. Soc. J.,* **10**(5), 12–18.
Reimnitz, E. & Plafker, G., 1976. *U.S. geol. Surv. Bull.,* No. 1415, 16 pp.
Reimnitz, E., von Huene, R. & Wright, F. F., 1970. *Prof. Pap. U.S. geol. Surv.,* 700C, 35–42.
Richards, H. G. & Strens, M. R., 1985. *Sci. Progr.,* **69,** 341–358.
Robbins, E. I., 1983. *Tectonophysics,* **94,** 633–658.
Roe, K. K. & Burnett, W. C., 1985. *Geochim. Cosmochim. Acta,* **49,** 1581–1592.
Rogers, J. K. & Crase, N. J., 1980. In, *Fertilizer Mineral Potential in Asia and the Pacific,* edited by R. P. Sheldon & W. C. Burnett, East-West Center, Honolulu, pp. 307–330.
Rona, P. A., 1973a. *Scient. Am.,* **229**(1), 86–95.
Rona, P. A., 1973b. *Ocean Mgmt,* **1,** 145–159.
Rona, P. A., 1976. *Nat. Resourc. Forum,* **1,** 17–28.
Rona, P. A, 1977. *EOS Trans. Am. geophys. Un.,* **58**(8), 629–639.
Rona, P. A., 1978. *Econ. Geol.,* **73,** 135–160.
Rona, P. A., 1982. *Mar. Technol. Soc. J.,* **16**(3), 81–86.
Rona, P. A., 1983a. *Nat. Resourc. Forum,* **7,** 329–338.
Rona, P. A., 1983b. *Mar. Mining,* **4,** 7–38.
Rona, P. A., 1984. *Earth-Sci. Rev.,* **20,** 1–104.
Rona, P. A., Boström, K., Laubier, L. & Smith, K. L., 1983. Editors. *Hydrothermal Processes at Seafloor Spreading Centers.* Plenum Press. New York, 796 pp.
Rona, P. A. & Lowell, R. P., 1980. Editors. *Seafloor Spreading Centers Hydrothermal Systems.* Dowden, Hutchinson & Ross, Inc., Stroudsburg, Penn., 424 pp.

Rona, P. A. & Neuman, L. D., 1976a. *Ocean Mgmt,* **3,** 57–78.
Rona, P. A. & Neuman, L. D., 1976b. *Mem. Am. Assoc. Petrol. Geol.,* **25,** 48–57.
Rowland, R. W., Goud, M. R. & McGregor, B. A., 1983. *U.S. geol. Surv. Circ.,* No. 912, 29 pp.
Rowland, T. J. & Cruickshank, M. J., 1983. *Oceans '83,* 703–707.
Sawkins, F. J., 1972. *J. Geol.,* **80,** 377–397.
Sawkins, F. J., 1984. *Metal Deposits in Relation to Plate Tectonics.* Springer Verlag, Berlin, 325 pp.
Sawyer, D. L., Smyres, G. A., Sjoberg, J. J. & Carnahan, T. G., 1983. *U.S. Bur. Mines Tech. Rep.,* TPR 122, 8 pp.
Schott, W., 1976. In, *Handbook of Strata-Bound and Stratiform Ore Deposits,* edited by K. H. Wolf, Elsevier, Amsterdam, pp. 245–294.
Schott, W., 1980. Editor. *Geol. Jb.,* **D38,** 204 pp.
Scolari, G., 1982. *Ann. Mines,* **189** (11–12), 65–70.
Scott, S. D., 1985. *Mar. Mining,* **5,** 191–212.
Sharma, G. D., 1977. In, *Proc. 4th Internat. Conf. on Port and Ocean Engineering under Arctic Conditions,* pp. 885–891.
Sheldon, R. P., 1981a. In, *Energy Resources of the Pacific Region,* edited by M. T. Halbouty, American Association of Petroleum Geologists, Tulsa, Oklahoma, pp. 537–539.
Sheldon, R. P., 1981b. *Ann. Rev. Earth Planet. Sci.,* **9,** 251–284.
Sheldon, R. P., 1982. *Scient. Am.,* **246**(6), 31–37.
Sheldon, R. P. & Burnett, W. C., 1980. Editors. *Fertilizer Mineral Potential in Asia and the Pacific.* East-West Center, Honolulu, 481 pp.
Shelf Working Group, 1980. *A Tentative Methodology for the Appraisal of Offshore Non-fuel Mineral Resource Potential.* Departmental Coordinating Committee on Ocean Mining, Department of Energy, Mines and Resources, Ottawa, Ontario, 30 pp. (Internal Report.)
Shigley, C. M., 1968a. *Occ. Publ. Grad. School Oceanogr. Univ. Rhode Is.*, No. 4, 45–50.
Shigley, C. M., 1968b. *Ocean Ind.,* **3**(11), 43–46.
Shinn, D. W., Sedwick, P. N. & Zeitlin, H., 1984. *Mar. Mining,* **5,** 57–73.
Shusterich, K. M., 1982. *Oceans '82,* 1316–1321.
Siddayo, C. M., 1984. *Ocean Mgmt,* **9,** 73–100.
Sillitoe, R. H., 1972a. *Econ. Geol.,* **167,** 184–197.
Sillitoe, R. H., 1972b. *Bull. geol. Soc. Am.,* **83,** 813–818.
Sillitoe, R. H., 1972c. *Trans. Instn Min. Metall.,* **81B,** 141–148.
Skornyakova, N. S., Gordeev, V. V. & Kuzmina, T. G., 1981. *Litologiya i poleznie iskoloemie,* No. 5, 79–90 (in Russian).
Skornyakova, N. S., Kurnosov, V. V., Mukhina, V. V., Kruglikova, S. B., Rudakova, L. N. & Ushakova, M. G., 1983. *Litologiyaipoleznie iskoloemie,* No. 1, 121–134 (in Russian).
Sleep, N. H. & Morton, J. L., 1983. *Oceans '83,* 782–786.
Smale-Adams, K. B. & Jackson, G. O., 1978. *Phil. Trans. R. Soc. Lond. Ser. A,* **290,** 125–133.
Smith, J. B., Holt, B. R. & Paul, R. G., 1985. *Proc. Offshore Technol. Conf.,* OTC 4899, 9–17.
Soong, K.-L. 1978. *Acta oceanogr. Taiwanica,* No. 8, 43–62.
Sorensen, P. E. & Mead, W. J., 1968. *Am. J. agric. Econ.,* **50,** 1611–1620.
Sorensen, P. E. & Mead, W. J., 1969. In, *Trans. 5th Mar. Technol. Soc. Conf. Exhibit, Miami,* pp. 491–500.
Spagni, D. & Ford, G., 1984. *Far Eastern Economic Rev.,* 29th March.
Spiess, F. N., Macdonald, K. C., Atwater, T., Ballard, R., Carranza, A., Cordoba, D., Cox, C., Diaz Garcia, V. M., Francheteau, J., Guerrero, J., Hawkins, J., Haymon, R., Hessler, R., Juteau, T., Kastner, M., Larson, R., Luyendyk, B.,

Macdougall, J. D., Miller, S., Normark, W., Orcutt, J. & Rangin, C., 1980. *Science,* **207**, 1421–1433.
Stavridis, J., 1985. *Proc. U.S. Naval War Inst,* Jan., 72–77.
Stephen-Hassard, Q. D., Chave, K. E., Fernando, Q., Keith, K. M., Meylan, M. A. & Miklius, W., 1978. *The Feasibility and Potential Impact of Manganese Nodule Processing in Hawaii.* State of Hawaii Department of Planning and Economic Development, Honolulu, 7 chapters + 11 appendices.
Stevens, J. F., 1970. *Ocean. Ind.,* **5**(11), 47–51.
Stowasser, W. F., 1980. *U.S. Bureau Mines Bull.,* No. 671, 663–682.
Stowasser, W. F., 1981. In, *Minerals Yearbook 1980, Vol. 1, Metals and Minerals.* U.S. Bureau of Mines, Washington, D.C., 950 pp.
Strens, M. R. & Cann, J. R., 1982. *Geophys. J. R. astron. Soc.,* **71**, 225–240.
Styrt, M. M., Brackmann, A. J., Holland, H. D., Clark, B.C., Pisutha-Arnond, V., Eldridge, C. S. & Ohmoto, H., 1981. *Earth Planet. Sci. Lett.,* **53**, 382–390.
Sujitno, S., 1977. *CCOP Tech. Bull.,* No. 11, 169–182.
Summerhayes, C. P., 1967. *N.Z. Jl mar. freshw. Res.,* **1**, 267–282.
Sutherland, D. G., 1985. *J. geol. Soc. Lond.,* **142**, 727–737.
Swan, D. A., 1974. *Mar. Technol.,* Jan., 9–18.
Tagg, A. R., 1979. *Offshore Heavy Metals Resources of Alaska.* NTIS No. PB81-192585, 42 pp.
Takahara, H., Handa, K., Ishii, K. & Kuboki, E., 1984. In, *Proc. Offshore Technol. Conf., OTC 4782,* pp. 69–72.
Takeuchi, T. K., 1979. In, *The Deep Seabed and its Mineral Resources, Proc. 3rd Internat. Ocean Symp.* 1978. Tokyo, pp. 105–109.
Tepordei, V. V., 1980. *U.S. Bur. Mines Bull.,* No. 671, 781–791.
Tepordei, V. V., 1981. In, *Minerals Yearbook 1980, Vol. 1, Metals and Minerals,* U.S. Bur. Mines, Washington, D.C., pp. 695–720.
The Delegation of China, 1981. *Proc. 17th Session CCOP,* 462–463.
The Delegation of the Republic of Korea, 1981. *Proc. 17th Session CCOP,* 192–205.
The Philippine Delegation, 1981. *Proc. 17th Session CCOP,* 218–220.
The Sedimentation Laboratory, South China Sea Institute of Oceanology, 1983. *Tropic Oceanol.,* **2**(3), 250–256 (in Chinese).
Thiede, J., 1983. Editor. *Report of the III. International Workshop on Marine Geosciences Heidelberg, F.R. Germany July 19–24,* 1982, 163 pp.
Thompson, G., 1983. In, *Chemical Oceanography, Vol. 8,* edited by J. P. Riley & R. Chester, Academic Press, London, 2nd edition, pp. 271–337.
Thompson, R. M. & Smith, K.G., 1970. *Prepr. Offshore Technol. Conf.,* **2**, 819–826.
Thompson, T. L., 1976. *Bull. Am. Assoc. Petrol. Geol.,* **60**, 1463–1501.
Tivey, M. K. & Delaney, J. R., 1985. *Mar. Mining,* **5**, 165–179.
Toenges, A. L., Kelly, L. W., Davis, J. D., Reynolds, D. A., Fraser, T., Crentz, W. L. & Abernethy, R. F., 1948. *U.S. Bur. Mines Bull.,* No. 474, 106 pp.
Tokunaga, S., 1967. *Bull. Japanese geol. Surv,* **18**(9), 65–72.
Tokunaga, S., 1969. *CCOP Tech. Bull.,* **2**, 117–122.
Towner, R. R., 1984a. In, *Australian Mineral Industry Annual Review for 1982,* Australian Government Publishing Service, Canberra, pp. 211–213.
Towner, R. R., 1984b. In, *Australian Mineral Industry Annual Review for 1982, Australian Government Publishing Service,* Canberra, pp. 244–252.
Towner, R. R., 1984c. In, *Australian Mineral Industry Annual Review for 1982, Australian Government Publishing Service,* Canberra, pp. 276–278.
Trondsen, E. & Mead, W. J., 1977. *Sea Grant Publ.,* No. 59, 188 pp.
Tufar, W., Gundlach, H. & Marchig, V., 1984. *Mitt. österr. geol. Ges.,* **77**, 185–245.
U.N. Ocean Economics and Technology Branch, 1981. *The Future of Offshore Petroleum.* McGraw-Hill, New York, 272 pp.

U.N. Ocean Economics and Technology Branch, 1982. *Assessment of Manganese Nodule Resources: The Data and the Methodologies.* Graham & Trotman Ltd, London, 79 pp.
U.N. Ocean Economics and Technology Branch, 1984. *Analysis of Exploration and Mining Technology for Manganese Nodules.* Graham & Trotman Ltd, London, 140 pp.
U.N. Ocean Economics and Technology Office, 1979. *Manganese Nodules Dimensions and Perspectives.* D. Reidel, Dordrecht, Holland, 194 pp.
U.S. Department of the Interior, 1979. *Program feasibility document OCS hard minerals leasing (with 23 appendices).* NTIS No. PB81-192551, 180 pp.
Usami, T., Saito, T. & Oba, S., 1982. In, *Proc. Offshore Technol. Conf.,* OTC 4297, pp. 763-768.
van Dyke, J. M., 1985. Editor. *Consensus and Confrontation: The United States and the Law of the Sea Convention.* The Law of the Sea Institute, University of Hawaii, Honolulu, 576 pp.
Varnavas, S. P. & Panagos, A. G., 1984. *Chem. Geol.,* **42,** 227-242.
Veeh, H. H., Burnett, W. C. & Soutar, A., 1973. *Science,* **181,** 844-845.
Vhay, J. S., Brobst, D. A. & Heyl, A. V., 1973. *Prof. Pap. U.S. geol. Surv.,* No. 820, 143-155.
von Rad, U. & Kudrass, H.-R., 1984. Compilers. *Geol. Jb.,* **D65,** 252 pp.
Von Stackelberg, U., 1982. Compiler. *Geol. Jb.,* **D56,** 215 pp.
Walter, P. & Stoffers, P., 1985. *Mar. Geol.,* **65,** 271-287.
Wang, C., 1982. *Manganese Nodules in the Ocean.* Ocean Press, Beijing, 244 pp. (in Chinese).
Wang, F. F. H. & McKelvey, V. E., 1976. In, *World Mineral Supplies Assessment and Perspective,* edited by G. J. S. Govett & M. H. Govett, Elsevier, Amsterdam, pp. 221-286.
Ware, T. M., 1966. *Trans. 2nd Ann. Mar. Technol. Soc.* Suppl., 7-13.
Wedepohl, K. H., 1980. In, *Geology and Geochemistry of Manganese, Vol. 1,* edited by I. M. Varentsov & Gy. Grasselly, Hungarian Academy of Sciences, Budapest, pp. 335-351.
Welling, C. G., 1977. *Oceans '77,* 26E-1-26E-2.
Welling, C. G., 1982a. *Mar. Technol. Soc. J.,* **16**(3), 5-7.
Welling, C. G., 1982b. In, *Economics of Ocean Resources A Research Agenda,* edited by G. M. Brown & J. A. Crutchfield, Washington Sea Grant Publication, University of Washington, Press, Seattle, pp. 90-98.
Welling, C. G. 1983. In, *The Law of the Sea and Ocean Development Issues in the Pacific Basin,* edited by E. L. Miles & S. Allen. Proc. Law of the Sea Institute Fifteenth Ann. Conf. 5-8th October, 1981, Honolulu, Hawaii, pp. 284-286.
Wells, S. M., 1981. *Proc. 4th Internat. Coral Reef Symp.,* **1,** 323-330.
Wells, S. M., 1982. *Oceans,* **15**(6), 65-67.
Wenk, E., 1969. *Scient. Am.,* **221**(3), 166-176.
White, P. M. T., 1982. *Trans. Instn Min. Metall.,* **91A,** 184-198.
White, W. C., 1964. *Bull. Aust. Bur. Mineral Resourc.,* **69,** 98-99.
White, W. C. & Warin, O. N., 1964. *Bull. Aust. Bur. Mineral Resourc.,* **69,** 173 pp.
Wijkman, P. M., 1982. *J. World Trade Law,* **16,** 27-48.
Wilcox, S. M., Mead, W. J. & Sorensen, P. E., 1972. *Ocean Ind.,* **7**(8), 27-28.
Williams, D. L., 1976. *J. Volcanol. Geothermal Res.,* **1,** 85-100.
Wilson, T. A. & Mero, J. L., 1966. *Bull. Calif. Div. Mines Geol.,* **190,** 343-353.
Wiltshire, J., 1984. *Proc. Pacific Congr. on Marine Technology,* PACON 84, MRM3/22-28.
Wollard, G. P., 1981. In, *Energy Resources of the Pacific Region,* edited by M. T. Halbouty, American Association of Petroleum Geologists, Tulsa, Oklahoma, pp. 507-516.

Xu, Y. & Qinchen, C., 1983. *J. Shandong College Oceanol.*, **13**(4), 71–74 (in Chinese, English Abstr.).
Yamakado, N., Handa, K. & Usami, T., 1978. *Proc. Offshore Technol. Conf.*, OTC 3136, 725–728.
Yeend, W., 1973. *Prof. Pap. U.S. geol. Surv.*, No. 820, 561–565.
Yuasa, M. & Yokota, S., 1982. *CCOP Tech. Bull.*, No. 15, 51–64.
Zablocki, C. J., 1978. Chairman. *Deep seabed minerals: Resources, diplomacy, and strategic interest.* U.S. 95th Congress 2d Session U.S. Congress Subcommittee on International Organisations. U.S. Government Printing Office, Washington, D.C., 123 pp.
Zierenberg, R. A., Shanks, W. C. & Bischoff, J. L., 1984. *Bull. geol. Soc. Am.*, **95**, 922–929.

ADDITIONAL REFERENCES ADDED IN PROOF

Bäcker, H., 1985. *Geowissenschaften*, 43–55.
Bischoff, J. L. & Pitzer, K. S., 1985. *Earth Planet. Sci. Lett.*, **75**, 327–338.
Clark, A. L., Humphrey, P., Johnson, C. J. & Pak, D. K., 1985. U.S. Minerals Management Serv., OCS Study, MMS 85–0006, 35 pp. + 6 plates.
Eade, J. V., 1984. Complier. *Proc. 13th. Session CCOP/SOPAC*, 126 pp.
Ergunalp, D. & Weber, H., 1985. *Erzmetall*, **38**, 238–242.
Hale, P. B. & McLaren, P., 1984. *Can. Min. Metall. Bull.*, **77**(869), 51–61.
Haynes, B. W., Law, S. C. & Barron, D. C., 1982. *U.S. Bur. Mines Inf. Circ.*, IC 8906, 60 pp.
Ligozat, H., 1985. *Underwater Technol.*, **11**(1), 29–30.
Liu, Y., 1985. *Mar. Geol. Quatern. Geol.*, **5**(3), 91–100 (in Chinese, English summary).
Morton, J., 1985. EOS *Trans. Am. geophys. UN.*, **66**(45), 756 only.
Odunton, N. A., 1984. *Proc. Offshore Technol. Conf.*, OTC 4781, 45–68.
Pasho, D. W., 1973. M.Sc. thesis, University of Southern California, 188 pp.
Pasho, D. W. & King, D. E. C., 1980. *Energy, Mines and Resources*, Canada Mineral Policy Sector, Int. Rept MRI 82/7, 37 pp.
Renard, V., Hekinian, R., Francheteau, J., Ballard, R. D. & Bäcker, H., 1985. *Earth Planet. Sci. Lett.*, **75**, 339–353.
Samson, J., 1984. *Energy, Mines and Resources*. Canada Ocean Mining Division Document 84–3, 162 pp + 2 app.
Skorniakova, N. S., Gordeev, V. V., Anikeeva, L. I., Chudaev, O. V. & Kholodkevich, I. V., 1985. *Okeanologiya*, **25**, 630–637 (in Russian).
Tan, Q., Sun, Y., Wang, Z., Liu, Q., Liu, H. & Jiang, Y., 1985. *Mar. Geol. Quatern. Geol.*, **5**(4), 40–47 (in Chinese, English summary).
Usami. T., Tsurusaki, K., Hirota, T. & Padan, J. W., 1979. *Mar. Technol.* 176–189.
Von Stackelberg, U. & the Shipboard Scientific Party, 1985. *Bundesanstalt für Geowissenschaften und Rohstoffe*, Circ. 2, 14 pp.

Oceanogr. Mar. Biol. Ann. Rev., 1986, **24**, 65-170
Margaret Barnes, Ed.
Aberdeen University Press

THE BENGUELA ECOSYSTEM
PART III. PLANKTON

L. V. SHANNON and S. C. PILLAR
Sea Fisheries Research Institute, Private Bag X2, Rogge Bay 8012, Cape Town, South Africa

INTRODUCTION

Although the living resources of the Benguela have been exploited for several centuries, it is only recently that any significant progress has been made as regards understanding the biological processes in the ecosystem. Collections of marine animals and plants were made by naturalists in the eighteenth and nineteenth centuries, perhaps the best known of which were described in Challenger reports. The South African demersal fishing industry developed in the early part of the twentieth century as a direct result of the pioneering studies by Dr J. D. F. Gilchrist, who is generally accepted as the father of marine science in southern Africa. Throughout the first half of the century research emphasis continued to be placed on the floristic studies. Indeed even with the advent of a major research programme after World War II to investigate the biology of the South African pilchard, *Sardinops ocellatus*, the work remained largely descriptive for the next two decades. Much of the effort was devoted to monitoring the system, the dynamics of which were not well understood. 1968 saw the start of process orientated biological oceanography in the Benguela region and since then substantial advances have been made in understanding the system's complex food web.

In this review, the third in a series on the Benguela, we have attempted a synthesis of information about the plankton and its dynamics. In keeping with the two published companion reviews which dealt with the physics and the chemistry of the Benguela (Shannon, 1985; Chapman & Shannon, 1985), we have tried to emphasize the processes, for example, primary production and zooplankton dynamics. It has not been possible to review all the available floristic literature in detail, and we have generally excluded from our discussion publications dealing with the subtidal and intertidal regions, except where they are relevant to the biological processes in the broader Benguela system. While the list of references is fairly comprehensive, it is by no means exhaustive and it is probable that we have missed some of the less readily available works *e.g.* reports in Russian published in the U.S.S.R. and possibly some recent papers by scientists in the German Democratic Republic. Nevertheless, the papers cited will serve to lead into other literature on the plankton of the Benguela. The review should be read in conjunction with reviews by Shannon (1985) and

Chapman & Shannon (1985) of the physical and chemical oceanography of the Benguela system.

PHYTOPLANKTON: DISTRIBUTION AND DYNAMICS

The Benguela system, which is bounded in the north by the Angola system and in the south by the Agulhas Retroflection area (see Shannon, 1985), lies within both tropical and temperate latitudes. Accordingly, the northern Benguela receives more insolation than the southern part of the system, although the meridional effect is moderated by the increased cloud cover in the north, in particular inshore where the frequency of fog is high. The wind field largely determines upwelling, and accordingly the supply of nutrients to the photic zone, as well as mixing in the upper 50 m. The seasonal effects are more noticeable in the southern Benguela which is also subject to frequent shorter perturbations due to the passage of easterly moving cyclones south of Africa during the austral summer. Moreover, the upwelling and quiescent—or downwelling—seasons in the southern Benguela tend to be out of phase with those in the north (refer to Shannon, 1985), while in the central part of the system upwelling occurs over most of the year. In view of the aforegoing, the spatial and temporal distribution of phytoplankton in the Benguela will be discussed on a regional basis, following the scheme adopted in Part I by Shannon (1985), where three areas were delineated *viz.* the northern Benguela, the central (Namaqua-Lüderitz) area, and the southern Benguela.

The first large-scale investigation of phytoplankton in the South Atlantic Ocean was that undertaken during the 1925–1927 expedition of the METEOR. The treatises of Hentschel (1928, 1933) and Hentschel & Wattenberg (1930) provided the first quantitative information about the high

Fig. 1.—R.R.S. WILLIAM SCORESBY in Table Bay harbour.

plankton biomass in the Benguela system and its general distribution. Indeed Hentschel (1928) regarded the waters around Walvis Bay as being among the most productive in the world. The voyages of the R.R.S. WILLIAM SCORESBY (Fig. 1) during 1950 enabled Hart & Currie (1960) to give a really detailed account of the planktonic flora in the Benguela. Although there have been several subsequent studies, that of Hart & Currie (1960) is still probably the primary reference work on the region. Recent theses by Austin (1980) and Schuette (1980) have included useful bibliographies relating to the phytoplankton of the northern Benguela, while Brown & Hutchings (1985) have provided an insight into the complex dynamics of phytoplankton around the South Western Cape.

In the following sections we shall attempt to review briefly the literature on species distribution and to synthesize knowledge about the dynamics of phytoplankton in the Benguela system.

FLORISTIC STUDIES

The spatial dichotomy (Fig. 2) in the phytoplankton between the northern and southern Benguela regions was first demonstrated by Hentschel (1928, 1933). His studies showed that Bacillariophyceae comprised the dominant microplanktonic group in the neritic zone, both off the Cape and off Namibia, while offshore Dinophyceae were relatively more important in the northern Benguela than in the south. From the literature it would appear that the Benguela is generally a diatom-dominated system. Certainly the importance of diatoms and, to a lesser degree, dinoflagellates has been

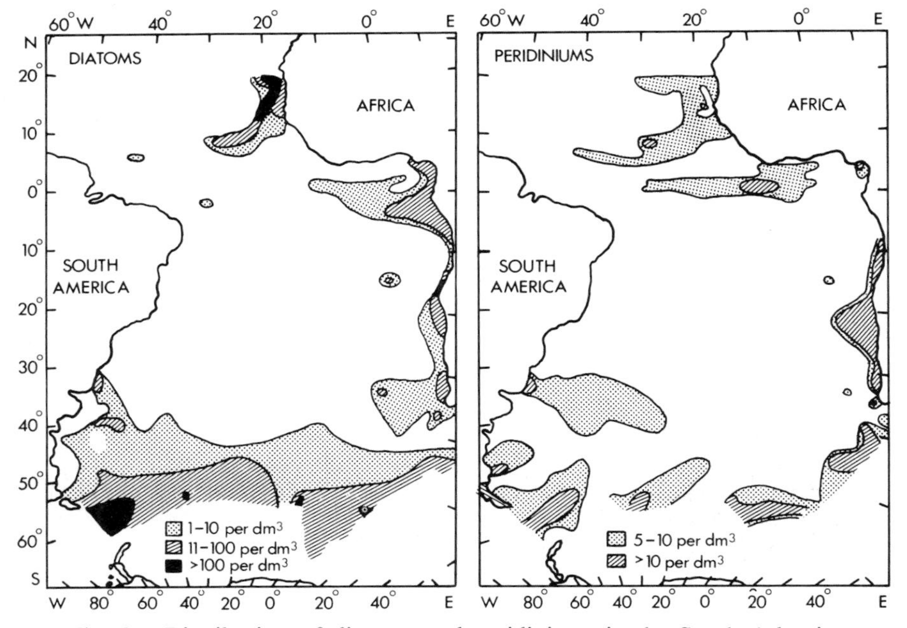

Fig. 2.—Distribution of diatoms and peridiniums in the South Atlantic Ocean (redrawn from Hentschel, 1933).

confirmed by a number of workers, *inter alia* Gilchrist (1914), Copenhagen & Copenhagen (1949), De Jager (1954, 1957), Davies (1957), De Decker (1973), Andrews & Hutchings (1980), Olivieri (1983a), and D. A. Horstman (pers. comm.) in the Cape, and Hart & Currie (1960), Kollmer (1962, 1963), Pieterse & Van der Post (1967), Reyssac (1973), Zernova (1974), Hulburt (1976), Austin (1980), Kruger (1980), and Kruger & Cruickshank (1982) in respect of the northern Benguela. Even in the Angolan neritic system the literature suggests that diatoms tend to dominate (Silva, 1953a, 1955), particularly in the south in the area bordering on the Benguela regime, although dinoflagellates evidently become increasingly important off central and northern Angola.

It is quite possible, however, that the apparent dominance of diatoms (and dinoflagellates) is an artefact of the methods of sample collection and examination used in the past. In some studies net tows were used and in others, where bottle samples were examined, high magnifications were not employed, with the result that the nanoplankton were totally overlooked. Recent studies (mainly unpublished) have, however, shown that the nanoplankton form relatively stable (relative to diatoms) and persistent populations, which become relatively more important under conditions not favourable for diatom growth. There also may be blooms, however, of small flagellates—*e.g.* during March 1983 in the southern Benguela Hutchings, Holden & Mitchell-Innes (1984) found an abundance of small green flagellates close inshore in upwelled water, and Norris (1983) has suggested that flagellates, which may be able to form somewhat stratified populations in upwelling systems, could be of utmost importance as primary producers in this environment. (The work of Norris, 1983, on delicate nanoplankton species in the southern Benguela revealed types that appear to be unusual in most coastal phytoplankton populations.) Hutchings *et. al.* (1984) recorded large numbers of coccolithophorids (2×10^6 cells/litre) at a station just inshore of the oceanic (thermal) front during their cruise in March 1983, while Austin (1980) also reported high coccolithophorid populations in the northern Benguela. Hobson (1971) found that microflagellates contributed a major fraction of phytoplankton carbon at most of his stations in the northern Benguela. In a recent study of nitrogen uptake by size-fractionated phytoplankton populations in the southern Benguela, Probyn (1985) found that picoplankton (<1 μm) and nanoplankton (<10 μm) chlorophyll *a* accounted for 2–49% and 13–99%, respectively, of the whole community chlorophyll *a* during December 1983. Thus, there is evidence that nanoplankton populations are more important in the Benguela than is generally appreciated. While it may not be unusual to find high microflagellate populations in the Benguela system during quiescent (non-upwelling) conditions, it may be unusual to find flagellates dominating in recently upwelled water.

In view of the paucity of literature on the nanoplankton, we shall, however, confine most of our discussion to diatoms and dinoflagellates, with the emphasis on the former group.

Northern Benguela area

The floristic studies of Pinto (1953) on radiolarians, Silva (1954) on tintinnids and Silva (1953a,b, 1955, 1957) provide a comprehensive account of the microplankton in the Angolan and extreme northern Benguela regions during 1951 and 1952. Silva (1953a) stressed that the work was qualitative rather than quantitative. In the south, off Baia dos Tigres, she found that there was a preponderance of disc-shaped diatoms at 10–20 m depth, in particular *Paralia sulcata, Coscinodiscus*, and *Arachnoidiscus ehrenbergii*, while off central Angola *Rhizosolenia* spp. were common. The families of Coscinodiscaceae, Chaetoceraceae and Biddulphiaceae were well represented. Silva (1955) did not record very high numbers of dinoflagellates, but it must be emphasized that most of her sampling was done within 50 km of the coast, and perhaps the dinoflagellate fraction would have been greater, had more stations been situated in oceanic water. The most commonly encountered species were *Protoperidinium depressum, Ceratium furca, Noctiluca scintillans*, and *Protoperidinium diabolus*.

Reyssac (1973) recorded 58 species of diatoms (of which there were five main species) and 32 species of dinoflagellates off central Namibia and he considered that the high diversity suggested high plankton activity in the region throughout the year. Kruger (1980) has recorded 537 microplankton species (including zooplankton) for the northern Benguela region, of which diatoms (61 genera) outnumbered the other goups. His view is, however, that the area is floristically undistinguished, sharing 73% of the species with the Mediterranean and 72% with the southwestern Indian Ocean (refer to Table I). Only a few diatoms, of which the principal one is *Delphineis karstenii (recorded as Fragilaria karstenii* by Hart & Currie, 1960, and other early workers), appear to be restricted to this region. On the other hand, *Skeletonema costatum*, recorded by Hart & Currie (1960) at a few stations is commonly encountered in the southern Benguela (D. A. Horstman, pers. comm.) but is generally conspicuous by its absence in collections from the Namibian region (Austin, 1980; I. Kruger, pers. comm.). The dominant species found during the two surveys of R.R.S. WILLIAM SCORESBY in 1950 are listed in Table II (from Hart & Currie, 1960).

From Table II it can be seen that *Chaetoceros* spp., *Rhizosolenia* spp., *Planktoniella sol*, *Nitzschia* spp., and *Asterionella glacialis* (*A. japonica*) were among the most common diatoms recorded by Hart & Currie (1960), while *Peridinium* (sic) spp. were important in the dinoflagellate group. Schuette (1980) observed that, in the sediment samples collected in water depths of less than 130 m between 19 and 24° S, *Chaetoceros* resting spores and *Delphineis karstenii* dominated, the latter being characteristically found in newly upwelled water (Austin, 1980). Schuette found that there was a significant contrast between the inshore and offshore sediment flora, with the latter being characterized by *Chaetoceros* resting spores, *Thalassiosira eccentrica, Coscinodiscus* spp., *Nitzschia kerguelensis*, and *Actinoptychus* spp. together with various oceanic species. From the available literature it seems that a wide diversity of diatoms can dominate in the neritic zone in the northern Benguela. Blooms of the following were recorded by Kollmer (1962, 1963) in the coastal waters off central Namibia between 1958 and 1960: *Leptocylindrus* spp., *Chaetoceros* spp., *Coscinosira polychorda*, *Thalassiosira* spp., *Delphineis karstenii*, *Thalassionema nitzschioides*, and *Asterionella glacialis*. Further offshore the key oceanic species were, in order of prevalence, *Rhizosolenia*

TABLE I

Frequency of occurrence of South West African higher taxonomic groups and main genera in the Mediterranean Sea and the southwestern Indian Ocean: percentages are only given for the number of species of South West African higher taxonomic groups that also occur in the Mediterranean Sea and the southwestern Indian Ocean; from Kruger (1980)

Taxonomic groups and genera	South West Africa		Mediterranean Sea		Southwestern Indian Ocean	
	Number of species of higher groups	Number of species of main genera	Number of species of higher groups co-occurring	Number of species of main genera co-occurring	Number of species of higher groups co-occurring	Number of species of main genera co-occurring
Diatoms	184		142 (77%)		152 (83%)	
Bacteriastrum		8		6		7
Biddulphia		8		7		6
Chaetoceros		46		37		40
Coscinodiscus		14		13		12
Nitzschia		6		6		6
Rhizosolenia		20		18		20
Thalassiosira		8		7		6
Thalassiothrix		6		3		6
Silicoflagellates	10		7 (70%)		5 (50%)	
Dictyocha		8		6		4
Coccolithophorids	15		10 (67%)		6 (40%)	
Dinoflagellates	158		118 (75%)		113 (72%)	
Ceratium		56		49		53
Dinophysis		19		13		10
Gonyaulax		6		5		4
Peridinium (sic)		35		26		28
Foraminifera	25		21 (84%)		20 (80%)	
Radiolarians	40		31 (78%)		21 (53%)	

TABLE I—*continued*

Taxonomic groups and genera	South West Africa		Mediterranean Sea		Southwestern Indian Ocean	
	Number of species of higher groups	Number of species of main genera	Number of species of higher groups co-occurring	Number of species of main genera co-occurring	Number of species of higher groups co-occurring	Number of species of main genera co-occurring
Tintinnids	95		57 (60%)		64 (67%)	
Codonellopsis		5		2		3
Dictyocysta		8		5		6
Epiplocylis		6		1		3
Eutintinnus		10		7		9
Parundella		7		6		6
Rhabdonella		8		3		8
Salpingella		8		6		6
Other	10		7 (70%)		4 (40%)	
Total number of species	537					
Total number of co-occurring species			393 (73%)		385 (72%)	

TABLE II

*Abridged frequency data: categories occurring as dominants at one or more of the thirty-nine repeated stations: the frequency data have been summarized in the form of double fractions—thus the entry "*Stephanopyxis turris *10/3 and 11/1" signifies that this species was present at ten out of the thirty-nine stations of the first survey and dominant at three of those ten and present at eleven of the second survey stations, but dominant at only one of them; from Hart & Currie (1960)*

During both surveys		During the first survey only		During the second survey only	
Stephanopyxix turris	10/3 and 11/1	*Eucampia zoodiacus*	11/4 and 0/0	*Skeletonema costatum*	1/0 and 3/1
Thalassiosira eccentrica	14/3 and 3/1	*Chaetoceros costatum*	4/4 and 1/0	*Thalassiosira* spp. non det.	2/0 and 14/5
T. subtilis	12/2 and 14/3	*C. didymum* (resting spores)	13/6 and 3/0	*Coscinodiscus* spp. non det.	8/0 and 28/1
Planktoniella sol	19/6 and 24/9	*C. pseudocrinitum*	2/1 and 0/0	*Asteromphalus heptactis*	5/0 and 6/1
Chaetoceros affine	2/1 and 4/1	*C. subsecundum*	11/3 and 3/0	*Chaetoceros atlanticum*	0/0 and 15/5
C. compressum	17/13 and 24/9	*C. teres*	15/7 and 2/0	*C. lorenzianum*	2/0 and 14/9
C. constrictum	22/12 and 26/19	*C. van heurckii*	9/5 and 1/0	*C. peruvianum*	8/0 and 10/1
C. convolutum	17/7 and 22/5	*Chaetoceros* spp. non det.	3/2 and 0/0	*C. sociale*	1/0 and 5/3
C. curvisetum	15/11 and 18/12	*Rhizosolenia simplex*	8/2 and 8/0	*Rhizosolenia setigera*	0/0 and 10/3
C. debile	4/3 and 4/2	*Leptocylindrus danicus*	5/2 and 2/0	*R. styliformis*	7/0 and 17/11
C. decipiens	5/1 and 14/8	*Pleurosigma capense*	3/1 and 1/0	*Thalassionema nitzschioides*	1/0 and 7/1
C. difficile	8/5 and 14/2	*Nitzschia closterium*	10/2 and 6/0	*Nitzschia longissima*	2/0 and 6/1
C. didymum (vegetative phases)	12/6 and 10/2	*Peridinium*[+] *crassipes*	3/1 and 0/0	*Codonella* spp.	1/0 and 17/1
C. strictum	11/7 and 10/6	*Gonyaulax spinifera*	14/6 and 2/0		
C. tetras	4/1 and 2/2	*Ceratium candelabrum*	13/1 and 4/0		
Rhizosolenia alata	23/6 and 22/5	*C. lineatum*	11/1 and 4/0		
R. hebetata	26/12 and 18/7	*C. fusus*	24/2 and 19/0		
R. imbricata	7/1 and 13/2	*C. tripos*	9/1 and 6/0		
Dactyliosolen mediterraneus	7/1 and 10/6	*C. arietinum*	19/1 and 12/0		
Fragilaria granulata	1/1 and 14/5	Acanthometridae	4/4 and 5/0		
*F. karstenii**	5/3 and 9/7	Radiolaria (other)	22/1 and 21/0		
Asterionella japonica†	17/9 and 14/7	*Rhabdonella brandtii*	9/2 and 1/0		
Nitzschia delicatissima	19/14 and 21/19	*Rhabdonella* spp.	10/1 and 4/0		
N. seriata	22/9 and 28/18	*Dictyocysta* spp.	20/2 and 13/0		
Peridinium spp. non det.	36/8 and 39/16	Ova, ? molluscan	2/1 and 0/0		

TABLE II—*continued*

During both surveys		During the first survey only		During the second survey only
Trichodesmium thiebautii	4/1 and 8/3	Pteropoda: *Limacina* juv.	4/3 and 6/0	
Foraminifera	24/3 and 33/7	Ova: euphausian	2/1 and 0/0	
Ova, mainly copepodan	21/10 and 28/3	Appendicularia	17/1 and 5/0	
Egg packets, mainly harpacticid	23/3 and 16/1			
Nauplii	37/16 and 37/14			
Copepoda	38/15 and 35/14			
Cast skins, etc.	9/4 and 32/10			
Faecal pellets	15/7 and 27/4			

**Delphineis karstenii*
†Asterionella glacialis
+Protoperidinium?

styliformis, *R. alata*, *Thalassiothrix longissima*, *Chaetoceros* spp., *Peridinium* (sic) and *Ceratium* species, and *Planktoniella sol*. The principal genera of diatoms recorded by Pieterse & Van der Post (1967) in the vicinity of Walvis Bay during their two-year study were, in order of importance *Thalassiosira* (the biogeography of various *Thalassiosira* species was discussed by Hasle, 1976), *Nitzschia*, *Leptocylindrus*, *Chaetoceros*, and *Thalassionema*. During autumn 1968, Hulburt (1976), who considered the northern Benguela system as not being nutrient-limited, recorded *Chaetoceros* spp., *Thalassiosira* spp., *Nitzschia* spp., *Rhizosolenia stolterfothii*, and *Asterionella glacialis* (*japonica*) as being the most common diatoms.

Hart & Currie (1960) noted that dinoflagellates were dominated by diatoms by around one to two orders of magnitude, and Kruger & Cruickshank (1982) found that the dinoflagellates comprised only 4% by number of the phytoplankton during their fish shoal ecology experiment in the northern Benguela. Dinoflagellates can dominate during quiescent conditions when there is a well-developed thermocline, such as is sometimes *encountered during summer (Austin, 1980) e.g.* red tides (Copenhagen, 1953; Brongersma-Sanders, 1947; Hart & Currie 1960; Pieterse & Van der Post, 1967). Hobson (1971) noted that microflagellates and *Gymnodinium* spp. contributed a major fraction to the phytoplankton carbon at all but two of his stations. The most commonly encountered dinoflagellates near Walvis Bay during the 1964–1966 red tides were *Heterocapsa triquetra*, *Gymnodinium galatheanum*, *Scrippsiella trochoidea* (*Peridinium trochoideum*), and *Gonyaulax tamarensis* (Pieterse & Van der Post, 1967). Hart & Currie (1960) also observed *Peridinium* (sic) spp. as being fairly common (see Table II). The dinoflagellate, *Brachydinium taylorii*, previously only found in the Mozambique Channel was recorded 100 km offshore at 19° S in August 1972 by Kruger (1979).

The average percentages of the main diatom groups in the area north of 28° S, as recorded by Hart & Currie (1960), are summarized in the self explanatory Table III. Chaetoceraceae was the main group in the neritic zone, while Soleniineae and Pennatae were relatively more important further offshore. Reyssac (1973) obtained very similar results, and in Fig. 3 the dominance of *Thalassiosira* and *Chaetoceros* species inshore and of *Rhizosolenia* and *Nitzschia* species at his outermost station (situated just west of the shelf-break) is evident. This figure illustrates the relative variability of the various dominant diatoms off central Namibia during 1965–1966. Readers are referred to Kollmer (1962, 1963), Pieterse & Van der Post (1967) and Austin (1980) for further information about the species monthly distribution in the region during 1958, 1959, 1964–1966, and 1971, respectively.

Southern Benguela area, north of Cape Point

In spite of the extensive monthly sampling in the southern Benguela during the 1950s and 1960s as part of the study of the pilchard, *Sardinops ocellatus*, surprisingly little has been published on the biogeography of the phytoplankton. Apart from the early work of *e.g.* Hendey (1937) whose analysis of material from six stations around 33° S indicated that

TABLE III

The average percentages of the main diatom groups in the Benguela north of 28° S during autumn 1950 and spring 1950 according to Hart & Currie (1960)

	Discineae	Biddulphiineae	Chaetoceraceae	Soleniineae	Pennatae
Inshore stations (<74 km from coast)					
Survey I (autumn)	3	2	80	1	14
Survey II (spring)	1	0	89	0	10
Outer shelf stations (74–185 km from coast)					
Survey I (autumn)	1	0	93	1	5
Survey II (spring)	4	0	38	34	25
Offshore stations (>185 km from coast)					
Survey I (autumn)	13	0	14	20	52
Survey II (spring)	2	0	18	66	14

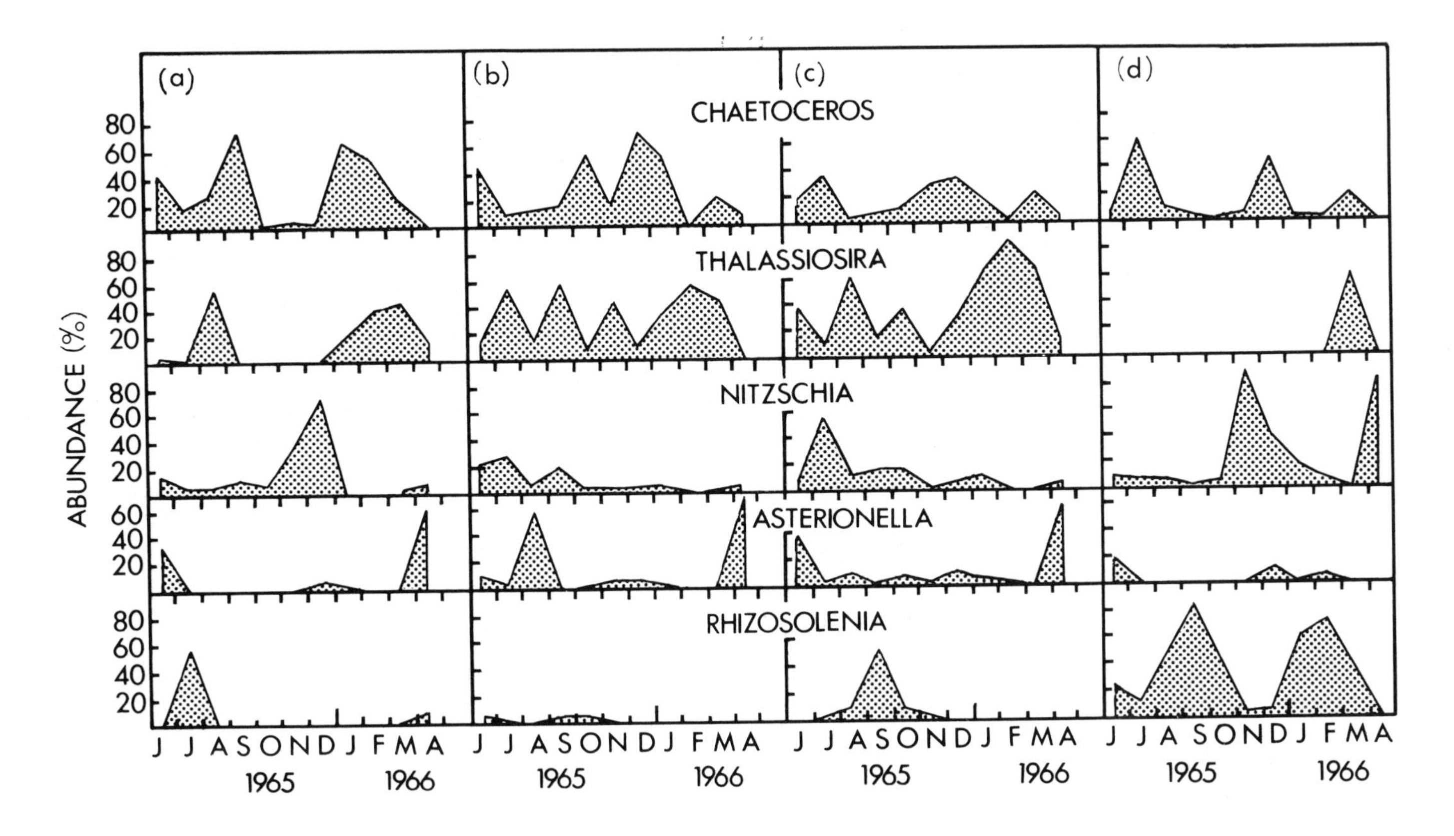

Fig. 3.—Relative abundance of the five principal diatom genera during 1965–1966 at stations 7 km (a), 28 km (b), 67 km (c), and 159 km (d) west of Walvis Bay (after Reyssac, 1973).

Planktoniella sol (present at all stations), and species of *Coscinodiscus*, *Rhizosolenia*, *Thalassiosira*, and *Fragilaria* were the main diatoms, and the study of Copenhagen & Copenhagen (1949) off the Cape Peninsula—*Chaetoceros* spp., *Thalassionema nitzschioides*, and *Melosira* sp. (*Skeletonema costatum*?) were the principal microplankton recorded—the only published taxonomic text of any substance is the major work of Boden (1950). Nevertheless, the studies by De Jager (1954, 1957) on the association between the environment and the phytoplankton in St Helena Bay, and of Davies (1957) on the feeding of pilchards in the area are useful. The most common diatom genera recorded by De Jager were *Chaetoceros* (most abundant), *Nitzschia*, *Thalassiosira*, *Skeletonema*, *Rhizosolenia*, *Coscinodiscus*, and *Asterionella*. Davies (1957) observed that phytoplankton contributed 70% of the diet of the pilchard (21% diatoms, 4% dinoflagellates, 45% unrecognizable phystoplankton) and that species of *Thalassiosira, Coscinodiscus*, *Rhizosolenia*, *Chaetoceros*, *Nitzschia*, *Ceratium*, and *Peridinium* (sic) were frequently encountered. Both he and De Jager noted that diatoms appeared to be the dominant group. Dinoflagellates commonly encountered by De Jager (1957) included species of *Prorocentrum*, *Ceratium*, and *Peridinium* (sic). Red-tide organisms are often present in the phytoplankton from the South Western Cape and several major blooms have been documented (e.g. Grindley & Nel, 1970; Brown, Hutchings & Horstman, 1979; Popkiss, Horstman & Harpur, 1979; Horstman, 1981). These will be discussed in the section on red tides.

In studies on marine alpha-radioactivity, Shannon (1969, 1972) listed the principal bloom causing phytoplankton species in samples collected in the southern Benguela, as analysed by Mrs E. Coghlan (née Nel). *Skeletonema costatum*, *Chaetoceros* spp., and *Nitzschia* spp. were the diatoms most commonly encountered. Although the detailed treatise of Mrs Coghlan and her co-workers on the phytoplankton around the South Western Cape during the mid 1960s has not been published, D. A. Horstman (pers. comm.) has indicated that *Chaetoceros* spp., *Skeletonema costatum*, and *Thalassiosira* spp. were the dominant diatoms, with *Skeletonema costatum* being the most abundant single species. In a study of the phytoplankton communities in relation to upwelling off the Cape Peninsula during the austral summer of 1972–1973, Olivieri (1983a) noted that *Nitzschia* spp. dominated, although *Thalassionema nitzschioides*, *Chaetoceros* spp., and *Skeletonema costatum* were significant during some months. From her work and previous records she concluded that a wide diversity can dominate coastal waters at various stages after upwelling. D. A. Horstman (pers. comm.) has observed that *S. costatum* has been very common off the Cape Peninsula during recent years *i.e.* analogous to its dominance during the mid- and late 1960s. It may be pure coincidence that there was a minimum in the equatorward wind stress in the southern Benguela region in 1965 and again more recently, and that the early 1970s were generally warmer years (refer to Shannon, 1985). Lower equatorward wind stress during summer would result in increased statification and a tendency towards nutrient depletion in the surface layer. As pennate diatoms have a lower sinking rate than centric diatoms in nutrient-depleted water (Bienfang & Harrison, 1984), species such as *Nitzschia* may be better adapted than *Chaetoceros*, *Skeletonema*, and *Thalassiosira* for survival and reseeding during "warm" periods (reduced nutrient supply, increased stratification)—refer also to Parsons (1979). Whether or not cyclical changes in the dominant diatoms have occurred in the southern Benguela on a time scale of decades is, however, a matter for speculation. Nevertheless, the dominance of *Nitzschia* spp. during 1972–1973 suggests that these changes do occur, with obvious implications for secondary producers.

Agulhas Bank region

In a study in the western Agulhas Bank, covering the nearshore region between Cape Point and Danger Point during 1975, Tromp, Lazarus & Horstman (1975) reported that *Chaetoceros costatum* and *Hemiaulus haukii* were the dominant "cosmopolitan" species encountered while *Chaetoceros compressum*, *C socialis*, *C. curvisetum*, *Nitzschia pacifica*, *N. seriata*, *Leptocylindrus danicus*, *Hemiaulus sinensis*, *Rhizosolenia delicatula*, and *Asterionella glacialis* (*A. japonica*) were the dominant "temperate" diatoms. The plankton of the Agulhas Bank proper has been discussed by De Decker (1973) and Carter & De Decker (1983). De Decker (1973) regarded this area as contiguous with the Benguela system. The most common diatoms on the bank according to De Decker were *Chaetoceros socialis*, *Leptocylindrus danicus*, *Nitzschia seriata*, and *N. delicatissima*, the last being common throughout the southwestern Indian Ocean (Nel, 1968).

In concluding this floristic section it is perhaps appropriate to mention the diatom *Anaulus birostratus*, mainly confined to the surf zone, which frequently occurs along beaches on the south coast and in False Bay, causing a localized brown discolouration of the water. The morphology of this species has been documented recently by Kruger & Wilson (1984).

RED TIDES

Outbreaks of red water are common in both northern and southern *Benguela regimes. They tend to be recorded close inshore (e.g.* Shannon, 1965), although they have been reported up to 30 km from the coast on occasions (Hutchings, Nelson, Horstman & Tarr, 1983), and appear to be most prevalent during quiescent periods which follow upwelling events. Brongersma-Sanders (1947) intimated that the presence of upwelled water was necessary for red-tide blooms, although obviously red tides also occur in non-upwelling areas. Causative organisms in the southern Benguela are generally dinoflagellates and sometimes ciliates, the majority of which are non-toxic (Horstman, 1981). The most common red-tide organism in the southern Benguela is *Noctiluca scintilans* (*N. miliaris)*, a non-toxic dinoflagellate (Brown *et al.*, 1979), which has also been reported near Walvis Bay by Marchand (1928) and others. Red-tide blooms often co-exist together with diatom blooms.

Red tides in the Benguela have been associated with mortalities of fish, mussels and crustaceans. Gilchrist (1914) recorded several early outbreaks which he linked to fish mortalities. Brongersma-Sanders (1948) first suggested that the mass mortalities in the Walvis Bay area were due to a dinoglagellate, a *Gynmodinium* sp., and Copenhagen (1953) cited a private communication from Professors Spärk and Nielsen of an intense bloom of a *Glenodinium* sp. in the area which coincided with a mass mortality of fish during December 1950. It was present in the top metre (1% light penetration depth was at 0·6 m) and appeared as a brown soup. This organism was subsequently recorded off Walvis Bay by Pieterse & Van der Post (1967) and Reyssac (1973) in substantial numbers. A bloom of *Heterocapsa triquetra* (''khaki'' coloured water) was recorded off Sandwich Harbour by Hart & Currie (1960) during March 1950 and they discussed its possible implication in fish mortalities. These authors also recorded a bloom of the ciliate *Mesodinium rubrum* inshore off southern Angola. In an intensive set of surveys in Walvis Bay during 1964–1966, Pieterse & Van der Post (1967) recorded blooms of four main species, *Heterocapsa triquetra* (the most common bloom organism), *Gymnodinium galatheanum*, *Gonyaulax tamarensis*, and *Scrippsiella trochoidea*. Blooms were associated with mortalities of juvenile fish on two occasions. These authors linked the blooms to pollution from fish factories, and although they found that neither temperature nor salinity was critical, blooms only occurred during calm periods.

Mortalities of marine life in the southern Benguela have generally been linked with a few species of *Gonyaulax* and *Mesodinium rubrum*. Blooms of the latter were first recorded off the Cape Peninsula by Hart (1934). Grindley & Taylor (1962, 1964) noted that a mortality in False Bay in March 1962 was associated with the decay of dense concentrations of the non-toxic *Gonyaulax polygramma*, while Grindley & Nel (1970) documented the mortality of hundreds of thousands of white sand mussels, *Donax serra*, and numbers of black mussels, *Choromytilus meridionalis*, near Elands Bay (north of St Helena Bay) during December 1966 due to *Gonyaulax grindleyi* (a new species). Popkiss *et al*. (1979) reported 17 cases of paralytic shellfish poisoning along the western coast during May 1978 as a result of an extensive bloom of *G. catanella*. Several fish mortalities have been caused by gill clogging, for example during the bloom of a *Gymnodinium* species in False Bay at the end of August 1976 (Brown *et al*., 1979). Horstman (1981) has reviewed all the reported red-tide occurrences in the southern Benguela region between 1959 and 1980, and listed 14 species responsible of which only two *viz*. *Gonyaulax catanella* and *G. grindleyi* were consistently toxic. The bloom species were: *Noctiluca miliaris*, *Mesodinium rubrum*, *Prorocentrum micans*, *Scrippsiella trochoidea*, *Ceratium furca* v. *berghii*, *Gonyaulax spinifera*, *G. polygramma*, *G. catanella*, *G. grindleyi*, *Gymnodinium splendens*, *G. arcuatum*, *Dinophysis acuminata*, *Protoperidinium minutum*, and an unidentified species of *Gymnodinium*. Huge mortalities of black and white mussels occurred at sites along the western coast during December 1967, September 1974, May 1978, and March 1980 due to *Gonyaulax catanella* and *G. grindleyi* (Horstman, 1981). Numerous red tides were reported during the first quarter of 1983, in particular *G. cantanella* by Hutchings, Holden & Mitchell-Innes (1984) during the warm event in the southern Benguela (Shannon, Crawford & Duffy, 1984). According to Horstman (1981) red-water blooms occur most frequently in the southern Benguela during autumn, although they are common in spring and summer.

That red tides occur regularly in the Benguela is not unexpected. Diatoms, which dominate the phytoplankton in upwelling ecosystems, have high nutrient requirements and are adapted to turbulent conditions (*cf.* Parsons 1979). Diatom blooms, resulting from the high nutrients in newly-upwelled water, quickly strip nutrients from the stabilizing water mass, thereby (together with the calm sunny weather which generally succeeds a bout of upwelling) cause the environment to change into one in which dinoflagellates can out-compete diatoms, due to the dinoflagellates preference for a stratified, high light environment and an ability to grow relatively more efficiently at low nutrient concentrations.

DISTRIBUTION OF BIOMASS AND SEASONAL CHANGE

The Namaqua–Lüderitz area

The principal centre of upwelling in the Benguela system is in the vicinity of Lüderitz (27° S), while another important upwelling centre is situated about 400 km further south near Hondeklip Bay in Namaqualand (Shannon, 1985). Upwelling favourable winds blow in the central Benguela throughout the year but with a maximum during the last quarter and a minimum between May and July (Bailey, 1979). Thus, the central Benguela region forms an effective 'buffer' between the northern and southern regions.

Relatively few stations have been occupied in the Lüderitz and Namaqua zones and consequently not much has been published on the phytoplankton in the area. The essential features of the horizontal distribution of microplankton (mainly diatoms) in the Benguela region off Namibia (*i.e.* north of the Orange River), during quasi-quiescent and active upwelling phases are illustrated in Figure 4, redrawn from diagrams in Hart & Currie (1960). During the first survey (quasi-quiescent conditions) in March 1950, Hart & Currie (1960) noted that the rich zone of coastal phytoplankton (largely *Chaetoceros* spp.) was narrow in the south near the Orange River, but broadened northwards. During their second survey in active upwelling conditions in September–October 1950 the zone of rich coastal phytoplankton was smaller, but offshore it was richer (*Planktoniella sol* was common) than on the first survey, but still more scanty than inshore. Maximum concentrations were recorded off Sylvia Hill, and small zooplankton were observed feeding voraciously. Relatively high concentrations were recorded offshore west of the Orange River, a feature noted subsequently also by Shannon, Schlittenhardt & Mostert (1984) in the chlorophyll distribution as deduced from NIMBUS-7 Coastal Zone Colour Scanner (CZCS) imagery of the region (Fig. 5). In this figure the progressive broadening of the productive neritic zone between 31 and 29° S is evident, and Shannon *et al.* (1984) speculated whether the S-shaped offshore feature was due to the proximity of the Orange Banks (150 km offshore) or whether it was an (advected) extension of the Namaqua upwelling tongue. What is significant about Figure 5 and other CZCS pigment maps is that the colour front tends to follow the shelf-break and that the productive area narrows appreciably at about 28° S *i.e.* immediately south of Lüderitz. The small zones of low "chlorophyll" adjacent to the coast in the Orange Bight may have been due to the presence of localized freshly upwelled water or to atmospheric contamination.

Ship studies of chlorophyll in the central Benguela indicated that

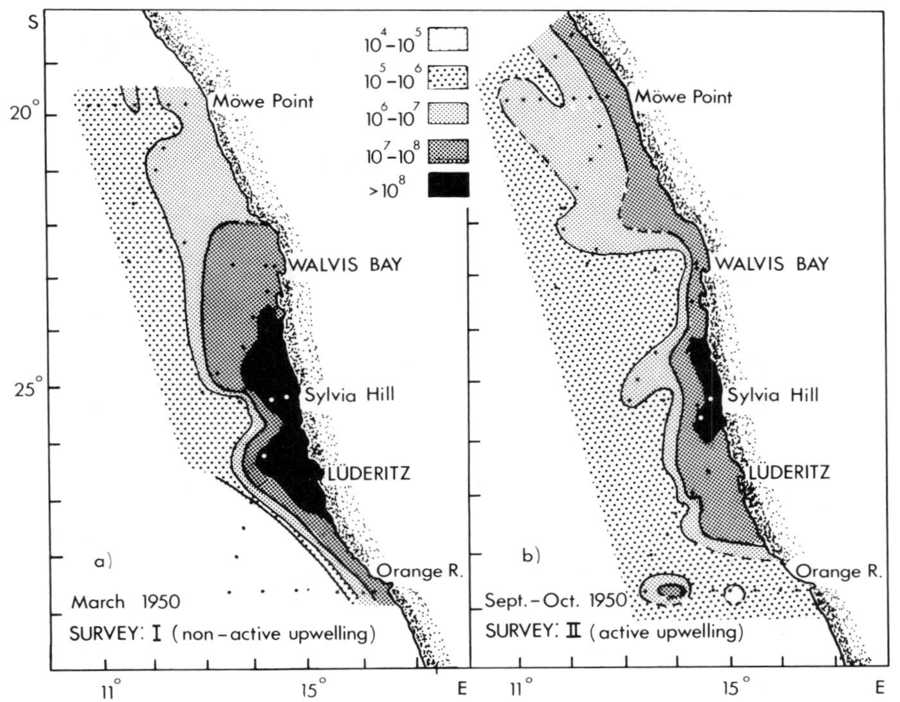

Fig. 4.—Distribution of the microplankton, estimated totals per net haul, off Namibia (from Hart & Currie, 1960).

relatively low (<3 mg·m^{-3}) chlorophyll *a* concentrations found off Lüderitz were associated with strong southerly winds and active upwelling (Fig. 6b)—Shannon, Hutchings, Bailey & Shelton (1984). High surface chlorophyll *a* concentrations (>10 mg·m^{-3}) were recorded in a 60 km wide coastal band during the intermediate and quiescent phases (Figs 5a and 5c, respectively; Shannon and Hutchings *et al.* 1984).

Little is known about the vertical distribution of plankton in the central part of Benguela. Bailey (1979) suggests that the top 20–50 m layer is well mixed, except during quiescent conditions when the high chlorophyll *a* concentrations are confined to the upper 10–15 m layer.

The northern Benguela

The northern Benguela, the area between approximately 17 and 25° S, is characterized by upwelling throughout most of the year (less intense than in the Namaqua-Lüderitz area) with a maximum during late winter and spring and a minimum during late summer (Shannon, 1985).

The typical standing crop of phytoplankton recorded by Hart & Currie (1960) in the area north of 24° S was in the range 10^6–10^7 cells·l^{-1} which is a concentration of the same order as that noted subsequently by Kollmer (1963), Kruger & Cruickshank (1982), and others. The distribution of chlorophyll *a* in the region has been discussed by Shannon & Hutchings *et*

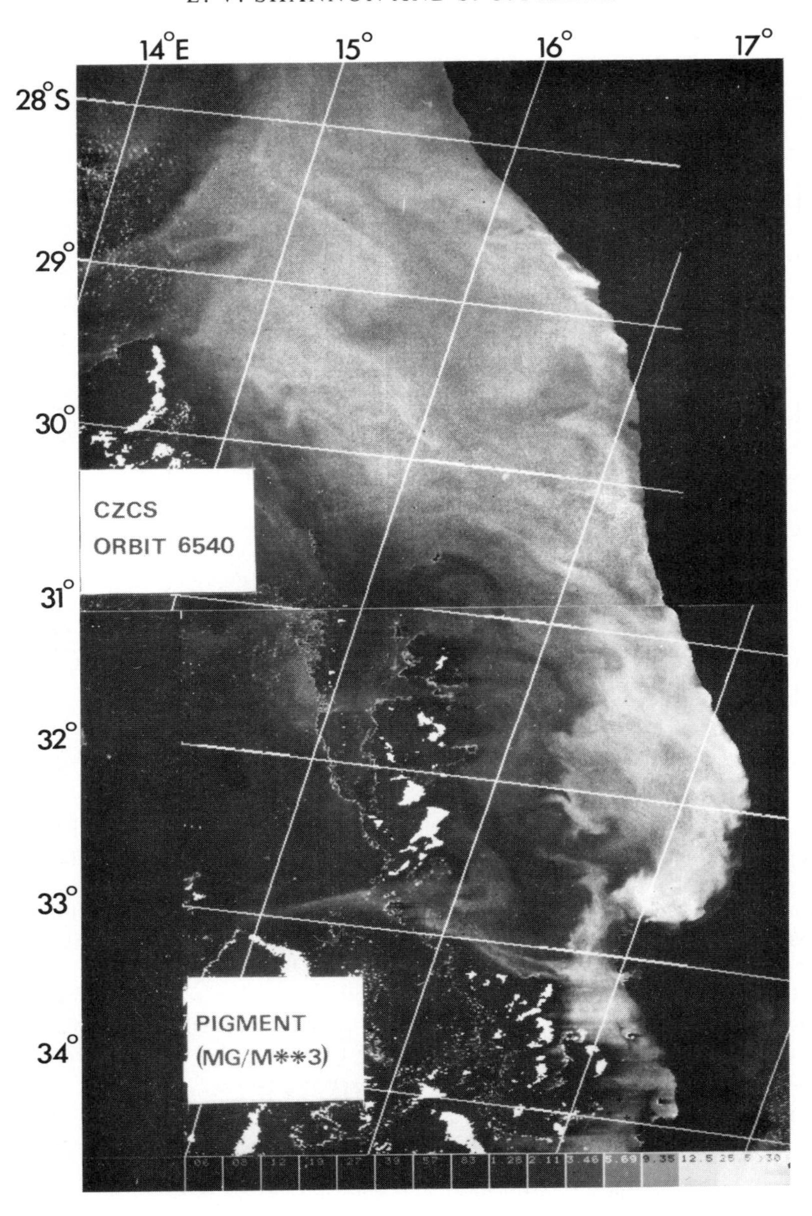

Fig. 5.—The distribution of estimated average euphotic chlorophyll in the southern Benguela region, CZCS orbit 6540, 9th February, 1980 (from Shannon, Schlittenhardt & Mostert, 1984).

al. (1984). They noted that vertical gradients of chlorophyll were closely associated with the stratification of the water column, and that, although subsurface maxima at depths up to 20 m were common, the concentrations in these maxima were seldom more than double the surface value, and

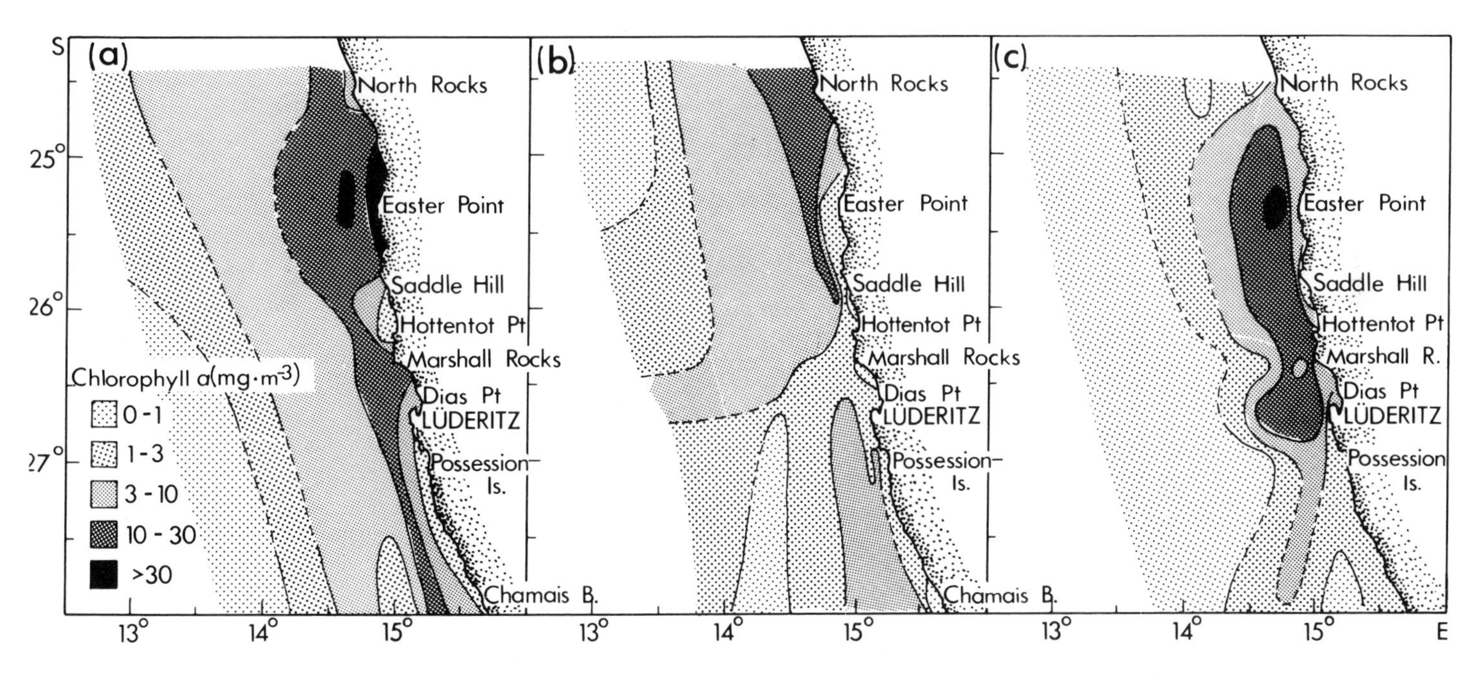

Fig. 6.—Surface chlorophyll *a* (mg·m^{-3}) in the Lüderitz region during various phases (a) intermediate—May 1976, (b) upwelling—November 1976, and (c) quiescent—February 1977 based on Bailey (1979) as redrawn by Shannon & Hutchings *et al*. (1984).

usually only less than 20% higher. The vertical distribution of phytoplankton off central Namibia from Kollmer's (1962) paper is shown in Figure 7 and from this it is evident that it is fairly uniform in the top 10–30 m. The surface layer is well mixed during much of the year and surface phytoplankton or chlorophyll *a* concentrations serve as fair indictors of activity in the euphotic layer, with obvious applications for the satellite remote sensing of the area (Shannon & Hutchings *et al.*, 1984). If the typical surface chlorophyll *a* concentration over the shelf is taken as being between 3 and 10 $mg \cdot m^{-1}$ on average (see Fig. 9), and if a mixed surface layer depth of 25 m and a chlorophyll *a*:carbon conversion ratio of 50 are assumed (the latter appears to be reasonable from the work of Hobson, 1971), then the standing stock of phytoplankton carbon in the neritic zone of the northern Benguela would be, on average, in the range 4–12 $g\ C \cdot m^{-2}$—a value of the same order as given by Shannon & Field (1985) for the southern Benguela region. Zernova (1974) indicated a phytoplankton biomass for the northern Benguela neritic zone of $0{\cdot}5$–$1{\cdot}3\ g \cdot m^{-3}$. If a mixed euphotic zone of 25 m and a dry mass:carbon ratio of 2·5 are assumed, Zernova's biomass estimate is equivalent to 5–13 $gC \cdot m^{-2}$ *i.e.* virtually identical to the biomass estimated from chlorophyll data.

The work of Kruger (1983) suggests that, during the austral summer, the phytoplankton concentrations reach a maximum about 15–20 km offshore, and then decline to about 10% of the maximum value about 100 km offshore (Fig. 8). Shannon & Hutchings *et al.* (1984) selected three months, May and September 1971 and January 1972 as representatives of an approaching upwelling season (intermediate situation), maximum upwelling, and quiescent conditions, respectively, and the surface temperature and chlorophyll *a* distributions for these months are shown in Figure 9. The zones of moderate to high chlorophyll *a* concentrations (3 to >10 $mg \cdot m^{-3}$) corresponding to a surface temperature of 13–15 °C during the intermediate phase are evident in Figure 9a. During active upwelling (Fig. 9b) both the newly upwelled water and the oceanic water were depleted in chlorophyll *a* with the result that the maximum occurred as a band situated offshore. During the quiescent phase (Fig. 9c) highest chlorophyll *a* concentrations were recorded in a narrow coastal band. The changes in phytoplankton and zooplankton distribution during September 1982 to March 1983 have been discussed by Kruger & Boyd (1984), and a selected set of maps is given in Figure 10. These adequately show the seasonal changes in the plankton volumes north of 26° S, with the dramatic increase in both phytoplankton and zooplankton abundance in the Walvis Bay area and further south with the approach of autumn. Although we are not aware of any published work on the effect of grazing of phytoplankton by zooplankton off Namibia, the impact implicit in Fig. 10c,d (north of 22° S) appears to be substantial.

When Figures 9 and 10 are viewed together with Figures 4 and 6, some striking features and similarities emerge. During quiescent and reduced upwelling phases, the maximum phytoplankton concentrations appear to occur south of 22–23° S, within 80–120 km of the coast. (The shape of the 10 $mg \cdot m^{-3}$ contour in Figure 6 is very similar to the shape of the boundary of the rich microplankton zone in Figure 4a.) Also, the high chlorophyll *a* within 60 km of the coast during active upwelling (Fig. 9b) is similar to

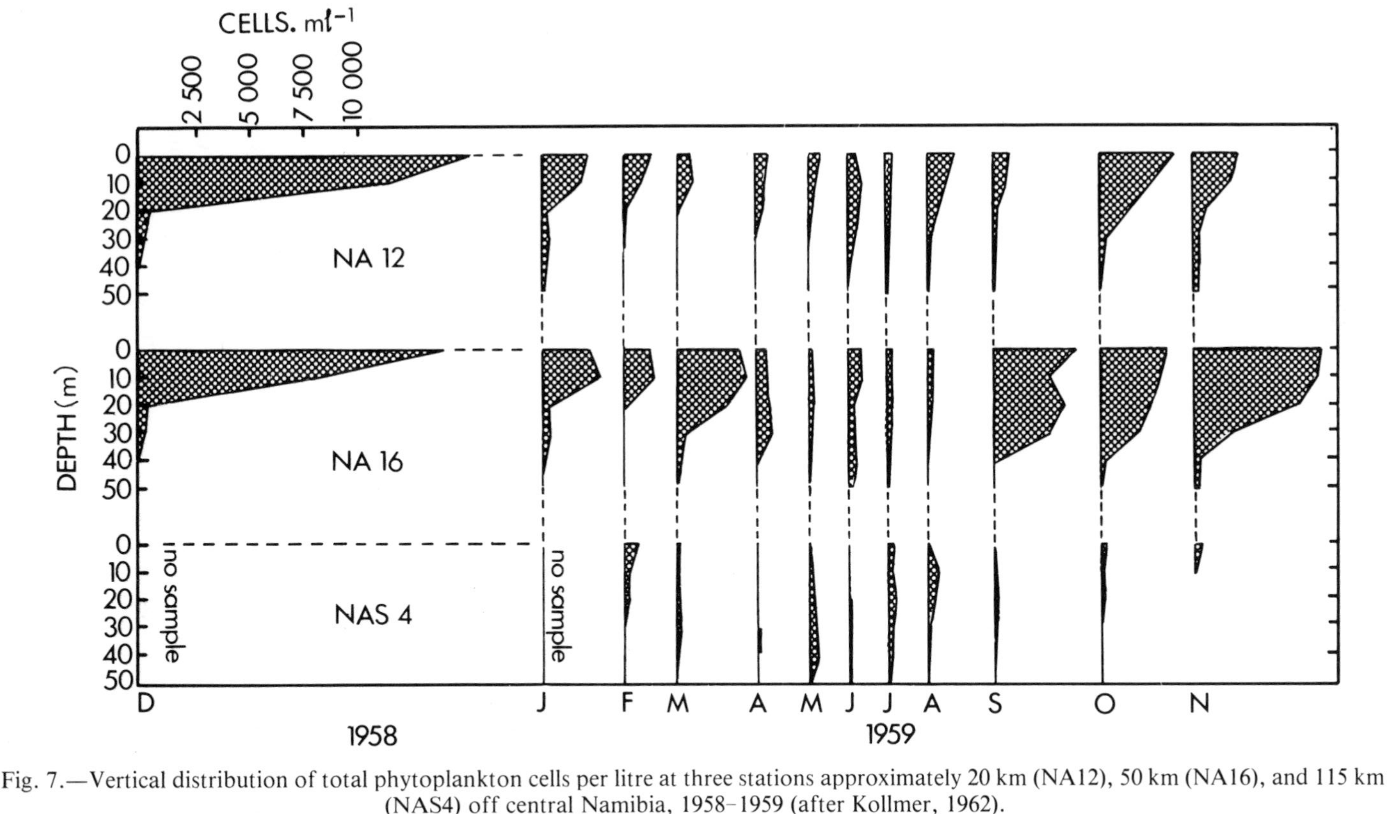

Fig. 7.—Vertical distribution of total phytoplankton cells per litre at three stations approximately 20 km (NA12), 50 km (NA16), and 115 km (NAS4) off central Namibia, 1958–1959 (after Kollmer, 1962).

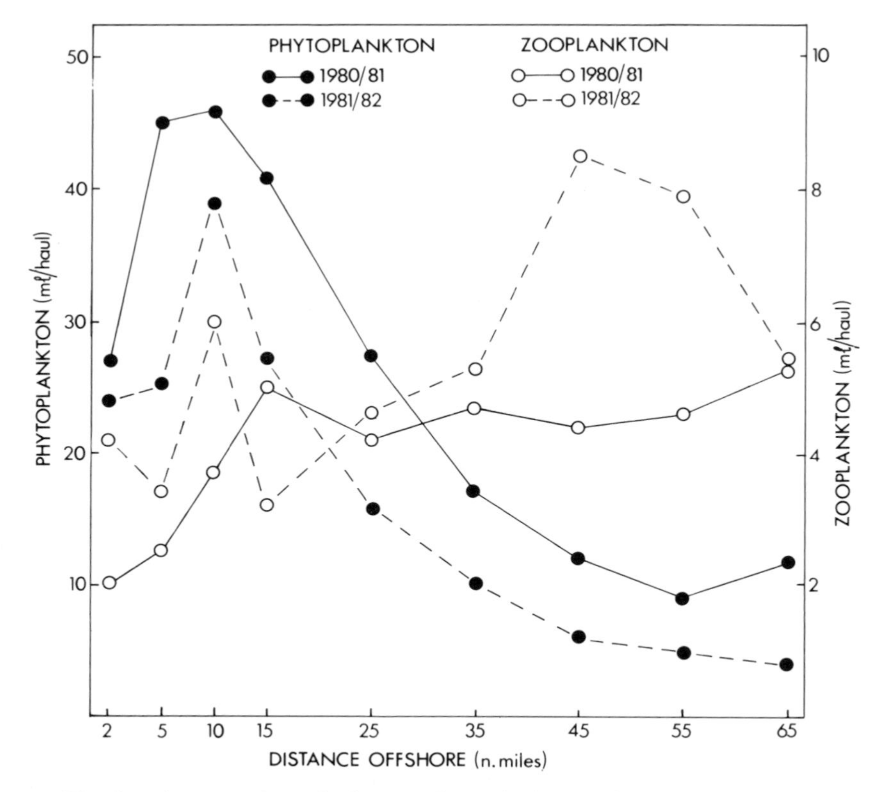

Fig. 8.—Average phytoplankton and zooplankton volumes plotted against distance offshore off Namibia during the period September–March in 1980–1981 and 1981–1982 (after Kruger, 1983).

Figures 4b and 10a. Furthermore, there appears to exist a broad zone of lowish chlorophyll (1–3 $mg \cdot m^{-3}$) at the approximate latitude of Walvis Bay during active upwelling, possibly associated with the zone of convergence or reduced upwelling between 22 and 23° S noted by Shannon (1985). A consistent feature of the data in Kruger & Boyd (1984) and Shannon & Hutchings *et al.* (1984), but not evident in Hart & Currie (1960) owing to station-line spacing, is the narrow zone of high phytoplankton biomass between Cape Cross and Möwe Point. In general, however, Hart & Currie (1960) seem to have had an uncanny foresight in their selection of station positions and in the timing of their cruises.

The studies of Kollmer (1962, 1963), Pieterse & Van der Post (1967) and Reyssac (1973) provide some insight into the seasonal changes in the dinoflagellate populations off central Namibia. The paper of Pieterse & Van der Post (1967) represents the most comprehensive account of dinoflagellates in the Benguela system, although it was limited to the general area near Walvis Bay. Aspects of work of these authors relating to red tides are discussed elsewhere. Gross changes in dinoflagellate abundance near Walvis Bay during 1964–1966 (Fig. 11) suggest that the calm conditions associated with periods of reduced upwelling, in particular during mid-summer when the water column is more strongly stratified, are a pre-requisite for dinoflagellate blooms.

The southern Benguela

The upwelling season in the southern Benguela extends between September and March, and is modulated or pulsed due to the passage of easterly moving cyclones south of the continent during the summer which result in wind reversals on a time scale of about one week (Shannon, 1985). During winter in the area south of 31° S, westerly winds tend to predominate, and winter storms plus reduced insolation result in the formation of a deep, well-mixed surface layer.

The first study of the seasonal changes in phytoplankton abundance in the southern Benguela was made during 1934–1935 by Copenhagen & Copenhagen (1949). These authors sampled within a radius of 30 km of Cape Town using towed nets, and subsequently estimated pigments colourimetrically following a method similar to that of Harvey (1934). They compared their findings with records of sea temperature and insolation. Phytoplankton blooms were noted during spring and autumn. It was not until the 1950s, however, that more comprehensive surveys of phytoplankton were undertaken (De Jager 1954, 1957), with samples being collected monthly at a number of stations in the Cape Columbine–St Helena area. During 1954 De Jager (1957) observed that phytoplankton was most abundant during spring and summer, less abundant in winter and scarce during the autumn. Densest blooms were confined to the upper 20 m. Hoy (1970) found highest chlorophyll concentrations on the west coast during late summer and early autumn, while east of Cape Point there were no obvious trends. De Decker (1973) noted that the cell numbers on the west coast were about ten times higher than on the Agulhas Bank. He found considerable month-to-month variability, but with lowest concentrations during winter and the highest abundances during summer (Fig. 12). Agulhas Bank settled volumes were characteristically about one third of those on the west coast.

Following the routine monthly surveys of the 1950s and 1960s, a programme of process related studies off the Cape Peninsula was started in 1968. A line of stations positioned along the inferred axis of the Cape Peninsula upwelling tongue—the upwelling monitoring line— was occupied monthly during 1971–1973, and spatial changes were equated with temporal changes in terms of the development of phytoplankton in upwelled water (Andrews, 1974, Andrews & Hutchings, 1980; Hutchings, 1981; Olivieri, 1983a; Brown & Hutchings, 1985). The monthly changes in the vertical distribution of chlorophyll *a* at stations 4, 22 and 50 km along this line are illustrated in Figure 13, which should be compared with a corresponding temperature graph (see Fig. 21 in Shannon, 1985 taken from Andrews & Hutchings, 1980). Distinct biomodality in the surface layer is suggested, with the spring and summer blooms corresponding to the cycles in wind and upwelling (refer to Shannon, 1985). The upwelling water off the Cape Peninsula is poor in chlorophyll *a* ($0{\cdot}4$–$0{\cdot}9\ \mathrm{mg{\cdot}m^{-3}}$) but rich in nutrients and accordingly chlorophyll maxima in the euphotic zone are characteristically recorded 20–80 km offshore

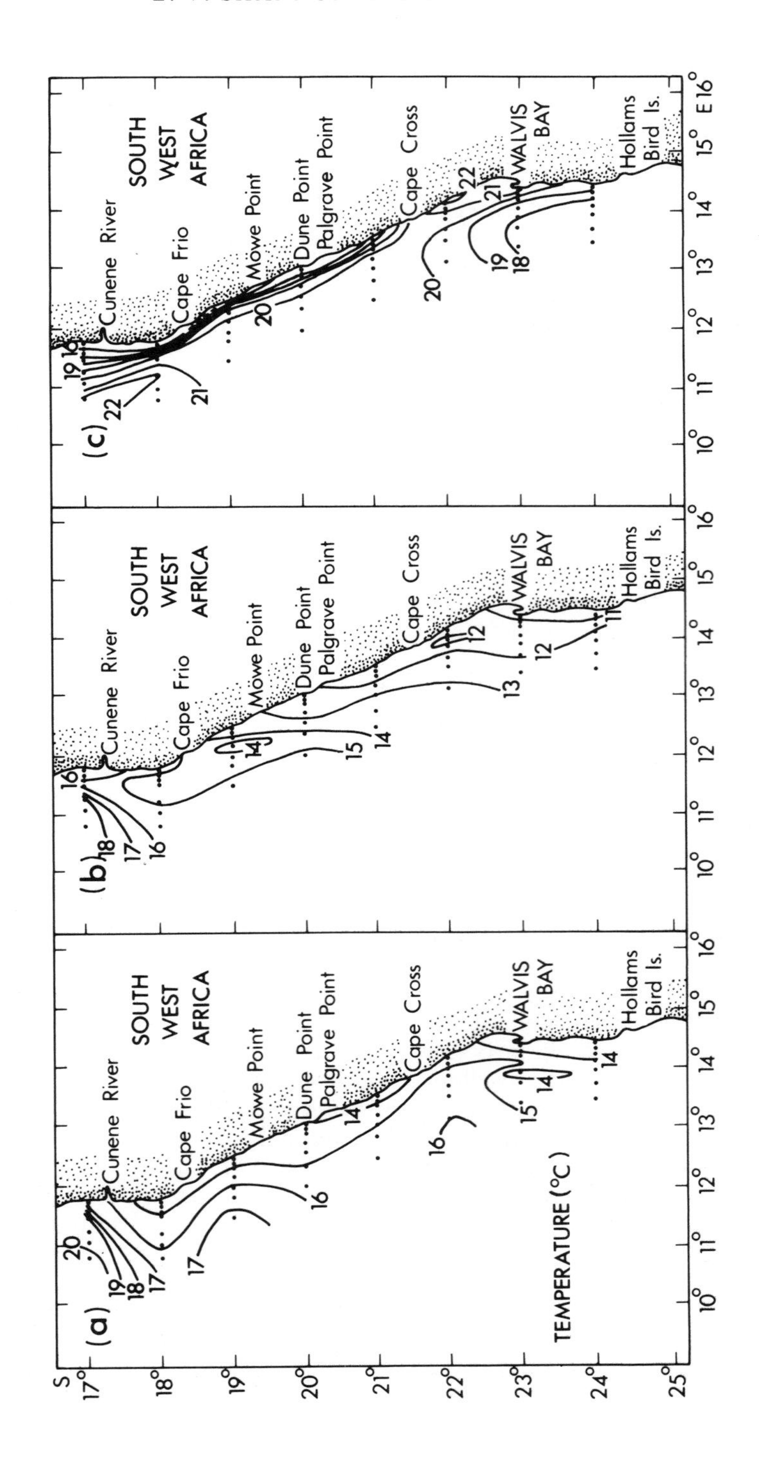
(a)
(b)
(c)
SOUTH WEST AFRICA
Cunene River
Cape Frio
Möwe Point
Dune Point
Palgrave Point
Cape Cross
WALVIS BAY
Hollams Bird Is.
TEMPERATURE (°C)
S
17°
18°
19°
20°
21°
22°
23°
24°
25°
10°
11°
12°
13°
14°
15°
E 16°

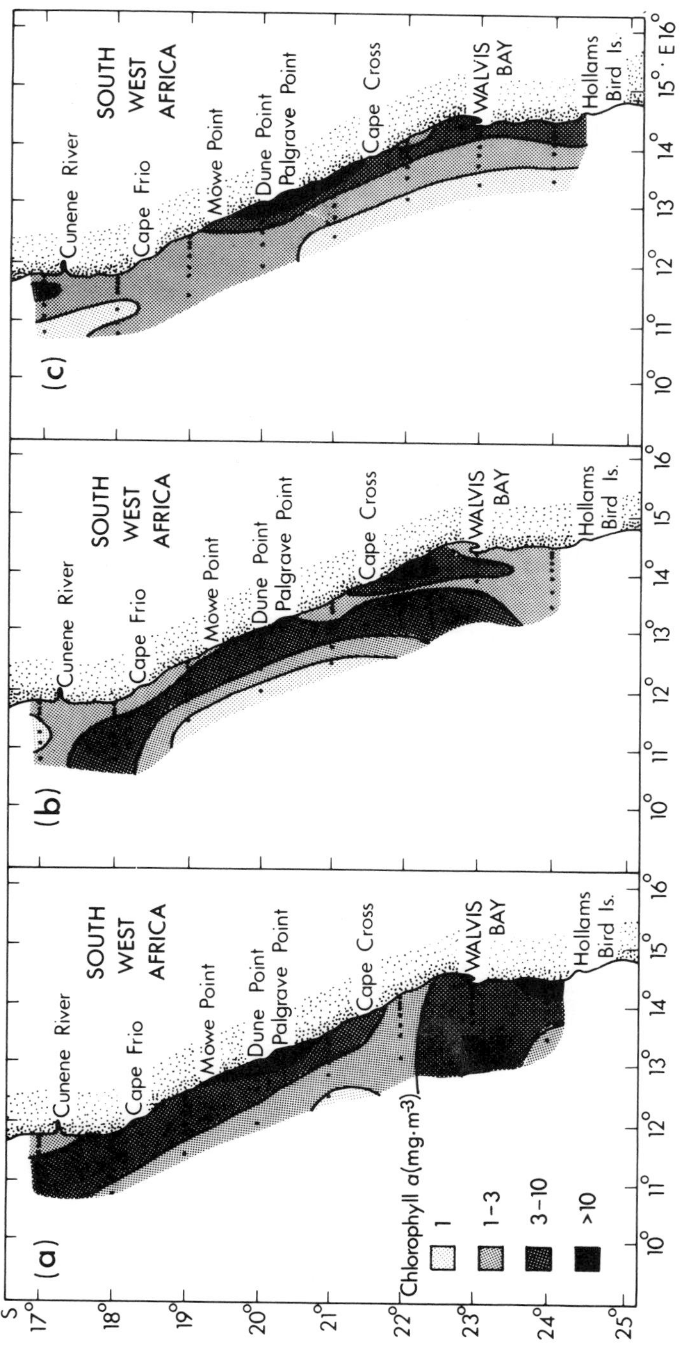

Fig. 9.—Surface temperature (°C) and surface chlorophyll *a* (mg·m^{-3}) between 17° and 24° S during various phases (a) intermediate—May 1971, (b) upwelling—September 1971, and (c) quiescent—January 1972 (after Shannon & Hutchings *et al.*, 1984).

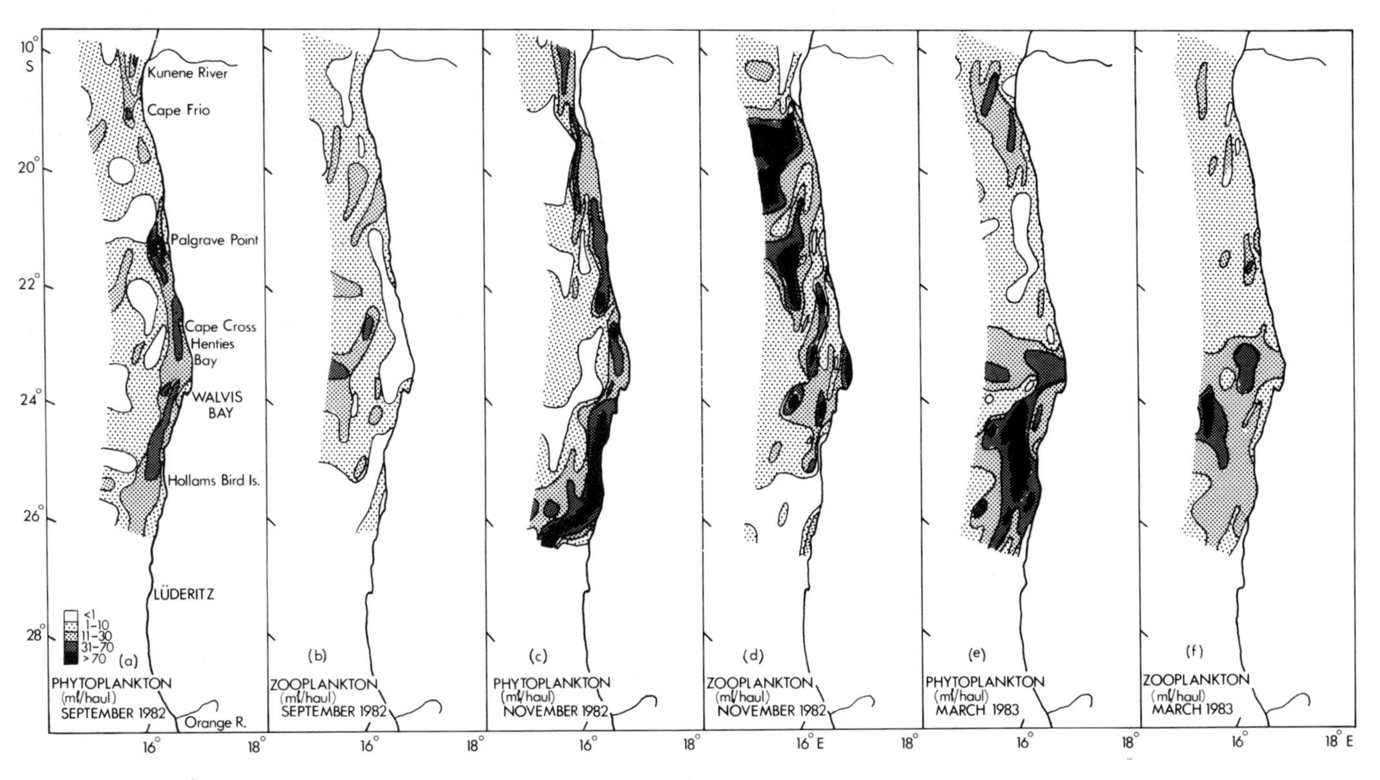

Fig. 10.—Distribution of phyto- and zooplankton off Namibia 1982–1983 (after Kruger & Boyd, 1984).

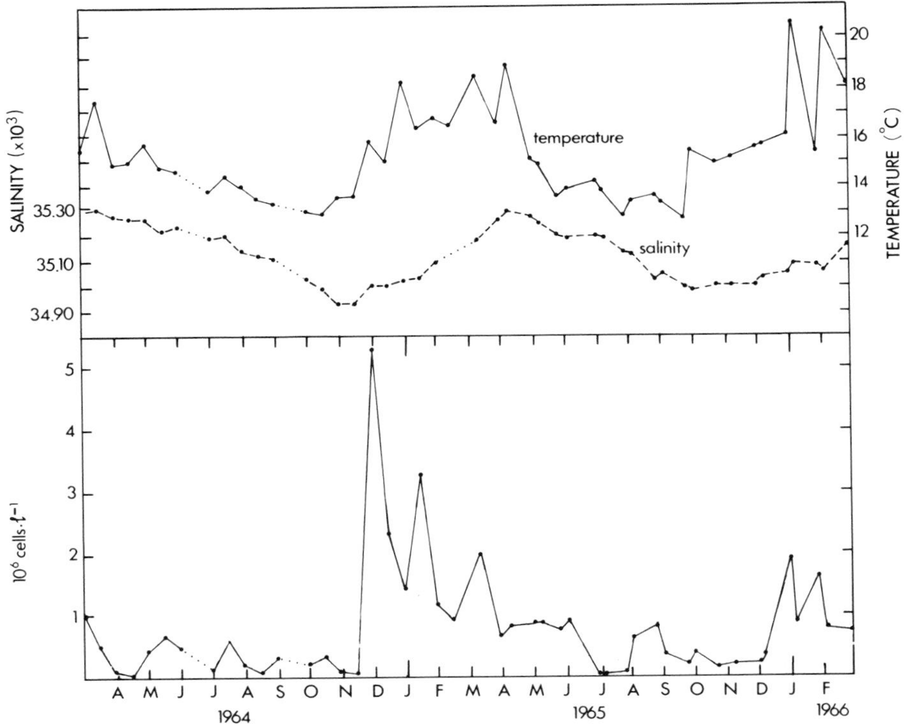

Fig. 11.—Bi-monthly averages of surface temperature and salinity, and surface and 5 m mean total dinoflagellates at stations near Walvis Bay, 1964–1966 (from Pieterse & van der Post, 1967).

(Andrews & Hutchings, 1980). According to these authors oceanic water and shelf water have chlorophyll *a* concentrations of about 0·4 and 0·6 $mg \cdot m^{-3}$, respectively. The seasonal variation in the phytoplankton standing stock in the upper 50 m is shown in Figure 14 (after Andrews & Hutchings, 1980). The chlorophyll *a*:carbon conversion ratio measured by them was characteristically 140 during winter and 70–90 during the upwelling season, and the mean phytoplankton biomass in terms of carbon was $11 \cdot 2$ $g \cdot m^{-2}$. During the upwelling season the biomass was higher and more variable (pulsed upwelling) than in winter (reduced upwelling, insolation and water column stability, increased turbulence). At a nearshore site Brown (1980, 1984), however, observed a similar mean phytoplankton biomass during summer and winter, although concentrations were far more variable in the upwelling season than in winter. Nitrate was noted by Andrews & Hutchings (1980) as being the limiting nutrient in the southern Benguela (refer also to Chapman & Shannon, 1985), and in Figure 15 (from Olivieri 1983a) the reduction in nutrients in the upper 50 m with increasing distance offshore, and the corresponding increase in the phytoplankton biomass is illustrated quite dramatically for the 1972–1973 upwelling season. (Recent work by Probyn, 1985, in the area strongly suggests that net plankton productivity is largely nitrate controlled, whereas nanoplankton

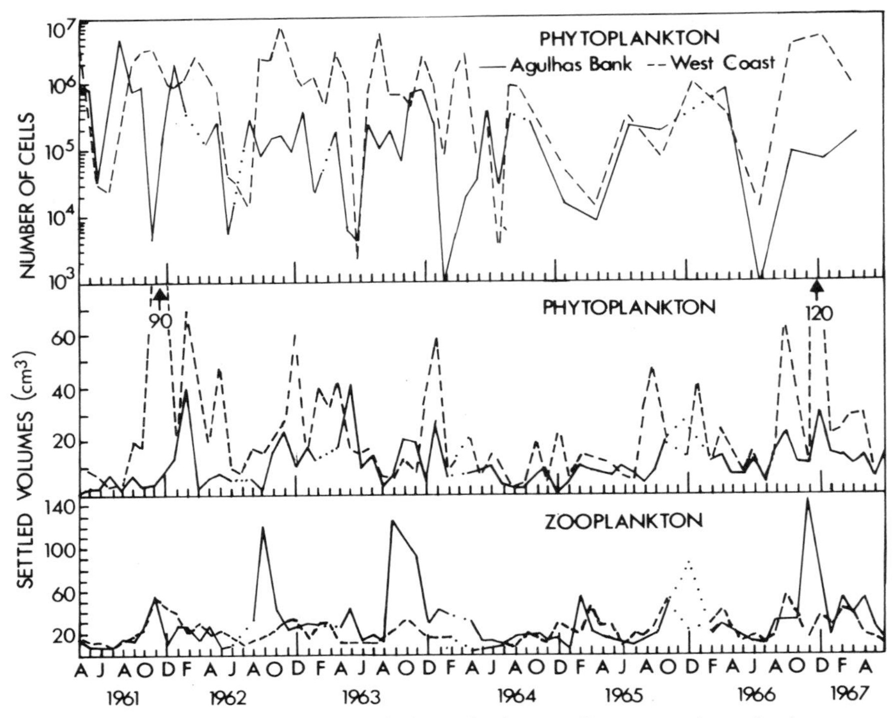

Fig. 12.—Monthly averages of phytoplankton cell counts, phytoplankton settled volumes and zooplankton settled volumes off the west coast of South Africa and the Agulhas Bank (from De Decker, 1973).

productivity is regulated by regenerated nitrogen.) Olivieri (1983a) identified three characteristic phytoplankton-rich zone widths.

(1) Limited zones following a few days of gentle southeasterly winds when the oceanic front was relatively close to the coast.
(2) Extensive zones characterized by mature blooms which followed periods of highly upwelling favourable winds when the front was approximately 100 km offshore.
(3) Suppressed zones which occurred after wind reversals or after a reduction in the upwelling favourable wind stress.

Although the results of the monthly sampling in the area east of Cape Point during 1975 have yet to be published, Tromp, Lazarus & Horstman (1975) have shown that the monthly changes in the nearshore chlorophyll *a* concentrations were closely related to upwelling, associated with easterly winds.

Monthly egg and larval surveys around the South Western Cape between 31° S on the western coast and 22° E on the southern coast during 1977–1978 facilitated the collection of a substantial set of chlorophyll *a* data. The mean chlorophyll *a* averaged over the top 20 m at the inshore stations showed a progressive decrease from about 7 $mg \cdot m^{-3}$ in St Helena Bay to about 2–3 $mg \cdot m^{-3}$ east of Cape Agulhas (Hutchings & Nelson *et al.*, 1983)—

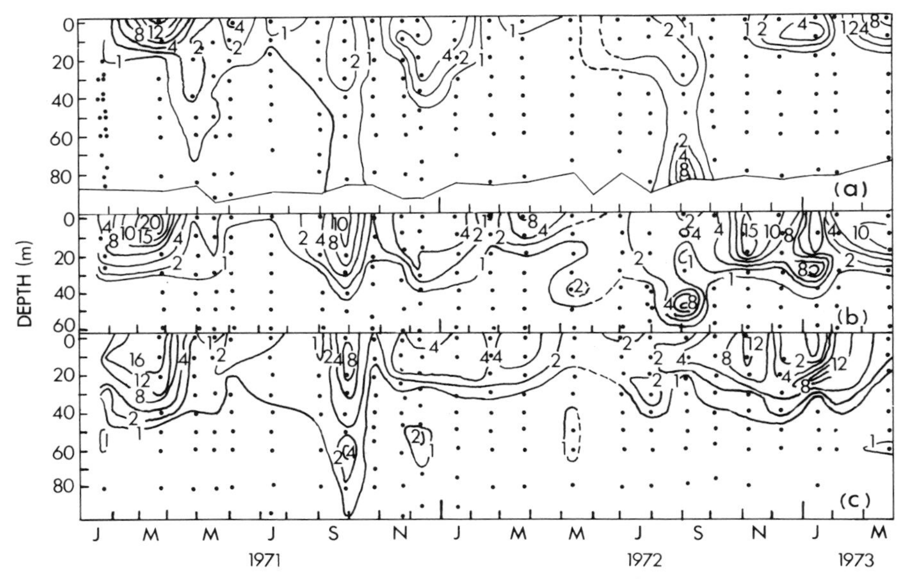

Fig. 13.—Seasonal changes of chlorophyll *a* ($mg \cdot m^{-3}$) approximately at stations 4 km (a), 22 km (b) and 50 km (c) along a line running north-west from the Cape Peninsula (after Andrews & Hutchings, 1980).

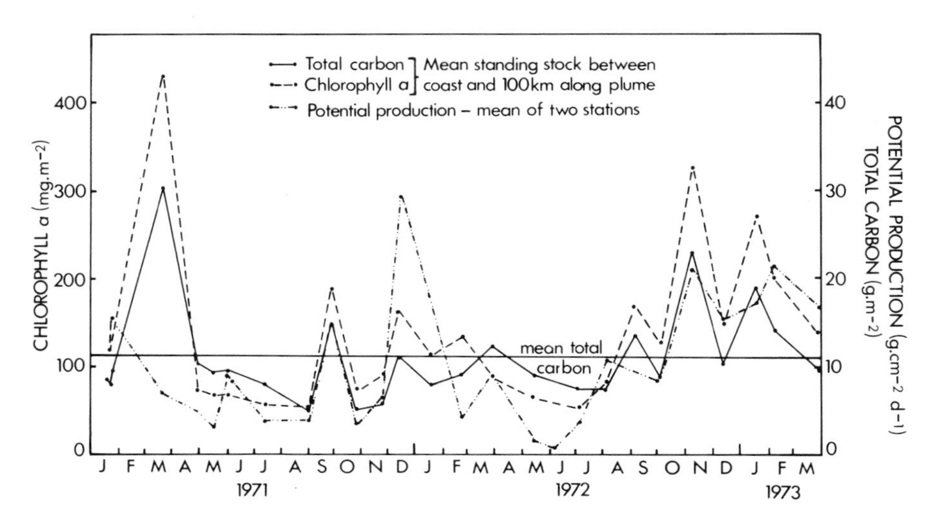

Fig. 14.—The seasonal variation of the standing stock and potential production of the phytoplankton along a line of stations running north-west from the Cape Peninsula (after Andrews & Hutchings, 1980).

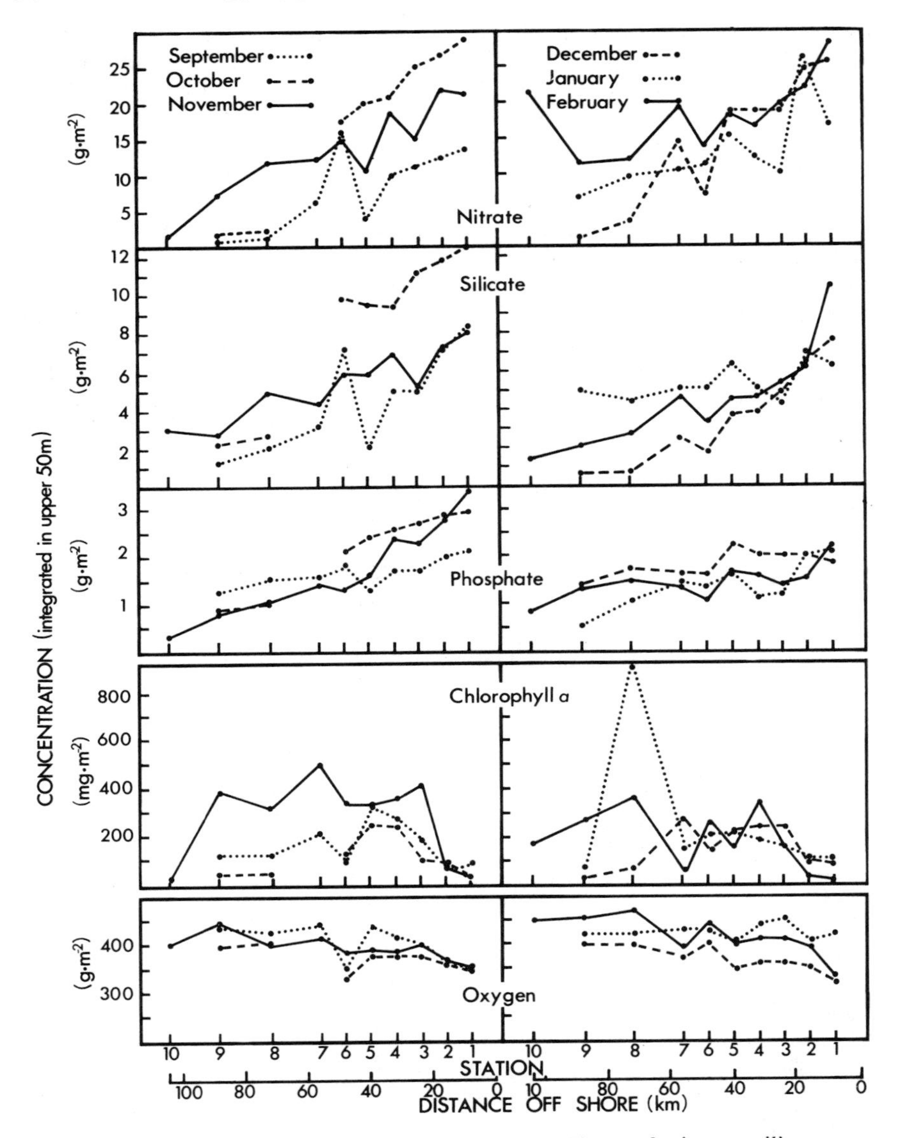

Fig. 15.—Monthly changes within the upper 50 m of nitrate, silicate, phosphate, chlorophyll *a* and dissolved oxygen along a line of stations running north-west from the Cape Peninsula (after Olivieri, 1983a).

a similar relative difference to that evident in De Decker's (1973) phytoplankton settled volume data (see Fig. 12). Relatively low concentations were recorded at stations at the bases of the Cape Columbine and Cape Peninsula upwelling tongues in recently upwelled water. Hutchings & Nelson *et al.* (1983) related their findings to the different wind-upwelling regimes on the southern and western coasts in an attempt to explain the distribution of the white sand mussel, *Donax serra*. The distribution of chlorophyll *a* from the 120-monthly stations was discussed in some detail by Shannon & Hutchings *et al.* (1984). These authors selected four months to show the seasonal trends in the distribution of chlorophyll *a* integrated over the upper 50 m (Fig. 16). In this figure the change from low chlorophyll (minimum upwelling) to high concentrations in spring following the onset of upwelling is evident. With the approach of summer the frequent upwelling events created a band of chlorophyll-rich water along the western coast, extending as far south as Cape Agulhas, while in autumn the relaxation of upwelling and reduction in insolation resulted in less intense fronts. Shannon & Hutchings *et al.* (1984) also examined the vertical distribution of chlorophyll *a* on selected lines during summer and winter (Fig. 17). In the area north of St Helena Bay (line 12) the distributions were similar during summer and winter, although during the latter season the mixed layer was about twice as deep. Off Cape Columbine (line 28) the strong frontal

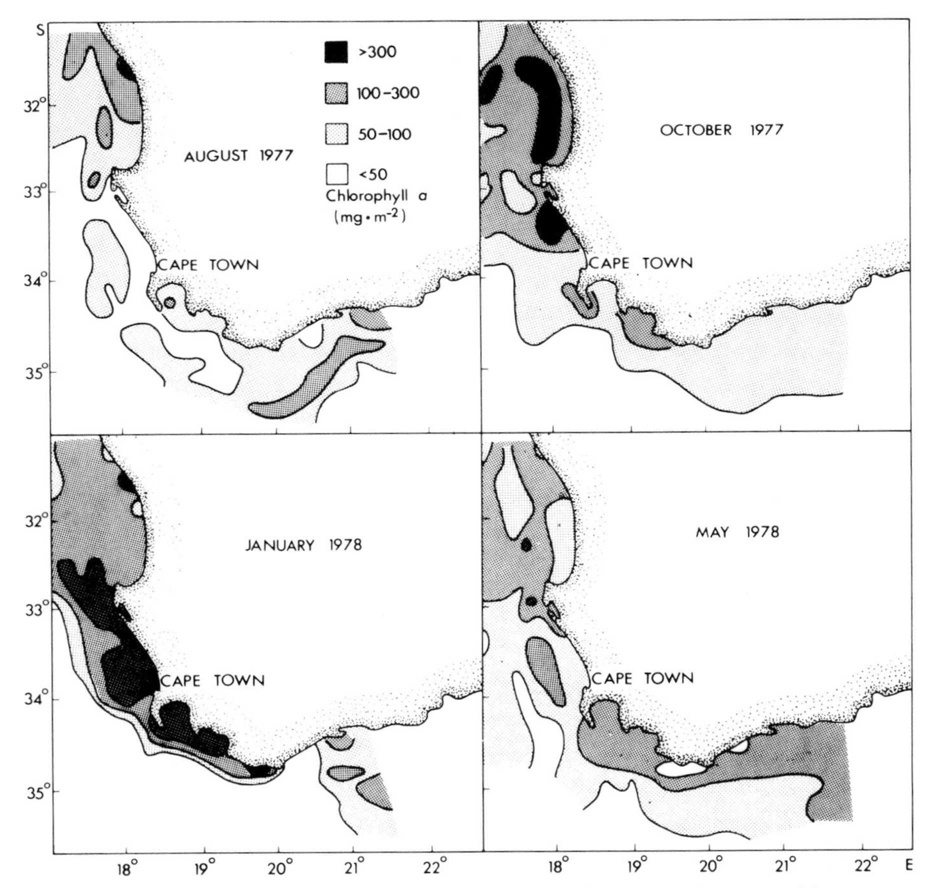

Fig. 16.—Distribution of chlorophyll *a* integrated through the upper 50 m ($mg \cdot m^{-3}$) in winter (August, 1977), spring (October 1977), summer (January 1978) and autumn (May 1978) off the South Western Cape (after Shannon & Hutchings *et al.*, 1984).

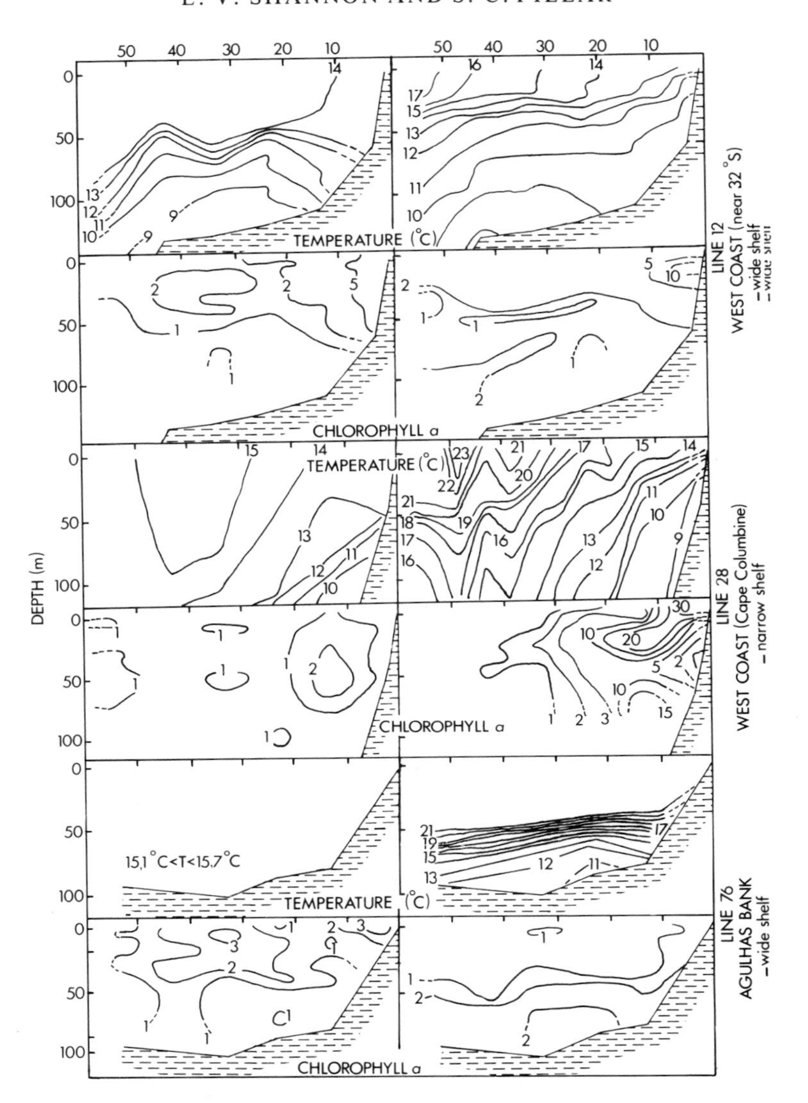

Fig. 17.—Vertical sections of chlorophyll *a* ($mg \cdot m^{-3}$) and temperature (°C) in winter (August 1977) and summer (January 1978) on the Agulhas Bank on the western coast (after Shannon, Hutchings, Bailey & Shelton, 1984).

features (subsurface chlorophyll maximum and sinking at the front—similar to that noted by Andrews & Hutchings 1980 off the Cape Peninsula) were evident during summer, but during winter well-mixed chlorophyll-poor water intruded close inshore. On the Agulhas Bank (line 76) extreme variations between summer and winter existed. The effect of winter storm mixing to the bottom is evident in Figure 17, and Shannon & Hutchings *et al*. (1984) suggested that the relatively high chlorophyll concentrations in the upper

40 m resulted from nutrients supplied by sediments. During summer they noted strong stratification with enhanced chlorophyll *a* concentrations at the thermocline (50–70 m) and suggested that the phytoplankton was maintained by a balance between light limitation and nutrient diffusion.

The extensive set of pigment maps generated from NIMBUS-7 CZCS data have been discussed by a number of authors *inter alia*, Anderson *et al.* (1981), Shannon & Anderson (1982), Shannon, Mostert, Walters & Anderson (1983), Shannon & Mostert (1983), Shannon & Lutjeharms (1983), Shannon, Schlittenhardt & Mostert (1984), Shannon & Hutchings *et al.* (1984), Shannon & Field (1985), and Shannon, Walters & Mostert (1985). The utility of satellite ocean colour scanners for measuring the mesoscale chlorophyll distribution in the case I optical waters (cf. Morel, 1980) of the South Western Cape, synoptically, has been clearly established. Shannon *et al.* (1983) examined five CZCS scenes of the area and found that the mesoscale feature expected in view of winds, upwelling and insolation, as well as the seasonal trends were evident. Distinct blooms were noted downstream of the upwelling tongues off the Cape Peninsula and Cape Columbine. The mesoscale frontal and tongue features during February 1980 were examined in some detail by Shannon, Schlittenhardt & Mostert (1984). The chlorophyll-poor newly-upwelled water in the Cape Columbine tongue is clearly visible in their map for 13th February 1980 (Fig. 18), as was the coastal zone of chlorophyll-rich water which extended as far as Cape Agulhas. East of this Cape low concentrations were indicated, but the authors warned that the subsurface chlorophyll maximum (described earlier), which is characteristic of the area during summer, would not have been visible to the satellite. More recently Shannon *et al.* (1985) examined the full set of CZCS ocean colour and thermal images for the period 1978–1980 and they noted that there was generally good agreement between the distributions and wind-induced upwelling, and excellent agreement between the positions of the oceanic thermal and the chlorophyll fronts in the southern Benguela, while Shannon & Field (1985) used the CZCS data in conjunction with regressions of surface-euphotic zone chlorophyll to estimate the standing stock of phytoplankton. Expressed per unit area their mean standing stock was highest in St Helena Bay (14 g $C \cdot m^{-2}$) and 9–10 g $C \cdot m^{-2}$ off the Cape Peninsula (*cf.* the value of 11·2 g $C \cdot m^{2}$ for the latter area recorded by Andrews & Hutchings, 1980, who used a slightly different carbon:chlorophyll *a* ratio).

What both the satellite and *in situ* studies have shown, however, is that in the southern Benguela the productive zone is much broader during winter than in summer, although biomass levels are lower and biological fronts are less pronounced.

PHYTOPLANKTON PRODUCTION

In a review of the primary productivity of various upwelling regimes, Cushing (1969) estimated the total annual fixation in the Benguela system to be 274×10^6 tonnes C, compared with 156×10^6 and 30×10^6 tonnes C for the Peruvian–Chilean and Californian systems, respectively. He made assumptions about areas of the systems and the duration of the productive (*i.e.* upwelling) seasons, and of course there were relatively few estimates of

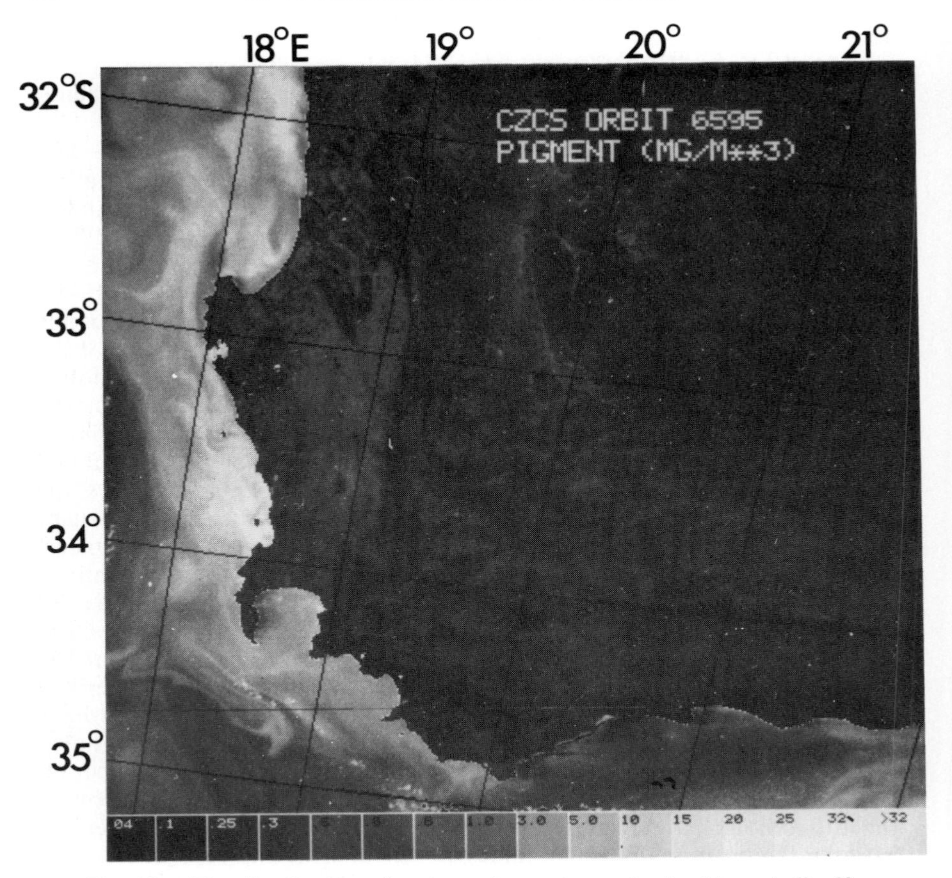

Fig. 18.—The distribution of estimated average euphotic chlorophyll off the western and southern coasts of South Africa, CZCS orbit 6595, 13th February, 1980.

primary productivity available at the time. Notwithstanding the apparent crudeness of Cushing's (1969) estimates, the fact is that the total annual carbon fixation in the Benguela region is of the same order of magnitude as that in the Peruvian–Chilean system, *i.e.* very substantial.

The first measurements of the rates of primary production off south-western Africa were made during the Danish expedition (1950–1952) of GALATHEA by Steemann Nielsen & Jensen (1957) who employed the carbon-14 uptake technique. Since then over two hundred measurements of depth-integrated rates of production have been made in the Benguela system. The majority of these were made close to the Cape Peninsula in the southern Benguela by Brown (1980, 1983a,b, 1984). Apart from the work by scientists from the German Democratic Republic (Schultz, 1982), few observations off Namibia and in the central region of the Benguela have been reported. The available information is summarized in Table IV. We have not included in this table data from Andrews, Cram & Visser (1970), Andrews (1974), and Andrews & Hutchings (1980) as their results were, at best, estimates of potential production under light saturation conditions (typically 10 g $C \cdot m^{-2} \cdot day^{-1}$).

Certain comments on the available data are necessary. The results of Henry, Mostert & Christie (1977), S. A. Mostert (pers. comm.), Brown, Hutchings & Horstman (1979), Borchers & Field (1981), and Carter (1982, 1983) relate to work undertaken in a semi-enclosed bay, or otherwise very close inshore *e.g.* in a kelp-bed. The rates of production in Hobson (1971) were expressed volumetrically, and these have been recalculated in Table IV, assuming a uniform surface mixed layer depth of 20 m. (From Hobson's density profiles and from other published physical data for the northern Benguela—refer to Shannon 1985—this is not unrealistic.) The estimates of Allanson, Hart & Lutjeharms (1981) refer to the offshore region south of Africa, and are possibly more typical of Agulhas or oceanic water, rather than of the Benguela *per se*. Few measurements of primary production have been made in the southern Benguela during the non-upwelling (winter) season (Henry, 1979; Brown, 1980, 1984; Brown & Henry, 1985). Similarly there are no published data in the northern Benguela during winter—a season of strong upwelling there.

Brown (1980, 1984) has contrasted two sites off the Cape Peninsula, one at Oudekraal, situated at the base of an upwelling tongue, the other near Robben Island, downstream of this upwelling centre. She found that the phytoplankton distribution and production at these inshore sites were highly variable during the upwelling season, but in the winter were more uniform due to reduced light levels, sinking and increased turbulence and consequently a deeper mean euphotic zone. Brown (1984) observed that the mean daily integrated production at the active upwelling site was lower than at the site further downstream (ratio 1·9) where the phytoplankton had had more time to reproduce. The rates of production at the downstream site were, on average, double those at the base of the Cape Peninsula upwelling tongue.

Brown (1983a,b) and Brown & Hutchings (1985) have reported the preliminary results of the extensive measurements made during the series of five plankton dynamics cruises around the South Western Cape between 1979 and 1981. Their findings will be discussed in more detail in the next section, but it should be noted that these cruises were located and timed specifically to study the dynamics of plankton, *viz.* seeding, blooming, grazing, and decay, off the Cape Peninsula, and accordingly the results may not necessarily be typical of the whole of the southern Benguela. Most of Schultz's (1982) work off Namibia was undertaken near to the Henties Bay area (22° S). This area may be a convergence zone (Shannon, 1985)—at least upwelling there is less intense than further north or south (Boyd, pers. comm.)—and because of this the production estimates may not be representative of the whole northern Benguela. In support of this, the rates of primary production at 20° S recorded by Schultz (1982) were somewhat higher ($2 \cdot 1 \pm 0 \cdot 6$ g $C \cdot m^{-2} \cdot day^{-1}$). Apart from the two measurements of Hobson (1971) there do not appear to be any estimates of primary productivity in the area around Lüderitz, the principal upwelling site in the Benguela system!

A number of factors can influence the reliability of production estimates and, in this respect, some useful work on the methodology has been done

TABLE IV

Summary of available data on phytoplankton primary production in the Benguela system (standard deviations in brackets): a, results corrected in accordance with Steemann Nielsen (1964); b, results adjusted for night respiration (see text); c, estimated assuming a constant surface mixed layer depth of 20 m

Area	Season	Technique used	Number of observations	Primary production: Range of values ($gC \cdot m^{-2} \cdot day^{-1}$)	Primary production: Mean ($gC \cdot m^{-2} \cdot day^{-1}$)	Reference
Cape west coast	Summer	C^{14}	4	[a]0·7–3·6	[a]2·2 (±1·2)	Steemann Nielsen & Jensen, 1957
Orange River	Summer	C^{14}	1		[a]0·8	
Walvis Bay	Summer	C^{14}	3	[a]0·7–5·5	[a]2·5	
Southern Angola	Summer	C^{14}	5	[a]0·5–2·8	[a]1·3 (±1·0)	
Walvis Bay area	Summer/autumn	C^{14}	1		[a]5·1	Bessonov & Fedosov, 1965
Southeast Trade Wind	Summer/autumn	C^{14}	2	[a]0·7–2·6	[a]1·6	
Drift Trade Wind	Autumn/spring	O_2	2	0–1·5	0·7	
Orange River area	Autumn	C^{14}	1		[b]0·3	Hobson, 1971
Lüderitz area	Autumn	C^{14}	2	[b]0·5–1·4	[b]0·9	
Sylvia Hill area	Autumn	C^{14}	2	[b]1·8–6	[b]3·9	
Walvis Bay area	Autumn	C^{14}	1		[b]1·2	
21° S	Autumn	C^{14}	1		[b]0·6	
Cape Frio area	Autumn	C^{14}	1		[b]3·2	
Southern Angola	Autumn	C^{14}	2	[b]0·9–2·6	[b]1·7	
Saldanha Bay	Year	O_2(P:B)	10	0·9–3·3	1·7 (±0·6)	Henry *et al.*, 1977; Henry, 1979
Oudekraal/Table Bay/	Winter	C^{14}	9	0·9–2·4	0·8 (±0·7)	
St Helena Bay	Spring/summer	C^{14}	17	0·5–11	5·1 (±2·6)	
False Bay (red tide inshore, Gordons Bay)	Spring	O_2	1		3·7	Brown *et al.*, 1979

Oudekraal	Spring/autumn	O_2	35		2·4 (±2·4)	Brown, 1980, 1984
	Winter	O_2	10		1·9 (±1·4)	
Table Bay (Robben Is.)	Spring/autumn	O_2	13		4·0 (±2·8)	
Southern Agulhas Bank	Summer	C^{14}	2		0·1	Allanson *et al.*, 1981
Southern zone, offshore	Summer	C^{14}	1		0·2	
Oudekraal (inshore)	Autumn	C^{14}	2		1·1	Borchers & Field, 1981
Dune Point (20° S)	Spring	C^{14}	6	1·1–2·8	2·1 (±0·6)	Schultz, 1982
Henties Bay (22° S)	Spring	C^{14}	31	0·3–5·5	1·4 (±0·9)	
Oudekraal (inshore, 20 m depth)	Year	O_2	29	[c]0·6–4·2	[c]2·6 (±1·76)	Carter, 1982, 1983
Cape Peninsula system	Spring	C^{14}	12	[c]1·3–10·8	[c]4·0 (±2·4)	Brown, 1983a;
	Late summer	C^{14}	25	[c]1·1–5·4	[c]2·9 (±1·4)	Brown & Hutchings, 1985
Cape Peninsula to Cape Columbine	Summer	C^{14}	12	[c]1·6–11·7	[c]3·5 (±2·8)	
Agulhas Bank	Late spring	C^{14}	5	[c]0·31–2·5	[c]1·07 (±0·9)	Brown, 1983b; Brown & Hutchings, 1985
Cape Columbine/ St Helena Bay	Summer	C^{14} CZCS	N/A	2–4	approx. 4	Shannon & Henry, 1983
Cape Peninsula system	Summer	C^{14}/ CZCS	N/A	2–4	approx. 3	
Southern zone	Summer	C^{14}/ CZCS	N/A	2–4	approx. 2–4	
Cape Columbine/ St Helena Bay	Mean annual	P:B(0·25)	N/A		3·5	Shannon & Field, 1985
Cape Peninsula	Mean annual	P:B(0·25)	N/A		2·2	
Southern zone	Mean annual	P:B(0·25)	N/A		2·4	
Total "active" system	Mean annual	P:B(0·25)	N/A		2·8	
Saldanha Bay	Winter	O_2	1		1·9	
	Spring/summer	O_2	4	3·6–1·0	6·3 (±2·8)	Mostert, pers comm.

off the Cape Peninsula by Brown (1980, 1982a,b,c, 1983c). This author conducted experiments on the effects of the size of the bottles used in the oxygen method and found no significant difference when using the 125, 250, 500, and 1000 cm^3 bottles, but that the 60 cm^3 bottle gave consistently lower results. Brown (1928b) compared the results from simulated *in situ* production experiments using a deck incubator and actual *in situ* production, and while the former yielded results slightly higher than the latter, the error was small. In a series of light-shock experiments Brown (1982c) found that short exposure of samples to high light levels did not make much difference. Prolonged exposure, however, stimulated production in surface samples but inhibited production in samples from near the bottom of the euphotic zone. Brown & Field (1985a) investigated diel fluctuations in phytoplankton production in nearshore water off the Cape Peninsula in order to improve estimates of daily production calculated from short-term incubations. Measurements during different times of the year showed diel fluctuations in production both under natural day–night and artificial light regimes, with greatest fluctuations being apparent when day and night were of equal length. These authors concluded that mid-day incubations, adjusted for the length of the day gave a fairly reliable indication of daily production. The changes in phytoplankton biomass and production in relation to the physical and chemical status of the water column at two nearshore sites in the southern Benguela were investigated by Brown & Field (1985b). They found that the low phytoplankton biomass in upwelling water, coupled with the frequency of upwelling, was more important in determining overall productivity than were the light levels and nutrient concentrations.

It is difficult to extrapolate the results of production measurements at specific sites in the Benguela to the broader system, and in this respect the work of Shannon & Henry (1983), who used ocean colour imagery in conjunction with discrete measurements of primary production and biomass, might have a useful application. Brown & Henry (1985) have generated a series of regression equations of chlorophyll *a* at the surface compared with chlorophyll *a* in the euphotic zone and compared with primary production for different areas in the southern Benguela in different seasons. They concluded that primary productivity estimates based on satellite observations of ocean colour are justified by greatly improved data coverage of biomass distribution, which they felt more than compensates for the variance associated with the regressions of primary production and biomass, so long as the physical and biological processes of the area under consideration are reasonably well understood.

From the available data in Table IV it is evident that the average rates of primary production in the Benguela are spatially and temporally variable. Brown & Hutchings (1985) have cited a value of 3 g $C \cdot m^{-2} \cdot day^{-1}$ for the annual average production in the inshore region of the southern Benguela. Shannon & Field (1985) considered that the best estimate of the annual average production in the "productively active" part of the southern Benguela system was 2·8 g $C \cdot m^{-2} \cdot day^{-1}$. For the northern Benguela we suggest that a slightly lower mean annual production rate would be appropriate, probably around 2 g $C \cdot m^{-2} \cdot day^{-1}$. The above rates should be compared with values of around 1·9 g $C \cdot m^{-2} \cdot day^{-1}$ off Peru and northwestern Africa cited by Barber & Smith (1981).

DYNAMICS OF PHYTOPLANKTON BLOOMS

Relatively little has been published on the dynamics of phytoplankton in the broader Benguela system, although the results of a series of experiments in the southern Benguela are at present being prepared for publication.

The study of Pieterse & Van der Post (1967) and that undertaken by scientists from the German Democratic Republic appear to be the sum total of work on the short term response of phytoplankton to changes in upwelling and nutrient supply in the central part of the northern Benguela. (The account of Wood & Corcoran, 1966, on the diurnal changes in phytoplankton in the "Benguela region" appears in fact to relate to the equatorial Atlantic off Gabon.) In Walvis Bay, Pieterse & Van der Post (1967) undertook intermittent daily sampling during the summer of 1964–1965, while during the third week of December 1965, they made measurements at the surface and at the bottom (about 5 m) at two-hourly intervals between 08·00 and 18·00 h. Their study showed clear evidence of photosynthesis and respiration during a red-water bloom, with a large diurnal change in dissolved oxygen content (between 6 $ml \cdot l^{-1}$ and 0 $ml \cdot l^{-1}$) being apparent in the bottom water. (Part of the change may have been due to advection.) These authors considered that the rapid growth of plankton was initiated by an increase in the nutrient supply.

The change in several conditions off Henties Bay, north of Walvis Bay, over a 15-day period during November 1976 (Fig. 19) evident from data in Schultz, Schemainda & Nehring (1979) has been discussed briefly by Chapman & Shannon (1985). The wind during the first 11 days of the study was upwelling-favourable, typically southerly, with maximum wind speeds on Day 1 and Days 6 and 7 and a minimum on Day 2. During the last four days winds were generally not favourable for upwelling. The changes in the primary production and chlorophyll *a* integrated over the top 10 m during the period are shown in Figure 20. A number of features are evident in Figures 19 and 20. First, the production measured at 06.00 h was on average about two-thirds of the noon value, but as expected this was not reflected in the chlorophyll values. (A similar trend was observed by Brown & Field, 1985a, off the Cape Peninsula in the southern Benguela.) Secondly, levels of nitrate and phosphate in the euphotic zone increased during Day 2, and a bloom occurred on Day 3, following which nutrient concentrations declined. Thirdly, increased upwelling and nutrient supply which occurred on Days 6 and 7 appears to have triggered a second bloom, with a maximum primary production of 2·5 $g\ C \cdot m^{-2} \cdot day^{-1}$ being recorded at noon on Day 10. Production declined thereafter due to reduced upwelling and probable nitrate depletion, although concentrations of chlorophyll *a* remained high (200 $mg \cdot m^{-2}$) for two further days before also declining. The level of primary production during the period between the third and twelfth days when the wind speed was typically 6 $m \cdot s^{-1}$or greater, was on average about 1·5 $g\ C \cdot m^{-2} \cdot day^{-1}$. Fourthly, the decay of the first bloom and possible sinking of phytoplankton can be followed in Figure 19. The rapid decrease in nitrate with a corresponding increase in nitrite during Day 9 in the bottom 40 m, was followed by an increase in the deeper phosphate concentrations.

The first studies of the development of phytoplankton populations in

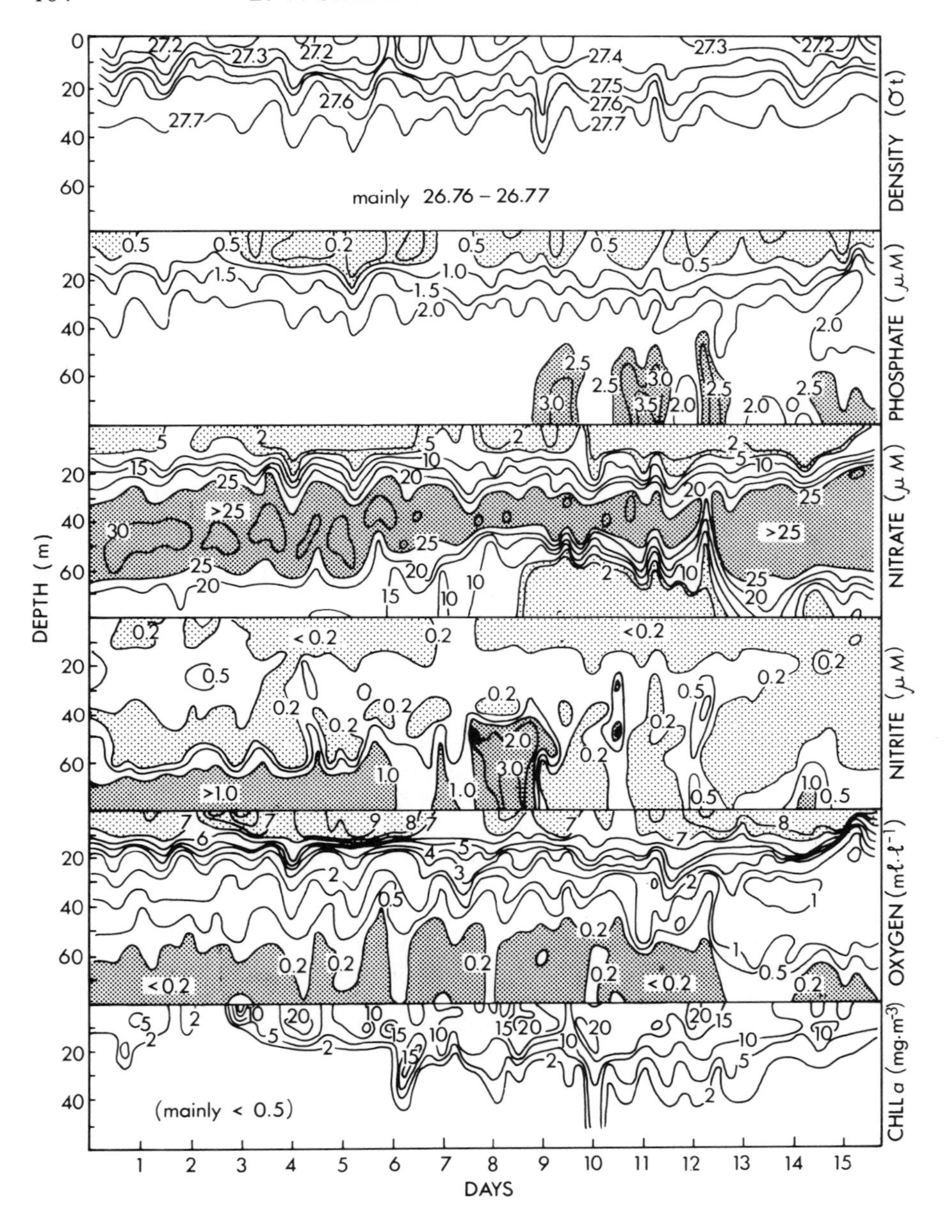

Fig. 19.—Changes in the water column at 22°16′ S: 14°05′ E near Henties Bay between 2nd and 17th November, 1976 (from Chapman & Shannon, 1985).

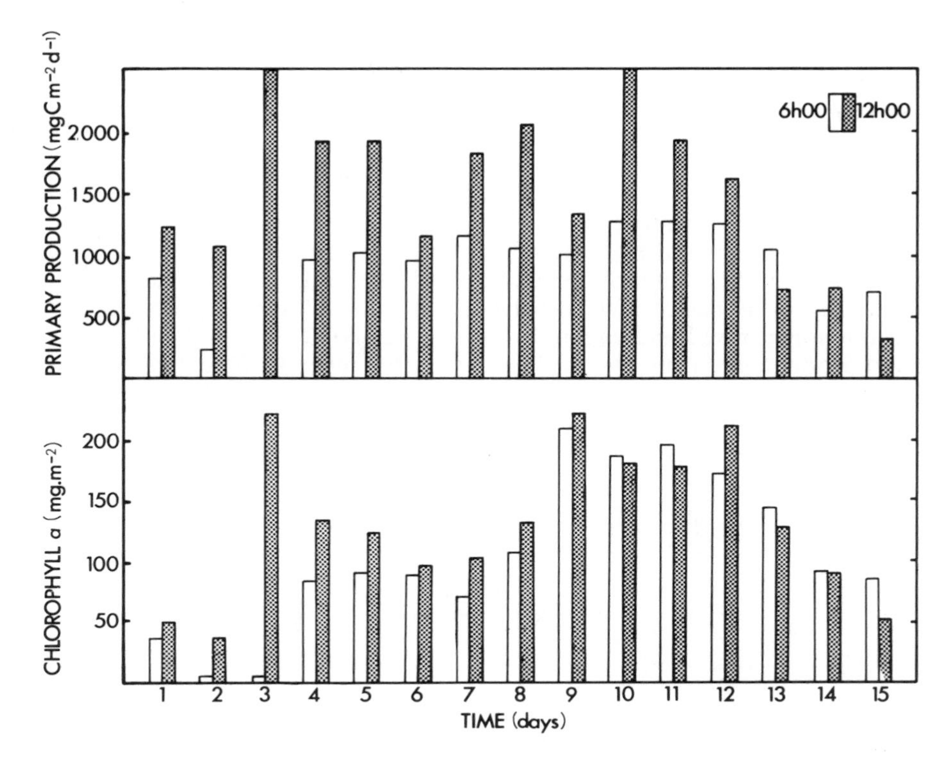

Fig. 20.—Primary production and chlorophyll *a* integrated over the top 10 m at a station at 22°16′ S; 14°05′ E (near Henties Bay) between 2nd and 17th November, 1976 (calculated using data from Schultz, Schemainda & Nehring, 1979).

response to upwelling events in the southern Benguela region were attempted during 1968 and 1969 by Andrews and his co-workers (Andrews & Cram, 1969; Andrews, Cram & Visser, 1970; Andrews 1974; Andrews & Hutchings, 1980). The growth of a phytoplankton bloom, the development of a subsurface biomass maximum followed by sinking near the oceanic front during a three day upwelling–relaxation–downwelling event was graphically illustrated by Andrews & Hutchings (1980). Intensive measurements made on six cruises off the Cape Peninsula between 1979 and 1983 have provided a good understanding of the dynamics of phytoplankton populations. The findings have been adequately reviewed in a recent paper by Brown & Hutchings (1985). Biplanar tetrahedra drogues were deployed at 10 m in large patches of newly upwelled water during five of the cruises, and these were followed for periods of between 5 and 10 days, sampling taking place three times per day. Phytoplankton development was estimated from *in situ* changes in chlorophyll *a*, protein, carbohydrate and particle volume concentrations and from *in situ* and on-deck incubations where carbon-14 uptake, including partitioning into photosynthetic products, and changes in particle volumes were used to estimate production (Hutchings,

Barlow, Brown & Olivieri, 1983). On the sixth cruise (undertaken during March 1983, following a summer of anomalously light winds, when the stratified conditions encountered were not ideal for diatom growth) the phytoplankton was dominated by small flagellates (Hutchings, Holden & Mitchell-Innes, 1984).

Brown (1983a) discussed the changes in phytoplankton biomass and production evident during four of the above-mentioned cruises, and observed that production increased initially as the phytoplankton populations increased near the sea surface (the generation time of phytoplankton in the southern Benguela is of the order of one day), then stabilized or decreased as the penetration of light and nitrate concentrations in the upper mixed and euphotic zones decreased. The change in biomass and production during one of the cruises (December 1980) is illustrated in Figure 21, and from this Brown & Hutchings (1985) concluded that the daily production, as calculated from carbon-14 uptake experiments, could not account for the observed changes in biomass. They suggested that other

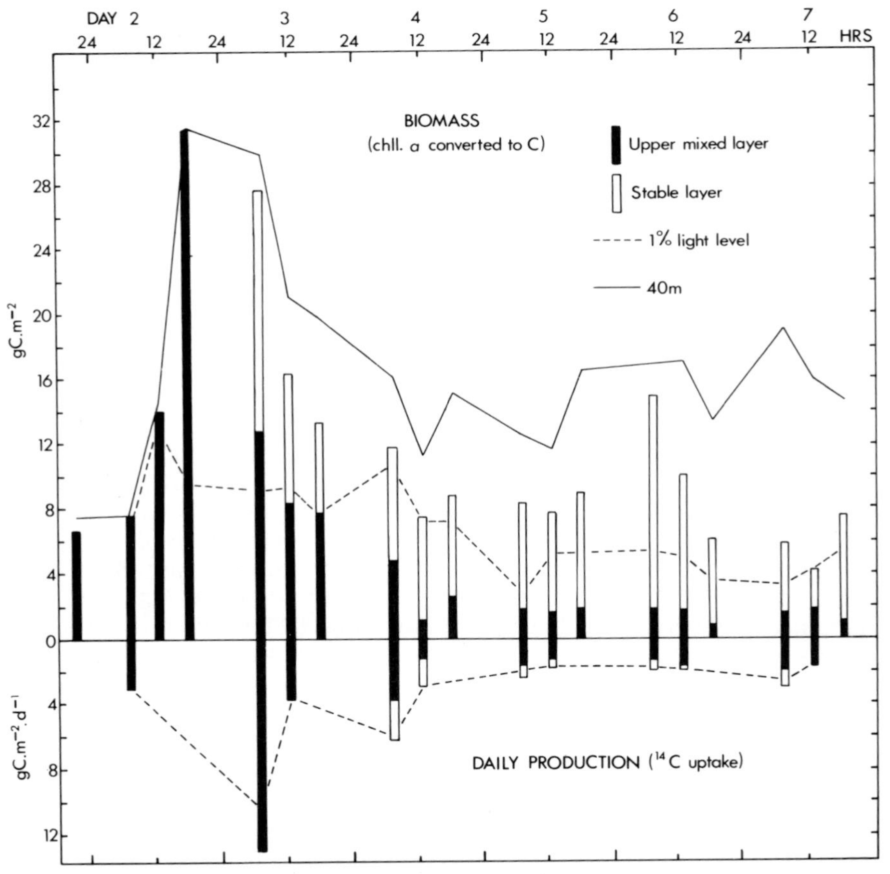

Fig. 21.—Phytoplankton biomass and production during a plankton dynamics cruise off the Cape Peninsula in December 1980 (from Brown & Hutchings, 1985).

processes such as dispersion, sedimentation, and zooplankton grazing would be important in regulating phytoplankton biomass. The impact of grazing was assessed by Olivieri & Hutchings (1983) from a series of 24-h feeding experiments undertaken during the plankton dynamics cruises. Their work showed that herbivorous copepods consumed on average about one third of the daily phytoplankton production, although individual estimates varied widely. It appears from their work that the grazing of phytoplankton by zooplankton might be relatively less important in the newly upwelled water off the Cape Peninsula than further north in the Benguela.

The depletion of nutrients, in particular nitrate, during a period of rapid diatom growth (Fig. 22) was demonstrated by Barlow (1982c) and also by Olivieri (1983b) in her discussion of the colonization and temporal changes in phytoplankton species diversity and abundance in the Cape Peninsula upwelling plume during five days in December 1979. The bloom reached its peak after three days and Barlow (1982c) showed that chlorophyll *a* concentrations increased by 19 $mg \cdot m^{-3}$ in the euphotic zone with a concomitant decrease in nitrate–nitrogen of 19 m $mol \cdot m^{-3}$. No species succession was noted by Olivieri (1983b), and the three principal seed species *viz. Chaetoceros compressum*, *C. debile*, and *Skeletonema costatum* remained dominant over the five days. Brown & Hutchings (1985), however, suggested that changes in species composition may well have occurred after the termination of the experiment. Olivieri (1983b) speculated that, as the cell sizes increased, the rates of nutrient absorption decreased. The progressive increase in cell size as the upwelled water aged

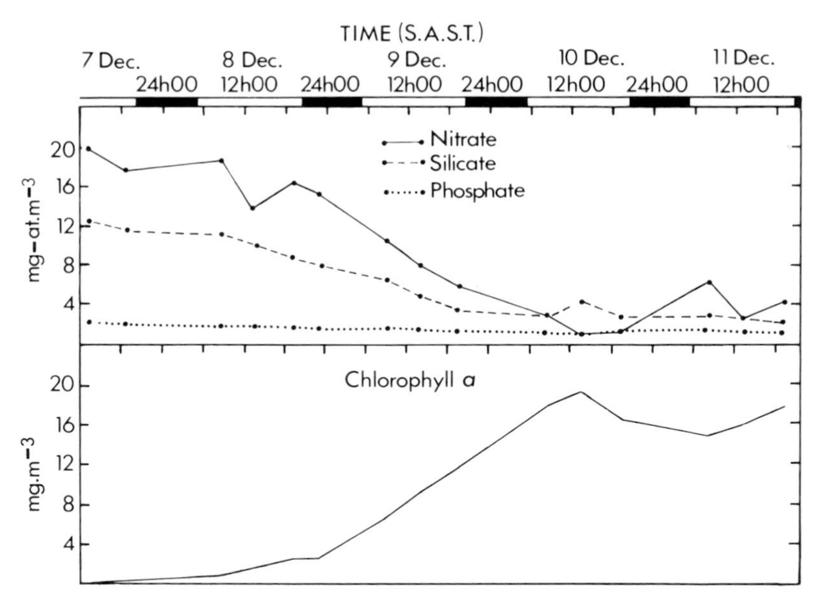

Fig. 22.—Changes in the concentrations of nutrients and chlorophyll *a* in the euphotic zone: results are mean concentrations integrated between the surface and the depth of the 1% light level; black bars indicated dark period (from Barlow, 1982c).

was interpreted by Olivieri as an adaptive method for successful colonization and maintenance as large cell sizes are advantageous for storage of intracellular nitrogen. She felt that, by increasing cell size, the diatoms would increase their sinking rate, thereby enabling them to utilize the higher levels of nutrients available at subsurface depths after the surface nutrients had been exhausted. She also considered that the dominants were able to adjust physiologically to progressively decreasing light levels.

Following experimental work off the Cape Peninsula during which biochemical analytical techniques were perfected (*e.g.* Barlow, 1981; Barlow & Swart, 1981) studies on the biochemical composition and physiology of phytoplankton were undertaken by Barlow (1980, 1982a,b,c, 1984a,b,c). Barlow (1980) noted that the ratio of protein to carbohydrate decreased as blooms aged and he concluded that this ratio was a suitable indicator of the physiological state of phytoplankton communities in an upwelling area. Three water types (or stages of maturity) were identified by Barlow (1982a) in the southern Benguela, the essentials of which are summarized in Table V. This author estimated that chlorophyll *a* existed in a ratio of about 1:1 between living phytoplankton and bacteria and detritus and he noted that the concentration of protein in detritus was higher than in living phytoplankton. Studies on the assimilation of carbon-14 into the photosynthetic end-products led Barlow (1982b) to suggest that, at high light intensities and high nutrient concentrations a certain amount of assimilated carbon and energy is channelled into protein for growth and cell maintenance, while excess is stored in the polysaccharide fraction. At low light intensities assimilation rates are reduced and less excess carbon and energy are available to produce polysaccharides, but protein synthesis is maintained. Decreased nitrate availability was shown by Barlow as leading to continued protein synthesis at the expense of carbohydrate synthesis at low light intensity, but enhanced the carbohydrate relative to protein at high light intensity. Barlow (1982c) showed that maximum protein concentrations occurred just prior to the bloom peak during the five-day

TABLE V

Characteristics of the three types of upwelled water off the Cape Peninsula according to Barlow (1982c)

	Type 1	Type 2	Type 3
Temperature (°C)	8–10	10–15	12–16
"Age" of water	recently upwelled	maturing upwelled	aged upwelled
Nitrate-N ($mmol \cdot m^{-3}$)	15–30	2–15	<2
Silicate-Si ($mmol \cdot m^{-3}$)	10–30	2–22	2–12
Phosphate-P ($mmol \cdot m^{-3}$)	1·2–3·5	0·5–2·5	0·5–1·5
ATP ($mg \cdot m^{-3}$)	0·14±0·03	0·57±0·03	0·73±0·04
Chlorophyll *a* ($mg \cdot m^{-3}$)	1	1–20	5–30
Protein ($mg \cdot m^{-3}$)	70±5	424±21	580±31
Total carbohydrate ($mg \cdot m^{-3}$)	32±3	189±11	929±88
Protein:carbohydrate ratio	2·7±0·2	2·5±0·1	0·7±0·03
Phytoplankton growth phase	active	active	slow

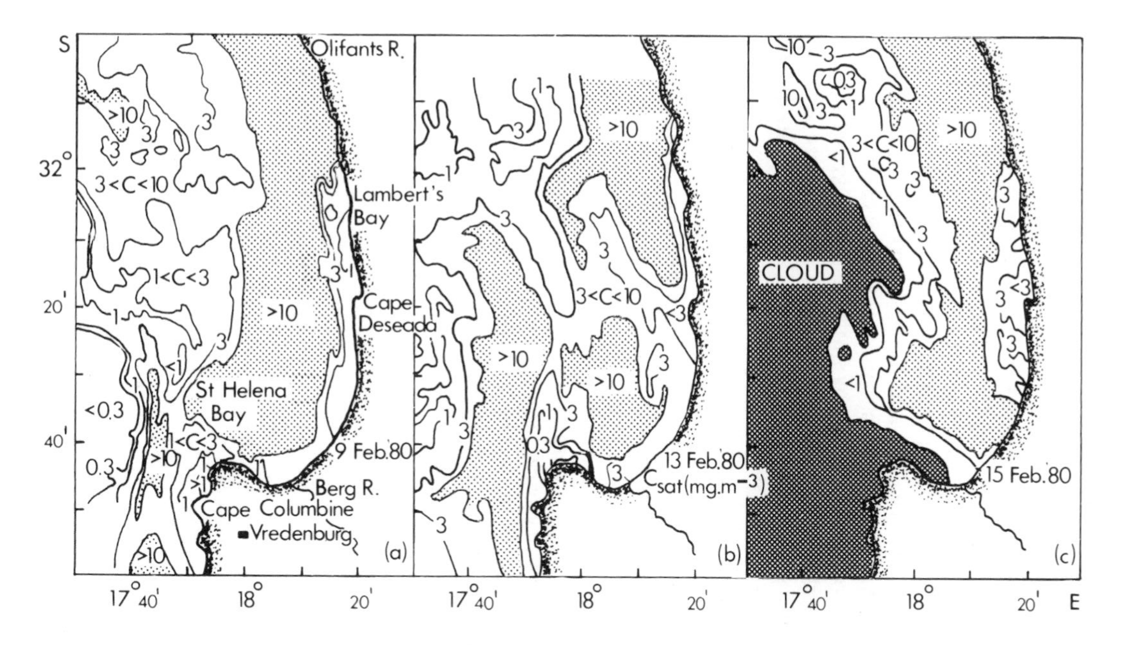

Fig. 23.—The distribution of estimated surface chlorphyll in the Cape Columbine and St Helena Bay area (a) CZCS orbit 6540, 9th February, 1980, (b) CZCS orbit 6595, 13th February, 1980, (c) CZCS orbit 6623, 15th February, 1980 (Shannon, Schlittenhardt & Mostert, 1984).

experiment in December 1979 (*cf.* Fig. 22), and that carbohydrate concentrations increased rapidly during the day but decreased at night. The percentage carbon incorporated into protein was, however, greater at night than during the day. In his study during a declining phase of a bloom during December 1980, Barlow (1984a) found that a large proportion of assimilated carbon was detected in the ethanol-soluble fraction at the 50% light level, and that as cells sank out of the mixed layer, chlorophyll *a* and protein concentrations increased in the stable and bottom layers. Cells were still found to be viable at the 1% light level. In a series of laboratory experiments Barlow (1984c) showed that the largest proportion of carbon was assimilated into polysaccharides during active growth, but that during no growth, synthesis of protein was the dominant metabolic process. The response of phytoplankton to turbulent and stable environment was investigated by Barlow (1984b), and he observed that total assimilation of carbon-14 photosynthetic products under mixing conditions was greater than under stable conditions, thereby implying faster growth in a turbulent environment. In stable conditions a greater percentage of carbon was incorporated into protein. Barlow, however, concluded that, since algal growth is linked to protein synthesis, growth in a stable environment will be faster than in a turbulent one.

The utility of satellite ocean colour scanners in studies of the dynamics of phytoplankton blooms has been demonstrated by Shannon & Mostert (1983) and Shannon, Schlittenhardt & Mostert (1984), and an indication of the mesoscale changes in near-surface "chlorophyll" which can occur in an upwelling region are illustrated in Figure 23. (Readers are referred to Figure 40 in Shannon, 1985, for details about the relevant upwelling history.) Except on an exploratory basis satellite studies on their own will not be very useful for investigating the complex dynamics of phytoplankton in the Benguela system. As a support to *in situ* studies, however, the application of satellite ocean colour imagery possesses considerable potential.

ZOOPLANKTON

The most important contributions to the systematics of South African zooplankton are those by Cleve (1904) and Stebbing (1905, 1910). Their general catalogues of the zooplankton fauna were later supplemented by studies on individual taxa such as Wolfenden (1911) on copepods, Tattersall (1925) and Illig (1930) on euphausiids and Barnard (1932, 1940) on amphipods. The two surveys of the Benguela Current by the R.S.S. WILLIAM SCORESBY between 20° S and 35° S during 1950 (Hart & Currie, 1960) provided the first detailed studies of individual zooplankton taxa from the Benguela system. Some of the findings stemming from these surveys were reported by Iles (1953) on ostracods, Boden (1954, 1955) on euphausiids, Tatersall (1955) on mysids, and Jones (1955) on cumaceans. With the development of the Cape and Namibian pelagic fisheries during the early 1950s regular zooplankton sampling began, first in the St Helena Bay area where fishing activities were more concentrated. The first account of the local zooplankton groups from this region was provided by De Jager (1954) from monthly sampling of an area bounded by Saldanha Bay (33° S) in the south and Lamberts Bay (32° S) in the north. This was later extended southwards to Table Bay and is referred to in the text as the "Cape routine area" (Fig. 24a). Regular collections from this area formed the basis of several distributional studies which include Nepgen (1957) on euphausiids, Van Zyl (1960) on tunicates, Heydorn (1959) on chaetognaths, and Siegfried (1963) on hyperiid amphipods.

During the early 1960s the southern limit of the "Cape routine area" was extended southwards to encompass the Agulhas Bank regions (38° S). The eastern limit of this area, referred to as the "Southern routine area" (Fig. 24b) was at San Sebastian Bay (21° E). Between 1963 and 1967, this area was sampled monthly and later on a quarterly basis which yielded further contributions to our knowledge of the major zooplankton groups. De Decker (1973) summarized most of the research effort in his account of the

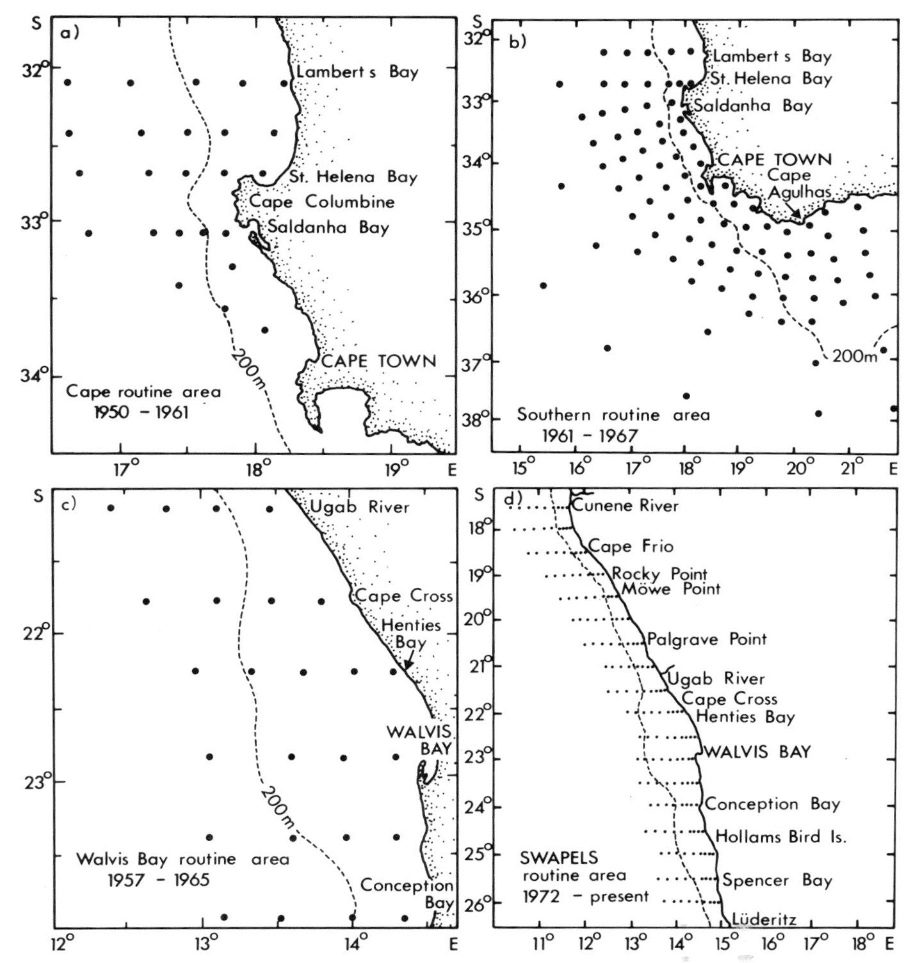

Fig. 24.—Zooplankton routine and research areas in the Benguela region between 1950 and 1977.

Agulhas Bank plankton. Other works include Lazarus (1967) and Lazarus & Dowler (1980) on the distribution of phyllosoma larvae and pelagic tunicates, respectively.

In the Namibian region of the northern Benguela system, material collected from an area stretching from Conception Bay in the south (24° S) to the Ugab River in the north (21° S), referred to as the "Walvis Bay routine area" (Fig. 24c), provided the first account of the centres of abundance of the major zooplankton groups (Kollmer, 1963). Later studies in this area by Unterüberbacher (1964) and Venter (1969) provided accounts of the copepod and chaetognath fauna. During the early 1970s, this area was extended northwards to the Kunene River (18° S) and southward to just north of Luderitz (26° S), by the initiation of the South West African Pelagic Egg and Larval Survey (SWAPELS). Regular ongoing sampling of the "SWAPELS routine area" (Fig. 24d) has yielded vast amounts of zooplankton material. No attempt has yet been made, however, to analyse individual zooplankton groups. Zooplankton biomass has been monitored regularly, but few published data are available (Kruger, 1983; Kruger & Boyd, 1984).

During the late 1960s, zooplankton sampling in the southern Benguela changed from monthly routine surveys over large areas to more intensive and discrete studies in time and space. These changes were initiated to provide better insight into the changes in zooplankton communities during local upwelling events. During December 1969 samples were collected over a 13-day period at discrete depth intervals using a centrifugal pump (Hutchings, Robertson & Allan, 1970) and WP-2 nets (Currie & Foxton, 1957) equipped with flowmeters. This net was considered more quantitative than the 'Discovery'-type gear which had been used previously on routine surveys. From 1971–1973 a transect of stations running northwest of the Cape Peninsula was monitored monthly to observe seasonal changes in zooplankton standing stock. In 1974, this line of stations was extended northwards by three lines to Cape Columbine which were monitored quarterly over a one-year period. The results of these surveys were reported by Hutchings (1979, 1981) and Andrews & Hutchings (1980). Between 1977 and 1978 monthly surveys, principally aimed at mapping the distribution of pelagic fish eggs and larvae around the South African western and southern coasts (Cape Egg and Larval Programme: CELP), provided material for further studies on local zooplankton communities. Collections from along one transect off Lamberts Bay, formed the basis of a study on the inshore–offshore and temporal distribution of the copepod community of this area, with notes on other major taxonomic groups (Hopson, 1983).

During nine deep-sea and two inshore cruises by the R.S. AFRICANA II, between 1961–1968, a pump situated at keel depth of about 3 m, was used to collect zooplankton while steaming between hydrological stations. The cruises encompassed an area between the Indian Ocean (26° S), the South Atlantic Convergence (47° S) and the southeastern Atlantic Ocean bordering the western coast mainland to as far north as Walvis Bay (23° S). Data collected from these cruises were recently presented by De Decker (1984) in his comprehensive account of the copepod distribution in the southern African seaboard.

The following paragraphs are not intended as an account of the progress in zooplankton faunistic studies of the Benguela system. They should serve as a summary of pertinent work on taxonomic and distributional studies, rather than an attempt to address zooplankton on a functional basis. This account should be viewed as a reference text to augment our understanding of the distribution of local zooplankton taxa. In recent years the direction of research has changed and more emphasis has been placed on the investigation of functional aspects of zooplankton ecology, rather than descriptive distributional studies. An outline of ongoing and future studies is included in the text to elaborate on this change in approach.

DISTRIBUTION OF BIOMASS AND SEASONAL CHANGES

Northern Benguela

Earlier studies by Kollmer (1963) and Unterüberbacher (1964) provided some insight into the distribution and seasonal changes in zooplankton abundance in the northern Benguela. From data collected in the "Walvis Bay routine area" between 1958 and 1962, they noted consistently high biomasses of zooplankton in an area west and northwest of Walvis Bay along the inferred upwelling plume which was characterized by low temperatures and salinities. Unterüberbacher (1964) observed two annual zooplankton peaks, one during the late spring and early summer (November–December) and the other mainly during the autumn (March–May). These findings agreed with the peaks of phytoplankton abundances recorded by Kollmer (1963). Later studies by Visser, Kruger, Coetzee & Cram (1973) and Wessels, Coetzee & Kruger (1974) showed that zooplankton and phytoplankton maxima occurred in belts parallel with the coastline. The phytoplankton maxima occurred inshore in cool upwelled water while the zooplankton peaked further offshore. Data shown from the South West African Pelagic Egg and Larval Surveys (SWAPELS) has provided some information on the distribution of zooplankton in the northern Benguela. Kruger (1983) has shown the inshore-offshore distribution of zooplankton in relation to phytoplankton abundance during the period from September to March 1980–1981 and 1981–1982 (see Fig. 8, p. 86). The pattern of high inshore phytoplankton concentrations and offshore peaks of zooplankton biomass estimates from the SWAPELS material, however, are of a semi-quantitative nature due to the imprecise method of measuring plankton volume, *i.e.* relative (zooplankton/phytoplankton ratios), rather than direct measurement. Furthermore, the type of gear used *i.e.* 80 μm mesh 'Discovery'-type N50 net, is inefficient as regards the capture of the larger zooplankton and the small mesh would be susceptible to clogging which would confound biomass estimates. Consequently, limited conclusions on biomass distribution and seasonal changes can be drawn from these data. Recent analysis of zooplankton from Bongo net collections taken during the SWAPELS surveys is providing more quantitative data on zooplankton biomass.

Southern Benguela

The first study of seasonal changes of zooplankton in the southern Benguela was by De Jager (1954) from his collections from the "Cape Routine area" between 1950 and 1951. He observed that biomass was lowest during winter when upwelling was minimal and highest during spring and summer when upwelling sustained high phytoplankton concentrations in his study area. He noted a three to four-fold decrease in zooplankton settled volume from inshore to offshore. De Decker (1973) showed higher zooplankton settled volumes over the Agulhas Bank than off the western coast from data collected between 1961 and 1967 in the "Southern routine area". This observation he noted as paradoxical considering the comparatively lower phytoplankton production over the Agulhas Bank. The author explained this observation as a bias in the settled volume estimates caused by large numbers of pelagic tunicates from the Bank collections which far exceeded those collected off the western coast.

Settled volume biomass estimates from the above studies were based on collections using the 'Discovery'-type N70 net which was limited quantitatively due to the absence of flow monitoring or depth recording instrumentation and no cognisance of vertical migration. Consequently, the data indicated little more than seasonal trends in zooplankton biomass. Data collected from October 1970 to March 1973 off the Cape Peninsula, using the more quantitative WP-2 net, enabled Hutchings (1979, 1981) and Andrews & Hutchings (1980) to define more clearly local seasonal cycles of zooplankton biomass. A transect of seven to ten stations running northwest from the Cape Peninsula, referred to as the "Upwelling monitoring line", was sampled monthly to observe the physical and biotic changes along the axis of the inferred upwelling plume. The authors used dry weight per unit area (g $DW \cdot m^{-2}$) which they considered to be a more suitable measure of zooplankton standing stock than the previously used settled volume biomass estimates ($ml \cdot 1000\ m^{-3}$).

The monthly changes in standing stock along the "Upwelling monitoring line", as illustrated by Andrews & Hutchings (1980), are shown in Figure 25. Considerable variability between successive months is evident but when a three-month running mean is applied to the data, distinct seasonal patterns are apparent. Highest values were recorded in the upwelling season between December to February and lowest values in the non-upwelling season between June to August. A two- to three-fold difference in standing stock was noted between these periods with an overall mean of 2·3 g $DW \cdot m^{-2}$. At a nearshore site in a kelp bed zone off the Cape Peninsula, Carter (1983) observed similar seasonal fluctuations in zooplankton biomass. Highest biomasses were noted in summer with reduced levels in autumn and lowest in winter with an overall mean biomass of 2·14 g $DW \cdot m^{-2}$. By comparison, St Helena Bay, some 150 km to the north was characterized by higher and more consistent standing stock than off the Cape Peninsula (Hutchings, 1981). The author relates his findings to the comparative homogeneity of the St Helena Bay region. Based on long term wind data, he found that upwelling-favourable winds were more common in St Helena Bay (45%) than off the Cape Peninsula (33%). Also the cyclonic water circulation within the St Helena Bay area favoured water retention

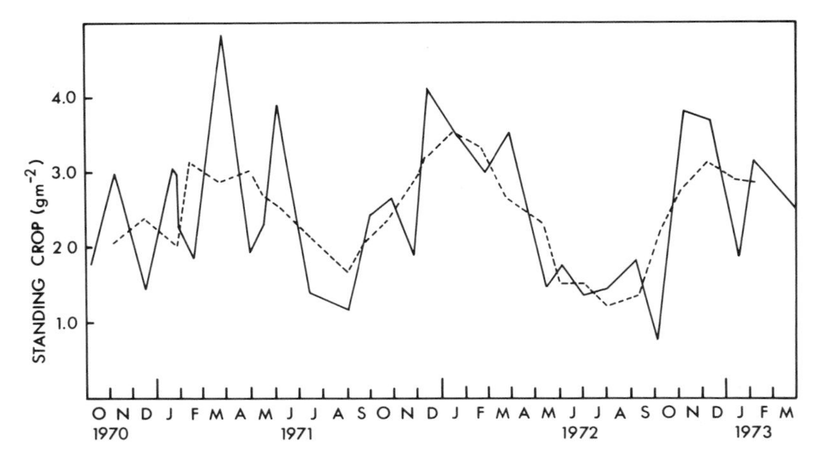

Fig. 25.—Seasonal variations of the mean zooplankton standing stock (g dry wt·m^{-2}) along the Upwelling Monitoring Line running north-west off the Cape Peninsula between 1970 and 1973 (from Andrews & Hutchings, 1980).

with mild but prolonged upwelling events, thus contributing further to the stability of this area relative to the more variable seasonal upwelling fluctuations associated with the Cape Peninsula. Hopson (1983) noted that upwelling was active during most of her monthly zooplankton collections in an area just north of St Helena Bay (Lamberts Bay) (32° S) between August 1977 and August 1978. Supporting the finding by Hutchings (1981) she noted uniformity and a lack of seasonal variation in biomass throughout the year.

The seasonal inshore-offshore fluctuations of zooplankton biomass along the Cape Peninsula upwelling plume, as illustrated by Andrews & Hutchings (1980) are shown in Figure 26. During the spring and summer multiple peaks of zooplankton occurred offshore as a result of offshore displacement during active upwelling. In winter when upwelling was minimal zooplankton peaks were smaller and closer inshore. The zones of high concentrations corresponded to temperatures between 10° and 16 °C. Intrusions of warmer water towards the coast, especially during winter months limited zooplankton to the inshore regions. The authors pointed out that the transect was occupied only once a month and between each collection four to five wind reversals could occur with subsequent marked changes in the relative distribution of the zooplankton. From data collected during December 1969 off the Cape Peninsula, Hutchings (1981) was able to describe the short-term changes in zooplankton standing stock during a brief wind reversal. As the wind blew from the west after a period of upwelling, the inshore surface water warmed rapidly and the zooplankton was more concentrated near the coast during this period of downwelling. As the wind backed to the southeast and increased in speed, upwelling resumed, and cold water reappeared which rapidly spread offshore displacing the zooplankton in a northwesterly direction. The author concludes that the duration of the southeasterly winds would influence the

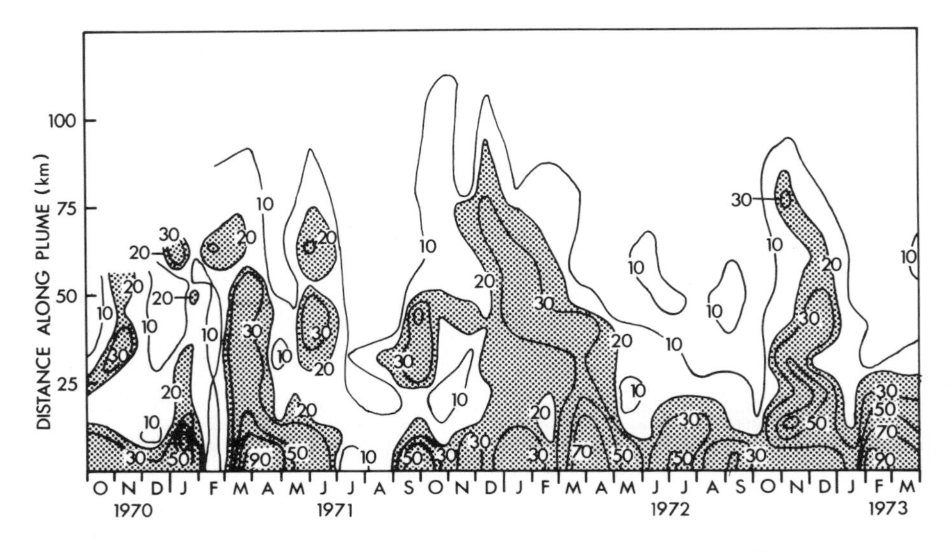

Fig. 26.—Seasonal and spatial changes of zooplankton standing stock (mg dry wt$\cdot$m^{-3}) along the Upwelling Monitoring Line between 1970 and 1973 (from Andrews & Hutchings, 1980).

displacement of zooplankton. Typically, these winds seldom persist for more than four to six days, consequently it is estimated that patches of zooplankton could be displaced some 40–90 km during this period.

Andrews & Hutchings (1980) attempted to relate zooplankton biomass to phytoplankton on a seasonal basis. By expressing the biomasses in terms of carbon, they found that zooplankton biomass formed 4–25% of the phytoplankton biomass with maximum values occurring during the summer and lowest during the winter. These correspond to periods with highest zooplankton biomass and phytoplankton biomass, respectively. Carter (1983) also noted similar seasonal cycles in zooplankton/phytoplankton ratios which form a 5% minimum during winter to a 64% maximum during summer. Using the mean zooplankton/phytoplankton biomass ratio (12%) from the data of Andrews & Hutchings (1980) and a higher ratio of 20% estimated from data compiled by Olivieri & Hutchings (1983), Shannon & Field (1985) estimated the lower and upper limits of the annual zooplankton production of the active upwelling area of the southern Benguela (40 000 km^2), to be $2{\cdot}0$–$6{\cdot}7\times10^6$ tonnes carbon, respectively. The estimated mean annual zooplankton production of $3{\cdot}6\times10^6$ tonnes carbon was equivalent to 9% of the phytoplankton production. Andrews & Hutchings (1980) reported spatial and temporal mismatch along the upwelling plume between zooplankton and phytoplankton, yet found a correlation ($0{\cdot}01<P<0{\cdot}05$) between zooplankton standing stock and gross phytoplankton production. No seasonal pattern was evident. Carter (1983) found no relationship between either zooplankton and phytoplankton biomass or gross phytoplankton production. The author, however, pointed out that his findings were not conclusive owing to the spatial and temporal limitations of his data. Hutchings (1981) noted that because of the more rapid response of phytoplankton than zooplankton to upwelling and downwelling events, it is expected that there would be inconsistent patterns of zooplankton and phytoplankton distribution of the Cape Peninsula.

Hutchings (1979) and Hopson (1983) compared their zooplankton standing stock estimates with those of previous records from the Benguela system as well as those from other upwelling areas *e.g.* California, Peru, Northwest Africa. Carter (1983) confined his comparisons to those of Andrews & Hutchings (1980). These authors, however, have stressed that the data must be viewed with several restrictions placed on the validity of the comparisons. The effect of spatial, seasonal, and diel variations are important sources of bias. Also those associated with sampling gear such as mesh size and avoidance of the sampler, particularly by the larger zooplankton, can confound valid comparisons when different samplers are used. Pillar (1984c) has shown that towing bridles similar to those in front of the 'Discovery'-type N70 nets an the WP-2 nets, increased the avoidance response of the larger zooplankton, such as euphausiids, to the net. This resulted in lower biomass estimates using these nets than when using Bongo nets which were free from bridle obstructions. The author showed that mesh size was of critical importance in biomass of local copepod fauna as a mesh of 200 μm retained two to three times the biomass of a 300-μm meshed net.

Considering the aforementioned, the standing stock values listed in Table VI should be considered reliable for regional comparative purposes only when similar sampling gear is used. We feel that meaningful regional comparisons are, however, premature until data on the temporal variability of the zooplankton are available from collections covering the whole depth range of the fauna. Despite these restrictions, the data presented should provide a useful reference as an overview of the zooplankton standing stock of the Benguela system.

Copepoda

Copepods are numerically the most abundant and diverse group of zooplankton in South African water. De Decker (1964) gave a systematic review of earlier contributions to copepod faunal studies of South Africa, citing Cleve (1904), Stebbing (1910), Wolfenden (1911), and Tanaka (1960) as the major works. He identified 92 copepod species of which 28 were first records for South African waters (Table VII). He lists three species which he regarded as typical members of the Agulhas Current, i.e. *Clausocalanus furcatus*, *Corycella concinna*, and *Temora discaudata* and five species, *Calanus finmarchicus*, *Euterpina acutifrons*, *Temora turbinata*, *Pseudodiaptomus nudus*, and *Calocalanus tenuis* as being specific to the Agulhas Bank fauna, while eight species were regarded as typical members of the cold water community of the Benguela Current, *i.e. Centropages brachiatus*, *Calanoides carinatus*, *Metridia lucens*, *Nannocalanus minor*, *Clausocalanus arcuicornis*, *Paracalanus parvus*, *Paracalanus crassirostris*, and *Ctenocalanus vanus*. Subsequent studies by De Decker (1973) have confirmed the dominance of these species in the Benguela system. The more recent and comprehensive account by this author (De Decker, 1984) on the

TABLE VI

Zooplankton standing stock estimates from the Southern and Northern Benguela: the estimated denotes by an asterisk were taken from the original source, and others were taken from Hutchings (1979); all data were converted to gC·m^{-2} using a conversion factor of 6·5% of wet weight (Cushing, 1971); dry weight is considered 10% of wet weight (Hutchings, 1979)

Area	Source	Mean depth sampled (m)	Gear	Mesh width (μm)	Standing stock (gC·m^{-2})	Comments
Southern Benguela	1 De Jager, 1954	100	N70V	200–460	1·59	Summer samples only 1950/51; Cape routine area
	2 De Decker, 1973	100	N70V	200–460	1·02	Annual mean 1961/67, Southern routine area
	3 Andrews & Hutchings, 1980	180	WP-2	200	1·49	Annual mean 1970/73; Cape Peninsula upwelling plume
	4 Hopson, 1983	100	Bongo	300	1·29*	Annual mean 1977/78; Lamberts Bay area
	5 Carter, 1983	20	Pump	44	1·39*	January, May, August and October/
	Carter, 1983	20	Pump	149	1·11*	November 1979 (mean); kelp bed zone off the Cape Peninsula
Northern Benguela	1 Kollmer, 1963	100	N70V	200–460	2·04	Annual mean 1958/59; Walvis Bay routine area
	2 Unterüberbacher, 1964	100	N70V	200–460	1·34	Annual mean 1959/62; Walvis Bay routine area
	3 Visser *et al.*, 1974	116	WP-2	200	0·96*	May-November 1971, April-October 1972 (mean) 19–23° S
	4 Wessels *et al.*, 1975	116	WP-2	200	0·63*	June and July (mean) 19–23° S

TABLE VII

Copepod species recorded from South African waters (from De Decker, 1964)

- *Acartia amboinesis*
  - *danae*
  - *negligens*
- *Acrocalanus gracilis*
  - *monachus*
- *Calanoides carinatus*
- *Calanopia minor*
- *Calanus finmarchicus*
  - *tenuicornis*
- *Calocalanus contractus*
  - *pavo*
  - *plumulosus*
  - *styliremis*
  - *tenuis*
- *Candacia armata (?)*
  - *bipinnata*
  - *catula*
  - *curta*
  - *truncata*
- *Canthocalanus pauper*
- *Centropages brachiatus*
  - *calaninus*
  - *elongatus*
  - *furcatus*
  - *gracilis*
  - *pacificus*
- *Clausocalanus arcuicornis*
  - *furcatus*
  - *paululus*
- *Clytemnestra rostrata*
  - *scutellata*
- *Copilia mediterranea*
  - *mirabilis*
- *Corycaeus africanus*
  - *agilis*
  - *asiaticus*
  - *crassiusculus*
  - *dubius*
  - *latus*
  - *longistylis*
  - *pacificus*
  - *speciosus*
  - *subtilis*
- *Corycella concinna*
  - *curta*
  - *gibbula*
- *Ctenoclanus vanus*
- *Eucalanus attenuatus*
  - *elongatus*
  - *monachus*
  - *mucronatus*
  - *pileatus*
  - *subcrassus*
- *Euchaeta wolfendeni*
- *Euterpina acutifrons*
- *Labidocera acutum*
  - *minutum*
- *Macrosetella gracilis*
- *Mecynocera clausi*
- *Metridia lucens*
- *Microsetella norvegica*
  - *rosea*
- *Nannocalanus minor*
- *Oithona fallax*
  - *nana*
  - *plumifera*
  - *rigida*
  - *tenuis*
- *Oncaea clevei*
  - *media*
  - *mediterranea*
  - *subtilis*
  - *venusta*
- *Pachos tuberosum*
- *Paracalanus aculeatus*
  - *crassirostris*
  - *denudatus*
  - *parvus*
- *Pleuromamma abdominalis*
  - *gracilis*
  - *robusta*
- *Pontellina plumata*
- *Pseudodiaptomus nudus*
- *Rhincalanus cornutus*
  - *nasutus*
- *Sapphirina gastrica*
  - *nigromaculata*
  - *ovatolanceolata-gemma*
- *Scolecithrix danae*
- *Temora discaudata*
  - *turbinata*
- *Undinula darwinii*
  - *vulgaris*

copepod fauna of southern African waters is based on widespread collections in the area between the Indian Ocean (26° S) and the southeastern Atlantic Ocean (23–47° S). The author provided distributional maps of 80 species which he considered occurred in sufficient numbers to warrant description. His cumulative data illustrate the low species diversity in the Benguela Current system compared with the Agulhas Current and Agulhas Bank regions (Fig. 27). The decrease in diversiy from the Bank northwards along the west coast supports the findings that the cool Benguela inshore waters limit the dispersal of a number of warm-water species of Agulhas origin while supporting a larger but less diverse population of cool-water species. This is adequately reflected in the distribution maps of three species of the genus *Centropages* (Fig. 28). *C. furcatus*, an Indo-Pacific species, is shown to be indicative of the Agulhas Current with little penetration into the Benguela system. *C. chierchiae*, a member of the Agulhas Bank community, has a wider distribution which reflects the advection of Agulhas Bank water into the Benguela system. Contrary to *C. chierchiae*, *C. brachiatus* thrives in the inshore waters in the Cape and the Namibian regions of the Benguela. *C. brachiatus* is shown to be less able to withstand advection than *C. chierchiae*. The southward penetration shown by *C. chierchiae* is only faintly displayed by *C. brachiatus*. To the north *C. brachiatus* is considered not to extend further than Namibia while *C. chierchiae* has been recorded along the African Atlantic coast (Thiriot, 1978).

De Decker (1984) also discussed the distribution of *Metridia lucens* and *Calanoides carinatus*, in relation to that of *Centropages brachiatus* (Fig. 29). Although they proliferate and co-occur in the Benguela, their distributions differ in the eastern Atlantic. *Metridia lucens* extends northwards from the Subantarctic to northern Namibia and it has not been recorded further north until approximately 15° N which is regarded as the southern limit of its northern population (Thiriot, 1978). De Decker (1984) noted that *Calanoides carinatus* was more eurythermic than *Centropagus brachiatus* which explains its more extensive southward and northeasterly

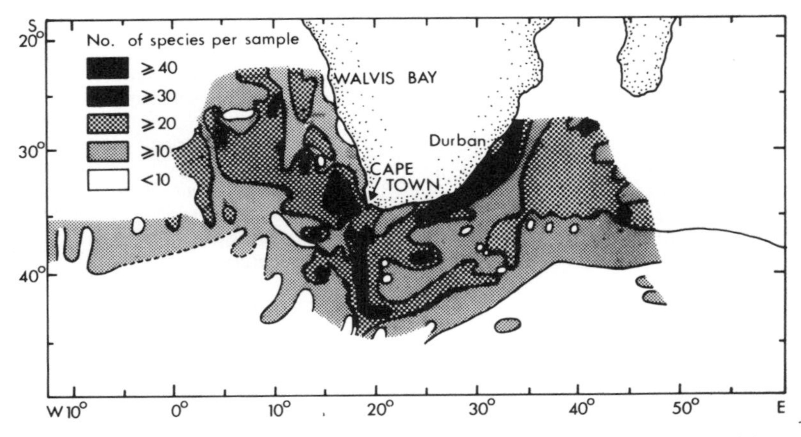

Fig. 27.—Diversity distribution of copepods collected off southern Africa, from cumulative data drawn from eleven cruises between 1961 and 1968 (from De Decker, 1984).

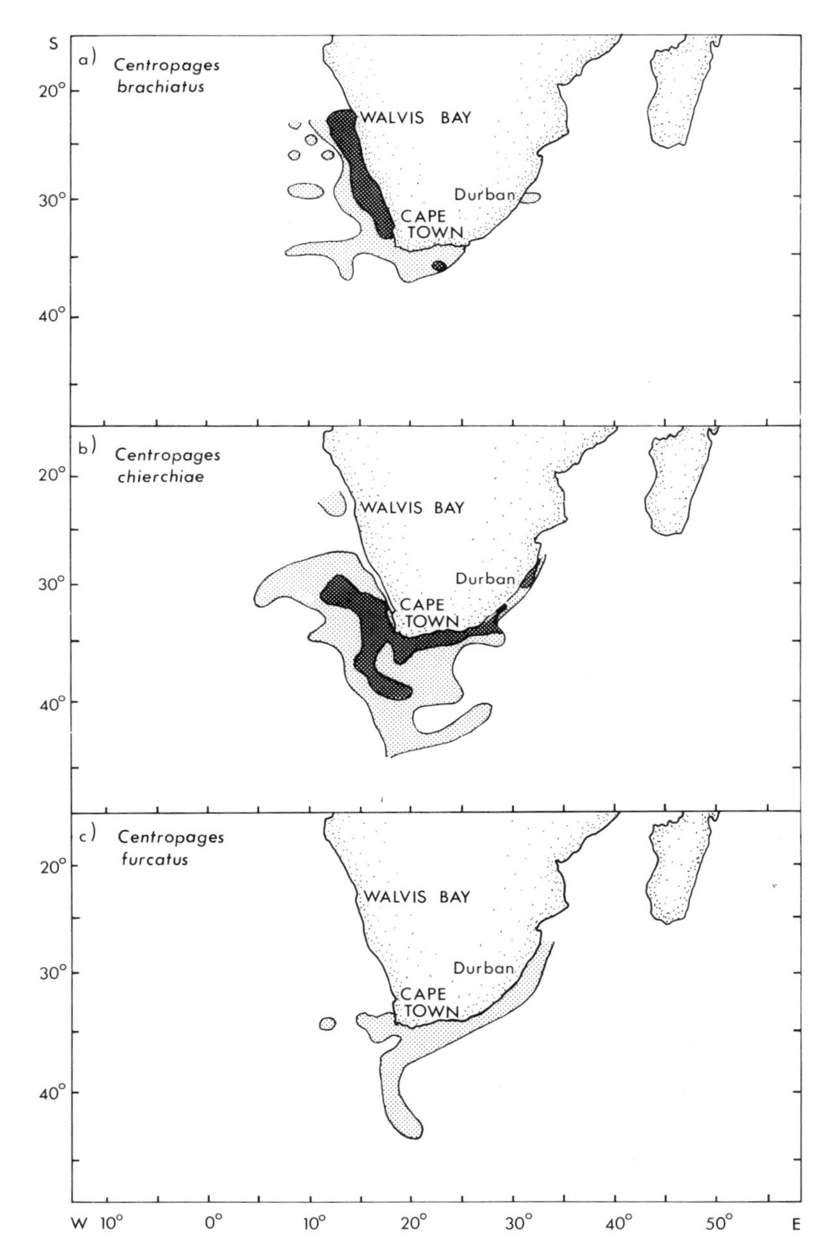

Fig. 28.—Distribution of (a) *Centropages brachiatus*, (b) *Centropages chierchiae*, and (c) *Centropages furcatus*: shaded areas suggest distribution patterns; darker shading indicates areas of higher abundance; from De Decker (1984).

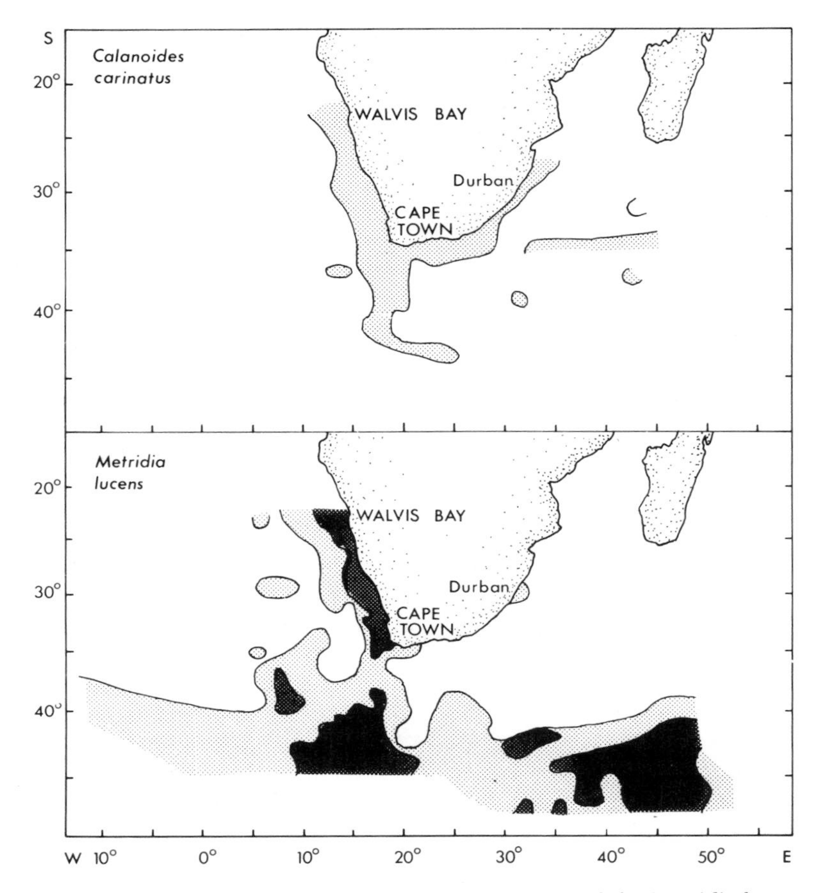

Fig. 29.—Distribution of (a) *Calanoides carinatus* and (b) *Metridia lucens*: shading as for Figure 28; from De Decker (1984).

distribution in his study area. Thiriot (1978) has recorded *Calanoides carinatus* as being a dominant copepod species on the eastern Atlantic seaboard.

Working in the "Walvis Bay routine area", Kollmer (1963) noted the importance of the genus *Paracartia* as a nearshore copepod and *Centropages brachiatus*, *Calanoides carinatus*, and *Metridia lucens* as being numerically dominant over the continental shelf, while the boundary area was typified by the presence of *Rhincalanus nasutus*. Unterüberbacher (1964) provided a more comprehensive account of the copepod fauna of this area. He found *Paracalanus crassirostris* and *Paracartia africana* to be a strictly neritic species while the cool-water community was represented mainly by *Centropages brachiatus*, *Calanoides carinatus*, and *Metridia lucens* which together with *Paracalanus parvus* and *Oithona similis* comprise about 80% of the copepods in the area studied. The author listed a number of copepod species which he considered to be members of the offshore warm-water community, with *Nannocalanus minor* being numerically the most important.

The data of Unterüberbacher (1964) provided a base line for studying the effect on the biotic fauna during the Benguela "El Nino" in 1963 when water with anomalously high temperature and salinity and unusual dissolved oxygen concentrations moved southwards into the northern Benguela from the tropical Atlantic (Stander & De Decker, 1969). The notable increase in numbers of *Nannocalanus minor* and the marked decline of some cool-water copepods, principally *Paracalanus parvus*, was suggested by the authors as a consequence of the warm-water intrusion. Furthermore, the occurrence of several species of copepods in the area, which were only known to occur further north, contributed towards their conclusion that the observed anomaly was not a result of local conditions, but due to a southward advance of warm surface water from Angola. The use of copepods as an indicator of water masses was also demonstrated by De Decker & Coetzee (1979) in their studies of the oil yield fluctuations from pelagic fish in the Namibian fishing grounds. They showed an inverse correlation between oil yield with the southward extension of a warm saline water mass of Angolan Current origin into the Namibian shelf region. This influx of water was marked by the presence of the neritic copepods *Temora turbinata* and *Euterpina acutifrons* which were regarded by the authors as of practical value in the interpretation and the identification of different water masses in the northern Benguela regions.

Coetzee (1974) discussed the vertical distribution of *Calanoides carinatus*, *Centropages brachiatus*, and *Metridia lucens* in the upper 50 m of the waters in the vicinity of Walvis Bay during April and August 1972. Despite gross environmental changes as observed from the stable highly stratified conditions during April to strong coastal upwelling with minimal stratification later in the year, there were no apparent changes in the vertical distribution of *Calanoides carinatus* or *Centropages brachiatus*. A strong diurnal vertical migration was displayed by *Metridia lucens* during both months with maximum numbers encountered in the surface layers during the night.

Lazarus (1975) provided a systematic account of the zooplankton communities from four sheltered bays in the southern Benguela system between the Cape Peninsula and St Helena Bay. Typical members of the cool nearshore copepod community were *Calanoides carinatus*, *Centropages brachiatus*, *Paracalanus crassirostris*, and *Oithona similis* which dominated the zooplankton throughout three years of observations (1964–1966). Other dominant Benguela species, such as *Nannocalanus minor* and *Metridia lucens* were relatively scarce in his collections, indicating their affinity for warmer offshore waters.

From an ecological rather than a systematic viewpoint, Hutchings (1979) produced a comprehensive study of the vertical and horizontal distribution of zooplankton in the southern Benguela upwelling region. Using more quantitative sampling methods than used on previous studies, his work was the first detailed investigation of the coupling between the biological and physical processes which act to control local zooplankton communities. Copepods collected at 36-h intervals, over a ten-day period from oceanic, newly upwelled and mature upwelled water, showed that temperature and chlorophyll abundance had an effect on the vertical structure of different species. Species were separated into three groups using association indices and typical members of each group were described in relation to hydrological condition. The largest group of copepods were typical of the cool upwelled water consisting mainly of those species described by earlier workers (*e.g.* De Decker, 1964, 1973) as preferring this habitat. With the exception of *M. lucens* there was very little indication of any extensive vertical migration in this group with the majority remaining in the mixed upper layer where chlorophyll concentrations were highest. A second group typified the frontal zone which borders the cool upwelled water and were represented by *Paracalanus crassirostris* and *Calocalanus tenius*. Strong vertical temperature gradients exerted a considerable influence on these species, as evidenced by their close association with the thermocline. Those species inhabiting the warm waters beyond the front formed the third group, which the author considered to be species characteristic of the Agulhas Bank, such as *Calanus finmarchicus*.

From the above data and from those drawn from additional material Hutchings (1979) was able to postulate several possible mechanisms whereby the zooplankton community was maintained in the southern Benguela system. He suggested that sinking at the oceanic front, combined with the periodic shoreward movement when upwelling relaxed and downwelling at the coast prevailed, could be sufficient to allow the zooplankton to return to the inshore regions. The southward surface currents and low oxygen undercurrents such as those described by Duncan & Nell (1969) and De Decker (1970) were also suggested by the author as a system of currents which could be responsible for replenishing the local zooplankton population.

In her studies on the temporal and inshore–offshore variations of zooplankton in the pelagic fish nursery and recruitment grounds in the Lamberts Bay area, Hopson (1983) singled out the copepod community for special attention because of their importance as potential food items of young fish. Only six species comprised 86% of the copepod community and there were sufficient variations in their distribution to suggest that different species had different life histories. Attention was drawn by the author to the fact that young and mature members of larger species such as *Calanoides carinatus* were found closer inshore than the smaller species *Paracalanus parvus* and *Ctenocalanus vanus*, the young of which were found further offshore than the adults. This size-determined life history was discussed by the author in relation to changes in the physical and biotic environment.

The zooplankton community of a nearshore southern Benguela kelp bed was found by Carter (1983) to be dominated by copepods typical of the cool-water groups of the Benguela system. Their vertical distributions were consistent with the community structure of those defined by Hutchings (1979), supporting the conclusions by Field *et al.* (1980) that little captive water is maintained in the west coast kelp beds. Respiration rates were determined by the author for *Centropages brachiatus* at three different temperatures (8°, 9·5°, and 13·5 °C). Rates ranged from 3·41–5·11 $\mu l\ O_2 \cdot mg\ DW^{-1} \cdot h^{-1}$ and the low Q_{10} values found for this species showed that it was well equipped to withstand frequent large temperature variations which characterized its normal habitat. In laboratory studies Borchers & Hutchings (1983) found the copepod *Calanoides carinatus* was highly adapted to the patchy food regime typical of the southern Benguela region. This tolerance of starvation was ascribed to an exceptionally large lipid reserve. Newly hatched and young copepodite stages were most vulnerable to starvation, depending on the temperature.

Lazarus (1975) found that certain of the smaller copepod species such as *Oithona similis*, *Paracalanus parvus*, and *Paracalanus crassirostris* were entirely herbivorous while several of the larger species such as *Calanoides carinatus*, *Centropages brachiatus*, and *Metridia lucens* had remains of both phytoplankton and zooplankton, principally copepods, in their gut. As these larger species do not occur in abundance in the relatively warm waters of the Agulhas Bank nor near the oceanic front, it is unlikely that they are important predators on young fish larvae in the southern Benguela (refer to spawning areas of pelagic fish shown in Figs 34 and 35).

Euphausiacea

The study of Boden (1954) provides the most comprehensive account of the euphausiids of southern African waters. He described and illustrated 42 species and constructed keys to the euphausiids from both the eastern and western seaboards of South Africa. He subsequently reported on 14 species drawn from material collected off Namibia by the R.S.S. WILLIAM SCORESBY (Boden, 1955). Nepgen (1957) reported on 18 species of euphausiids from collections in the "Cape routine area" (Table VIII) of which *Euphausia lucens* and *Nyctiphanes capensis* were numerically the dominant species, constituting 76 and 23%, respectively, of his total collections from two years of monthly sampling (1954–1956). Boden (1955) first described the larval developmental stages of both these species, but later Bary (1956) and Talbot (1974) concluded that Boden's description of *Euphausia lucens* was incorrect. Rearing studies by Pillar (1984b, 1985) confirmed both Bary's and Talbot's descriptions of *Euphausia lucens* as being correct, while supporting Boden's original description of *Nyctiphanes capensis* larvae.

The distribution of euphausiid eggs was first documented by De Jager (1954) in the "Cape routine area", where he found maximum concentra-

TABLE VIII

Euphausiid species recorded off the west coast of South Africa (from Nepgen, 1957)

Euphausia lucens	*Nematoscelis megalops*
Euphausia recurva	*Nematoscelis microps*
Euphausia similis var. *armata*	*Stylocheiron carinatum*
Euphausia similis	*Stylocheiron longicorne*
Euphausia hanseni	*Stylocheiron elongatum*
Euphausia mutica	*Stylocheiron abbreviatum*
Nyctiphanes capensis	*Stylocheiron affine*
Thysanoëssa gregaria	*Stylocheiron maximum*
Thysanoëssa parva	*Nematobrachion flexipes*

tions during winter and spring and a decline during summer and autumn. He showed maximum occurrence of eggs in waters approximately 37 km from the coast where the temperature was generally higher than 14 °C. Nepgen (1957) working in a similar area recorded maximum egg production during the spring and summer at a similar distance offshore in waters with temperatures ranging from 11–14 °C (integrated over 0–50 m depth). More recent work by Hopson (1983) in Lamberts Bay showed a late winter to spring peak in egg production with maximum during summer, again offshore in the relatively warmer water. As *Nyctiphanes capensis* do not release their eggs directly into the water it appears that the above authors were reporting on the distribution of the eggs of *Euphausia lucens*. Although peaks were recorded by these authors, egg production was noted throughout the year as were the occurrence of euphausiid larvae. Larval peaks were in accordance with the seasons of maximum egg production.

In the southern Benguela, Nepgen (1957) and Gow (unpubl. data) cited by De Decker (1973), found that swarms of *Euphausia lucens* and *Nyctiphanes capensis* occurred closer inshore than those of *Euphausia recurva*. Less abundant species such as *E. similis*, *E. spinifera*, *Thysanoëssa gregaria*, and *Nematoscelis megalops* occurred further offshore in warm oceanic waters (Fig. 30). *Euphausia lucens* was dominant in Gow's material

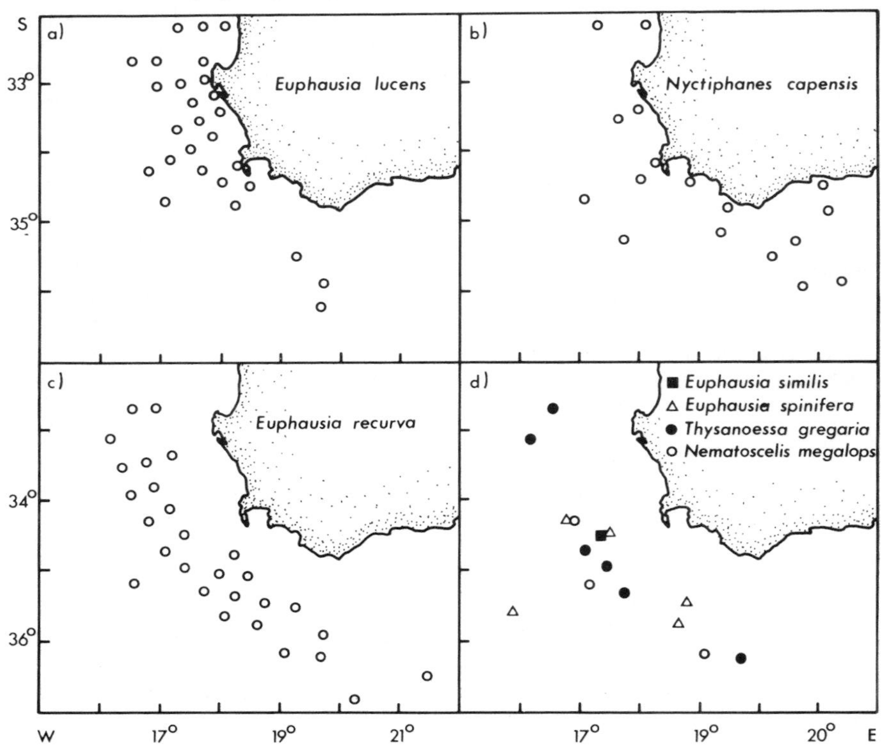

Fig. 30. The occurrence of euphausiid swarms off the western and southwestern coast of South Africa from cumulative data from 1964–1966 (from De Decker, 1973).

and were caught in the Lamberts Bay area in the winter, while summer swarms were located further south as far as 75 km west of Cape Point. Winter and spring concentrations of *Nyctiphanes capensis* extended from Lamberts Bay southwards as far as 160 km south of Cape Agulhas, while very few swarms were recorded during the summer months. Hopson (1983) also found winter and spring euphausiid maxima in the Lamberts Bay area. Further south, Talbot (1974) found that *Euphausia lucens* and *Nyctiphanes capensis* were only found in the cooler coastal waters of the western Agulhas Bank during the winter when temperatures were lowest and the influence of the Agulhas Current was weakest. She concluded that during the summer *Euphausia lucens* was shifted westward into the Benguela system due to the influx of warm tropical waters from the east coast. Talbot (1974) concluded that both *Nyctiphanes capensis* and *Euphausia lucens* were resident species of the Benguela system and not indicative of the Agulhas zooplankton community.

Hutchings (1979) discussed the vertical distribution of the larval stages of *E. lucens* during a period of active upwelling off the Cape Peninsula. He found that the early stage larvae (calyptopis) were concentrated higher in the water column than the older furcilia larvae. Pillar (1984a) examined the vertical distribution and diel movement of *E. lucens* from material collected in an area 65 km west of St Helena Bay (33° S) during a comparative study of the sampling performance of zooplankton nets (Pillar, 1984c). He showed evidence of ontogenetic layering in the vertical range of this species. While the calyptopis larvae remained near the surface and did not migrate, the older stages did, with the juveniles and adults moving through a greater depth range than the younger furcilia larvae (Fig. 31).

The euphausiids of the northern Benguela current have received less attention than those further south in the Cape waters. Boden (1955) recorded *Nyctiphanes capensis* as the dominant species in inshore Namibian waters. He concluded that *Euphausia lucens*, which was scarce in his collections, had approached the northern limits of its range in the Benguela system. While *Nyctiphanes capensis* has been recorded further north, as far as 23° N, *Euphausia lucens* has not been recorded north of the Namibian border (17° S) (Thiriot, 1978). Unterüberbacher (1964) reported euphausiid larvae to be abundant beyond the shelf off Walvis Bay and D'Arcangues (1977) described the vertical distribution of larvae which she observed during February and May 1974 and February 1975. She showed that two distinct layers existed in the water column, one concentrated in the upper layers in temperatures ranging from 14–18 °C and the other deeper in 12–13 °C. She also found the larvae to be tolerant of a wide range of dissolved oxygen levels ($1 \cdot 2$–$7 \cdot 3$ $ml \cdot l^{-1}$). During November in inshore waters off Walvis Bay, Thomas (1980) observed extensive day-time surface swarms of early larval stages of *Nyctiphanes capensis*. Night-time surface swarms of adults were observed by Cram & Schülein (1974) during an unusual occurrence of surface-shoaling Cape hake (*Merluccius capensis*). *Nyctiphanes capensis* have been encountered in sound-scattering layers off Namibia by D'Arcangues (1977).

The feeding habits of the local euphausiid fauna was first reported by Nepgen (1957) who found them to be omnivorous. Observations on *N. capensis* by Lazarus (1975) showed that the larval stages contained only

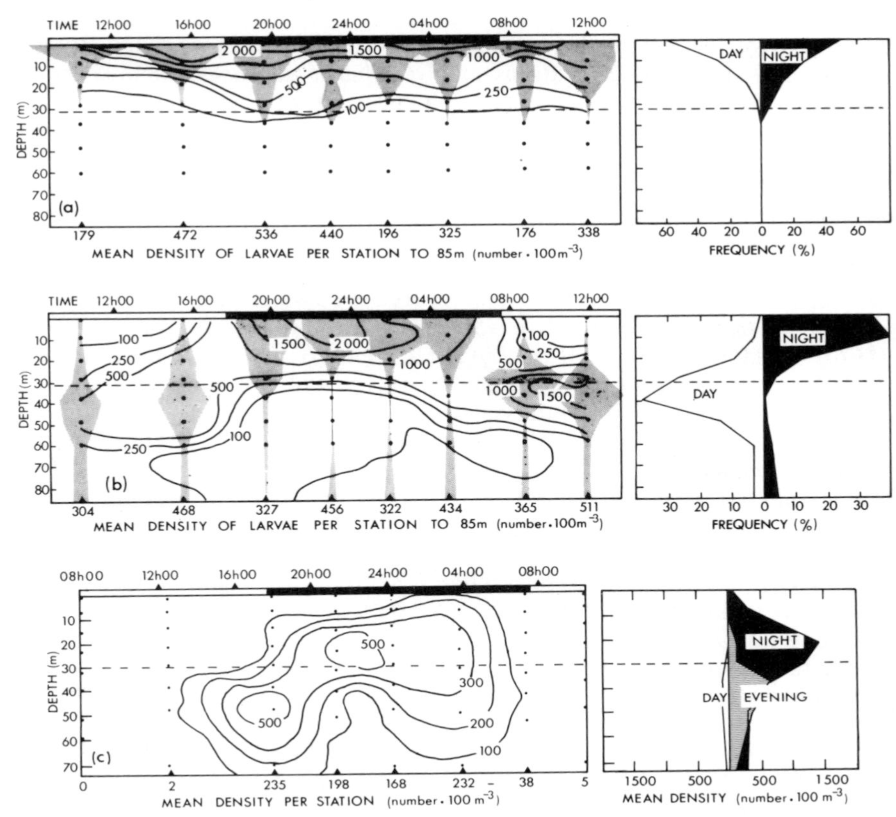

Fig. 31.—Vertical distribution of (a) calyptopis larvae and (b) furcilia larvae of *Euphausia lucens* based on pump collections and (c) juvenile and adults based on Miller net collections: the thermocline is denoted by a broken line, the sampling depths as dots and the proportion of animals at each depth by a shaded area; from Pillar (1984a).

diatoms whereas the juvenile and adults contained, in addition to diatoms, copepod and amphipod remains. Larval stages of euphausiids in the gut indicated them to be cannibalistic. Pillar (1984b, 1985) carried out selective feeding experiments with larval stages of *Euphausia lucens* and *Nyctiphanes capensis* using mixtures of *Artemia* nauplii, copepod nauplii, and phytoplankton as food. He showed that a mixed or pure diet of phytoplankton (*Tetraselmis chuii*) and *Artermia* nauplii, but not a diet of copepod nauplii, was sufficient for proper larval development. He pointed out, however, that *Artemia* nauplii is a poor indicator of carnivorous feeding behaviour since it is captured as a passive particle, whereas the larvae would have more difficulty in capturing the more active copepod nauplii. Nevertheless, he provided evidence that they can capture and utilize copepod nauplii at later stages of their larval development.

Field observations by Pillar (1983) have shown the co-occurrence both horizontally and vertically of *Euphausia lucens* and anchovy larvae in the southern Benguela system. Laboratory studies indicate that juvenile and adult stages of this euphausiid are capable of capturing and ingesting 1 to 2-day-old anchovy larvae, which increased significantly with increasing age of predator.

Chaetognatha

Heydorn (1959), in a survey of the works by Cleve (1905), Ritter-Zahony (1909, 1911), Gray (1923), Thomson (1947), and Fraser (1952), recorded 19 species of chaetognaths in South African waters, 12 species of which were present in his own material collected in the "Cape routine area". The chaetognath community of the Agulhas Current has been investigated by Stone (1969) who recorded 18 species, the majority being cosmopolitan warm-water forms of Indo-Pacific origin. From material collected in the "Southern routine area", Masson (1973) recorded 24 species of chaetognaths (Table IX) which included all those reported by both Heydorn (1959) and Stone (1969). Venter (1969) reported on the occurrence of 10 species from collections in the "Walvis Bay routine area", which were also included in Masson's material. As evidenced in the data obtained by the above authors, there is a close relationship of different chaetognath species with specific water masses in the Benguela system. *Sagitta friderici*, which is the most commonly occurring chaetognath, is indicative of cold neritic water while *S. minima*, the second most common species, prefers the warmer offshore waters of mixed origin. Heydorn (1959) describes *S. enflata* and *S. regularis* as useful indicators of the penetration of water of Agulhas Current origin into the Benguela system. Other less common species are also considered to be useful in labelling water masses and determining their origin, even after there has been substantial mixing. *S. serratodentata* and *S. tasmanica* are indicative of the influx of South Atlantic subtropical surface water in the region. The occurrence of bathypelagic species such as *Eukrohnia hamata* and *Sagitta decipiens* in coastal regions and the offshore displacement of *S. friderici* is considered as evidence of recent upwelling (Masson, 1973; Thiriot, 1978).

TABLE IX

Chaetognath species recorded off the west and southwest coast of South Africa (from Masson 1973)

Sagitta friderici	*Sagitta enflata*
Sagitta lyra	*Sagitta tasmanica*
Sagitta robusta	*Sagitta regularis*
Sagitta decipiens	*Sagitta serratodentata*
Sagitta macrocephala	*Sagitta neglecta*
Sagitta pacifica	*Sagitta bipunctata*
Sagitta hexaptera	*Eukrohnia bathypelagica*
Sagitta pulchra	*Eukrohnia fowleri*
Sagitta bedoti	*Eukrohnia hamata*
Sagitta minima	*Krohnitta pacifica*
Sagitta ferox	*Krohnitta subtilis*
Sagitta planctonis	*Pterosagitta draco*

The degree of penetration of chaetognaths into the Benguela system either from Agulhas Current water or from South Atlantic subtropical surface water is dependent on the amount of upwelling in process. The cold upwelled waters of the western coast act as a barrier to the northern extension of some species of chaetognaths. When this barrier is relaxed, many chaetognath species are transported northwards by the Benguela Current. The scarcity, however, of some typically Indo-Pacific species such as *S. robusta* and *S. neglecta* indicates the restricted movement of these forms into the Benguela system. Venter (1969) recorded no Indo-Pacific species in his material collected in the northern Benguela region which seems to suggest that the Lüderitz upwelling zone is an effective environmental barrier to certain species in the Benguela.

Masson (1973) found that both *S. friderici* and *S. minima* spawned continuously throughout the year. There were, however, notable peaks in spawning in winter and spring in accordance with the maximum occurrence of the adult population. He postulated that there were at least five generations of *S. friderici* per year in the southern regions and noted a tendency for spawning to occur inshore in the winter and offshore during the summer. For certain species, such as *S. lyra* and *Eukrohnia hamata*, immature stages were found to inhabit shallower water than the adults. This pattern has been documented for other species in other areas (Alvariño, 1965), hence any surface current will have a pronounced effect on the juveniles and less effect on the adult distribution. The fact that the juveniles are more widely distributed than the adults of a particular species can be explained by their susceptibility to surface water movement. Masson (1974) also attributes this observation to the possibility that the juveniles have a greater tolerance to changing conditions when transported out of their preferred habitat.

Venter (1969) noted the neritic preference of *Sagitta friderici* in Namibian waters. It constituted numerically 90% of all the species recorded in coastal waters there while forming only 24% of the chaetognath population in the offshore regions. Although it was described as being associated with cold water it was noted that its maximum occurrence was not coincidental with periods of upwelling. Its seasonal occurrence in this region is similar to that found in the southern Benguela, being maximum during the winter and spring. In the northern Benguela Current *S. tasmanica* occupies the place held by *S. minima* in the warm offshore waters of the southern Benguela regions. The former species represented 70% of the chaetognath population in the offshore waters while the latter species only constituted 3% of the whole chaetognath community (Venter, 1969). Although both *S. friderici* and *S. tasmanica* are reported by Thiriot (1978) as being particularly dominant in upwelled areas along the African Atlantic coast, *S. friderici* is often located far beyond the centre of upwelling along the west African seaboard.

The carnivorous feeding behaviour of chaetognaths was supported by Lazarus (1975) who identified 19 types of prey in the gut of *S. friderici*. Only one species of phytoplankton was recorded, and occurred only in 1% of the specimens. Copepods, cirripede nauplii, decapod larvae, and hyperiid amphipods all formed important parts of their diet, also the cannibalistic nature of this species was noted by the author. No fish larvae remains were noted in the gut contents. Consumption of fish larvae by chaetognaths has, however, been frequently inferred elsewhere from both examination of field specimens and from laboratory experiments (Hunter, 1984), but its rôle as a predator in the Benguela system is not well documented.

Amphipoda, Hyperiidae

Using material drawn mainly from the reports of Barnard (1932, 1940), Siegfried (1963) compiled a list of 77 hyperiid species found off the western coasts of South African and Namibia (Table X). From material collected from an area between the Agulhas Bank on the southern coast (38° S), to the northern border of Namibia (17° S), he found most species occurred north of the Cape of Good Hope with 27 occurring off the southern coast. Dick (1970) examined material covering the southwestern and eastern coast of South Africa and recorded 105 species, of which 20 were additional to the western coast fauna. The numerical importance of *Themisto gaudichaudi* (recorded as *Parathemisto gaudichaudi* by Siegfried, 1963) in the amphipod fauna of the Benguela system is confirmed by the above authors and by Siegfried (1965), who reported it as being one of the most characteristic and abundant members of the neritic plankton off the western coast of South Africa.

Siegfried (1963) found ovigerous females of *T. gaudichaudi* throughout the year. Further observations by Siegfried (1965) from the "Cape routine area", showed an increase in their abundance during the spring and a low occurrence during the winter. The maximum reproductive activity displayed during the spring was reflected in his data by an increase in the abundance of mature individuals during the summer and autumn. This early attainment of maturity allowed for the production of a number of successive generations each spring and summer, which he suggested maintained an abundant population in the southern Benguela region. Hopson (1983) found a progression of growth stages of *T. gaudichaudi* in her collections off the Lamberts Bay area. In agreement with Siegfried (1965), her data suggested continuous reproduction throughout the year with an overwintering group of adults which produced a spring spawning, which in turn produced another generation in late summer.

The cool neritic preference of *T. gaudichaudi* (Fig. 32) was shown by Siegfried (1963, 1965), while Hutchings (1979) and Hopson (1983) also found it more abundant in cool inshore waters off the western coast during periods of active upwelling. It has also been reported in appreciable numbers over the Agulhas Bank where small populations were found centred over an area of upwelling (De Decker, 1973).

Siegfried (1965) found *T. gaudichaudi* to be a voracious feeder on plankton and to be unselective in their feeding habits. In most cases he found the frequency of occurrence of prey in their guts reflected the frequency with which the prey occurred in the water. Zooplankton items ranged from small copepodite stages to juvenile euphausiids, decapod larvae, and fish larvae. He identified five genera of phytoplankton which were consumed to a greater degree by the juveniles than by the adults. Lazarus (1975) found 61 different kinds of food in the guts of this species of

TABLE X

Hyperiid amphipod species recorded off the western coast of South Africa and Namibia (from Siegfried, 1963)

Lanceolidae

Lanceola serrata
Lanceola pacifica
Scypholanceola vanhoeffeni

Scinidae

Scina crassicornis
S. curvidactyla
S. incerta
S. langhansi
S. borealis
S. uncipes f. *affinis*
S. oedicarpus
S. woltereckí
S. rattrayi
S. tullbergi
S. nana
S. exisa
S. stenopus

Vibiliidae

Vibilia viatrix
V. propinqua
V. antarctica
V. armata
V. cultripes
V. chuni
Cyllopus magellanicus

Paraphronimidae

Paraphronima gracilis
P. crassipes

Cystisomatidae

Cystisoma coalitum

Hyperiidae

Hyperia galba
H. promontorii
Hyperioides longipes
Hyperoche medusarum
*Parathemisto (Euthemisto) gaudichaudi**
Phronimopsis spinifera

Dairellidae

Dairella latissima

Phronimidae

Phronima sedentaria
P. atlantica
P. atlantica var. *solitaria*
P. pacifica
P. colletti
P. stebbingi
Phronimella elongata

Phrosinidae

Phrosina semilunata
Primno macropa
Anchylomera blossevillei

Lycaeopsidae

Lycaeopsis themistoides

Pronoidae

Eupronoe minuta
E. maculata
E. armata
Parapronoe crustulum
P. campbelli
Sympronoe parva

Lycaeidae

Tryphana malmi
Lycaea nasuta

Brachyscelidae

Brachyscelus crusculum
B. rapax
Thamneus platyrhynchus

Oxycephalidae

Simorhynchotus antennarius
Oxycephalus clausi
O. latirostris
Streetsia pronoides
S. porcella
S. steenstrupi
Cranocephalus scleroticus
Leptocotis tenuirostris
Rhabdosoma whitei

Parascelidae

Parascelus edwardsi
Schizoscelus ornatus
Thyropus sphaeroma

Platyscelidae

Platyscelus ovoides
P. armatus
P. serratulus
Hemityphis rapax
Amphithyrus bispinosus
A. sculpturatus
Paratyphis maculatus
Tetrathyrus forcipatus

**Themisto gaudichaudi*

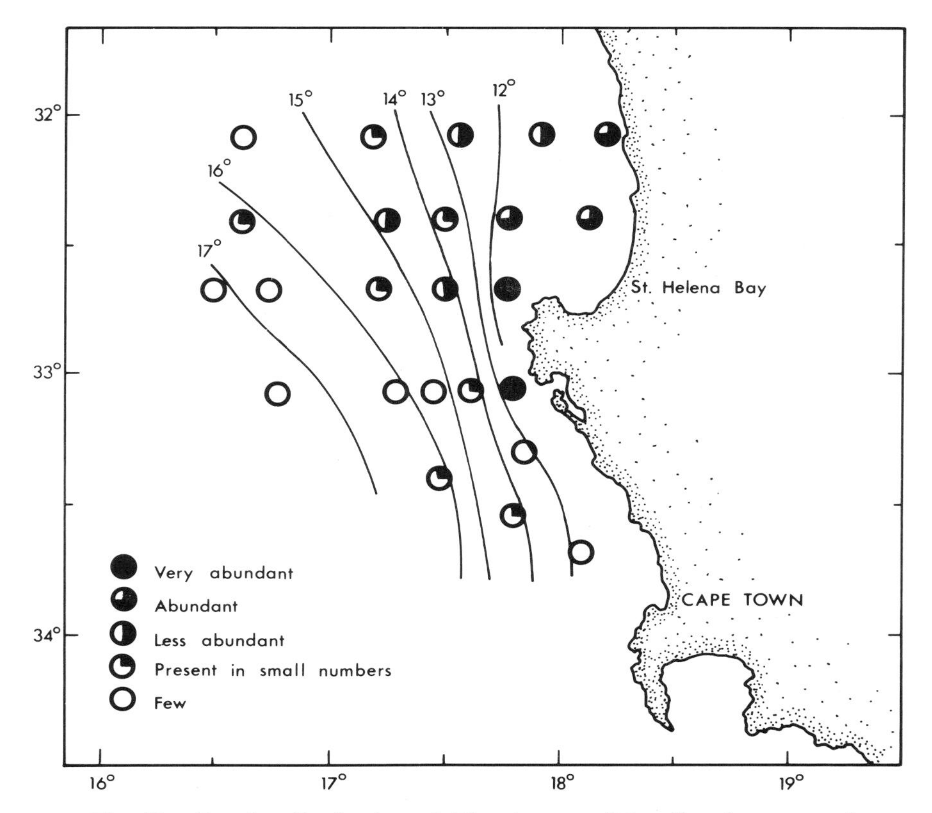

Fig. 32.—Density distribution of *Themisto gaudichaudi* and average of 0–50 m mean integral temperatures from April 1954–March 1955 (after Siegfried, 1965).

which 12% consisted of zooplankton remains. Field evidence indicates this species to be strongly predatory on fish larvae (Hunter, 1984).

Tunicata

Van Zyl (1960) investigated the distribution of doliolids and salps in the "Cape routine area" and listed 11 species of thaliaceans. A follow-up of this work by Lazarus & Dowler (1980) reported on 23 species of thaliaceans and 26 species of appendicularians from material collected monthly over a two-year period (1964–1965), from the "Southern routine area" (Table XI). A brief summary of unpublished data by Dowler was incorporated into an earlier report by De Decker (1973) in his account of the Agulhas Bank plankton. The seven most commonly occurring western coast tunicates recorded by Lazarus & Dowler (1980) were three thaliaceans, *Thalia democratica*, *Salpa fusiformis*, and *Doliolium denticulatum* and four appendicularians, *Oikopleura longicauda*, *O. dioica*, *Fritillaria formica*, and *F. pellucida*. The use of several of these species as indicators

TABLE XI

Thaliacea and Appendicularia species recorded off the western and southwestern coast of South Africa (from Lazarus & Dowler, 1980)

Oikopleura intermedia	*Fritillaria* sp.
Oikopleura fusiformis	*Folia* sp.
Oikopleura longicauda	*Oikopleura* sp.
Oikopleura cophocerca	*Kowalevskaia oceanica?*
Oikopleura rufescens	*Thalia democratica*
Oikopleura dioica	*Thalia longicauda*
Oikopleura parva	*Salpa fusiformis*
Oikopleura cornutogastra	*Salpa maxima*
Oikopleura albicans	*Salpa cylindrica*
Megalocercus huxleyi	*Ihlea magalhanica*
Stegosoma magnum	*Iasis zonaria*
Althoffia tumida	*Traustedtia multitentaculata*
Tectillaria fertilis	*Pegea confoederata*
Fritillaria formica	*Cyclosalpa pinnata*
Fritillaria borealis	*Doliolum denticulatum*
Fritillaria pellucida	*Doliolum nationalis*
Fritillaria megachile	*Doliolum denticulatum* var. *ehrenbergii*
Fritillaria haplostoma	*Doliolum mirabilis*
Fritillaria fraudax	*Dolioletta gegenbauri*
Fritillaria bicornis	*Doiloiletta tritonis*
Fritillaria taeniogona	*Pyrosoma* sp.

of different water masses was demonstrated by De Decker (1973) and Lazarus & Dowler (1980).

Of the thaliaceans, *Thalia democratica* and *Doliolum denticulatum* are indicative of warm water and good indicators of waters of Agulhas origin. In the typical forms of *D. denticulatum*, there is a strong tendency for the oozoids (nurse stages), the phorozoids (solitary stages), and the gonozoids (aggregate stages) to avoid the cool inshore waters. The restriction placed by temperature on the distribution of *D. denticulatum* in the southern Benguela region was illustrated by Lazarus & Dowler (1979) during their October 1965 collections when areas of cool water in the Walker Bay, Cape Peninsula and St Helena Bay regions excluded this species (Fig. 33a). They noted that this species was abundant in the warmer well-mixed waters further off the western coast and over the Agulhas Bank. Only when the warmer waters (>14 °C) penetrated inshore on the western coast did the authors observe *D. denticulatum* near the coast. This distributional pattern was also observed in *Thalia democratica* and to a lesser extend in *Salpa fusiformis*. They were not found in areas associated with cool water and were only observed offshore in mixed warm water.

Among the less abundant species of thaliaceans reported by De Decker (1973) only *Doliolum nationalis* showed a preference for areas of cool water. On the rare occasions when it was observed it occurred inshore in upwelling areas off the western coast. The affinity of this species for low temperatures was also noted by Van Zyl (1960). *D. nationalis* is exceptional among local thaliacean species in having its maximum occurrence during winter when the other species show a decline during this period. The peak abundances of the other species were reported by the above authors as occurring during the spring and summer.

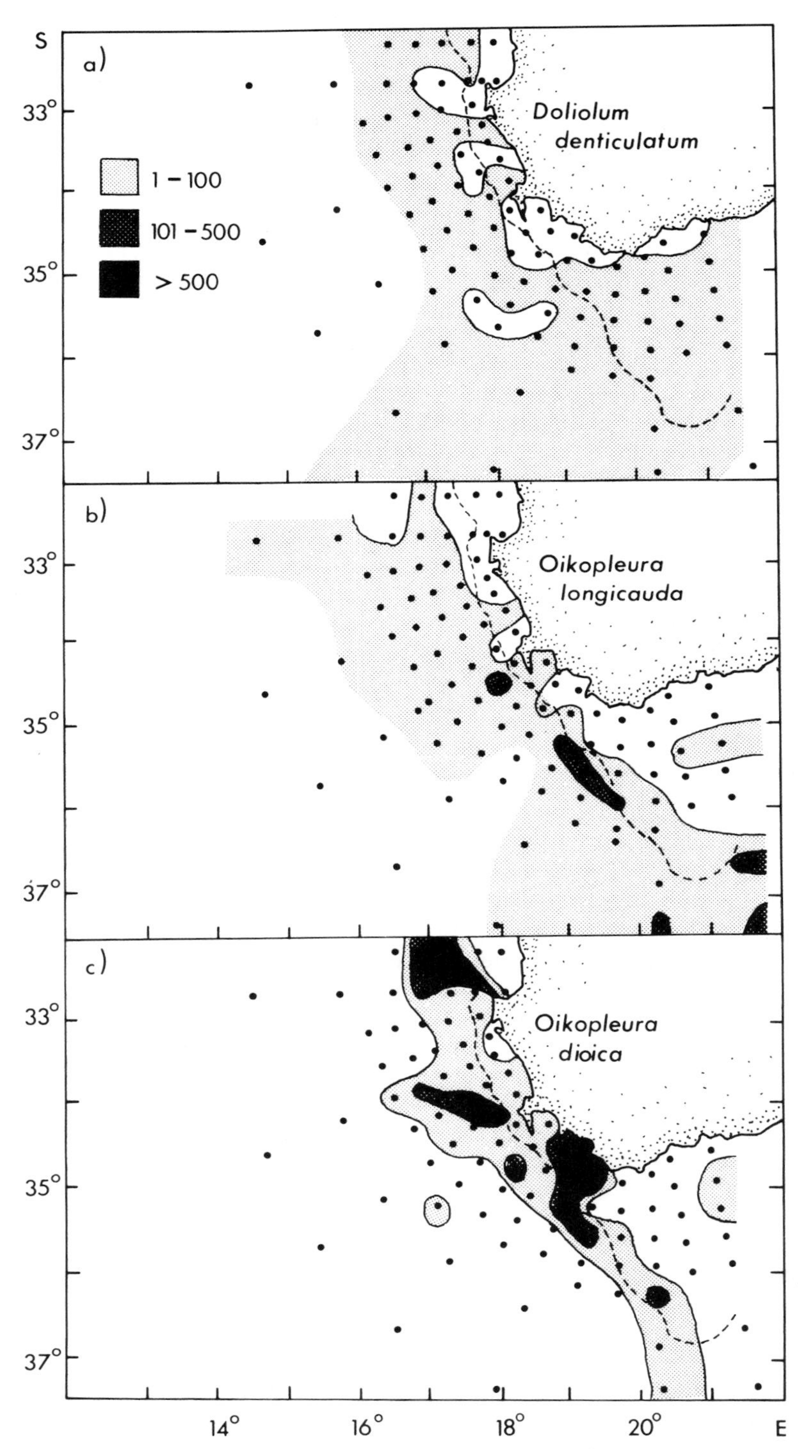

Fig. 33.—Distribution of (a) *Thalia democratica*, (b) *Doliolum denticulatum*, and (c) *Oikopleura dioica* during the summer and winter period of 1964 (after Lazarus & Dowler, 1980).

Of the common appendicularian species reported by Lazarus & Dowler (1980), *Oikopleura dioica* and *Fritillaria formica* showed a preference for cool water, whereas *Oikopleura longicauda* and *Fritillaria pellucida* showed a preference for warmer waters. This pattern was clearly indicated in their July 1964 distribution maps showing the two species of the genera *Oikopleura* (Fig. 33b and c). During this period, inshore upwelling was noted east of the Cape Peninsula. There was very little penetration into this area of *O. longicauda*; further offshore in warmer waters the species was, however, well represented. Contrary to this, *O. dioica* occurred in abundance close inshore in the cooler waters while progressively decreasingly in numbers offshore. The former species is shown to be indicative of the warm waters of Agulhas origin, while the latter proved to be indicative of cooler upwelled water. A comparison of the abundance of these two species by the above author showed that over the entire area studied *O. dioica* was by far the less abundant species. It, however, was predominant in the appendicularian community of the western coast. Similar findings were also noted by Lazarus (1975) from inshore collections taken between Table Bay and St Helena Bay.

Analysis of gut contents by Van Zyl (1960) showed that thaliaceans fed mainly on phytoplankton in inshore areas and zooplankton in offshore areas. Zooplankton remains consisted mainly of young stages of copepods, amphipods, and euphausiids. Large salps such as *Salpa maxima* had traces of fish eggs together with phytoplankton remains in their stomachs. Lazarus (1975) found inshore specimens of appendicularians to be entirely herbivorous.

Decapod larvae

Decapod larval studies in the Benguela system have been principally concerned with the development and distribution of the phyllosoma stages of the commercially important rock lobster, *Jasus lalandii*. Lazarus (1967) examined phyllosoma larvae from material collected over a three-year period (1961–1964) in the "Southern routine area" (see Fig. 24) and identified 13 larval stages. He estimated that these larvae spent between 9–11 months in the plankton after hatching from eggs in September to November each year. The early stage larvae were found in large numbers in nearshore waters during the summer but thereafter middle and late stage larvae were rarely encountered in his inshore collections. Late stage larvae were only found well offshore as far as 300 km west during the winter. Later studies by Pollock & Goosen (1983) in the southeastern Atlantic confirmed the offshore distribution of the late stage phyllosoma larve, although relatively few individuals were caught. They found their distribution extended westward over the shelf break to as far as 13°30′ E. The contention of both Lazarus (1967) and Pollock & Goosen (1983), that some of the larvae encountered further west are those of *Jasus tristani* of Tristan da Cunha origin is still inconclusive owing to the taxonomic difficulties in distinguishing between these two species at the phyllosoma stage.

Since phyllosoma larvae are unable to swim against currents of any appreciable magnitude, return transport mechanisms are thought to play an important rôle in preventing all the larvae from being advected out of the

Benguela system. The rôle of vertical migratory behaviour coupled with underlying inshore countercurrents and eddies have been proposed by Lazarus (1967), Duncan & Nell (1969), and Pollock & Goosen (1983) as being possible return mechanisms. Pollock & Goosen also reported late larval stages occurring in the top 11 m at night and deeper than 200 m during the day. No significant data on the distribution of the older puerulus stages could be gleaned from the above collections due to their ability to avoid the sampling gear. Pollock &, Goosen speculated that once metamorphosis from phyllosoma into puerulus occurs offshore, their progressively stronger swimming capabilities enables them to return to the shallow coastal waters.

Lazarus (1975) found only early phyllosoma larvae and the occasional puerulus stage of *J. lalandii* in his nearshore collections. Gut contents of the former yielded only phytoplankton remains, whereas the latter were principally carnivorous, feeding mainly on crustaceans and larval bivalves.

TABLE XII

Synopsis of work on six other taxa

Taxa	Details of findings	References
Cladocera	Dominant species in the southern Benguela are *Evadne spinifera* and *Podon* sp. They prefer thermoclines or warm surface water. Minimum during winter in nearshore water. *Penilia arirostris* is found off the western coast but is considered a typical member of the Agulhas Bank community.	De Decker, 1973; Hutchings, 1979; Carter, 1983
	Dominant species in the northern Benguela are *Podon polyphemoides* and *Evadne noidmanni*. Both species are abundant throughout the year. The former found inshore of the shelf while the latter species extends beyond the shelf.	Unterüberbacher, 1964
Mysidacea	The dominant species is *Mysidopsis major* with *Mysidopsis similis*, *Mysidopsis schultzei*, *Anchialina truncata* and *Gastrosaccus psammodytes* being present in appreciable numbers. Breeding is continuous throughout the year with a spring maxima and a winter low.	Lazarus, 1975
	The mean biomass of *Mysidopsis major* in a southern Benguela kelp bed was estimated to be 116 mg dry wt$\cdot$m^{-2} being an order of magnitude less than the smaller zooplankton biomass (principally copepods). Their mean daily ration requirements represented only 0·08% of the total primary production. Production of this species in the study area was estimated as 246 mgC$\cdot$m$^{-2}\cdot$yr^{-1} with an annual P/B estimate of 5·7.	Carter, 1983

TABLE XII—*continued*

Taxa	Details of findings	References
Ctenophora	Two species, *Pleurobrachia pileus* and *Beroe forskali* recorded in nearshore Cape waters. Maximum occurrence of the former species during winter and the latter during summer and autumn. Noted as being carnivorous and a potential predator of fish larvae.	Lazarus, 1975
Cnidaria	Eleven genera recorded in nearshore Cape waters. Sizes and seasonality of different genera discussed. Zooplankton remains were observed in the stomachs of a number of species.	Lazarus, 1975
	Stranding of *Physalia* on the beaches in the South Western Cape during January 1983 is considered to be a consequence of the anomalous meteorological oceanographical conditions during the 1982–1983 summer period.	Shannon & Chapman 1983
Cirripedia	Abundance and distribution of cypris nauplii and larval stages of the western coast are described for *Balanus algicola* (the most dominant species) and for *Balanus amphitrite* var. *denticulata*, *Tetraclita serrata* and *Octomeris angulosa*. Their abundance declined during the winter months.	Lazarus, 1975
	Cypris nauplii and larvae recorded as preferring warm water.	Hutchings, 1979; Carter, 1983
Isopoda	Seven species are recorded of which *Eurydice longicornis* and *Paridotea ungulata* comprised numerically 87 and 9%, respectively, of the population. Peaks of abundance and maximum breeding occurs during the spring months.	Lazarus, 1975

Other species

Table XII lists six other zooplankton taxa that have been studied to a lesser degree than those discussed earlier. The major contributions to the study of these groups was by Lazarus (1975) from inshore collections between the Cape Peninsula and St Helena Bay.

PRESENT AND FUTURE ZOOPLANKTON RESEARCH IN THE BENGUELA

The main direction of zooplankton research in recent years has been towards a better understanding of the functional aspects of zooplankton populations, not as a single entity, but on dominant species to which the concept of population dynamics may best be applied. Research in this direction has lagged behind that of the physical, chemical, and phytoplankton work, largely because of the lack of intensive short-term surveys such as those first reported by Hutchings (1979) in the Cape Peninsula. With the exception of the work by Carter (1983) on the energetics of zooplankton in the kelp-bed zone, when time scales in the order of hours to days were used, the only attempts at elucidating the dynamics of local zooplankton are from surveys of at least one month apart. Under such sampling programmes documentation is limited to large scale phenomena, such as faunal changes associated with seasonal hydrographic changes and faunal differences between nearshore and offshore regions. Nevertheless, surveys such as The Cape Egg and Larval Programme (CELP) have provided material which has added considerably to our knowledge of the spatial and temporal distribution of local zooplankton communities, *e.g.* Hopson (1983). These collections, using Bongo nets, are considerably more quantitative than earlier investigations. In addition, the superior catching ability of these nets has facilitated more reliable mapping of the distribution and abundance of larger zooplankton such as euphausiids. The mesh size of these nets (300 μm) has allowed for a more quantitative evaluation of all developing stages of local euphausiid species, principally *Euphausia lucens* and *Nyctiphanes capensis*. These data are at present being documented in relation to concomitant physical–chemical data as a basis towards a better understanding of the life history adaptations of these species in the southern Benguela system.

Recent developments in the techniques for sampling zooplankton have resulted in the construction of a small meshed (200 μm) large-mouth multiple net system (RMT 1+6), which can sample up to six different depths in one operation using electronically operated opening and closing devices. This net has allowed for the examination of the vertical distribution of a wide size range of zooplankton organisms. In addition, it allows sampling of the entire water column so that organisms may not migrate beyond the tracking range of the net. On-going hydroacoustic surveys for mapping pelagic fish stocks off the South African seaboard have incorporated the RMT 1+6 in their sampling programme in order to address questions regarding the importance of zooplankton in the feeding ecology of economically important fish species. These collections have provided useful material for examining the depth distribution of key zooplankton species in relation to environmental factors such that their transport and maintenance mechanisms can now be quantitatively evaluated.

In considering future research, short-term studies will be conducted primarily in the Cape Columbine and St Helena Bay area. In this important fishery recruitment area environmental conditions, such as wind and current patterns, have been closely monitored over a number of years, hence some physical understanding of the processes is available and could be used in the study of the dynamics of zooplankton in this region. A major emphasis will be placed on the formation of patches of suitable zooplankton food organisms and the rates of consumption of young and adult fish in these patches. Given that reasonable estimates of zooplankton production can be derived from biomass estimates in the field and laboratory and field-derived estimates of grazing, growth and generation times from ongoing research, the relative rates of growth versus feeding and dispersion need to be examined in order to derive better estimates of the carrying capacity of zooplankton in important recruitment and spawning areas of the Benguela system. Central to this approach will be the deployment of a continuous depth profiling sampler such as the Batfish (Herman & Denman, 1977) which may soon be operational. The high vertical and horizontal resolution capabilities of this instrument offer tremendous scope for refining our present broad-scale studies so that critical relationship concepts between zooplankton and environmental variables can be formulated and tested more efficiently. This instrument will form the basis of a monitoring programme which will be initially deployed between Lamberts Bay and the Cape Peninsula to include the important St Helena Bay nursery area and will later be extended to include the Agulhas Bank spawning area.

The interest in quantitative analysis of zooplankton dynamics has resulted in the development of culturing and rearing facilities which was initiated in the early 1980s. Two main categories of study are at present being investigated. First, continuous cultures have been used as an aid in taxonomic work. The young stages of many of the commoner species in the Southern Hemisphere are poorly described and this information is vital to any work using a population dynamics approach. In rearing the euphausiids *Euphausia lucens* and *Nyctiphanes capensis*, Pillar (1984b, 1985) studied their larval development under different trophic conditions. He showed that morphological characters were infuenced by feeding condition, with similar morphological variations existing in wild populations. Culturing of dominant local copepod species such as *Centropages brachiatus* and *Calanoides carinatus* has contributed significantly to an improved understanding of their successive larval stages. To date nearly 60 species of planktonic marine and estuarine organisms, mainly fish larvae and copepods, have been successfully reared or cultured in the laboratory.

From a consideration of standing stock and laboratory determined respiration, feeding, and excretion rates, Carter (1983) estimated the production and daily ration requirements of the zooplankton community of a kelp-bed zone off the Cape Peninsula. This approach is at present being extended towards the construction of energy and nitrogen budgets of the dominant euphausiid, *Euphausia lucens*. Factors influencing the physiological processes such as temperature, swimming speed, feeding behaviour, food type, and feeding rates are being investigated in this study. As techniques are improved, further studies on other major zooplankton groups are envisaged.

To investigate the idea that invertebrate predation on the commercially important anchovy larvae is of significance, extensive observations are being made on the food preferences and the effectiveness of various zooplankton groups as predators on these larvae. Having anchovy larvae of various sizes as well as juveniles and adults in captivity, has also permitted studies on cannibalistic behaviour in the laboratory.

The emphasis for future culture work lies in estimating quantitative values for the tropho-population dynamics of selected zooplankton groups. Olivieri (1985) and Carter (1983) have estimated that local zooplankton populations exert little impact on the phytoplankton community. It is suggested that the important factor limiting local zooplankton populations may be the transient nature of food availability and their inability to return to sites of high primary production. Laboratory studies by Borchers & Hutchings (1983) have shown that the adults of the common west coast copepod, *Calanoides carinatus* are resistant to periods of starvation (8–12 days) but the juvenile stages are less so. There is clearly room for more laboratory studies on the feeding, growth, starvation tolerance, and survival of zooplankton, such processes which cannot be successfully monitored in the ocean. These studies should be combined with field work for studying patchiness and its general significance to the trophodynamics of the zooplankton community. Knowledge of their general vertical and horizontal distributions and the system of currents and counter-currents need to be drawn together to provide a picture of the life history adaptations of representative species.

ICHTHYOPLANKTON

Although Gilchrist (1916) described the eggs and larvae of 38 species of Cape fishes including pilchard, anchovy, hake, sole, saury, horse mackerel, and snoek, little was known about the distribution of the ichthyoplankton in time and space until comparatively recently[1]. The spawning area of the pilchard off Namibia was first documented by Hart & Marshall (1951) following the two surveys of R.R.S. WILLIAM SCORESBY, while eggs and young stages of commercially important fish species, notably hake (Hart & Marshall, 1951; Hart & Currie, 1960) and anchovy (Hart & Currie, 1960) were also collected during the surveys. With the development of the Cape and Namibian pilchard fisheries during the 1950s regular egg surveys were undertaken in the areas which were considered to be the principal spawning grounds. Although some of the findings were reported by Davies (1954, 1956) for the Cape and Stander (1963) and Matthews (1964) for Namibia, the results of the surveys undertaken between 1950 and 1967 remain largely unpublished. Crawford (1980), however, summarized the essential details of the distribution of pilchard and anchovy eggs collected around the South Western Cape.

As a result of the sharp decline in the Namibian pilchard catch in 1970 a comprehensive research programme, the Cape Cross Programme (Cram & Visser, 1973) was initiated. The quantitative South West African Pelagic Egg and Larval Surveys (SWAPELS) were undertaken monthly in the area between 18° S and 24° S (O'Toole, 1974) and at the same time the development of eggs and larvae of key species were investigated in a series of comprehensive laboratory studies. The on-going SWAPELS programme has been augmented during recent years by investigations by Spanish and other overseas workers under the auspices of the International Commission for the South East Atlantic Fisheries (ICSEAF).

It is perhaps worth commenting at this stage as to the reasons why ichthyoplankton, which is after all a minor component of the plankton, has

[1]Subsequent to writing this section a doctoral dissertation by Olivar (1985) has become available. The emphasis of the work is on taxonomy, distribution, and abundance of the ichthyoplankton collected between 1979 and 1984.

received so much attention (although this is not reflected in the prime scientific literature). There are two reasons for a fishery organization to conduct regular ichthyoplankton surveys: the first is to attempt to relate the data back to spawner stock size or forward to recruitment; the second is to investigate the ecological processes. Much of work done to date has unfortunately been justified on the basis of the first approach, which is of dubious worth. In fact the first useful step towards estimating the spawner stock size of anchovy from egg production data was only made in November 1984 (in southern Benguela). The second (*i.e.* ecological) approach was adopted during 1977 and 1978 when quantitative surveys of pelagic eggs and larvae were undertaken around the South Western Cape—the Cape Egg and Larval Programme (CELP)—and which were aimed at answering questions such as what environmental factors control spawning and recruitment. The detailed findings of this research are at present being prepared for publication by Dr P. A. Shelton of the Sea Fisheries Research Institute.

In the following paragraphs the relevant literature on the ichthyoplankton of the Benguela is reviewed. The principal spawning areas and seasons of nine important species, as inferred from published work by Shannon (in press), are shown in Figures 34 and 35. Taxonomic studies are mentioned, but are not discussed in any detail. Readers are referred to Brownell (1979) which is a useful reference text on the development of eggs and larvae of 40 species of local fishes. The biology of the post-larval stages is not discussed as this will be included in the section dealing with commercial fisheries in Part IV of this series of reviews. The taxonomic names of species are the latest as prescribed by the J. L. B. Smith, Institute of Ichthyology; the older names are given in parentheses.

PILCHARD, SARDINOPS OCELLATUS

The egg of the southern African pilchard, *Sardinops ocellatus* (*S. sagax, S. ocellata*), is round—the typical diameter is 1·7 mm (Gilchrist, 1916)—and has a single oil globule of about 0·18 mm. It is slightly larger than the egg of the Californian sardine *S. caerulea* (Davies, 1954) and can be readily identified in net ichthyoplankton. Blastodisc-stage eggs collected off central Namibia in water of 16·6–19·5 °C, when incubated, hatched and maintained at 13–22 °C, yielded larvae that could, in terms of a functional jaw and fully pigmented eyes, commence feeding (King 1975, 1977a). At temperatures below 13 °C (King, 1975, 1977a) hatching was successful but subsequent development was retarded. The optimum temperature for development of "warm water" spawned eggs appears to be about 18 °C. The development of the yolk-sac, larval and metamorphic stages of *S. ocellatus* has been described by Louw & O'Toole (1977). Young stages were noted as being very similar to other pilchard species. Growth rate estimates range from 0·3–0·6 $\mathrm{mm \cdot day^{-1}}$ for small larvae to 0·2–0·3 $\mathrm{mm \cdot day^{-1}}$ for larger larvae (O'Toole 1977a; Boyd & Badenhorst 1981).

During the 1950s and 1960s most of the Benguela ichthyoplankton samples were collected using relatively large towed nets (N100H). From 1972 the more quantitative towed Bongo nets with 0·50- and 0·94-mm mesh sizes were used (King & Robertson, 1978), and accordingly only qualitative comparisons should be made between the pre- and post-1971 collections.

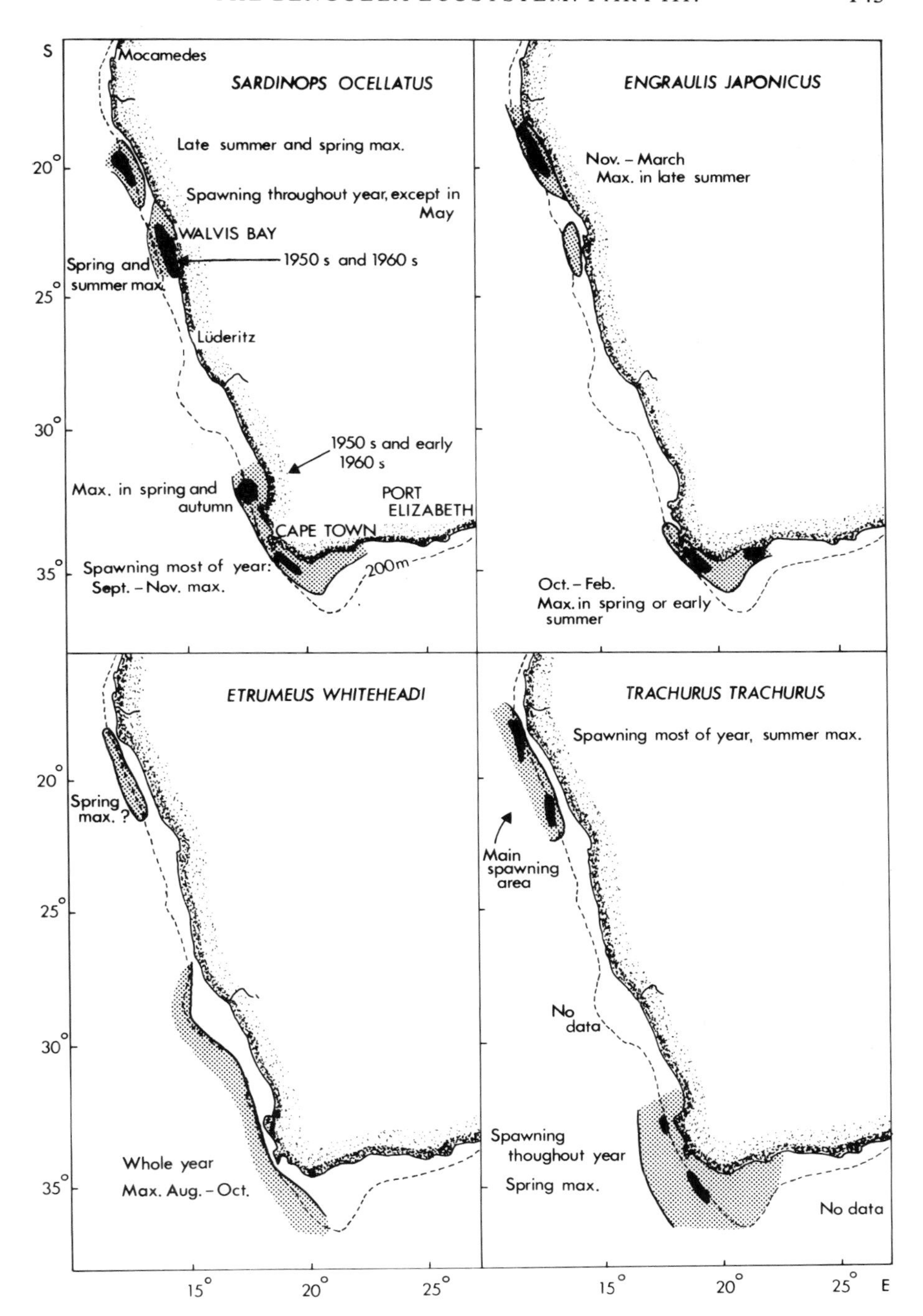

Fig. 34.—Distribution of eggs and larvae of pilchard, anchovy, round herring, and horse mackerel in the Benguela region as inferred from published work.

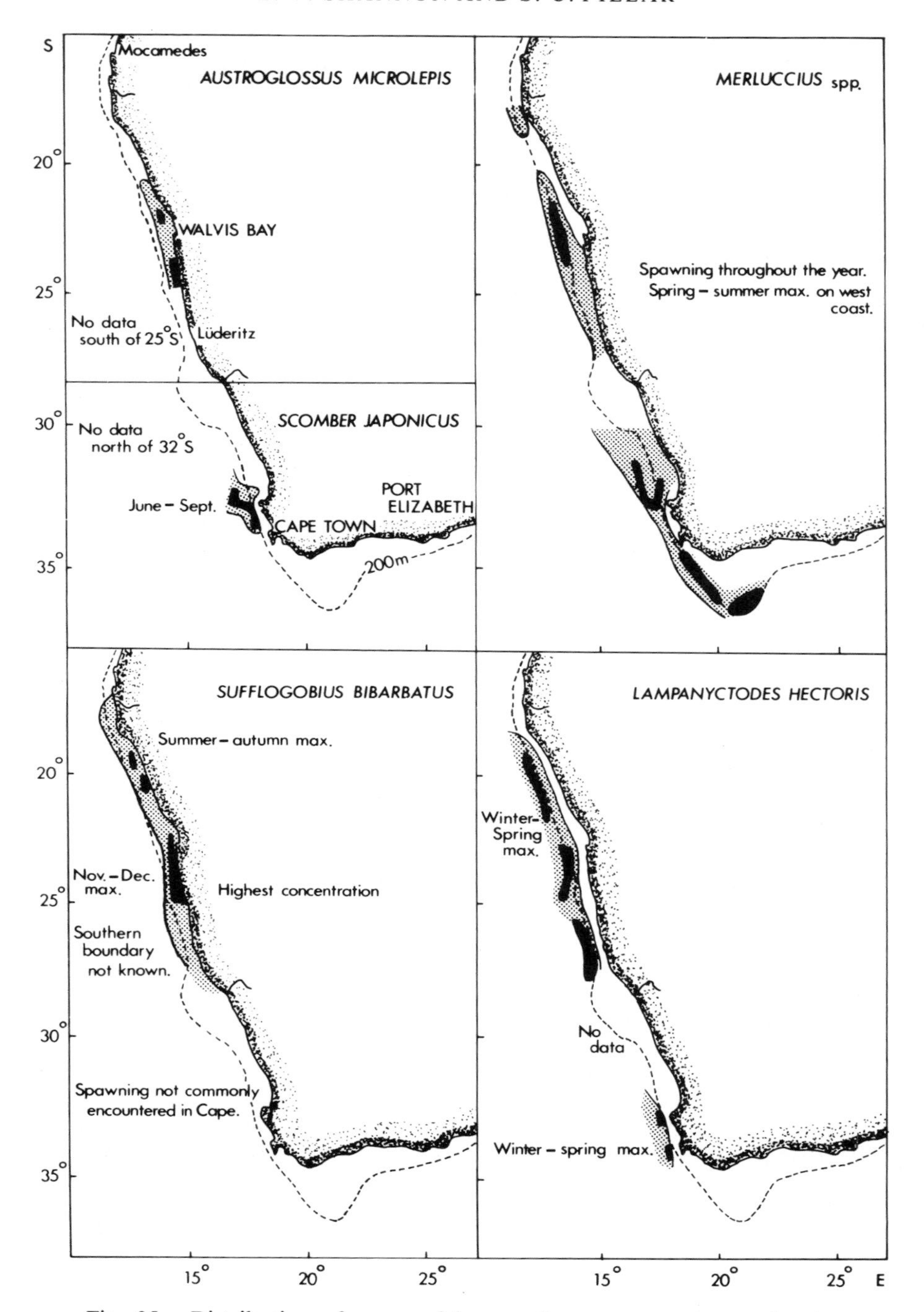

Fig. 35.—Distribution of eggs and larvae of west coast sole, mackerel, hake, goby, and lanternfish on the Benguela region as inferred from published work.

Separate spawning areas of the pilchard exist off the South Western Cape and Namibia. Although genetic studies (Thompson & Mostert, 1974) have yielded inconclusive results, it is generally accepted that the Cape and Namibian stocks are separate populations (*e.g.* Newman, 1970) separated by the major upwelling area (a potential environmental barrier) which exists in the central Benguela (refer to Boyd & Cruickshank, 1983; Shannon, 1985). Extensive spawning of the pilchard off Namibia was first documented by Hart & Marshall (1951) and Hart & Currie (1960). These authors recorded pilchard eggs at the inshore stations along most of the Namibian coast with the highest concentrations within 50 km of the shore off central Namibia (at 23° S and 25° S) being associated with high concentrations of diatoms (*Delphineis karstenii* and *Chaetoceros* spp.). In a report on the reproduction of the pilchard, Matthews (1964) described the distribution of eggs between 1957 and 1960 in the area between Cape Cross (21°50′ S) and Conception Bay (24° S). Subsequent studies (*e.g.* Cram & Visser, 1973; O'Toole, 1974, 1977a; King, 1977b) have shown that there are two principal spawning areas off Namibia, one inshore off central Namibia south of 22° S (as located by Hart & Marshall 1951; Matthews, 1964) and a second area situated further offshore between about 19 and 21° S (Figs 34 and 36). Hart & Currie (1960) did not comment on the existence of the offshore spawning area but its existence prior to the collapse of the Namibian pilchard stocks cannot be ruled out. Nevertheless, there is a possibility that this offshore area might have become more important following the stock collapse and the reduction in the size of pilchards at maturity. In the South Western Cape, Davies (1954, 1956) documented a spawning area near St Helena Bay (32–33° S) and noted that considerable spawning also took place south of the area he studied, the larvae being more widely distributed than the eggs. The existence of the second large spawning ground on the Agulhas Bank has been documented by Crawford (1980), Davies, Newman & Shelton (1981), Crawford, Shelton & Hutchings (1983), and Shelton & Armstrong (1983). There is some evidence to suggest that spawning on the Agulhas Bank increased following the altered size composition of the Cape pilchard stock and coincidental with a reduced size in maturity (Crawford, 1981). Thus, two spawning areas exist (or existed!) off Namibia and two off the South Western Cape—one in each region in close proximity to upwelling and one in each in the more stratified and warmer waters near the boundaries of the Benguela. We shall comment further on this later.

Both in the Cape and in Namibia pilchards spawn in most months judging from the work of a number of authors, although Stander (1963) and Matthews (1964) recorded virtually no eggs off central Namibia during May between 1957 and 1960. Two distinct spawning maxima were noted by Matthews (1964), one in late winter–spring, the other in late summer–autumn, a finding which has been confirmed by King (1977b), O'Toole (1977a), Le Clus (1983), and others. Matthews (1964) viewed the bimodality as indicating the commencement and termination of spawning by a single Namibian stock and studies by Le Clus (1979a) on the oocyte development and spawning frequency of pilchard have shown that the fish has the potential to spawn at least twice. O'Toole (1977a), however, viewed the two spawning maxima as being linked to environmental signals (O'Toole, 1977b). He noted that the autumn spawning was preceded by the

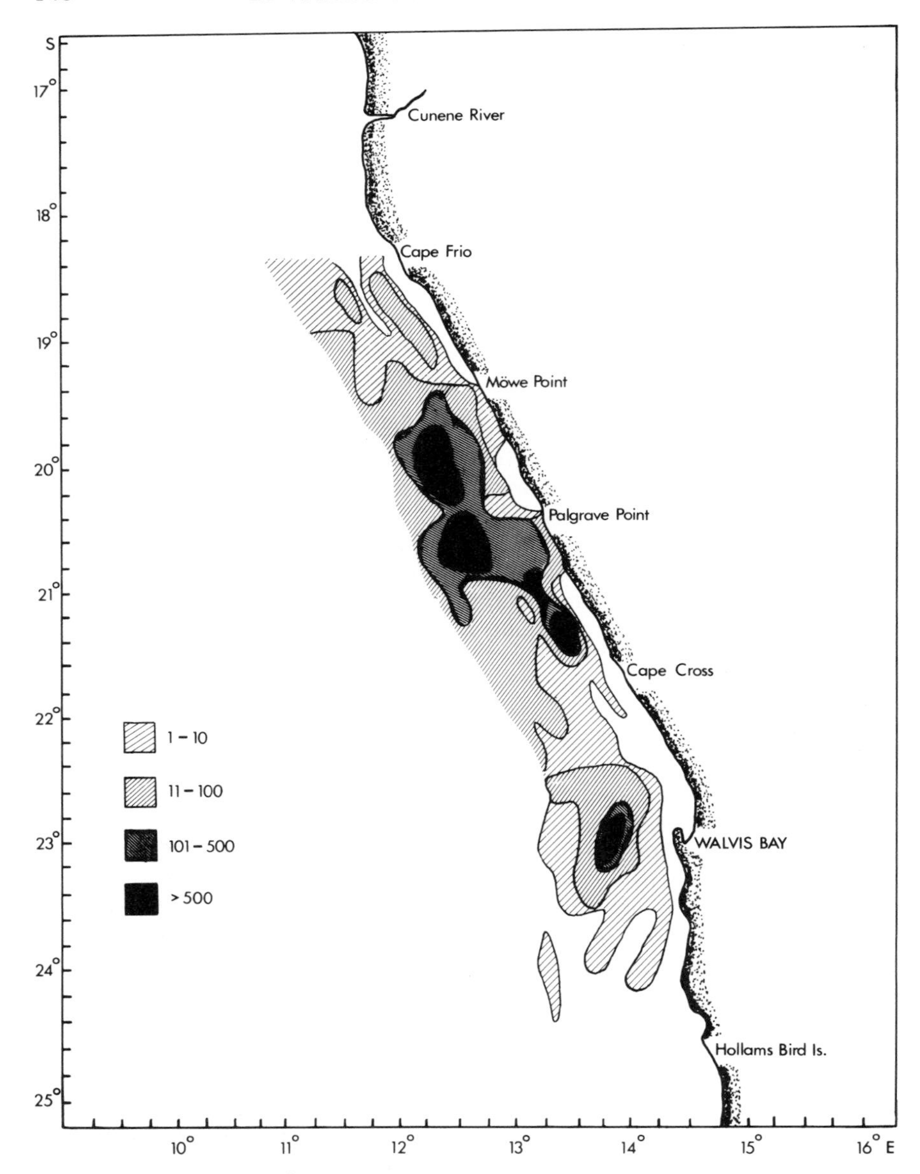

Fig. 36.—Distribution and abundance (cumulative totals) of pilchard larvae during the period August 1973–April 1974 (after O'Toole, 1977a).

intrusion of warm oceanic water while spring spawning followed intense upwelling which suggested that temperature and food were the respective triggers. In the Cape, Davies (1954, 1956) and Crawford (1981) showed that egg production was maximal in September and late summer and reached a minimum during May–June. Results from CELP suggested a maximum egg abundance on the Agulhas Bank during October (Davies *et al.*, 1981).

Davies (1956) considered that 13 °C was the threshold temperature for

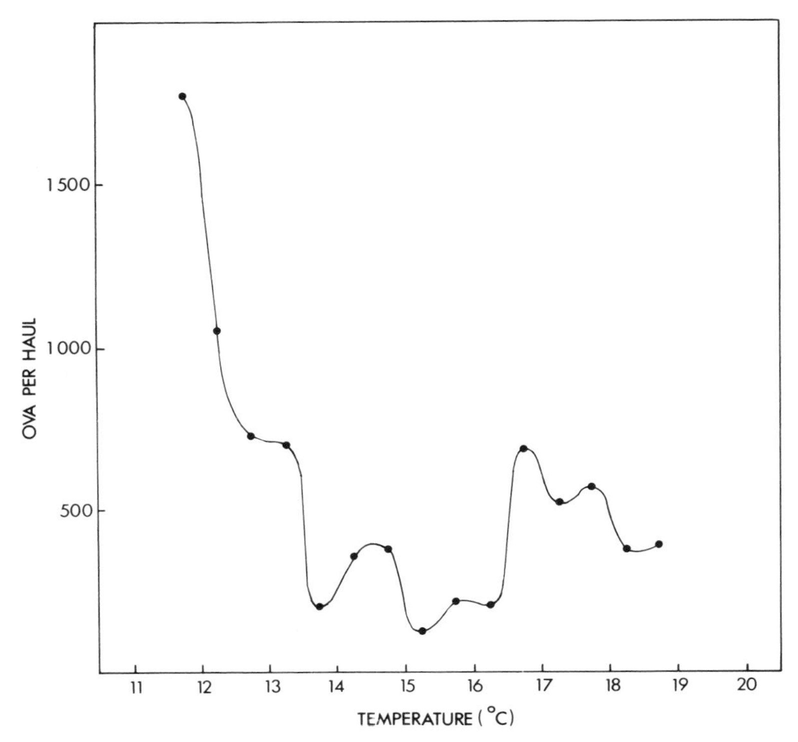

Fig. 37.—The relation between abundance of pilchard ova and surface temperature during the period 1957–1961 as given by Stander (1963).

major spawning in the Cape, maximum numbers of eggs per haul having been collected within the surface temperature range 13–16 °C. (These were the temperatures commonly encountered in their survey grid.) Off central Namibia, Stander (1963) and Matthews (1964) recorded maximum egg abundance (Fig. 37) in water with a surface temperature of 12–13 °C (*i.e.* below the threshold noted by King, 1977a) during August-September and 14–19 °C during February-March. These ranges should be compared with the range of 13–14·5 °C in which Hart & Marshall (1951) collected most of their pilchard eggs. King (1977b) and O'Toole (1977a) recorded maximum concentrations of eggs and larvae south of 22° S during late winter and spring within a surface temperature range 13–17 °C (mainly 14–16 °C) and off northern Namibia during summer and autumn within a surface temperature range 16–22 °C (mainly 17–20 °C). According to O'Toole (1977a) 75% of the eggs are found within the upper 20-m layer, while maximum concentrations of larvae are usually found near the thermocline (20 m). As a result, the larval temperature "preferences" would be about 1 °C lower than the surface temperatures ranges cited. King (1977b) found that 16·5 °C was a division between the northern and southern spawning off Namibia with no summer–autumn spawning offshore occurring at temperatures below 16·5 °C and no spring–summer spawning occurring inshore when the water was warmer than 16·5 °C. These findings led King

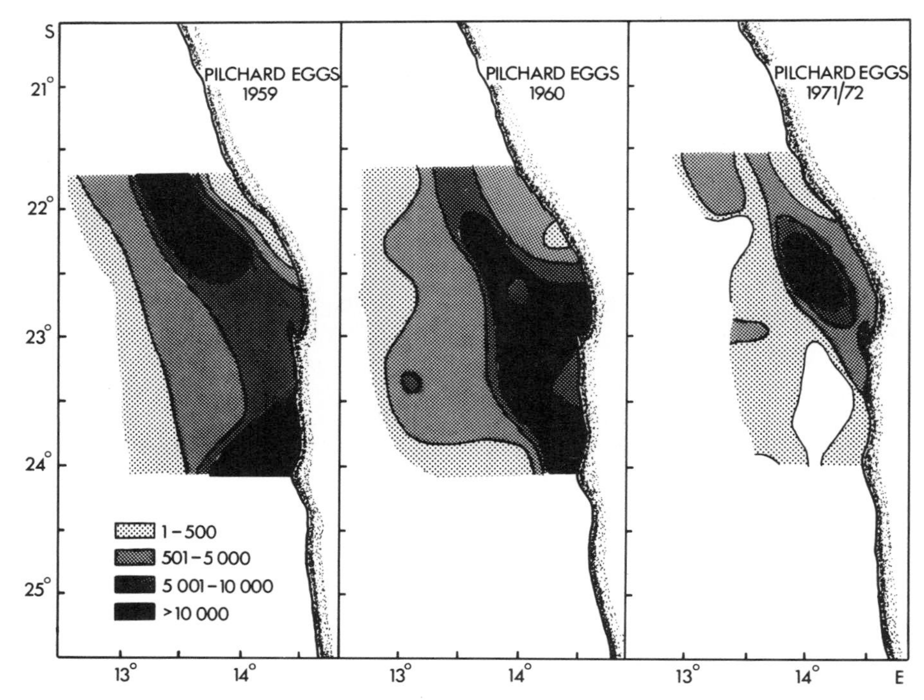

Fig. 38.—The distribution and abundance (cumulative totals) of pilchard eggs in 1959, 1960, and 1971–1972 (after King, 1977b).

(1977b) and O'Toole (1977a) to conclude that two separate breeding stocks of pilchard existed off Namibia. Work by Le Clus (1977) on the methodology of fecundity determination enabled her (Le Clus, 1979b) to show that there was an active response of oocyte mass to temperature, with the winter-spring oocyte mass being 20% higher than in summer-autumn. Although her work did not contradict King's or O'Toole's findings, it does suggest that the eggs spawned early in the season are perhaps physiologically better adapted to a colder environment and may indicate that King's (1977a) critical temperature of 13 °C may only apply to eggs spawned in warm water. Le Clus (1979b) found no difference in the mass of eggs spawned by two- and three-year old fish.

The results of egg and larval monitoring during recent years off Namibia have been reported by Le Clus & Kruger (1982), Le Clus (1983), Agenbag, Kruger & Le Clus (1984), Boyd, Hewitson, Kruger & Le Clus (1985), and others. Although a direct comparison with earlier studies is tenuous, *e.g.* Figure 38 there is an indication that the pilchard component of the ichthyoplankton has declined both in the Cape and in Namibia—symptomatic of the collapse of the pilchard fisheries in both regions. Since 1971 pilchard appears to have contributed less than 4% to the Namibian neritic ichthyoplankton (O'Toole, 1974; Olivar, 1982, 1983), with gobies currently being dominant. That wide fluctuations in ichthyoplankton have occurred over the last 30 years is well documented. Stander & De Decker (1969) reported a six-fold decline in the average numbers of eggs per haul in 1963 during a warm-water intrusion (refer to Shannon, 1985) and, although this decline may have been due, in part, to a decline in the parent stock, pilchard eggs were virtually absent during the normal spawning months. To what extent patchiness influenced the results is not known. In 1982–1983 pilchard egg production was comparable with that in 1981–1982 and much higher than in the preceding three years (Boyd *et al.*, 1985). These authors reported a sharp decline in 1983–1984, however, probably as a result of very intense upwelling in November 1983 followed by a warm-water intrusion early in 1984.

There now appears to be relatively little pilchard spawning close inshore south of Walvis Bay during the upwelling season. Similarly, eggs were abundant offshore of St Helena Bay in the early 1950s but the intensity of spawning declined to negligible levels by 1966. The sharp contrast between December 1961 and November–December 1977 in the spatial distribution of pilchard eggs around the South Western Cape was demonstrated by Shelton & Armstrong (1983)—Figure 39. Crawford (1980) attributed the decline in spawning off St Helena Bay to the elimination of the older pilchards from the population as a result of over-fishing. Referring to the Namibian temperature-spawning curve of Stander (1963)—Figure 37—three maxima are evident, one at 12 °C, the second between 14 and 15 °C, and the third above 16·5 °C. It is tempting to speculate whether the first maximum was not perhaps associated with spawning of older pilchards, for by the time of the studies of King and O'Toole, only the second and third maxima were detected, and there had been a northward contraction in spawning range towards the nursery areas (Crawford *et al.*, 1983). R. J. M. Crawford (pers. comm.) has suggested that the older pilchards may have spawned eggs which were better adapted physiologically (*e.g.* larger yolks would permit slower development) to colder conditions, and as a result could spawn successfully (in the Cape at least) close to the recruitment area—this seems to make ecological sense because loss caused by dispersal would be reduced. Eggs and larvae from younger fish might not survive this harsher

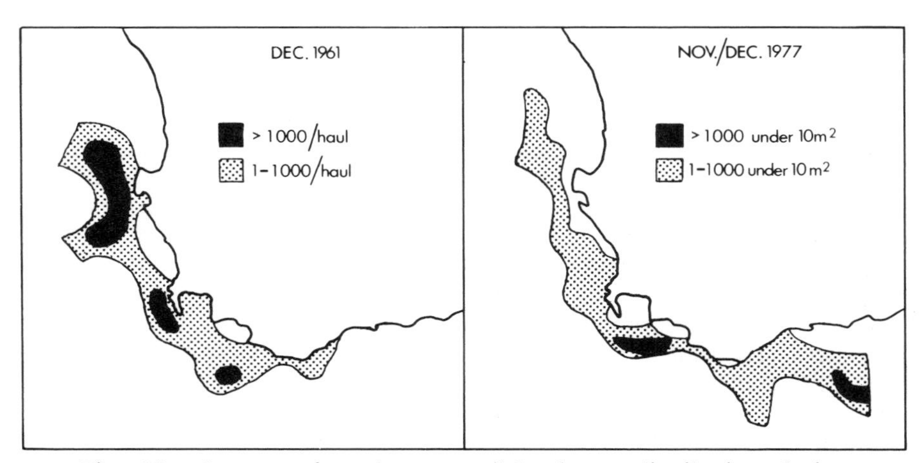

Fig. 39.—A comparison between pilchard egg distribution during December 1961 and November to December 1977 (after Shelton & Armstrong 1983).

environment, with the result that the warmer, more stratified waters of the Agulhas Bank and offshore of northern Namibia would present preferred habitats. This rather interesting hypothesis, if correct, could have important consequences for the fishery. Reductions in the geographic range of egg production, as have been observed for both pilchard populations, would adversely affect the probability of larvae encountering favourable environmental conditions, increase changes of recruitment failure (Saville, 1980), and decrease likelihood of favourable eruptions (Newman & Crawford, 1980). Moreover it may imply that, in the Cape, the spawning-jet transport hypothesis of Shelton & Hutchings (1981, 1982)—discussed on page 151—may not be the most efficient transport mechanism to ensure good recruitment for pilchard, and that a capability to spawn near to upwelling may be a prerequisite for good recruitment. The surface currents around the South Western Cape (*e.g.* Shannon, 1985) are such that eggs spawned in the south and transported around the Cape of Good Hope may be lost to the system, whereas those spawned close to St Helena Bay may be expected to be retained more efficiently within the system.

ANCHOVY, *ENGRAULIS JAPONICUS*

The egg of the southern African anchovy *Engraulis japonicus* (*capensis*) is smaller than that of the pilchard and is oval, typically $1{\cdot}3 \times 0{\cdot}55$ mm (Anders, 1965). It can readily be identified in the net ichthyoplankton. Anchovy eggs were first collected at sea off the southern coast 70 years ago by means of a plankton tow (Gilchrist, 1916), but inspite of the existence of large known anchovy populations, the species was not actively fished or researched until after 1964. The N100H tows commonly used before this date had a coarse mesh and collected few anchovy eggs. Davies (1954) reported very few eggs in collections made of St Helena although he did find some larvae during summer and concentrations of post-larvae and juveniles in the area from late summer to winter. Regular quantitative Bongo net surveys only commenced in the 1970s (*e.g.* O'Toole, 1974), King, Robertson & Shelton (1978) have described the development of anchovy eggs and larvae. Samples were collected off the Cape Peninsula in 18 °C water. Eye pigmentation and jaw development was complete in two days after hatching at 22 °C and in three days at 18 °C. They concluded that 14 °C was the minimum temperature for successful spawning as larvae reared below this threshold failed to develop pigmented eyes and a functional jaw. The anchovy has been reared successfully from the egg stage through metamorphasis in the laboratory by Brownwell (1983a) in a series of feeding experiments. Larval growth rates are in the range $0{\cdot}4$–$0{\cdot}9$ mm$\cdot$day^{-1} (O'Toole, 1977a; Badenhorst & Boyd, 1980) but the characteristic rate at 19–20 °C is $0{\cdot}5$–0.6 mm$\cdot$day^{-1} (O'Toole, 1977a).

Separate spawning areas of the anchovy exist off the South Western Cape and Namibia (Fig. 34). Although genetic studies have yielded inconclusive results (Grant, 1983) it is generally accepted (*e.g.* Report on a Symposium on Anchovy Distribution and Migration held in June 1983 in Cape Town under the auspices of the Benguela Ecology Programme, unpubl. document, 70 pp.) that the Cape and Namibian stocks represent separate populations. Some interchange is, however, probable. Badenhorst & Boyd

(1980) advanced the hypothesis that the large anchovy larvae found on occasions off southern Namibia were spawned off the Cape, consistent with a mean northward drift of 13 $cm \cdot s^{-1}$ and a growth rate of 0·5 $mm \cdot day^{-1}$. This idea was supported by the relationship between larval size and abundance and latitude between Cape Columbine and Lüderitz (Boyd & Hewitson, 1983), larvae being larger but less abundant north of the Cape spawning area.

The spawning area off the Cape was first mapped by Anders (1965). Most of the spawning takes place east of Cape Point on the Agulhas Bank, with maximum egg concentrations 40–100 km offshore in 16–19 °C water. Anders (1965) did, however, record eggs over a wide temperature range (9.5–22 °C) and concluded that the threshold temperature for anchovy spawning was 4 °C lower than for the pilchard. (Her chart for the summer of 1964–1965 showed high concentrations of anchovy eggs in St Helena Bay.) This appears to contradict the threshold level of 14 °C given by King *et al.* (1978) for anchovy spawning. Their study was, however, done on eggs collected in warm (18 °C) water and the question arises as to whether eggs spawned in water colder than 14° C, perhaps by older anchovy (largely exterminated after the 1960s), could develop normally. The eastward extent of anchovy spawning is not certain, and large concentrations of eggs have been collected off East London and Durban on occasions, *e.g.* December 1973 (Anders, 1975). Drogue experiments conducted off the Cape Peninsula and on the Agulhas Bank during the spring of 1976 led Shelton & Hutchings (1981, 1982) to propose that a shelf-edge frontal jet current entrains eggs and developing larvae and transports them around the Cape Peninsula and northwards to the recruitment area north of 32°30′ S. Pulses of early stage eggs in the plankton during the drogue study were related to nocturnal spawning and the authors noted that the thermocline acted as a barrier to vertical migration of larvae close to the core of the 0·55 $m \cdot s^{-1}$ jet. The frontal system presumably provides a stable area with a suitable balance between temperature and food for larval survival. The distribution of eggs and larvae during 1977–1978 and the inferred rôle of jet-transport is shown in Figure 40. Anchovies are serial spawners and the spawning season in the Cape extends from October to February (Crawford, 1980; Davies, Newman & Shelton, 1981; Shannon, Hutchings, Bailey & Shelton, 1984) with maximum numbers of eggs present during spring or early summer (Crawford, 1981). The Agulhas Bank is evidently a suitable spawning area, the temperature of the surface layer (*e.g.* Shannon, 1966; Christensen, 1980) seldom falling below the 'critical' 14 °C level. Spawning coincides with the increasing stratification of the water column and the formation of a deep chlorophyll maximum layer (Shannon & Hutchings *et al.*, 1984). Although the planktonic biomass per unit area is lower than on the western coast, the availability of food in strata must be energetically attractive for spawners and first-feeding larvae. The surface discontinuity at 21° E (refer to Shannon, 1985) may be important in determining whether or not larvae recruit on the western coast, and Shannon & Hutchings *et al.* (1984) considered that larvae found east of this meridian would probably recruit into the fishery on the southern coast.

The spawning area of anchovy off Namibia was first mapped by O'Toole (1974). Most of the spawning takes place north of 22° S (King, 1977b;

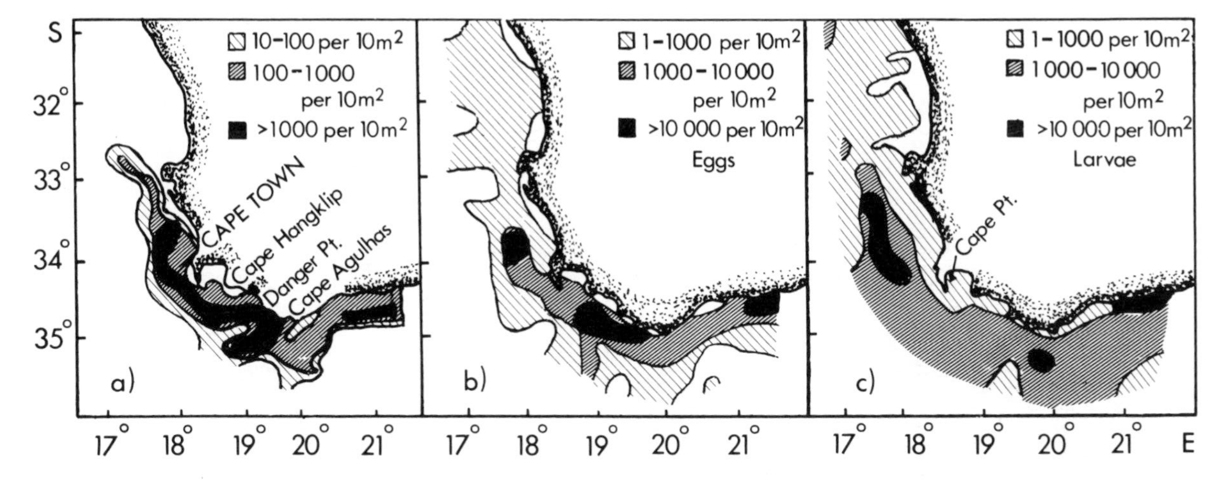

Fig. 40.—a, distribution of anchovy eggs during November to December 1977 around the South Western Cape (after Shelton & Armstrong, 1983). b, cumulative distribution of anchovy eggs, August 1977–August 1978 (after Shannon, Hutchings, Bailey & Shelton, 1984). c, cumulative distribution of anchovy larvae, August 1977–February 1978, April and June 1978 (after Shannon & Hutchings *et al.*, 1984).

O'Toole, 1977a; Le Clus, 1985), although less intense spawning does occur at times between Walvis Bay and Hollams Bird Island (25° S) and possibly even further south (Le Clus, 1985, Le Clus & Kruger, 1982). The northernmost extent of spawning is not known. Eggs have been found 120 km from the coast, but maximum concentrations usually occur about 20 km offshore (Le Clus, 1985). O'Toole (1977a) noted that the maximum concentrations of larvae tend to be west of the spawning area, typically 60–110 km offshore in the area between Cape Frio (18°20′ S) and Palgrave Point (20°20′ S), illustrating the net westward drift of the surface water. The monthly distribution of eggs during the summer of 1980–1981 and cumulative totals of larvae for the period 1972–1973 and 1973–1974 (Fig. 41) shows this feature. The Namibian spawning season is continuous between spring and autumn (King, 1977b; O'Toole, 1977a) but most of the spawning is between November and March with a maximum in late summer (Agenbag *et al.*, 1984; Le Clus, 1985). King (1977b) reported spawning activity within the temperature range 13·2–21 °C, usually above 16 °C, while O'Toole (1977a) noted that spawning as inferred from egg and larval distributions was most frequent when the surface temperature was between 18 and 20 °C (equivalent to a temperature at 20 m of 16–19 °C). O'Toole (1977a) found that three times as many larvae were caught at night as in the day, suggesting vertical migration, and Badenhorst & Boyd (1980) attributed this discrepancy, in part, to the avoidance of nets by larger larvae during the day. Le Clus & Kruger (1982) and Agenbag *et al.* (1984) noted that intensive spawning during late summer lagged one month behind the maximum phytoplankton concentrations, which could result in spawning

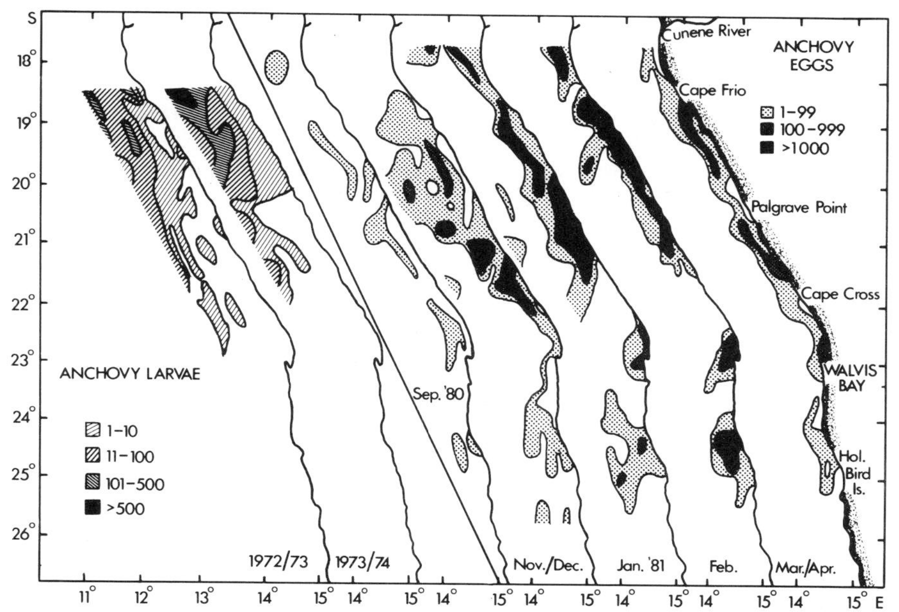

Fig. 41.—Distribution and abundance of anchovy larvae (cumulative totals) for 1972–1973 and 1973–1974 (after O'Toole, 1977a) and monthly anchovy egg distribution during 1980–1981 (after Le Clus, 1985).

coinciding with a maximum zooplankton biomass. It is, however, difficult to explain why spawning ceases abruptly at the end of March on the basis of this hypothesis. There is, however, usually a fairly sharp intensification of upwelling during late March or April off northern Namibia which may effectively terminate spawning. Le Clus (1985) has examined possible reasons for the preferred timing and siting of spawning of anchovy off Namibia and concluded that these were related to the period of minimum upwelling, maximum influx of warm Angolan and oceanic water, maximum stratification and minimum turbulence and offshore advection (refer to Shannon, 1985). Le Clus (1985) suggested that reduced upwelling could favour larval survival. Thus, the spawning areas off both the Cape and Namibia are characterized by warm water, strong stratification and the absence of, or reduction in, upwelling. Spawning off Namibia lags behind that in the Cape by a few months, however, and the larval transport mechanisms appear to be different (*cf.* frontal jet-transport in the Cape), although those on the eastern Agulhas Bank may be analogous to the processes off Namibia.

Sampling of anchovy eggs and larvae has been reasonably regular off Namibia since the early 1970s and intermittent off the Cape since 1964. The results of the monitoring have been discussed by *inter alia* Badenhorst & Boyd (1980), Le Clus & Kruger (1982), Agenbag *et al.* (1984), Boyd *et al.* (1985), Le Clus (1985), and Shelton (1984). Although widely variable, anchovy larvae appear to comprise about 15% of the ichthyoplankton (O'Toole, 1974; Olivar, 1983) and there is some evidence that egg abundance has lagged recruitment by one year, reflecting trends in the adult stock (Crawford, Shelton & Hutchings, 1983). Using data from a single cruise in November 1982, Shelton (1984) found no evidence of spawning failure just prior to the environmental perturbation in the southern Benguela during the summer of 1982–1983. Boyd *et al.* (1985) showed, however, that total anchovy egg production during 1983–1984 was one twentieth the long-term mean in the northern Benguela, and attributed this spawning failure to abnormally strong upwelling in November 1983 which was followed by a Benguela "El Niño" type event early in 1984.

HORSE MACKEREL, *TRACHURUS TRACHURUS*

The egg of the horse mackerel or marsbanker *Trachurus trachurus* is smooth and spherical (diameter about 0·9 mm) and has a single fairly large (0·22 mm) oil globule (O'Toole, 1977a). The larval stages of this species have been described by Haig (1972a), while King, O'Toole & Robertson (1977) have described the *in vivo* development of eggs and larvae. The work of the last authors indicated that the yolk was fully absorbed after 27 h and 80 h at temperatures of 22 and 13 °C, respectively.

The eggs and larvae of horse mackerel are fairly widely distributed around South Africa. Davies (1954) recorded eggs and larvae in the offshore waters west of St Helena Bay during summer, and Haig (1972a) showed that the larvae occurred in the South Western Cape research area (32° S–22° E) from the coast as far as 350 km offshore between 1951 and 1965, but most frequently at the western edge of the Agulhas Bank (southwest of Cape Point) and west of Cape Columbine. Crawford (1980) indicated that the fish spawned in the Cape during 1977–1978 offshore in fairly deep water. Fish spawn throughout the year, with a maximum in spring (Haig, 1972a; P. A. Shelton, pers. comm.). It is probable that the abundance of horse mackerel eggs in the southern Benguela has decreased with the declining catch of adults during the last three decades.

The spawning area of horse mackerel off Namibia was first documented by O'Toole (1974), who showed that maximum abundance of larvae occurred on the outer shelf between 18 and 23° S, and Komarov (1964a,b) who indicated the preferred areas as 17–20° S, in particular the interaction zone between the Benguela and Angolan systems near Cape Frio. The southern boundary of spawning off Namibia seems to be around 22–23° S, while the northern and offshore boundaries have yet to be established (Olivar & Rubiés, 1983). Maximum larval abundance occurs 50–100 km offshore (O'Toole, 1977a; Olivar & Rubiés, 1983). Off northern Namibia horse mackerel spawn between spring and autumn (at least) with a summer maximum, most of the spawning taking place in mixed water of 16–19 °C (O'Toole, 1977a). O'Toole (1977c) showed from experimental work that most larvae were near the thermocline (25 m, 18 °C), with the majority of newly hatched individuals being near to the chlorophyll maximum layer. Estimates of the per cent contribution of horse mackerel to all larvae collected off Namibia ranges from 0·4% in November 1979 (Olivar, 1981), 3% during 1972–1973 (O'Toole, 1974) to 41% during March-April 1981 (Olivar & Rubiés, 1983). In this respect a dramatic increase in parent stock off Namibia has been suggested during recent years, *e.g.* see Crawford *et al.* (1983).

ROUND HERRING, *ETRUMEUS WHITEHEADI*

The egg of the round herring or red-eye *Etrumeus whiteheadi* (*E. micropus, E. teres*) is smooth, spherical (diameter about 1·4 mm) with a narrow perivitelline space and has a large lightly segmented yolk and no oil globule (O'Toole & King, 1974). These authors have described the development of the eggs and early larval stages. Hatching of eggs occurred after 135 h at 11 °C and after 36 h at 20·5 °C. Developmental stages of round herring were reported in the St Helena Bay samples by Davies (1954), and subsequent investigations have shown that the eggs and larvae are abundant, perhaps the second most abundant species, over an extensive offshore area around the South Western Cape (Crawford, 1980; Davies *et al.*, 1981; Olivar 1984). Spawning in the southern Benguela is continuous throughout the year but with a distinct late winter–spring maximum (Davies *et al.*, 1981). The eggs and larvae of round herring are less abundant off Namibia with estimates ranging between 1% (Olivar, 1983) and 6% (Olivar, 1982) of the total larval ichthyoplankton. Highest concentrations of larvae in the northern Benguela are over the shelf-break between 18 and 22° S. Little is known about the distribution of round herring eggs and larvae between 25 and 30° S except for the study during July 1983 by Olivar & Rubiés (1985) who found concentrations of eggs and larvae near the shelf-break between 27 S and 29° S. The inferred main spawning areas in the Benguela are shown in Figure 34 (p. 143). It seems possible that the much higher abundance of eggs and larvae of the species off the Cape may be associated with the intense sub-surface fronts which are characteristic of the southern Benguela (refer to Shannon, 1985). It is beyond the scope of this review, however, to examine this hypothesis.

MACKEREL, *SCOMBER JAPONICUS*

Little is known about the distribution of mackerel eggs and larvae in the Benguela system. The spawning in the Cape was defined by Baird (1974, 1977) as being between June and September and he showed that maximum numbers of eggs were near the shelf-breaks between 32 and 34° S in water with a surface temperature of 13–17 °C. Few larvae were caught. Mackerel eggs were reported in Benguela collections by Porebski & Koronkiewicz (1975), but no details were given. The spawning area is shown in Figure 35 (see p. 144) which suggests that the Columbine divide or divergence and jets associated with the bathymetry of the region (*cf.* proximity of the Cape Canyon) may be important (refer to Shannon, 1985).

HAKE, *MERLUCCIUS* SPP.

Two species of hake *Merluccius capensis* and *Merluccius paradoxus* exist in the Benguela and it has not yet been possible to distinguish between their eggs and larvae (O'Toole, 1978a) on the basis of microscopic analysis. The egg is spherical (diameter about 0·95 mm) and has a relatively large (0·22 mm) oil globule (O'Toole, 1978a). It is difficult to distinguish between the eggs of hake and those of species such as mackerel and horse mackerel (Porebski & Koronkiewicz, 1975) and Porebski (1975) proposed a simple surface adhesion test to assist egg separation. The development of eggs and early larvae was described by Matthews & De Jager (1951), and Haig (1972b) documented the larval stages of *Merluccius* spp.

Hake eggs and larvae are widely distributed throughout the Benguela. Davies (1954) reported the frequent occurrence of post larvae near St Helena Bay, mainly during summer. Results of the cruises of the R.S. PROFESSOR SIEDLECKI during the summer of 1972–1973 (Porebski & Koronkiewicz, 1975) and R. S. AFRICANA during winter 1983 (Olivar, 1984) showed that, in the southern Benguela, spawning occurs between 37 and 30° S (at least), while the distribution of larvae given by Haig (1972b) suggests that they are most common near the shelf-breaks around 33° S (refer to the interpretation in Fig. 35) and along the western edge of the Agulhas Bank. Although Hart & Marshall (1951) and Hart & Currie (1960) reported the presence of hake larvae off Namibia, the distribution of larvae was first mapped by O'Toole (1974). The maximum abundance of hake eggs and larvae there appears to be in the vicinity of the inner shelf-break (typically 50–100 km offshore) and the distribution extends from the Angolan border to at least 28° S as suggested by Porebski & Koronkiewicz (1975), O'Toole (1978a), Olivar (1981), and Olivar, Rubiés & Salat (1982). According to O'Toole (1978a) the area between Henties Bay (22° S) and Hollams Bird Island (25° S) is the principal spawning ground in the northern Benguela (Fig. 42). Hake spawning occurs throughout the year in midwater (Botha, 1973), but spawning is at a maximum between October and February in the Benguela system (O'Toole, 1978a; Botha, 1980). O'Toole (1978) noted

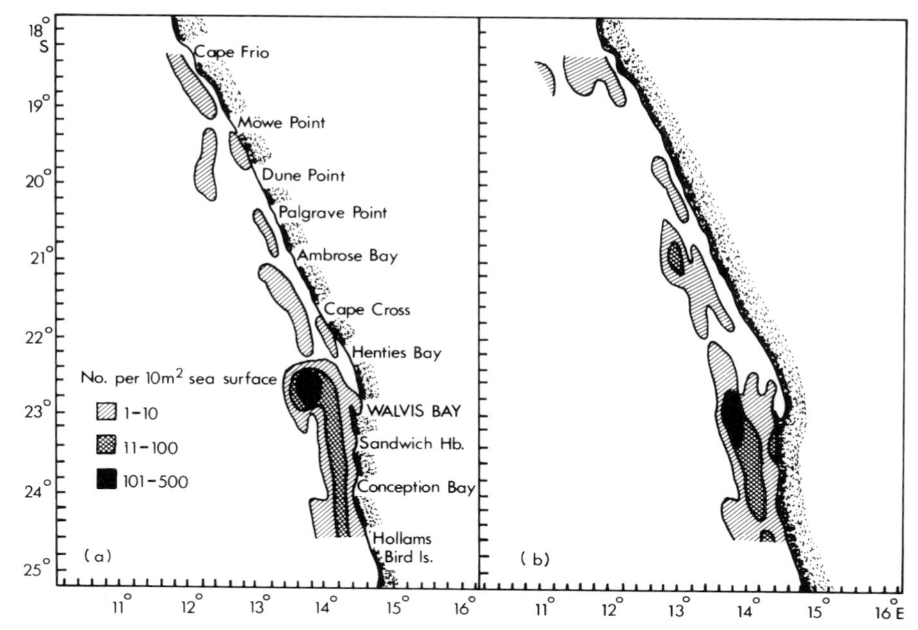

Fig. 42.—Distribution and abundance of hake larvae (cumulative totals) off Namibia during 1972–1973 and 1973–1974 (after O'Toole, 1978a).

significant diurnal vertical migration of the larvae and, by considering the temperature at 20 m and catch success, suggested that the range of 12–16 °C was the preferred spawning temperature. O'Toole (1978a) agreed with Botha (1973) that there was little evidence of a horizontal spawning migration of hake. Hake larvae comprise between 1 and 8% of the larval ichthyoplankton north of Walvis Bay (O'Toole, 1974; Olivar, Salat & Rubiés, 1981; Olivar, 1982).

The egg and larval distribution data (Fig. 35, see p. 144) seem to suggest the existence of at least two stocks, one off the Cape, the other off Namibia. From a study of adult hake physiological conditions in the whole Benguela region Assorov & Berenbeim (1983) documented the presence of two spawning stocks of *Merluccius capensis* off Namibia (one off Cape Frio and the other off Walvis Bay) and a spawning stock of each hake species on the Agulhas Bank. To some extent this last paper confirms the conclusions reached from the distribution of eggs and larvae.

GOBY, *SUFFLOGOBIUS BIBARBATUS*

The larvae of the pelagic or bearded goby, *Sufflogobius bibarbatus*, comprise a major fraction—typically more than one half—of the Namibian larval ichthyoplankton with survey abundance estimates ranging between 20 and 67% (O'Toole, 1974, 1978b; Olivar, 1981, 1982). As such this species is an important component of the northern Benguela ecosystem. The larval and early juvenile stages of goby were first described by O'Toole (1978b). Although adult goby are present in Cape waters, significant catches of eggs and larvae have not been reported in the literature on the southern Benguela. The spatial distribution of the larvae is shown in Figure 35, and the abrupt termination at 25° S evident in this figure reflects the extremity of the sampling grid. The larvae of the species are distributed along much of the Namibian coast between winter and late summer, but the southern boundary of spawning has yet to be established. During July 1983 Olivar & Rubiés (1985) recorded small numbers of goby larvae as far south as 28°30′ S. Spawning is most intense during spring and early summer in the coastal waters south of the Walvis Bay bight (O'Toole, 1978b). This author indicated that the larvae were caught in this southern area in upwelled water (11–16 °C) during late winter and spring and off northern Namibia in the intermediate zone between the warm oceanic and colder coastal waters. Comparison of Figures 34 and 35 (see pp. 143 and 144) shows that the spawning areas of the pilchard and goby are remarkably similar, as is the timing of spawning. Like pilchard, the goby larvae are coincident with the distribution of the high phytoplankton biomass south of Walvis Bay (*cf.* distribution of *Delphineis karstenii* and *Chaetoceros* spp.). Recent studies by Crawford, Cruickshank, Shelton & Kruger (1985) have suggested that the stock of older pilchard that historically occurred near Walvis Bay has largely been replaced by the goby, and if so, then this may imply that the present biomass of goby larvae may be much higher than was perhaps the case in the 1950s.

LANTERNFISH, *LAMPANYCTODES HECTORIS*

The larvae of *Lampanyctodes hectoris*, the most common species of myctophid or lanternfish in the Benguela, were described by Ahlstrom, Moser & O'Toole (1976). The eggs and larvae of this species appear to be widely distributed in the region and are very abundant both in the Cape (Davies, Newman & Shelton, 1981; Prosch & Shelton, 1983; Olivar, 1984) and off Namibia where it comprises 8–51% of the ichthyoplankton (Ahlstrom *et al.* 1976; Olivar, 1981, 1982). Recently Olivar & Rubiés (1985) reported the occurrence of large concentrations of eggs and larvae of lanternfish near the shelf break off southern Namibia during July 1983. Off the Cape large numbers of eggs were recorded west of Cape Columbine near the Cape Canyon and near the Cape Point Valley (off Cape Point) by Prosch & Shelton (1983) who associated the distribution with the frontal jet currents at these sites. Crawford (1980) suggested that the main Cape spawning may take place further offshore and that the presence of eggs near the canyons was a result of the outcropping of the thermocline there. The main spawning season in the Cape during 1977–1978 was July-October (Davies *et al.*, 1981; Prosch & Shelton, 1983) while larvae were most abundant during August–October and were fairly widely distributed.

Off Namibia Ahlstrom *et al.* (1976) and Olivar (1981, 1982) found that larvae were widely distributed south of 19° S. Maximum abundance was noted in offshore waters (>60 km from the coast), particularly in the area 19–22° S and south of 23° S. Ahlstrom *et al.* (1976) indicated that, although myctophid larvae were caught at a surface temperature between 14 and 21 °C, 60% were caught within a narrower range *viz.* 14–15·5 °C.

TABLE XIII

Synopsis of work on eight other species

Species	Details of work and findings	References
West coast sole, *Austroglossus microlepis*	Distribution of larvae off Namibia during 1972–1973 described (3% of ichthyoplankton)	O'Toole, 1974
	Larvae occurred in coastal waters (within 30 km of coast) off Namibia during 1972–1974 south of Cape Cross, with heaviest concentrations in vicinity of Conception Bay and Hollams Bird Island and between Cape Cross and Walvis Bay (Fig. 35). Spawning season early spring–early summer. Small larvae (<4 mm) occurred south of 23° S, larger larvae in the north.	O'Toole, 1977c
Snoek, *Thyrsites atun*	Description of embryonic and larval development at 18·5 °C. Eggs hatched 50 h after fertilization.	De Jager, 1955
	Description of the larval stages of snoek and the distribution of larvae around South Western Cape. Larvae were widely distributed, mainly further than 50 km from the coast.	Haig, 1972b
	Readers are also referred to information on the distribution of eggs and larvae contained in the *Colln scient. Pap. int. Commn. SE. Atl. Fish* series.	
Lightfish, *Maurolicus muelleri*	Eggs of this species were more abundant around South Western Cape in 1977–1978 than pilchard, anchovy and round herring. Mainly distributed off the shelf, below thermocline.	Shelton & Davies, 1979 P. A. Shelton, pers. comm.
	Eggs and larvae were plentiful offshore off southern Namibia during July 1983	Olivar & Rubiés, 1985
	Eggs and larvae plentiful between 30° and 36° S during winter of 1983.	Olivar, 1984
	Distribution and abundance of eggs and larvae off Namibia. Mainly occurred offshore south of 20° S.	Olivar, 1981, 1982
Jacopever, *Helicolenus dactylopterus*	Description of larval stages and distribution around South Western Cape. Larvae occurred mainly offshore between Cape Columbine and Danger Point.	Haig, 1972b
Frostfish, *Lepidopus caudatus*	Eggs and larvae occurred just west of shelf break (surface temperature 16–18 °C) during summer of 1972–1973. More abundant near Lüderitz than in Cape or off northern Namibia.	Porebski & Bielaszewska, 1975

TABLE XIII—*continued*

Species	Details of work and findings	References
Dragonet, *Paracallionymus costatus*	Larvae plentiful between 30° and 36° S during winter of 1983.	Olivar, 1984
	Eggs and larvae abundant around the South Western Cape.	P. A. Shelton, pers comm.
Saury, *Scomberesox saurus scomberioides*	Growth rate measured (0·6 mm·day^{-1}) during 47 days after hatching at 18–19 °C. Feeding behaviour was observed.	Brownell, 1983b
	Distribution of eggs and late larvae around the South Western Cape during 1977–1978 discussed.	Dudley, Field & Shelton, 1985
Other	Description of development of sciaenid eggs collected off Lüderitz in June 1967. Species not positively identified, although suggested it was geelbek, *Atractoscion aequidens*.	Meyer-Rochow, 1972

OTHER SPECIES

A summary of the available literature on the ichthyoplankton of eight further Benguela species is given in Table XIII. Readers are also referred to the Collection of Scientific Papers of the International Commission for the South-east Atlantic Fisheries for details of abundance and distribution of a number of lesser species (*e.g.* the listings in Olivar 1981, 1982, 1983 and Olivar & Rubiés, 1985), and to the taxonomic text of Brownell (1979).

SYNTHESIS AND DISCUSSION

The following paragraphs are based on a working paper submitted to ICSEAF (Shannon, in press).

From the aforegoing text and diagrams it is evident that, for a number of species, preferred spawning areas exist in both the northern and the southern Benguela regions. While there are relatively few data on the ichthyoplankton of the central region of the Benguela, the indications are that, with the possible exception of the goby, this part of the system is not an important spawning area, being characterized by strong southerly winds (intense upwelling, cold water, high turbulence, low stratification) for much of the year. This environmental barrier effectively divides the Benguela in terms of spawning, which in turn implies a strong probability of discrete populations of adults of some neritic species existing off the Cape and off Namibia.

There are several evident similarities between the preferred spawning habitats off the Cape and off Namibia, but there are also some important differences. In both the Cape and Namibia maximum numbers of larvae occur downstream (*i.e.* alongshore or offshore) of the areas of maximum egg abundance. In the case of the anchovy the preferred habitats in both regions are close to warm-water regimes (Angolan and Agulhas systems), and spawning coincides with the increasing stratification of the water column. Off northern Namibia the peak spawning is after the main upwelling season, while on the Agulhas Bank it is after the cessation of strong winter mixing. In the one case nutrients will have been supplied by upwelling, in the other by mixing (storm, internal waves breaking) and so

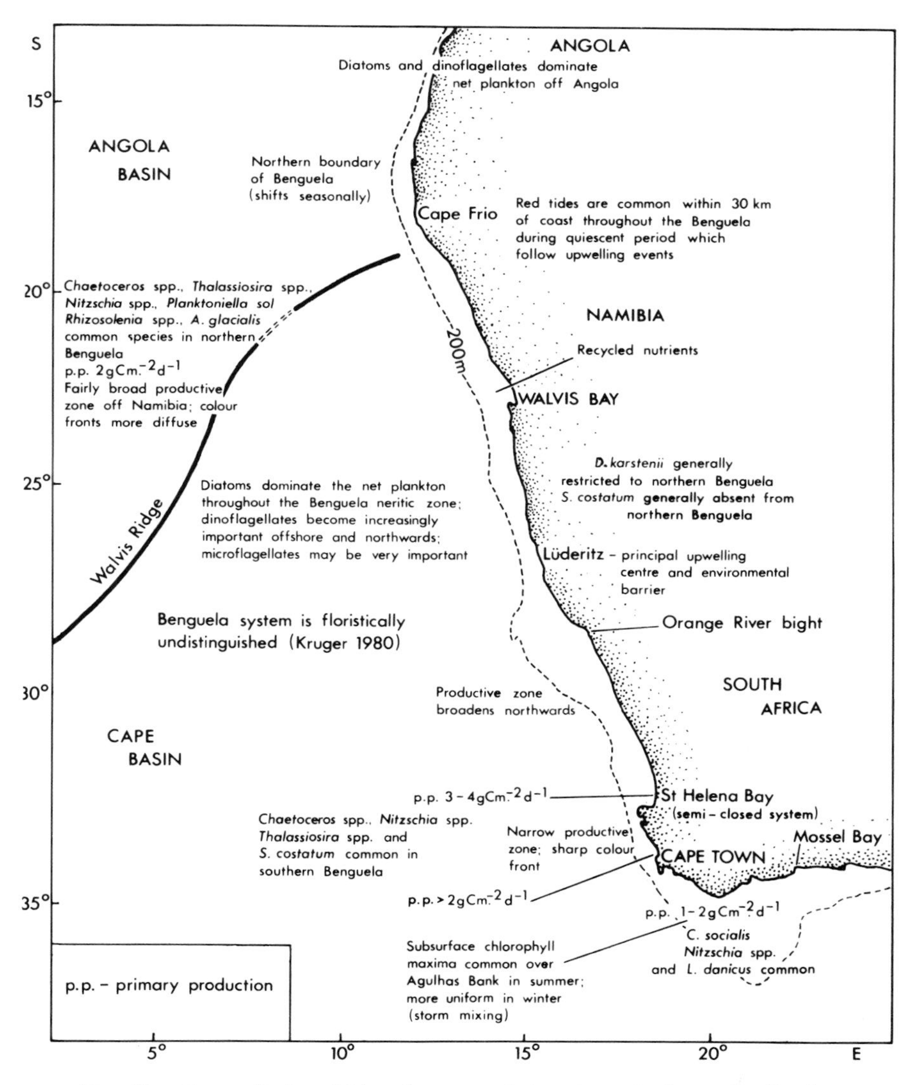

Fig. 43.—Some distinguishing features of the phytoplankton distribution in the Benguela system.

effectively 'primed' the system. Off northern Namibia the egg and larvae of anchovy occur close to the probable nursery areas (the same may be true for the eastern Agulhas Bank although this still has to be shown), while eggs spawned on the western Agulhas Bank will have to rely on a transport mechanism (frontal jet) in order to reach the recruitment area. In the case of the pilchard the important spawning areas close to major upwelling centres off both Namibia and the Cape, evident in the 1950s have effectively disappeared, this disappearance coinciding with the extermination of older adult fish. The remaining adult age classes at present spawn in similar areas

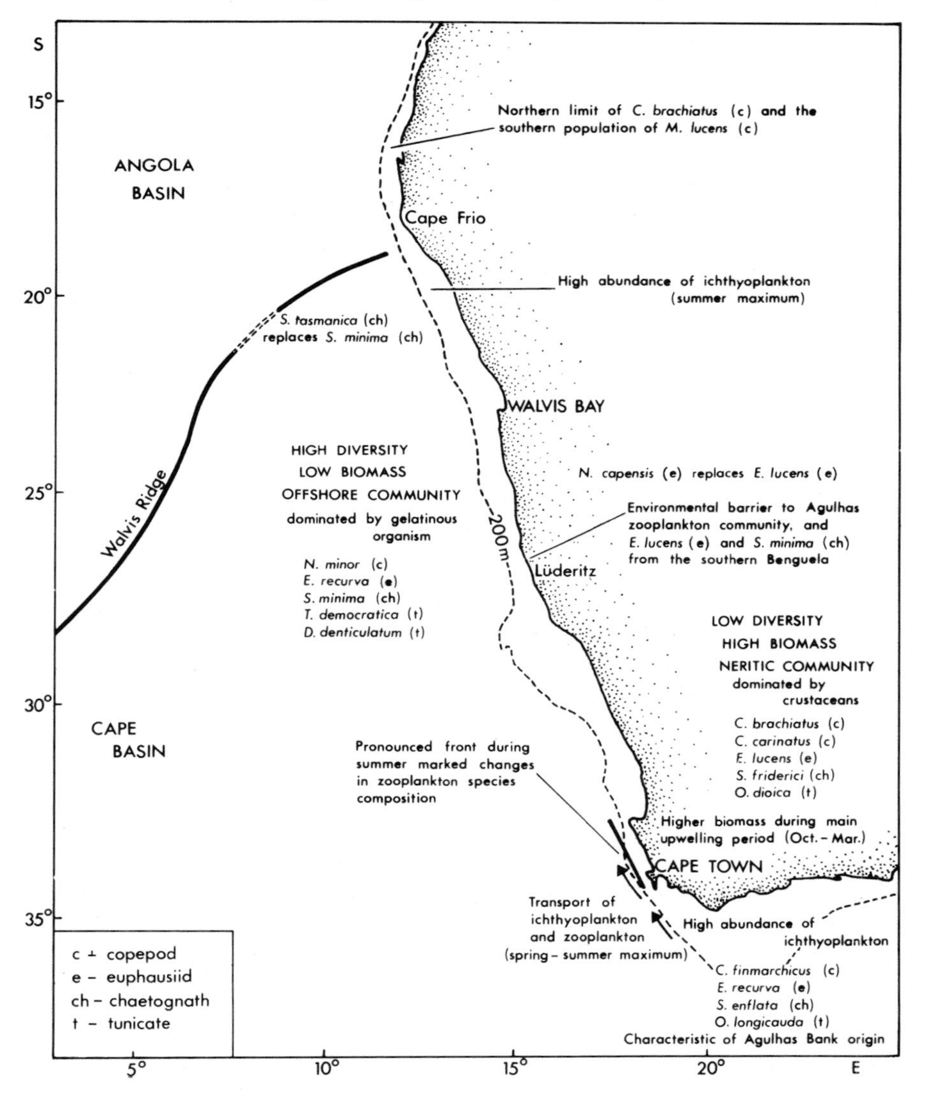

Fig. 44.—Some distinguishing features of the zooplankton and ichthyoplankton distribution in the Benguela system.

as the anchovy *i.e.* close to warm-water regimes. This may suggest that the pilchard spawning in anchovy-type areas assumed greater importance following the collapse of the respective pilchard parent stocks and compensatory reductions in sizes at maturity. The goby which generally shares the same spawning habitat and season with the Namibian pilchard (and may have partially replaced this species in that ecosystem) is evidently not abundant in the Cape (P. A. Shelton, pers. comm.). There may be a link between the abundance of goby larvae south of Walvis Bay and the indemic diatom, *Delphineis karstenii*, which is eaten by the larvae. Whereas the eggs and larvae of hake, and probably lanternfish also, appear to be fairly widely distributed in the northern and southern Benguela, the abundance of horse mackerel (and by implication probably also the eggs and larvae of this species) is substantially higher off northern Namibia than off the Cape. Conversely, round herring is abundant in Cape ichthyoplankton but less so off Namibia. A possible explanation for the different distribution of round herring in the southern and northern Benguela may be related to the different subsurface frontal features in the two parts of the system. In the Cape the subsurface front is intense and is maintained throughout most of the year. Off northern Namibia the front is more diffuse. It does seem that energy exported offshore from the Namibian upwelling region is available at present to horse mackerel and lanternfish, while that exported out of the Cape upwelling system is available to round herring, lanternfish, and lightfish. This concept will be discussed in some detail in Part IV of this review series.

A CONCEPTUAL IMAGE OF THE DISTRIBUTION OF PLANKTON IN THE BENGUELA

Some distinguishing features of the plankton distribution in the Benguela are highlighted in Figures 43 and 44. It is hoped that it may prove useful for scientists in other disciplines and to those not familiar with the plankton of the Benguela. It should be viewed by readers in conjunction with the appropriate text and not in isolation.

ACKNOWLEDGEMENTS

The authors wish to thank the following for their useful comments on the manuscript and advice in its preparation: Penny Brown, Betty Mitchell-Innes, and Sakkie Kruger (phytoplankton); Dot Armstrong, Larry Hutchings, and Robin Carter (zooplankton); Robert Crawford, Peter Shelton, and Frances Le Clus (ichthyoplankton). We are also appreciative of the assistance received from Tony van Dalsen, whose group prepared all the artwork, Mariana van Niekerk who typed the manuscript, and the Institute's ever-helpful librarians Ben Wessels and Adrienne Melzer. The synthesis was undertaken under the auspices of the Benguela Ecology Programme and we gratefully acknowledge the rôle of the participants in this programme, whose interaction facilitated the development of many of the concepts. The work was funded by the Sea Fisheries Research Institute.

REFERENCES

Agenbag, J. J., Kruger, I. & Le Clus, F., 1984. *S. Afr. J. mar. Sci.,* **2,** 93–107.

Ahlstrom, E. H., Moser, H. G. & O'Toole, M. J., 1976. *Bull. 5th Calif. Acad. Sci.,* No. 75(2), 138–152.

Allanson, B. R., Hart, R. C. & Lutjeharms, J. R. E., 1981. *S. Afr. J. Antarct. Res.,* **10,** 3–14.

Alvariño, A., 1965. *Oceanogr. Mar. Biol. Ann. Rev.,* **3,** 115–194.

Anders, A. S., 1965. *S. Afr. Shipp. News Fish. Ind. Rev.,* **20**(11), 103–107.

Anders, A. S., 1975. *S. Afr. Shipp. News Fish. Ind. Rev.,* **30**(9), 53–57.

Anderson, F. P., Shannon, L. V., Mostert, S. A., Walters, N. M. & Malan, O. G., 1981. In, *Oceanography from Space,* edited by J. F. R. Gower, Plenum Publishing Corp., pp. 381–386.

Andrews, W. R. H., 1974. *Téthys,* **6,** 327–340.

Andrews, W. R. H. & Cram, D. L., 1969. *Nature, Lond.,* **24,** 902–904.

Andrews, W. R. H., Cram, D. L. & Visser, G. A., 1970. *Proceedings of symposium Oceanography in South Africa,* Durban, August 1970, South African National Committee for Oceanographic Research, CSIR, Pretoria, Paper B3, 13 pp. (unpubl.).

Andrews, W. R. H. & Hutchings, L., 1980. *Prog. Oceanogr.,* **9,** 1–81.

Assorov, V. V. & Berenbeim, D. Y., 1983. *Colln scient. Pap. int. Commn SE Atl. Fish.* **10**(1), 27–30.

Austin, N. E. H., 1980. M.Sc. thesis, University of Natal (Pietermaritzburg), South Africa, 127 pp.

Badenhorst, A. & Boyd, A. J., 1980. *Fish. Bull. S. Afr.,* **13,** 83–106.

Bailey, G. W., 1979. M.Sc. thesis, University of Cape Town, South Africa, 225 pp.

Baird, D., 1974. Ph.D. thesis, University of Stellenbosch, South Africa, 241 pp.

Baird, D., 1977. *Zool. Afr.,* **12,** 347–362.

Barber, R. T. & Smith, R. L., 1981. In, *Analysis of Marine Ecosystems,* edited by A. R. Longhurst, Academic Press, London, pp. 31–68.

Barlow, R. G., 1980. *J. exp. mar. Biol. Ecol.,* **45,** 83–93.

Barlow, R. G., 1981. *Fish. Bull. S. Afr.,* **14,** 99–101.

Barlow, R. G., 1982a. *J. exp. mar. Biol. Ecol.,* **63,** 209–227.

Barlow, R. G., 1982b. *J. exp. mar. Biol. Ecol.,* **63,** 229–237.

Barlow, R. G., 1982c. *J. exp. mar. Biol. Ecol.,* **63,** 239–248.

Barlow, R. G., 1984a. *Mar. Ecol. Prog. Ser.,* **16,** 121–126.

Barlow, R. G., 1984b. *J. Plankt. Res.,* **6,** 385–397.

Barlow, R. G., 1984c. *J. Plankt. Res.,* **6,** 435–442.

Barlow, R. G. & Swart, M. J., 1981. *Fish. Bull. S. Afr.,* **15,** 89–93.

Barnard, K. H., 1932. *'Discovery' Rep.,* **5,** 1–326.

Barnard, K. H., 1940. *Ann. S. Afr. Mus.,* **32,** 381–543.

Bary, B. M., 1956. *Pacif. Sci.,* **10,** 431–467.

Bessonov, N. M. & Fedosov, M. V., 1965. *Oceanology,* **5,** 88–93.

Bienfang, P. K. & Harrison, P. J., 1984. *Mar. Biol.* **83,** 293–300.

Boden, B. P., 1950. *Trans. R. Soc. S. Afr.,* **32,** 321–434.

Boden, B. P., 1954. *Trans. R. Soc. S. Afr.,* **34,** 181–243.

Boden, B. P., 1955. *'Discovery' Rep.,* **27,** 337–376.

Borchers, P. L. & Field, J. G., 1981. *Botanica Mar.,* **24,** *89*–91.

Borchers, P. L. & Hutchings, L., 1983. *Proceedings of Fifth National Oceanographic Symposium,* Grahamstown, South Africa, January 1983. S288: paper B21 (abstract).

Botha, L., 1973. *S. Afr. Shipp. News Fish. Ind. Rev.,* **28**(4), 62–67.

Botha, L., 1980. Ph.D. thesis, University of Stellenbosch, South Africa, 182 pp.

Boyd, A. J. & Badenhorst, A., 1981. *Fish. Bull. S. Afr.,* **15,** 43–48.

Boyd, A. J. & Cruickshank, R. A., 1983. *S. Afr. J. Sci.,* **79,** 150–151.
Boyd, A. J. & Hewitson, J. D., 1983. *S. Afr. J. mar. Sci.,* **1,** 71–75.
Boyd, A. J., Hewitson, J. D., Kruger, I. & Le Clus, F., 1985. *Colln scient. Pap. int. Commn SE. Atl. Fish.,* **12** (in press).
Brongersma-Sanders, M., 1947. *Proceedings Koninklijke Nederlandsche Akademie van Wetenschappen,* Vol. 50(6) pp. 660–665.
Brongersma-Sanders, M., 1948. *Verh. Akad. Wet. Amst.* Sect. 2, **45**(4), pp. 1–112.
Brown, P. C., 1980. M.Sc. thesis, University of Cape Town, South Africa, 98 pp.
Brown, P. C., 1982a. *Fish. Bull. S. Afr.,* **16,** 25–29.
Brown, P. C., 1982b. *Fish. Bull. S. Afr.,* **16,** 31–37.
Brown, P. C., 1982c. *Fish. Bull. S. Afr.,* **16,** 39–44.
Brown, P. C., 1983a. *S. Afr. J. Sci.,* **79,** 144 only.
Brown, P. C., 1983b. *Proceedings of the First Annual Congress of the Phycological Society of Southern Africa,* Johannesburg, South Africa, January 1983, p. 7 only.
Brown, P. C., 1983c. *S. Afr. J. mar. Sci.,* **1,** 139–143.
Brown, P. C., 1984. *S. Afr. J. mar. Sci.,* **2,** 205–215.
Brown, P. C. & Field, J. G., 1985a. *Botanica Mar.,* **28,** 201–208.
Brown, P. C. & Field, J. G., 1985b. *J. Plankt. Res.,* **8,** 55–68.
Brown, P. C. & Henry, J. L., 1985. In, *South African Ocean Colour and Upwelling Experiment*, edited by L. V. Shannon, Sea Fisheries Research Institute, Cape Town, pp. 221–218.
Brown, P. C. & Hutchings, L., 1985. *International Symposium on the Most Important Upwelling Areas off Western Africa (Cape Blanco and Benguela),* Inst. Invest. pesq. Barcelona, Vol. 1, pp. 319–344.
Brown, P. C., Hutchings, L. & Horstman, D., 1979. *Fish Bull. S. Afr.,* **11,** 46–52.
Brownell, C. L., 1979. *Ichthyol. Bull. J. L. B. Smith Inst. Ichthyol. S. Afr.,* **40,** 84 pp.
Brownell. C. L., 1983a. *S. Afr. J. mar. Sci.,* **1,** 181–188.
Brownell, C. L., 1983b. *S. Afr. J. mar. Sci.,* **1,** 245–248.
Carter, R. A., 1982. *Mar. Ecol. Prog. Ser.,* **8,** 9–14.
Carter, R. A., 1983. Ph.D. thesis, University of Cape Town, South Africa, 203 pp.
Carter, R. A. & De Decker, A. H. B., 1983. Review paper H7, presented at Fifth National Oceanographic Symposium, Grahamstown, South Africa, January 1983.
Chapman, P. & Shannon, L. V., 1985. *Oceanogr. Mar. Biol. Ann. Rev.,* **23,** 183–251.
Christensen, M. S., 1980. *S. Afr. J. Sci.,* **76,** 541–546.
Cleve, P. T., 1904. *Mar. Invest. S. Afr.,* No. 3, 177–210.
Cleve, P. T., 1905. *Mar. Invest. S. Afr.* No. 4, 125–128.
Coetzee, D. J., 1974. M.Sc. thesis, University of Stellenbosch, South Africa, 107 pp.
Copenhagen, W. J., 1953. *Investl Rep. Div. Fish. S. Afr.,* No. 14, 35 pp.
Copenhagen, W. J. & Copenhagen, L. D., 1949. *Trans. R. Soc. S. Afr.,* **32,** 113–123.
Cram, D. L. & Schülein, F. H., 1974. *J. Cons. perm. int. Explor. Mer,* **35,** 272–275.
Cram, D. L. & Visser, G. A., 1973. *S. Afr. Shipp. News Fish. Ind. Rev.,* **28**(3), 56–63.
Crawford, R. J. M., 1980. *J. Fish Biol.,* **16,** 649–664.
Crawford, R. J. M., 1981. *Fish. Bull. S. Afr.,* **14,** 1–46.
Crawford, R. J. M., Cruickshank, R. A., Shelton, P. A. & Kruger, I., 1985. *S. Afr. J. mar. Sci.,* **3,** 215–228.
Crawford, R. J. M., Shelton, P. A. & Hutchings, L., 1983. In, *Proceedings of the Expert Consultation to Examine Changes in Abundance and Species Composition of Neritic Fish Stocks,* San José, Costa Rica, April 1983, edited by G. D. Sharp & J. Csirke, *FAO Fish. Rep.,* No. 291(2), 407–448.

Currie, R. I. & Foxton, P., 1957. *J. mar. biol. Ass. U.K.*, **36**, 17–32.
Cushing, D. H., 1969. FAO Fish. Tech. Paper No. 84, FAO, Rome, 40 pp.
Cushing, D. H., 1971. *Adv. Mar. Biol.*, **9**, *255*–334.
D'Arcanques, C., 1977. *Ann. Inst. Océanogr. Paris*, **53**, 87–104.
Davies, D. H., 1954. *Investl Rep. Div. Fish. S. Afr.*, No. 15, 28 pp.
Davies, D. H., 1956. *Investl Rep. Div. Fish. S. Afr., No. 22, 155 pp.*
Davies, D. H., 1957. *Investl Rep. Div. Fish. S. Afr.*, No. 30, 40 pp.
Davies, S. L., Newman, G. G. & Shelton, P. A., 1981. *Colln scient. Pap. int. Commn SE. Atl. Fish.*, **8**(II), 51–74.
De Decker, A. H. B., 1964. *Investl Rep. Div. Sea Fish. S. Afr.*, No. 49, 33 pp.
De Decker, A. H. B., 1970. *Investl Rep. Div. Sea Fish. S. Afr.*, No. 84, 24 pp.
De Decker, A. H. B., 1973. In, *Ecological Studies, Analysis and Synthesis, Vol. 3*, edited by B. Zeitzschel, Springer Verlag, Berlin, pp. 189–219.
De Decker, A. H. B., 1984. *Ann. S. Afr. Mus.*, **93**(5), 303–370.
De Decker, A. H. B. & Coetzee, D. J., 1979. *Ann. S. Afr. Mus.*, **78**, 69–79.
De Jager, B.v.D., 1954. *Investl Rep. Div. Fish. S. Afr.*, No. 17, 26 pp.
De Jager, B.v.D., 1955. *Investl Rep. Div. Fish. S. Afr.*, No. 19, 16 pp.
De Jager, B.v.D., 1957. *Investl Rep. Div. Fish. S. Afr.*, No. 25, 78 pp.
Dick, R. I., 1970. *Ann. S. Afr. Mus.*, **57**(3), 25–86.
Dudley, S. F. J., Field, J. G. & Shelton, P. A., 1985. *S. Afr. J. mar. Sci.*, **3**, 229–237.
Duncan, C. P. & Nell, J. H., 1969. *Investl Rep. Div. Sea Fish. S. Afr.*, No. 76, 19 pp.
Field, J. G., Griffiths, C. L., Linley, E. A., Carter, R. A. & Zoutendyk, P., 1980. *Estuar. cstl mar. Sci.*, **11**, 133–150.
Fraser, J. H., 1952. *Mar. Res.*, No. 2, 52 pp.
Gilchrist, J. D. F., 1914. *Mar. Biol. Rep., Cape Town,* **2**, 8–35.
Gilchrist, J. D. F., 1916. *Mar. Biol. Rep., Cape Town (for the year ending 30 June 1916),* **3**, 1–26.
Grant, W. S., 1983. *Proceedings of Fifth National Oceanographic Symposium*, Grahamstown, South Africa, January 1983: S288: paper B10 (abstract).
Gray, B. B., 1923. *Proc. R. Soc. Qd*, **34**, 171–180.
Grindley, J. R. & Nel, E. A., 1970. *Fish. Bull. S. Afr.*, **6**, 36–55.
Grindley, J. R. & Taylor, F. J. R., 1962. *Nature, Lond.*, **195**, 1324 only.
Grindley, J. R. & Taylor, F. J. R., 1964. *Trans. R. Soc. S. Afr.*, **37**(2), 111–130.
Haig, E. H., 1972a. *Ann. S. Afr. Mus.*, **59**(8), 139–150.
Haig, E. H., 1972b. *Ann. S. Afr. Mus.*, **59**(3), 47–70.
Hart, T. J., 1934. *Nature, Lond.*, **134**, *459*–460.
Hart, T. J. & Currie, R. I., 1960. *'Discovery' Rep.*, **31**, 123–298.
Hart, T. J. & Marshall, N. B., 1951. *Nature, Lond.*, **168**, 272–273.
Harvey, H. W., 1934. *J. mar. biol. Ass. U.K.*, **19**, 761–773.
Hasle, G. R., 1976. *Deep-Sea Res.*, **23**, 319–338.
Hendey, N. I., 1937. *'Discovery' Rep.*, **16**, 151–364 plus 8 plates.
Henry, J. L., 1979. Poster presentation, Fourth South African National Oceanographic Symposium (S189), Cape Town, South Africa, July 1979.
Henry, J. L., Mostert, S. A. & Christie, N. D., 1977. *Trans. R. Soc. S. Afr.*, **42**, 383–398.
Hentschel, E., 1928. *Int. Revue ges. Hydrobiol. Hydrogr.*, **21**, 1–16.
Hentschel, E., 1933. *Wiss. Ergebn. dt. Atlant. Exped. 'Meteor'*, **11**, Parts 1 and 2, 344 pp.
Hentschel, E. & Wattenberg, H., 1930. *Annln Hydrogr. Berl.*, **58**, 273–277.
Herman, A. W. & Denman, K. L., 1977. *Deep-Sea Res., 385*–397.
Heydorn, A. E. F., 1959. *Investl Rep. Div. Fish. S. Afr.*, No. 36, 56 pp.
Hobson, L. A., 1971. *Invest. pesq.*, **35**, 195–208.
Hopson, S. D., 1983. M.Sc. thesis, University of Cape Town, South Africa, 124 pp.
Horstman, D. A., 1981. *Fish. Bull. S. Afr.*, **15**, 71–88.

Hoy, H., 1970. *Fish. Bull. S. Afr.*, **6**, 1–9.
Hulburt, E. M., 1976. *Limnol. Oceanogr.*, **21**, 193–211.
Hunter, J. R., 1984. *Flodvigen rapportser,* **1**, 533–562.
Hutchings, L., 1979. Ph.D. thesis, University of Cape Town, South Africa, 206 pp.
Hutchings, L., 1981. In, *Coastal Upwelling,* edited by F. A. Richards, *Coastal and Estuarine Sciences* 1, AGU, Washington, D.C., pp. 496–506.
Hutchings, L., Barlow, R., Brown, P. C. & Olivieri, E. T., 1983. Proceedings of Fifth National Oceanographic Symposium, Grahamstown, South Africa, January 1983, paper B16 (abstract).
Hutchings, L., Holden, C. & Mitchell-Innes, B., 1984. *S. Afr. J. Sci.*, **80**, 83–89.
Hutchings, L., Nelson, G., Horstman, D. A. & Tarr, R., 1983. In, *Sandy Beaches as Ecosystems,* edited by A. MacLachlan & T. Erasmus, Dr W. Junk Publishers, The Hague, Boston, Lancaster, pp. 481–500.
Hutchings, L., Robertson, A. A. & Allan, T. S., 1970. *Proceedings of Symposium Oceanography in South Africa,* Durban, August 1970, 8 pp.
Iles, E. J., 1953. *'Discovery' Rep.*, **26**, 259–280.
Illig, G., 1930. *Wiss. Eregbn. dt. Tiefsee-Exped. 'Valdivia'*, **22**(6), 400–625.
Jones, N. S., 1955. *'Discovery' Rep.*, **27**, 279–292.
King, D. P. F., 1975. M.Sc. thesis, University Stellenbosch, South Africa, 54 pp.
King, D. P. F., 1977a. *Investl Rep. Sea Fish, Brch S. Afr.*, No. 114, 35 pp.
King, D. P. F., 1977b. *Fish. Bull. S. Afr.*, **9**, 23–31.
King, D. P. F., O'Toole, M. J. & Robertson, A. A., 1977. *Fish. Bull. S. Afr.*, **9**, 16–22.
King, D. P. F. & Robertson, A. A., 1978. *Fish. Bull. S. Afr.*, **10**, 15–19.
King, D. P. F., Robertson, A. A. & Shelton, P. A., 1978. *Fish. Bull. S. Afr.*, **10**, 37–45.
Kollmer, W. E., 1962. *Investl Rep. mar. Res. Lab. S.W. Afr.*, No. 4, 44 pp.
Kollmer, W. E., 1963. *Investl Rep. mar. Res. Lab. S.W. Afr.*, No. 8, 78 pp.
Komarov, Y. A., 1964a. Contribution No. 94. Atlantic Committee, International Council for the Exploration of the Sea, C. M. 1964, 5 pp.
Komarov, Y. A., 1964b. *Trudy Atlant. NIRO,* **11**, 87–99.
Kruger, I., 1979. *Fish Bull. S. Afr.*, **11**, 23–25.
Kruger, I., 1980. *Fish Bull. S. Afr.*, **13**, 31–53.
Kruger, I., 1983. *Colln scient. Pap. int. Commn SE. Atl. Fish.*, **10**(2), 121–138.
Kruger, I. & Boyd, A. J., 1984. *Colln scient. Pap. int. Commn S. E. Atl. Fish.*, **11**(1), 109–133.
Kruger, I. & Cruickshank, R. A., 1982. *Fish. Bull. S. Afr.*, **16**, 99–114.
Kruger, I. & Wilson, E. G., 1984. *S. Afr. J. mar. Sci.*, **2**, 163–194.
Lazarus, B. I., 1967. *Investl Rep. Div. Sea Fish. S. Afr.*, No. 63, 38 pp.
Lazarus, B. I., 1975. Ph.D. thesis, University of Stellenbosch, South Africa, 372 pp.
Lazarus, B. I. & Dowler, D., 1980. *Fish. Bull. S. Afr.*, **12**, 93–119.
Le Clus, F., 1977. *Fish. Bull. S. Afr.*, **9**, 11–15.
Le Clus, F., 1979a. *Fish Bull. S. Afr.*, **12**, 53–68.
Le Clus, F., 1979b. *Fish. Bull. S. Afr.*, **11**, 26–38.
Le Clus, F., 1983. *Colln scient. P. int. Commn S.E. Atl. Fish.*, **10**(II), 139–145.
Le Clus, F., 1985. M.Sc. thesis, University Port Elizabeth, South Africa, 202 pp.
Le Clus, F. & Kruger, I., 1982. *Colln scient. Pap. int. Commn S.E. Atl. Fish.* **9**(II), 121–145.
Louw, E. & O'Toole, M. J., 1977. *Ann. S. Afr. Mus.*, **72**(7), 125–145.
Marchand, J. M., 1928. *Fish. Mar. Biol. Surv., Un. S. Afr.*, Spec. Rep. No. 5, pp. 1–11 of Annual Report No. 6 for 1928.
Masson, C., 1973. M.Sc. thesis, University of Stellenbosch, South Africa, 231 pp.
Matthews, J. P., 1964. *Investl Rep mar. Res. Lab. S.W.Afr.*, No. 10, 96 pp.
Matthews, J. P. & De Jager, B.v.D., 1951. *Investl Rep. Fish. mar. Biol. Surv. S. Afr.*, No. 13, 8 pp.

Meyer-Rochow, V.B., 1972. *Vie Milieu,* **23**, sér. A, 11–19.
Morel, A., 1980. *Boundary-layer Met.,* **18**(2), 177–201.
Nel, E. A., 1968. *Investl Rep. Div. Sea Fish. S. Afr.,* No. 62, 178 pp.
Nepgen, C. S. de V., 1957. *Investl Rep. Div. Sea Fish. S. Afr.,* No. 28, 30 pp.
Newman, G. G., 1970. *Investl Rep. Div. Sea Fish. S. Afr.,* No. 86, 6 pp.
Newman, G. G. & Crawford, R. J. M., 1980. *Rapp. P.-v. Réun. Cons. int. Explor. Mer,* **177**, 279–291.
Norris, R. E., 1983. *Proceedings of Fifth National Oceanographic Symposium* Grahamstown, South Africa, January 1983, S288, paper B17 (abstract).
Olivar, M. P., 1981. *Colln scient. Pap. int. Commn SE. Atl. Fish.,* **8**(II), 161–173.
Olivar, M. P., 1982. *Colln scient. Pap. int. Commn SE. Atl. Fish.,* **9**(II), 221–240.
Olivar, M. P., 1983. *Colln scient. Pap. int. Commn SE. Atl. Fish.,* **10**(II), 147–160.
Olivar, M. P., 1984. *Colln scient. Pap. int. Commn SE. Atl. Fish.,* **11**(II), 13–21.
Olivar, M. P., 1985. Doctoral thesis, University of Barcelona, 710 pp.
Olivar, M. P., Salat, J. & Rubiés, P. 1981. *Colln scient. Pap. int. Commn SE. Atl. Fish.,* **8**(II), 175–183.
Olivar, M. P. & Rubiés, P., 1983. *Colln scient. Pap. int. Commn SE. Atl. Fish.,* **10**(II), 161–172.
Olivar, M. P. & Rubiés, P., 1985. *Colln scient. Pap. int. Commn SE. Atl. Fish.,* **12** (in press).
Olivar, M. P., Rubiés, P. & Salat, 1982. *Colln scient. Pap. int. Commn SE. Atl. Fish.,* **9**(II), 241–248.
Olivieri, E. T., 1983a. *S. Afr. J. mar. Sci.,* **1**, 199–229.
Olivieri, E. T., 1983b. *S. Afr. J. mar. Sci.,* **1**, 77–109.
Olivieri, E. T., 1985. Ph.D. thesis, University of Cape Town, South Africa, 214 pp.
Olivieri, E. T. & Hutchings, L., 1983. *S. Afr. J. Sci.,* **79**, 145 only.
O'Toole, M. J., 1974. *S. Afr. Shipp. News Fish. Ind. Rev.,* **29**(11), 53–59.
O'Toole, M. J., 1977a. Ph.D. thesis, University Cape Town, South Africa, 272 pp plus appended publications.
O'Toole, M. J., 1977b. *Fish. Bull. S. Afr.,* **9**, 46–47.
O'Toole, M. J., 1977c. *Fish. Bull. S. Afr.,* **9**, 32–45.
O'Toole, M. J., 1978a. *Fish. Bull. S. Afr.,* **10**, 20–36.
O'Toole, M. J., 1978b. *Investl Rep. Sea Fish. Brch S. Afr.,* No. 116, 28 pp.
O'Toole, M. J. & King. D. P. F., 1974. *Vie Milieu,* **24**, sér A, 443–452.
Parsons, T. R., 1979. *S. Afr. J. Sci.,* **75**, *536*–540.
Pieterse, F. & Van der Post, D. C., 1967. *Investl Rep. mar. Res. Lab. S.W. Afr.,* No. 14, 125 pp.
Pillar, S. C., 1983. *Proceedings of Fifth National Oceanographic Symposium,* Grahamstown, South Africa, January 1983. S288, paper B24 (abstract).
Pillar, S. C., 1984a. *S. Afr. J. mar. Sci.,* **2**, 71–80.
Pillar, S. C., 1984b. *S. Afr. J. mar. Sci.,* **2**, 43–48.
Pillar, S. C., 1984c. *S. Afr. J. mar. Sci., 2,* 1–18.
Pillar, S. C., 1985. *J. Plankt. Res.* **7**, 223–240.
Pinto, J. de S., 1953. *Trab. Missā Biol. Marit.,* No. 6, 129–151.
Pollock, D. E. & Goosen, P. C., 1983. *Proceedings of Fifth National Oceanographic Symposium,* Grahamstown, South Africa, January 1983, S288, paper B3 (abstract).
Popkiss, M. E. E., Horstman, D. A. & Harpur, D., 1979. *S. A. Med. J.,* **55**, 1017–1023.
Porebski, J., 1975. *Colln scient. Pap. int. Commn SE. Atl. Fish.,* **2**, 102–106.
Porebski, J. & Bielaszewska, E., 1975. *Colln scient. Pap. int. Commn SE. Atl. Fish.,* **2**, 107–112.
Porebski, J. & Koronkiewicz, A., 1975. *Colln Scient. Pap. int. Commn SE. Atl. Fish.,* **2**, 92–101.
Probyn, T. A., 1985. *Mar. Ecol. Prog. Ser.,* **22**, 249–258.

Prosch, R. M. & Shelton, P. A., 1983. *S. Afr. Shipp. News Fish. Ind. Rev.*, **38**(12), 47–49.
Reyssac, J., 1973. *Bull. Inst. Fond. Afr. Noire,* **35**, ser. A. (2), 273–298.
Ritter-Zahony, R. V., 1909. *Zool. Anz.,* **34**, 787–793.
Ritter-Zahony, R. V., 1911. *Dt. Südpol.-Exped.,* **13**(5), 1–71.
Saville, A., 1980. *Rapp. P.-v. Réun. Cons. int. Explor. Mer,* **177**, 513–517.
Schuette, G., 1980. Ph.D. thesis, Oregon State University, U.S.A., 115 pp.
Schultz, S., 1982. *Rapp. R.-v. Réun. Cons. int. Explor. Mer,* **180**, 202–204.
Schultz, S., Schemainda, R. & Nehring, D., 1979. *Geodat. Geophys. Veröff., Reihe IV,* **28**, i–vii + 1–43.
Shannon, L. V., 1965. *S. Afr. Ship. News Fish. Ind. Rev.,* **20**(11), 115 only.
Shannon, L. V., 1966. *Investl Rep. Div. Sea Fish, S. Afr.,* No. 58, 22 pp., plus 30 pp. figures.
Shannon, L. V., 1969. *Investl Rep. Div. Sea Fish. S. Afr.,* No. 68, 38 pp.
Shannon, L. V., 1972. *Investl Rep. Div. Sea Fish. S. Afr.,* No. 98, 80 pp.
Shannon, L. V., 1985. *Oceanogr. Mar. Biol. Ann. Rev.,* **23**, 105–182.
Shannon, L. V. (in press). *Colln scient. Pap. int. Commn SE. Atl. Fish.*
Shannon, L. V. & Anderson, F. P., 1982. *S. Afr. J. Photogram., Remote Sensing Cartogr.* **13**, 153–169.
Shannon, L. V. & Chapman, P., 1983. *S. Afr. J. Sci.,* **79**, 454–458.
Shannon, L. V., Crawford, R. J. M. & Duffy, D. C., 1984. *S. Afr. J. Sci.,* **80**, 51–60.
Shannon, L. V. & Field, J. G., 1985. *Mar. Ecol. Prog. Ser.,* **22**, 7–19.
Shannon, L. V. & Henry, J. L., 1983. *S. Afr. J. Sci.,* **79**, 144 only.
Shannon, L. V., Hutchings, L., Bailey, G. W. & Shelton, P. A., 1984. *S. Afr. J. mar. Sci.,* **2**, 109–130.
Shannon, L. V. & Lutjeharms, J. R. E., 1983. Proc. of *Earth Data Information Systems Symp.,* Pretoria, South Africa, September 1983, edited by C. Martin, S. Afr. Soc. Photogrammetry, Remote Sensing and Cartography, 14 pp.
Shannon, L. V. & Mostert, S. A., 1983. *S. Afr. J. Sci.* **79**, 45 only.
Shannon, L. V., Mostert, S. A., Walters, N. M. & Anderson, F. P., 1983. *J. Plankt. Res.,* **5**, 565–583.
Shannon, L. V., Schlittenhardt, P. & Mostert, S. A. 1984. *J. geophys. Res.,* **89**, 4968–4976.
Shannon, L. V., Walters, N. M. & Mostert, S. A., 1985. In, *South African Ocean Colour and Upwelling Experiment,* edited by L. V. Shannon, Sea Fisheries Research Institute, Cape Town, pp. 183–210.
Shelton, P. A., 1984. *S. Afr. J. Sci.,* **80**, 69–71.
Shelton, P.A. & Armstrong, M., 1983. In, *Proceedings of the Expert Consultation to Examine Changes in Abundance and Species Composition of Neritic Fish Stocks,* San José, Costa Rica, April 1983, edited by G. D. Sharp, & J. Csirke, *FAO Fish Rep.* 291(3), 1113–1132.
Shelton, P. A. & Davies, S. L., 1979. *S. Afr. Shipp. News Fish. Ind. Rev.,* **34**(6), 28–29.
Shelton, P. A. & Hutchings, L., 1981. *Rapp. P.-v Réun. Cons. int. Explor. Mer,* **178**, 202–203.
Shelton, P. A. & Hutchings, L., 1982. *J. Cons. int. Explor. Mer,* **40**, 185–198.
Siegfried, W. R., 1963. *Investl. Rep. Div. Sea Fish S. Afr.,* No. 48, 12 pp.
Siegfried, W. R., 1965. *Zool. Afr.,* **1**, 339–352.
Silva, S. E., 1953a. *Trab. Missã Biol. Marit.,* No. 3, 1–72.
Silva, S. E., 1953b. *Trab. Missã Biol. Marit.,* No 4, 75–86.
Silva, S. E., 1954. *Trab. Missã Biol. Marit.,* No. 11, 181–243.
Silva, S. E., 1955. *Trab. Missã Biol. Marit.,* No. 15, 107–141.
Silva, S. E., 1957. *Trab. Missã Biol. Marit.,* No. 18, 27–85.
Stander, G. H., 1963. *Investl Rep. mar. Res. Lab. S.W.Afr.,* No. 9, 57 pp.

Stander, G. H. & De Decker, A. H. B., 1969. *Investl Rep. Div. Sea Fish. S. Afr.,* No. 81, 46 pp.
Steeman Nielsen, E., 1964. *ICES C.M. 1964—Plankton Comm.,* Doc. 105, pp. 1–2.
Steemann Nielsen, E. & Jensen, E. A., 1957. *Galathea Rep.,* **1,** 49–137.
Stebbing, T. R. R., 1910. *Ann. S. Afr. Mus.,* **6,** 281–593.
Stone, J. H., 1969. *Ecol. Monogr.,* **39,** 433–463.
Talbot, M. S. 1974. *Zool. Afr.,* **9,** 93–145.
Tanaka, O., 1960. *Special Publ. Seto Marine Biol. Laboratory,* 1–95.
Tattersall, O. S., 1955. *'Discovery' Rep.,* **28,** 1–190.
Tattersall, W. M., 1925. *Fish. Mar. Biol. Surv. Cape Town,* No. 4(5), 12 pp.
Thiriot, A., 1978. In, *Upwelling Ecosystems,* edited by R. Boje & M. Tomczak, Springer-Verlag, Berlin, pp 31–61.
Thomas, R. M., 1980. *Fish. Bull. S. Afr.,* **13,** 21–23.
Thompson, D. & Mostert, S. A., 1974. *J. Cons. int. Explor. Mer,* **36,** 50–53.
Thomson, J. M., 1947. *Counc. sci. ind. Res. (Australia) Bull.,* No. 222, 1–43.
Tromp, B. B. S., Lazarus, B. I. & Horstman, D. A., 1975. *Gross Features of the S.W. Cape coastal waters.* Sea Fish. Res. Inst. S. Afr. (unpubl. rep.) 43 pp.
Unterüberbacher, H. K., 1964. *Investl. Rep. Mar. Res. Lab. S.W. Afr.,* No. 11, 78 pp.
Van Zyl, R. P., 1960. *Investl Rep. Div. Fish. S. Afr.,* No. 40, 31 pp.
Venter, G. E., 1969. *Investl Rep. mar. Res. Lab. S. W. Afr.,* No. 16, 73 pp.
Visser, G. A., Kruger, I., Coetzee, D. L. & Cram, D. L., 1973. *Cape Cross Programme, Phase III report.* Sea Fish. Res. Inst. S. Afr. (unpubl. rep.), pp. 1–9.
Wessels, G. F. S., Coetzee, D. L. & Kruger, I., 1974. *Cape Cross Programme, Phase IV report.* Sea Fish. Res. Inst. S. Afr. (unpubl. rep.), pp 1–8.
Wolfenden, R. N., 1911. *Dt. Südpol.-Exped.,* **12,** 181–380.
Wood, E. J. F. & Corcoran, E. F., 1966. *Bull. mar. Sci.,* **16,** 383–403.
Zernova, V. V., 1974. *Oceanology,* **14,** 882–887.

Oceanogr. Mar. Biol. Ann. Rev., 1986, **24**, 171–263
Margaret Barnes, Ed.
Aberdeen University Press

JAPANESE GELIDIALES (RHODOPHYTA), ESPECIALLY *GELIDIUM*

ISAMU AKATSUKA
Department of Biology, School of Huzoku Kôtôgakkô, Chûô University, 22–1, Kita-mati-3, Nukui, Koganei-si, Tôkyô-to, 184 Japan

INTRODUCTION

Recently, macroscopic seaweeds have attracted attention as one of the world's natural resources and their marine agronomy or mariculture is becoming important (Michanek, 1971, 1972, 1975, 1978, 1983). Since early days the Japanese people have been using seaweeds as a food and in recent decades they have developed cultivation techniques for *Porphyra*, *Undaria*, and *Laminaria*. Unfortunately, the mariculture of *Gelidium* and related species has not yet been successful and, indeed, little research is being carried out on this aspect. Therefore, to stimulate these studies I propose to review all the available literature on the Gelidiales, especially *Gelidium* which is the most important and most frequently investigated agarophyte in Japan in fields other than mariculture.

Although *G. amansii* has been studied for many decades, it is quite apparent that a comprehensive knowledge of this plant is lacking. There are several monographs on *G. amansii*, *e.g.* Okamura (1934a) on the taxonomy, Katada (1955) on the biology with special reference to the early development, Yamasaki (1962) on the marine agronomical surveys employing fisheries statistics, and Yamada (1967) on manuring in the sea. *G. amansii* has been frequently employed for investigations on various topics that are of interest in relation to its economic importance. Many of the reports are in Japanese and, therefore, not readily accessible to non-Japanese scientists and technologists.

Three decades ago, Suto (1954) wrote a review, also in Japanese, of *Gelidium*. Since then, only one review (Santelices, 1974) on the Gelidiales has appeared dealing with morphology, ecology, and marine agronomy. In these two reviews, however, only the biological and marine agronomical aspects were considered. Suto (1954) when dealing with the latter aspects mainly referred to reports of the Japanese Prefectural Fisheries Experimental Station (FES). Various FES reports concerning the propagation of *G. amansii* have referred to similar methods used in great number since the year 1800. These surveys and experiments were made without rigid scientific control and experimental design and without statistical analyses of the results obtained. In addition, FES reports have seldom been based on sequential surveys or experiments of two years or more duration. I have, therefore, selected only those FES reports giving reliable data.

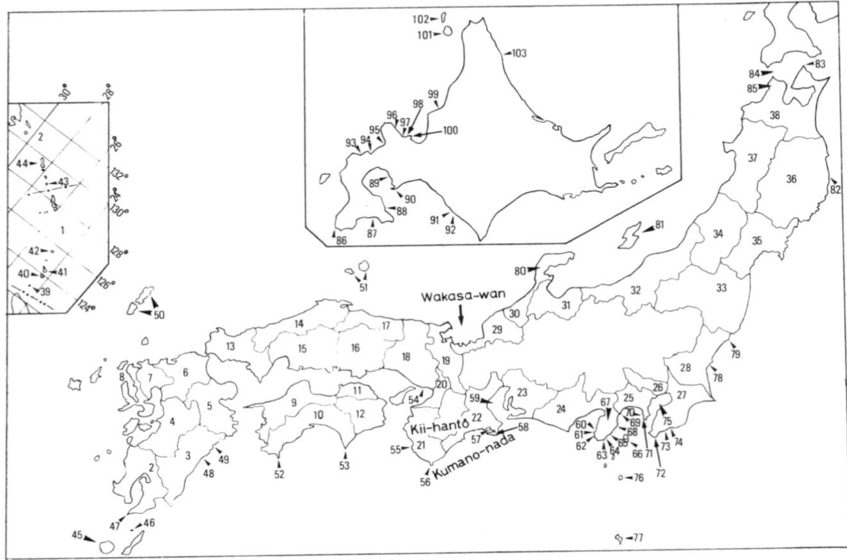

Fig. 1.—Map of geographic and political names of Japan. Names of maritime provinces (=ken) of Japan: Aiti (23), Akita (37), Aomori (38), Ehime (9), Hirosima (15), Hukui (29), Hukuoka (6), Hukusima (33), Hyôgo (18), Ibaraki (28), Isikawa (30), Iwate (36), Kagawa (11), Kagosima (2), Kanagawa (25), Kôti (10), Kumamoto (4), Kyôto (19), Mie or Miye (22), Miyagi (35), Miyazaki (3), Nagasaki (8), Niigata (32), Ôita (5), Okayama (16), Okinawa (1), Ôsaka (20), Saga (7), Simane (14), Sizuoka (24), Tiba (27), Tokusima (12), Tôkyô (26), Tottori (17), Toyama (31), Wakayama (21), Yamagata (34), Yamaguti (13). Geographic names appearing in the text and Tables, other than those indicated in Fig. 2: Aburatubo (71), Amami-ôsima Island (44), Asizuri-misaki Cape (52), Awa-Kominato (74), Hakodate (87), Hamazima (57), Harutati (91), Hatizyô-zima I. (77), Hurubira (96), Hutomi (73), Inatori (65), Iriomote-zima I. (40), Ise-wan Bay (59), Isigaki-zima I. (41), Izu-Ôsima I. (66), Izu-hantô Peninsula (67), Izura (79), Iwanai (95), Kitami-Esasi (103), Kôbe (54), Kominato (74), Komoto (82), Kumomi (61), Mage-sima I. (46), Manazuru and Manazuru-misaki C. (69), Masike (99), Matumae (86), Mera (62), Mimitu (48), Mituisi (92), Miyake-zima I. (76), Miyako-zima I. (42), Muroran (90), Muroto-misaki C. (53), Nagai (70), Nisina (60), Nitiren-zaki C. (68), Nobeoka (49), Noto-hantô C. (80), Ôarai (78), Oki Islands (51), Ôma-zaki C. (83), Osyoro (98), Rebun-tô I. (102), Risiri-tô I. (101), Sado I. (81), Sata-misaki C. (47), Simamaki (93), Sima-hantô Pen. (58), Sirahama and Simoda (Sizuoka-ken, 64), Sirahama (Tiba-ken, 72), Sirahama (Wakayama-ken, 55), Siwono-misaki C. (56), Susaki (63), Takasima-misaki C. (100), Tôkyô-wan B. (75), Tugaru-hantô Pen. (85), Tugaru-kaikyô Straits (84), Tusima Is. (50), Usu (89), Usuziri (88), Utasutu (94), Yaku-sima I. (45), Yoiti (97), Yonaguni-zima I. (39), Yoron-zima I. (43). Single (—·—·—) and double (—··—··—) chain lines in the appended map show borders between Kagosima- and Okinawa-ken and Japan and Taiwan, respectively.

Literature dealing with chemical aspects not included in Suto's review (1954) has been noted. The very early papers published many decades ago, written with old Japanese letters and style and without punctuation and English summary, have been reviewed because even few Japanese researchers will have read them.

Wherever necessary, *G. pacificum*, *G. subfastigiatum*, and other *Gelidium* species are collectively treated as the closely related species *G. amansii*; other members of the Gelidiales have been added to descriptions and discussions as required.

The literature survey for this review was first completed at the end of March 1979 for a review on the Japanese Gelidiales. Articles published from that date up to December 1984 have now been added and this review is, therefore, an up-dated version of the one written in Japanese by Akatsuka (1982a).

The following terminology has been used throughout the review. Erect axis, frond, or thallus: frond, used in comparison with the prostrate axis; main axis: erect axis directly issuing from the prostrate axis, not including branches and pinnules of any orders; branch of first order: branch disposed directly to main axis; ultimate pinnules: branches or proliferations in the last order; tengusa: Japanese collective noun for the species of *Gelidium* and frequently *Pterocladia*; Gelidiaceae: *sensu* Fan (1961). Japanese geographical names and the names of journals using Japanese language are Romanized following to the Kunrei-siki system, even if the names of journals have been previously Romanized by another system; the names of Japanese authors also follow this system when Romanized names are not given in the published papers. Japanese geographical and political names used in connection with the descriptions in this review are presented in Figure 1; sea currents and names of the major islands are found in Figure 2. Tables presenting chemical analyses may be regarded as results obtained using the whole thallus of adult plants, despite the lack of these descriptions in the reports cited.

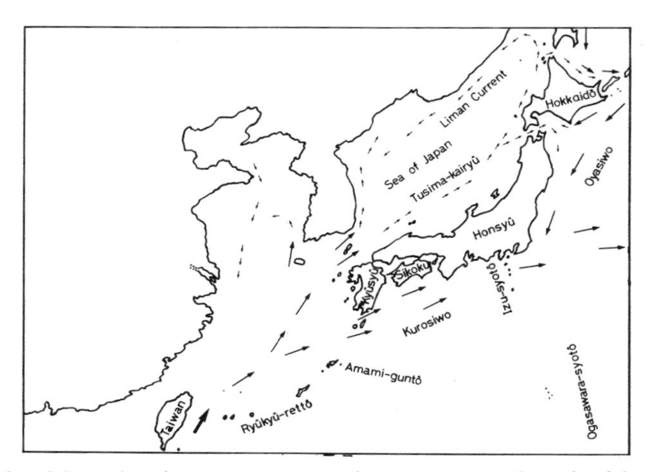

Fig. 2.—Map showing sea currents in summer and main islands of Japanese archipelago.

MORPHOLOGY AND TAXONOMY

This section is divided into parts dealing with the morphology of the mature thallus, and the morphology of spore germination and the early stages of development.

THALLUS MORPHOLOGY AND ITS VARIABILITY

It is generally known that the habits of seaweeds largely vary according to environmental conditions. It is particularly so in *Gelidium*. Okamura (1934a) and Segi (1959) made re-description of *G. amansii* based on Japanese specimens and holotype, respectively.

External morphology and taxonomy

Since Kylin (1923) established the Order Gelidiales, Fan (1961) proposed a new family Gelidiellaceae including the single genus *Gelidiella* on the basis of the absence of internal hyphae and lack of a sexual generation. In another family, Gelidiaceae, nine genera were recognized, *viz., Acanthopeltis, Acropeltis, Beckerella, Gelidium, Porphyroglossum, Pterocladia, Ptilophora, Suhria*, and *Yatabella*. Recently, Santelices & Montalva (1983) advanced an opinion that *Acropeltis* should be merged with *Gelidium*, and the new combination *Gelidium chilensis* (Montagne) Santelices et Montalva (incorporating the only species, *Acropeltis chilensis* Montagne, of the former genus) was proposed. Among the genera of the family, Fan (1961) distinguished *Gelidium* on the basis of its monopodial growth, the absence of an inner cortex in the thallus as with *Beckerella*, the marginal ramuli having smooth surfaces not imbricated on axis, the base of plant forming prostrate axes, cystocarps always bilocular with one or more ostioles on each surface, and the carposporangia not in short chains. Dixon (1973) placed the family Gelidiaceae (*sensu* Kylin, 1956) in the Nemaliales (= Nemalionales). Santelices (1974) has fully discussed the validity of the Order Gelidiales.

As a first step in identification, several books in colour have been published for Japanese seaweeds, *e.g.* Higashi (1934), Okada (1934), Segawa (1956), Arasaki (1964), Chihara (1970), and Gakken (1975). In addition, Asahi News Paper Co. (1978a,b,c) issued three numbers of an ecological coloured photo-atlas. Okamura 1900–1902, 1907–1909, 1909–1912, 1913–1915, 1916–1923, 1923–1928, 1929–1932, 1933–1937) published eight volumes of the icones of Japanese algae and these volumes are recognized as being of great value for scientists. The names of species and varieties of Japanese Gelidiales which have so far been described as Japanese records are listed in Table I. Of these, four (*Gelidium johnstonii, G. polystichum, G. pulchrum*, and *G. purpurascens*) whose occurrence was reported for Japan by Segi (1955, 1957) are those originally described from California as new species by Gardner (1927) and Setchell & Gardner (1937). Segi (1977) examined the type or authentic specimens of these species from California and claimed coincidence with his specimens. The similarity between Japanese and Pacific North American *Gelidium* species has been recently pointed out by Hommersand (1972). Dixon (1966) stated that

TABLE I

Names of species, varieties, and forms of the Gelidiales recorded from Japan to date: only the names given as new records and new species by Segi (1954) without any descriptions are omitted

---

Acanthopeltis japonica. Okamura, 1892.
Beckerella irregularis. Akatsuka et Masaki, 1983.
B. subcostata (Okamura). Kylin, 1956.
Echinocaulon rigidum (Vahl) Kützing (= *Gelidiella acerosa*). Okamura, 1932a.
Gelidiella acerosa (Forsskål) Feldmann et Hamel. Okamura, 1936.
G. ramellosa (Kützing) Feldmann et Hamel. Ichiki, 1958 (found on back of a turtle).
Gelidiopsis rigidum (Vahl) Weber-van Bosse (= *Gelidiella acerosa*). Okamura, 1912.
Gelidium amamiense. Tanaka, 1965.
G. amansii (Lamouroux) Lamouroux. Von Martens, 1866.
G. amansii f. *elatum*. Okamura, 1934a.
G. amansii f. *elegans*. Okamura, 1934a.
G. amansii f. *teretiusculum*. Okamura, 1934a.
G. amansii f. *typicum*. Okamura, 1934a.
G. capillaceum Kützing (= *Pterocladia capillacea*). Von Martens, 1866.
G. cartilagineum (L.) Greville. De Toni, 1895.
G. clavatum. Okamura, 1934a (= *Gelidium kintaroi*).
G. corneum (Hudson) Lamouroux. Von Martens, 1866.
G. corneum var. *capillaceum* Gmelin. Von Martens, 1866.
G. corneum var. *pinnatum* Turner. Von Martens, 1866.
G. corneum var. *pulchellum* Greville. Segi, 1957.
G. crinale (Turner) Lamouroux. Okamura, 1914.
G. decumbensum. Okamura, 1934a.
G. densum. Okamura, 1935b (non Gardner; = *Gelidium yamadae*).
G. divaricatum Von Martens. Okamura, 1900.
G. elegans. Kützing, 1868 (= *Gelidium amansii*).
G. isabelae Taylor. Tanaka, 1965.
G. japonicum (Harvey), Okamura, 1901 (= *Onikusa japonica*).
G. johnstonii Setchell et Gardner. Segi, 1954.
G. kintaroi (Okamura). Yamada, 1941.
G. latifolium (Greville) Bornet. Higashi, 1936.
G. linoides Kützing. Yendo, 1909.
G. nanum. Inagaki, 1950.
G. pacificum. Okamura, 1914.
G. polycladum Kützing. Kützing, 1869.
G. polystichum Gardner. Segi, 1955.
G. pristoides Turner. Von Martens, 1866.
G. pulchrum Gardner. Segi, 1957.
G. purpurascens Gardner. Segi, 1955.
G. pusillum (Stackhouse) Le Jolis. Okamura, 1902.
G. pusillum f. *foliaceum*. Okamura, 1934a.
G. pusillum var. *conchicola* Piccone et Grunow. Okamura, 1931.
G. pyramidale Gardner (= *Pterocladia pyramidale* Dawson). Segi, 1957.
G. repens. Okamura (non Kützing), 1899 (= *Gelidium pusillum*).
G. rigens Greville. Kützing, 1849.
G. rigidum (Vahl) Greville (= *Gelidiella acerosa*). Okamura, 1893.
G. subcostatum Okamura (= *Beckerella subcostata*). Okamura, 1894 (in Schmitz, 1894).
G. subfastigiatum. Okamura, 1934a.
G. tenue. Okamura, 1934a.
G. vagum. Okamura, 1934a.
G. yamadae (Okamura). Fan, 1951.
Onikusa japonica (Harvey). Akatsuka, 1986.
Porphyroglossum japonicum (Harvey). Schmitz, 1894 (= *Onikusa japonica*).
Pterocladia capillacea (S.G. Gmelin) Bornet et Thuret. De Toni, 1895.
P. densa. Okamura, 1934a.
P. nana. Okamura, 1934a.
P. tenuis. Okamura, 1934a.
Suhria japonica. Harvey, 1860 (= *Onikusa japonica*).
Yatabella hirsuta. Okamura, 1900.

Gelidium is one of the most polymorphic genera of the Florideophyceae and identification is notoriously difficult. He also claimed that a complete revision of the genus is highly desirable and such a revision demands a knowledge, on a global scale, of the range of form and of the intergrades between currently accepted taxa. Segi (1955, 1957) did not notice the plasticity of species of this genus and reported the four species previously mentioned from Japan without careful identification. Dixon & Irvine (1977a) have claimed that the work of Segi (1963) has almost no meaning.

Lamouroux (1813) transferred *Fucus amansii*, which had been described previously by himself (Lamouroux, 1805), to *Gelidium*. Von Martens (1866) recorded specimens collected from Simoda, Yokohama, and Nagasaki. Descriptions of *G. amansii* given by Kützing (1849) and Lamouroux (1805) were too brief to define fully the entity. Therefore, Okamura (1934a) made a new diagnosis in order to re-define *G. amansii* after the first description in the icones (Okamura, 1913–1915). In his diagnosis he listed four forms within the species. These were f. *typicum*, f. *elatum*, f. *elegans*, and f. *teretiusculum*. Afterwards, var. *latioris* was added from a collection in northern Taiwan (Okamura, 1935b).

Okamura (1934a, 1936) stated that *G. amansii* is closely related to *G. subfastigiatum* and *G. tenue* and that to distinguish from these two species is frequently difficult. A form of *G. amansii*, f. *teretiusculum*, growing in the deep water habitats also resembles *G. linoides* Kützing, but the latter can be distinguished on the bases of its alternate-dichotomous ramification. De Toni (1897), however, listed *G. linoides* as a doubtful species. *G. pacificum* is another species frequently difficult to separate from *G. amansii*. Dixon (1963) stated that "it can be argued that a species is a pattern of development. The study of speciation requires information at two levels, the morphological and the cytological. Although morphological studies are of secondary importance, they must not be dismissed as being of no significance." *G. japonicum* and *G. pristoides* are very similar in appearance, but they show some differences in microscopic characters (Akatsuka, 1983).

Notwithstanding, the morphological delimitation of species must be first established in the study of speciation. The criteria adequate throughout the species of the genus, especially measurable characters, must be discovered. The presence of the taxonomic criteria, with which even an expert on the Gelidiales finds it difficult to identify the species, is the problem. The delimitation of species of *Gelidium* and *Pterocladia* is being studied by Akatsuka; Japanese representatives of *P. tenuis* and *P. densa* were described as being conspecific with *P. capillacea* by Stewart (1968) but her samples were few and small.

Many of the species enumerated as Japanese by Levring, Hoppe & Schmid (1969) are not found in the country; their bibliographical sources were industrial or chemical literature.

Akatsuka is at present using suitable criteria, *e.g.* the width, thickness of the middle region of the main axis and their ratio in his study of Japanese *Gelidium*. These criteria were chosen to be of significance together with various other characters including surface cell morphology as described in the literature (Akatsuka, 1982d).

Light microscopic structures

Algae are the Thallophyta and have not any special organs. Therefore, only a few special structures have been known. In this subsection, the special structures are only mentioned.

Vegetative structures. Dixon (1958b) and Dixon & Irvine (1977b) have described the species of *Gelidium* and *Pterocladia* occurring in the British Isles. In British species, cystocarps are rare in *Gelidium* and have never been found in *Pterocladia*, so that the use of cystocarpic structure for the discrimination is meaningless. Faced with sterile or tetrasporangial specimens, one can do little more than examine the external form. Although Segi (1955, 1957) illustrated many transverse sections of the Japanese species of *Gelidium*, no discussion was given on the comparative aspects and their application to the taxonomy. Several authors (Okamura, 1934a; Feldmann & Hamel, 1936; Loomis, 1949; Dixon, 1958b; Fan, 1961) examined the distribution of internal hyphae and the compactness of the medullary region, but the instability of these characters rendered them of little use in distinguishing *Gelidium* and *Pterocladia*. The distribution of the hyphae also varies among plants from a given locality (Stewart, 1968) and its use as a mean of distinguishing these two genera is no longer accepted (Santelices, 1974). Sohn & Kang (1978) have, however, claimed the usefulness of the distribution of hyphae in distinguishing *Gelidium divaricatum* and *G. microphysa* from other Gelidiaceae. The distribution of pterocladian hyphae was studied by Felicini & Perrone (1986).

Akatsuka (1970a, 1981) found the shape and orientation of surface (=outermost cortical) cells useful in distinguishing between *Gelidium* and *Pterocladia*. These surface cell characters also make it possible to distinguish or group all genera of the Gelidiaceae (Akatsuka, 1986). Akatsuka (1983) confirmed the use of surface cell characters in distinguishing a number of samples from various countries of the world, so that it is useful not only for Japanese but also for British and material from other regions. Very juvenile thalli of *Gelidium* are difficult to identify, but not so those of *Pterocladia* by the use of surface cell characters.

Although internal hyphae occur in all the Gelidiaceae, their function is unknown. Feldmann & Hamel (1936) assumed their rôle to be to provide structural support for the thallus, but no critical study of the ultrastructure or function has yet been made. That each apical cell of the thallus cuts off only two pericentral cells has been found by Papenfuss (1966) as a characteristic feature of the Gelidiales. Dixon (1963) commented that as in *Pterocladia*, the tips of the axes of *Gelidium* are cylindrical in summer and spatulate in winter. The length and extension of the prostrate axis and the number of the erect axes produced from *G. amansii* are unknown.

Akatsuka (1978) briefly described the cuticle and hairs of seven species of the Gelidiales including *G. amansii* and claimed their potential usefulness for both identification and systematics. He assumed that cuticular types may be useful, but further surveys are necessary to provide statistical significance; electron microscopic studies are also required. The idioblast of *Gelidiella* is the mother cell of a hair (Akatsuka, 1982b) Dromgoole & Booth (1985) noted the hairs in *Gelidium caulacantheum* J.Ag.

Reproductive structures. No comparison of the cystocarpic or tetrasporic structures of Japanese Gelidiaceae has yet been made. The male reproductive cells of *G. amansii* and other species of the Gelidiales have been described (Akatsuka, 1970b, 1973b, 1979). For comparison, the insufficient data on width and height of the male frond, size of antheridia, and thickness of mucilage layer (see Fig. 20) are summarized in Tables II and III. It is uncertain whether the biometry of these characters of the male plant shows intergeneric differences, due to the sparse data.

Although Tazawa (1975) distinguished several types in the developmental processes of the Floridean antheridial mother cells, his literature survey was insufficient; he attempted to classify the Florideae on the basis of the male reproductive structures. The fundamental structure of the reproductive organs of *G. amansii*, except for numerical aspects, is the same as for other species of *Gelidium*.

TABLE II

Measurements of antheridia and male thalli (from Akatsuka, 1970b, 1973b, 1979): a, antheridia are shown in Fig. 20; b, middle region; c, basal region; d, portion outside the antheridial layer, shown in Fig. 20; e, correction of Akatsuka (1979); f, dried specimen

	Antheridial length in sectional view (μm)[a]	Male thallus (erect axis)			Thickness of mucilage layer (μm)[d]
		Height (mm)	Width (mm)		
			Maximum[b]	Minimum[c]	
G. amansii	2–3 × 3–5	60	2	0·8	4
G. pacificum	2–3 × 2–5	30–90	3(1·8[f])	0·8	7
G. pusillum	2–3 × 3	2	—	—	5
G. japonicum	2–3 × 3–5[e]	30–85	4·0	2·5	—
P. tenuis	2 × 4	20–100	—	—	—
B. subcostata	2–3 × 2–5[e]	—	—	—	—

TABLE III

Remarks on male ramulus and ramification (from Akatsuka, 1970b, 1973b, 1979): a, correction of the value presented in Akatsuka (1979)

G. amansii:	dicho- to tetrachotomous, occasionally single
G. pacificum:	extremely ramified from single base
G. pusillum:	poorly ramified, male patches also on the main axis and branches
G. japonicum:	up to 1·5 mm length, 150–250 μm width
P. tenuis:	single ramulus, not specially disposed, but warped to one side, and larger than sterile ramuli; male patches are also formed on main axis and primary branches
B. subcostata:	length 3 mm; width 250–350 μm[a]

Ultrastructure

Hara (1972) and Hara & Chihara (1974) stated that the chloroplast of *G. amansii*, *G. linoides*, *G. japonicum*, *Pterocladia capillacea* (=*P. tenuis*) and *Gelidiella acerosa* is a 'P-type'. The features of P-type chloroplasts are pyrenoid absent, outermost lamellae present, bundles of tubular lamellae absent, shape of chloroplasts discoid or parietal, numbers of the chloroplasts in each cell many, site of the chloroplasts within a cell parietal. Compared with the ultrastructure of the chloroplast of the Bangiophycidae and Nemaliales of Florideophycidae, it was suggested by these authors that the fine structure of advanced Orders (including the Gelidiales) are systematically stable.

Electron microscope studies have not been presented regarding the relationship between the cell wall and the chemical constituents of the intercellular matrix in members of the Gelidiales.

Histochemistry

Okamura & Ôisi (1905) histochemically tested *Gelidium amansii* using classical staining with Methyl Blue (non-Methylene Blue) and other dyes. After boiling, *G. amansii* lost its intercellular mucilage and they concluded that the intercellular substance was agar; their staining method was, however, non-specific for agar. Hayashi & Okazaki (1970) suggested that the agar is extracted from the outer layer of the cell wall and that this layer has a rôle as a dialysis membrane which serves as an ion exchanger. Akatsuka & Iwamoto (1979) applied staining methods of animal histochemistry to develop a specific test for agar. They employed Toluidine Blue (TB) O, a metachromatic dye, and Alcian Blue for histochemical detection of sulphate groups bound to the galactan of agar. They found that in *G. pacificum*, a closely related species of *G. amansii*, the agar-binding sulphates are localized in the intercellular spaces and the outer layers of parenchymatous cells, except for internal hyphae. By employing the periodic acid-Schiff (PAS) reaction and $ZnCl_2$–I method, they concluded that cellulose is present in the inner layer of parenchymatous cells and in almost the entire layer of internal hyphae. From these facts, it has been recognized that the walls of parenchymatous cells of *G. pacificum* are histochemically composed of two layers. In their study, it was assumed that PAS and TB do not interfere with each other, because the agar does not react with periodic acid (Barry, Dillon & McGettrick, 1942). These methods which give consistent results with *G. pacificum* could be applied to *G. amansii*. It would be interesting to determine whether or not the localization of agar and cellulose varies in the various developmental stages of *G. amansii* and in the related taxa of the Gelidiales. Akatsuka (unpubl. data) found histochemical peculiarity in the cell walls of *Gelidiella* (*G. acerosa*) and scarcity of the intercellular matrix.

Floridean starch is known to be an important storage substance in the Rhodophyta. If the gradual disappearance of the floridean starch occurs in the tissue of *Gelidium amansii*, it would be of interest to study sugar metabolism throughout the life of this species. No report on the variation of the amounts of floridean starch and on histochemical methods for the detection of floridoside have been presented. Ueda (1972) studies fluorochromes for the detection of the nucleus, chromosomes, DNA or chloroplasts in various cells of *G. amansii* as tabulated (Table IV) and treated the results taxonomically. "Fluorescence taxonomy" was proposed on the basis of correspondence of his results with present brown and red algal systematics.

TABLE IV

Results of fluorescence examination of G. amansii *(from Ueda, 1972)*

	Cell wall	Protoplasm
Primary fluorescence in:		
distilled water	Negative	Pale orange
Secondary fluorescence in:		
Acridine orange	Pale red or orange	Yellow
Coriphosphin	Orange, including cuticle	Yellow
Al-Morin	Blue, including cuticle	Green
Rhodamine B	Blue in wall; red in cuticle	Orange

Cytology

The average size of nuclei in the Rhodophyta is of the order of 3 μm (Dixon, 1973). The diameter of nuclei of tetrasporangial mother cell of *Acanthopletis japonica* measured 3·5 μm (Kaneko, 1968). Yabu (1975) compiled cytological literature on Japanese algae. Although chromosome numbers of two Japanese species of the Gelidiales were counted (Kaneko, 1966, 1968), that of *Gelidium amansii* is unknown; those known in the Gelidiales are summarized in Table V. Kaneko (1968) expected that the basic chromosome number of the members of the Gelidiales is five on the basis of these literature values, but to verify this more accurate counts of chromosomes are required.

TABLE V

Chromosome numbers of gelidialean algae

Species	Chromosome number	Reference
G. *corneum*	$n=4$–5, $2n?=9$–10	Dixon, 1954
G. latifolium	$n=4$–5, $2n=?9$–10	Dixon, 1954
G. latifolium (tetraspores)	$n? \approx 18$	Boillot, 1963
G. latifolium var. *luxurians*	n or $2n=25$–30	Magne, 1964
G. vagum	$n=7$–10	Kaneko, 1966
A. japonica	$n=15$, $2n=30$	Kaneko, 1968
Gelidiella acerosa	$n=4$, $2n=8$	Rao, 1974

Rao (1974) stated that on the basis of his counting of chromosome numbers, *i.e.* four in tetrasporangium and eight in vegetative cell, tetrasporangia of *Gelidiella acerosa* do normal meiosis. He concluded that there was a haploid sexual phase which has been overlooked. However, to conclude on the life cycle of *Gelidiella*, the amount of DNA should be determined in vegetative cell and tetraspore. For example, Goff (1984) has revealed that for any vegetative cell type of *Wrangelia plumosa*, an increase in cell size is linearly correlated with an increase in cell ploidy; there is no difference in the amount of nuclear DNA in similarly-sized cells of gametophyte and sporophyte plants.

DEVELOPMENTAL MORPHOLOGY

In the morphological studies of the developmental process of the gelidialean algae, the stages from the germination of tetra- or carpospores to early development are known, but there has been lack of success in studies on completing their life cycle in laboratory culture. Moreover, long-term observations on the sea bottom have not been successful because of technical difficulties.

Germination

Studies of the germination of spores of the Gelidiales were first carried out by Killian (1914) and Kylin (1917). Those of *Gelidium amansii* and other species of Gelidiales have been repeatedly studied by many investigators, such as Ôno (1927, 1932), Inoh (1941, 1947), Ueda & Katada (1943), Takamatsu (1944), Katada (1949, 1950, 1955), Katada *et al.* (1953), Katada & Matsui (1954), Yamasaki (1960, 1962), Chihara & Kamura (1963), Yoshida & Yoshida (1965), and Kaneko (1966). As stated in Umezaki's review (1966), all the species of Japanese Gelidiales examined show the "Gelidial type" germination (the term after Inoh, 1947); but the germination of spores of *Beckerella subcostata* have not been reported. The feature of the spore germination or early development of the Gelidial type is that the first unequal-sized division of the sporelings results in the formation of upper small and lower larger cells. Germination shows the same sequence in both tetraspores and carpospores (Katada, 1955; Yamasaki, 1962). Ôno (1927, 1932) made detailed observations on the release and germination of *Gelidium amansii* under the conditions of laboratory culture.

Both kinds of spores are polygonal inside the sporangia but after release become globular measuring 30–(35)–40 μm (see Umezaki, 1966, p. 54 for a comparison of the spore diameters of various species of the Gelidiales). Sedimented spores are filled with contents and have a single nucleus. Although spores show a red colour in their centres, their margins are colourless. Experiments have shown that in autumn most spores die after release, but those released in summer show both germination and development. Carpospores after release show amoeboid movement and have no cell wall before adhesion to the substratum (Suto, 1954). The spores seemed to be released after the destruction of the cortical layer of sporangial stichidia and settle at a rate of 1 mm/20 s. Suto (1954) also observed that the number of spores counted in the natural dense population of *G. amansii* was 10^2/l and the time used for fixing was 5–20 min ($\bar{x}=10$) in *G. amansii*, as with *Pterocladia tenuis*, *Gelidium pacificum*, and *Acanthopeltis japonica*. Okuda & Neushul (1981) measured the size and sinking rate of 11 species of red algal spores. From their data, it has been concluded that the larger spores sink faster than smaller ones, as would be expected, and that the variability of pore sizes is species specific. Ogata (1952) experimented on the attachment potentiality of spores and showed that those of coralline algae became attached to slide glasses, but those of *Gelidium* did not. These results are of interest in relation to "the isoyake" phenomenon discussed on page 246.

Development of the sporelings

Ôno (1927, 1932) first described the development of the erect axis from sporelings of *G. amansii*. In the natural habitat the sporelings form several haptera from which prostrate axes develop to give lawn-like networks. Following this development, the prostrate axes bear younger tips here and there which become the erect axes. Therefore, it is commonly found that a thallus clump is composed of many individual fronds originating from a single spore. In contrast to this, direct development of the erect axes from the sporelings was occasionally observed. Ueda & Katada (1943) cultured the sporelings of *G. amansii* under natural conditions. When the length of the prostrate axes had reached about 4 mm and was fairly well ramified, the axes bore the erect branches to a height of 2 mm. Their observations agree with those of Ôno (1927) in that the erect axes occur after the establishment of networks of prostrate axes. Takamatsu (1944) described the development of the sporelings from fragmented information obtained from observations carried out on the sea-bed. Contrary to the previous authors, he stated that the sporelings of *G. amansii* elongated erectly and, on most occasions, produce lateral axes. Generally, the lateral axes elongate and become the erect axes, which are the fronds of *G. amansii*. Otherwise, the thinner lateral axes elongate horizontally and become the prostrate axes. When circumstances are unsuitable, the sporelings may first produce the prostrate axes and the erect axes then develop after haptera formation. Ueda & Katada (1949) and Katada (1955) carried out an experimental study on the development of the sporelings, employing laboratory cultures, artificial tide-pools, and submerging of rocks within the natural population of *G. amansii*. They confirmed that the lateral branches (this term after Takamatsu, 1944) which developed from the basal part of the sporelings develop either erectly ("erect type") to form erect axes or horizontally ("prostrate type") to form the prostrate axes. The prostrate type is frequently found on steep or overhanging surfaces of substrata or embayed areas. Yamasaki (1960, 1962) pointed out that the lateral branches were derived from the larger cell produced by the first unequal-sized division of spores. He also agreed with Takamatsu (1944) and Katada (1955) in that the sporelings of *G. amansii* usually show the "erect type" development. He claimed that the type of the development of sporelings may be related to the depth at which the *G. amansii* is growing.

AUTOECOLOGY

Autoecology of seaweeds is an important discipline giving a basis for marine agronomy or their mariculture. Among Japanese seaweeds, the gelidiacean algae as well as *Porphyra* and some brown algae have been the subjects of experiments, observations and field surveys.

DISTRIBUTION

There are many reports related to the distribution of the Gelidiales, including those of the Japanese Prefectural Fisheries Experimental Station.

Geographical and topographical distribution

Akatsuka (1982c) collected new and existing information of the geographical distribution of each species of the Gelidiales in Japan and its vicinity. From his analysis, some features of the distribution of each species have been found. Geographical records of species of the Gelidiales are compiled in maps (Figs 3–13). The records presented in these maps have been selected from literature listed in Akatsuka (1982c) and my own collections with the exception of doubtful records and possibly extinct occurrences at the present day; the records given only in the names of ancient Provinces by Okamura (1934a, 1936) are plotted around the centres of coastlines of the Provinces.

Gelidium species occur very commonly around the coasts of Japan, except for Okhotsk and the Pacific coast of Hokkaidô (Ôno, 1929; Okamura, 1934a, 1935). *G. amansii* is absent on the east coast of Hokkaidô between Kitmai-Esasi and Harutati (Fig. 3). Kang (1966) reported the distribution of *G. amansii* on the entire coast of the Korean peninsula. *G. amansii* also occurs on the Yellow Sea coast of China and in southern China (Tseng & Chang, 1959). According to L. P. Perestenko (pers. comm.), "the Gelidiales were not usually found in U.S.S.R. waters outside

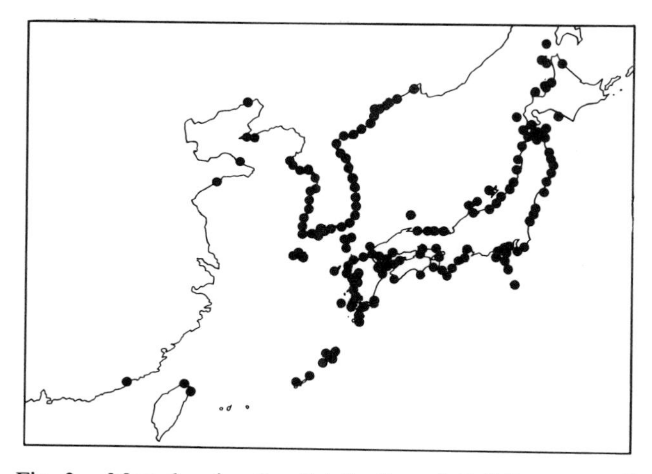

Fig. 3.—Map showing the distribution of *Gelidium amansii*.

the Bay; *G. vagum* is a common species in the Bay; *G. amansii* and *G. pacificum* have only been once collected there; it is possible that *G. pacificum* collected in 1923 was brought accidentally from somewhere into the Bay." There are some questions about the identification of this *G. vagum* on the basis of my own examination of specimens provided by Dr Perestenko. This limited distribution of the Gelidiales on the U.S.S.R. coast could be because the cold Liman Current washes the coast. Although Segi (1957) recorded *G. corneum* var. *pulchellum*, which is at present treated as *G. pulchellum*, this species in the British Isles was referred to *G. pusillum* by Dixon & Irvine (1977b). I refer Segi's specimen to *G. amansii* on the basis of his figure of its habitat. Although *G. amansii*, the commonest species of Japan, is known from warm seas in the world, it has also been recorded from subtropical and tropical seas showing the maximum surface temperatures (28–29 °C) in summer, *e.g.* Ryûkyû, northern Taiwan, Penang Island of the Malay peninsula, Indonesia, the Philippines, Thailand, Kalimantan, and Sulawesi (Table VI). *G. elegans* Kützing (=*G. amansii*, Okamura, 1934a) listed by Okamura (1932b) from the Bering Sea is incorrect. On the assumption that the identification of the specimens described in the reports cited in the Table is correct, light intensity rather than surface temperature seems to affect strongly the distribution of *G. amansii* in subtropical and tropical seas, especially in coral reefs.

G. subfastigiatum, a species very closely related to *G. amansii*, occurs in the region from the Sea of Japan coast of Hokkaidô to the Miyagi-ken coast of Pacific Honsyû (Okamura 1934a, 1936). Okamura suggested that the origin of *G. subfastigiatum* was in northern Japan, but Akatsuka (1982d) concluded that *G. subfastigiatum* is conspecific with *G. amansii*.

Nakaniwa (1975) claimed that Izura (Ibaraki-ken) is the northern limit of the distribution of *G. pacificum*. Investigations should re-examine the existing records of this species (Fig. 4) in the localities outside the regions between Wakayama- and Tiba-ken described by Okamura (1934a,b, 1936). Especially, the records from southern Korea and the Bay of Peter the Great

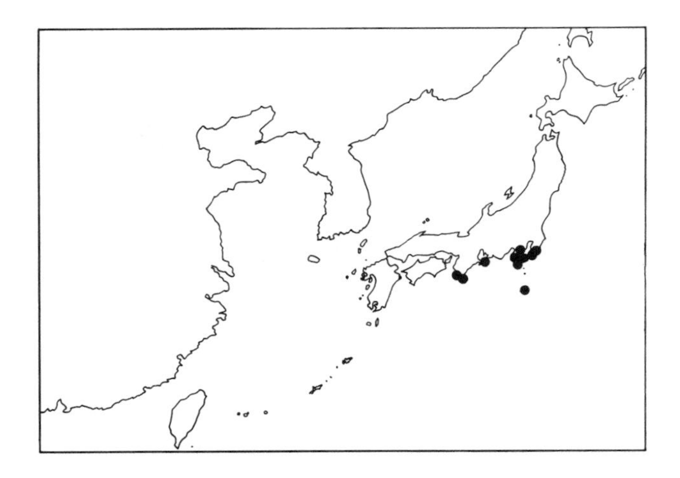

Fig. 4.—Map showing the distribution of *Gelidium pacificum*.

are doubtful. I saw specimens collected in southern Korea and sent from Dr J.-W. Kang; these specimens are intergrades between *G. pacificum* and *G. amansii* (unpubl. data). In contrast, the occurrence of the species in Muroto- and Asizuri-misaki Capes, Miyazaki- and Kagosima-ken can be expected on the basis of their surface temperature. A report (Miyazaki FES, 1939) has described the occurrence of the species along the entire coast of Miyazaki-ken, but my visit to the coast gave negative results and it was probably a form of *G. amansii*. The morphological continuity between *G. pacificum* and *G. amansii* should be studied, on the basis that some intergrade specimens of these two species have been collected (Akatsuka, 1982d). It is interesting that some Venezuelan specimens of *G. serratum* I have seen strongly resemble typical specimens of *G. pacificum* (unpubl. data).

Okamura (1936) described the range of distribution of *G. vagum* from the western coast of Hokkaidô to the Bay of Wakasa-wan on the coast of the Sea of Japan and to Hukusima-ken on the Pacific coast. In addition, dispersed records have been presented from central Japan, western Kyûsyû, Yellow Sea, and southern Korea (Fig. 5). In my opinion this species has a close relationship to Californian *G. johnstonii* Setchell et Gardner.

G. pusillum has been recorded from the northern part of Taiwan to Aomori-ken on both Pacific and the Sea of Japan coasts (Fig. 6). No record of this species seems to be reported from Hokkaidô, China other than the Yellow Sea area, and the coast of the U.S.S.R. But I have heard of its occurrence in Hokkaidô from Mr Fujita of Hokkaidô University. The U.S.S.R. coast of the Sea of Japan showing below 2 °C surface temperature in February is inadequate to the life of *Gelidium* on the basis of the conclusion by Kinosita (1942, 1944, 1949) that in Hokkaidô the *Gelidium* is absent in regions where for some months the monthly mean surface temperatures calculated from values of ten years are lower than 2 °C (in other words, annual mean surface temperature lower than 10 °C for ten years). But this theory can be applied to the case of harvestable amounts only, on the basis of records of *Gelidium* sp. from Kitami-Esasi on the Okhotsk

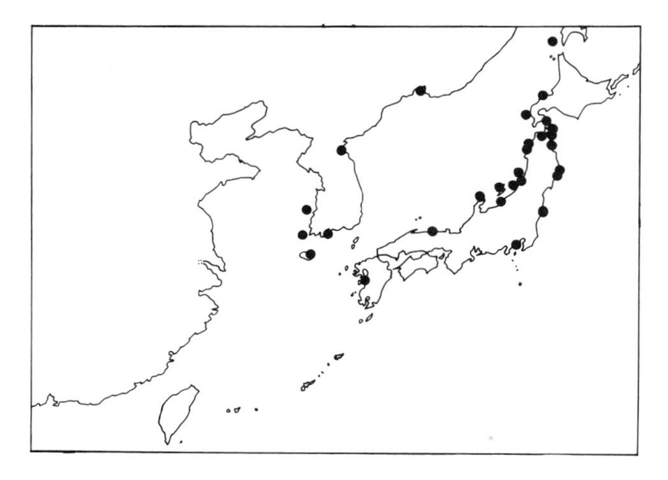

Fig. 5.—Map showing the distribution of *Gelidium vagum*.

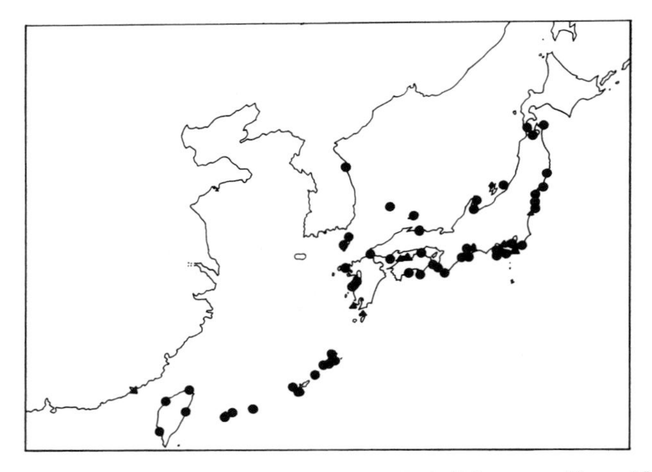

Fig. 6.—Map showing the distribution of *Gelidium pusillum* (●) and *G. crinale* (▲).

coast of Hokkaidô. The absence of the species in the Bay of Peter the Great is because the land-ice limit in the Bay in January and February is the same as that of Okhotsk coast of Hokkaidô. But it is interesting that in these two regions occur some species (*G. amansii*, *G. vagum*, *G. pacificum*?, *Gelidium* sp.) of *Gelidium* other than this species. The problem is that *G. pusillum* has less tolerance to low temperature in comparison with *G. amansii* and related species recorded from Okhotsk and the Bay of Peter the Great.

G. crinale of the British Isles was treated as referable to *G. pusillum* on the basis of continual observations at the same place (Dixon & Irvine, 1977b). It can, however, not be said that the same treatment is justified in Japanese populations of this species, because of the scarcity of records of its collections (Fig. 6).

Okamura (1936) stated that Japanese *G. divaricatum* is distributed from Taiwan to Mutu-ôma-zaki of Aomori-ken along the Pacific coast. Records from other regions have now been added (Fig. 7).

Okamura's (1936) descriptions on the distribution of *Onikusa japonica* (= *Gelidium japonicum*) is added to the records from the west coast of Kyûsyû (Fig. 8). Nakaniwa (1975) indicated that the northern limit of distribution of the species is Ôarai, Ibaraki-ken. It is a question that in spite of the obvious warm-water adaptation of this species showing similar range of distribution to *Acanthopeltis japonica*, there is no record from Tusima Island located offshore of the northern Kyûsyû and Cheju-do Island of southern Korea. In consideration of both its distribution in northern Kyûsyû and the direction of the Tusima (warm) Current, the occurrence of *Onikusa japonica* in Tusima, Cheju-do and even in the southern coast of Korean Peninsula can be expected. No record of the species has been reported from the Ryûkyû Rettô Archipelago beyond south of the Amami-Guntô Islands.

Okamura (1936) stated that *Pterocladia tenuis* has spread from Ryûkyû to the western coast of Hokkaidô. A record from the surface of the coral

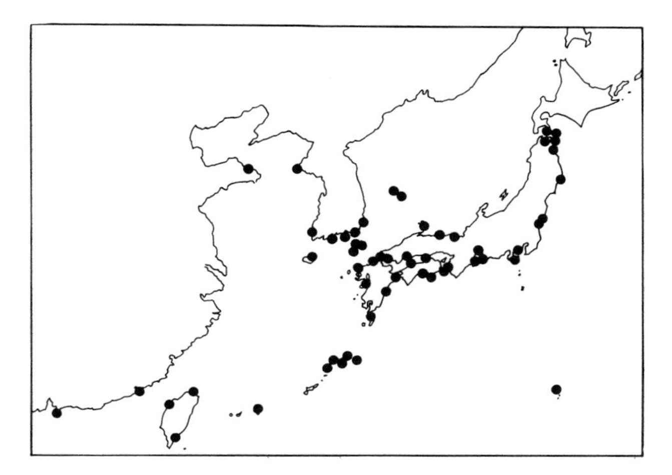

Fig. 7.—Map showing the distribution of *Gelidium divaricatum*.

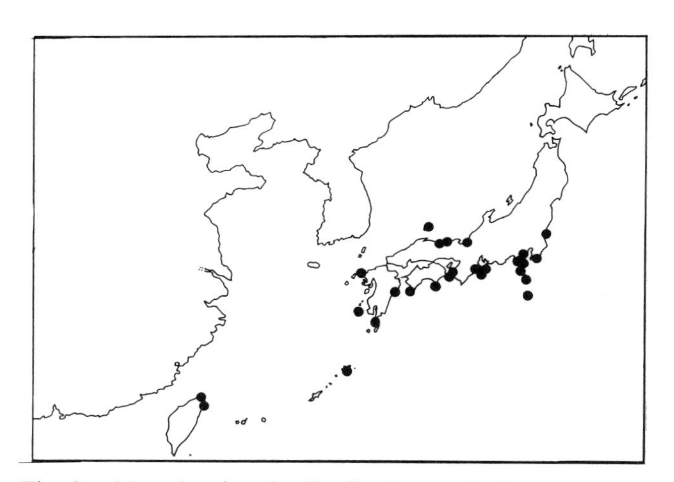

Fig. 8.—Map showing the distribution of *Onikusa japonica*.

reef in Isigaki-zima Island of the southernmost groups of Ryûkyû has been added (Fig. 9). The records of this species and *Gelidium amansii* overlap each other on the western coast of Hokkaidô. Although the latter has been recorded from the Bay of Peter the Great, the former species lacks any record in the Bay. These facts mean that the difference between these two species is distribution in the Bay is because of the adaptability of *Pterocladia tenuis* to warmer waters and the lower (below 2 °C) surface temperature in winter in the Bay than that of west of Hokkaidô.

The records of *Pterocladia nana* are few (Fig. 10). This is because of the difficulty of the delimitation between this species and *P. tenuis*.

The distribution records of *Beckerella subcostata* (Fig. 11) are few because of its deep-water habitats. Okamura (1936) described the Pacific

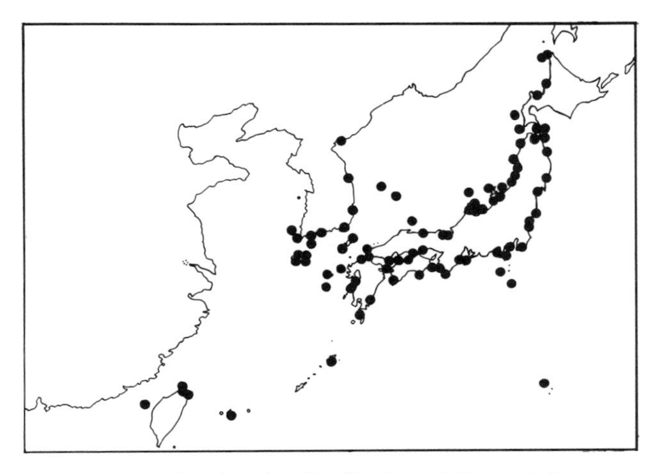

Fig. 9.—Map showing the distribution of *Pterocladia tenuis*.

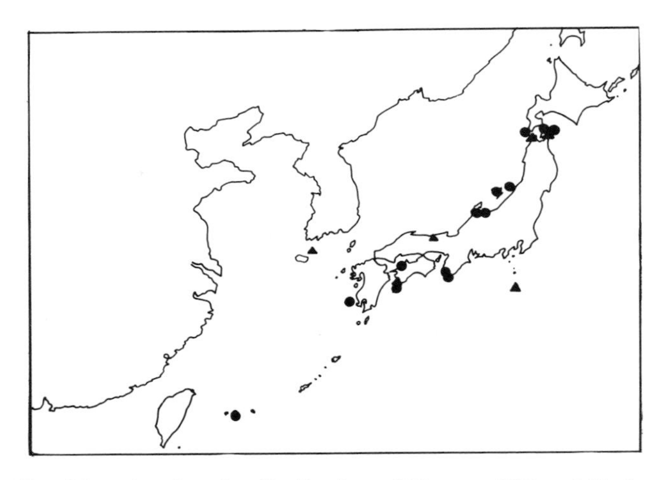

Fig. 10.—Map showing the distribution of *P. nana* (●) and *P. densa* (▲).

coast between Kyûsyû and Tiba-ken, Hatizyô-zima Island and Hakodate as its range of distribution. The occurrence in Hakodate is, however, doubtful. There are additional records from western Kyûsyû (Migita, 1985, pers. comm.) and Tugaru-hantô Peninsula (Nanao, 1974); the isolated record of the latter must be confirmed by re-collection. This species shows apparent restriction of warm waters. Further collecting activities using SCUBA diving are interesting on the north coast of Kyûsyû and Cheju-do Island located at the mouth of the Sea of Japan, in which the warm Tusima-Kairyû Current is flowing.

Okamura (1900–1902) described the distribution of *Acanthopeltis japonica* in the regions between Miyazaki-ken and Tiba-ken. This species is found in warm seas as seen in his and other additional records (Fig. 12). The existing record from Cheju-do suggests the possibility of the occurrence of

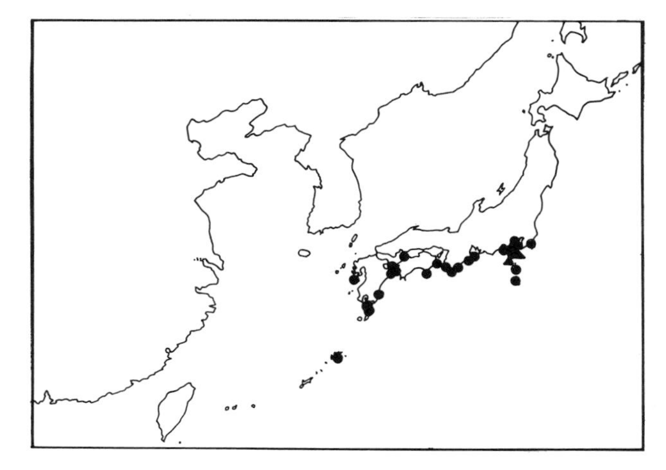

Fig. 11.—Map showing the distribution of *Beckerella subcostata* (●) and *B. irregularis* (▲).

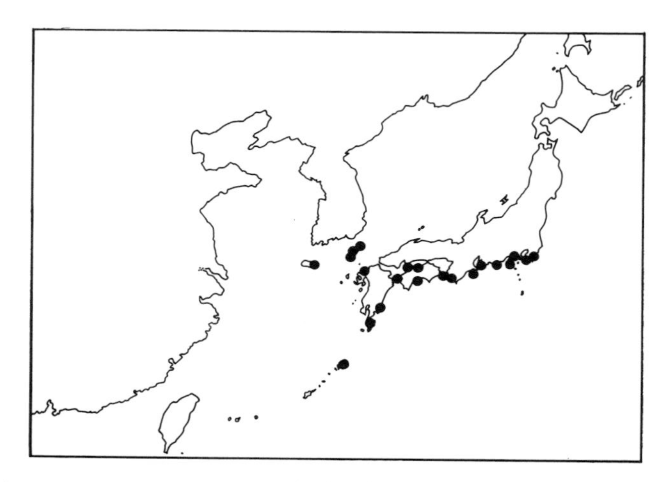

Fig. 12.—Map showing the distribution of *Acanthopeltis japonica*.

this species in the Sea of Japan, especially on Oki Island, Noto-hantô Peninsula, and Sado Island. Careful collections using SCUBA diving are needed for a better understanding of the distribution in the Sea of Japan. The species shows a range of distribution almost overlapping with that of *Onikusa japonica*, but no record from the northern part of Taiwan. The absence of a record of *Acanthopeltis japonica* from Taiwan is possibly because this species has sublittoral habitats, differing from *Onikusa japonica*. Careful surveys in the sublittoral zone are needed on the northern coast to confirm the absence of the species in Taiwan.

Gelidiella acerosa being apparently a warm-sea species (Fig. 13) has been widely recorded from the islands beyond the south of Kyûsyû and the islands and capes washed by Kurosiwo. The occurrence on Sima-hantô

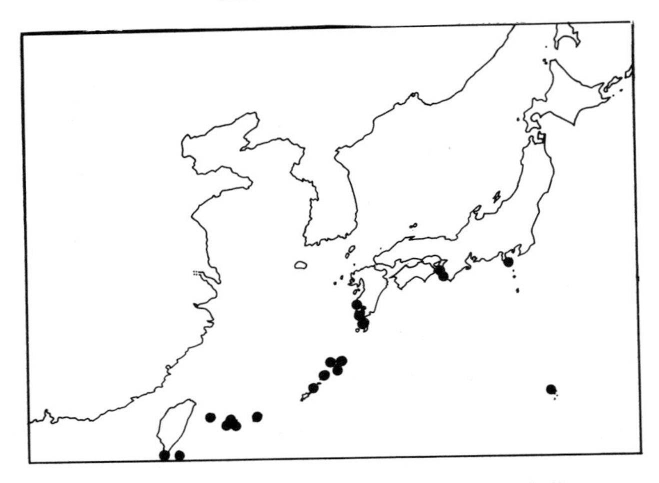

Fig. 13.—Map showing the distribution of *Gelidiella acerosa*.

Peninsula, Asizuri- and Muroto-misaki Capes, Miyazaki- and Kagosima-ken, and each island of Izu-syotô Islands are likely on the basis of the record from Kii-sirahama which is away from the axis of flow of Kurosiwo. Distinct differences in marine algal distribution have been found between northern and southern Taiwan. In northern Taiwan, *Gelidium* and *Pterocladia* having considerable sizes grow abundantly (Fan, 1951), but in southern Taiwan *Gelidiella acerosa* has only been recorded among genera or species of the Gelidiales (Chiang, 1973).

The distribution of species not plotted in Figures 3–13 cannot be discussed in detail, because the records are too few. Although *Gelidium corneum* has been recorded from Nagasaki-ken (Ichiki, 1958), it has been omitted on distribution maps in this review on the basis of the complexity of its entities.

Okamura (1934a,b, 1936) described the range of distribution of *G. linoides* between Wakayama- and Tiba-ken. De Toni (1897) doubted the independency of *G. linoides*. Okamura (1934a) recognized the similarity of this species to *G. nudifrons* Gardner and *G. arborescens* Gardner, but he described distinguishing criteria *e.g.* shape of transection and substance. This species is difficult to distinguish from the plants of *G. amansii* collected in deep habitats.

G. kintaroi appears to be a species distributed in southern waters on the basis of the collection by Okamura (1934a) from Taiwan and Amoy and by Fan (1951) from the northern Taiwan; but Takamatsu (1938) reported this species from Iwate-ken, northern Honsyû of Japan. Iwate-ken washed by Oyasiwo is the coldest region in Honsyû of Japan. In spite of the presence of a regular arrangement of the ultimate pinnules resembling that of *G. tenue* and *G. vagum*, the way in which Takamatsu (1938) identified his specimens was not mentioned by him.

G. yamadae has been recorded by Okamura (1935b) and Fan (1951) from three localities of Japan and one of northern Taiwan. Many more collections are needed in order to discuss the distribution of this species.

The above discussion on a species level has been presented chiefly on the basis of literature. Therefore, precise conclusions on the distribution of *Gelidium* and *Pterocladia* become possible after the completion of studies on the delimitation of species and the re-identification of specimens used in the descriptions in the existing reports. On the basis of Figures 3–13, all Japanese species of the Gelidiaceae (*sensu* Fan, 1961) are distributed along the coasts from the second to fifth regions of Japanese marine algal distribution distinguished by Okamura (1926); they do not occur in the first, *i.e.* subarctic, region. The Gelidiellaceae is found in his third, *i.e.* subtropical, region.

The distribution of *Gelidium* in Hokkaidô has been described and discussed by several authors (Ôno, 1929; Kinosita, 1939, 1942, 1943, 1944; Kinosita & Sibuya, 1941a,b). According to them, *Gelidium* was abundantly harvested in Risiri-tô Island, Masike, Osyoro, Yoiti, Hurubira, Iwanai, Utasutu, Simamaki, Matumae, and Hakodate. *Pterocladia* in Hokkaidô has a dwarf thallus and an agar content of low quality. It also interferes with the propagation of *Gelidium* because of its occurrence in the same habitats (Ôno, 1927). *Gelidium* species including *G. subfastigiatum*, *G. amansii*, and *G. vagum* in Hokkaidô are distributed in the region from Usu, Itanki-hama and Ottyoku-hama of Muroran (Nakamura & Ito, 1956) to Risiri- and Rebun-tô islands *via* the Tugaru-kaikyô strait. Kinosita (1942, 1944) claimed that the distribution of *Gelidium* in Hokkaidô corresponds to the region along the Tusima (warm) Current; he assumed this temperature to be necessary to *Gelidium*, because no *Gelidium* occurred in the region where water temperature drops to 2 °C and below. The lowest temperature rather than the yearly average was stressed as a limiting factor for the distribution of *Gelidium*. As seen in Figure 2 and the illustration of Kinosita (1942), the warm current reaches to offshore of the Okhotsk and Pacific coast of Hokkaidô during the summer and autumn. Therefore small quantities of dwarf *Gelidium* possibly grow in these regions. In fact, Chihara (1972) collected *G. amansii* from Harutati on the Pacific coast of Hokkaidô and suggested that the northern limit of the species in Pacific Hokkaidô is in the region east of the Hidaka district. Also recently, Fukuhara (1977) reported on the northernmost record of several dwarf plants of *Gelidium* sp. from Kitami-Esasi of Okhotsk. These were smaller (5–7 cm in maximum height) than those collected from the Sea of Japan coast of Hokkaidô and led him to the following conclusion on the distribution of *Gelidium* in Hokkaidô. The northern limit of the distribution which has been accepted to date is the limit of the amount harvestable by fishermen; *Gelidium* may be widely found on the coast of Hokkaidô.

Kawana (1956) discussed the relationships between cold and warm currents and the *G. amansii* growing along the coast of the Pacific peninsula or islands in Japan. In the regions west of the Kii-hantô peninsula, vigorously growing *Gelidium* is found on the southeastern side on substrata to which a strong warm current (Kurosiwo) flows. In contrast, the region of the Izu-syotô islands with the force of the Kurosiwo stronger than in the western regions shows best growth on substrata on northern, northeastern, and eastern sides. In these last regions, the southwestern to west sides which the Kurosiwo attacks strongly have poor *Gelidium*. From these facts he concluded that mixing of the cold and warm currents is important to obtain a rich growth of *G. amansii*. From the literature survey, Endo & Matsudaira (1960) found that *G. amansii*, *G. japonicum* and *Acanthopeltis japonica* occur in waters where the annual means of water temperature are higher than 10 °C, 16 °C, and 17 °C, respectively. They proposed that the upper limit of the annual mean temperature allowing the occurrence of *Gelidium amansii* was at 26 °C or higher, which is the annual mean temperature of the southern end of Taiwan. This supposition is, however, inadequate, because *Gelidium* may not grow in the region (Chiang, 1973).

World records of the occurrence of *G. amansii* are given in Table VI. With regard to the distribution of *Gelidium* around Hokkaidô, the Karahuto (=Sakhalin or Saghalien) coast should be mentioned. Tokida (1954) recorded the occurrence of *G. amansii* and *G. vagum* from Kaiba-tô (Moneron Island) located offshore of the western coast of Karahuto. In shallow water of the region these species were growing in small amounts. In contrast to this island, Risiri- and Rebun-tô islands located near Karahuto produce large quantities of *Gelidium* on a scale suitable for harvesting. *Gelidium* on the Karahuto mainland has not been recorded; therefore, *Gelidium* has not yet spread to the Karahuto coast. In the world, *Gelidium* spp. occurring in the coldest seas are *G. crinale* from the Falkland Islands (Cotton, 1915) and *G. latifolium* from the south of Bergen (Jorde, 1966).

TABLE VI

The records of distribution of G. amansii *in the world: 1, incorrect; 2, now separated into several species by Gardner (1927); 3, original literature not shown, but is Kützing (1843)?, the description "In mari Kamtschatico" was given for* G. crinitum *by him*

Places	References
Mediterranean Sea[1]	Zaneveld, 1959
From California to British Columbia[2]	Setchell & Gardner, 1903
California and Mexico	Zaneveld, 1959
Korean Peninsula	Kang, 1966
Kaiba-tô (=Moneron) Island (Karahuto=Sakhalin)	Tokida, 1954
Mage-sima Island (Kagosima-ken, Japan)	Tanaka, 1950
Usyuku (Amami-ôsima, Okinawa-ken)	Kida & Kitamura, 1964
Yoron-zima I. (Okinawa-ken)	Tanaka, 1964
Yellow Sea and East China Sea (China)	Tseng & Chang, 1959
North Taiwan	Fan, 1951
Bashi (=Batan) Islands	von Martens, 1866
The Philippines	von Martens, 1866
Kalimantan and Sulawesi Islands (Indonesia)	Johnston, 1966
Molukka Islands (Indonesia)	Kützing, 1849
Samut Prakan (=Paknam) (Thailand)	von Martens, 1866
Penang Island (Malaysia)	Sivalingam, 1978
Southern Madagascar	Michanek, 1975
From Madagascar to Mauritius Islands	Lamouroux, 1805
From Agulhas Cape to St Lucia Bay (South Africa)	Seagrief, 1967
From Natal to Zululand (South Africa)	Jackson, 1976
Bering Sea[1]	Okamura, 1932b[3]

The former species in the British Isles was regarded as conspecific with *G. pusillum* by Dixon & Irvine (1977a,b). *Gelidium* is absent from the Arctic and Antarctic Seas.

Okamura (1934a, 1935b) stated that *Gelidium* does not occur in the Ryûkyû-rettô archipelago to the south of Yakusima island. Suto (1954) gave the southern limit of *G. amansii* and *Pterocladia tenuis* as on the Sata-misaki cape of Kagosima-ken where the monthly water temperature on average ranged between 16–17 °C. Table VII summarizes the records of *Gelidium* of large sizes from Yoron-zima and Okinawa-zima islands at the southern end of the Japanese distribution. The southern limit of large *Gelidium* is located at Okinawa-zima. Suto (1954) pointed out that a systematized survey of geographic variation in the quantity of light energy is needed in addition to plant distribution studies. Okamura (1935b) discussed the rôle of water temperature and the specific gravity of sea water with respect to discontinuous distribution between southern Kyûsyû and northern Taiwan. The temperature around Cheelung, where abundant gelidioids occur, nearly reaches that of southern Kagosima of Kyûsyû and Siwono-misaki of Honsyû rather than that of the Ryûkyû-rettô archipelago where gelidioids are of scanty occurrence. The salinity of the coastal streams along Cheelung differs only slightly from those along Ryûkyû. The

TABLE VII

Geographic records of gelidialean algae in southern and western regions around Kyûsyû (for literature, see the text): A. = Acanthopletis; B. = Beckerella; G. = Gelidium; Gla. = Gelidiella; P. = Pterocladia

Ryûkyû Rettô Archipelago

Mage-sima Island: *A. japonica, B. subscostata, Gla. acerosa, G. amansii, G. divaricatum, G. japonicum, G. crinale, G. pusillum, P. tenuis*

Amami-ôsima Islands: *A. japonica, B. subcostata, Gla. acerosa, G. amansii, G. divaricatum, G. japonicum, G. pusillum, P. tenuis*

Yoron-zima Island: *Gla. acerosa, G. amansii*

Okinawa-zima Island and Okinawa-syotô Islands: *G. pusillum, G. corneum* var. *pulchellum* (*G. amansii*), *P. nana*

Miyako-zima Island: *G. pusillum, Gla. acerosa*

Ikema-zima Island: *Gla. acerosa, G. pusillum*

Isigaki-zima Island: *Gla. acerosa, G. pusillum, P. tenuis, P. nana*

Kuro-, Aragusuku-, Kayama-, Kohama- and Taketomi-zima Islands: *Gla. acerosa, G. pusillum*

Iriomote-zima Island: *Gla. acerosa, G. divaricatum, G. pusillum*

Yonaguni-zima Island: *Gla. acerosa*

Taiwan

P'eng-chia Hsü (= Agincourt Islands): *G. planiusculum, G. latiusculum*

Chilung (= Cheelung, Keelung) to Tansyui (= Tansui) areas: *G. amansii, G. grubbae, G. japonicum, G. kintaroi, G. latiusculum, G. planiusculum, G. pusillum, G. tsengii, G. yamadae, P. tenuis, P. nana*

P'eng-hu Tao Island: *G. kintaroi*

Hung-t'ou-hsü and Tung-sha Tao Islands: *Gla. acerosa*

China Mainland

East China Sea coast: *G. pacificum*

Entire Chinese coast: *G. divaricatum, G. amansii*

compilation of the distribution of gelidialean algae based on literature surveys (*e.g.* Okamura, 1915, 1931, 1935b; Tseng, 1938; Yamada & Tanaka, 1938; Fan, 1951; Tanaka, 1956b, 1964; Segi, 1957; Tseng & Chang, 1959; Segawa & Kamura, 1960; Chiang, 1962; Kida, 1964, 1974; Taniguti, 1970, 1979; Akatsuka, 1973a) with reference to the occurrence of various species of the Gelidiales in southern and western regions around Kyûsyû is given in Table VII. Chiang (1973) surveyed the seaweed vegetation in the intertidal zone of the southern end of Taiwan which is an emerged coast of lime stone. He compared the vegetation in the region with that of the northern region of Taiwan and concluded that southern Taiwan has an affinity with Ryûkyû rather than northern Taiwan. But it is obvious that an examination by a survey of sublittoral zone in southern region is necessary on the basis of the discussion of the distribution of *Gelidium* in Ryûkyû, because his conclusion was made on the basis of the observation of intertidal zone only.

No *G. amansii* occurs in Indonesia and the Philippines, according to Weber-van Bosse (1921, 1932) and Eisses (1953), respectively. Although Zaneveld (1959) described this species from the Mediterranean Sea, Gayral (1966) has not described the occurrence of *G. amansii* on French coast in the Mediterranean and Atlantic. Records of *G. amansii* from California and Mexico are cited in old literature and the upper and lower Californian *G. amansii* is now separated into several species. For example, several species were established by Gardner (1927). As stated by Hommersand (1972), comparative studies of the species of Pacific North America and Japan are required. He pointed out that typical specimens of *G. amansii* have a habit similar to *G. purpurascens* and simple reproductive branches like these of *G. pulchrum*; some deep-water forms resemble *G. nudifrons*. *G. amansii* and its relatives are the species distributed in the region between the Indian and North Pacific Oceans.

Topographic observations have been rarely described. From the Figure 3 and my own experience, it can be concluded that *G. amansii* occurs on almost all rocky coasts of Japan (in regions other than the eastern half of Hokkaidô), excluding the innermost parts of bays.

Vertical distribution

The vertical distribution of all genera of Japanese Gelidiaceae is presented in Table VIII. Of the Japanese Gelidiaceae, *G. amansii* most widely occurs between the intertidal zone and a depth of 40 m. In the sublittoral zone, *Beckerella subcostata* was recorded between the depths of 5 and 128 m. In Izu-syotô islands of Pacific Japan the occurrence of *Gelidium* reaches its greatest depth, over 15 m (Kawana, 1956; Kawana & Ebata, 1959). In distant regions from the Izu islands the growing layer of *Gelidium* gradually shifts to shallower layers and ultimately no vegetation appears in very distant regions of the north and south (*e.g.* from Nagai, Miura-hantô peninsula to Mimitu, Miyazaki-ken). Embayed areas show similar vegetation to that in the shallow water.

Ôsuga & Yamasaki (1960) attempted a quantitative analysis of the vertical distribution using three concrete blocks as substratum for the *Gelidium*. These blocks were tetrehedral and trapezoid and were set on the

TABLE VIII

Vertical distribution of Japanese Gelidiaceae: IT, intertidal zone; LTL, low tide level; HTL, high tide level

Species	Depth (m)	Locality	Reference
Gelidium amamiense	60	Amami-ôsima, Kagosima-ken	Tanaka, 1965
G. amansii	IT–1·0	Rebun-tô, Hokkaidô	Yamada, 1980
	0·5–2·0	Otaru, Hokkaidô	Ôno, 1932
	0–8	Risiri-tô, Hokkaidô	Kaneko & Niihara, 1977
	0·9–1·9	Usuziri, Hokkaidô	Saito & Atobe, 1970
	1·5–15	Hamazima, Mie-ken	Takayama, 1939b
	4–15	Nobeoka, Miyazaki-ken	Miyazaki FES, 1939
	2–30	Wakayama-ken	Kawana, 1940
	20–40	west off Izu-hantô Pen.	Ueda & Okada, 1940
G. crinale	below LTL	Aiti-ken	Okamura, 1934a
G. divaricatum	around HTL	not specified	Okamura, 1934a
G. linoides	15–30	Wakayama- to Tiba-ken	Okamura, 1936
G. nanum	around LTL	Irago-zaki C., Aiti-ken	Inagaki, 1950
G. pacificum	4·5–16·5	Wakayama- to Tiba-ken	Okamura, 1936
G. pusillum	around HTL	not specified	Okamura, 1934a
G. tenue	65; 128	west off Izu-hantô Pen.	Ueda & Okada, 1938
G. vagum	IT–1·0	Rebun-tô, Hokkaidô	Yamada, 1980
Pterocladia densa	IT	not specified	Okamura, 1934a
P. nana	around HTL	not specified	Okamura, 1934a
P. tenuis	IT- deep	not specified	Okamura, 1934a
Beckerella irregularis	12–23	Kôzu-sima, Tôkyô-to	Akatsuka & Masaki, 1983
B. subcostata	5–30	Sikine-zima, Tôkyô-to	Kida & Nakamura, 1961
	64; 128	west off Izu-hantô Pen.	Ueda & Okada, 1938
Onikusa japonica	IT–22·5	not specified	Okamura, 1934a
Acanthopeltis japonica	below LTL	not specified	Okamura, 1936
Yatabella hirsuta	12–13·5	Oryûzako, Miyazaki-ken	Okamura, 1936

sea-bed at depths of 6, 10, and 14 m. They counted the individual plants on each of the northern, eastern, southern, and western surfaces and analysed the relation between the amount of *G. amansii* and the light intensity on each surface. Their explanation of the data and the per cent occurrence is difficult to understand. Kawana (1956) described the relationship between the vertical and geographic distribution of *G. amansii* with regard to light intensity. In the Oyasiwo Current, *G. amansii* occurs abundantly on the top of the substratum rather than on the side. Otherwise, in the region along Kurosiwo *G. amansii* grows abundantly on the side rather than the top in depths at less than 8 m. He explained the cause affecting these results; in the northern region the light does not reach to the deeper layers, because the water is very turbid whereas in southern regions there is higher transparency of the sea water and light reaches a greater depth. Therefore, *G. amansii* occurs in the shallow layer (*i.e.* top of substratum) in the northern region in order to gain more light energy. Alternatively, *G. amansii* in the southern grows in deeper water to avoid strong light intensity. He generally regarded regions shallower than 15 m as optimal depths for the growth of *G. amansii* on the Pacific coast of Japan, because *G. amansii* are found more often from regions shallower than 15 m depth all around the coast.

Hellebust (1970) stated that *G. crinale* (= *G. pusillum*, after Dixon & Irvine, 1977b) with wide light intensity tolerances (euryphotic) is found in the Bay of Naples near the surface in bright sunlight, at considerable depths, or in shadowed areas near the surface. This is interesting in relation to its world-wide distribution.

LIFE HISTORIES OF THE NATURAL POPULATIONS

The 'theoretical' life cycle of the Gelidiaceae (*sensu* Fan, 1961) generally accepted is as follows. The members of this family show triphasic life cycle including tetrasporophyte (Fig. 14), male (Fig. 15) and female (Fig. 16) gametophytes, and carposporophyte. The plants of the first two phases are isomorphic and macroscopic. The tetrasporophyte is an asexual generation and produces tetraspores which are asexual reproductive cells with haploid nucleus through reduction division of mother cell (Fig. 17) transformed from the cortical cell of ultimate pinnule (Fig. 18). The tetraspores develop to male or female plants through somatic mitosis, without nuclear fusion into any other reproductive cell. When surface water temperature rises, the male and female plants form male and female gametes, *i.e.* spermatia and eggs, respectively, on the ultimate pinnules. Cells outside near the pericentral cells of the female plants transform into eggs followed to the elongation of trichogyne. The trichogyne projects outside the thallus surface (Fig. 19) for fertilization. The egg of Rhodophyta fitted with the trichogyne is called carpogonium. The spermatium (Fig. 20) is formed through the transformation of outermost cortical cell of ultimate pinnule into antheridial mother cell which divides periclinally to cut off the antheridium. The 'cuticle' over the surface of thallus of the mature male plant transforms into a thick mucilaginous layer to release easily non-motile spermatia (Fig. 20). After a released spermatium attaches to the trichogyne of a carpogonium, the fusion of their nuclei takes place to lead the nucleus of the carpogonium to the diploid phase. The diploid carpogonium makes some processes accompanying cell divisions to form carposporophyte (Fig. 21). The carposporophyte is the aggregated body of cells called gonimoblast and locates inside an ultimate pinnule (Fig. 22) of a female gametophytic plant. This diploid carposporophyte bears diploid carpospores (Fig. 21) by the division of the gonimoblast cells. The carposporophyte is generally regarded as a kind of asexual generation parasitic within the tissue of the female gametophyte. The portions of mature pinnules bearing mature carpospores project roundly and an open pit(s) called carpostome(s) or ostiole(s) (Fig. 22) on one or both surfaces of the pinnule releases the mature carpospores. The projected portion of the pinnule is called cystocarp. The diploid carpospores released into the sea develop to tetrasporophytic plants through somatic mitosis. The process of the germination of the carpospores is identical with that of tetraspores showing "gelidial type" germination (the term after Inoh, 1947) as its peculiarity is in beginning unequal cell-division of spores and the following unequal aggregation of cells produced shown in Figure 23.

In the British Isles, *Gelidium* and *Pterocladia* are sterile or rare (Dixon & Irvine, 1977b). This fact suggests that each phase in the life cycle of *Gelidium* and *Pterocladia* may be incapable of expression under certain environmental conditions.

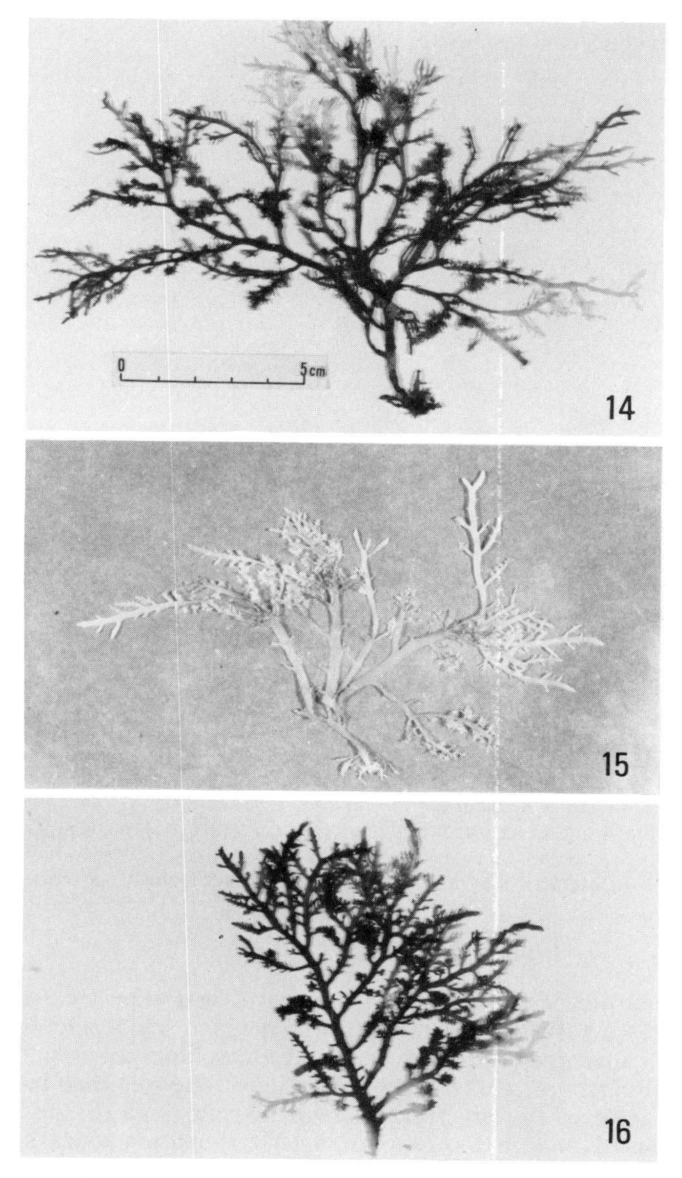

Fig. 14.—*Gelidium pacificum*: tetrasporangial plant.

Fig. 15.—*Gelidium pacificum*: male plant; scale as in Fig. 14.

Fig. 16.—*Gelidium pacificum*: female/cystocarpic plant; scale as in Fig. 14.

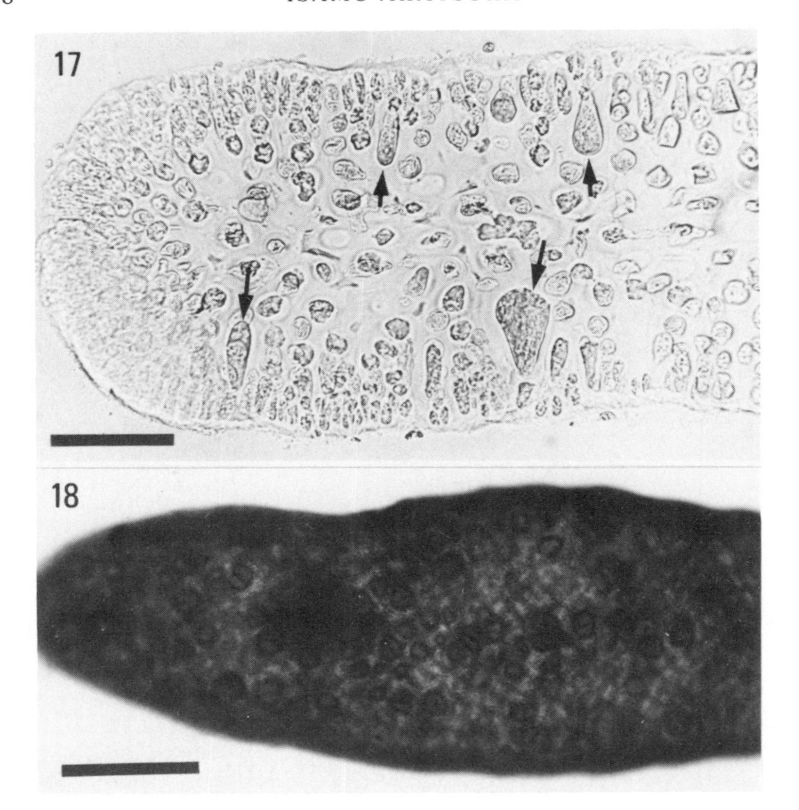

Fig. 17.—*Gelidium pacificum*: transverse section of a tetrasporangial pinnule showing tetrasporangial mother cells (arrows); scale = 40 μm.

Fig. 18.—*Gelidium pacificum*: an ultimate pinnule bearing tetraspores; scale = 200 μm.

Phases in the life history

The germination of *Gelidium amansii* spores was expected to be most abundant from July to August, because of the high frequencies of juvenile thalli which appeared on newly submerged stones observed from August to December (Okamura, 1911e, 1918). Experiments showed that germination in August achieved the largest numbers of juvenile thalli. In the Izu-hantô peninsula and Tiba-ken 10–12 months after germination when their height is 2–10 cm the juvenile thalli form reproductive cells. Continuous observations on the same individual thalli are, however, necessary for a survey of the phases of the life cycle. Okamura assumed that the thallus newly developed from the spores produces reproductive cells only once a year; after the first release of spores, thalli survived for a further 2–3 years, producing reproductive cells once a year. During the life of *G. amansii*, the reproductive ramuli or main axes may occasionally be lost. Matumoto (1922) found the maximum occurrence of spore germination of *G. amansii* during July from the observations on the sea-bed over 24 months. Ôno

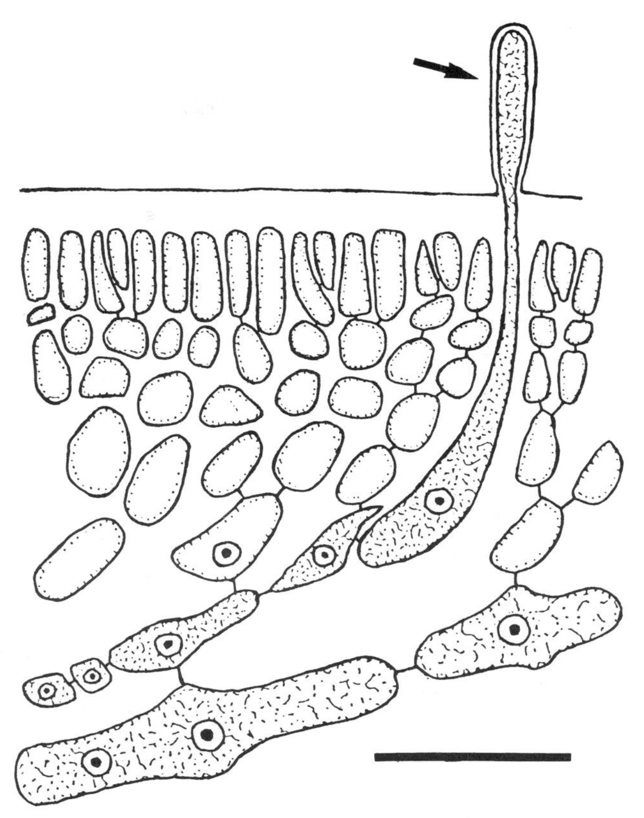

Fig. 19.—*Gelidium cartilagineum*: carpogonium with trichogyne (arrow); scale = 20 μm; after Kylin (1928).

(1927, 1929, 1932) studied the life history of *G. amansii* in the coast of Otaru, Hokkaidô. He found that the onset of spore formation is ruled by sea-water temperature and cystocarp formation was delayed in comparison with that of tetrasporangia (Table IX). Takayama (1939a,b) reported the fruiting season of *G. amansii* on the Pacific coast of Mie-ken as from April to December. The tetrasporophytes were first recognized early in April and gradually became abundant with two major peaks. These peaks were from mid June to mid July and mid September to late October. The maximum period of the spore formation was in the season when the water temperature reached 20 °C. When the temperatures were above 25 °C, the maximum period was interrupted because tetrasporangial stichidia were lost. Occasional findings of tetrasporophytes were made in November and March. On the other hand, cystocarps were found from early August to early October; the maximum occurrence was in August when the maximum water temperature was 25–26 °C. This maximum period of the cystocarp occurrence terminated when the sea-water temperature fell below 25 °C. Cystocarps were occasionally found in April. Determination of reproductive season in the life cycle is obtained by counting the fertile plants in

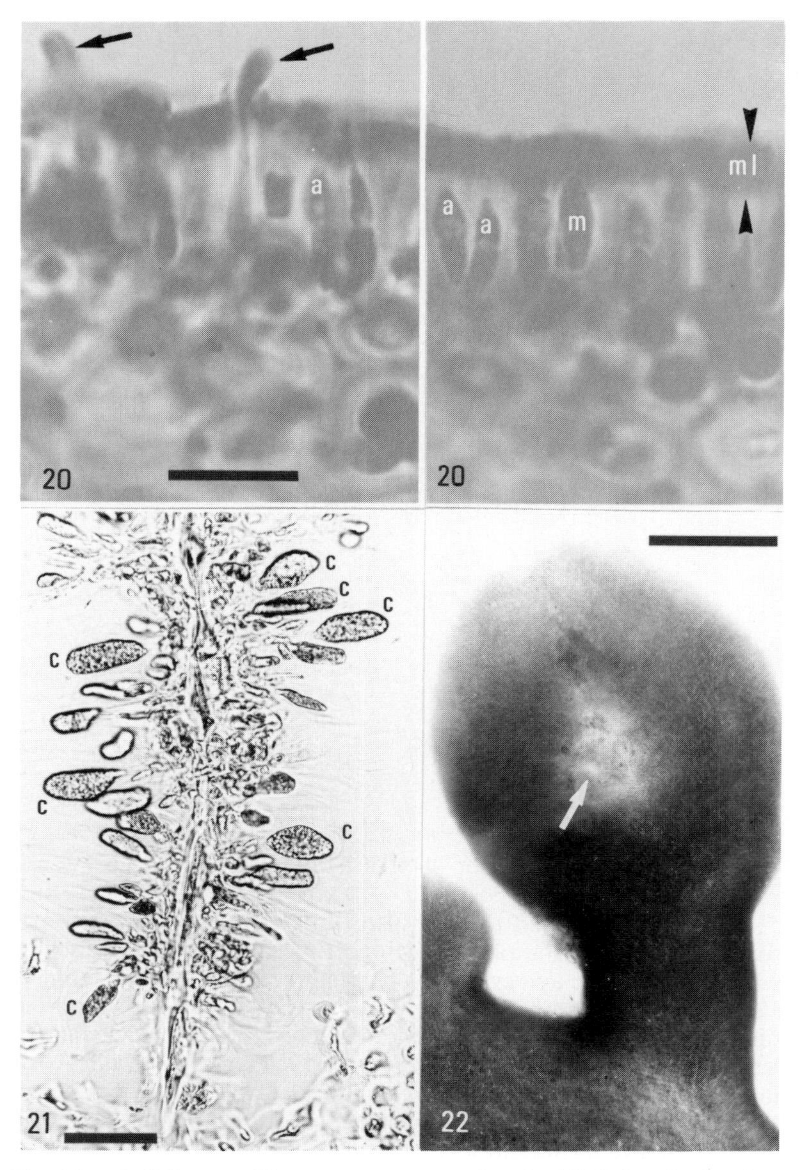

Fig. 20.—*Gelidium pacificum*: two sections of antheridial pinnule showing a releasing spermatia (arrows), antheridia (a), antheridial mother cell (m); mucilaginous layer (ml) between two arrowheads is also shown by metachromatic positive staining with Toluidine Blue O; scale = 10 μm.

Fig. 21.—*Gelidium pacificum*: transverse section of cystocarp showing carposporophyte with carposporangia (c); scale = 40 μm.

Fig. 22.—*Gelidium pacificum*: a cystocarp showing an ostiole (arrow) for release of carpospores; scale = 200 μm.

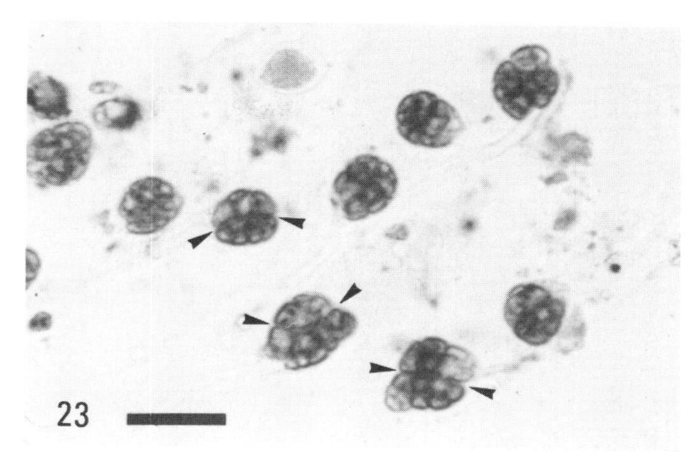

Fig. 23.—*Gelidium pacificum*: sporelings showing unequal aggregation of cells derived from the first division into one small and the other large daughter cells; arrows show borders between large and small aggregations of cells; rhizoids are also found; scale = 50 μm.

random samples and plotting their frequency distribution. Takayama (1939a), however, gave no description of his sampling method or the number of plants. His conclusion that there are two peaks of tetrasporangial formation, therefore, may be doubted; modern survey methods are required to test his conclusion. A comparison of the reproductive season of *G. amansii* on various coasts from north to south compiled from the data of several workers is presented in Table IX.

Male plants of the Gelidiaceae were collected in the following months: May (*Acanthopeltis japonica*, Simoda); July (*Gelidium japonica*, *Beckerella subcostata*, both Miyake-zima island); August (*Pterocladia tenuis*, Awa-Kominato; *Gelidium pacificum*, Nitiren-zaki, Izu-hantô peninsula); September (*G. pusillum*, Awa-Kominato; *G. amansii*, Izu-Mera) (Kaneko, 1968; Akatsuka, 1970b, 1973b, 1979)).

For a complete understanding of the life history of gelidioid algae, surveys of growing and defoliation seasons must be simultaneously carried out in addition to the studies of the seasons of ripening, formation and release of spores. Reliable data can be obtained when these surveys are carried out monthly for at least 12 months, but preferably for 24 or more months. Ideally, geographical variability should be also surveyed.

As an example of a survey in the sea, Ôno (1929, 1932) observed the yearly variation of fronds of *G. amansii* on Otaru, Hokkaidô. According to him, in June, when the water temperature rises rapidly, ultimate pinnules and prostrate axes are formed and elongate, and new erect axes begin to develop from the prostrate axes. The luxuriant growth season can be divided into a primary (July to August) and a secondary (September to November) season, but he stated that the transition period between them is difficult to distinguish because of its shortness. The defoliation season was also divided into a primary (from August to September) and secondary (from November to May) one. Kinosita (1949) and Kinosita & Sibuya

TABLE IX

Phenology of tetrasporic and cystocarpic plants of G. amansii *at some localities in Japan: A, abundant; F, fair; R, rare; N, absent; 1, month divided into three sections of ten days; empty columns indicate no data; Matsuura (1958) and Miyazaki FES (1939) did not distinguish 'abundant' and 'fair'*

	Month[1]										
	Mar.	Apr.	May	June	July	Aug.	Sept.	Oct.	Nov.	Dec.	References
Tetrasporic											
Otaru			R	FFF	FFA	AAA	AAA	FFF	RRR	RN	Ôno, 1927, 1929, 1932
Manazuru	FFF	FFF	FFF	FFF	FFF						Matsuura, 1958
Kominato			FF	FAA	FFF	FFF	FFF	NNN			Ueda, 1936
Inatori			FF	FAA	FFF	FFF	FFF	NNN			Ueda, 1936
Hamazima	FFF	FFF	FFF	FAA	AAF	FFF	FAA	AAA	FFR		Takayama, 1939a
Wakayama			FFF	FFF	FFF	FAF	FFF	FFF	A		Onodera *et al.*, 1954
Miyazaki		FFF	FFF	FFF	FFF	FFF	F				Miyazaki FES, 1939
Cystocarpic											
Otaru					F	FFF	FFF	FFF	FFF	RR	Ôno, 1927, 1929, 1932
Manazuru	FFF	FFF	FFF	FFF	FFF						Matsuura, 1958
Kominato					F	FAA	AAA	AFF			Ueda, 1936
Inatori					F	FAA	AAA	AFF			Ueda, 1936
Hamazima					F	AAA	AAA	AFF	FFR		Takayama, 1939a
Wakayama				FFF	FFF	AAA	FFF	FFF	A		Onodera *et al.*, 1954
Miyazaki		FFF	FFF	FFF	FFF	FFF	R				Miyazaki FES, 1939

(1941a,b) also reported the relationships between sea surface temperature and the reproductive season of *G. subfastigiatum* (= *G. amansii*) at three places in western Hokkaidô. Field observations and sample collections were carried out by them on the 1st and 15th of each month over one year and the number of reproductive ramuli counted. The beginning and end of the formation of reproducive cells were determined by examining sections under a microscope. The species began to form tetra- and carpospores when the surface temperature rose to near 12 °C (early June and Risiri-tô island) or near 15 °C (at Matumae and Takasima-misaki cape). As shown in Table IX, the formation of tetraspores showed two peaks within 12 months in Mie-ken and Wakayama-ken, but only a single peak was recognized on the western coast of Hokkaidô. With regard to this difference in the fruiting season of the tetraspores between northern and southern Japan, they pointed out that the seasons showing the peaks correspond to the peroids with water temperatures over 20 °C. When the temperature remained above 20 °C for six months, the secondary peak appeared in Mie-ken. This temperature, however, lasts for less than three months in Hokkaidô.

Tetraspores are released in the months when the sea temperature falls below 17–18 °C. On the other hand, the release of carpospores and decaying of cystocarps takes place when the surface temperature falls below 13–14 °C, and follows the months of the release of tetraspores in Hokkaidô (Kinosita & Sibuya, 1941a,b). Ôno (1927, 1929) stated that the release of carpospores was observed at temperatures lower than that of tetraspores. Ôno (1932) claimed that carpospores play a part in propagation at times in the year of low temperature, because they tolerate low temperatures better than do tetraspores. Otherwise, the most optimal temperature for the germination and early development were 21° and 16 °C for the tetraspores and carpospores of *G. subfastigiatum* (= *G. amansii*), respectively (Kinosita, Hirano & Takahasi, 1935a,b). They stated that the germinating carpospore has a greater tolerance of lower temperatures than the tetraspores. Suto (1950b) estimated spore numbers produced in a day as 10^4 to 10^6 per g of thallus. Onodera, Matuoka & Ogino (1954) studied the fruiting season of *G. amansii* at two sites on Wakayama-ken by means of frond sectioning (Table IX). Both the tetraspores and carpospores showed two peaks in their formation like those of tetraspores in Hamazima on Mie-ken reported by Takayama (1939a). Their sampling methods and sample sizes have not been described in detail. Suto (1954) compiled data fragmentally reported by various FES reports on the spore releasing season in the region from Hokkaidô to Miyazaki-ken, but the surveys were not obtained by continuous observations over 12 or more months. Not only spore release, but other seasons, such as the growing one, must be investigated at monthly intervals for 24 months and in places from the north to the south for a better understanding of the life history of *G. amansii*.

Matsuura (1958) roughly illustrated and described the annual reproductive cycle at Manazuru-misaki Cape (Table IX). It is odd that no records of fertile plants are described by him from August onwards, the maximum season of spore formation and release. This can possibly be attributed to his sampling which was twice a month, but no description of sample sizes was given. The absence of fertile plants in some months may be due to the harvest by fishermen or to the small sample sizes and number of collections.

Suto (1954) generalized on the relationships between water temperature and sequence in each stage of the reproductive cycle of *G. amansii* growing in central Pacific Honsyû as follows: (1) in water temperatures of 10–20 °C (winter to spring), thallus growth season; (2) near 20 ° C, male reproductive cells formed; (3) 20–21 °C, tetraspores formed; (4) 24–25 °C, carpospores formed; (5) declining of the growth rate; (6) decaying and loss of reproductive ramuli after the release of reproductive cells; (7) decaying and loss of the tips of main axes; and (8) termination of thallus growth, allowing growth of epiphytes.

Table IX shows the possibility that both carpo- and tetraspores have the beginning at their formation at a later time and have a shorter duration of the reproductive season for both kinds of spores in northern than in southern regions. Suto (1950a) reported that the peak release of reproductive cells in 1947 was delayed by about 20 days compared with other years; in 1947 water temperatures were considerably lower than the mean monthly values for other years. From these data, it is obvious that the formation and release of reproductive cells relate directly to the coastal water temperature.

Ôno (1929) stated that tetrasporangial stichidia are borne on new ramuli formed in the season when the frond has just ceased its growth and that they begin their formation immediately after the emergence of erect axes from prostrate axes. He also observed the frequent occurrence of tetrasporangial stichidia on young fronds of about 5 cm height. Otherwise, carposporangial ramuli are usually found on the fronds showing maximum growth.

Proportions of each generation

In his review, Umezaki (1966) briefly commented on the proportions of generations of the Gelidiales and their relationship to the life history. He felt that Kylin's (1923) upgrading of the Gelidiaceae to the order, Gelidiales, on the basis of the idea of a *Polysiphonia*-type life history without experimental evidence was not at all acceptable. Umezaki agreed with Dixon (1961). Suto (1954) reviewed literature regarding the proportions, but the data cited by him was not systematic, mixing frequency and weight of thalli. The proportions of *Gelidium amansii* were first described by Ôno (1927). He neglected the proportion of males, as did Takayama (1938, 1939a) (Table X). If the neglected males were equal to the frequency of cystocarpic plants, then the sexual and asexual proportions would be nearly equal. In fact, the double cystocarp proportion presented by him showed equality with that of tetrasporangia. Therefore, male plants were probably included by him in the tetrasporangial group. He proposed that the sexual and asexual proportions did not vary with seasons. The number of tetrasporangial stichidia formed on a frond varied with change of sea-water temperature. In his field observations the number of cystocarpic ramuli was proportional to the height of the frond. He assumed that the dominance of the tetrasporophyte over the cystocarpic plants in untouched regions was caused by the higher potentialities of the propagation of prostrate axes of tetrasporophytes compared with those of the cystocarpic plants. Takayama (1938, 1939a) described the proportions of the tetrasporophytic and

TABLE X

Data on the percentage of each generation of Japanese Gelidium *excluding the reports using the expression (weight per weight, g): data for* G. amansii, *excluding* for* G. pacificum; *T, tetrasporic; C, carposporic; M, male; S, sterile;—, no data*

Place	Date	Frequency (%)				Reference
		T	C	M	S	
Hamazima (Mie-ken)	—	80	20	—	—	Takayama, 1938
Hamazima (Mie-ken)	—	70	30	—	—	Takayama, 1939a
Hutomi (Tiba-ken)	—	49	25	20	6	Takamatsu, 1946c*
Otaru (Hokkaidô)	Aug. or Oct.	70	30	—	—	Ôno, 1927, 1932

cystocarpic plants in Mie-ken (Table X) but the date of observation was not given. From the proportions calculated he supposed that the tetraspores played the more important rôle in propagation. Kinosita & Sibuya (1941a,b) and Kinosita (1949) surveyed twice monthly the proportions by weight of carposporangial and tetrasporangial thalli from March 1938 to February 1939. The proportions were calculated from the wet weights of the samples. Their method was, however, not reliable because topographic life-history studies require the counting of frequencies of individuals of each generation, not only the wet weight measurements. Although Kinosita & Sibuya (1941a,b) also compared the number of cystocarpic with that of tetrasporangial pinnules in the same plant, their method and analysis of the results were not statistical. Tokyo FES (1941, 1942) also made surveys of the proportions employing weight measurements on *G. pacificum*, but neglected the male plants. They also discussed the frequencies of the reproductive ramuli per thallus. This counting is useful in determining the fertility of the population. In addition, they counted the numbers of spores per tetra- and carposporangia; more tetraspores than carpospores were released. Not all the tetraspores observed in the sporangia were released. Onodera, Matuoka & Ogino (1954) also reported that tetrasporic plants dominated over the carposporic ones in *G. amansii* collected at two sites at Wakayama-ken from May to November (no sample in October), as well as those described in the reports of the earlier authors. Differences in the proportions between two collection sites were also found. Differences of 10–30% in weight between the carposporic and tetrasporic plants were described. Yamasaki (1962) investigated the proportion of cystocarpic plants to tetrasporic on substrata newly built with stones. Two years after building up the substrata the carposporic plants greatly dominated over the tetrasporic. The proportion decreased after three and more years, approaching the value of the plants collected from untouched regions near the experimental site. They supposed that the higher ratio of the tetrasporophytes in untouched regions was caused by the higher potentialities of propagation of prostrate axis compared with the ratio of cystocarpic plants. In the study of Newfoundland Ceramiaceae, Whittick (1984) stated that the tetrasporophyes apparently perennate while the gametophytes appear to be annual.

Yamasaki (1962) concluded that the alternation of generations of thalli with cystocarps was the same as or more than that of tetrasporic ones, but his deductions are not readily acceptable. First, he did not mention the proportion of male plants and secondly, he did not discuss the meaning shown by the proportion of each generation at a certain time. Factors affecting the proportion of generations are: released and surviving numbers of reproductive cells including spermatia produced by each generation; the ratio of success of fertilization for the next generation; and the production of new erect axes from prostrate axes. It is nonsense for a life history, especially the alternation of generations and especially the alternation of nuclear phases, to be discussed only on the basis of the proportion of generations at one particular time during the life cycle. Surveys of the proportions need to be carried out at least over 12 months. To claim a conclusion about the regularity of the alternation of generations on the basis of the proportion of tetrasporic and cystocarpic plants which has been misunderstood as the proportion of sexual and asexual generations must be avoided. Yamasaki (1962) also claimed that the high proportion of asexual (tetrasporic) plants to sexual (excluding male) ones is because of the high potentiality of propagation by the prostrate axes of the asexual plants; this is probable. Furthermore, he supposed the greater longevity of the sexual (meaning cystocarpic) plants than that of the asexual ones to be significant.

According to Takamatsu (1946c), there is scarcely any difference between the number of tetrasporic plants and the summed numbers of cystocarpic and male plants; but what does this mean? A certain portion of plants of each generation persists to the next year, because *Gelidium* is a perennial. The state of affairs becomes complex by the addition of environmental changes during a year. Therefore, surveys of the alternation of generations (including males) must be carried out for some years in an area where harvesting of *Gelidium* is prohibited.

Takamatsu (1946c) discussed the proportion of sexual to asexual generations of *G. pacificum*. To date, no one else has investigated the true proportions, including that of male plants. From his values (Table X), he pointed out the nearly equal frequencies of cystocarpic and male plants. The frequencies of sexual plants, including both male and carposporophytic generations was approximately equal to those of the asexual plants or tetrasporophytes. Thus, the asexual plants did not dominate over the sexual. In addition, he claimed that in the studies by the earlier workers the male plants had been neglected because their external appearance resembled the tetrasporophytes. Few publications on the true proportions, including males, have been presented since then. Takayama (1939a) described the proportions (Table X) in which male plants were lacking, but the sexual to asexual ratio was near to one, if the males showed equal proportion to cystocarpic plants (30%). Notwithstanding, he (Takayama, 1938) recorded the proportions of sexual to asexual as 1:1.

Dixon & Irvine (1977b) recognized problems with respect to the life histories of British Gelidiaceae, because *Pterocladia pinnata* (Hudson) Papenfuss (= *P. capillacea*) in the British Isles are completely sterile on the basis of eight years investigation (Dixon, 1958a) and sexual plants of

Gelidium latifolium, *G. pusillum*, and *G. sesquipedale* are absent or relatively rare in Great Britain. In contrast, gametophytes (cystocarpic) of many species of the Gelidiaceae have been reported from Spain, Portugal, and France (Funk, 1927; Feldmann & Hamel, 1936; Seoane-Camba, 1965).

The earlier Japanese investigators proposed that *G. amansii* had an 'irregular' life history, because of the much larger proportion of tetrasporangial plants compared with the carposporangial. But Yamasaki (1962) described the complete life-history of *G. amansii* in Japan; over half of the fertile plants were cystocarpic after two years on substrata newly built with stones.

I recognize two points which have to be discussed on the life history and the proportion of each generation. First, up to now two ways of the expression of the proportion of the generations have existed—on the grounds of the numbers or the weight of plants. I judge the frequencies of individual plants of each generation as being strictly necessary for life-history studies because of the importance of frequencies rather than the weight in each given population. Secondly, the proportions to date have included two types which are tetrasporophyte to cystocarpic plants or asexual to sexual plants. In most reports (except for Takamatsu, 1946c) the sexual plants have not usually included the males. The males appear to have been confused with the tetrasporophytes or sterile plants due to their featureless appearance resembling these two phases of life history. The reason why are there so many tetrasporophytes, has been discussed for many localities of the world. The abundance of the tetrasporophytes has been simultaneously explained by two investigators. Whittick (1984) presented field, culture, and cytological evidence to support a hypothesis that meiosis in the tetrasporangia of some species of Ceramiaceae is facultative and under environmental control and that this accounts for the abundance of the tetrasporophyte phases in Newfoundland. Kain (1984) gave six reasons why 99% of fertile plants of *Plocamium cartilagineum* in the Isle of Man are tetrasporophytes. She suggested that the sixth reason discussed, that the gameto- and tetrasporophytes might occupy different niches, is important.

The proportion of the different phases is an important part of ecological life-history studies. Such studies should last at least 24 months and should cover places of various latitude and depth. In the case of the absence of gametophytes (for example in *Gelidiella* species in some regions of the world) the cytological approach is also required to support nuclear alternation in the life cycle.

Longevity of erect fronds and prostrate axes

For four years Okamura (1918) made field observations on a *Gelidium amansii* population growing on substrata newly built with rocks on the sea bed. After one year the growing thalli showed no constrictions and were smooth and soft to touch. After two years he observed that scars from which reproductive ramuli had been lost after the release of spores in the previous year regenerated adventitious reproductive ramuli. After three years, these regenerated reproductive ramuli produced more new ramuli at the points of constrictions which occurred in the upper region regenerated in the previous year. He assumed the longevity of *G. amansii* to be three or more years, because no regenerated fourth-year ramuli were found. The constriction was also used for the determination of perennation by Dickinson (1949) in the description of *G. helenae* (=*Beckerella heleniae*) and *Gelidium rumpii* (*Beckerella rumpii*). *Gelidium sesquipedale* in Spain leaves a mark each year on a branch which was used for population studies of growth and longevity (Seoane-Camba, 1969). The mark so-called by him is probably the constriction of Okamura (1918). Matumoto (1922) also determined the longevity of *G. amansii* as 36 months on the basis of the direct observations of natural populations.

Dixon (1958b) studied the perennation of the erect fronds and the prostrate axes of British Gelidiaceae. The erect frond may persist for four or more years in *Pterocladia capillacea* and *Gelidium sesquipedale*. Dixon (1973) also stated that evidence had been obtained indicating the persistence of populations of *Pterocladia capillacea* for at least 40 years.

The prostrate axes are remarkably persistent in all British species. The prostrate branches of *Gelidium pulchellum* survive in crevices in rock, beneath a covering of *Lithothamnion*. From observations on seven species of the Gelidiaceae (Akatsuka, 1982d), there is evidence of the long-term persistence of the prostrate axis and that their cell sizes are generally greatest in the transitional part between main axis and prostrate (=creeping) axis. In *Gelidium latifolium* the production of erect fronds over a period of years from the same prostrate axis has been demonstrated by Dixon (1958b). Similar observations have been made on the Japanese coast (Ôno, 1927; Miyazaki FES, 1939; Takamatsu, 1946a). *G. crinale* (=*G. pusillum*: Dixon & Irvine, 1977b) shows greater longevity at the lower levels of vertical distribution (Dixon, 1963), but this tendency has not been investigated for Japanese *Gelidium* spp.

Vegetative propagation and regeneration

Regeneration is commonly recognized in the Gelidiales. Okamura (1911a,b) made field observations on the adhesive potentialities of fragments of *G. amansii* to substrata in Sirahama of Tiba-ken. The fragments produced haptera and fixed after ten days, when caught in the crevices of rocks. Scattering the fragments over the sea bed appeared to be meaningless for propagation, because most of the fragments were lost by wave action. Instead of a fragment-scattering technique he found twining many fruiting thalli into a rope and tying this tightly to a rock followed by submerging the rock in the sea for spore dispersion as effective. Okamura (1918) examined the rate of regeneration of fragments of *G. amansii* on the Pacific coast of Tiba-ken and in laboratory culture. The growth rate of fragments (original length unknown) in the sea gave lengths of 10 cm after 43 days of submerging in the sea, and 15–18 cm and 24–27 cm after 50 and 71 days, respectively. Laboratory culture of fragments of the tetrasporophytes and cystocarpic plants failed, probably because water movement and nutrients were not provided. These field and culture studies showed that *G. amansii* fragments have the potentiality to survive even at the shortest length, 15 mm, when the fragment can fix to the substratum. Haptera can project from any part of the thallus and can develop prostrate axes. These may cause the formation of clumps. In laboratory culture, Miyazaki FES (1939) found that the fragmented thalli of *G. amansii* produce many ramuli composed of single cell-rows. When the thallus dissected from basal regions was planted in holes bored in dead corals in the sea, haptera were formed after 24 days. The erect axis and branches appear to form haptera due to a contact stimulus. From these observations the erect and prostrate fronds in the experiments may not be clearly differentiated. It has been stated that the vegetative propagation by the prostrate axis is more important than that by spores. The comparison of propagation potentialities of the prostrate axes of tetrasporic plants with sexual ones has not been made. Katada (1954, 1955) also confirmed the significant potentialities of the regeneration of *G. amansii*. As the fronds of *G. amansii* were kept attached by rope to concrete blocks placed on the sea bed, these fronds produce haptera on branches the next spring with the later development of many juvenile fronds. He stated that these events were frequently found on plants detached by wave action or the harvesting activities by fishermen in natural habitats. Adventitious lateral branches of the Gelidiaceae develop in a similar manner to the normal branches of ultimate growth (Dixon, 1958b).

Recently, Felicini (1970) and his co-workers (Felicini & Arrigoni, 1967; Felicini & Perrone, 1972) published a series of experimental studies on the regeneration of *Pterocladia capillacea*.

VARIOUS RELATIONSHIPS

To elucidate the relationships between the Gelidiales and other marine organisms in the occupation of their habitats, light energy and nutrients are important in marine agronomy as also in ecology. Systematic studies on these aspects are absent from field surveys because of technical difficulties; only fragmentary knowledge has been obtained to date.

Yendo (1911) enumerated the interrelationships between seaweeds and other organisms as follows. (1) Parasitism by other algae: generally, many of the parasitic algae invading the intercellular space or matrix of the host algae have chloroplasts and they are semi-parasitic and able to produce carbohydrates. (2) Common epizoa attached to useful seaweeds are Hydrozoa and Bryozoa: these animals attach to full-grown portions of laminariacean algae. (3) Abnormality caused by perforation drilled by crustacean animals. (4) Decrease of assimilatory pigments caused by the decrease of utilizable light caused by the attachment of eggs of fishes and other animals to the surface of seaweeds. (5) Attaching of other algae, especially diatoms: Yendo (1911) pointed out that the identical species never attach to the same species of the algae; of the diatoms, *Cocconeis*, *Isthmia*, *Arachnoidiscus*, and *Licmophora* are the species frequently attaching to seaweeds; meristematic portion is very difficult to be attached to by other algal species.

Competition

Santelices (1974) reviewed the competition of the Gelidiaceae citing two papers and commenting that no clear conclusions have yet been reached.

Ôsuga & Yamasaki (1960) could not complete a study of competition between tetrasporic and sexual plants of *Gelidium amansii* over five years at sea, because of their sampling errors and drastic changes in environmental conditions. In relation to "isoyake" (see p. 246), Masaki (1984) reported that the spores of *Gelidium* cannot settle on living crustose coralline algae having two and more epithallial layers; and that *Lithophyllum yessoense* in culture showed sloughing off of its outer epithallial cells and discarded anything attached to its surface.

Miyazaki FES (1939) recognized periodism between the biomass of *Gelidium* and other seaweeds, especially coralline algae. They also reported that young erect axes emerging from the prostrate axis are persistent over a considerable period under the cover of unarticulated coralline algae. Suto (1954) suggested that *G. amansii* is difficult to grow among a population of coralline algae. Takamatsu (1946a) demonstrated the close relationships between *Gelidium* including *G. amansii* and melobesioidean corallines in Simoda of Sizuoka-ken. The prostrate axes of matured *Gelidium* survive under the hard cover of the melobesioidean algae, so that the frond of *Gelidium* can be mechanically supported by the Melobesioideae. It is difficult for sporelings of *Gelidium* to develop under this cover and they cease development and die; the rate of development of the sporelings is also considerably lower than that of the Melobesioideae. In contrast, dead melobesioidean algae are of value to the settlement of *Gelidium* spores and afford protection for the spores against wave action owing to their coarse surfaces caused by the erosion by sea water.

According to the literature (Ôno, 1932; Suto, 1954) plants competing with *Gelidium* are as follows: *Eisenia bicyclis* (Kjellm.) Setch., *Ecklonia cava* Kjellm., *Eckloniopsis radicosa* (Kjellm.) Okam., *Padina arborescens* Holmes, sargassoids, *Cladophora wrightiana* Harvey, coralline algae, *Phyllospadix iwatensis* Makino, *Polysiphonia morrowii* Harvey, *Rhodymenia palmata* (L.) Grev., *Laurencia nipponica* Yam., *Symphyocladia latiuscula* (Harv.) Yam., *S. gracilis* (Mart.) Falkenb., *Dictyopteris divaricata* (Okam.) Okam., *Laminaria religiosa* Miyabe., *Pterocladia tenuis* Okam., *Halyseria divaricata* Okam., and *Scytosiphon lomentaria* (Lyngb.) J. Ag. Within these species, large plants shade the fronds of *Gelidium*.

Epibionts

The available data on epibionts of *Gelidium* in Japan are summarized in Table XI. *Arachnoidiscus* can cause different damage to *Gelidium* depending on its age; *Licmophora* attaches heavily to the *Gelidium amansii* frond remaining after the spore release (Suto, 1954); and *Harposiphonia* can kill sporelings of *Gelidium* by covering them. According to Ôno (1929, 1932), in Hokkaidô *Arachnoidiscus ornatus*, a diatom, covers heavily the frond surfaces of *Gelidium*. He also stated that young shells of bivalves frequently attach to *Gelidium* thallus, but that these can fall off on drying. The cover of diatoms is frequently found in periods of calm water on the coast. These diatoms and shells are gradually lost in autumn. As an alternative, coralline algae and bryozoans appear in autumn. He also found that the epibionts decrease in the periods before and after the growing season of *Gelidium*. Ohmi (1963) reported *Porphyra atropurpurea* (Olivi)

TABLE XI

Epibionts of Gelidium *recorded in Japan*

Epiphytes

Chlorophyta: no literature, but I found *Enteromorpha* sp. and *Monostroma* sp. during culture studies on *Gelidium pacificum*.

Rhodophyta: *Lithothamnion* (Miyazaki FES, 1939), *Harposiphonia tenella* (Katada, 1955), *Porphyra atropurpurea*, *P. kuniedai* (Ohmi, 1963).

Bacillariophyta: *Arachnoidiscus ornatus*, *Biddulphia pulchella*, *Climacosphenia moniligera*, *Cocconeis scutellum*, *Gomphonema kamtschaticum* var. *californicum*, *Licmophora juergensii* var. *elongata*, *Skeletonema costatum*, *Trachyneis aspera* (these were dominants) (Takano, 1961), *Entophyla*, *Licmophora* (Okamura, 1909).

Epizoa

Porifera: Halichondrida (Miyazaki FES, 1939).

Tentaculata: Bryozoa (in March) (Miyazaki FES, 1939).

Mollusca: shells (12 spp.) (Yokota, 1939), shells (147 spp.) (Kanamaru, 1932), *Cantharidus japonicus*, *Batillus cornatus* (Katada, 1955).

Annelida: Nereidae (as nest constructor; much in August) (Miyazaki FES, 1939).

Arthropoda: *Gammarus* (nest constructor; much in August) (Miyazaki FES, 1939), crabs (13 spp.), puerulus of *Panulirus japonicus* (Yokota, 1939).

De Toni and *P. kuniedai* Kurogi as epiphytes on *Gelidium* sp. and *G. Divaricatum*, respectively.

Katada (1955) observed that the thallus of *G. amansii* changed to yellow because of a covering by microalgae on its thallus. It is known that the Gelidiaceae can be epiphytes on algae, for example non-articulate corallines.

Substrata suitable for Gelidium

Miyazaki FES (1939) enumerated the substrata adhered to by *G. amansii* (Table XII).

Katada (1955) compiled data from the literature and his own experience on the quality of such substrata for Gelidiaceae. More instances of the settlement of spores were observed on fresh rocks carried from mountains for the experiments than on those cobbles or boulders cast ashore; on stones having coarse surfaces than on smooth ones; and on soft stones more than on hard ones. All these have commonly coarse surfaces. Rather than the kind of rocks used as substrata, it is the coarseness of surfaces that is of primary importance for the settlement of spores and sporelings of *Gelidium* (Miyazaki FES, 1939).

Phycophagous animals

Ino (1958) studied the colour changes of shells of *Turbo cornutus* (= *Batillus cornutus*) when several seaweeds were used as food. The species made white shells when it was fed *Eisenia bicyclis* (a brown alga), brown with *Sargassum ringgoldianum*, dark greenish-brown with *Gelidium amansii*, and striped black in white with a mixture of *G. amansii* and

TABLE XII

Enumeration of substrata adhered to by Gelidium: *reference not shown in the Table is Miyazaki FES (1939)*

Abiotic substrata
- Stone, iron bolts, concrete blocks set for lobsters.

Biotic substrata
- Shells, living: *Pinna bicolor*, abalones, byssus of *Pteria* (*Austropteria*) *brevilata*.
- Shells, dead: Ostreidae, Pectinidae.
- Balanomorpha's shell, dead.
- Madreporaria, living: Acroporidae.
- Madreporaria, dead: Acroporidae, Faviidae.
- Plants, living: rhizoids of *Sargassum ringgoldianum*, rhizoids of *Ecklonia bicyclis* (my own finding), rhizoids of *Laminaria* sp. Fukuhara (1977).
- Plants, dead: pines, cedars, oak fragments.

Eisenia as diet. Growth rate was maximal with *Eisenia* but it fell when *Gelidium amansii* and *Sargassum* were used as food. Ueda & Katada (1949) found many juvenile plants of *Gelidium* grazed by gastropods in tide-pools of Tiba-ken. Katada (1955) mentioned *Cantharidus japonicus*, *Homalopoma nocturnum*, *Hiloa megastoma*, *Fossarina picta*, and *Batillus cornutus* as grazers of *Gelidium amansii*. He stated that juvenile *Gelidium* is frequently injured by grazing animals or by a covering of epiphytic algae. Santelices (1974) reviewed the grazing of the Gelidiaceae by phycophagous fishes, citing five reports. Harada & Kawasaki (1982) statistically analysed the attraction index of seaweeds as food of abalone. The indexes of *G. amansii* and *G. divaricatum* were the highest among 14 species of red algae tested. They (Harada & Hirano, 1983) also studied the relationship between the attraction index and some other factors influencing that index. The index increased with the increase of day the seaweeds to be tested had been stored. They concluded that the index was not positively correlated with changes in the chemical constituents analysed; the correlation coefficient between the index and the chemical constituents of various seaweeds was not significant. Some combinations of chemical constituents of seaweeds attracted the abalones of their food.

BIOMASS STUDIES

To elucidate the change in biomass of Gelidiales in field and laboratory culture studies is of primary significance in relation to resource recruitment and its management. The measurement of the biomass with respect to weight, height, and counting of the gelidialean plants, and also the variability in appearance is conveniently treated in this section.

DEVELOPMENT OF SPORELINGS IN LABORATORY CULTURE

As generally recognized, to culture sporelings of *Gelidium* through to the mature plants in the laboratory is very difficult and has not so far been successful. Katada (1955) reported the growth of sporelings of *G. amansii* in laboratory culture using slides as substrata. After three months, adventitious ramuli occurred on various portions of the sporelings. The growth rate of the adventitious ramuli was considerably higher than that of the main axis of sporelings which reached less than 1 mm at the end of the experiment. Rao & Suto (1970) cultured *G. amansii* in an artificial culture medium. After transference to the artificial medium sporelings, germinated in natural sea water, were suitable for the development of unialgal bacterized cultures. A light and dark regime of 13 h L to 11 h D in the medium supported normal growth of *G. amansii*; the authors supplied a photograph without a scale bar in their paper.

VARIATIONS IN BIOMASS OF NATURAL POPULATIONS

No sequential studies have been made on the variation of *G. amansii* biomass between the season after harvest in the summer and next spring before harvest. Suto (1954) pointed out that inadequate methods have been employed in surveys of natural populations by various prefectural FES and their data are possibly biased.

Weight, height, and frequency

The data available on growth measurements are given in Table XIII. Suto (1954) stated that biomass studies using quadrats are rare. From data compiled by him the height of *G. amansii* plants after 12 months on newly submerged stones varied between 1–18 cm following the reports by various prefectural FES. These data are, however, barely meaningful, because *Gelidium* spores did no always germinate and develop simultaneously after submerging of the experimental stones. In addition, monthly and yearly sequential observations were not made by the FES and the reproducibility of most of their data is doubtful. Okamura (1918) cut 3 cm from the base of several plants of *G. amansii* and *G. pacificum* growing on natural rocks and marked them with hemp fibres. The rocks were then submerged again on the sea bed. Twelve months later he measured the length of branchlets regenerated from the cut points. Growth rate was calculated as 9–12 cm and 18–21 cm a year in *G. amansii* and *G. pacificum*, respectively. From the growth rate calculated, he suggested that "niban-kusa" (= secondary re-harvestable *Gelidium* after the closed season) so-called by fishermen is not the plants regrown from the remaining portion of the thalli of *Gelidium* cut for harvest, but grew from the small plants which remained after the harvest activities. Matumoto (1922), employing a helmet-type diving suit, observed the growth of *G. amansii* on the sea bed. In his report no descriptions on the sample size and the kind of marking method used have been given. Takayama (1939b) also studied the growth rate of *G. amansii* on the newly submerged stones. The values obtained correspond closely to those given by Okamura (1918), Matumoto (1922), and Sizuoka FES (1951). Aoki & Yasuda (1941) investigated the variations of the dry weight of *G. amansii*, each month in Korea; the weight decreased from September to May. Sizuoka FES (1951) surveyed the biomass of *G. amansii* and *G. pacificum* growing on the eastern coast of the Izu-hantô peninsula. *Gelidium* gave 0·37–20 kg/m^2 in collections at 45 sites and in depths of

TABLE XIII

Compiled list of measurements of growth of Gelidium amansii *on substrata newly built up with stones or concrete blocks: a, other than these data, Suto (1954) compiled some data from seven FES located from Akita-ken to Nagasaki-ken as a mean value of 10 cm/12 months; b, the unit "kg" is wet weight*

Months after building substrata	Measurement[b]	Locality	Reference
6–10	12 cm	Hamazima, Mie-ken	Takayama, 1939b
8	4·1 cm (mean)	Sirahama, Sizuoka-ken	Yamasaki, 1962
9	10 cm	Nisina, Sizuoka-ken	Okamura, 1911c, 1917
10	6·6 cm (mean)	Sirahama, Sizuoka-ken	Yamasaki, 1962
12[a]	3–10 cm	Hamazima, Mie-ken	Matumoto, 1922
12[a]	9–11 cm (mean)	Sirahama, Sizuoka-ken	Yamasaki, 1962
12[a]	0·28 kg/m[b]	Sirahama, Sizuoka-ken	Yamasaki, 1962
16	12–14 cm (mean)	Sirahama, Sizuoka-ken	Yamasaki, 1962
20	9–12 cm (mode)	Hamazima, Mie-ken	Takayama, 1939b
24	15–17 cm (mean)	Sirahama, Sizuoka-ken	Yamasaki, 1962
24	12–18 cm	Hamazima, Mie-ken	Matumoto, 1922
24	0·78 kg/m[b]	Sirahama, Sizuoka-ken	Yamasaki, 1962
30	18 cm (mode)	Hamazima, Mie-ken	Takayama, 1939b
36	21–27 cm	Hamazima, Mie-ken	Matumoto, 1922
36	18–20 cm (mean)	Sirahama, Sizuoka-ken	Yamasaki, 1962
36	1·08 kg/m[b]	Sirahama, Sizuoka-ken	Yamasaki, 1962
36	1·23 kg/m[b] (mean)	Hamazima, Mie-ken,	Takayama, 1939b
48	1·33 kg/m[b]	Sirahama, Sizuoka-ken	Yamasaki, 1962
60	1·32 kg/m[b]	Sirahama, Sizuoka-ken	Yamasaki, 1962

2–13 m but, unfortunately, the distribution of sample statistics is unknown. Katayama & Inaba (1953) investigated the frequency distribution of the fresh weight/m^2 of attached *G. amansii* at various depths; their results were very briefly given without any statistical backing. Nonaka, Ôsuga & Sasaki (1962) analysed the differences between values of the standing crop of *G. amansii* on the eastern coast of Izu-hantô peninsula. The maximum was 2·3 kg fresh weight/m^2 and the minimum 0·9 kg/m^2; these weights were measured during the growing season of *G. amansii*. Yamasaki (1962) studied the yearly changes of the growth rate of *G. amansii* on concrete blocks submerged in late August. Three years after the submerging of the stones the population showed fairly similar standing crops to that of a natural reef. He noted the small increment of *G. amansii* after three years and he mentioned that the areas of "isoyake" barely showed any increment of *G. amansii* even on newly submerged concrete blocks.

Nonaka, Ôsuga & Sasaki (1962) surveyed the growth and recovery of natural populations of *G. amansii* in Sirahama of Sizuoka-ken. Harvesting of l m^2 quadrats was carried out 11 times at irregular intervals between May and September of the next year. The population showed maximum height and weight corresponding to daylength. One to two months after harvest

(June to July) the population gave maximum recovery with a minimum from November to February.

Tokyo FES (1941) reported on the growth rate of *G. pacificum* in the ocean. A collection was made three times a month every month for a year and the heights of 100 plants were measured. The minimum height (monthly mean) was found in June (97 mm) and the maximum in January and February (125·5 mm). The FES commented that the differences of means between winter and summer samples are barely recognizable, and can be almost neglected presumably because of an error in experimental design.

Katada & Satomi (1975) described the seasonal variation of the standing crop of *Gelidium (G. amansii*?). In the communities of "stem- or frond-survivors" involving *Gelidium*, the difference between the maximum and minimum crops was usable as an estimate of annual net production, when tissue losses were negligible. They concluded that there was an alternation in each six months of the maximum and minimum standing crops of *Gelidium*, but this was on the basis of the trend line drawn using non-monthly and insufficient data.

Coverage

Saito & Atobe (1970) described the seasonal changes of coverage and frequency of *G. amansii* in the southern Hokkaidô. *G. amansii* showed the maximum frequency in June and Septemer, and maximum coverag was found in June. They concluded that *G. amansii* belongs to the 'spring type' of category of seaweed community on the basis of the vegetative growth period. Recently, Lee, Kim, Lee & Hong (1975) surveyed bimonthly the coverage using quadrats set in Kwang-yang Bay in southern Korea and described *G. amansii* occurring in March and May with the lowest coverage.

GROWTH OF THALLUS IN THE NATURAL POPULATION

Weight and height

Ueda & Katada (1949) and Katada (1955) observed that the juvenile plants of *G. amansii* grew most rapidly in February and March and very slowly in the season from summer to winter. The increase in the growth rate of seaweeds in early spring should be related to the increase of the water temperature and the light regime. For careful analysis monthly measurements of thallus growth for at least 24 months are required. Matsuura (1958) found that the season showing increase of growth rate of seven species of the Gelidiaceae in Manazuru coincided with the seasonal rise in temperature from 13° to 20 °C. These species disappeared, however, when temperatures rose above 25 °C during the season investigated. Yamasaki measured the height of thalli grown on concrete blocks submerged for five years (Fig. 33 in Yamasaki, 1962).

It is interesting that *Pterocladia capillacea* in the intertidal zone of southern California showed peak development in the late fall to winter months when sand is being removed from the beaches by storm surf (Stewart, 1983). It is well known that the growth of *Gelidium* is vigorous in places with appropriate movement of sea water, but poor in calm water.

Polymorphism

Yendo (1911) described the *G. amansii* of Masike (Hokkaidô) as very small; I confirmed this and also observed the dwarf *Pterocladia tenuis* during a visit to Masike. Okamura (1934a) and Fan (1951) distinguished five forms of polymorphism within *Gelidium amansii* (Table XIV) but these forms strongly resembled each other.

Dixon and Irvine (1977b) and Dixon (1958b) described the polymorphism of British species of *Gelidium*. Identification of members of the genus *Gelidium* has always been notoriously difficult because of the absence of clearly circumscribed entities. From sequential field observations of marked plants in Britain and a more detailed understanding of growth and behaviour, most of the species accepted by Feldmann & Hamel (1936) which occur in Britain can be reduced to two aggregate groups, *G. pusillum* and *G. latifolium*. The difference of frond size between each generation of *G. amansii* were briefly described by Miyazaki FES (1939) as the cystocarpic plants having a smaller size and thickness compared with those of the tetrasporic ones. Dixon (1963) claimed that the form of the mature rhodophytan thallus is determined by the interaction of three factors: (1) disposition of the axes; (2) shape of the axes; and (3) life-span of the frond. The maximum size and frequency of branching increases from the upper to the lower limit of the distribution in the intertidal region.

Colour of thallus

Yanagawa (1943) recorded that Gelidium was bleached to yellow in shallow water under the high intensity of light. Yamada (1960, 1961, 1964, 1967, 1972, 1976) studied mainly the natural bleaching of *G. amansii* on the Izuhantô peninsula. It is caused by the decrease of the phycobilin contents, including phycoerythrin and phycocyanin. The decrease of phycobilin relates to a rise in sea-water temperature or of light intensity. When nitrogenous and phosphate fertilizers were applied, there was a recovery of the phycobilin content. When the temperature fell by 2–3 °C or clouds covered the sky, the plants bleached in summer recovered their colour and when a plant recovered its colour the chlorophyll content decreased.

CHEMICAL CONSTITUENTS

The chemical composition and content of seaweeds depends on the season of the year, habitat, depths of habitat, the state of development, and the portion of thallus. Variations in the composition among a number of plants collected in the same area at the same time were reported by Black, Richardson & Walker (1959). There has been little systematic work on variations in the composition of other classes than the Fucaceae (Percival & McDowell, 1967).

Trouble is caused in the earlier reports because the following details are frequently omitted: the name of species analysed, date, place, and depth of collection, the status and period of storage of the material before chemical analysis, the source and condition of material used.

TABLE XIV

Comparison of variants of G. amansii (*from Okamura, 1934a, 1935b; Fan, 1951*)

Character	f. *typicum*	f.*elegans*	f. *elatum*	f. *teretiusculum*	var. *latioris*
Thallus height (cm)	10–15	10–15	3 or more	10–15	14
Ramification	Irregular	2–3 orders	With large distances	Irregular; large distance	Irregular
Ramuli	Sharp terminal, irregular length	Uniform length	—	Bend back	Often short, filiform, curved
Geographical distribution	Very common	Very common	From Kie-ken to Chiba-ken; deep habitats	Japan Sea, Seto-naikai	North Taiwan

Murata (1963) described various free-amino acids, peptides, organic bases, organic acids, and monosaccharides in his review on the extractable constituents of marine algae.

AGAR

Seaweeds grow in sea water containing considerable amounts of various salts. They should, therefore, accumulate or absorb ions derived from these salts. In absorption of ions, free sulphate groups are incorporated into the cell wall and the intercellular matrix may play a part (Nisizawa, 1970). There seems little doubt that a sulphated polysaccharide, agar, helps to build up a flexible structure suited to the environment of seaweeds (Percival & McDowell, 1967). Stiffer gels are present in plants exposed to more severe wave action (Rees & Conway, 1962).

Chemical properties of agar

Payen (1859) first named the component "gélose", extracted directly from *G. corneum* and *Plocaria lichenoides*. The gélose lacked nitrogen, was soluble in boiling water, formed a colourless gel after cooling, contained sulphate, and gave a yellow-orange colour with iodine. His report was the first description of a sulphated polysaccharide. Yanagawa (1933) analysed constituents of agar prepared from various agarophytes including *Gelidium*, *Acanthopeltis*, *Campyraephora*, and *Gracilaria*. Aoki (1940) and Aoki & Yasuda (1941) studied relationships between the season of collection and agar content (%) of *Gelidium amansii* on the Korean coast. The agar contents slightly decreased from April to September and increased during the period following to October. They also reported the relationship between depth of habitat and agar content. Their collections were made at depths of 0·1, 2·3, and 4·0 m below low tide mark in April, June, August, and October at two sampling sites. They concluded that no clear connection existed between the depth and agar content, but they employed no statistical test for significance. Chen & Wu (1953) investigated the agar content of several species of the Gelidiaceae collected at one to nine sites in Taiwan. They found 22–36% in *G. amansii* and 30–42% in other species, but the date of collection and sample sizes were not given in their report.

Araki (1937, 1956) found that agar prepared from *G. amansii* is composed of at least two polysaccharides, agarose (AG) and agaropectin (AP). AP and its derivatives contained the greater part of the sulphates, uronic acids, pyruvic acids, and ash; AG and its derivatives rarely contained these acids and ash. Izumi (1971) reported that agar prepared from *G. amansii* consists of polydisperse polysaccharides having a similar macromolecular structure. Chemical heterogeneity of polysaccharides in algal cell wall matrices presumably shows species- and tissue-specific differences.

The amount of 3,6-anhydride such as agar and porphyran in algal tissues may be important as an adaptive ecological character (Rees & Conway, 1962). Eppley & Cyrus (1960) suggested that one of the functions of porphyran might be to act as a cushion against physical buffeting. These ideas on porphyran are also applicable to agar in *Gelidium*. More detailed investigations of algal polysaccharides, particularly during cell differentiation and thallus development, will provide an important clue in the elucidation of their biological functions (Percival & McDowell, 1967).

The polysaccharide composition of agar from *G. amansii* changed with the conditions of extraction, even if the collection date and sites were the same (Tagawa & Kojima, 1973). Largely sulphated polysaccharides were mostly obtained when extracted at 100 °C for 1 h and decreased at 120 °C for 3 h, presumably because of the degradation of polysaccharide structure in the latter conditions.

Although Murakami (1955, 1960) analysed agars prepared from *G. amansii*, *Beckerella subcostata*, *Acanthopeltis japonica*, *Gracilaria confervoides*, and *Ceramium hypnaeoides*, no description was given about the date of collection of these species.

Physical properties of agar

Watase, Akahane & Arakawa (1975) concluded that the effect of alkali concentration in pre-treatment on the ability of gelation of agar gels can be explained in terms of the conformation of agar molecules in water. Watase & Nishinari (1981) rheologically compared the samples of *Gelidium amansii* collected in May and August at the same place namely Susaki, Izu-hantô peninsula. The molecular weight and the amount of sulphate residues were larger in the August than in the May sample. Sulphate groups are responsible for the gelling ability of agar (Takahasi, 1941), and it seems related to the mechanism of polymerization rather than the dissociated radicals (Hayashi & Okazaki, 1970). Watase & Nishinari (1981) suggested that gel formation is related to the molecular weight at high concentration and to the amount of sulphate residues at low concentration. The mechanism of gelation by agarose could be similar to that for carrageenans; computer modelling shows the possibility of double helices formation (Rees, 1969). Since gel rigidity is related to the proportion of cross links, helix formation would appear to be progressively inhibited by an increasing number of sulphuric ester groups and by their interactions with cations. Gel strength decreases in the following order: AG, agar, AP (Hayashi & Okazaki, 1970). Akahane *et al.* (1982) studied the ion selectivity of some extracellular viscous polysaccharides from *G. amansii* and another three species of red algae. They concluded that the affinity of K^+ ions to agar and carrageenan studied was higher than that of Na^+ ions. Sequential studies on the physical properties of agar are in progress by Akahane and Watase.

Ohta & Tanaka (1964) studied the utility of *Gracilaria verrucosa* (Huds.) Papenf., *Hypnea charoides* Lam., and *Grateloupia okamurai* Yam. as agarophytes; admixture of material of *Grateloupia okamurai* to that of *Gelidium amansii* reduced the gel-strength of agar, leading to an increase of its yield.

OTHER SUBSTANCES

Tsuchiya (1962) pointed out that carbohydrates and inorganic substances were major components of seaweeds. Systematic studies of the geographic

TABLE XV

General analysis of G. amansii *(from Yanagawa & Nisida, 1930): percentage on dry weight basis; *further information not available*

Constituents	Place						
	Izu-hantô Peninsula*			Wakayama*	Korea*	Korea*	Tsing-tao
Water	14·00	17·06	19·93	20·27	14·24	16·04	20·42
Crude fat	—	0·04	—	0·04	0·02	0·14	0·09
Crude protein	16·14	17·25	22·64	9·70	13·85	11·60	11·55
Crude fibre	13·49	10·49	11·81	12·47	10·91	11·15	15·08
Ash	3·56	4·05	7·34	6·94	5·14	5·48	12·60
Crude agar	60·06	43·31	58·51	43·90	51·80	47·39	38·01
Galactose	—	22·32	17·06	21·26	26·00	—	19·38

TABLE XVI

Vertical variation of analysis in G. amansii *of Wakayama-ken (from Onodera* et al., *1954)*

Depth habitat	Ash (% dry wt basis)		Water (% dry wt basis)		Specific gravity of hot water extracts	
	Max.	Min.	Max.	Min.	Max.	Min.
1·5 m (Mio-wan Bay)	5·9 (June)	3·6 (Apr.)	7·7 (Nov.)	2·6 (June)	2·2 (Apr.)	1·2 (Nov.)
9·0 m (Hiki-ura)	6·0 (Dec.)	3·5 (Oct.)	5·0 (Dec.)	1·9 (Oct.)	1·8 (Apr.)	1·2 (June, July, Sept., Dec.)

variations in the chemical constituents of the Gelidiaceae have rarely been made. Yanagawa & Nisida (1930) analysed the thallus of *G. amansii* collected from various places in Japan and its adjacent areas (Table XV). In their report the date and exact site of collection of materials and sample size were not recorded. Onodera, Matuoka & Ogino (1954) reported the geographical, vertical, and seasonal variations of *G. amansii* samples collected at Wakayama-ken (Table XVI); they found no significant difference of constitutents in the algae from these two sites. Huzikawa (pers. comm.) indicated that the gel strength of the *Gelidium* extract tends to increase in samples collected in northern compared with southern regions. Viscosity tends to increase in samples from the southern region. Fragments of descriptions on seasonal variations of analysis of *G. amansii* described by Ôtani & Huzikawa (1935) are shown in Table XVII. No comparison, however, was made of sample size, place, and depth of collection with which the values obtained varied. Asano, Kuroda & Asahi (1951) studied the seasonal variations of chemical composition of *G. amansii* in samples collected in each of 12 months for a year at Muroran of Hokkaidô, but the sampling method employed was not described. Their data are simply compiled in Table XVIII. Tsuchiya (1962) commented on the seasonal variation of total nitrogen and protein-N contents of seaweeds. Seaweeds generally show the maximum amount of nitrogen during January to March.

Tables XVII and XVIII show that protein or nitrogen reaches maximum values in winter, but the difference between maximum and minimum is small (2% in both Tables). Ash also shows small differences (2–3%) between the maximum and minimum value, but Table XVII lacks data for July and October.

No studies on the variations of chemical constituents of *G. amansii* with depth have been made.

TABLE XVII

Tendency for seasonal variation in some chemical constituents of G. amansii *(entire thallus) compiled from Ôtani & Huzikawa (1935): a, values read from graphs*

	Collection date	% on dry wt basis
Nitrogen[a]	Nov. to June, monthly	Maximum: 4 (Dec.) Minimum: 2 (Nov. and June) Remarks: constant (3·3) from Jan. to May
Ash	Nov. to June, monthly	Maximum: 15·38 (Apr.) Minimum: 12·38 (June), 12·65 (Nov.) Remarks: other taxa maximum in April were *Ulva*, *Undaria pinnatifida*, *Hizikia*, *Sargassum thunbergii*, *Gloiopeltis*
Reduced sugars[a]	Nov. to June, monthly	Maximum: 27 (June) Minimum: 20 (Jan. and Apr.)

TABLE XVIII

Seasonal variation in general analysis of entire thallus of G. amansii *(from Asano* et al., *1951)*

	% on dry wt basis	Trend
Ash	Max.: 18 (Aug.) Min.: 14 (Feb.)	Increased from winter to summer
Fibre	7–8	Constant
Fat	0·25–0·28	Constant
Protein	Max.: 18 (Feb.) Min.: 16 (Sept.)	Decreased from winter to summer
Carbohydrate	60	Constant

ORGANIC SUBSTANCES

Sugars

Although fucose was detected in agar, it has not been found in plants of *G. amansii*, *G. vagum* or *Beckerella subcostata* (Nakamura, 1977). Nakamura (1971) and Nagashima (1976) reported the distribution of alcohol-stable sugars and sugar alcohols in the Gelidiaceae (Table XIX). Ross (1953) semi-quantitatively analysed minor sugars in five species of the Gelidiaceae (Table XX). His results on fucose do not agree with those for several other species of the Gelidiaceae reported by Nakamura (1977). Tables XIX and XX show an absence and presence of trehalose, respectively in some *Gelidium* and *Pterocladia* species.

Storage polysaccharides

Floridean starch is a glycogen-like polysaccharide and stains reddish brown with I–KI solution. This substance is specifically contained in species of the Rhodophyta in the form of grains which measure 0·5–25 μm in diameter

TABLE XIX

Sugar alcohols of three species of Gelidiaceae: A, Nagashima (1976); B, Nakamura (1971); —, not detected; +, scarce; + + +, abundant; t, trace; empty column, doubtful result

	Floridoside		Isofloridoside		Trehalose		Laminitol	
	A	B	A	B	A	B	A	B
G. amansii	+ + +	+ + +	+	t	—	—		+
G. japonicum	+ + +	+ + +	+	t	—	—	+	+
P. tenuis	+ + +	+ + +	+	t	—	—	+	+

TABLE XX

Minor sugars of five species of Gelidiaceae in per cent dry wt basis (from Ross, 1953): RSA, Republic of South Africa; CAL, California; w, 5–10%; t, less than 5%; —, negative

Species	Site; Date	Glucose	Xylose	Mannose	Fucose
Gelidium pristoides	RSA; June	t	w	t	w
G. cartilagineum	RSA; Jan.	t	w	t	w
G. coulteri	CAL; Aug.	w	w	t	t
Pterocladia pyramidale	CAL; Aug.	w	w	t	t
Suhria vittata	RSA; Apr.	t	w	—	w

and show birefringence under polarized light (Nisizawa, 1970). The mean length of chain of floridean starch is unknown. The content of the substance in the Gelidiaceae (*G. cartilagineum*, *G. coulteri*, *G. pristoides*, *Pterocladia pyramidale*, and *Suhria vittata*) was reported as 1–2% (Ross, 1953), but variations with the season, locality, developmental stages of thallus, portion of thallus, and depth of habitats are all unknown.

Structural polysaccharides

Cellulose plays a part in the mechanical support of the algal thallus. Alpha-cellulose isolated from *Gelidium amansii* was identified as true cellulose by Araki & Hashi (1948). The fibre fraction of the residues of agar extraction of *G. amansii* showed the curve of IR-spectra of cellulose pulp (Katsuura, 1962). Table XXI shows the amounts and physico-chemical properties of *G. amansii* cellulose reported to date. The molecular weight and degree of polymerization of the cellulose of *G. amansii* are considerably larger than those of terrestrial plants, presumably because the plant must have thallus flexibility to tolerate severe wave action (Akahane & Fuse, 1968). Holocellulose from *G. amansii* contains D-galactose, D-glucose, xylose, and L-fucose as shown by paper chromatographic analysis. In these sugars, fucose seems to make polyfucose sulphate. The degree of crystallization of *G. amansii* cellulose calculated from X-ray diffraction analysis was lower than those (70–80%) from ramie (a fibre from the stem of a woody Asian plant). The cellulose isolated from *G. amansii* holocellulose which showed a lower degree of orientation was judged to be cellulose-I, the same as that known in other members of the Rhodophyta.

Structural polysaccharides of the water-soluble polysaccharides of *G. pacificum* were galactan (abundant), mannan (doubtful), xylan (doubtful) and those of crude fibre (beta-cellulose) were glucan (abundant) and mannan (present) (Iriki, 1971).

Fats and oils

Algal lipids generally equal 1% on a dry weight basis (Tsuchiya, 1962). Certain seaweeds contain essential oils giving a turpene-like smell.

TABLE XXI

Yield and nature of cellulose isolated from Gelidiaceae (% dry wt basis): a, Percival method; b, A.O.A.C. method; c, South African material; d, Californian material; —, means no data

Species	Yield (% dry wt)	Physical property	Reference
G. amansii	11·3[a], 9·8[b]	Average degree of polymerization (D.P.) = 330	Hachiga & Hayashi, 1966
G. amansii	14·12	D.P. = 72 M.W. = 12000 Degree of crystallinity = 50–55%	Akahane & Fuse, 1968
G. japonicum	13·3[a], 8·8[b]	D.P = 300	Hachiga & Hayashi, 1966
G. pacificum	11·7[a], 10·1[b]	D.P. = 340	Hachiga & Hayashi, 1966
G. pristoides	6·5 (June)[c]	—	Ross, 1953
G. cartilagineum	9·0 (Jan.)[c]	—	Ross, 1953
G. coulteri	8·5 (Aug.)[d]	—	Ross, 1953
P. tenuis	10·7[a], 9·7[b]	D.P. = 380	Hachiga & Hayashi, 1966
P. pyramidale	8·2 (Aug.)[d]	—	Ross, 1953
B. subcostata	9·0[a], 6·5[b]	D.P. = 260	Hachiga & Hayashi, 1966
Suhria vittata	7·9 (Apr.)[c]	—	Ross, 1953

G. amansii showed a trace of the oil (Ando, 1953). Ito, Nagashima & Matsumoto (1956) isolated chalinasterol from *G. amansii* and *Pterocladia*. Sterols of this kind have not been found in any plants. Tsuda, Akagi & Kishida (1957, 1958) and Tsuda *et al.* (1958) found cholesterol in *Gelidium amansii*, *G. japonicum*, *Beckerella subcostata*, *Pterocladia tenuis*, and *Acanthopeltis japonica* and stressed the first record of cholesterol from the plant kingdom. They also found chalinasterol in *Gelidium amansii* and *Pterocladia tenuis* as did Ito, Nagashima & Matsumoto (1956). Tsuchiya (1962) pointed out that most species of the Rhodophyta have cholesterol, but Chlorophyta and Phaeophyta have not. In relation to the existing reports concerning the effect of marine algal sterols to lower hypercholesterolemic level, Kato & Ariga (1983, 1984) analysed 39 species of green, brown, and red marine algae and demonstrated 20–200 mg of various sterols/100 g dry wt basis; of these species, *Gelidium amansii* and *Meristotheca papulosa* showed minimum amounts and *Codium fragile* and *C. intricatum* showed maximum amounts. The main sterols were cholesterol in red algae, fucosterol in browns, and fuco- and isofucosterols in greens.

Amino acids and their derivatives

Kuriyama (1961a,b, 1965) studied sulphur-containing amino acids and a new one found widely in Rhodophyta including *Gelidium amansii*, was named rhodoic acid (Kuriyama, 1961c). Ito (1963) studied the distribution of D-cysteinolic acids in seaweeds. *G. amansii* exhibited a considerable yield of taurine, but no detectable amount of D-cysteinolic acid. Ito, Miyazawa & Matsumoto (1977) performed screening studies on amino acids and aminosulphonic acids of seaweeds. The amino acids and their derivatives known from *G. amansii* are given in Table XXII. Ito (1969) published a review on free amino acids of marine algae. Takagi (1970) reviewed briefly the low molecular nitrogen compounds of marine algae including *Gelidium* species.

Amines

Hamana & Matsuzaki (1982) analysed di- and polyamines of many species of seven phyla of algae; of the algae analysed, *G. amansii* showed 0·147 (putrescine), 0·001 (spermidine, spermine, norspermidine), and 0·033 (cadaverine) in μmol/g wet wt. They suggested the uselessness of the polyamines in algal phylogenetics.

Organic acids

Creac'h (1953) reported citric acid contents of 24–747 mg/100 g dry matter in red algae including *G. attenuatum*. He noted that the amount was small in young plants and larger in matured ones. Watanabe (1937) detected pyruvic acid in *Gelidium* (*amansii* ?), *Ulva* of the Chlorophyta and Myelophycus of the Phaeophyta. Hayashi & Okazaki (1970) described the

TABLE XXII

Amino acids and their derivatives of G. amansii: *a, μM/100 g-fresh material (Ito* et al., *1977) unless specially noted; b, Miyazawa, Ito & Matsumoto (1967, 1969) on dry wt basis (%); c, Ogino (1955), % on dry wt basis using material from Wakayama; d, Ito* et al. *(1966) and Ito & Hashimoto (1966) reported the occurrence of gigartinine in* G. amansii; *–, means not detected; +, occurrence (not quantitatively expressed)*

Amino acids	Yield[a]	Amino acids	Yield[a]
Taurine	314	Methionine	—
N-dimethyl taurine	1–2[b]	Methionine sulphoxide	44
Aspartate	39	Isoleucine	59
Threonine	72	Leucine	—
Serine	30	Tyrosine	25
Asparagine	37	Phenylalanine	92
Glutamine	124	Lysine + Ornithine	32
Glutamate	336	Histidine	trace
Citruline	—	Cit-Arg	—
Proline	154	Arginine	27(0.494[c])
Glycine	141	Gigartinine	+[d]
Alanine	238	(NH_3)	(806)
Valine	64		

occurrence of pyruvic acid in the residue of agaropectin in agar prepared from *Gelidium amansii*. They stated that carboxyl groups in agaropectin seem to be responsible for the gel-forming potentialities of *G. amansii* agar. D-gluculonic acid was determined as 5% in agar from *G. amansii* (Araki, 1932, 1937; Yanagawa, 1936). Ross (1953) reported the yield of uronic anhydride from five species of the Gelidiaceae as a percentage of the dry weight (Table XXIII). Kato & Ariga (1983, 1984) analysed unsaturated fatty acids and found that they were distributed widely in the analysed species and that they were more abundant than saturated fatty acids. The composition of the acids of brown algae was intermediate between green and red algae including *G. amansii*.

TABLE XXIII

Yield of uronic anhydride of the Gelidiaceae (% dry wt basis; from Ross, 1953): RSA, Republic of South Africa; CAL, California

Species	Place	Month	% uronic anhydride
G. pristoides	RSA	June	2·6
G. cartilagineum	RSA	Jan.	2·4
G. coulteri	CAL	Aug.	4·1
P. pyramidale	CAL	Aug.	4·6
Suhria vittata	RSA	Apr.	2·9

Assimilatory pigments

The Rhodophyta incorporates water-soluble chromoproteins as assimilatory pigments. The ecological variations of the chromoprotein contents are unknown. Phycoerythrin, phycocyanin, and chlorophyll are common assimilatory pigments occurring in the Rhodophyta. Since being detected in *G. amansii* (Yakushiji, 1935), cytochrome-*c* has been studied for its spectral and oxidoreductive characteristics (Katoh, 1959). Absorbing bands of cytochrome-*c* of *G. amansii* coincided with those of two red, one brown, two green, and one blue-green algae. *Pterocladia tenuis* showed the absorption peaks of reduced cytochrome resembling those of *Gelidium amansii* (Sugimura, Toda, Murata & Yakushiji, 1968).

Vitamins

Watanabe (1937) determined the vitamin B_2 (=riboflavin) content as 0·324, 0·14, and 0·14 μg/g fresh wt in representatives of red, brown, and green algae, respectively. *G. amansii* contained 0·05 and 0·03 μg/g fresh wt of riboflavin and lumiflavin (=photolytic riboflavin), respectively. Hashimoto & Sato (1954) determined vitamin B_{12} (=cyanocobalamin) in *G. amansii* as 0·4 μg/100 g fresh wt in the material collected at Aburatubo of Kanagawa-ken in December. Screening investigations on the vitamin B complex in 40 species were reported by Kanazawa (1961). His analysis of the complex in *G. amansii* is given in Table XXIV. He recommended *G. amansii* as a resource material for folic acid, riboflavin, and cholin. Also according to him, vitamin B_{12} is widely distributed in the animal world, but in the plant world presumably only in marine algae.

Kanazawa (1963) reviewed the distribution and structure of algal vitamins, algal physiology in relation to vitamins, and biosynthesis of vitamins in algae.

TABLE XXIV

Vitamin B complex in G. amansii *(from Kanazawa, 1961)*

	μg/g (dry basis)
B_1 (thiamine)	1·60
B_2 (riboflavin)	17·95
Niacin (nicotinic acid)	20·10
Pantothenic acid	1·22
Choline	4885
Inositol	443
	ng/g (dry basis)
Lipoic acid	570
B_{12} (cyanocobalamine)	35·9
Folic acid	782
Biotin	61

Volatile and flavour constituents

Katayama (1960) investigated the distribution of volatile constituents in the marine algae. He described that dimethylsulphide and acrylic acid were not found in Phaeophyta; furfuryl alcohol was not found in Rhodophyta. Fujiwara (1969) reviewed the flavour of seaweeds in relation to various chemical constituents including rhodoic acid of *Gelidium*.

Unknown substance

Tsujino & Saito (1961) found a new compound with an absorption maximum in the 320–330 nm range, which is specific to red algae including *G. vagum* of the Gelidiaceae. Sivalingam, Ikawa & Nisizawa (1974) and Sivalingam, Ikawa, Yokohama & Nisizawa (1974) reported a new UV-absorbing substance having an absorption maximum at 334 nm contained in *G. amansii*. Its content varies according to the depth of habitat of the material and the chlorophyll and phycoerythrin contents. They suggested that this substance might have some rôle in algal metabolism.

Katayama (1961) studied the volatile constituents of seaweeds and detected one of them from *Gelidium* sp., but it was found not to be dimethyl sulphide.

INORGANIC CONSTITUENTS

Tsuchiya (1962) reviewed the inorganic constituents of seaweeds; they contain 15–40% ash on a dry wt basis and have potentialities as a resource material for important inorganic nutrients if used for human food. Juvenile thalli contain greater amounts of inorganic elements than those of aged ones. The school of Yamamoto and Ishibashi issued many reports relating to the selective absorption of inorganic elements by seaweeds (Ishibashi & Yamamoto, 1958a,b, 1959, 1960a,b,c; Yamamoto, 1960a,b,c; Morii, 1961, 1962; Ishibashi, Yamamoto & Morii, 1962; Ishibashi, Yamamoto & Fujita, 1964a,b; Ishibashi *et al.*, 1965; Yamamoto, Fujita & Ishibashi, 1965; Yamamoto, Fujita, Shigematsu & Ishibashi, 1968; Yamamoto, Fujita & Shigematsu, 1969; Yamamoto, Fujita & Ishibashi, 1970; Yamamoto, Yamaoka, Fujita & Isoda, 1971; Yamamoto, 1972; Yamamoto & Ishibashi, 1972; Yamamoto, Otsuka & Uemura, 1976; Yamamoto, Otsuka, Okazaki & Okamoto, 1979). Table XXV shows the inogranic constituents of *G. amansii* citing these articles together with several by other workers. Of the inorganic constituents, sulphate is an important one bound to mucilaginous polysaccharides of the red seaweed *Gelidium*. Therefore, many investigators have made analyses of the sulphate or sulphur contents of the gelidiacean plants (see Hayashi & Okazaki, 1970, pp. 248 and 400). Some inorganic substances showed differences in the amounts contained in the Gelidiales. Notwithstanding, a systematic study on the variation of the amount or component of inorganic substances in relation to the following: season, geographic locality, taxonomic group, stage in growth or development, and vertical distribution is necessary.

TABLE XXV

Inorganic elements detected in G. amansii: *a, mean values are expressed in μg/g dry matter, unless stated otherwise; b, CF means concentration factor with respect to natural sea water*

Elements	Mean values[a]	CF[b]	Sample size	Remarks: Reference
Al	142	13 200	3	Yamamoto *et al.*, 1979
As	3	—	—	Tagawa & Kojima, 1976
B	177	9·21	3	H_3BO_3: 450–1500 ppm, Kobayasi, 1962
Br	(0·1612–0·1327%)		—	Sirahama *et al.*, 1944
Ca	7000	4·38	4	Yamamoto *et al.*, 1979
Ca	5 g/kg fresh material			Ueda *et al.*, 1973
Co	0·34	850	3	Yamamoto *et al.*, 1979
Cr	0·50	2500	1	Yamamoto *et al.*, 1979
Cu	7·9	653	2	Yamamoto *et al.*, 1979
Fe	356	10700	8	Yamamoto *et al.*, 1979
Ga	0·07	—	3	Yamamoto *et al.*, 1979
I	(0·0151–0·0609%)		—	Sirahama *et al.*, 1944
	(0·4302, 0·4662, 0·7262%)			Itano, 1933
K	300	0·20	1	Yamamoto *et al.*, 1979
Mg	5340	1·20	5	Yamamoto *et al.*, 1979
Mn	43	5400	3	Yamamoto *et al.*, 1979
Mo	0·27	6·68	3	Yamamoto *et al.*, 1979
Na	200	0·005	1	Yamamoto *et al.*, 1979
Ni	1·22	153	3	Yamamoto *et al.*, 1979
P	(1·06%/fresh material)		—	Tadokoro & Ugazin, 1933
Se	(0·67 ppm)	—	—	Horiguchi *et al.*, 1971
Si	4540	273	3	Yamamoto *et al.*, 1979
Sr	90	2·81	3	Yamamoto *et al.*, 1979
Sr	42 mg/kg fresh weight			Ueda *et al.*, 1973
Ti	23·1	5780	3	Yamamoto *et al.*, 1979
V	2·48	310	3	Yamamoto *et al.*, 1979
Zn	155	3170	5	Yamamoto *et al.*, 1979

Natural radio activity: 3±2 cpm/0·2 g/April in Kagosima (Tanaka, 1956a)

PHYSIOLOGICAL STUDIES IN LABORATORY CULTURE

For physiological studies, culturing in controlled experimental conditions should be realized. The development from sporelings to mature thalli has not yet been achieved in laboratory culture of gelidialean plants. Existing literature treats only with early development. The complete life history sequence is not required for physiological studies. In the growing season of spring and early summer, controlled cultures can be achieved, because of the high activity of the Gelidiales thalli at these times.

There may be differences between early development and thallus growth in individual physiological activity, and optima and tolerance of conditions. Therefore, both stages are treated separately in this section.

SPORE RELEASE, GERMINATION AND EARLY DEVELOPMENT

Many reports have described morphological aspects during spore release and early development, as mentioned before (p. 181–182). Thus, the physiological aspects of these stages related to various conditions of laboratory culture are surveyed in this subsection.

Temperature

Water temperature is a significant factor affecting spore ripening, release, germination, and development. As regards the periodicity of spore release of *G. amansii* in relation to temperature, Katada and his co-workers published a series of papers (Ueda & Katada, 1943, 1949; Katada, 1949, 1955; Katada *et al.*, 1953; Katada & Matsui, 1954). Tetra- and carpospores were released for several hours in a day (Katada, 1955). Yamasaki (1962) recorded 1600–1800 h as the time in which tetraspores were released in his experiments carried out from February to June. The commencement of carpospore release preceded that of the tetraspores. The time of release of carpospores showed seasonal variation as with that of tetraspores (Katada, 1955). The range of the hours of release varied with change of water temperature. He also found that a temperature over 25 °C promoted spore release, and that the temperature during the night affected the hour of release on the following day.

Ôno (1927) claimed that during the reproductive season the release of spores is promoted by a rise in water temperature, but delayed by a sudden fall in the temperature. The release of carpospores took place in a period of lower temperatures than that of tetraspores. The colour of stichidia immediately before spore release changes to pale yellow. I regard this colouration as the result of a secretion of a thick mucilaginous layer.

The effect of water temperature on the early development of tetra- and carpospores has never been systematically studied. Katada (1955) found no difference between both types of spore in relation to the optimal temperature effective in the early development of sporelings of *G. amansii*. Kinosita, Hirano & Takahasi (1935a,b), however, reported differences between both types of spores of *G. amansii* collected and cultured in Otaru, Hokkaidô as follows: 21 °C for the tetrasporic sporelings; 16 °C for carposporic ones. Konisita's results may be related to the fact that tetrasporic germination was made in the middle of July and carposporic germinations were throughout October. The maximum growth rate of the early developmental stages of sporelings was observed at 25 °C by Ohno (1969). Lethal temperatures for sporelings was found to be 28–29 °C (Katada, 1955). The sporelings in culture survived for 24 h at 34 °C, and for 3 h and less in 36 °C and 38 °C, respectively (Ohno, 1969). Furthermore, Yamasaki (1962) recorded 31 °C as the survival limit for six-day cultures made at the end of June.

Light quality and intensity

Katada (1949) made a preliminary investigation of the effect of light quality on the growth of *G. amansii* employing coloured bottles. The data obtained are incomplete and rigid experimental design and suitable tools were not applied. When sporelings were kept in weak light intensity, chloroplasts increased and rhizoids actively emerged. Yamasaki (1962) studied the light factors affecting the sporelings but his experimental design, data analysis, data presentation and discussion were inadequate. Ohno (1969) reported that the saturated light intensity for photosynthesis of *G. amansii* was lower than that of the species of *Ulva*, *Monostroma*, and *Sargassum* growing in the intertidal zone. Otherwise, he suggested that the saturated light intensity for the development of *Gelidium amansii* is presumably at a lower level than that for photosynthesis. In addition, it was claimed that the optimal light intensity for early development is possibly higher than that for post development and growth.

Salinity (specific gravity) and pH

Katada (1955) reported that the specific gravity (s.g.) between 1·0103 and 1·0350 in the culture at 15 °C temperature gave no difference in the rates of the survival and lengths of *G. amansii* sporelings but it promoted the elongation of rhizoids. Surviving sporelings greatly decreased in media of s.g. 1·0067 compared with that in 1·0103. No lethal s.g. has been determined. Ohno (1969) reported that the sporelings of *G. amansii* in cultures at 25 °C were very sensitive to change of salinity. Sea water in which there was maximum elongation of the sporelings had a salinity of 33·78‰ and a pH value of 8·5. All sporelings died in a culture with twice the sea water concentration. Identical results were obtained by Ogata & Matsui (1965b) in experiments on the photosynthesis of *G. amansii*. After the treatment with low concentrations of sea water, the number of tetraspores released and the germination rate of tetraspores decreased compared with those cultured in normal sea water (Yamasaki, 1962). Although the sporelings immediately after the germination tolerated low s.g. of sea water, those sporelings at the early developing stages did not.

Desiccation

Although Takamatsu (1946b) investigated the rate of spore release under conditions of desiccation, his experiment lacked controls. Katada (1955) concluded from his experiments that the desiccation of tetrasporic and carposporic plants delayed spore release, and did not induce it. Ohno (1969) investigated the tolerance to desiccation of spores immediately after their release. The absolute humidity in his experiment was not measured, but relative humidity was employed. The maximum numbers survived for four hours in conditions of 80% humidity. From these results he concluded that *G. amansii* is more sensitive to desiccation than are other intertidal species.

Darkness

Suto (1954) described the germination of spores and survival of sporelings for considerable times in darkness. The sporelings grew when re-illuminated. Ohno & Arasaki (1969) and Ohno (1969) cultured the sporelings of *G. amansii* in the darkness. After two weeks in the dark, every sporeling remained healthy in various temperatures. But over three weeks the sporelings that died increased and none were alive after four weeks. Their data show low reliability, because no statistical test for significance with regard to the relationships to temperature and light intensity was given.

Nutrition

With regard to the recovery from natural bleaching, Yamada (1967) made culture studies on the absorption of nitrogen by *G. amansii*. Nitrate and ammonium nitrate were considerably absorbed, but nitrite appeared not to be so. Urea has been utilized by *G. amansii*, but may give pharmaceutically bad effects. In the culture with ammonium sulphate, protein-N contained in *G. amansii* was maximum in comparison with that using other nitrogen sources.

Water movement

The effect of wave action has been studied by Miyazaki FES (1939). In intertidal regions, *G. amansii* occurred abundantly on substrata washed by waves. Katada (1955) studied the effect of water movement on the position of adhesion of released spores. In his experiments the spores aggregated on a line made by water movements. He claimed that the coarse surfaces of rocks and the plants and sessile animals living there would contribute to the occurrence of various micro-movements of sea water in nature.

Suto (1950a) claimed that tide has no direct relationship to spore release. Large amounts of *G. amansii* grow in places of wide vertical ranges of water movement, but small amounts of *G. amansii* occur when there is severe mixing of water. No *G. amansii* is found in places covered with mud. High rates of growth are found on the sides of rocks projecting from the sea surface where the water moves over a wide range on the vertical face.

Mechanism of spore release

The mechanism of spore release of the gelidiacean species has not been explained. Suto (1954) supposed that the spores were released by the increase of osmotic pressure of matured spores. Takamatsu (1946b) observed that drying in the shade enhanced spore release in *G. japonicum*, but no induction of release by drying was recognized with *G. amansii* and *G. pacificum* (Suto, 1950a; Katada, 1955). Artificial techniques for the enhancement or induction of spore release have not been devised.

GROWTH AND TOLERANCE OF THALLUS

Many decades ago, Okamura (1909, 1911c, 1911e) examined the tolerance of *G. amansii* thallus to changes in air temperature, desiccation, changes in salinity, and direct exposure to the sun; mature plants were found to tolerate fresh water for 0·5 to 1·5 h. These tests were, however, made using non-quantitative methods without careful experimental design.

Fukuhara (1977) collected *G. amansii* at Kitami-Esasi on Hokkaidô where the water temperature falls below 1 °C in February. Suto (1954) assumed that the optimal temperature for the growth of *Gelidium* lies in the range from 10° to 20 °C on the basis of the comparison of water temperature with yield of harvest. He suggested a decrease of the growth rate at temperatures over 20 °C in the spring in southern regions of Japan.

The addition of sodium nitrate (5 mg/l) for nitrogen source resulted in the healthy growth of sporelings of *G. amansii* during 30-days culture (Rao & Suto, 1970). Their experiments were done at 20 °C, in light of 3000 lx, and sodium phosphate (25 mg) was added. They claimed that the growth of sporelings of *G. amansii* was enhanced when unialgal bacterized cultures were transferred from sea water to artificial media. In the cultures of Yamada (1967), maximum growth rate was observed in shaking vessels kept at 22 °C with the addition of ammonium nitrate (5 mg/l). He suggested that the optimal concentration of ammonium nitrate required by *G. amansii* differed with the season. He also listed the amount of micro-nutrients necessary for the culture of *G. amansii*: chemicals promoting the growth were iron (0·05–0·1 mg/l), manganese (0·01–0·03 mg/l), vitamin B_{12} (1·0–10 μg/l), adenine, uric acid, and xanthine (0·1–1·0 ng/l. Copper (0·01–0·1 mg/l) inhibited growth, and boron (5·0–20 mg/l) barely affected it. For definite conclusions to be drawn from his studies a statistical analysis of the results is necessary.

Hirose (1978) stated that *Pterocladia tenuis* observed in 1953 perished after 22 years, *i.e.* in 1975, differing from *Gelidium amansii* which survived on the coast of Kôbe, at the bottom of Ôsaka Bay. He concluded that *G. amansii* was better able to tolerate the increase of COD (chemical oxygen demand) to above 3 mg/l.

PHOTOSYNTHESIS

Although photosynthesis and respiration studies are important to understand the gelidialean taxa physiologically and ecologically, few records have been presented on these aspects.

Influential factors

Experimental studies on photosynthesis of the Gelidiaceae were first made by Tseng & Sweeney (1946) using *G. cartilagineum*. Ogata & Matsui (1965a,b) reported on the photosynthetic activity of *G. amansii* under various conditions. Their data are summarized in Table XXVI together with those of Yokohama (1971, 1972, 1973a,b). Yamada (1967) recorded oxygen evolution values for the photosynthesis of *G. amansii* employing Winkler's method. No description of any control or replicate experiments were given in his report, in addition to an evident error in the description of the results, nor was the water temperature in the culture recorded. Yokohama (1973a) stated that the plants growing in deeper habitats are "shade-plants" on the basis that *G. amansii* collected from 10 m depth showed evidently low net assimilation (▲20 μl $O_2 \cdot cm^{-2} \cdot h^{-1}$) than that from 1 m depth. Although Yokohama (1971, 1972, 1973b) studied the effect of water temperature on the photosynthetic rate of various seaweeds, the entire data (Figs 1–3) of the first publication have been reproduced as Figures 1–3 of the last two publications without reference. In this section, I refer to Yokohama (1973b) only. The examined species including *G. amansii* showed photosynthetic optimal temperature higher in summer plants than that in winter ones. Both summer and winter plants of *G. amansii* collected in Simoda, Sizuoka-ken of central Japan showed a maximum photosynthetic rate of 30 °C but an intense decrease at 35 °C. The optimal photosynthetic temperature of

TABLE XXVI

Experiments on factors influencing photosynthesis of G. amansii: **compensation point calculated was 580 lx and $Q_{O_2} = 0{\cdot}85$ at 20 °C*

Factor	Condition	Observation	Reference
CO_2	30 °C, 10 klx	Max. photosynthetic rate at; $NaHCO_3 = 7{\cdot}5 \times 10^{-3}$M	Ogata & Matsui, 1965a
Light intensity	30 °C, $NaHCO_3$ ($7{\cdot}5 \times 10^{-3}$M) 20 °C, KOH (20%)	Max. photosynthetic rate at: 15 klx Max. photosynthetic rate at: 10 klx	Ogata & Matsui, 1965a* Yokohama, 1973a
Salinity	30 °C, 10 klx	Max. photosynthetic rate in normal concentration	Ogata & Matsui, 1965b
pH	30 °C, 10 klx	Max. photosynthetic rate at pH 7·7	Ogata & Matsui, 1965b
Desiccation	30 °C, 10 klx	Photosynthetic acitivity was inhibited proportionally	Ogata & Matsui, 1965b
Temperature	30klx, KOH (20%)	Max. photosynthetic rate at 30 °C	Yokohama, 1971, 1972, 1973a,b

G. amansii and other plants of the same species collected in winter in Komoto, Iwate-ken of northern Japan showed differences of 2–7 °C lower than the optimal temperatures of winter plants from Simoda (Hata & Yokohama, 1976).

Effects of inhibitors

Under conditions of 30 °C, 10klx and with the addition of sodium carbonate ($2 \cdot 5 \times 10^{-3}$ M), Ogata & Matsui (1965a) tested the inhibition of photosynthesis of *G. amansii* by several chemicals (10^{-6} M to 10^{-2} M of potassium cyanide, hydroxylamine, and sodium azide). Interspecific comparisons were made with regard to sensitivity to these chemicals. Yamada & Kubota (1972) reported that NH_4^+–N (1–10 mg/l), NO_2^-–N (0·25–1·0 mg/l) and chlorine (1–10 mg/l) inhibited the photosynthetic activity of *G. amansii*.

RESPIRATION

The respiration experiments are frequently made at the same time as those on photosynthesis; the former are treated below.

Influential factors

Watanabe (1937) investigated oxygen consumption of *G. amansii* as affected by the addition of various respiratory substrates to the culture medium (Table XXVII). His experiments were carried out with a Warburg

TABLE XXVII

Respiratory rates in 5 mm long thalli of G. amansii *in solution of various substances (from Watanabe, 1937): a, ranges of the differences of respiratory rates (%) calculated as increased (+) and decreased (−) volumes of O_2 in the solution of various substances compared with Q_{O_2} (0·85) in the conditions of no addition of the solutions (value, 0%, calculated means no effect); b, amino acids include a peptide (leucil-glycine) and a peptone; c, sodium salts of fatty acids were used; d, oleic and linoleic acids were used at 0·01 M concentration; e, acetone was used; f, mono-, di-, tri-, tetra-, and polysaccharides were used*

Substance added	Number of compounds used	Concentration	Range of respiratory rates[a] at 25 °C in the dark	
			Minimum	Maximum
Amino acids	16[b]	0·05 M	−13 (*d*-arginin)	+34 (*d*-isoleucin)
Fatty acids[c]	16	0·01 M	−5 (isovaleric acid)	+19 (butyric acid)
Alcohols				
Monohydric	6	0·025 M	−11 (butylalcohol)	+7 (propylalcohol)
Polyhydric	6	0·05 M	+1 (adonit)	+6 (mannit)
Carbonic acids	11	0·05 M[d]	−11 (malonic acid)	+32 (malic acid)
Aldehydes	5	0·025M	−94 (valeraldehyde)	+7 (propylaldehyde)
Ketone	1[e]	0·025 M	−16 (acetone)	
Carbohydrates[f]	15	0·05 M	0 (raffinose)	+16 (laevulose)

manometer at 25 °C in darkness. Respiratory quotient ($RQ = CO_2/O_2$ in volume) of *G. cartilagineum* ($\bar{x} = 1 \cdot 11$) was calculated by Tseng & Sweeney (1946). Addition of sodium hypochlorite at a concentration adjusted to evolve 30 mg/l of free chlorine inhibited the respiration of *G. amansii* (Yamada, 1960); he did not test the significance of the differences nor did he make replicate tests. Ogata (1968) found that when desiccated, the rate of respiration of *G. amansii* fell and when rehydrated, it slightly recovered; he concluded that *G. amansii* and *Ulva pertusa* barely tolerated desiccation. Ogata & Takada (1968) investigated the influence of salinity on oxygen consumption. Their experiments were carried out at 25 °C and $Qo_2 = 1 \cdot 99$ in natural sea water. The oxygen consumption of all the species examined increased under hypotonic conditions; in *Gelidium amansii* it was increased at half the concentration of natural sea water. Akagawa, Ikawa & Nisizawa (1972) and Nisizawa, Akagawa & Ikawa (1972) examined red, including *G. amansii*, and green algae in the course of studies on the fixation of $^{14}CO_2$ cultures of brown algae in the dark.

Effects of inhibitors

No report on systematic studies of inhibitors to respiration is available to date.

ENZYMES, HORMONES, ANTIBIOTICS AND OTHER BIOLOGICALLY ACTIVE SUBSTANCES

Takagi (1953a,b,c) and Takagi & Murata (1954a,b,c) reported optimal conditions for catalase and nitrate reductase of *G. amansii* (Table XXVIII)

TABLE XXVIII

Optimal conditions of catalase and nitrate reductase (from Takagi, 1953a,b,c; Takagi & Murata, 1954a,b,c): — means no data; a, catalase activity (f) = [1/t·log (a/a − x)] × 100; in this, a = titration value (ml) of 0·02 N $Na_2S_2O_3$ at zero min, a − x = titration value (ml) of 0·02 N $Na_2S_2O_3$ after reaction time, t = reaction time (min); b, nitrate reductase activity is expressed by the amount of NO_2^- (µg) produced

	Species	Collection date	Catalase	Nitrate reductase
Activity (*f*)	*G. amansii*	June	3·37[a]	—
		July	—	2·55[b]
	G. vagum	June	13·02[a]	—
		July	—	1·18[b]
Optimal pH	*G. amansii*	July	7·17–7·38	—
		Sept.	—	7·17–8·30
Optimal temperature (°C)	*G. amansii*	July	10–15	—
		May	—	10–12·5

and many other seaweeds. Ikawa, Asami & Nisizawa (1972) and Nisizawa (1978) recorded the activity of fucose diphosphate aldolase extracted from various marine algae including *G. amansii* during the course of an investigation on the biosynthetic pathways of mannitol contained in brown algae. Abe, Uchiyama & Sato (1972) and Abe, Uchiyama, Sato & Muto (1974) detected an auxin-like substance in extracts of *G. amansii*. The *Rf* value of this substance coincided with that of indole acetic acid. The substance enhanced the growth of *Avena* coleoptile section, but showed low activity on growth regulation. On the basis of the results of analysis by chromatograph, UV-absorption spectra, and mass spectra indole-auxins seem to be widely destributed in marine algae. Gibberellins and anti-algal substances have been studied in marine algae other than *G. amansii.* Kamimoto (1955, 1956) studied antimicrobial activities of extracts from various algae. The results of his bacteriological tests of extracts from *G. amansii* are listed in Table XXIX. Recently, Ohta (1977) named beckerelide, a new antimicrobial substance ioslated from *Beckerella subcostata* of the Gelidiaceae. During his screening investigations on marine

TABLE XXIX

Bacteriological tests of extracts from powder made of entire thallus of G. amansii *(from Kamimoto, 1955, 1956):* B. = Bacillus; M. = Mycobacterium; S. = Staphylococcus; T. = Typhoid; −, *no growth of bacteria;* ±, ⊥, +, + +, + + +, *increasing degree of bacterial growth; a, each extracting solution was initially made by adding ten times the volume of each solvent to powder of* G. amansii, *by extracting with Soxhlet apparatus, and adjusting pH to 7·0–7·2; the initial extracts (control) were diluted for bacteriological tests;* 10^{-2}, etc. *means 1/100,* etc. *dilution; b, non-germicidal or bacteriostatic; c,* M. tuberculosis *var.* hominis

Solvent for extraction	Bacterial species	Concentration of extract[a]					
		10^{-2}	5×10^{-2}	10^{-3}	5×10^{-3}	10^{-4}	Control
Ether	*B. subtillis*	+	+ +	+ +	+ +	+ +	+ +
	M. tuberculosis						
	after 4 wk	−	−	−	−	+	+
	after 8 wk	+	+	+ +	+ +	+ + +	+ + +
Acetone	*S. aureus*	+ + +	+ + +	+ + +	+ + +	+ + +	+ +
	B. subtillis	+ +	+ +	+ +	+ +	+ +	+ + +
	Escherichia coli	+	+	+	+	+	+ +
	T. bacillus	−	−	−	+	+	+ +
Alcohol	*B. subtillis*	−[b]	−[b]	−[b]	⊥	⊥	+
	B. anthracis	−	+	+	+	+	+
	Shigaella sp.	−[b]	−[b]	−[b]	+	+	+
	M. tuberculosis[c]						
	after 4 wk	+	+	+ +	+ +	+ +	+ +
	after 8 wk	+ +	+ +	+ +	+ +	+ +	±

algal antimicrobial substances, the methanol extract of this species exhibited considerable antimicrobial activity against *Bacillus subtillis*. He elucidated the structure of new halogenated lactones as antimicrobially active constituents, and named bromobeckerelide, chlorobeckerelide, and γ-pyroglutamic acid (Ohta, 1978). Very recently, Ochi (1983) studied antibiotic actvities of methanol and methylene chloride extracts of numerous marine algae by the "paper disc (assay) method" using an 8-mm disc of filter-paper. He gave the results including those of some Gelidiaceae (*Gelidium amansii*, *G. pacificum*, *Beckerella subcostata*, and *Onikusa japonica*) as all negative against *Escherichia coli*, *Bacillus subtillis*, *Saccharomyces cerevisiae*, and *Penicillium crustosum* during 48-h culture. He also presented the results that Chlorophyta showed no activity; Dictyotaceae in Phaeophyta and Bonnemaisoniaceae and Rhodomelaceae in Rhodophyta showed many species of prominent activity.

Hori, Yamamoto, Miyazawa & Ito (1979) analysed quaternary ammonium bases in seven species of seaweeds on the basis that many biologically active substances in marine algae contain nitrogen. In their analyses, *Gelidium amansii* showed the presence of beta-homobetaine, glycine betaine, gamma-butyrobetaine and stachydrine. There was a no specific difference in the distribution of these and other ammonium bases. The same authors (Hori, Miyazawa & Ito, 1981) studied the presence of hemagglutinins in 53 species of green, brown, and red marine algae with erythrocytes of rabbits, horses, sheep, ducks, chickens, and humans. They found components with strong activity in 14 species, but *G. amansii*, *Beckerella subcostata* and *Pterocladia capillacea* (=*P. tenuis*) of Gelidiaceae gave negative reactions.

NUTRITIONAL VALUE

On the basis of chemical studies of *Gelidium amansii* and another six taxa of red and brown marine algae, Mori, Kusima, Iwasaki & Omiya (1981) concluded that seaweeds are an excellent source of dietary fibre.

Kisu, Goto, Aizawa & Kimura (1980) compared the growth among four groups of rats fed various combinations of diets. The highest weight was observed in a group fed the diet composed of protein (16%), fat (10%), thalli of *G. amansii* (0·4 g), and vinegar (1·2 g) per rat per day. They concluded that this result is mainly due to the increase of absorption of inorganic elements contained largely in *G. amansii* by vinegar in the diet; they stressed the digestive importance of cellulose of *G. amansii* and agar.

MARICULTURE OR MARINE AGRONOMY OF *GELIDIUM*

Ecological observations and experiments on spores and thalli have already been described. The culture in the sea, however, has never been studied on a large scale beginning with either sporelings or thalli. One of the reasons is that various culture techniques in the sea are difficult, because thalli of *Gelidium* of considerable size grow in subtidal regions. Wildman (1974) very briefly described some aspects of the mariculture techniques and their problems with *Gelidium* and other agarophytes (Tengusa).

General textbooks at least partly dealing with the mariculture or utilization of the Gelidiales are as follows: Yendo (1911); Hoppe & Schmid (1962); Ueda, Iwamoto & Miura (1963); Levring, Hoppe & Schmid (1969); Chapman (1970); Okazaki (1971); Venkataraman, Goyal, Kaushik & Roy-choudhury (1974). The following reviews deal with various useful seaweeds: Tseng (1947) for North America; Simons (1971) for South Africa; Boney (1965) and Michanek (1975) for world-wide scope. Although Chapman's (1970) book deals with the uses of seaweeds, the numerous descriptions and figures quoted for Japan are rather ancient, for example the drawings of the working and processing of agar were reproduced from those published in an old article from the beginning of this century. In the review on the distribution and uses of seaweeds of southeast Asia published by Zaneveld (1955), the descriptions on distribution of *Gelidium amansii* and *G. latifolium* are unusable, because the literature used is taxonomically old. Suto (1965) claimed that economically, a consideration of the method of harvesting is more important than the techniques of propagation.

HARVEST AND YIELD

Ôno (1932) claimed that the optimal season for harvesting *Gelidium* was July and August, because May and June were the months when diatoms, *e.g. Arachnoidiscus* settled, and because from late autumn to the following spring is the destructive season for the thallus. The report of Miyazaki FES (1939) pointed out several facts and problems as follows. Due to its potentialities to form haptera, prostrate axes and ultimate pinnules, these probably play more important rôles in the propagation of *Gelidium* than the spores. If harvest activities were carried more often *Gelidium* beds would be long-lasting; in contrast, if the activities were not carried out at least for one year, the bed would not last as long. Recently, Cuyvers (1978) described the harvesting of *Gelidium* on Miyake-zima Island.

Variations in Gelidium *yield*

On the basis of the analysis of harvesting statistics and of the data on water temperature observed over ten years, Kinosita & Simizu (1935) and Kinosita (1944) reported on the relationship of sea-water temperature and the yield of *Gelidium* along the coast of Matumae on Hokkaidô. The yield increased over that in a previous year in the period between the next spring and summer, when the higher temperatures were between December and February. Otherwise, they claimed that the lowest temperature during December to February affects the development and growth of the sporelings of the *Gelidium* in the forthcoming spring and summer. Takayama (1938) divided the region of *Gelidium* harvesting in Mie-ken into three sub-regions on the basis of the yearly variations of the yield. The largest variations of the yield were commonly found in the region near the mouth of Ise-wan Bay which opens to the Pacific Ocean. Kinosita (1939) stated that the yield of the *Gelidium* in 1939 was poor in comparison with the average value over ten years in Hokkaidô. In 1939 there were considerably lower (4·1 °C) water temperatures in late December compared with the 10-year average (9·2 °C). He claimed the relationship between yield and temperature to be the same in the past years as in 1939. Takayama (1939b) analysed the relationship between the yield and temperature in the region along the coast of Kumano-nada on Mie-ken; when the temperature during November to January was higher, the yield was poor in the following summer. The same was described by Yamasaki (1962). On the other hand, a higher yield was gained even after lower temperatures occurred during April and May immediately before the harvesting season. But Yamada (1967) stated that the yield of the year in which the temperature in June and July was higher than that of the previous year was increased compared with that of the previous year. Contradictory statements are contained in these reports on yield, but the reports are consistent in treating temperature as a single factor which resulted in the yearly variations in yield. That temperature has been regarded as the only factor causing variations in yield may be because no thought has been given to other possible factors causing variations, such as the yearly stages of the life history, occurrence of "isoyake", light energy, nutrient salts, dissolved CO_2, and even social factors. The last of these, the social factor, can be considered because yield is a social phenomenon and not purely biological.

Short or long periods of decrease of yield have been distinguished by Miyazaki FES, but the data underlying these have not been statistically analysed. Kawana & Ebata (1959) also claimed the existence of very long-term periodicities of variation of *Gelidium* yield, as much as 16–30 years.

Seaweed fishermen call the thalli of harvestable size which has regrown about one month after the closing of the first harvest of *Gelidium*, "niban-kusa", meaning "second harvestable *Gelidium*". On the basis of experiment, Okamura (1918) explained that the dwarf plants of the underbrush, growth of which had been suppressed under "ichiban-kusa" (meaning "first harvestable *Gelidium*"), fully grows into the "niban-kusa" after a short period following the first harvest season. He proposed this hypothesis on the basis of fishermen's experiences that a vigorous recovery of biomass of *Gelidium* is to be expected when the first harvest has been heavily performed. On the contrary, Miyazaki FES (1939) regarded "niban-kusa" as regenerated fronds growing from the cut points of first harvesting activity. Okamura (1918) concluded on the basis of the observations on the sea bed that the *Gelidium* underbrush does not reach to harvestable size during only one month. In comparing these two hypotheses, it has been decided that Okamura's conclusion is valid. The marking of individual plants would be required to test completely these two hypotheses.

During the years when the harvesting activity ceased, the *Gelidium* population was destroyed (Miyazaki FES, 1939). Adequate frequency of harvesting is, therefore, important for the recovery of the harvestable biomass of *Gelidium*. In relation to the harvesting frequencies, the Miyazaki FES report indicates the important rôle of prostrate axes for the propagation of *Gelidium* on the basis of the following observations. The prostrate axes continue to grow not only during the early stages of development but also in the period after the erect axes have grown considerably. Suto (1954) recommended that the thallus base and prostrate axis be retained during the harvesting and the harvesting of one-year fronds should be prohibited; the former is difficult in practice. Compared with larger plants like the Laminariaceae, in *Gelidium* it is difficult to cut the erect frond only because of its relatively small size, and the whole thallus of *Gelidium* is apt to be torn, including the prostrate axes. Yamasaki (1962) investigated the recovery and seasonal change of the wet weight and frond height of *Gelidium* remaining after the harvest in Sirahama, Sizuoka-ken. In operations over 22 months, 1 m^2 quadrats were used for semi-random sampling with the systematic arrangement of 6 to 36 sampling sites. The sampling was done 12 times during the study. The biomass of *Gelidium* recovered by the harvest season of the next year after a heavy harvest during the late spring and summer of the previous year. He concluded that over-harvesting does not badly affect the biomass of the next year and that the great decrease of the biomass of useful seaweed called "isoyake" may not be attributed to over-harvesting. He recommended three intermittent harvests in the same place during a harvest season to obtain maximum yield from the *Gelidium* population.

In contrast, Barilotti & Silverthorne (1972) recommended August through November on Santa Cruz Island as the best harvesting period, because the agar content is maximal, reproduction has already occurred, and regeneration is at its highest rate. This optimal harvesting period seems to vary between different geographical areas as well as between subtidal and intertidal populations. Katada & Satomi (1975) calculated the mean daily rate of recovery based on the data of Nonaka, Ôsuga & Sasaki (1962). The maximum rate of the recovery was 20 g fresh $wt \cdot m^{-2} \cdot day^{-1}$. The annual amount of recovery was calculated 1·3kg dry $wt \cdot m^{-2}$. No simple comparison can be made between these figures because the units are different. Boney (1965), quoting Okamura (1911a) and Yamasaki (1960), stated that the best time for harvesting *G. amansii* in Japan appeared to be between August and October, probably coinciding with the period of maximum photosynthetic activity. Nevertheless, no direct evidence for his speculation has been presented. It is empirically said that *Gelidium* shows maximum growth in June, and the photosynthetic activity should be in maximum in this month. Besides, as mentioned before, Ôno (1932) concluded that the seasons from the end of August to the spring of the next year are unsuitable to harvest *Gelidium* as this is the time of thallus defoliation.

Resource management of Gelidium

In terms of resource management, marine algal biomass should be obtained not by traditional harvesting, but by cultivation. Biomass studies of the Gelidiales should be carried out to serve as a basis for this. Experiments on the mariculture of this group have, however, never been continuously performed. Thus, this section can treat only some attempts in the past to obtain the conservation of natural populations of this group of plants.

Conservation of natural biomass

Although propagation programmes including building up of substrata with stones or concrete blocks, weed removal and other miscellaneous activities have been frequently carried out, their results have not been statistically analysed in the following reports except for that of Kato (1955) of *Gelidium* populations occurring on built substrata. Ôisi (1917a) enumerated many technical suggestions and information for *Gelidium* harvesting as follows. Good propagation is found in areas where algae with similar forms to *G. amansii* such as *Polysiphonia*, *Scinaia*, and *Carpopeltis* are growing, and in areas which are enclosed rather than open because of the aggregation of released spores. The fronds regenerated from prostrate axes have higher potentialities for propagation than the fronds developed from spores. Sand and mud inhibit the development and growth of *Gelidium amansii* germlings by covering them. The upper portion of the rocky substratum immediately under the water surface is preferable for the propagation of Melobesioideae rather than *Gelidium*. In the places where river water stagnates, no vigorous growth of *Gelidium* is found. When the water temperature during July falls, no vigorous propagation occurs in the following year, because reproductive cells cannot mature after that month. For the purpose of breeding Ôisi recommended removing poorly grown thalli and the transplanting of strains showing rapid growth. Overharvesting must be prohibited. In places with continuous movement of sea water, other seaweeds cover the *Gelidium* population; subsequently, the growth of *Gelidium* decreases. For transplantations it is important to select excellent strains and a suitable fruiting season and to submerge fresh stones as substrata. As a rule, plants showing rapid growth give large biomasses. *Gelidium* growing in shallow water shows variable biomass from year to year, because the plants are easily detached and undergo damage. Removing sargassoid algae is recommended at least twice a year.

Ôno (1932) stated that *Gelidium* preferably grows in places showing the movement of wave rather than that of strong flow. He also claimed that for transplantation, water temperature, specific gravity, depth of growth and other conditions should be the same as those in the original place where the plant was grown. Okamura (1911d) observed the maximum occurrence of *G. amansii* in the next early summer following substrata building with newly submerged stones in August, but sequential decreases of the frequencies of plants was observed on the stones newly submerged in September to November, falling to nothing on stones submerged in December. Suto (1965) stated that the vegetation of gelidioids is strictly restricted by topography, but it shows stability after its establishment.

Silverthorne (1977) studied the rational policy of the management of *G. robustum* (Gardder) Hollenberg et Abbott, commercially valuable in California, through computational procedures. He also indicated the guidelines for further research.

Artificial substrata

Suto (1965) claimed that the primary technique for marine agronomy of the gelidioid algae is the submerging of substrata for extension of the substrata in areas preferable to the gelidioids.

Artificial propagation has consisted largely of increasing the available substrata by throwing large rocks into the coastal waters in areas near existing beds where spores will be sufficiently abundant to seed the rocks (Wildman, 1974). Ôsima (1936) put forward ideas to increase the harvestable yield of *Gelidium*. If the reef was poorly developed, it was more effective to submerge large numbers of new stones, for example as many as

32×10^4. In addition, he recommended the construction of artificial reefs and transplantation of *Gelidium* thalli from regions having no natural reefs. This idea is too difficult to realize, because newly submerged stones are very easily buried in sand. Ôno (1932) and Takayama (1939b) recommended July for submerging the stones, because this month is immediately before the season of spore release. The most effective practice is to fill crevices in the reef by rocks and this will give abundant biomass of *Gelidium*. The stones to be used should have a coarse surface and tetrahedral shape for settlement of spores. The data of Takayama (1939b) are compiled together with other peoples in Table XIII (p. 214). Miyosi (1958) reported the results of experiments on stone-submerging and manuring in relation to the propagation of *Gelidium* and recommended the seasons for the stone-submerging between May and August on the basis of achievement of harvestable sizes. Ôno (1932) recommended andesite and basalt as rocks for substrata to be submerged. Matumoto (1922) recorded the maximum numbers of spores adhering to substrata and maximum growth of *G. amansii* plants which appeared on the sides of the submerged stones. Katayama, Inaba, Hiyosi & Ôsuga (1952) transplanted *G. amansii* growing on rocks by tying the rocks with concrete blocks nearby. This experiment was done in late July on the eastern coast of the Izu-hantô peninsula. They failed to propagate *G. amansii*, and concluded that their failure was due to doing the experiment in July which is the reproductive or degradation season. For success the transplantation should be done during and before the growing season. The optimal season for submerging of stones was determined in one of two ways. One was the estimation of the month of spore release by microscopical examination of plants sampled in the sea, and the other was the measurement of *G. amansii* biomass on the stones submerged each month. Suto (1954) listed results on the biomass of *G. amansii* appearing on submerged rocks as reported by nine prefectural FES of the region from Akita-ken to Miyazaki-ken. These results were obtained during a period of less than 12 months. It is now well recognized that the selection of the season and site is important for the submerging stones. Kato (1955) first analysed statistically the factors relating to stones. He performed a two-way analysis of variance with season, locality, and quality of stones. The wet weight of *G. amansii* growing on the same stone showed significant differences between months (July, September and November) and between four localities along the coast of Tokusima-ken in September. In conclusion, at $P = 0{\cdot}95$ level, the biomass produced on the stones dominated in July and in western and southern localities, but no significant difference was found in the quality of stones. As he stated the test of significance at the $P = 0{\cdot}95$ level could not be realized, because no random sampling was done on many newly submerged stones. Ôsuga & Yamasaki (1960) observed a number of *G. amansii* plants occurring on the tetrahedral concrete blocks newly submerged at various depths. The dominant occurrence was found on the surfaces facing to the west and north of the blocks submerged at 6 m and 10 m depths. Occurrence was minimum on blocks submerged at 14 m depth. The three surfaces made an angle of 60 ° with the bottom of the tetrahedron. Their observations were not satisfactorily quantitative and the significance of differences was not statistically tested. For several years Yamasaki (1962) surveyed the variation of the wet weight of a *G. amansii* population grown on newly submerged concrete blocks (see Table XIII, p. 214). The increment of biomass (wet weight) of the population ceased three years after block-submerging. He expected that a few years after submerging of the blocks the *G. amansii* on the blocks would reach to the same density as that of the natural population. The submerged blocks in the region with "isoyake" resulted in extremely low increments of *G. amansii* biomass. Tokyo FES (1941) experimented on the propagation of *G. pacificum*. Stones measuring 43 × 32 × 12 cm were submerged monthly and the frequencies of the species growing on them were counted monthly; maximum frequencies were found on the stones submerged from June to September.

Closed season

Yamasaki (1962) graphically analysed the yield of the *G. amansii* harvest from the data accumulated during 60 years by the Sirahama Fishermen's Guild of the Izu-hantô peninsula. In this region, the 'closed season' had been established during 20 to 30 days in June. The yield rapidly decreased by the harvest activities had considerably recovered after the closed season. The yield obtained after the closed season was usually higher than that immediately before the season. On the basis of these facts he proposed that unless the closed season is established, the yield should decrease from year to year. There are critical problems about his interpretations of the effects of the closed season, because no controlled area was set up in his design of field observations. In addition, no definition of the term 'closed season' was given. The definition of the term is that time in which harvesting is prohibited in spite of harvestable quantities of the fully grown plants of *Gelidium* remaining. On the other hand, the 'actual closed season' is usually set at the time when *G. amansii* populations barely remain due to the heavy harvesting activities. Thus, the word "closed" has no meaning, because there are practically bare areas immediately before the "actual" closed season. This discussion of the result in Sirahama is supported by that in Kumomi (western coast of Izu-hantô), in the reverse sense. In the latter area, the closed season as strictly defined above is practised for conservation of the *Gelidium* resource (Yamasaki, 1962); *i.e.* the closed season is set during one month at the time when there remain considerable amounts of plants of harvestable size. In this case, the recovery of biomass of *Gelidium* rarely occurred. The practice of correct 'closed season' had no effect. This closing is not effective because of almost a complete harvest before the onset of closing and the short duration of the closed period. To replace closed periods of this kind, I recommend yearly regional closing, as is the case with the large Phaeophyta in Japan.

Ôisi (1917b) reported that at Nisina, Sizuoka-ken, success was realized by prohibiting the harvesting of *Gelidium* until the mid-season of spore-releasing. Although he stated that this success is because of conservation of the following generation, as stated before, the propagation of *Gelidium* is strongly related to vegetative propagation of prostrate axes.

Removing algal weeds, and denudation of surface of substratum covered by coralline algae

Okamura (1911c,e, 1917) experimented on ways of increasing the propagation of *G. amansii* involving the removing algal weeds. After one year of removing sargassoid algae, the biomass of *G. amansii* increased, but *Cladophora wrightiana* dominated over *Gelidium* after two years. Removing Melobesioideae, non-articulated coralline algae, by crushing with a large iron ball gave no effective result for the propagation of *G. amansii* population. Instead of this technique he recommended cracking the substratum itself to produce fresh surfaces. Tokyo FES (1941) determined the suitable season for the removing of algal weeds as being in July to September on the basis of the monthly observations of stones submerged monthly. Summarized results of weed-removing were reported by various prefectural FES, but the data were obtained using an unreliable method (Suto, 1954).

Culture techniques in the sea

Yendo (1911) enumerated the factors affecting to the transplantation of gelidioid algae as follows: light quality, water temperature, salinity, depth of the substrata, quality of the substrata, the age of plant to be transplanted or cultured. He pointed out that of these factors, the first five should be identical with those of original place of life of the plants; on the last factor, the plants to be transplanted should be young, because the fertile thalli are easy to decompose. Ueda (1936) experimented by suspending *G. amansii* fragments (10–12 cm long) from the sea surface. No growth was observed during August and November, but rapid growth was found in early December. Tôyama (1939) reported that good growth was achieved in the following March after mature plants were twisted into rope and tied to earthen pots, which were then submerged on a sandy bottom of the sea in the middle of August. Huzimori (1940) devised two techniques for the mariculture of *G. amansii*. In the first, the "fragment twining method", thallus fragments were twined into rope and suspended from the sea surface. The minimum length of the fragments sufficient to obtain growth was 5 mm. The fragments suspended in the sea developed prostrate axes and were then fixed to the rope by them. A calm, embayed area is preferable for the culture of *Gelidium* using this technique. The result of the experiment practised in Tokyo-wan Bay from March to April was that the fragments of 6 cm long grew up to 10 cm in 30 days after suspension and bore branchlets. The wet weight also increased 5–8 times after 30 days. Except for the experiment in summer the fragments 5–10 cm length grew to harvestable size after two months. After harvest the remaining basal portions of fragments continued to grow. If bimonthly harvest is practised the method can be economically established. October has been recommended as the season for rope suspension, because in the coming winter *G. amansii* will reach the growing season and epibionts will be lost. The second method, "sporeling transplanting method" involved the following procedures. Thirty days after released spores had adhered to slides, the sporelings which had developed were scraped off the slide and transplanted into a large vessel followed to the adhesion of the sporelings to thin rope by mixing them with sea water. After 4–5 days, the sporelings developed several rhizoids, so fixing them to the rope. The thin rope was then twined into a thick rope of 1 cm diameter and suspended in the sea. The experiment unfortunately failed, because epibionts heavily attacked the space among the sporelings of *G. amansii*. Field culture using any techniques has not been practised up to the present in Japan because of economic aspects. MacFarlane (1968) introduced a field culture experiment of *Gelidium* employing fragments at Hamazima, Mie-ken FES. Katada (1955) noticed that the regeneration potentiality of *Gelidium* sporelings rather than culture can be applied to the propagation technique. Gibor (1976) recently published on an artificial flotation method using Perlite, a generic name applied to some volcanic rocks, as a most promising flotation substance for the cultivation of benthic algae, including *Gelidium*. Cyano-acrylate contact glue proved very effective and non-toxic for fixing *Gelidium* species to Perlite particles. Growth studies using this method are in progress. Ôsima (1936) studied a manuring technique employing nitrate and phosphate salts to enhance the growth of *G. amansii* plants growing on the sea bed. Yamada (1967) developed Ôsima's work on manuring of the natural population on the coast of Izu-hantô peninsula. He devised a fertilizer composed of nitrate and phosphate salts with guano manure as a binder, because of the low cost and little dissolution of the effective constituents. He claimed that the height and weight of thalli in the manured area was increased in comparison with those in the controlled area except for a few cases but he did not test the significance of the differences observed.

A mass seeding technique at the relevant time has not been devised, because the control of spore release has not been established (Suto, 1950a). Suto (1954) pointed out that the indoor seeding and seed stock transplantation of *Gelidium* are difficult at the mass level. A comparison of tetrasporophyte and carposporic plants was attempted by him to assess the suitability as seed stock but reliable reports are not available.

Neish (1979) has reviewed on the developments in the culture of algae and seaweeds and the future of the industry.

DISADVANTAGES TO *GELIDIUM* POPULATION

At the present day *Gelidium* is harvested from natural populations, therefore, the variation, especially decrease and extinction, of the biomass of *Gelidium* directly effects the yield.

Yendo (1911) discussed the decrease of the vegetation of (useful) sea-weeds: (1) the decrease by "isoyake"; (2) the decrease caused by the competition of useful and unuseful algae; and (3) the decrease by inappropriate ways of harvesting, *i.e.* exhaustive harvesting and too long duration of harvest activity.

The "isoyake"

The term "isoyake" was given to the great decrease or disappearance of useful seaweeds by Japanese seaweed fishermen. Tamura (1951) briefly reviewed "isoyake" with respect to the shallow-water culture of marine organisms. No scientific explanation has been given, but a few practical comments have been made. Yendo (1903a) found the following three stages within the course of occurrence of the "isoyake": (1) the decrease of *Gelidium*, *Eisenia bicyclis*, lobsters, abalones; (2) extinction of useful seaweeds, but sargassoid algae remain; and (3) substratum becomes white or whitish-yellow because of the death of coralline algae, organisms become absent except for small quantities of the sargassoids. He also described that the damage of the "isoyake" was not found at depths of 30 m or more but was great in shallow-water areas. He noticed that a large river opened near the damaged area of "isoyake", and concluded that the "isoyake" is caused by dilution of sea water by river water. Okamura (1908, 1909) thought that the "isoyake" of *Gelidium* occurred on the eastern coast of the Izu-hantô peninsula probably because of the over-harvesting of *Gelidium* which results in an increase of non-articulated coralline algae. Thus, the "isoyake" of *Gelidium* may occur due to the occupation of the substratum by spores of the corallines. Erect fronds of *Gelidium* are frequently covered by non-articulated coralline algae (Miyazaki FES, 1939), but *Gelidium* has potentialities of vegetative propagation even under the cover of such algae (Ôno, 1927). The potentiality is due to the ability of high propagation or regeneration by the prostrate axes. Miyazaki FES (1939) described a "rock-surface scraping machine" fitted with small gears and iron brushes devised to crush coralline algae and to aid *Gelidium* propagation. The younger buds of *Gelidium* are frequently found surviving burial under the cover of non-articulated corallines. This machine can be used during any stage of the life cycle of *Gelidium*.

Okamura & Tago (1915) investigated the "isoyake" of *Laminaria* in Aomori-ken. The great decrease of *Laminaria* possibly took place because of the approach to the Tusima (warm) Current in the Sea of Japan to the Aomori coast and because of the mixing of a large volume of river water with sea water. The former cause might result in unsuitable changes of temperature of coastal water for *Laminaria* which is adapted to colder water. Alternatively, the latter might result in a lower salinity of the coastal water and a difficulty of adjusting osmotic pressure by the Laminariales. They also surveyed the same case and the "isoyake" has been defined by them as the state showing seaweed populations heavily destroyed by certain circumstantial conditions followed by changes of the surface of the sea bed into a denuded state except for the occurrence of non-articulated corallines. Ohmi (1951) studied the "isoyake" of *Laminaria japonica* and *Undaria pinnatifida* occurring in the region around the Tugaru-kaikyô strait. The population of *Gelidium* also became poor and its substratum was covered by Melobesioideae. *Sargassum thunbergii* had grown in small quanitities. He concluded that the "isoyake" in this case was directly attributable to the sedimentation of volcanic ash from Mount Komaga-take. It is questionable whether or not he neglected common environmental factors such as the competition with other algae and the succession of vegetation. He supposed that the case of Aomori-ken was principally due to the shift of Tusima Current by which the temperature and chemical nature of the sea water changed, and the attachment of floating ice. The basis of his reasoning appears to be the heavy damage of *Laminaria* adapted to the cold current region. Other than *Laminaria*, a small decrease was observed in *Gelidium* and other algal populations adapted to the warm current. Kawana (1956) examined depths of 15 m or more; he claimed that the "isokaye" of *Gelidium* in Izu-hantô peninsula extended out to the neighbouring region due to the effect of current and water temperature. He explained that the "isoyake" is caused by the change of sea-water temperature considerably shifted from the optimal temperature for the germination and growth of *Gelidium*. If the shift of the axes of Kurosiwo or Oyasiwo Currents along the Pacific coast is the exclusive factor of the "isoyake", the "isoyake" can be forecast. Such forecasting has not been practised. Mie-ken FES (1931) reported the "isoyake" in spring, when the specific gravity of the sea water in late winter increases to over 1·0257. This development of "isoyake" is presumably because of the non-occurrence of juvenile fronds in the more concentrated sea water. Very recently, Masaki & Akioka (1980) investigated the "isoyake" that occurred in Hokkaidô. According to them, a shift of axis of the sea current leading to the "isoyake" following the rise of water temperature did not apply to the occurrence on the Sea of Japan coast of Hokkaidô. In the oligotrophic state, especially involving phosphate salts, of the Sea of Japan which is in the warm Tusima Current region, the "isoyake" may occur on the Sea of Japan coast of Hokkaidô. General factors related to all the regions damaged by "isoyake" were not defined and specific factors on each occasion in each region should be considered. As seen in the above descriptions, the results and conclusions of the study of the "isoyake" are many and various. The cause of the "isoyake" is not known. The "isoyake" has also not been clearly defined. Therefore, I present the definition as : the state showing that, no matter what the cause, although the melobesioidean and/or sargassoid algae are growing, other algae with fronds of considerable size are greatly reduced or completely lost even if they are in their growing season. This state becomes critical when the lost algae are economically useful ones, *e.g. Laminaria* and *Gelidium*. I have attempted to analyse each of the causes of "isoyake" so far claimed by several authors.

Decrease of water temperature (Okamura & Tago, 1915; Kawana, 1956; Yamasaki, 1962). The range of optimum temperature for the life of mature and young fronds of *Gelidium* may not be so narrow that the "isoyake" occurs when the temperature of coastal water suddenly falls because of approach of a cold current or cold mass of sea water to the coast. Water temperature becomes critical for the ripening, releasing, and germination of spores. Because of the perennial nature of *Gelidium*, persisting portions of the *Gelidium* frond can reproduce by bearing spores again in the next year, even if a decrease of temperature occurs during the ripening and releasing of spores, and this can be actually ignored as the cause of "isoyake". In addition, low temperature which badly affects the stages of life cycle of *Gelidium* is probably not maintained for 12 months. In view of the competition of *Gelidium* with other algae, especially non-articulated corallines, the decrease or "isoyake" of the *Gelidium* population can, however, occur after a few years in which the fall in water temperature reaches a level inappropriate for the growth development of *Gelidium* during summer or from winter to spring. On the other hand, it is generally noted that vegetative propagation of prostrate axes rather than sporelings derived from spores is more effective for the propagation of *Gelidium* plants. This suggests a limited effect of temperature on the *Gelidium* "isoyake", because it is to be expected that the temperature does not vary between wide enough ranges in which lowest extreme kills matured tissue such as the prostrate axes. Consequently, temperature can be ignored as a singly cause of the "isoyake".

The above discussion of lower temperatures can also be applied in the case of higher ones. According to Akatsuka (in prep.) the upper limit of water temperature on coasts where *Gelidium* (including *G. amansii*) grows is 28–29°C. Temperatures above this value do not really exist on coasts except in tide-pools. Temperature was also given as the upper limit of tolerance of *Gelidium* plants by Katada (1955). Yamasaki (1962) reported the interpretation of harvesting statistics through graphical presentation, and in spite of showing no data on coastal water temperature, he stated the approach of the warm Kurosiwo Current to be the cause of "isoyake" on the Izu-hantô peninsula.

Large volume of river water run into the coastal sea water as a result of heavy rain (Yendo, 1903a,b; Okamura & Tago, 1915). According to Yendo (1911), the characteristics of places showing the "isoyake" are as follows: the presence of strong current offshore of the coast; the presence of a large river running into the sea near the place; if the sea bottom is shallower than that surrounding the area; no decrease is observed at the intertidal zone. He concluded that the primary factor for the occurrence of the "isoyake" is the running of a large volume of fresh water from rivers into the sea. The large volume of fresh water lies on the bottom in front of the coast, because the mixing of fresh and sea waters does not occur by the disturbance of strong current flowing offshore of the coast. Fresh water stays at the depth of about 8 m because of its lower specific gravitiy than that of sea water. Therefore, when the sea bottom is at a depth of about 8 m, sublittoral algae die due to the change of osmotic pressure.

Over-harvesting of the Gelidium *population* (Okamura, 1908, 1909). This hypothesis is based on the idea that the substratum for the adhesion of spores of melobesioidean corallines increases because of over-harvesting of gelidioid algae, and finally the propagation of the corallines leads to the "isoyake". But Miyazaki FES (1939) reported that when the *Gelidium* population remained without harvesting, the population was damaged by other algae; when harvested satisfactorily, development of the population was observed.

Oligotrophy of coastal sea water (Masaki & Akioka, 1980). Based on the idea that the growth of articulated coralline algae is accelerated by decreases in phosphate salts (Brown, Ducker & Rowan, 1977), Masaki & Akioka (1980) stated that the decrease of common seaweeds and the increase of corallines in the western coast of Hokkaidô may be caused by the decrease of dissolved nutrient salts in the Sea of Japan.

Grazing by phycophagous animals (Kito, Kikuchi & Uki, 1980). When the number of phycovorous animals using gelidioid algae as their food increases among the gelidioid population, young plants are also eaten. These authors concluded that the extinction of seaweeds on sea bed of Iwate-ken takes place due to over-grazing by animals such as abalones and sea-urchins. In this case, unless being eaten by the animals, non-articulated coralline algae can spread so occupying substratum, and this can become a cause of the "isoyake". According to Masaki (pers. comm.), however, there are observations that molluscs, *e.g.* limpets and perhaps sea-urchins, feed on non-articulated corallines, especially those with several layers of cover cells. On the contrary, Ôno (1927) and Miyazaki FES (1939) stated that young thalli of *Gelidium* survived under the cover of the corallines. Studies of competition and grazing by these animals under controlled conditions is needed. Suto (1965) stated that almost all attempts to recover the vegetation of gelidioid algae from the "isoyake" state have failed.

Experiments to demonstrate the occurrence of the "isoyake", are almost impossible, because in such experiments each of the above factors must be artificially controlled over a wide area, and they can be easily upset by the effect of environmental conditions.

SUMMARY

Throughout the world, species of *Gelidium* and *Pterocladia* are obviously numerous but the real existence of some is doubtful. In early times many taxonomists of gelidioid algae seem to have been oblivious of the high variability of the external morphology or to have temporarily described many species without regard of it even if they were aware of it.

The shape, arrangement, and orientation of surface cells are useful taxonomic criteria for genera of the Gelidiales. Cystocarpic structure cannot be used with certainty as a valuable criterion for distinguishing *Pterocladia* from *Gelidium*. Among the genera of the Order, it is speculated that the genus *Gelidiella* (Gelidiellaceae) is most primitive, and *Yatabella* and *Acanthopeltis* (Gelidiaceae) are most differentiated.

Differences among the higher ranks of algae are found in chloroplast structure by transmission electron microscopic (TEM) observations. But differences at the species level have not been found with TEM. TEM or scanning electron microscope (SEM) studies of the cell walls and cuticle of the Gelidiales have not been made. These structures can, however, be expected to be of interest particularly in relation to their chemical analysis.

Histochemical study of the Gelidiales has rarely been reported. Investigations of polysaccharides and sulphates, being primary components of the gelidialean plants, on a seasonal and geographical basis, as well as of the thallus during development, would be valuable.

Chromosome counts are necessary for the phylogenetic speculation of the members of the Gelidiaceae; so far the chromosome numbers of few species of the family have been determined with certainty. Because the Gelidiaceae possesses small nuclei and not a few numbers of the chromosomes, obtaining precise and accurate data is difficult.

In species of the genera, except for *Beckerella*, it has been established that the germination of spores is "gelidial-type". Morphological studies of early development have been completed in detail, but observations during post-developmental stages are lacking, because culture techniques to develop the sporeling into the mature thallus have not been perfected. In the course of the development of erect and prostrate axes, two types have been described; these types must be studied experimentally.

A satisfactory knowledge of geographical distribution of the Gelidiales in Japan exists, except for *Beckerella* and *Yatabella*. Of these two genera, the former has species growing in deep habitats and the latter has not been reported in Japan from localities other than Oryûzako, Miyazaki-ken. The distribution of *Gelidium amansii*, the most common species of *Gelidium* in Japan, is not quite satisfactory in the region between Madagascar and Taiwan. On the other hand, many species of *Gelidium* which were made with dividing the populations of *G. amansii* of Pacific northwest America closely resemble the *G. amansii* of Japan. Therefore, these species must be carefully examined and their numerical characters tested for comparison with the Japanese *G. amansii*.

General descriptions of the vertical distribution are needed for all four seasons at many sites extending over the entire Japanese coasts. Such systematic studies do not exist.

A survey of each phase in life cycle of the gelidioid algae should be made at least bimonthly for 12 or 24 months at many sites on the entire Japanese coasts. The variability of proportion of each generation existing at the same time and in the same place should be studied in this way. Three years is the minimum length of life of *G. amansii*; to survey the maximum longevity has less meaning, except for its rôle as a basis for the selection of good strains for mass culturing. The variability of longevity of *G. amansii* needs to be investigated geographically and in relation to the depth of habitat.

There is no report on the competition and succession of the gelidioid algae; as the "isoyake" should relate to these phenomena, a continual survey is required. Physiological studies on regeneration will provide important basic knowledge for the artificial culture of gelidioid plants. Few studies are available on variability of biomass of the gelidioids in the sea. Also there is no systematic study on the difference of growth of individual plants in indoor culture, because culture techniques for *Gelidium* have not been established.

Long-term observations on gelidioid populations in the sea to estimate their biomass variation are difficult because of the disturbance by wave action and the difficulty of establishing an area where harvesting of the gelidioids is prohibited. This kind of observation can, however, succeed with intertidal species, *e.g. G. pusillum* and *G. japonicum*. Because the gelidioids are perennial, their precise ecological study needs to be made continuously over at least 24 months.

Systematic reports with high reliability are not available concerning the variation of agar content. In many early reports dealing with the general analysis of the chemical composition of seaweeds, the identification of the species used was doubtful or completely mistaken. Furthermore, a number of reports on this kind give neither date nor place of collection, nor sample size. Chemical analysis of thallus without any attention to the accuracy of identification of species and the variability of contents is not scientifically acceptable. The composition and content of polysaccharides show less distinct difference among taxonomic groups than those of inorganic constituents. The biochemical study of sulphates contained in gelidioids is interesting in relation to their metabolism.

Most physiological studies of the Gelidiales are known only for the early developmental stages of the thallus. Although there are some reports dealing with several physical and chemical factors, they were made in relation to the taxonomic aspects or the depth of habitats. Pure physiological studies of the gelidioids dealing with topics other than taxonomic and ecological aspects are scarce.

Artificial culture techniques for seaweeds, especially larger ones, will be further developed in the near future. Technology for the successful mariculture of seaweeds and the establishment of them as natural resources are being developed in some research institutes of the world at present. Mariculture of *Gelidium* is also being experimented on by some investigators. Difficulty in the harvesting technique is the biggest problem in *Gelidium* mariculture, because of its small size compared with *Laminaria* and *Undaria* in Japan. Therefore, indoor or outdoor mass culture techniques are essential for *Gelidium*. Preliminary experiments of indoor culture using *G. pusillum*, *G. divaricatum*, *G. japonicum* and *Pterocladia* species, which are intertidal species, can be expected to yield valuable data, on the basis of their high tolerance to environmental changes. Indoor-, outdoor-, or mariculture will be of great value if investigated properly, on the basis of speculation that (sulphated) galactan contained in gelidioid algae may not be artificially synthesized in the near future. At present the artificial culture of *Gelidium* should be restricted to methods using vegetative propagation through regeneration capability, because the culture initiated from spores is known to be unsuccessful because of the low tolerance of sporelings to environmental factors.

It can be concluded about the studies of each field of gelidioid biology that most of them are insufficient, fragmentary, and not systematic. Investigation, especially related to ecological aspects, must be published as the work is completed, with continuation over at least 12 months and preferably for 24 or more. Finally, modern biological sciences require the random sampling of material, careful experimental design, and statistical analysis of the results.

ACKNOWLEDGEMENTS

Dr Margaret Barnes, the editor and Professor A. D. Boney of the University of Glasgow improved the English of the manuscript. Dr E. Nishide of the Faculty of Agriculture and Veterinary Science of Nihon University kindly read the section on chemical constituents.

REFERENCES

*Japanese, Chinese and Korean articles with English or German summary.
**The above languages without summary in the last two languages.

Abe, H., Uchiyama, M. & Sato, R., 1972. *Agric. biol. Chem.*, **36**, 2259-2260.

Abe, H., Uchiyama, M., Sato, R. & Muto, S., 1974. *Proc. int. Conf. Plant Growth Substances*, **8**, 201-206.

Akagawa, H., Ikawa, T. & Nisizawa, K., 1972. *Bontanica mar.*, **15**, 126-132.

*Akahane, T. & Fuse, T., 1968. *Nippon Nôgei Kagaku Kaisi*, **42**, 740-742.

Akahane, T., Kawashima, S., Hirao, I., Shimizu, T. & Minakata, A., 1982. *Polymer J.*, **14**, 181-188.

*Akatsuka, I., 1970a. *Bull. Jap. Soc. Phycol.*, **18**, 72-76.

*Akatsuka, I., 1970b. *Bull. Jap. Soc. Phycol.*, **18**, *112*-115.

*Akatsuka, I., 1973a. *Bull. Jap. Soc. Phycol.*, **21**, 39-42.

Akatsuka, I., 1973b. *J. Jap. Bot.*, **48**, 52-54.

Akatsuka, I., 1978. *Revue algol.*, n.s., **13**, 349-358.

Akatsuka, I., 1979. *Revue algol.*, n.s., **14**, 17-20.

Akatsuka, I., 1981. *Nova Hedwigia*, **35**, 453-463.

**Akatsuka, I., 1982a. *Makusa wo Tyûsin to Sita Tengusa-rui Kenkyû no Genzyô: Sôrui no Kyôzai-ka ni Kansuru Kenkyû I.* Nippon Sigaku Kyôiku Kenkyûzyo Tyôsa Siryô, No. 86, 72 pp.

*Akatsuka, I., 1982b. *Jap. J. Phycol.*, **30**, 235-236.

Akatsuka, I., 1982c. *Proc. Education Inst. Private Sch. Japan*, **18, *Ser. 2*, 371-396

Akatsuka, I., 1982d. *Nova Hedwigia*, **36**, 759-774.

Akatsuka, I., 1983. *Nova Hedwigia*, **38**, 197-207.

Akatsuka, I., 1986. *Botanica mar.*, **29**, 59-68.

Akatsuka, I. & Iwamoto, K., 1979. *Botanica mar.*, **22**, 367-370.

Akatsuka, I. & Masaki, T., 1983. *Bull. Fac. Fish. Hokkaidô Univ.*, **34**, 11-19.

*Ando, Y., 1953. *Bull. Jap. Soc. scient. Fish.*, **19**, 713-716.

Aoki, K., 1940. *Bull. Jap. Soc. scient. Fish.*, **8, 223-226.

Aoki, K. & Yasuda, K., 1941. *Tyôsen Zenra Nandô Suisan Sikenzyô Hôkoku, Mokpo (Korea)*, **14, 81-95.

**Araki, C., 1932. In, *Kyôto Kôtô Kôgei Senmon Gakkô 30 Syûnen Kinen Hô*, pp. 131-164.

Araki, C., 1937. *Nippon Kagaku Kaisi*, **58, 1214-1234.

Araki, C., 1956. *Mem. Fac. ind. Arts Kyôto tech. Univ., Sci. Technol.*, **5**, 21-25.

**Araki, C. & Hashi, Y., 1948. In, *Colln of Treatises in Commem. of 45 years Anniv. of Kyôto tech. Coll.*, pp. 64-68.

**Arasaki, S., 1964. *How to Know the Seaweeds of Japan and Its Vicinity, Fully Illustrated in Colours.* Hokuryûkan, Tokyo, 217 pp.

**Asahi Newspaper, 1978a. *Sekai no Syokubutu: Syûkan Asahi Hyakka*, No. 113. Asahi Newspaper, Tokyo, 28 pp.

**Asahi Newspaper, 1978b. *Sekai no Syokubutu: Syûkan Asahi Hyakka*, No. 114. Asahi Newspaper, Tokyo, 28 pp.

**Asahi Newspaper, 1978c *Sekai no Syokubutu: Syûkan Asahi Hyakka*, No. 115. Asahi Newspaper, Tokyo, 28 pp.

*Asano, M., Kuroda, K. & Asahi, T., 1951. *Bull. Hokkaidô reg. Fish. Res. Lab.*, **1**, 28-35.

Barilotti, C. D. & Silverthorne, W., 1972. *Proc. int. Seaweed Symp.*, **7**, 255-261.

Barry, V. C., Dillon, T. & McGettrick, W., 1942. *J. chem. Soc.*, 183-185.

Black, W. A. P., Richardson, W. D. & Walker, F. T., 1959. *Econ. Proc. r. Dublin Soc.*, **4**, 137-149.

Boillot, A, 1963. *Revue gén. Bot.*, **70**, 129-137.

Boney, A. D., 1965. *Adv. mar. Biol.,* **3,** 105-253.
Brown, V., Ducker, S. C. & Rowan, K. S., 1977. *Phycologia,* **16,** 125-131.
Chapman, V. J., 1970. *Seaweeds and Their Uses,* Methuen, London, 2nd edition, 304 pp.
**Chen, X. -H. & Wu, Z. -S., 1953. *Rep. Taiwan Fish. Res. Inst.,* for 1953, 77-79.
Chiang, Y. -M., 1962. *Taiwania,* **8,** 131-153.
Chiang, Y. -M., 1973. *Bull. Jap. Soc. Phycol.,* **21,** 97-102.
**Chihara, M., 1970. *Common Seaweeds of Japan in Colour.* Hoikusha, Ôsaka, 173 pp., 64 pls.
*Chihara, M., 1972. *Mem. natn. Sci. Mus.,* **5,** 151-162.
Chihara, M. & Kamura, S., 1963. *Phycologia,* **3,** 69-74.
Cotton, A. D., 1915. *J. Linn. Soc., Bot.,* **43,** 137-231, pls. 4-10.
Creac'h, P. V., 1953. *Proc. int. Seaweed Symp.,* **1,** 42-43.
Cuyvers, L., 1978. *Sea Frontiers,* 285-293.
De Toni, G. B., 1895. *Memorie R. Ist. veneto Sci.,* **25,** 1-78, pls. 1-2.
De Toni, G. B., 1897. *Sylloge Algarum, Vol. 4, Sect. 1.* Seminary, Padua, 386 pp.
Dickinson, C. I., 1949. *Kew Bull.,* 565-567, pls. 3-4.
Dixon, P. S., 1954. *Br. phycol. Bull.,* **1,** p. 4 only.
Dixon, P. S., 1958a. *Proc. int. Seaweed Symp.,* **3,** 15-16.
Dixon, P. S., 1958b. *Ann. Bot.,* n.s., **22,** 353-368.
Dixon, P. S., 1961. *Botanica mar.,* **3,** 1-16.
Dixon, P. S., 1963. *Syst. Ass. Publ.,* No. 5, 51-62.
Dixon, P. S., 1966. In, *Trends in Plant Morphogenesis,* edited by E. G. Cutter, Longmans & Green, London, pp. 45-63.
Dixon, P. S., 1973. *Biology of the Rhodophyta.* Oliver & Boyd, Edinburgh, 285 pp.
Dixon, P. S. & Irvine, L. M., 1977a. *Bot. Notiser,* **130,** 137-141.
Dixon, P. S. & Irvine, L. M., 1977b. *Seaweeds of the British Isles. Vol. I. Part 1.* Br. Mus. (Nat. Hist.), London, xi + 252 pp.
Dromogoole, F. I. & Booth, W. E., 1985. *N.Z.J. mar. freshw. Res.,* **19,** 43-48.
Eisses, J., 1953. *Indones. J. nat. Sci.,* **109,** 41-56.
*Endo, T. & Matsudaira, Y., 1960. *Bull. Jap. Soc. scient. Fish.,* **26,** 871-876.
Eppley, R. W. & Cyrus, C. C., 1960. *Biol. Bull. mar. biol. Lab., Woods Hole,* **118,** 55-65.
Fan, K.-C., 1951. *Rep. Lab. Biol. Taiwan Fish. Res. Inst.,* **2,** 1-22, pls. 1-5.
Fan, K.-C., 1961. *Univ. Calif. Publs Bot.,* **32,** 315-368.
Feldmann, J. & Hamel, G., 1936. *Revue algol.,* **9,** 85-140, pls. 2-6.
Felicini, G. P., 1970. *G. Bot. Ital.,* **104,** 35-47.
Felicini, G. P. & Arrigoni, O., 1967. *G. Bot. Ital.,* **101,** 199-217.
Felicini, G. P. & Perrone, C., 1972. *G. Bot. Ital.,* **106,** 351-358.
Felicini, G. P. & Perrone, C., 1986. *Phycologia,* **25,** 37-46.
*Fujiwara, T., 1969. *Kôryô,* No. 92, 35-44.
*Fukuhara, E., 1977. *Bull. Jap. Soc. Phycol.,* **25,** Suppl., 43-44.
Funk, G., 1927. *Pubbl. Staz. zool. Napoli,* **7,** Suppl., 507 pp. Tav. I-II.
**Gakken, 1975. *Gakken Tyûkôsei Zukan. Vol. 12: Kaisô.* Gakken, Tokyo, 290 pp.
Gardner, N. L., 1927. *Univ. Calif. Publs Bot.,* **13,** 273-318, pls. 36-54.
Gayral, P., 1966. *Les Algues des Côtes Françaises.* Deren, Paris, 2nd edition, 633 pp.
Gibor, A., 1976. *Botanica mar.,* **19,** 397-399.
Goff, L. J., 1984. *J. Phycol.,* **20,** Suppl., p. 22 only.
*Hachiga, M. & Hayashi, K., 1966. *Hakkô Kôgaku Zassi,* **44,** 753-757.
Hamana, K. & Matsuzaki, S., 1982. *J. Biochem.,* **91,** 1321-1328.
Hara, Y., 1972. *Proc. int. Seaweed Symp.,* **7,** 153-158.
Hara, Y. & Chihara, M., 1974. *Sci. Rep. Tokyo Kyôiku Daig., Sect. B,* **15,** 209-235, pls. I-XV.

Harada, K. & Hirano, M., 1983. *Bull. Jap. Soc. scient. Fish.*, **49**, 1547–1551.
Harada, K. & Kawasaki, O., 1982. *Bull. Jap. Soc. scient. Fish.* **48**, 617–621.
Harvey, W. H., 1860. *Proc. Am. Acad. Arts Sci.*, **4**, 327–335.
*Hashimoto, Y. & Sato, T., 1954. *Bull. Jap. Soc. scient. Fish.*, **19**, 987–990.
*Hata, M. & Yokohama, Y., 1976. *Bull. Jap. Soc. Phycol.*, **24**, 1–7.
**Hayashi, K. & Okazaki, A., 1970. *Kanten Handobukku.* Kôrin Syoin, Tokyo, 534 pp.
Hellebust, J. A., 1970. In, *Marine Ecology, Vol. I, Part 1,* edited by O. Kinne, Wiley-Interscience, London, pp. 125–158.
**Higashi, M., 1934. *Gensyoku Nippon Kaisô Zuhu.* Seibundô, Tokyo, 80+5+4 pp., 80 pls.
Higashi, M., 1936. *Suisan Kenkyû Si,* **31, 1–9.
Hirose, H., 1978. *Proc. int. Seaweed Symp.*, **9**, 173–179.
Hommersand, M. H., 1972. *Proc. int. Seaweed Symp.*, **7**, 66–71.
Hoppe, H. A. & Schmid, O. J., 1962. *Botanica mar.*, **3**, Suppl., 9–119.
Hori, K., Miyazawa, K. & Ito, K., 1981. *Bull. Jap. Soc. scient. Fish.*, **47**, 793–798.
Hori, K., Yamamoto, T., Miyazawa, K. & Ito, K., 1979. *J. Fac. appl. biol. Sci. Hiroshima Univ.*, **18**, 65–73.
*Horiguchi, Y., Noda, H. & Naka, M., 1971. *Bull. Jap. Soc. scient. Fish.*, **37**, 996–1001.
**Huzimori, S., 1940. *Suisan Kai,* No. 693, 36–44.
*Ichiki, M., 1958. *Bull. Jap. Soc. Phycol.*, **6**, 34–37.
Ikawa, T., Asami, S. & Nisizawa, K., 1972. *Proc. int. Seaweed Symp.*, **7**, 526–531.
Inagaki, K., 1950. *J. Jap. Bot.*, **25**, 20–26.
Ino, T., 1958. *Bull. Tôkai reg. Fish. Res. Lab.*, **22**, 33–36, pl. I.
Inoh, S., 1941. *Syokubutu oyobi Dôbutu,* **9, 877–880.
**Inoh, S., 1947. *Kaisô no Hassei.* Hokuryûkan, Tokyo, 4+4+255 pp.
**Iriki, Y., 1971. In, *Hikaku Seiri Seika Gaku Yori Mita Sôrui no Keitô Bunrui.* Monbusyô Kagaku Kenkyûhi Hozyokin (Dai 4 Bu, Syokubutu Gaku) ni Yoru Sôgô Kenkyû Hôkoku (Syôwa 44–45 Nendo), pp. 13–17, Tabs 1–2.
Ishibashi, M., Fujinaga, T., Yamamoto, T., Fujita, T. & Watanabe, K., 1965. *Nippon Kagaku Zassi,* **86, 728–733.
Ishibashi, M. & Yamamoto, T., 1958a. *Nippon Kagaku Zassi,* **79, 1179–1183.
Ishibashi, M. & Yamamoto, T., 1958b. *Nippon Kagaku Zassi,* **79, 1187–1190.
Ishibashi, M. & Yamamoto, T., 1959. *Rec. oceanogr. Wks Japan,* Spec. No. 3, 109–115.
Ishibashi, M. & Yamamoto, T., 1960a. *Rec. oceanogr. Wks Japan,* Spec. No. 4, 73–78.
Ishibashi, M. & Yamamoto, T., 1960b. *Rec. oceanogr. Wks Japan,* Spec. No. 4, 79–85.
Ishbashi, M. & Yamamoto, T., 1960c. *Rec. oceanogr. Wks Japan,* **5**, 55–62.
Ishibashi, M., Yamamoto, T. & Fujita, T., 1964a. *Rec. oceanogr. Wks Japan,* **7**, 17–24.
Ishibashi, M., Yamamoto, T. & Fujita, T., 1964b. *Rec. oceanogr. Wks Japan,* **7**, 25–32.
Ishibashi, M., Yamamoto, T. & Morii, F., 1962. *Rec. oceanogr. Wks Japan,* **6**, 157–162.
Itano, A., 1933. *Ber. Ôhara Inst. landw. Forsch.*, **6**, 59–72, pls. 4–6.
Ito, K., 1963. *Bull. Jap. Soc. scient. Fish.*, **29**, 771–775.
Ito, K., 1969. *Bull. Jap. Soc. scient. Fish.*, **35, 116–129.
Ito, K. & Hashimoto, Y., 1966. *Nature, Lond.*, **211**, 417 only.
Ito, K., Miyazawa, K. & Hashimoto, Y., 1966. *Bull. Jap. Soc. scient. Fish.*, **32**, 727–729.

*Ito, K., Miyazawa, K. & Matsumoto, F., 1977. *J. Fac. Fish. Anim. Husb. Hiroshima Univ.*, **16**, 77–90.
Ito, S., Nagashima, H. & Matsumoto, T., 1956. *Nippon Kagaku Zassi,* **77, 1119–1121.
Izumi, K., 1971. *Carbohydrate Res.*, **17**, 227–230.
Jackson, L. F., 1976. *Inv. Rep. Oceanogr. Res. Inst. S. Afr. Ass. mar. biol. Res.*, No. 46, 72 pp.
Johnston, H. W., 1966. *Tuatara,* **14**, 30–63.
Jorde, I., 1966. *Sarsia,* **23**, 1–52.
Kain, J. M., 1984. *Br. Phycol. J.*, **19**, 195 only.
Kamimoto, K., 1955. *Jap. J. Bact.*, **10, 897–902.
Kamimoto, K., 1956. *Jap. J. Bact.*, **11, 307–313.
Kanamaru, T., 1932. *Venus,* Kyôto, **3, 271–281.
*Kanazawa, A., 1961. *Kagosima Daig. Suisan Gakubu Kiyô,* **10**, 38–69.
Kanazawa, A., 1963. *Bull. Jap. Soc. scient. Fish.*, **29, 713–731.
*Kaneko, T., 1966. *Bull. Jap. Soc. Phycol.*, **14**, 62–70.
Kaneko, T., 1968. *Bull. Fac. Fish. Hokkaidô Univ.*, **19**, 165–172, pls. I–VI.
**Kaneko, T. & Niihara, Y., 1977. In, *Recent Studies on the Cultivation of Laminaria in Hokkaidô,* Japanese Society of Phycology, Koganei (Tokyo), pp. 21–38.
Kang, J.-W., 1966. *Bull. Pusan Fish. Coll., Part nat. Sci.* **7**, 1–125, pls. I–XII.
*Katada, M., 1949. *Bull. Jap. Soc. scient. Fish.*, **15**, 359–362.
*Katada, M., 1950. *Suisan Kenkyû Si,* **40**, 93–97.
**Katada, M., 1954. In, *Ann. Meet. Jap. Soc. scient. Fish.*, for 1954, Abstracts, 10–11.
*Katada, M., 1955. *J. Shimonoseki Coll. Fish.*, **5**, 1–87, pls. I–VII.
*Katada, M. & Matsui, T., 1954. *J. Shimonoseki Coll. Fish.*, **3**, 229–233.
*Katada, M., Matsui, T., Nakatsukasa, T., Kojo, J. & Miura, A., 1953. *Bull. Jap. Soc. scient. Fish.*, **19**, 471–473.
Katada, M. & Satomi, M., 1975. In, *Advance of Phycology in Japan,* edited by J. Tokida & H. Hirose, Dr W. Junk, The Hague & Gustav Fischer, Jena, pp. 211–239.
Katayama, K. & Inaba, S., 1953. *Suisan Zôsyoku,* **1 [1], 29 only.
**Katayama, K., Inaba, S., Hiyosi, S. & Ôsuga, H., 1952. *Sizuoka-ken Suisan Sikenzyô Gyômu Hôkoku,* 186–192.
*Katayama, T., 1960. *Bull. Jap. Soc. Phycol.*, **8**, 79–84.
*Katayama, T., 1961. *Bull. Jap. Soc. scient. Fish.*, **27**, 75–84.
*Kato, M. & Ariga, N., 1983. *Gihu Daig. Kyôyôbu Kenkyû Hôkoku,* **18**, 53–62.
*Kato, M. & Ariga, N., 1984. *Gihu Daig. Kyôyôbu Kenkyû Hôkoku,* **19**, 57–64.
*Kato, T., 1955. *Bull. Jap. Soc. scient. Fish.*, **21**, 88–91.
Katoh, S., 1959. *J. Biochem.*, **46**, 629–632.
**Katsuura, K., 1962. In, *Ann. Meet. Jap. Soc. Chem., Tyûgoku Sikoku Sibu,* for 1962, Abstracts, 15–16.
Kawana, T., 1940. *Teisui,* **19 [7], 26–28.
Kawana, T., 1956. *Suisan Zôsyoku,* **3 [3], 1–11.
Kawana, T. & Ebata, R., 1959. *Suisan Zôsyoku,* **7 [2], 20–23.
Kida, W., 1964. *Rep. Fac. Fish. prefect. Univ. Mie,* **5**, 217–235.
**Kida, W., 1974. In, *Okinawa-ken Kaityû Kôen Keikaku Tyôsasyo*. Okinawa-ken, Naha-si, pp. 63–84, 3 pls., 2 maps.
**Kida, W. & Kitamura, S., 1964. In, *Amami-ôsima Kaiyô Seibutu Tyôsa Hôko kusyo,* Toba Suizokukan, Toba-si, pp. 71–90.
**Kida, W. & Nakamura, I., 1961. In, *Izu Sikine-zima Kaiyô Seibutu Tyôsa Hôko kusyo*. Toba Suizoku kan, Toba-si, pp. 35–52.
Killian, K., 1914. *Z. Bot.*, **6**, 209–278.
**Kinosita, T., 1939. *Hokkaidô Suisan Sikenzyô Zigyô Zyunpô,* No. 444, 228–229.

Kinosita, T., 1942. *Kaiyô no Kagaku,* **2, 410–417.
**Kinosita, T., 1943. *Hokkaidô Suisan Sikenzyô Zigyô Zyunpô,* No. 575, 131–139.
Kinosita, T., 1944. *Hokusuisi Geppô,* **1, 193–202.
**Kinosita, T., 1949. *Hokkaidô Senkai Suizoku no Zôsyoku ni Kansuru Kenkyû. Vol. 2. Nori, Tengusa, Hunori, oyobi Ginnansô no Zôsyoku ni Kansuru Kenkyû.* Hoppô Syuppansya, Sapporo, 109 pp.
**Kinosita, T., 1951. *Suisan Kagaku Sôsyo, Vol. 7. Hokkaidô Senkai Zôsyoku Gaiteki Seibutu Hen.* Samonzi Syoten, Otaru, 64 pp.
**Kinosita, T., Hirano, Y. & Takahasi, T., 1935a. *Hokkaidô Suisan Sikenzyô Zigyô Zyunpô,* No. 300, 993–994.
**Kinosita, T., Hirano, Y. & Takahasi, T., 1935b. *Hokkaidô Suisan Sikenzyô Zigyô Zyunpô,* No. 301, 1002–1005.
**Kinosita, T. & Sibuya, S., 1941a. *Hokkaidô Suisan Sikenzyô Zigyô Zyunpô,* No. 495, 1–7.
*Kinosita, T. & Sibuya, S., 1941b. *Suisan Kenkyûsi,* **36,** 69–79. (This is a reproduction of 1941a.)
**Kinosita, T. & Simizu, Z., 1935. *Hokkaidô Suisan Sikenzyô Zigyô Zyunpô,* No. 298, 971 only.
*Kisu, Y., Goto, T., Aizawa, Y. & Kimura, F., 1980. *Seikatu Kagaku Kenkyûzyo (Miyagi Gakuin Zyosi Daig.),* **13,** 9–17.
Kito, H., Kikuchi, S. & Uki, N., 1980. In, *Proc. int. Symp. Coastal Pacific mar. Life,* edited by Western Washington Univ. & Office of Sea Grant NOAA, Western Washington Univ. Press, Bellingham, Washington, 55–66.
**Kobayasi, M., 1962. *Nagano-ken Eisei Kenkyûzyo Tyôsa Kenkyû Hôkoku,* 41, 20 pp.
Kuriyama, M., 1961a. *Nature, Lond.,* **192,** 969 only.
Kuriyama, M., 1961b. *Bull. Jap. Soc. scient. Fish.,* **27,** 694–698.
Kuriyama, M., 1961c. *Bull. Jap. Soc. scient. Fish.,* **27,** 699–702.
Kuriyama, M., 1965. *Free Amino Acids in Marine Algae.* Doctoral Dissertation Hokkaidô Univ., 66 pp.
Kützing, F. T., 1843. *Phycologia Generalis.* Brockhaus, Leipzig, xxxii + 459 pp.
Kützing, F. T., 1849. *Species Algarum.* Brockhaus, Leipzig, vi + 922 pp.
Kützing, F. T., 1868. *Tabulae Phycologicae, Bd 18.* Nordhausen, 35 pp, Taf. 1–100.
Kützing, F. T., 1869. *Tabulae Phycologicae, Bd 19.* Nordhausen, 36 pp, Taf. 1–100.
Kylin, H., 1917. *Ark. Bot.,* **14** [22], 1–25.
Kylin, H., 1923. *K. Svenska VetenskAkad. Handl.,* **63,** [11], 1–139.
Kylin, H., 1928. *Lunds Univ. Årsskr.* N. F. Avd. 2, **24,** [4] 127 pp.
Kylin, H., 1956. *Die Gattungen der Rhodophyceen.* Gleerups, Lund, xv + 673 pp.
Lamouroux, J. V. F., 1805. *Dissertations sur plusieurs espèces de* Fucus, *peu Connues ou Nouvelles,* Fasc. 1, Noubel, Agen, 85 pp., Tab. 1–36.
Lamouroux, J. V. F., 1813. *Annls Mus. Hist. nat. Paris,* **20,** 21–47, 115–139, 267–293, pls. 7–13.
Lee, I.-K., Kim, Y.-H., Lee, J.-H. & Hong, S.-W., 1975. *Korean J. Bot.,* **18,** 109–121.
Levring, T., Hoppe, H. A. & Schmid, O. J., 1969. *Marine Algae: A Survey of Research and Utilization.* Cram de Gruyter, Hamburg, 421 pp.
Loomis, N. H., 1949. *Occ. Pap. Allan Hancock Fdn,* No. 6, 29 pp.
MacFarlane, C. I., 1968. *Prof. Rep. Ind. Dev. Serv. Dep. Fish.* Canada, No. 20, 96 pp., 1 map.
Magne, F., 1964. *Cah. Biol. mar.,* **5,** 461–671.
Martens, G., von, 1866. *Die Tange.* K. Geheime ober-hofbuchdruckerei, Berlin, 152 pp., Taf. 1–8.
*Masaki, T., 1984. *Jap. J. Phycol.,* **32,** 71–85.

**Masaki, T. & Akioka, H., 1980. In, *Hokkaidô ni Okeru Isoyake no Genzyô ni Tuite,* Hokkaidô Saibai Gyogyô Sinkô Kôsya, Sapporo, pp. 4–19.

Matsuura, S., 1958. *Bot. Mag., Tokyo,* **71,** 93–109.

Matumoto, T., 1922. *Suisan Kenkyû Si,* **17, 229–230.

Michanek, G., 1971. *A Preliminary Appraisal of World Seaweed Resources.* FAO Fish. Circ., No. 128, 37 pp.

Michanek, G., 1972. *Proc. int. Seaweed Symp.,* **7,** 248–250.

Michanek, G., 1975. *Seaweed Resources of the Ocean. FAO Fish. Tech. Pap.,* No. 138, 127 pp., 1 map.

Michanek, G., 1978. *Botanica mar.,* **21,** 469–475.

Michanek, G., 1983, In, *Marine Ecology, Vol. V, Part 2,* edited by O. Kinne, John Wiley & Sons, Chichester, pp. 795–837.

**Mie-ken Suisan Sikenzyô (FES), 1931. *Mie-ken Suisan Sikenzyô Zigyô Hôkoku, Ser. Yôsyoku no Bu,* for 1931, 31–33.

**Miyazaki-ken Suisan Sikenzyô (FES), 1939. *Miyazaki-ken Suisan Sikenzyô Gyômu Gaiyô,* for 1938–1939, 100–113.

**Miyazawa, K., Ito, K. & Matsumoto, F., 1967. In, *Autumn Meet. Jap. Soc. scient. Fish.* 1967, Abstracts, p. 58 only.

Miyazawa, K., Ito, K. & Matsumoto, F., 1969. *Bull. Jap. Soc. scient. Fish.,* **35,** 1215–1219.

Miyosi, M., 1958. *Suisan Zôsyoku,* **6 [1], 59–62.

*Mori, B., Kusima, K., Iwasaki, T. & Omiya, H., 1981. *Nippon Nôgei kagaku Kaishi,* **55,** 787–791.

Morii, F., 1961. *Nippon Kagaku Zassi,* **82, 1510–1511.

Morii, F., 1962. *Nippon Kagaku Zassi,* **83, 77–81.

Murakami, S., 1955. *Sci. Rep. Saitama Univ., Ser. B,* **2,** 35–38.

Murakami, S., 1960. *Sci. Rep. Saitama Univ., Ser. B.,* **3,** 251–254.

Murata, K., 1963. *Bull. Jap. Soc. scient. Fish.,* **29, 189–197.

*Nagashima, H., 1976. *Bull. Jap. Soc. Phycol.,* **24,** 103–110.

**Nakamura, S., 1971. In, *Hikaku Seiri Seika Gaku yori Mita Sôrui no Keitô Bunrui,* Monbusyô Kagaku Kenkyûhi ni yoru Sôgô Kenkyû Hôkoku (Syôwa 44–45 Nendo), pp. 23–24, Tabs la-lc.

*Nakamura, T., 1977. *Kyûsyû Zyosi Daig. Kiyô,* **13,** 23–32.

**Nakamura, Y. & Ito, S., 1956. In, *Iburi Sityô Kan-nai Senkai Zôsyoku Tekiti Tyôsa Hôkoku,* Hokkaidô-tyô, Sapporo, pp. 1–18.

*Nakaniwa, M., 1975. *Bull. Jap. Soc. Phycol.,* **23,** 99–110.

Nanao, Y., 1974. *Bull. Jap. Soc. Phycol.,* **22, 29–38.

Neish, I. C., 1979. In, *Advances in Aquaculture,* edited by T. V. R. Pillay & W. A. Dill, Fishing News Books, Farnham, pp. 395–402.

**Nisizawa, K., 1970. In, *Tatô Seikagaku. Vol. 2. Seibutugaku Hen,* edited by H. Egami *et al.*, Kyôritu, Tokyo, pp. 605–635.

Nisizawa, K., 1978. In, *Handbook of Phycological Methods: Physiological and Biochemical Methods,* edited by J. A. Hellebust & J. S. Craigie, Cambridge University Press, Cambridge, pp. 239–244.

Nisizawa, K., Akagawa, H. & Ikawa, T., 1972. *Proc. int. Seaweed Symp.,* **7,** 532–536.

**Nonaka, T., Ôsuga, H. & Sasaki, T., 1962. *Sizuoka-ken Suisan Sikenzyô Izu Bunzyô Kenkyû Hôkoku,* No. 18, 3 pp., Tab. I.

**Ochi, M., 1983. In, *Suisangaku Siriizu. Vol. 45. Kaisô no Seikagaku to Riyô,* edited by Japanese Society of Scientific Fisheries, Kôseisya Kôseikaku, Tokyo, pp. 101–119.

Ogata, E., 1952. *Kagaku (Science), Tokyo,* **22, 364–365.

Ogata, E., 1968. *J. Shimonoseki Univ. Fish.,* **16,** 139–152.

Ogata, E. & Matsui, T., 1965a. *Jap. J. Bot.,* **19,** 83–98.

Ogata, E. & Matsui, T., 1965b. *Botanica mar.*, **8**, 199–217.
Ogata, E. & Takada, H., 1968. *J. Shimonoseki Univ. Fish.*, **16**, 117–138.
Ogino, C., 1955. *J. Tokyo Univ. Fish.*, **41**, 107–152.
Ohmi, H., 1951. *Bull. Fac. Fish. Hokkaidô Univ.*, **2**, 109–117.
*Ohmi, H., 1963. *Bull. Jap. Soc. Phycol.*, **11**, 38–44.
Ohno, M., 1969. *Rep. Usa mar. biol. Stn.* **16**, 1–46.
*Ohno, M. & Arasaki, S., 1969. *Bull. Jap. Soc. Phycol.*, **17**, 37–42.
*Ohta, F. & Tanaka, T., 1964. *Kagosima Daig. Suisan Gakubu Kiyô*, **13**, 38–44.
Ohta, K., 1977. *Agric. biol. Chem.*, **41**, 2105–2106.
Ohta, K., 1978. *Proc. int. Seaweed Symp.*, **9**, 401–411.
Ôisi, H., 1917a. *Suisan Kenkyû Si*, **12, 104–109.
Ôisi, H., 1917b. *Suisan Kenkyû Si*, **12, 289–291.
**Okada, Y., 1934. *Gensyoku Kaisô Zuhu.* Sansei-dô, Tokyo, 191+41 pp., 164 pls.
*Okamura, K., 1892. In, *Iconographia Florae Japonicae*, edited by R. Yatabe, Vol. I, Part 2. Maruya, Tokyo, pp. 157–158, pl. 39.
Okamura, K., 1893. *Bot. Mag., Tokyo*, **7, 369–376.
Okamura, K., 1899. *Bot. Mag., Tokyo*, **13**, 2–10, pl. I.
Okamura, K., 1900–1902. *Illustrations of the Marine Algae of Japan, Vol. I.* Okamura, Tokyo, pls. 1–30.
Okamura, K., 1907–1909. *Icones of Japanese Algae. Vol. I.* Kazama-syobô, Tokyo, pls. 1–50.
Okamura, K., 1909–1912. *Icones of Japanese Algae. Vol. II.* Kazama-syobô, Tokyo, pls. 51–100.
Okamura, K., 1913–1915. *Icones of Japanese Algae. Vol. III.* Kazama-syobô, Tokyo, pls. 101–150.
Okamura, K., 1916–1923. *Icones of Japanese Algae. Vol. IV.* Kazama-syobô, Tokyo, pls. 151–200.
Okamura, K., 1923–1928. *Icones of Japanese Algae. Vol. V.* Kazama-syobô, Tokyo, pls. 201–250.
Okamura, K., 1929–1932. *Icones of Japanese Algae. Vol. VI.* Kazama-syobô, Tokyo, pls. 251–300.
Okamura, K., 1933–1937. *Icones of Japanese Algae. Vol. VII.* Kazama-syobô, Tokyo, pls. 301–345.
Okamura, K., 1908. *Suisan Kenkyû Si*, **3, 131–136.
Okamura, K., 1909. *Suisan Kôsyûzyo Siken Hôkoku*, **5, 221–230.
Okamura, K., 1911a. *Bot. Mag., Tokyo*, **25, 373–378.
Okamura, K., 1911b. *Suisan Kôsyûzyo Siken Hôkoku*, **7, 72–73.
Okamura, K., 1911c. *Suisan Kôsyûzyo Siken Hôkoku*, **7, 230–236.
Okamura, K., 1911d. *Suisan Kôsyûzyo Siken Hôkoku*, **7, 236–238.
Okamura, K., 1911e. *Suisan Kôsyûzyo Siken Hôkoku*, **7, 238–246.
Okamura, K., 1915. *Bot. Mag.*, Tokyo, **29, 57–58.
Okamura, K., 1917. *Suisan Kenkyû Si*, **12, 267–268.
Okamura, K., 1918. *Suisan Kôsyûzyo Siken Hôkoku*, **13, 1–10, pls. 1–2.
Okamura, K., 1926. *Proc. Pan-Pacific Sci. Congr.*, **3**, 958–963.
Okamura, K., 1931. *Bull. biogeogr. Soc. Japan*, **2**, 95–122, pls. 10–12.
Okamura, K., 1932a. *J. Jap. Bot.*, **8, 174–178.
Okamura, K., 1932b. *Rec. oceanogr. Wks Japan*, **4**, 30–150.
Okamura, K., 1934a. *J. imp. Fish. Inst.*, **29**, 47–67, pls. 16–33.
Okamura, K., 1934b. *Rec. oceanogr. Wks Japan*, **6**, 13–18, pl. 7.
Okamura, K., 1935a. *Syokubutu oyobi Dôbutu*, **3, 1501–1504.
Okamura, K., 1935b. *Nippon Gakuzyutu Kyôkai Hôkoku*, **10, 441–443.
**Okamura, K., 1936. *Nippon Kaisô Si.* Uchida Rôkakuho, Tokyo, 975 pp.
Okamura, K. & Ôisi, H., 1905. *Suisan Kôsyûzyo Hôkoku*, **3, 93–104.
**Okamura, K., & Tago, K., 1915. *Aomori-ken Simokita-gun Isoyake Tyôsa.* Aomori-ken Naimubu, Aomori-si, 16 pp.

Okazaki, A., 1971. *Seaweeds and Their Uses in Japan.* Tôkai University Press, Tokyo, 165 pp.
Okuda, T. & Neushul, M., 1981. *J. Phycol.,* **17,** 113–118.
**Ôno, I., 1927. *Hassei-zyô yori Mitaru Tengusa no Hansyoku ni Tuite.* Hokkaidô Suisan Sikenzyô Tokubetu Hôkoku (mimeographed), 9 pp., 3 pls.
**Ôno, I., 1929. *Hokkaidô ni Okeru Tengusa Hansyoku Ziki. Engan Suizoku Hansyoku Hogo Siken Hôkoku* (mimeographed), No. 3, 7 pp.
**Ôno, I., 1932. *Hokkaidô ni okeru Senkai Riyô: Suisan Zôsyoku Kôwa.* Hokkaidô Suisan Kai, Sapporo, 124 pp.
Onodera, H., Matuoka, T. & Ogino, C., 1954. *Suisan Zôsyoku,* **2 [2], 10–16.
**Ôsima, K., 1936. *Toyama Wan no Yûyô Kaisô to Gyoson Sinkô no Iti Taisaku.* Toyama-ken Suisan Kai, Toyama-si, 46 pp.
Ôsuga, H. & Yamasaki, H., 1960. *Suisan Zôsyoku,* **8, 111–116.
**Ôtani, T. & Huzikawa, K., 1935. *Kaisô no Kagaku.* Kôsei Kaku, Tokyo, 357 pp.
Papenfuss, G. F., 1966. *Phycologia,* **5,** 247–255.
Payen, M., 1859. *C. r. hebd. Séanc. Acad. Sci., Paris,* **49,** 521–530.
Percival, E. & McDowell, R. H., 1967. *Chemistry and Enzymology of Marine Algal Polysaccharides.* Academic Press, London, 219 pp.
Rao, P. S., 1974. *Cytologia,* **39,** 391–395.
Rao, P. S. & Suto, S., 1970. *Bull. Tôkai reg. Fish. Res. Lab.,* **63,** 61–64.
Rees, D. A., 1969. *Adv. Carbohydr. Chem. Biochem.,* **24,** 267–332.
Rees, D. A. & Conway, E., 1962. *Biochem. J.,* **84,** 411–416.
Ross, A. G., 1953. *J. Sci. Food Agric.,* **4,** 333–335.
Saito, Y. & Atobe, S., 1970. *Bull. Fac. Fish. Hokkaidô Univ.,* **21,** 37–69.
Santelices, B., 1974. *Gelidioid Algae: a Brief Résumé of the Pertinent Literature.* Sea Grant Tech. Rep. Hawaii Univ., No. 1, 111 pp.
Santelices, B. & Montalva, S., 1983. *Phycologia,* **22,** 185–196.
Schmitz, F., 1894. *Hedwigia,* **33,** 190–201, Taf. 10.
Seagrief, S. C., 1967. *The Seaweeds of the Tsitsikama Coastal National Park.* National Parks Board of the Republic of South Africa, Pretoria, 147 pp.
**Segawa, S., 1956. *Coloured Illustrations of the Seaweeds of Japan.* Hoikusha, Ôsaka, 175 pp, pls. 72 + 12.
**Segawa, S. & Kamura, S., 1960. *Marine Flora of Ryûkyû Islands.* Biol. Inst. Ryûkyûs, Naha-si, Okinawa-Ken, 72 pp.
Segi, T., 1954. *Bull. Jap. Soc. Phycol.,* **2, 13–19.
Segi, T., 1955. *Rep. Fac. Fish. prefect. Univ. Mie,* **2,** 124–137, pls. 5–13.
Segi, T., 1957. *Rep. Fac. Fish. prefect. Univ. Mie,* **2,** 456–461, pl. 24.
Segi, T., 1959. *Rep. Fac. Fish. prefect. Univ. Mie,* **3,** 251–255, pls. 9–10.
Segi, T., 1963. *Rep. Fac. Fish. prefect. Univ. Mie,* **4,** 509–525, pls. 1–52.
*Segi, T., 1977. *Bull. Jap. Soc. Phycol.,* **25,** Suppl., 327–331.
Seoane-Camba, J., 1965. *Investigación pesq.,* **29,** 3–216.
Seoane-Camba, J., 1969. *Proc. int. Seaweed Symp.,* **6,** 365–374.
Setchell, W. A. & Gardner, N. L., 1903. *Univ. Calif. Publs Bot.,* **1,** 165–418, pls. 17–27.
Setchell, W. A. & Gardner, N. L., 1937. *Proc. Calif. Acad. Sci.,* Ser. 4, **22,** 65–98, pls. 3–25.
Silverthorne, W., 1977. *Botanica mar.,* **20,** 75–98.
Simons, R. H., 1971. In, *Proteins and Food Supply in the Republic of South Africa—A Symposium,* edited by J. W. Claassens & H. J. Potgieter, Balkema, Cape Town, pp. 151–160.
Sirahama, K., Sasa, S. & Uno, T., 1944. *Bull. Jap. Soc. scient. Fish.,* **12, 224–231.
Sivalingam, P. M., 1978. *Jap. J. Phycol.,* **26,** 161–164.
Sivalingam, P. M., Ikawa, T. & Nisizawa, K., 1974. *Pl. Cell Physiol., Tokyo,* **15,** 583–586.
Sivalingam, P. M., Ikawa, T., Yokohama, Y. & Nisizawa, K., 1974. *Botanica mar.,* **17,** 23–29.

**Sizuoka-ken Suisan Sikenzyô (FES), 1951. *Sizuoka-ken Suisan Sikenzyô Zigyô Hôkoku,* 1950–1951, 265–273.
*Sohn, C. -H. & Kang, J. -W., 1978. *Publs Inst. mar. Sci. natn. Fish. Univ. Busan,* **11,** 29–40.
Stewart, J. G., 1968. *J. Phycol.,* **4,** 76–84.
Stewart, J. G., 1983. *J. exp. mar. Biol. Ecol.,* **73,** 205–211.
Sugimura, Y., Toda, F., Murata, T. & Yakushiji, E., 1968. In, *Structure and Function of Cytochromes,* edited by K. Okunuki *et al.*, University of Tokyo Press, Tokyo, pp. 452–458.
*Suto, S., 1950a. *Bull. Jap. Soc. scient. Fish.,* **15,** 671–673.
*Suto, S., 1950b. *Bull. Jap. Soc. scient. Fish.,* **15,** 674–677.
**Suto, S., 1954. *The Agar Seaweed, "Tengusa"-harvesting and Propagating its Resources: Aquaculture Science Series, No. 8,* Dept. Fish. Tokyo University, 53 pp.
**Suto, S., 1965. *Engan Kaisôrui no Zôsyoku.* (Suisan Zôyôsyoku Sôsyo, No. 9). Nippon Suisan Sigen Hogo Kyôkai, Tôkyô, 36 pp.
Tadokoro, T. & Ugazin, H., 1933. *Scient. Pap. Inst. algol. Res. Hokkaidô Univ.,* **1, 40–45.
*Tagawa, S. & Kojima, Y., 1973. *J. Shimonoseki Univ. Fish.,* **22,** 67–75.
*Tagawa, S. & Kojima, Y., 1976. *J. Shimonoseki Univ. Fish.,* **25,** 67–74.
*Takagi, M., 1953a. *Bull. Jap. Soc. scient. Fish.,* **18,** 483–487.
*Takagi, M., 1953b. *Bull. Jap. Soc. scient. Fish.,* **19,** 798–802.
*Takagi, M., 1953c. *Bull. Jap. Soc. scient. Fish.,* **19,** 803–808.
Takagi, M., 1970. *Bull. Fac. Fish. Hokkaidô Univ.,* **21,** 227–233.
*Takagi, M. & Murata, K., 1954a. *Bull. Fac. Fish. Hokkaidô Univ.,* **4,** 306–309.
*Takagi, M. & Murata, K., 1954b. *Bull. Fac. Fish. Hokkaidô Univ.,* **4,** 310–313.
*Takagi, M. & Murata, K., 1954c. *Bull. Fac. Fish. Hokkaidô Univ.,* **5,** 173–175.
**Takahasi, T., 1941. *Kaisô Kôgyô.* Sangyô Tosyo, Tokyo, 417 pp.
Takamatsu, M., 1938. *Saitô Hô-onkai Mus. Res. Bull.,* **14,** 77–143.
**Takamatsu, M., 1944. *Sigen Kagaku Kenkyûzyo Ihô,* No. 6, 55–59, pls. 1–2.
**Takamatsu, M., 1946a. *Sigen Kagaku Kenkyûzyo Ihô,* No. 9, 35–36.
**Takamatsu, M., 1946b. *Sigen Kagaku Kenkyûzyo Ihô,* No. 10, 19–23.
**Takamatsu, M., 1946c. *Sigen Kagaku Kenkyûzyo Ihô,* No. 10, 24 only.
Takano, H., 1961. *Bull. Tôkai reg. Fish. Res. Lab.,* No. 31, 269–274, pls. 1–2.
**Takayama, K., 1938. *Mie-ken Suisan Sikenzyô Zihô,* No. 101, 205–213.
Takayama, K., 1939a. *Suisan Kenkyû Si,* **34, 211–213.
**Takayama, K., 1939b. *Mie-ken Suisan Sikenzyô Zihô,* No. 107, Suppl., 12 pp.
Tamura, T., 1951. *Hokusuisi Geppô,* **8, 28–36.
**Tanaka, T., 1950. In, *Kagosima Kokuritu Kôen Kôhoti Gakuzyutu Tyôsa Hôkoku (Kôhen),* Kagosima-ken, Kagosima-si, pp. 136–147.
*Tanaka, T., 1956a. *Mem. Fac. Fish. Kagosima Univ.,* **5,** 205–209.
Tanaka, T., 1956b. *Nanpô Sangyô Kagaku Kenkyûzyo Hôkoku,* **1, 13–22, pls. I–II.
Tanaka, T., 1964. *Proc. int. Seaweed Symp.,* **4,** 276–279.
Tanaka, T., 1965. *Mem. Fac. Fish. Kagosima Univ.,* **14,** 52–71.
Taniguti, M., 1970. *Syokubutu to Sizen,* **6 [8], 12–15.
*Taniguti, M., 1979. *Rep. envir. Sci. Mie Univ.,* **4,** 93–121.
Tazawa, N., 1975. *Scient. Pap. Inst. algol. Res. Hokkaidô Univ.,* **6,** 95–179, pls. 1–10.
Tokida, J., 1954. *Mem. Fac. Fish. Hokkaidô Univ.,* **2,** 1–264.
Tôkyô-hu Suisan Sikenzyô (FES), 1941. *Tôkyô-hu Suisan Sikenzyô Gyômu Kôtei, Yôsyoku bu,* **16, 101–106.
Tôkyô-hu Suisan Sikenzyô (FES), 1942. *Tôkyô-hu Suisan Sikenzyô Gyômu Kôtei, Yôsyoku Bu,* **17, 90–98.
Tôyama, N., 1939. *Teisui,* **18 [2], 37–39.

Tseng, C. -K., 1938. *Lingnan Sci. J.,* **17,** 591–604.
Tseng, C. -K., 1947. *Econ. Bot.,* **1,** 69–97.
*Tseng, C. -K. & Chang, C. -F., 1959. *Oceanologia Limnol. Sin.,* **2,** 43–52.
Tseng, C. -K. & Sweeney, B. M., 1946. *Am. J. Bot.,* **33,** 706–715.
**Tsuchiya, Y., 1962. *Suisan Kagaku.* Kôseisya Kôseikaku, Tokyo, 447 pp.
Tsuda, K., Akagi, S. & Kishida, Y., 1957. *Science,* **126,** 927 only.
Tsuda, K., Akagi, S. & Kishida, Y., 1958. *Chem. pharm. Bull., Tokyo,* **6,** 101–104.
Tsuda, K., Akagi, S., Kishida, Y., Hayatsu, R. & Sakai, K., 1958. *Chem. pharm. Bull., Tokyo,* **6,** 724–727.
*Tsujino, I. & Saito, T., 1961. *Bull. Fac. Fish. Hokkaidô Univ.,* **12,** 49–58.
Ueda, R., 1972. *Proc. int. Seaweed Symp.,* **7,** 198–200.
*Ueda, S., 1936. *Bull. Jap. Soc. scient. Fish.,* **5,** 183–186.
**Ueda, S., Iwamoto, K. & Miura, A., 1963. *Suisan Syokubutu Gaku. Kôseisya Kôseikaku, Tokyo,* 640 pp.
Ueda, S. & Katada, M., 1943. *Bull. Jap. Soc. scient. Fish.,* **11, 175–178.
*Ueda, S. & Katada, M., 1949. *Bull. Jap. Soc. scient. Fish.,* **15,** 354–358.
*Ueda, S. & Okada, Y., 1938. *Bull. Jap. Soc. scient. Fish.,* **7,** 229–236.
*Ueda, S. & Okada, Y., 1940. *Bull. Jap. Soc. scient. Fish.,* **8,** 244–246.
Ueda, T., Suzuki, Y. & Nakamura, R., 1973. *Bull. Jap. Soc. scient. Fish.,* **39,** 1253–1262.
Umezaki, I., 1966. *Acta phytotax. geobot., Kyôto,* **22, 49–63.
Venkataraman, G. S., Goyal, S. K., Kaushik, B. D. & Roychoudhury, P., 1974. *Algae: Form and Function.* Today & Tomorrow's Printers & Publishers, New Delhi, 562 pp.
Watanabe, A., 1937. *Acta phytochim., Tokyo,* **9,** 235–254.
*Watase, M., Akahane, T. & Arakawa, K., 1975. *Nippon Kagaku Kaisi,* 1975, 1564–1571.
*Watase, M. & Nishinari, K., 1981. *Nippon Syokuhin Kôgyô Gakkaisi,* **28,** 437–443.
Weber-van Bosse, A., 1921. *Siboga-Expeditie Monogr.,* Leiden, 59b, 126 pp., pls. 6–8.
Weber-van Bosse, A., 1932. *Mém. Mus. r. Hist. nat. Belg., Hors. Sér.,* **VI,** Fasc. 1, 1–27, pls. 1–5.
Whittick, A., 1984. *Br. Phycol. J.,* **19,** 201 only.
Wildman, R., 1974. *Proc. US-Japan Meet. Aquaculture,* **1,** 97–101.
Yabu, H., 1975. In, *Advance of Phycology in Japan,* edited by J. Tokida & H. Hirose, Dr W. Junk, the Hague & Gustav Fischer, Jena, pp. 125–135.
Yakushiji, E., 1935. *Acta phytochim., Tokyo,* **8,** 325–329.
Yamada, I., 1980. *J. Fac. Sci. Hokkaidô Univ., Ser. V,* **12,** 11–98.
Yamada, N., 1960. *Suisan Zôsyoku,* **7 [3], 24–28.
*Yamada, N., 1961. *Bull. Jap. Soc. scient. Fish.,* **27,** 953–957.
*Yamada, N., 1964. *Bull. Jap. Soc. scient. Fish.,* **30,** 908–911.
**Yamada, N., 1967. *Sizuoka-ken Suisan Sikenzyô Izu Bunzyô Kenkyû Hôkoku,* No. 32, 96 pp.
Yamada, N., 1972. *Proc. int. Seaweed Symp.,* **7,** 385–390.
Yamada, N., 1976. *J. Fish. Res. Bd Can.,* **33,** 1024–1030.
**Yamada, N. & Kubota, S., 1972. *Sizuoka-ken Suisan Sikenzyô Kenkyû Hôkoku,* No. 5, 79–87.
Yamada, Y., 1941. *Scient. Pap. Inst. algol. Res. Hokkaidô imp. Univ.,* **2,** 195–215, pls. 40–48.
Yamada, Y. & Tanaka, T., 1938. *Scient. Pap. Inst. algol. Res. Hokkaidô imp. Univ.,* **2,** 53–86.
Yamamoto, T., 1960a. *Nippon Kagaku Zassi,* **81, 381–384.
Yamamoto, T., 1960b. *Nippon Kagaku Zassi,* **81, 384–388.

Yamamoto, T., 1960c. *Nippon Kagaku Zassi,* **81, 388–391.
Yamamoto, T., 1972. *Rec. oceanogr. Wks Japan,* **11**, 65–72.
Yamamoto, T., Fujita, T. & Ishibashi, M., 1965. *Nippon Kagaku Zassi,* **86, 53–59.
Yamamoto, T., Fujita, T. & Ishibashi, M., 1970. *Rec. oceanogr. Wks Japan,* **10**, 125–135.
Yamamoto, T., Fujita, T. & Shigematsu, T., 1969. *Rec. oceanogr. Wks Japan,* **10**, 29–38.
Yamamoto, T., Fujita, T., Shigematsu, T. & Ishibashi, M., 1968. *Rec. oceanogr. Wks Japan,* **9**, 209–217.
Yamamoto, T. & Ishibashi, M., 1972. *Proc. int. Seaweed Symp.,* **7**, 511–514.
Yamamoto, T., Otsuka, Y., Okazaki, M. & Okamoto, K., 1979. In, *Marine Algae in Pharmaceutical Science,* edited by H. A. Hoppe *et al.*, Walter de Gruyter, Berlin, pp. 569–607.
Yamamoto, T., Otsuka, Y. & Uemura, K., 1976. *J. oceanogr. Soc. Jap.,* **32**, 182–186.
Yamamoto, T., Yamaoka, T., Fujita, T. & Isoda, C., 1971. *Rec. oceanogr. Wks Japan,* **11**, 7–13.
*Yamasaki, H., 1960. *Bull. Jap. Soc. scient. Fish.,* **26**, 116–122.
**Yamasaki, H., 1962. *Sizuoka-ken Suisan Sikenzyô Izu Bunzyô Kenkyû Hôkoku,* No. 19, 92 pp.
Yanagawa, T., 1933. *Ôsaka Kôgyô Sikenzyo Hôkoku,* **14 [5], 1–38.
Yanagawa, T., 1936. *Ôsaka Kôgyô Sikenzyo Hôkoku,* **17 [5], 14–22.
**Yanagawa, T., 1942. *Kanten.* Sangyô Tosyo, Tôkyo, 352 pp., 6 pls.
Yanagawa, T. & Nisida, Y., 1930. *Ôsaka Kôgyô Sikenzyo Hôkoku,* **11 [14], 1–32.
Yendo, K., 1903a. *Suisan Tyôsa Hôkoku,* **12, 1–33.
Yendo, K., 1903b. *Suisan Tyôsa Hôkoku,* **12, 34–38.
Yendo, K., 1909. *Bot. Mag., Tokyo,* **23**, 117–133.
**Yendo, K., 1911. *Kaisan Syokubutu Gaku.* Hakubun-kan, Tokyo, 748 + 84 pp., pls. I–VIII.
**Yokohama, Y., 1971. In, *Hikaku Seiria Seika Gaku yori Mita Sôrui no Keitô Bunrui.* Monbusyô Kagaku Kenkyûhi (Dai 4bu Syokubutu Gaku) ni yoru Sôgô Kenkyû Hôkoku (Syôwa 44–45 Nendo), pp. 45–46, pls. 1–3.
Yokohama, Y., 1972. *Proc. int. Seaweed Symp.,* **7**, 286–291.
*Yokohama, Y., 1973a. *Bull. Jap. Soc. Phycol.,* **21**, 119–124.
Yokohama, Y., 1973b. *Int. Revue ges. Hydrobiol.,* **58**, 463–472.
Yokota, T., 1939. *Suisan Kenkyû-Si,* **34, 133–134.
*Yoshida, T. & Yoshida, M., 1965. *Bull. Jap. Soc. Phycol.,* **13**, 92–97.
Zaneveld, J. S., 1955. *Economic Marine Algae of Tropical South and East Asia and Their Utilization.* FAO Indo-Pacific Fisheries Council Special Publ., No. 3, 55 pp.
Zaneveld, J. S., 1959. *Econ. Bot.,* **13**, 89–131.

Oceanogr. Mar. Biol. Ann. Rev., 1986, **24**, 265–307
Margaret Barnes, Ed.
Aberdeen University Press

THE STRUCTURE OF SUBTIDAL ALGAL STANDS IN TEMPERATE WATERS

DAVID R. SCHIEL*
and
MICHAEL S. FOSTER
Moss Landing Marine Laboratories, P.O. Box 450, Moss Landing, California, U.S.A. 95039

INTRODUCTION

The processes that affect the pattern and persistence of marine communities are receiving increasing attention in the ecological literature. In particular, intertidal studies on temperate rocky shores have been the major source of information, not only contributing assessments of interactions among species, but also serving as a focus for discussions about testing hypotheses and ecological experimentation (*cf* Strong, Simberloff, Abele & Thistle, 1984). This attention to intertidal areas is hardly surprising given their accessibility, the gradient of physical conditions in the relatively short distance between tide marks, and the large number of algae, herbivores, and predators of many phyla that inhabit most shores (Stephenson & Stephenson, 1949, 1972). In contrast, subtidal areas have been studied comparatively less, even though rocky reefs on most shores are usually much more extensive subtidally than intertidally. Only in recent years, however, has there been routine access to shallow subtidal habitats, allowing descriptive and experimental evaluation of some of the processes that structure communities there.

This comparative lack of attention to research in the shallow subtidal and relatively small contribution to our knowledge of marine systems generally are reflected in the fact that only 6–14% of the references cited in important recent reviews of marine community processes involved subtidal algal communities (Lubchenco & Gaines, 1981; Dayton, 1984; Underwood & Denley, 1984). Moreover, most of the studies cited were concerned with plant–herbivore interactions. The relationships between plants and abiotic factors, and plant–plant interactions have received far less attention, and the autecology of even prominent algal species is poorly understood. There has been an increasing amount of research done in subtidal algal communities, however, since the early 1970s. The goals of this paper are to review the literature for studies done in these communities, focusing on

*Present address: Fisheries Research Centre, Ministry of Agriculture & Fisheries, P.O. Box 297, Wellington, New Zealand

large brown algae, to examine critically the descriptive and experimental work, and to offer a perspective for future endeavours.

Large macroalgae of the Orders Laminariales and Fucales are the dominant organisms on rocky reefs in temperate and high-latitude regions of both hemispheres, in waters less than about 25 m in depth (Dawson, Neushul & Wildman, 1960; Druehl, 1967a, 1970; Shepherd & Womersley, 1970, 1971; North, 1971a,b; Barrales & Lobben, 1975; Velimirov, Field, Griffiths & Zoutendyk, 1977; Kain, 1979; Choat & Schiel, 1982). These algae provide the character of the habitats in which they occur, often forming dense stands which cover large areas of substratum. The vertical relief that macroalgae afford in the water column provides habitat for many species of invertebrates and fish, and may have major effects on the distribution and abundance of these organisms (Foster & Schiel, 1985). Major research programmes on subtidal, algal-dominated communities have been undertaken in many regions: the west coast of North America, Alaska, Nova Scotia, Great Britain, South America, South Africa, New Zealand, and Australia. Yet, there is surprisingly little in the way of consensus about the processes that structure algal stands. This is due to several factors: the recent emergence of subtidal research, the few localities in which research programmes have been done compared with the much broader regions where subtidal algal communities occur, and to the very few manipulative experiments that have addressed specific hypotheses. In many ways, the history of subtidal research mimics that of the intertidal. From early descriptions of floras and faunas, broad-scale correlations were sought (*e.g.*, Dawson *et al.*, 1960; Kain, 1962). Localized abundances of echinoid grazers in the 1960s drew attention to the fact that stands of large brown algae were vulnerable to wholesale removal (review in Lawrence, 1975). The focus on grazers has continued (*e.g.*, Mann, 1973; Estes, Smith & Palmisano, 1978; Lawrence & Sammarco, 1982), while experimental work to clarify their effects is more recent.

The effects of abiotic factors such as light, nutrients, and suitable substratum on the growth of algae have been addressed in many studies. These factors are considered to be the limiting resources, and the evidence for this is discussed. Large algae, particularly in dense stands, may alter the levels of these resources through interactions such as pre-emption of substratum and shading by canopies. The effects of plant–plant interactions are also discussed.

To a large extent our knowledge of the ecology of subtidal algal stands is a reflection of what has been amenable to study. Figure 1 emphasizes this problem by showing the entire life history of laminarian algae and suggesting how the abundance and distribution of each stage may be affected by different environmental variables. This complete life history is truly "the plant", and changes in the abundance and distribution of one stage can affect the others. Because of the difficulties of working with microscopic spores, gametophytes, and sporophytes in the field, however, most studies have concentrated on adult sporophytes, giving us a biased view of the ecology of these plants. Moreover, many of the factors that can affect "the plant" have probably not been measured at the relevant temporal and spatial scales (*e.g.*, nutrients; Zimmerman & Kremer, 1984), and factor interactions remain largely unknown. Most studies of subtidal

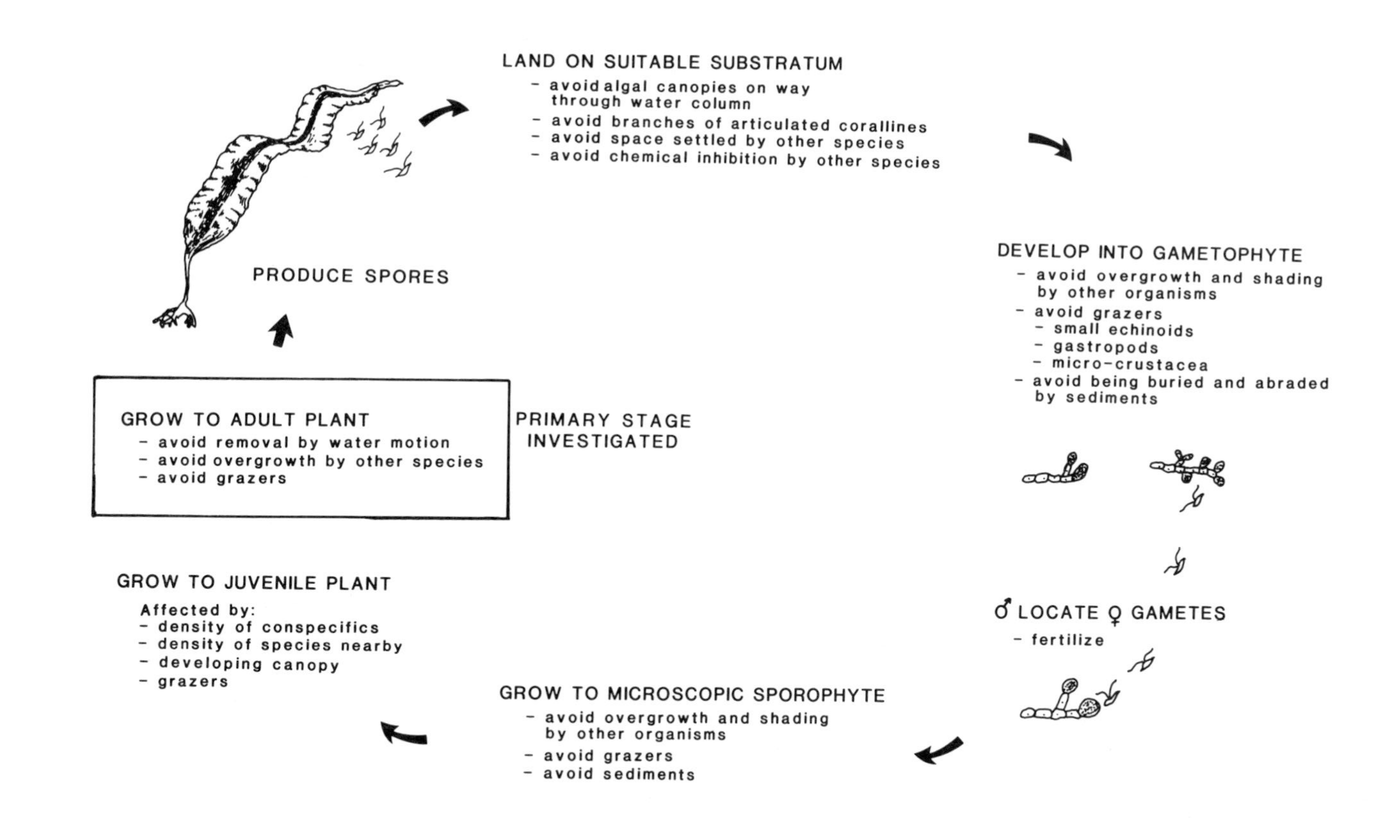

Fig. 1.—Summary of the life history stages of laminarian algae: some of the major hazards to development that must be overcome at each stage are listed; most studies in subtidal algal communities have been done on adult plants.

algae have been constrained to dealing with post-recruitment events. Relatively little is known from the field about fecundity, dispersal and events between settlement (when propagules arrive and attach) and what is recognized as recruitment (when they or their products are visible in the field). Recent advances in techniques for working with these microscopic stages and for measuring environmental variables (Foster, Dean & Deysher, 1985), and innovative approaches to the study of the entire life history (*e.g.*, Chapman, 1984) may significantly enhance our understanding of the ecology of subtidal plant communities.

We discuss the influence of biological and physical factors in a demographic framework. How do events affect the recruitment, growth, reproduction, and survival of algal species? This perspective is based on the premise that the structure of a community is essentially a numbers game, even though some species are larger and may have a greater effect than others, and that knowing how various factors affect numbers of plants is the key to understanding the patterns and persistence of algal stands. The categorization of the sections that follow reflects what we consider to be the necessary spatial and temporal context for examining the structure of subtidal algal stands, and the factors thought to have the major effects on this structure. The lengths of sections do not necessarily indicate the relative importance of the factors discussed to community structure; they merely reflect the amount of literature on these topics. We have generally confined this review to the widely available literature. Many reports and other local publications for *Macrocystis* are discussed by Foster & Schiel (1985), and for *Laminaria* by Kain (1979).

SCALES OF SPATIAL VARIABILITY

If a major goal of ecological studies is to explain how community patterns are produced and maintained, then it is clearly necessary to identify as accurately as possible the patterns themselves. The physical and biotic factors that affect community structure may vary considerably between localities, with consequent effects on the species composition, community organization and productivity of algal-dominated reefs. Within any one locality, changes in light, turbidity, sedimentation, nutrients, grazers, *etc*, occur with depth (as discussed below) and affect the composition of algal communities. Before assessing the importance of these effects and their variability, it is important to identify the patterns of spatial distribution and abundance of the algae and associated species to provide a framework within which these factors can be examined experimentally.

Although there are many studies that describe algal floras and associated organisms in one or more localities, few specifically address the scales of variability at which the major species of algae and herbivores occur. Even by limiting the consideration of major species to fucalean and laminarian algae and large herbivorous invertebrates, there are few studies that partition the variability in distribution and abundance into circumscribed components. Research efforts have tended to be concentrated at easily accessible sites of specific interest, often obscuring broader patterns of community structure. For example, studies on the effects of sea urchins on removal of large algae have been done at a few sites where sea urchins happen to very abundant (Jones & Kain, 1967; Breen & Mann, 1976a,b; Pearse & Hines, 1979; Duggins, 1980). Studies on plant growth, reproduction, and survival have been done mostly at sites where large grazers are not prevalent (Chapman & Craigie, 1977; Kain, 1979; Reed & Foster, 1984; Schiel, 1985a,b). As a result, it is often difficult to gain a perspective on whether the investigated processes are important only in relatively small patches, in nearby algal stands, or in the entire kelp bed. Major considerations, therefore, are not only whether these processes are important, but in providing the context in which to assess them. The levels of spatial variability we will discuss are (1) geographic, (2) among local areas, (3) within sites, and (4) between depths. We can find no study that encompasses all of these.

DISTRIBUTIONAL VARIATION AMONG LOCALITIES

Geographic

This section is concerned with spatial variation in abundance of large brown algae between widely separated localities. It does not deal with the extensive literature on phytogeographic boundaries, which has been reviewed elsewhere (Michanek, 1979; Druehl, 1981; van den Hoek, 1982).

One of the major functional differences between algal stands on a geographic scale is the canopy layering of vegetation. Stipitate laminarians, with a canopy height of less than about 3 m are abundant in the northern hemisphere (Druehi, 1969; Abbott & Hollenberg, 1976; Kain, 1979). Fewer laminarian species exist in the southern hemisphere where fucalean algae are abundant and can form dense stands subtidally (Lindauer, Chapman & Aiken, 1961; Shepherd & Womersley, 1970, 1971, 1976; Choat & Schiel, 1982). Considerably more vertical structuring in the water column occurs in both hemispheres where species that form floating surface canopies predominate. The giant kelp *Macrocystis pyrifera* has a broad distribution in both hemispheres (Womersley, 1954; Druehl, 1970). In the southern hemisphere it is the only species that forms significant surface canopies in deep water, while in the northern hemisphere the distribution of this species is slightly overlapped by *Nereocystis luetkeana*, which may form dense stands in localities from central California to Alaska (Druehl, 1970). Another kelp. *Alaria fistulosa*, can also form surface canopies in Alaska (Druehl, 1970; Dayton, 1975). Members of the fucalean genus *Sargassum* may form surface canopies seasonally in some areas (Ambrose & Nelson, 1982), as can *Cystoseira osmundacea* in central California (Schiel, 1985a). The surface canopies formed by these species can drastically reduce the amount of light available to organisms in the understorey (Neushul, 1971b; Gerard, 1984; Reed & Foster, 1984).

Along any coastline, the geographic limits of different species generally overlap considerably. Information on the geographical distribution of a species is usually recorded as the presence of that species in a given locality, rather than the abundance of individual plants in stands. Most commonly, temperature, insolation, and nutrient availability are cited as either imposing or being corrleated with distributional limits (*cf.* Gaines & Lubchenco, 1982). Kain (1979 reviews the distributional information for many species of the genus *Laminaria*. She found that particular species change in their depth distributions over widely separated localities and suggests this is due to competitive interactions with other laminarians, substratum availability (Kain, 1962), grazing (Jones & Kain, 1967) and irradiance (Kain, 1966; Kain, Drew & Jupp, 1976). For example, *L. digitata* occurs intertidally and in the shallow subtidal in the northeastern Atlantic (Kain, 1962), while it extends to depths of 7 m on the coast of Norway (Kain, 1971), and 20 m off Nova Scotia (Edelstein, Craigie & McLachlan, 1969). Where *L. hyperborea* occurs, *L. digitata* tends to be restricted to shallower water, possibly because *L. hyperborea* grows faster at intermediate depths and forms a canopy over other species (Kain, 1975a, 1976).

We can find no evidence that a species of laminarian or fucalean algae changes its size distribution or is incapable of forming dense stands at extremes of its recorded geographic limits. Despite co-occurring with many different species over its geographic limits, one much-studied species, *Macrocystis pyrifera*, forms dense stands of large plants at its northern limit near Santa Cruz in central California (Foster & Schiel, 1985), in an isolated population at its southern limit in the northern hemisphere off Baja California, Mexico (North, 1972), and off Chile (Santelices & Ojeda, 1984b), Argentina (Barrales & Lobban, 1975) and New Zealand (Moore, 1943; Gerard & Kirkman, 1984). The same is true for *Nereocystis luetkeana* which has dense populations at its southern limit in central California (Estes & Van Blaricom, in press) and various localities to its northern boundary in Alaska (Estes *et al.*, 1978; Duggins, 1980). Choat & Schiel (1982) looked at the distribution, abundance, and size-frequencies of a stipitate laminarian, *Ecklonia radiata*, over 7° latitude in northern New Zealand. They found that this species formed dense stands at all localities sampled and that local differences of size frequency with depth were more pronounced than differences between localities.

The distributional evidence points to geographic boundaries being imposed by the physical requirements of light, temperature, and nutrients for individual species, rather than grazing by larger invertebrates and interactions among the sporophytes of algal species. The large degree of variability on smaller spatial scales, between and within localities, is the result of physical requirements and biotic interactions. Gaines & Lubchenco (1982) discuss these possibilities as hypotheses with reference to the differing morphology and diversity of algal species between temperate and tropical areas.

Among local areas

All of the factors correlated with the broad-scale distribution of algal species have been implicated as causes of distributional variation among more localized areas (Michanek, 1979; Druehl, 1981). The availability of hard substrata may impose limits on the depth distribution of algae in areas where light penetration is still sufficient for algal growth (Kain, 1962; Neushul, 1967; Choat & Schiel, 1982; Foster & Schiel, 1985). Water motion is also correlated with differences in the abundances of species between localities (Kain, 1979; Cowen, Agegian & Foster, 1982; Dayton *et al.*, 1984) and with their morphology (Sundene, 1964; Larkum, 1972; Gerard & Mann, 1979). Localized upwelling may affect temperature and nutrients, and can support lush algal stands (Dawson, 1951; Velimirov *et al.*, 1977; Foster & Schiel, 1985). Most studies of these are subjectively descriptive, however, rather than quantitative. For example, studies on the west coast of North America have mentioned the associations of particular algal species with varying conditions of exposure, temperature, and light, although neither the physical conditions nor the algal abundances at each locality were quantified (McLean, 1962; Devinny & Kirkwood, 1974). Some studies have given an estimate of the distribution of algae within depths or along depth gradients and described the general physical characteristics of sites Shepherd & Womersley, 1970, 1971, 1976; Kain, 1971; Field *et al.*, 1980; Foster, 1982; Santelices & Ojeda, 1984b). These studies were undertaken for a variety of reasons, from general descriptions of coastlines to specific characterizations of particular localities. Other studies have described localities as either precursors to, or as part of, experimental work. Estes *et al.* (1978) described differences in the biota at several island localities in the Bering Sea. Stands of large brown algae were greatly reduced in size in localities where dense aggregations of sea urchins were abundant. Local differences in the sizes of algal stands due to sea urchins have also been noted in Nova Scotia (Breen & Mann, 1976a,b), Britain (Jones & Kain, 1967), California (Leighton, 1971; Harrold & Reed, 1985), and New Zealand (Choat & Schiel, 1982).

The general conclusion from all of these studies is that among localities along a coastline there is a great deal of variation in algal abundances, some of which appears to be correlated with obvious physical and biotic differences. These studies, however, did not sufficiently assess variation between localized sites to determine how much can be ascribed to smaller scale habitat differences.

DISTRIBUTIONAL VARIATION WITHIN SITES

Between depths

From the earliest underwater observations it was clear that subtidal vegetation is broadly zoned along a depth gradient (Kitching, Macan & Gilson, 1934; Kitching, 1941; Andrews, 1945) and virtually all descriptive subtidal studies centre on this point. Three broad subtidal zones have been recognized along the west coast of North America in localities where hard substrata are available (McLean, 1962; Neushul, 1965, 1967; Druehl, 1967a; Aleem, 1973; Foster, 1975b; Foster & Schiel, 1985). Species of kelp such as *Egregia menziesii* and *Eisenia arborea* usually occur inshore with fucalean algae such as *Cystoseira osmundacea* and *Sargassum* spp. Kelps that form surface canopies, such as *Macrocystis* and *Nereocystis,* are prominent at middle depths, with an understorey of stipitate laminarians such as *Pterygophora californica*. The zone in deepest water, seaward of surface canopies, is inhabited by sparse stands of understorey kelps such as *Agarum fimbriatum* and *Laminaria farlowii*, encrusting corallines and small foliose red algae. Each species may broadly overlap with others along a depth gradient. In Nova Scotia, *Laminaria* spp. occupy shallow water and *Agarum* occurs at depths to 30 m (Mann, 1972). A similar pattern is found in Alaska where *Alaria fistulosa* and *Laminaria longipes* may occupy shallow areas, with *Agarum cribrosum* and other laminarian species common in deeper water (Dayton, 1975). On British shores, laminarian species are broadly zoned with depth, with *Laminaria digitata* occurring in the shallow subtidal, followed by *L. hyperborea*, *L. saccharina*, and *Saccorhiza polyschides* along progressively deeper areas of reefs (Kain, 1962, 1979; Norton, 1978). New Zealand and Australia provide a contrast, where fucalean species are abundant in shallow water to several metres depth and laminarian species dominate areas of deeper reefs (Shepherd & Womersley, 1970, 1971, 1976; Choat & Schiel, 1982).

Of particular importance to algal distribution is the co-occurrence of grazers, especially sea urchins. In Alaska it appears that echinoids, particularly strongylocentrotids, can form dense aggregations at all depths to 20 m (Estes *et al.*, 1978). This may also occur in isolated localities elsewhere (Chapman, 1981). Other studies have recorded the dominance of sea urchins at particular depths, rather than being densely aggregated along an entire depth gradient. This occurs in California (Leighton, Jones & North, 1966; Mattison *et al.*, 1977; Dean, Schroeter & Dixon, 1984), Washington (Vadas, 1977), Britain (Prentice & Kain, 1976), southern Chile (Castilla & Moreno, 1982), and New Zealand (Choat & Schiel, 1982).

We conclude that broad zones of algae and sea urchins are a real and consistent feature of most shores where suitable physical conditions for macroalgal settlement and growth are present. Within any one locality, however, the distribution and abundances of species are broadly overlapping, while between localities the positions of the zones and identities of species present change from place to place. As on rocky intertidal shores, these zones serve only as descriptions, and as areas of focus for studies concerning the demography and interactions of species.

Within depths

Few published surveys discuss the distributional variation of plants or animals within a depth stratum, but qualitative observations and the high variances associated with abundance estimates (Rosenthal, Clarke & Dayton, 1974; Kain, 1977; Pearse & Hines, 1979; Choat & Schiel, 1982; Dayton *et al.*, 1984) suggest that distributions may be clumped at fairly small scales. This variability can result from a number of processes, including variation in the distribution of many abiotic factors, as well as environmental changes created by the organisms themselves, particularly by the larger kelps. Dense stands of macroalgae significantly reduce the amount of light available to organisms in the understorey, affecting the types and abundance of understorey vegetation (Neushul, 1971b; Kain, 1979; Kastendiek, 1982; Dayton *et al.*, 1984; Reed & Foster, 1984). Local patchiness in the abundances of sea urchins (Vance, 1979; Choat & Schiel, 1982; Cowen *et al.*, 1982; Dean *et al.*, 1984) and shifts in their behaviour (Harrold & Reed, 1985), predator–prey interactions (Bernstein & Jung, 1979; Schmitt, 1982), territorial behaviour and interactions among fish (Clarke, 1970; Hixon, 1980; Larson, 1980; Choat, 1982), physical disturbance (Cowen *et al.*, 1982; Velimirov & Griffiths, 1979), and limitations in spore dispersal (Anderson & North, 1966; Deysher & Norton, 1982; Dayton *et al.*, 1984; Schiel, 1985b) can also contribute to localized variation in community structure. Stochastic events are probably important, but the detailed descriptions necessary to detect them have not been done. In diverse and structurally complex algal communities, most of the patterns of within-depth distribution remain undescribed, and the mechanisms producing local patchiness are only beginning to be explored.

A methodological problem

Correlations between algal abundances and various physical and biological factors have been cited in dozens of studies, often with poor quantitative assessments. The existence of patterns and abundance of species constitutes evidence that these physical factors and biological interactions may affect the structure of these communities. They do not at the same time, however, demonstrate the importance or unimportance of these factors in producing observed patterns (*cf.* Underwood, in press). Presumably, these patterns will form the basis of experiments by using them to pose testable and refutable hypotheses. The question can validly be put, then, how good are the descriptions of pattern upon which experiments should be based?

A major problem concerns the variables that have been used to describe pattern. Perhaps because of the variety of reasons subtidal algal stands have been examined, there is a number of ways population estimates of algal abundances have been determined. They fall broadly into (1) standing crop estimates such as biomass and per cent cover, and (2) numerical abundances. The uses of each are not the same. Standing crop estimates have often been used in studies which seek to describe algal resources along a coast (*e.g.*, Shepherd & Womersley, 1970; Field *et al.*, 1980) or that lead to considerations of energetics (Mann, 1972). Biomass estimates have been used as an indication of productivity between sites and depths (*e.g.*, Jupp & Drew, 1974). Because size classes are lumped within a species and conceal numerical abundances, these measures cannot be used for any considerations of the population biology or demography of algal species and, therefore, have little use in assessing interactions. Numerical estimates of abundance, especially when used in tandem with size-frequency measurements, can form the basis of ecological considerations and experiments. Questions concerning the recruitment and longevity of a species and changes in size composition of algal communities can only be addressed by enumerating individuals. Biomass estimates can also be calculated from these by using weight–length equations (*cf.* Duggins, 1980; Choat & Schiel, 1982). It is not always clear from the literature why a particular method was chosen over another.

Regardless of the assessment used, sampling designs have generally been poor. Of 40 papers cited here that discuss distribution and abundance only 5 (13%) used random samples within defined limits, and only 9 (23%) of the studies mentioned any measure of variance about the mean at particular depths. Most often, variability has been recognized by schematic representations of quadrats or transects, illustrating differences in distribution between them (38% of the studies). We conclude that most of these studies were done for natural history descriptions and were not intended to lead to *in situ* experimental examination of processes and interactions.

EFFECTS OF ABIOTIC FACTORS

Subtidal algae require a particuar set of abiotic conditions for their establishment and persistence and, once established, alter these conditions for themselves and other species. This section reviews the former while the effects of alterations caused by plants are discussed later (see p. 289). This division is arbitrary, since physical and biotic factors interact.

The principal abiotic factors identified as affecting algal stands are salinity, substratum, sedimentation, temperature, nutrients, light, and water motion. Some of these factors, such as temperature, can be easily measured, and there is a long history of field and laboratory studies that correlate plant distributions with various abiotic conditions. Simple correlations, however, can be misleading because of factor covariance and interactions, changing relationships between abiotic variables and plant responses with changes in the scale of measurement, and differences in the responses of each life history stage. Some of these problems have been discussed by others (Jackson, 1977; Gagne, Mann & Chapman, 1982; Lüning, 1982; Zimmerman & Kremer, 1984; Foster & Schiel, 1985), and the difficulties of extrapolating laboratory data to the field are discussed by Connell (1974).

SALINITY

Large subtidal laminarian and fucalean algae are rare or absent in estuaries (*e.g.* Mathieson & Fralick, 1973), and it is generally assumed that these plants are open coast species intolerant of low salinities. An exception is *Laminaria saccharina* which does grow where salinity is reduced (Sundene, 1964; Druehl, 1967b). Druehl (1967b) used field correlations and transplant experiments, combined with laboratory cultures under different salinity and temperature regimes, to show that *L. saccharina* is tolerant of salinities down to 17‰ at low temperatures around Vancouver Island. This species occurred in areas of reduced salinities, and did not survive when transplanted to the open coast where wave action was high. Another species, *L. groenlandica*, was more sensitive to reduced salinity but more tolerant of wave action, and was restricted to open coast habitats. The causes of distributional differences in other estuarine habitats may be more complex as turbidity, siltation, temperature, and nutrients usually all increase as salinity decreases.

SUBSTRATUM AND SILTATION

Stands of large algae are usually restricted to hard substrata where plants can remain attached in moving water, and much of the patchiness found within a particular depth can be due to the presence of sandy areas unsuitable for algal attachment (*e.g.*, Rosenthal *et al.*, 1974; Foster, 1975a).

Moreover, subtidal algal stands are often terminated at their lower limits by continuous expanses of sand (*e.g.*, Kain, 1962). There are exceptions; viable drifting plants may occur in the confines of ocean eddies such as the Sargasso Sea or in sheltered coastal habitats (Moore, 1943; Burrows, 1958; Gerard & Kirkman, 1984). Sediment accumulation in and around *Macrocystis* holdfasts can enable these plants to grow attached in relatively calm water on sandy bottoms near Santa Barbara, California (Thompson, 1959; Neushul, 1971a). Subtidal *Pelagophycus porra* grows loosely attached to unconsolidated sand and shells in calm water at Santa Catalina Island, California (Parker & Bleck, 1966; Coyer & Zaugg-Haglund, 1982).

The hardness of a rocky substratum can indirectly affect the survival and age structure of stands. The stipitate laminarian *Pterygophora californica* forms extensive stands of large (and presumably old; see Reed & Foster, 1984) plants at a semi-exposed site in California with a hard conglomerate and sandstone substratum, while only small plants are found at nearby exposed sites with a soft mudstone substratum (Foster, 1982; Reed & Foster, 1984). The composition of red algae at the exposed site may reflect more frequent disturbances on soft rock. Age structure can also be related to cobble size on unconsolidated substrata, and the presence or absence of cobbles may affect species composition (Kain, 1962).

Algal stands are generally found on the tops rather than sides of rocky outcrops in areas of high relief, and Kain (1962) noted that *Laminaria hyperborea* was limited to rock with less than a 20% slope from horizontal near Port Erin, Isle of Man. More vertical slopes are often covered with sessile animals (Foster & Schiel, 1985). These differences could be due to a number of factors, including changes in light, spore dispersal and competition with sessile animals, and represent a fruitful area of investigation.

Sedimentation, with its associated effects on light, nutrients, and the ability of plants to attach, can drastically reduce the survival of spores and gametophytes in the laboratory (DeVinny & Volse, 1978; Norton, 1978), but there are no published experimental studies of the effects of this factor on algae in the field. Suspension and redeposition of patches of sediment within algal stands is commonly observed (*e.g.*, Rosenthal *et al.*, 1974; Foster, 1975a), and such disturbances may have important effects on community structure.

TEMPERATURE AND NUTRIENTS

There is now considerable evidence that low nitrogen levels are correlated with high temperatures (Jackson, 1977; Gagne *et al.*, 1982; North, Gerard & Kuwabara, 1982; Zimmerman & Kremer, 1984); the effect of low nutrients was not usually determined in past studies that measured only temperature. Field correlations and laboratory studies have most often suggested that temperature is the primary cause of the geographic boundaries of species (see review by Druehl, 1981), but experiments are needed to separate the effects of changes in nutrients and perhaps daylength.

In addition to correlations with geographic boundaries, mortality and declines in growth occur within local stands during periods of high temperatures and associated low nitrogen (Chapman & Craigie, 1977;

Jackson, 1977; Dayton & Tegner, 1984a; Zimmerman & Kremer, 1984). Chapman & Craigie (1977) added nutrients to a *L. longicruris* population during a period of high temperature and low ambient nutrients, and growth was improved relative to unfertilized controls. Gagne *et al.* (1982) reviewed prior studies on factors affecting growth in *Laminaria* spp. and, along with their own observations, suggested that different growth patterns at various sites in Nova Scotia were primarily controlled by local differences in nitrogen availability. Other factors, however, such as internal stores of nitrogen and carbon, and perhaps genetic differences in populations, may also be important.

The consequences of this natural variation in growth to stand structure have not been addressed. Changes in *Macrocystis pyrifera* abundance caused in part by high temperatures and low nutrients can favour the proliferation of other species, potentially causing long term changes in stand structure (Dayton & Tegner, 1984a).

LIGHT

The effects of varying light intensity and quality on subtidal seaweeds has been recently reviewed (Lüning, 1981) and will not be discussed at length here. In the absence of other possible limiting factors, it appears that stands of laminarian algae are restricted to areas where benthic irradiance is 1% of the surface irradiance or greater (around 50–100 $E \cdot m^{-2} \cdot yr^{-1}$; Lüning, 1981). Local variations in water quality and, therefore, light penetration can affect the depth distribution of algal stands (Lüning, 1981; Choat & Schiel, 1982; Foster, 1982). On a geographic scale, this rather low light requirement allows populations of *Laminaria solidungula* to persist in the high Arctic where ice forms for part of the year (Chapman & Lindley, 1980).

The growth rate, density, and biomass of algal stands generally decrease with depth (Kain, 1966; Velimirov *et al.*, 1977; Choat & Schiel, 1982), and patchiness may increase (Velimirov *et al.*, 1977). Since light attenuates with depth, it is tempting to suggest that these changes are caused by light. As pointed out by Jackson (1977), however, a number of other factors such as temperature, sedimentation, nutrients, and water motion also change along depth gradients. The relative effects of these factors on stand structure remain to be determined.

The influence of light may be further complicated by the ability of plants to store carbohydrates during times of high irradiance, and then use these reserves to maintain growth when light declines. For example, the irradiance necessary to saturate photosynthesis is similar in both *L. digitata* and *L. hyperborea* (Lüning, 1979). The storage capabilities of the former are, however, limited, so that high growth rates cannot be maintained in low light. Lüning (1979) suggests that this may be why *L. digitata* does not persist in deeper water while *L. hyperborea* does.

It is known that blue light is necessary for gametogenesis in all laminarian algae so far examined and that the total amount of blue light required varies with temperature (Lüning & Neushul, 1978; Lüning, 1980) and instantaneous irradiance (Deysher & Dean, 1984). Moreover, Lüning (1982) suggests that sporophyte growth may be related to photoperiod in some species. All of these observations suggest that the effects of light on stand structure in the field are quite complex.

WATER MOTION

The piles of drift plants on beaches after storms are a clear indication that water motion can directly affect subtidal algal stands, and wave action is one of the most frequently cited causes of adult plant mortality. Effects are particularly evident in stands of *Macrocystis pyrifera* because the stipes and lamine are subjected to wave forces throughout the water column (ZoBell, 1971; Rosenthal *et al.*, 1974; Barrales & Lobban, 1975; Foster, 1982; Dayton & Tegner, 1984a). Older plants appear to be more susceptible to removal, perhaps because of deterioriation of the holdfast (Ghelardi, 1971; Barrales & Lobban, 1975) combined with more biomass in the water column. Much of this deterioration is caused by small grazing animals that live in the holdfast (Ghelardi, 1971). Koehl & Wainwright (1977) found that undamaged *Nereocystis luetkeana* could sustain the drag caused by most waves. They found that up to 90% of detached plants, however, were broken at a flaw in the stipe due to abrasion or grazing. Limpets may cause similar damage to older *Laminaria hypoerborea* plants (Kain & Svendsen, 1969). This differential mortality may produce different population age or size structures in exposed compared with protected locations (Kain, 1963; DeWreede, 1984), and among patches within a stand (Rosenthal *et al.*, 1974). Survival may also differ between sites if exposure is the same but the substratum differs (see p. 275).

The depth distribution of particular species has often been correlated wth water motion. Kain (1962) correlated the distribution of three species of *Laminaria* with exposure to water motion, and noted that the upper limit of *L. hyperborea* is raised in more sheltered locations (Kain, 1971). The absence of *Macrocystis pyrifera* in shallow water may be due to increased water motion (North, 1971c) and increased wave action may be partly responsible for the northern limit of stands of this species on the Pacific coast of North America (Foster & Schiel, 1985).

Reductions in biomass or loss of entire plants during storms may trigger a variety of other changes in algal stands. Removal of overstorey plants can lead to increased recruitment and growth of understorey species (Cowen, Agegian & Foster, 1982; Dayton & Tegner, 1984a) and may stimulate sea urchins to graze attached plants because sufficient drift is no longer available (Ebeling, Laur & Rowley, 1985; Harrold & Reed, 1985). Water motion may directly affect sea-urchin behaviour (Lees & Carter, 1972), distribution (Lissner, 1980), and mortality (Ebeling *et al.*, 1985) leading to changes in the subtidal vegetation.

CONCLUSIONS

Field correlations and laboratory studies have provided information outlining the general abiotic environment of subtidal algal stands, and numerous correlations that may account for differences between stands. Few of these primarily single factor explanations, however, have been experimentally tested in the field, and the effects of interactions are largely unknown. Further descriptive and experimental work is needed to evaluate the importance of various factors and interactions to all the life history stages of plants in particular stands, and to examine the relationships between the abiotic environment, stand density, and plant growth and reproduction.

EFFECTS OF PLANT-HERBIVORE INTERACTIONS

Many studies mention that the activities of grazers affect the distribution and abundance of large brown algae in subtidal areas, but there have been relatively few experimental studies that assess the nature of their effects. Most of our present knowledge about the effects of grazers on algal assemblages comes from experiments in the intertidal, where gastropods are usually the most important grazers (see Underwood, 1979, for a review). Much of the small-scale spatial patchiness in intertidal algal assemblages is the result of grazing on filamentous and foliose plants, as well as on algal spores (*e.g.*, Dayton, 1971; Underwood & Jernakoff, 1981, 1984). Algal cover and diversity may be dependent on the density of grazers in a given area (Lubchenco, 1978; Lubchenco & Gaines, 1981; Underwood, Denley & Moran, 1983). The regime of grazing changes, however, in the boundary between intertidal and subtidal regions. In the subtidal, limpets may decline in abundance (Underwood & Jernakoff, 1981) while larger gastropods such as trochids and turbinids may be diverse and abundant (Lowry, McElroy & Pearse, 1974; Watanabe, 1984a), and sea urchins often become the major large invertebrate grazer (Lawrence, 1975). Herbivorous fish may also affect subtidal algal assemblages (Choat, 1982; Gaines & Lubchenco, 1982).

EFFECTS OF GRAZING BY ECHINOIDS

If any generalization has become prominent in the literature on kelp communities, it is the dominant effect of sea urchins on the distribution and abundance of large brown algae. Echinoids are known to cause abrupt changes in community organization by wholesale grazing of algal stands (Lawrence, 1975). Their effects can be so pronounced that algal-echinoid interactions are considered to be the main factor affecting community structure in some parts of the world (Breen & Mann, 1976b; Mann, 1977; Estes, Smith & Palmisano, 1978; Tegner & Dayton, 1981; Moreno & Sutherland, 1982). The words "control" and "regulating" are frequently used when discussing the effects of echinoids on algae, and "overgrazing" is often mentioned, evocative of an untoward shift from a natural community dominated by large brown algae to a modified and depauperate assemblage of grazers (*e.g.*, Estes *et al.*, 1978; Duggins, 1980; Tegner & Dayton, 1981; Bernstein, Williams & Mann, 1981). The general implication has been that the grazing activities of sea urchins may have a comprehensive effect on the character of biotic assemblages on rocky reefs, producing alternate stable states—one dominated by macroalgae, the other by echinoids (Simenstad *et al.*, 1978; Moreno & Sutherland, 1982; *cf.* Harrold & Reed, 1985). This argument has also been expanded to an evolutionary context, suggesting that the evolution of kelp life histories and competitive abilities is the result of responses to echinoid grazing activities (Vadas, 1977; Steinberg, 1984).

These generalizations, however, are often made without a spatial context and, in the case of stable states, without the necessary assessment of temporal scales (*cf* Connell & Sousa, 1983). Other studies show that echinoids may be restricted in their distribution to particular depths (Prentice & Kain, 1976; Andrew & Choat, 1982; Choat & Schiel, 1982) or subhabitats within depths (Cowen *et al.*, 1982; Foster, 1982). Their feeding activities, therefore, and consequent effects may be confined to particular depths or patches on different reefs (Lissner, 1980; Keats, South & Steele, 1982; Schiel, 1982; Harrold & Reed, 1985). The variability in the distribution and abundance of grazers within and between localities makes it difficult to evaluate the generality of sea urchin–algae interactions. There are also geographic differences in algal and echinoid types and distributions (Lawrence, 1975) which require evaluation before a broad understanding of their interactions can be achieved.

We now review observational and experimental studies examining direct effects of grazing on macroalgae and also the effects on sea urchins of predators, that may indirectly affect community structure. These effects have been generally documented in three categories: (1) the wholesale removal of algae and the provision of cleared substratum suitable for kelp recruitment, (2) alteration of the species composition of stands *via* feeding preferences and selective removal of species, and (3) the effects of predators on sea urchins.

The removal of stands of kelp by echinoids, especially strongylocentrotids and echinometrids, is a worldwide phenomenon (Lawrence, 1975). Although there is a general pattern to these grazing episodes, there are few studies that quantify the abundances of grazers and their affects on the numbers of kelp plants at specific sites. The large scale removal of *M. pyrifera* from stands in southern California was reported by Leighton, Jones & North (1966) and North & Pearse (1970). Leighton (1971) recorded the advance of a "feeding front" of sea urchins into a kelp stand at Point Loma. The echinoids in the vanguard were mostly large *Strongylocentrotus franciscanus*, tightly aggregated at 30 per m^2, while 18 m back smaller *S. purpuratus* were more abundant (39 per m^2). The aggregation moved 27 m into a kelp stand during a two-month period, consuming or detaching all *Macrocystis pyrifera* in its path by grazing through the holdfasts or lower fronds. The echinoids further back in the aggregation fed on this newly-freed plant material. There is no information on how large the feeding aggregation was, but a kelp stand of about 100×200 m was reportedly destroyed entirely after several months (Leighton, 1971).

The most thorough documentation of grazing effects was done further north, at the San Onofre kelp forest in southern California. Dean, Schroeter & Dixon (1984) used a series of observations and experiments to assess the effects of two species of sea urchins on *M. pyrifera*. Two different modes of feeding were seen for *Strongylocentrotus franciscanus* (a large echinoid averaging 107 mm test diameter in the kelp forest). Aggregations were either relatively small and stationary over three years, or else large and mobile, advancing at the rate of 2 m per month. Stationary aggregations were at maximum densities of 7 per m^2 in patches of 5–25 m width, feeding mainly on drift kelp, and having no significant effect on kelp recruitment and abundance. Mobile aggregations had densities of 47 per m^2 at the leading edge, and up to 14 per m^2 20 m behind, in patches of about 20 × 30 m. Most macroalgae in their path were grazed. Small *Macrocystis pyrifera* transplanted into the mobile aggregations were consumed over a two-day period, but remained intact amongst stationary echinoids. The results of a similar experiment with *Lytechinus anamesus*, a smaller echinoid (2–4 cm, test diameter) were equivocal, with small *Macrocystis pyrifera* being consumed in some trials and ignored in others. Of particular interest in this study was the careful 3-year observations of kelp and echinoid abundances along several transects through the kelp forest. Stationary and mobile aggregations of echinoids occurred within 100 m of each other, and feeding fronts were seen only three times during the course of the study. These quite different modes of feeding activity were very local-scale events, and apparently were caused by the paucity of drift algae, which produced a change of foraging behaviour of the sea urchins. Dean *et al.* (1984) also concluded that both types of aggregation appeared to be unrelated to predation pressure from lobsters and fishes, although density estimates of these predators were anecdotal.

A relationship between the availability of drift algae and the feeding behaviour of *Strongylocentrotus franciscanus* was found in two other studies in California. Harrold & Reed (1985) found that sea urchins moved greater distances and fed on benthic organisms in "barren" patches, while they remained relatively stationary and fed on drift plants in kelp stands. In central California, Mattison *et al.* (1977) found that sea urchins fed mainly on drift plants in areas devoid of macroalgae. Greater movement was exhibited by echinoids further away from the kelp forest. One of the major differences between actitivies of echinoid grazers in the eastern Pacific and those elsewhere may, therefore, be related to the preponderance of large kelps, and hence ample drift plants, compared with the smaller stipitate laminarians found in most other temperate areas of the world.

Descriptive studies on the removal of kelp by echinoids have also been given for the coastline of Nova Scotia. Breen & Mann (1976a,b) found that the widths of *Laminaria* beds were reduced considerably over a 5-year period, and suggested this was due mostly to an increase in the abundances of *Strongylocentrotus droebachiensis*. The pattern described was similar to areas elsewhere: the formation of mobile aggregations of echinoids, wholesale grazing of kelp, relatively stable border areas where invasion of kelp stands ceased, and extensive rock flat areas devoid of kelps. Anecdotal information describing a similar pattern is available from Newfoundland (Himmelman & Steele, 1971) and the British Isles (Kain, 1979).

In all cases for which there are data on the numbers of echinoids, large scale removal of kelps appears to be the result of high densities of echinoids. Breen & Mann (1976b) found that at low densities, *S. droebachiensis* did not remove kelp stands, and suggested a non-linear effect of grazing. They postulated that a threshold abundance of 2 kg per m^2 of echinoids was required for destructive grazing because of the weight necessary to hold down the kelp blades and prevent them from movements that would dislodge the urchins. In New Zealand, Schiel (1982) found a non-linear feeding effect for *Evechinus chloroticus* on seven species of algae in controlled field experiments. Echinoids clumped on some replicates and had significantly different effects among algal species on the amount of plant material removed. Aggregations of *Evechinus* are independent of sex, and may be associated with spawning, feeding, and aspects of the physical environment (Dix, 1969, 1970). Clumping around a food source has also been demonstrated for *Strongylocentrotus franciscanus* (Russo, 1979) and *S. droebachiensis* (Garnick, 1978; Vadas *et al.*, 1986). The evidence to date suggests an all-or-nothing effect of echinoids on the removal of plants from kelp stands. We can, however, find no study that experimentally manipulated echinoid densities in kelp stands and recorded their effects on the abundance of plants. An important step in understanding algal-echinoid interactions, therefore, is examining the factors that might affect echinoid aggregations. These are discussed below.

There are two persistent (i.e., several months to years) features of reef areas with kelp and high numbers of echinoids: (1) kelp-echinoid boundaries and (2) extensive areas devoid of fucalean and laminarian algae. After a grazing episode, dense aggregations of echinoids often form a distinct border between grazed areas and kelp stands. These borders may persist for several months or longer, and have been noted in California for *Macrocystis–Strongylocentrotus franciscanus* (Mattison *et al.*, 1977), in Britain for *Laminaria hyperborea–Echinus esculentus* (Kain, 1977), in Nova Scotia for *Laminaria* spp.–*Strongylocentrotus droebachiensis* (Lang & Mann, 1976), and in New Zealand for *Ecklonia–Evechinus chloroticus* (Choat & Schiel, 1982). Both Lang & Mann (1976) and Choat & Schiel (1982) found that echinoids were larger and more abundant at this boundary than they were on rock-flat areas several metres away from kelp stands. The reasons for the creation and persistence of these border areas are not clear, and several hypotheses have been proposed. Increased water motion may prevent the inshore movement of echinoids (Lissner, 1980); drift algae may be available to the larger echinoids on these borders causing them to cease active foraging (Mattison *et al.*, 1977); localized predators associated with kelp stands may prevent further impingement of echinoids into kelp areas (Breen & Mann, 1976b; Wharton & Mann, 1981); there may be longer-term settling and recruitment inhibition of both echinoids and algae in particular areas of reefs (*cf.*, Cameron & Schroeter, 1980). These are not mutually exclusive, and it is likely that there is an interaction of several factors.

Echinoid-dominated areas of reef have persisted for over 10 years in Nova Scotia (Chapman, 1981) and New Zealand (Ayling, 1981; Andrew & Choat, 1982). These have been termed "barren grounds" in most studies (Lawrence, 1975). We feel that this terminology is inappropriate and misleading (*cf.* Choat & Schiel, 1982). "Barren grounds" implies a detrimental shift from kelp stands to areas devoid of organisms. While this may have occurred in some instances, the term usually means an absence or severe reduction only in the numbers of fucalean and laminarian plants. Other organisms are usually not mentioned, and rarely quantified. These areas, however, may have a high per cent cover of coralline algae (Ayling, 1981; Noro, Masaki & Akioka, 1983), sponges and gastropods (Ayling, 1981). As feeding substrata these areas may also support higher abundances of some species of prominent reef fish than nearby kelp stands (*cf.* Andrew & Choat, 1982). Echinoid-dominated areas are, therefore, distinct and probably complex habitats, with some degree of persistence temporally and spatially, and are affected by a variety of physical and biotic factors.

The removal of high numbers of echinoids from areas of reefs usually results in the recruitment of macroscopic algae. The sequence of algal recruitment varies between studies. Fucalean or laminarian algae may be among the early colonizers and the transition to a fairly dense algal stand may occur in the relatively short time of about a year (Jones & Kain, 1967; Foreman, 1977; Pearse & Hines, 1979; Duggins, 1980; Andrew & Choat, 1982; Ebeling *et al.* 1985). In other situations, foliose, fleshy and crustose algae may recruit. Chapman (1981) pointed out the importance of nearby reproductive plants to algal recruitment in areas where echinoids are dominant. There is growing evidence that most recruits appear within a few metres of adult plants for large brown algae (Anderson & North, 1966; Dayton, 1973; Paine, 1979; Deysher & Norton, 1982; Schiel, 1985b; Dayton *et al.*, 1984), although Chapman (1981) reported dispersal of up to 600 m. Other factors that are potentially important, but rarely discussed in this context, are (1) the presence of sessile species such as sponges and coralline algae that may affect algal settlement (Breitberg, 1984; Reed & Foster, 1984), (2) sand or sediment cover that may inhibit gametophyte attachment and growth (Devinny & Volse, 1978; Norton, 1978), (3) the grazing activities of herbivorous gastropods that can be associated with high numbers of echinoids (Ayling, 1981; Choat & Schiel, 1982), and (4) seasonal and year-to-year variation in algal reproduction and recruitment (Foster, 1975a; Kain, 1975b Neushul *et al.*, 1976; Kennelly & Larkum, 1983; Schiel, 1985a,b) that suggests the timing of echinoid removals and the provision of free space may be critical to the recruitment of fucalean and laminarian algae.

EFFECTS OF NON-ECHINOID GRAZERS

In a report from California, Harris, Ebeling, Laur & Rowley (1984) suggested that fish, particularly *Medialuna californiensis* and *Girella nigricans*, can be important grazers of small *Macrocystis pyrifera* sporophytes on a local scale. On one reef, grazing affected 59% of small plants (<10 cm tall) that were concealed in a turf of ephemeral algae, while 94% of those in open reef quadrats were grazed. Larger plants were not grazed, suggesting a size refuge. The result of this grazing was a small-scale change in the dispersion pattern of juvenile *M. pyrifera* on the reef. They did not, however, report the abundances of fish present over the reef and no observations of fish feeding behaviour were mentioned. These herbivorous fish have been noted to destroy adult plants transplanted to an artificial reef where naturally occurring macroalgae were rare (Grant, Wilson, Grover & Togstad, 1982), suggesting grazing effects on adults may be related to plant density. Choat (1982) concluded that there were no studies that demonstrated extensive modification of the biota of temperate reefs by grazing fishes.

There are few published studies that have assessed the effects of gastropods on algae in subtidal regions. Watanabe (1983, 1984a,b) examined the effects of three species of *Tegula* in a shallow *Macrocystis pyrifera* forest. These gastropods normally live and feed on the fronds and laminae of *M. pyrifera*. During and after storms they are abundant on substrata below the plants, but quickly occupy fronds again when calmer conditions prevail. Their grazing activities, however, had no discernible effect on algal distribution and abundance.

Other invertebrate grazers such as abalone (*Haliotis* spp.) and sea stars (*Patiria miniata*) may have small-scale effects on algal abundances in *Macrocystis* forests (*cf*., Tegner & Levin, 1982), but their effects have not been assessed experimentally. Small crustaceans can be very abundant in algal turfs (Kennelly, 1983) and may be major grazers of algal spores. Experiments assessing their effects, and the interactions of grazers that co-occur on areas of substratum, have yet to be done.

Large areas of substratum in a kelp bed may be covered by sessile invertebrates (Foster & Schiel, 1985). The effects of these on algal distribution are largely unknown. Besides overgrowing plants and pre-empting space for attachment of algal propagules (Foster, 1972, 1975b; Neushul *et al*., 1976), their filtering activities could remove algal spores from the immediate area.

FEEDING PREFERENCES BY ECHINOIDS

Most sea urchins are considered to be generalist feeders, consuming a wide variety of plant and animals material (Lawrence, 1975). Feeding, however, is usually selective in some sense and a knowledge of which plant species are acceptable or preferred by echinoids may help to determine how they affect community structure. Many studies have shown that a preference hierarchy can be established for sea urchins consuming algal species in laboratory experiments (Leighton, 1971; Lawrence, 1975; Vadas, 1977; Larson, Vadas & Keser, 1980). An important question is whether these preferences reflect the manner in which algae are removed from natural stands by the same sea-urchin species. In a major study, Vadas (1977) found that *Strongylocentrotus droebachiensis* clearly preferred *Nerocystis luetkeana* to *Agarum cribrosum* in laboratory experiments. Sea-urchins grew faster and had a greater reproductive output when fed *Nereocystis* for long periods. He postulated optimal feeding methods for sea-urchins in nature, and argued for the co-evolution of algae and urchins based on selective removal, plant defences, and benefits to urchins. For these arguments to obtain, however, it must also be shown not only that echinoids can differentiate between algal species and feed selectively on them, but also that their grazing effects are of some consequence to the algal species. Rather than selective removal of algae from natural stands, Vadas (1977) indicated that echinoids fed mainly on drift *Nereocystis,* plants which had already been removed by other causes. In addition, the densities of sea-urchin aggregations were not mentioned as a factor important to plant removal, although this factor is associated with a change in feeding activities in nature (Lawrence, 1975; Breen & Mann, 1976a,b; Schiel, 1982).

In another study, Schiel (1982) used a combination of laboratory and field experiments and observations of algal removal in natural stands to demonstrate selective feeding by *Evechinus chloroticus*. For seven algal species, he found (1) no correlation between rankings of algal preferences for the laboratory and field experiments, (2) no correlation between rankings for the field experiment and the order in which algae were removed from natural stands, and (3) no overall correlation between availability rankings of algae in natural stands, in terms of numbers of plants, and the sequence in which species were removed. He suggested that some species were more vulnerable to removal, due to narrow or easily damaged stipes and holdfasts.

Many studies have concluded that availability is the major factor in determining which algae are consumed by echinoids (Paine & Vadas, 1969a,b; Himmelman & Steele, 1971; Ayling, 1978). It is often difficult to disentangle preference from availability. For example, *Strongylocentrotus droebachiensis* may co-occur with and consume *Desmarestia* in the field, but because of the algal morphology and movement in currents it may often be unobtainable to echinoids (Himmelman & Steele, 1971). More commonly, the presence of drift algae may confound the assessment of availability of food species (*e.g.*, Vance, 1979; Vance & Schmitt, 1979). Large laminarian algae, such as *Nereocystis* and *Macrocystis*, may be major components of drift algae in some habitats, even though they are not locally abundant as attached plants in the specific sites where dense aggregations of echinoids occur (Mattison *et al.*, 1977; Vadas, 1977; Duggins, 1981a).

We conclude that the available evidence does not support a major rôle for feeding preferences in the removal of algae from natural stands. While selective feeding undoubtedly has benefits in terms of growth and gonad production for echinoids, it appears to have, at most, a minor effect on algal stands.

EFFECTS OF PREDATORS ON HERBIVORES

A growing literature cites the effects of various predators, especially fish, lobsters, sea stars, and sea otters on the distribution, abundance and behaviour of sea urchins. These studies are of considerable importance because they argue for a keystone rôle of some predators in the maintenance and persistence of algal stands, based on (1) the demonstrated detrimental effects of high densities of echinoids on algal abundances, and (2) the fact that predators can "regulate" or "control" echinoid numbers. The first of these was discussed in the previous section, indicating that echinoid effects are temporally and spatially variable. The second merits consideration because of the potential importance of predators to the structure of algal communities.

Arguments concerning the importance of predators on echinoids and algal assemblages have been considered in terms of both recent or current effects and also extended back in time to historical effects. In this section, we discuss primarily the evidence for current effects. How do predators affect the distribution and abundance of herbivores?

Fish

Several studies report that predatory fishes affect the distribution of sea urchins, but few of them demonstrate the effects experimentally. Choat (1982) reviewed this subject, and we will mention only a few studies that have argued forcefully for fish effects on invertebrates and consequent effects on community structure.

In southern California the sheephead wrasse, *Semicossyphus pulcher*, can be locally abundant on kelp-dominated reefs, and may have a high per cent content of sea urchin remains in its gut (Quast, 1971; Cowen, 1983). Nelson & Vance (1979) observed that sheephead and the diademetid echinoid, *Centrostephanus coronatus*, have opposite activity patterns, with sea urchins becoming active at night at the time the wrasse is inactive. Because of this, and the fact that sheephead will eat *Centrostephanus*, they inferred that the evolution of behavioural patterns of *Centrostephanus* was the result of interactions with *Semicossyphus*. Cowen (1983) found that the removal of *Semicossyphus* from one site resulted in an increase in the proportion of *Strongylocentrotus franciscanus* that were exposed on the reef, relative to a non-manipulated site. He also found an increase in the number of echinoids present in the sheephead-removal site, although it was not clear if this was due to enhanced recruitment or to movement of echinoids from cryptic habitats. Tegner & Dayton (1981) used correlative information to argue for *Semicossyphus* having an important rôle in the local distribution of *Strongylocentrotus franciscanus* and *S. purpuratus*. They found a difference between depths in the numbers of sheephead wrasse, suggesting that urchins in the shallowest depth (12 m) were less often in cryptic habitats due to the absence of the wrasse. Alternative explanations are that sizes and behaviour of echinoids may vary locally, associated with substratum heterogeneity, depth and water motion (Lissner, 1980; Andrew & Choat, 1982; Choat & Schiel, 1982; Dean, Schroeter & Dixon, 1984).

In eastern Canada, Bernstein, Williams & Mann (1981) used gut content analysis and negative correlations between the numbers of *S. droebachiensis* and the numbers of fish (*Anarhichas lupus* and *Hippoglossoides platessoides*) to argue for significant modification of echinoid behaviour due to predatory fishes. A very complicated scenario involving the break-up of a large aggregation of echinoids was based partially on the observation that a feeding front dispersed when fishes became seasonally abundant. An alternative explanation is that the echinoids dispersed into nearby rock flat areas due to factors other than predators, such as an alteration in feeding behaviour, as noted in other studies (Lawrence, 1975; Mattison *et al.*, 1977). Choat (1982) pointed out some of the problems in interpretation of correlative studies, and also the lack of adequate controls for studies where echinoids were presented to fish in the field. Disturbances by divers may attract fish to prey items placed in the field, yielding an over-estimate of fish effects.

Andrew & Choat (1982) presented one of the few subtidal studies that specifically tested for effects of fish on echinoids and algae. In a rock flat area devoid of macroalgae, they used a factorial design and replicated 2 m^2 cages that included normal densities of the echinoid *Evechinus chloroticus* and excluded predatory fishes that were abundant in the study site. They found that fish exclusion enhanced the abundance of juvenile echinoids over a 16-month period, and that this effect was independent of the presence of adult urchins. The removal of echinoids from areas of reef resulted in a rapid recruitment of fucalean and laminarian algae. Even though significant fish effects were noted, they concluded that a consistent but low number of *Evechinus* recruits escaped predation on the rock flats, and that these were enough to maintain the urchin-dominated rock flats over several years.

Lobsters

Predation by lobsters is considered to have major effects on the distribution, abundance, size-frequencies and behaviour of echinoids in California (Tegner & Dayton, 1981; Tegner & Levin, 1983) and Nova Scotia (Breen & Mann, 1976a; Lang & Mann, 1976; Mann, 1977; Bernstein *et al.*, 1981, Wharton & Mann, 1981). Using feeding trials in the laboratory, Tegner & Levin (1983) found that *Panulirus interruptus* preferred *Strongylocentrotus purpuratus* to *S. franciscanus*, and that there were differences in how the echinoids were handled by lobsters. The longer spines of *S. franciscanus* apparently made them more difficult to consume. A bimodal size-frequency distribution of *S. franciscanus* was ascribed to the sheltering of juveniles beneath larger individuals (Tegner & Dayton, 1981), rather than to recruitment and year-class events (*cf.* Andrew & Choat, 1982; Himmelman *et al.*, 1983). Citing anecdotal and correlative information from records of lobster catches, Tegner & Dayton (1981) suggested that increases in echinoid abundances that occurred in areas of southern California in the 1950s were partly the result of declining lobster numbers.

Similar arguments have been used in Nova Scotia Mann (1977) found that declines in the commercial catch of lobsters in the early 1970s occurred at the time when increases in the numbers of *S. droebachiensis* were noted. The most forceful statement concerning current effects of lobsters and other predators was by Bernstein *et al.* (1981). Laboratory trials were used in which sea urchins were presented to crabs. Urchins tended to congregate (>2 individuals) more often in the presence of crabs. In addition, it was reported that crabs "attacked" both hidden and exposed (presumably solitary) echinoids, but not aggregated ones. From this it was argued that aggregations are an effective defence against predation by crabs, and presumably by lobsters, and that aggregations are a behavioural response to predation. Bernstein *et al.* (1981), however, present no details of experimental designs, such as numbers of crabs used and whether the trials were replicated, no information on the actual consumption of echinoids, and no data on the abundances of crabs or lobsters in or near kelp beds.

Wharton & Mann (1981) reviewed the correlative data on the increase of echinoids, the decline of kelp and lobsters and offered an overview of their putative relationships. They presented the following scenario (pp. 1343–1344): (1) an increase in echinoid numbers, (2) kelp beds were destroyed, leaving echinoid-dominated reefs, (3) the absence of kelp rendered young lobsters more vulnerable to predators, and lobster catches declined. Their historical data indicate a time lag between the grazing of kelp beds and the decline in lobster stocks. At the same time, however, they go on to argue that increases in sea-urchin densities, a prelude to kelp decline, must have been the result of decreased predation pressure by lobsters and other predators (pp. 1345, 1347). Pringle, Sharp & Caddy (1980) provided a more cautious interpretation of echinoid–kelp–lobster information. They cite alternative explanations, indicating that (1) data on echinoid increases have been gathered at only a few sites, (2) the main cause of lobster declines was overfishing rather than environmental causes or declines in habitat, and (3) there is insufficient evidence to support the argument that lobsters are the major predators of echinoids or that increases in echinoid abundances were due to release from predation by lobsters. Wharton & Mann (1981) rejected this interpretation, arguing for a reduction in the rejection level of hypotheses. They stated that because ecosystems are complex it is difficult to demonstrate the validity of interactions and that "one can always demonstrate that ecological evidence is insufficient and that a verdict of 'not proven' is justified". They plea for a change in consensus from "not proven" to "seems plausible, worth investigating further".

Sea otters

The hypothesis that sea otters (*Enhydra lutris*) enhance kelp abundance by removing echinoids has received considerable attention. Otters consume up to a fourth of their body weight per day, feeding on a wide variety of invertebrates and some fishes (Kenyon, 1969, 1969; Shimek, 1977; Stephenson, 1977; Hines & Loughlin, 1980; Estes, Jameson & Johnson, 1981; Estes, Jameson & Rhode, 1982; Ostfield, 1982). Attempts at experimentally assessing their effects on nearshore communities have been hampered by the obvious logistic constraints of dealing with a legally protected, mobile predator. To judge the effects of sea otters, observations on the aftermath of their moving into areas have been used and also "natural experiments" (*cf.* Connell, 1974; Hurlbert 1984), comparing sites with and without otters. The importance of otter–echinoid–kelp relationships is based on three types of evidence: (1) when dense aggregations of echinoids are present, kelp abundance at those sites may be low; (2) sea otters preferentially feed on large echinoids when these are available; in sites where otters are abundant, sea urchins tend to be scarce, small in size and occur in cryptic habitats (Lowry & Pearse, 1973; Estes *et al.*, 1978; Breen *et al.*, 1982) and kelp abundant (Dayton, 1975; Estes *et al.*, 1978); and (3) historical evidence that the otter's range once extended along the west coast of North America, suggesting they were important predators in kelp forests (Estes & Van Blaricom, in press).

Estes *et al.* (1978) examined the distribution and abundance of kelp and echinoids at several sites in the Bering Sea, some of which had populations of sea otters. In a site without otters, *Strongylocentrotus polyacanthus* were paricularly abundant and there was a sparse cover of macroalgae. Echinoids were larger at this site when compared with the site where otters actively foraged. Estes *et al.* (1981) found that in areas of the Aleutian Islands recently repopulated by otters, their diets consisted mainly of echinoids, whereas benthic fish were the most important prey to an established population. Duggins (1980) removed dense aggregations of sea urchins from a site in Alaska and, as with echinoid removals in other areas (Lawrence, 1975), found that a lush algal flora developed. It has been argued, therefore, that otters may be a "keystone species" in shallow communities in the north Pacific (Estes & Palmisano, 1974; Estes *et al.*, 1978; Duggins, 1980).

The evidence is not clear for the effects of sea otters in California. After otters expanded their range, dense echinoid populations were reduced along the Monterey Peninsula, leaving generally small, concealed individuals (McLean, 1962; Lowry & Pearse, 1973). Indeed, there are no known examples of dense aggregations of large sea urchins persisting where otters are present. It has been argued that *Nereocystis luetkeana*, an annual kelp, tends to persist in the presence of sea urchins, while *Macrocystis pyrifera* does not, and that there is a change from *Nereocystis* to *Macrocystis* after otters forage in a locality, removing echinoids (discussed in Estes *et al.*, 1981). While this may be the case for some sites in central California, the evidence is equivocal for others. Sea urchins are often quite patchy in their distribution and effects, and there is often not a straightforward relationship between their abundances and that of particular kelp species (Rosenthal, Clarke & Dayton, 1974; Lawrence, 1975; Duggins; 1981a; Harrold & Reed, 1985). In addition, most of the coast of central California is exposed. Winter storms and turbulent sea conditions can affect kelp abundance, with the annual *Nereocystis* tending to replace *Macrocystis* in exposed localities (*e.g.*, Cowen, Agegian & Foster, 1982). Moreover, both canopy species can persist outside the range of sea otters, sheephead wrasse, and lobsters, even though populations of large sea urchins may occur (Cowen *et al.*, 1982; Foster, 1982).

The historical evidence also indicates that the abundances and effects of sea otters are variable. Simenstad, Estes & Kenyon (1978) examined evidence from Aleut middens in Alaska and concluded that alternate stable states existed in nearshore communities. They argued that strata containing large quantities of echinoid tests and limpet shells coincided with the absence of otters, which were hunted by Aleuts. The presence of fish remains coincided with times when otters were present, suggesting that macroalgae predominated. The middens on San Nicolas Island off southern California also have evidence that prehistoric man hunted otters. Dayton & Tegner (1984b) pointed out the large numbers of abalone shells seen on the island, and suggested that aboriginal man had a significant impact on the nearshore community.

Estes & Van Blaricom (in press) reviewed data on the fluctuations of many shellfish populations and the possible effects of otters. The advent of otters and the decline of many shellfisheries in recent times were often coincident with increased fishing pressure, pointing to competition between modern man and otters for particular resources. They concluded that the near extinction of the sea otter permitted the development of shellfisheries in the first place. The interactions between otters, particular fisheries and kelp stands are complex. Both recent and past historical evidence tends to obscure the natural variability in the distribution and abundance of kelp forests, echinoids and otters. So far as the presence or absence of otters is concerned, three conclusions are clear: (1) otters do have a significant impact on the numbers of invertebrates, (2) factors other than predation of echinoids by otters (or other predators), such as winter storms and turbulent coastlines, regularly affect the abundance of kelp in central California, and (3) despite occasional occurrences of dense aggregations of echinoids in southern California where otters do not occur, algal forests still predominate where rocky reefs are present (Foster & Schiel, 1985). Various effects of sea otters on nearshore communities are discussed in Van Blaricom & Estes (in press).

Other predators

Sea stars may contribute to the local-scale variability in echinoid abundances. Schroeter, Dixon & Kastendiek (1983) found that *Patiria miniata* was an abundant predator of *Lytechinus anamesus* in a southern California kelp forest, affecting the local distribution of the echinoid. This sea urchin can be a major grazer of juvenile laminarians in local patches (Dean *et al.*, 1984), and an alteration in its dispersion patterns could contribute to algal patchiness. *Pycnopodia helianthoides* feeds on sea urchins (Moitoza & Phillips, 1979) and its presence can lead to local changes in echinoid abundance, allowing algal recruitment (Paine & Vadas, 1969a; Duggins, 1983). Bernstein *et al.* (1981) mention sea stars as one of a number of predators that may influence the abundance of echinoids in Nova Scotia.

Accumulating observations suggest that disease can have both local (Pearse, Costa, Yellin & Agegian, 1977) and large scale (Miller & Colodey, 1983) impacts on sea-urchin populations. This can lead to the expansion of existing kelp stands (Pearse & Hines, 1979; Miller, 1985) and a large scale switch from sea-urchin to algal domination (Miller, 1985).

METHODOLOGICAL PROBLEMS

Of the 31 experimental manipulations cited in 25 papers in this section, 15 (48%) had serious problems in design, and therefore in interpretation. Many of these studies (32%) did not give adequate information with which to judge exactly what was done, often obscuring (1) the number of replicates, (2) in the case of laboratory experiments, the number of trials and how many animals were used, or (3) how the results were analysed. Where removal of echinoids was done, virtually all studies had only one clearance site, which obscures natural variability between sites (Hurlbert, 1984).

An equally serious consideration concerns the degree of difference required to disprove a hypothesis. Poor correlative information has often been used to sustain arguments, especially those involving predator effects on echinoids. Where variability is obscured because of the sampling design used or measures of variability do not exist, it is often considered to be insignificant. We do not argue that many of the scenarios presented for predator–echinoid–kelp interactions are not "plausible", but rather that better conceived and designed experiments and sampling would result in better evidence, so that equally plausible alternative hypotheses can be rigorously tested.

EFFECTS OF INTERACTIONS AMONG PLANTS

INTERACTIONS BETWEEN SPECIES

Algal species may affect each other in a number of ways. Canopy layering may significantly reduce the light available to understorey plants and nutrients may be altered in the vicinity of dense algal forests (Jackson, 1977) with potential effects on the recruitment, growth, reproduction, and survival of understorey species. Other effects are the occupation of primary substratum by species of all canopy levels thus reducing the amount of free space for attachment by other species and also impeding spore fall from higher canopy levels. Not all of these factors have been examined experimentally.

The most common method used to determine the effects of one algal species on another is the selective removal of algal canopies and recording the subsequent recruitment or persistence of algae in the clearance areas relative to an unmanipulated area with the canopy intact. Dayton (1975) removed canopies of several species of laminarian algae at different depths at a site in Alaska. He found that when the species forming a surface canopy inshore, *Alaria fistulosa*, was removed other species did not invade the clearances, and he concluded that *Alaria* was not a competitive dominant. Canopies of *Laminaria* spp., however, appeared to preclude the recruitment of *Alaria*. The rhizoidal growth pattern of *Laminaria longipes* ensured that relatively rapid vegetative growth would allow this species to reform a canopy quickly after removal and so exclude other species from invasion. Two other canopy removals indicated that *Agarum cribrosum* had little effect on other species, but that its cover increased when *Laminaria* spp. were removed. These experiments showed that canopies of individual species can affect other species. The nature of these effects, however, was not clear. Presumably, the reduced recruitment beneath canopies was the result of light inhibition. Lower canopies may also impede the arrival of propagules to the substratum. In these experiments, per cent cover of species was assessed and it was not always clear if new recruitment had occurred or if small juveniles already present had grown to replace the canopy in clearances. One of the main problems, however, was the lack of information about reproduction for individual species. The apparent failure of some species to recruit successfully and the success of others could have been a reflection of the seasonal production of propagules (*cf.*, Foster, 1975a; Kain, 1975b, 1979) which may not have been available for some species at the time of canopy clearances.

The initiation of clearances at different times of the year may result in colonization by different species. Ambrose & Nelson (1982) found in southern California that a canopy of *Sargassum muticum* may prevent successful recruitment of *Macrocystis pyrifera* into localized patches. This may be partially due to the fact that the Sargassum canopy is densest during the early summer when *Macrocystis* recruits would normally appear.

Many canopy clearances by Dayton *et al.* (1984) showed that canopies of several laminarian species in southern California suppressed recruitment of algae. Proximity to reproductive adults appeared to be one of the major factors in determining if recruitment could occur in cleared areas. Dayton *et al.* (1984) also found that turfs of articulated coralline algae precluded the recruitment of large brown algae.

Reed & Foster (1984) examined the effects of three canopy levels on recruitment of algae using field experiments in which most (although not all) combinations of factors (*i.e.*, the presence and absence of each canopy) were used. Canopies of over- and understorey plants were removed: *Macrocystis pyrifera*, *Pterygophora californica*, articulated and encrusting corallines. Light measurements showed that either a dense *Macrocystis* or *Pterygophora* canopy could reduce light intensity to 0·1–0·3 of that found at the same depth outside of canopies. The combined effect of both canopies was not much greater at these low light levels (0·1–2·5% of surface light) that approach the lower limit needed for the growth of laminarian algae (Lüning, 1981). They found that a dense canopy of either *Macrocystis* or *Pterygophora* inhibited algal recruitment. Highest recruitment occurred in treatments that had the branches of articulated corallines removed, but left encrusting coralline algae in place, suggesting that the branches impede spore-fall, that they may reduce irradiance to the substratum, or that they harbour many micro-grazers. Clearances of coralline algae to bare rock did not result in high recruitment, but the target areas of clearances (25 × 25 cm) were much smaller than treatments with encrusting algae (1 × 1 m). Reed & Foster (1984) concluded that the major factor affecting recruitment was the reduction of irradiance caused by the *Pterygophora* canopy.

Other studies indicate that fucalean algae showed greater recruitment to cleared substratum than to areas where low-lying algae occupied the substratum. Crusts of coralline algae have been found to inhibit successful recruitment of many sessile species in kelp forests (Breitburg, 1984). Deysher & Norton (1982) found that *Sargassum muticum* recruited almost exclusively to patches of cleared substratum rather than to adjacent controls covered by various species of algae. The timing of clearances was also important. The longer the interval between the production of free space and inoculation by embryos, the fewer the recruits that appeared, presumably due to interactions with other species (*cf.* DeWreede, 1983). In central California *Cystoseira osmundacea* may require free primary substratum to recruit successfully (Schiel, 1985a). Pre-emption of space by low-growing algae is also important to the recruitment of laminarian and fucalean species in the low intertidal (Sousa, Schroeter & Gaines, 1981).

Few studies specifically address situations in which several species recruit into an area to determine if one or more has detrimental effects on others. Studies using small artificial substrata suggest that species which recruit early inhibit those recruiting later (Foster, 1975a). In a 20 × 10 m clearance of a *Macrocystis pyrifera* canopy, Pearse & Hines (1979) found significantly more recruitment of several laminarian species than in a nearby uncleared plot. This result was attributed to differences in light levels: 0·2% of surface light reached the bottom beneath the *M. pyrifera* canopy compared with 3·8% in the cleared area. In non-vegetated areas where large numbers of sea urchins had died, Pearse & Hines (1979) found that many species of laminarians recruited. As a *M. pyrifera* canopy became established over two years, *Laminaria dentigera*, *Pterygophora*, and *Nereocystis* became less abundant and were found only as isolated individuals in the understorey. This effect was attributed to shading by the *Macrocystis pyrifera* canopy.

The result was not so clear in other studies where sea urchins were removed. Several species of laminarians recruited into such areas in Britain (Kain & Jones, 1966; Jones & Kain, 1967), in British Columbia (Pace, 1981), and in Nova Scotia (Breen & Mann, 1976b), but no clear dominant, that had negative effects on other species, quickly emerged. This was probably due to the similar canopy heights of the stipitate laminarians in these studies producing a different result from areas which have large and abundant surface-canopy kelps.

Clearances in which there is recruitment of annual kelps appear to undergo the quickest changes in species composition. Paine & Vadas (1969a) reported that *Nereocystis leutkeana* became the dominant alga in the first year on subtidal reefs kept free of sea urchins, forming some 90% of the biomass of algae. In the following year, *Laminaria groenlandica* became the dominant alga. Duggins (1980) removed *L. groenlandica* from several small plots in Alaska, resulting in a high recruitment of annual kelp. By the second year, however, *L. groenlandica* was again dominant (x±S.D. = 53±27 plants/m^2 compared with 8±7 plants/m^2 for other species). Successful recruitment did not occur in the control plot where the canopy was left intact.

A different result of canopy clearance occurred in Chile; Santelices & Ojeda (1984a) cleared a *Macrocystis pyrifera* canopy from a 5×50 m transect. There was increased recruitment of *M. pyrifera* to the removal area but the dominant understorey plant, *Lessonia flavicans*, increased in biomass in the non-removal area. In addition, there was a change in the species covering the bottom after canopy removal. The method of canopy removal, however, was different from other studies, with *Macrocystis pyrifera* plants being cut 1 m below the sea surface rather than immediately above the holdfasts. The senescence of cut fronds could have negatively affected *Lessonia* and other species in the removal area. Relative to *Macrocystis* forests in the northern hemisphere, the changes appeared to be more subtle in this southern kelp forest depauperate in laminarian species.

One other study examined interactions among three species in the shallow subtidal, indicating that the distribution of an understorey species can be the result of interactions between two canopy species. Kastendiek (1982) found that a red alga, *Pterocladia capillacea*, occurred mostly beneath a canopy of the kelp *Eisenia arborea*. This kelp was sandwiched between two zones where *Halidrys dioeca* was abundant. The removal of *Eisenia* canopies allowed *Halidrys* to spread adventitiously and pre-empt space, while removal of both *Halidrys* and *Eisenia* resulted in the expansion of *Pterocladia*. Thus, *Eisenia* was the competitive dominant at the intermediate depth because its canopy excluded *Halidrys*, which in turn allowed *Pterocladia* to occupy the space beneath.

CONCLUSIONS

We conclude that the evidence clearly shows there are important effects among algal species, and that these are more pronounced between species of differing canopy heights. The number of plants required to exert an effect increases as canopies (and plants) decrease in size. For example, a closed canopy of *Macrocystis pyrifera* at 1 plant per 10 m^2, may exert a similar effect to a *Pterygophora californica* canopy at 8 plants per m^2 (*cf*., Dayton *et al*., 1984; Reed & Foster, 1984). The nature of these effects, however, is not always clear. Most studies cited here recorded recruitment of plants, fewer recorded subsequent survival, and none examined growth and reproduction of plants trapped beneath the canopies of others.

The most serious problems with these studies involved experimental and sampling designs. Most of these studies (75%) had pseudo-replication (Hurlbert, 1984) in at least one level, usually involving the clearance of only one site and leaving one unmanipulated site as a control, thus potentially obscuring localized effects unrelated to the experiments. In many experiments the designs were so obscure that it was not possible to decipher exactly what was done, how big the clearances or controls were, how many replicates there were, and what sampling procedure was used. We could find no study that used a balanced design in which interactions of factors could be assessed (*cf.* Underwood, 1981). If an understanding of interactions between species is to be achieved studies must assess these interactions and also determine whether a dominant species limits resources to others, physically or chemically interferes with them, or simply pre-empts the space required for their attachment.

INTERACTIONS WITHIN SPECIES

Effects of density

The hypothesis that the growth, reproduction, and survival of laminarian and fucalean algae are adversely affected by high population densities has rarely been examined. There is a large literature on within-species effects of terrestrial plants, showing that individual plants in dense stands usually exhibit lower growth, reproduction, and survival rates relative to plants in similar environments at lower densities (Harper, 1977). The fact that large, essentially monospecific aggregations of large brown algae commonly occur worldwide (Stephenson & Stephenson, 1949, 1972; Dawson, Neushul & Wildman, 1960; Mann, 1972, 1973; Dayton, 1975; Kain, 1979; Kirkman, 1981; Choat & Schiel, 1982) suggests that in some situations there might be advantages to plants growing in denser stands. There is evidence both for and against this. In New Zealand, Schiel & Choat (1980) found that a subtidal laminarian, *Ecklonia radiata*, and a subtidal fucalean species, *Sargassum sinclairii*, had the largest plants in dense, single-species stands on one semi-exposed reef. These plants were of the same age and in an apparently similar habitat, although differences among local sites (boulders) may have been important. They hypothesized that these results were due to plants of both species growing faster in high-density stands, and that there may be a cushioning effect for individuals in high-density stands, which affords some protection from battering by water motion. Cousens (1982) and Cousens & Hutchings (1983) reached different conclusions for stands of brown algae in Nova Scotia. They found that large plants of an intertidal alga, *Ascophyllum nodosum*, occurred both at low and high densities along a shoreline. They concluded that the most important factor in determining differences in size and morphology was not density but rather small-scale variation in water motion along the shore. In this study, however, ages of plants were not known which obscures growth histories, plants at different densities were located in different habitats on the shore, and the putatively important physical variable of water motion was not quantified.

Other studies have followed the histories of plants at different densities through time. Black (1974) found that survival was density-dependent for recruits of *Egregia laevigata* but that this was not the case after plants were three months old. In this study there were negative effects on growth rates and sizes of plants in the densest stands. Schiel (1985b) described the histories from recruitment of two fucalean species in low (1–40 plants/m^2) and high (>2000/m^2) densities. For both *Sargassum sinclairii* and *Carpophyllum maschalocarpum* he found that plants grew faster, attained a larger size, and had a greater amount of reproductive tissue at the higher density. Low-density plants exhibited greater survival, but these plants tended to be small and non-reproductive. Schiel (1985b) suggested these results could be expected if light and nutrients were not limited and if high densities dampened the effects of severe water motion. Although light can be reduced to plants trapped below canopies (Kain, 1979; Pearse & Hines, 1979; Kirkman, 1981; Reed & Foster, 1984), this effect may be reduced in stands of the same age and canopy height where water motion allows light flashes through the stands (Kitching, 1941; Kain, 1979; Gerard, 1984). Stands of large kelps such as *Macrocystis pyrifera* can also affect nutrient flow (Gerard, 1982a,b,c; Jackson, 1977; Wheeler, 1982) but this effect may not be seen in stands of smaller algae. Although Schiel (1985b) showed that different recruitment densities resulted in different plant responses, an alternative explanation of these results is that areas of boulders where dense recruitment occurred may have been favourable to plant growth, whereas low density areas were not. This factor could be examined by experimental thinning of stands of dense recruits.

Neushul & Harger (in press) planted adult *M. pyrifera* (mean size = 25 fronds per plant) at densities of 1 per m^2, 1 per 4 m^2, and 1 per 6 m^2 on a test farm for mariculture in southern California. Plants at the lowest density had the most fronds and greatest weights after a year, while those at the highest density were the smallest, suggesting that shading in the denser parts of the stand caused poorer growth. Possible complicating factors were: (1) the use of adult plants, which may already have adopted a particular growth regime, (2) the use of one small experimental plot (0·24 hectare) with different densities being contiguous, and (3) the placement of low and medium densities toward the outside of the stand where peripheral light may have enhanced their growth.

Density effects are probably not the same for all species of large brown algae. Density-dependent mortality may be greater for algae that form surface canopies, such as *Macrocystis* and *Nereocystis*. A major source of mortality for adult plants is large plants that have broken loose from the substratum and drift through kelp forests entangling other plants (Rosenthal, Clarke & Dayton, 1974; Dayton *et al.*, 1984). Another potential source of mortality is entwining of young plants as they grow through the water column (*cf.*, Dayton *et al.*, 1984). These effects may be exacerbated in dense stands during times of surgy conditions.

It has been suggested for some fucalean species that higher densities of adult plants may result in release *en masse* of gametes, which may be important for successful recruitment (Fletcher & Fletcher, 1975; Fletcher, 1980). Greater densities may also be important to effect successful fertilization of gametes for laminarian algae, although this has not been investigated.

Even if plants in some situations perform better at higher densities, there will obviously be an upper limit to the numbers of plants that can be crowded into an area without negative consequences. Intraspecific competition for light, nutrients, and space therefore, may prove to be an important structuring force in stands of large brown algae. Testing for these effects, however, requires experiments that control for the ages of plants, habitats, depths, and localized factors associated with sites. These experiments can take the form of thinning experiments, whereby areas with high recruitment have treatments thinned to lower densities. High recruitment into an area indicates that it is suitable, at least for the initial life stages of the alga. Site effects will be reduced if different density treatments are contained in the area of initially high recruitment, or if more than one site is used, using replicates within each (*e.g.*, randomized block design; Winer, 1971). Another way to approach this problem is to outplant sporophytes at differing densities and record subsequent growth and survival.

Effects of spore dispersal

The problem of determining whether or not spores have actually arrived to an area is important to arguments about competition among species (Underwood & Jernakoff, 1981, 1984; Underwood & Denley, 1984). There is increasing evidence that spore or germling dispersal is limited for many species of brown algae, with most recruits appearing within a few metres of reproductive adult plants (Anderson & North, 1966; Dayton, 1973; Paine, 1979; Deysher & Norton, 1982; Dayton *et al.*, 1984; Sousa, 1984; Schiel, 1985a,b). The problem of whether algal sporelings can survive and grow in sites away from adult plants has been addressed in a few studies.

Inoculating sites by tying down fertile plant tissue often results in a higher recruitment of algae to these treatments than to "unseeded" controls (Deysher & Norton, 1982; Dayton *et al.*, 1984). Because propagules themselves have no means for long range dispersal, this suggests that a proximate source of propagules, either through attached adult plants or drifting fertile tissue, is a requisite to high recruitment.

Adult distribution may be a reflection of spore distribution. Assessing this is a persistent problem in algal research and one which is only just beginning to be addressed for large brown algae (*e.g.*, Schonbeck & Norton, 1978, 1980; Kennelly, 1983; Schiel, in prep.). Spore-fall is probably not evenly distributed in natural situations (*cf.* Cook, 1980) and may be affected by current and surge conditions in much the same way as "seed shadows" occur in some terrestrial situations (Harper, 1977). Dense aggregations of spores may be important to the production of algal stands, but innovative experiments are required to determine this for marine plants. In such an experiment, Chapman (1984) examined the period from spore settlement to survival of adults for *Laminaria longicruris* and *L. digitata*. He found that the greatest proportional mortality for both species occurred between the time microscopic plants attached to the substratum and when they became visible, a period of about six weeks. Stands of *L. longicruris* produced 9×10^9 spores$\cdot m^{-2} \cdot yr^{-1}$ and *L. digitata* 20×10^9spores$\cdot m^{-2} \cdot yr^{-1}$. Of these, 1 in 1000 survived to microscopic recruits for *L. longicruris* and 1 in 20 000 for *L. digitata*. The eventual recruitment of visible plants was 1 per m^2 and 2 per m^2, respectively. More visible sporophytes appeared in areas where the red algal turf had been removed. Overall, Chapman (1984) found a good correlation between spore production and the numbers of microscopic recruits for *L. longicruris* and a poor correlation for *L. digitata*. This is a unique study in that it examined the numbers per area of every life history stage of two species, from spores to adult survival. Studies like it are essential to clarify the distributional histories of species, and to assess whether competition among algal species is an important structuring force in algal communities.

THE PERSISTENCE OF ALGAL STANDS

Studies done in several parts of the world document that kelp stands may occupy particular sites for at least several years (Mann, 1977; Kain, 1979; Dayton *et al.*, 1984; Foster & Schiel, 1985). Particular areas dominated by sea urchins have also been observed for several years (Wharton & Mann, 1981; Andrew & Choat, 1982) and, as mentioned above, have led some to consider kelp beds and echinoid-dominated areas as alternate stable states (Estes & Palmisano, 1974; Estes, Smith & Palmisano, 1978; Simenstad *et al.*, 1978; Moreno & Sutherland, 1982). A consideration of stability, however, requires a knowledge of several factors. Among them are the minimum spatial scales required for a community to become established, a time scale sufficient for a turnover of the individuals in populations, and some measurements of disturbing forces and how populations and communities respond to them (Connell & Sousa, 1983). These have not been examined in much detail for most kelp systems.

"Patches" of species have been referred to in many studies (*e.g.*, Dayton *et al.*, 1984; Reed & Foster, 1984), but there are few data on the sizes of patches, their spatial distribution in a kelp bed, and the species composition of individual patches through time. Algae may occur as isolated plants, as plants surrounded by other species or in dense aggregations, of various sizes, of conspecifics. The spatial distribution of plants in kelp stands has not been examined in most places, and yet can have major consequences on the composition of communities (*cf.* Hubbell, 1979). If plants do tend to occur in patches of predominantly single species (*cf.* Dayton *et al.*, 1984) then this fact should be revealed by an examination of the scales at which clumping is detected. How big an area is required for a kelp stand to exist? This will vary between species types (*e.g.*, those forming surface canopies and stipitate laminarians). For large kelps such as *Macrocystis* the problem of scale is probably exaggerated throughout their life histories. Although recruits may be very dense in a small area (1 m^2), spatial requirements change so that a density of adult plants of approximately 1 per 10 m^2 is a dense forest (Pearse & Hines, 1979; Reed & Foster, 1984). 8–20 plants per m^2 can form a closed canopy for various species of stipitate laminarians (Kain, 1979, 1979; Choat & Schiel, 1982; Reed & Foster, 1984). Clearly, these spatial requirements of different species are a necessary consideration in discussing stability. The minimum size necessary for a stand of a species to maintain itself is not known, however, and has not been examined.

Dayton *et al.* (1984) specifically examined patch dynamics in a *M. pyrifera* forest in southern California. They identified four distinct types of patches, two of single species (*M. pyrifera*, *Pterygophora californica*) and two of two species (*Laminaria* and *Cystoseira*, *Pterygophora* and *Eisenia*). Canopy clearances were used in parts of these patches (in most cases 1–4 m^2) and subsequent recruitment was recorded. The usual result was that the nearest fertile species recruited into canopy-free areas, while few recruits of any species appeared beneath mature canopies. The borders of the original patches apparently did not change significantly over several years. The sizes and abundancies of patches, and their spatial relationships in the kelp forest, were not specifically addressed.

The age distributions of prominent species in populations can also yield clues about stability and persistence. Unfortunately, it appears that there are few marine algae for which ages of plants can be determined independently if recruitment histories are unknown. Some stipitate laminarians have internal growth rings that may be annual (Kain, 1963; Novaczek, 1981; DeWreede, 1984; Reed & Foster, 1984) but this has not been verified in different habitats and growth conditions. Consequently, the full life histories of plants in different situations cannot be verified without long-term field observations.

There have been few studies which have produced survival curves for even a single species recruiting at one time. Rosenthal *et al.* (1974) followed a small population of *Macrocystis pyrifera* finding that most plants disappeared in the first year, while some adults lived for 7+ years. Dayton *et al.* (1984) used estimates of age based on short-term survival data and growth rings in some species to construct survival tables for several laminarian species. It is unlikely, however, that one survival curve for a single population adequately describes the performance of a species. Most of the factors discussed earlier will affect survival. For example, populations exposed to grazing will have different survival curves from those ungrazed.

The apparently limited dispersal of spores in large brown algae makes it likely that if relatively small areas are cleared in larger patches of one species that the area will be inoculated with spores of that species and it will replace itself. This form of self-replacement of many species could occur on a larger spatial scale, so that a kelp forest may exist in an area for longer than one generation of individuals. There are few studies, however, which examine the abundances of even a few of the most prominent species in a kelp forest over several years or a generation. The persistence of *M. pyrifera* populations along the coast of California has been documented for over 30 years from aerial photographs of the surface canopy (Wilson & North, 1983). Large fluctuations in the areal extent of canopies are evident, presumably reflecting in some manner the changes in numbers of plants and suggesting that these populations have not been stable.

Populations of sea urchins have also been documented as persistent, lasting about 10 years in Nova Scotia (Chapman, 1981) and New Zealand (Andrew & Choat, 1982). Andrew & Choat (1982) showed that recruitment of juvenile echinoids can occur into these areas, suggesting that the echinoid-dominated areas are maintained beyond the life span of individuals. The same may be the case in Nova Scotia (Miller, 1985). Evidence for other areas shows various patterns.

Sea urchins may recruit yearly in southern California (Dayton & Tegner, 1984b) and some protection of juveniles may be afforded by concealment beneath adults (Tegner & Dayton, 1977). The life history features of echinoid populations in dense aggregations have not been examined. There is evidence, however, that both physical and biotic factors may have a large effect on the dominance of reefs by echinoids. Ebeling, Laur & Rowley (1985), for example, found that when a severe storm removed most detached and drift *M. pyrifera* from a reef in southern California resident echinoids emerged from cracks and consumed other laminarian algae. A second storm removed many of the echinoids and provided free space, which was colonized by laminarian algae. Harrold & Reed (1985) found that particularly high algal recruitment with subsequent drift production resulted in a change in sea urchin behaviour. Duggins (1981b) recorded that echinoids on a reef in Alaska began feeding on a sudden influx of salps, which allowed algae to become established locally. The effects of sea urchin disease were mentioned above. These sorts of relatively small-scale, local events comprise most of our detailed knowledge of the dynamic nature of species interactions on reefs and suggest that kelp forest communities are not stable when the relative abundances of species are considered.

Dayton & Tegner (1984b) discussed broad-scale events that may have great effects on the long term persistence of algal communities. Broad climatic effects, such as the severe storms and high water temperatures associated with El Niños (Jackson, 1977; Dayton & Tegner, 1984a) may affect species composition of stands, nutrient levels, and larval transport and survival. Disease outbreaks may also occur over large areas (Miller, 1985). Larger time scales have seen effects: the high levels of exploitation by modern man of sea otters, shellfish, reef fish, and lobsters combined with stresses of human coastal populations have surely had effects on inshore algal systems, and prehistoric man also may have severely exploited predators such as sea otters as discussed above.

We conclude that what evidence exists for kelp communities indicates that they are not stable over a period as small as several years, when the relative abundances of the several prominent species that normally inhabit kelp beds are considered. Where recruitment rates and survival information is lacking, it cannot be assumed that there is constancy in these numbers.

CONCLUSIONS

Descriptive studies over the past 50 years have provided the broad outlines of the composition and distribution of subtidal algal stands, while correlative and manipulative investigations have suggested the importance of various factors and processes affecting them. Even though most of these studies have dealt only with adult phases in algal life histories, they have revealed that a number of processes and factors may interact to influence stand structure at various spatial and temporal scales (Fig. 2). Below we summarize some of the generalizations that have emerged.

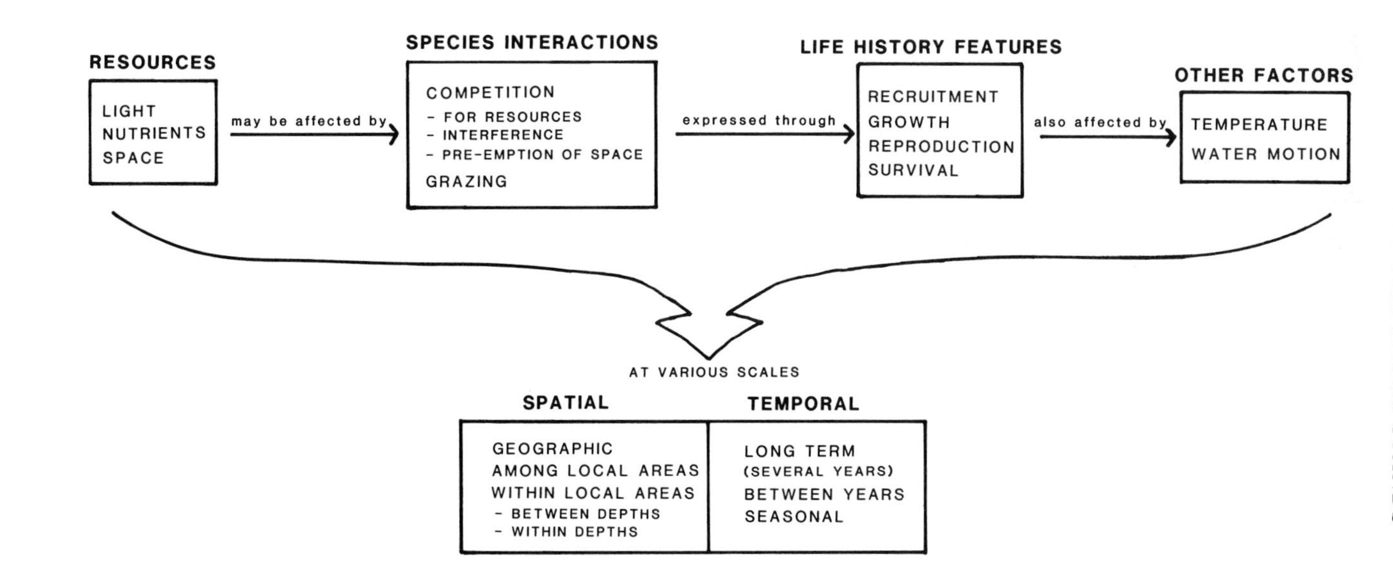

Fig. 2.—Summary of important factors in the dynamics of algal stands and the temporal and spatial scales at which they occur.

SPATIAL VARIATION

Species composition, distribution, and abundance change geographically and between and within local areas in response to broad environmental conditions and local site characteristics. Temperature is most often correlated with geographic distribution, although its direct influence is difficult to test in the field. Local differences between sites have been correlated with a number of factors, the more important of which appear to be differences in substratum, exposure to water motion, and grazing. Variation with depth has been correlated most often with changes in light, water motion, substratum, and grazing. Although patterns of distribution within depths have been recognized, particularly clumped distributions (or patchiness), the causes of pattern at this scale are only just beginning to be unravelled.

For kelp forests in California, Dayton *et al.* (1984) have suggested that within stands biological processes have the most important effects on persistence and stability, while between stands the main differences are due to physical effects. This can be the case where extreme examples of differential physical effects, such as water motion, may directly result in differences between stands. In other situations, however, differences between stands may be due to biological effects such as intense grazing. Moreover, as discussed in previous sections, a range of less extreme physical and biotic factors interact within any kelp stand. For example, the formation of a closed canopy by one species is a biological factor, but one of the most important results is a drastic reduction in light available to the understorey; differences in canopies between stands are not always due directly to physical factors and may involve biological factors such as spore availability and interactions with other species on the substratum. At this stage, the relative importance of many physical factors within stands, such as sediment distribution, substratum type and topography, nutrient availability, *etc.* is not known because most have not been examined in detail.

TEMPORAL VARIATION

With few exceptions (Rosenthal *et al.*, 1974; Kain, 1979; Pearse & Hines, 1979; Chapman, 1984; Dayton *et al.*, 1984; Schiel, 1985a,b), we know little about the recruitment, growth, and survival of subtidal algal populations. Seasonal and year-to-year variations are documented mostly for floating canopies of *Macrocystis* in California. These variations have been ascribed to water motion, nutrient availability and, most notoriously, to grazing by sea urchins. As discussed above, grazing by sea urchins has prompted the suggestion that dense algal stands and sea urchin-dominated areas represent alternate stable states. As pointed out by Connell & Sousa (1983), however, community changes at sites in Alaska have been attributed to human activities, as they have in southern California (Tegner & Dayton, 1981) and Nova Scotia (Wharton & Mann, 1981). Regardless of causes, we argue that it is inappropriate to infer a "normal equilibrium" state (*e.g.*, abundant kelp and few sea urchins) for shallow, rocky subtidal communities. In most cases, there is not enough information through time to judge whether algal stands are sufficiently persistent to be called "stable". The existence of different states may simply reflect how the community is defined.

As discussed previously, large temporal changes in the composition of algal stands can occur without the influence of sea urchins, and areas with high densities of sea urchins within kelp stands may be a regular feature of other reefs. In addition, a range of composition can be observed in algal stands separated by short distances, even in regions where there are predators which putatively control echinoid numbers. For example, in one kelp stand at San Nicolas Island, California, sea urchins are confined to crevices amongst abundant vegetation, apparently as a result of predation by fish (Cowen, 1983). Another stand on the same island is characterized by a mosaic of patches, some with abundant vegetation where echinoids feed on drift algae, and some with mostly encrusting coralline algae where sea urchins remove attached plants (Harrold & Reed, 1985). There are other sites on the same island where upright vegetation is almost entirely absent, and sea urchins occur at high densities over hundreds of square metres of substratum.

Although more information is needed for algal stands, these observations and others discussed previously indicate that community composition of shallow subtidal reefs is dynamic with relative abundances changing in space and time. We therefore suggest that such reefs have a natural dynamic range of composition and relative abundance and that this range should be included in the definition of the community. Great changes in composition do not necessarily represent alternative stable states or communities but simply end points in the range of one community. Perhaps using dynamic range as a working hypothesis will stimulate research on the entire spectrum of community composition and not just on the extremes.

THE RELATIVE IMPORTANCE OF VARIOUS FACTORS

Ecologists have focused on one factor after another in the search for generalizations concerning the regulation of population and community structure. As discussed above, the structure of many subtidal algal stands has been interpreted almost solely in terms of the interactions between grazers, their predators and the algae. Others have pointed out the important effects of exposure to water motion. Within stands these and other factors, such as dispersal, plant density, and competition for light, have demonstrated influences on the distribution and abundance of particular species.

As more studies are done at more sites, it is likely to be found that single factors and simple cause-and-effect relationships do not adequately explain or predict the structure of algal stands. A more useful concept may be typologies, where the relative influence of various factors are assigned to different 'types' of stands at different spatial and temporal scales. For example, on a relatively large spatial scale, water motion may be most important to the structure of stands growing on relatively soft rock exposed to high water motion. Sites under these conditions may form one site 'type' with a characteristic range of composition (*cf.* Foster, 1982; Foster & Schiel, 1985). At a smaller spatial scale, dispersal and light reduction may be the most important factors influencing localized stands or patches of understorey kelps. In these stands, extreme water motion or extensive grazing by echinoids may be important on the scale of tens of years, but not seasonally or year-to-year.

Such a classification scheme, emphasizing the relationships between sites, scales, and factors would be similar to that in an analysis of variance model which partitions the variance into different components. It might help to remove some of the confusion that results from mixing different scales, and from generalizing about "algal stands" from studies limited in space and time to one "algal stand". It should also provide a more rigorous context for future investigations.

ACKNOWLEDGEMENTS

This review covers literature published before August 1985. Many of the ideas expressed here, and some of the text, are expanded versions of subchapters from Foster & Schiel (1985). We thank the following who discussed various aspects of subtidal ecology with us, critically read the manuscript, and offered helpful suggestions: Drs J. H. Choat, J. H. Connell, S. D. Gaines, G. P. Jones, W. P. Sousa, and N. L. Andrew. We gratefully acknowledge the support of Moss Landing Marine Laboratories.

REFERENCES

Abbott, I. A. & Hollenberg, G. J., 1976. *Marine Algae of California.* Stanford University Press, Stanford, 827 pp.
Aleem, A. A., 1973. *Botanica Mar.,* **16,** 83–95.
Ambrose, R. F. & Nelson, B. V., 1982. *Botanica Mar.,* **25,** 265–267.
Anderson, E. K. & North, W. J., 1966. *Proc. Int. Seaweed Symp.,* **5,** 73–86.
Andrew, N. L. & Choat, J. H., 1982. *Oecologia (Berl.),* **54,** 80–87.
Andrews, H. L., 1945. *Ecology,* **26,** 24–37.
Ayling, A. L., 1978. *J. exp. mar. Biol. Ecol.,* **33,** 223–235.
Ayling, A. M., 1981. *Ecology,* **62,** 830–847.
Barrales, H. & Lobban, C. S., 1975. *J. Ecol.,* **63,** 657–677.
Bernstein, B. B. & Jung, H., 1979. *Ecol. Monogr.,* **49,** 335–355.
Bernstein, B. B., Williams, B. E. & Mann, K. H., 1981. *Mar. Biol.,* **63,** 39–49.
Black, R., 1974. *Mar. Biol.,* **28,** 189–198.
Breen, P. A. & Mann, K. H., 1976a. *Mar. Biol.,* **34,** 137–142.
Breen, P. A. & Mann, K. H., 1976b. *J. Fish. Res. Bd Canada,* **33,** 1278–1283.
Breen, P. A., Carson, T. A., Foster, J. B. & Stewart, E. A., 1982. *Mar. Ecol. Prog. Ser.,* **7,** 13–20.
Breitburg, D. L., 1984. *Ecology,* **65,** 1136–1143.
Burrows, E. M., 1958. *J. mar. biol. Ass. U.K.,* **37,** 687–703.
Cameron, R. A. & Schroeter, S. C., 1980. *Mar. Ecol. Prog. Ser.,* **2,** 243–247.
Castilla, J. C. & Moreno, C. A., 1982. In, *Echinoderms: Proceedings of the International Conference, Tampa Bay,* edited by J. H. Lawrence, A. A. Balkema, Rotterdam, pp. 257–263.
Chapman, A. R. O., 1981. *Mar. Biol.,* **62,** 307–311.
Chapman, A. R. O., 1984. *J. exp. mar. Biol. Ecol.,* **78,** 99–109.
Chapman, A. R. O. & Craigie, J. S., 1977. *Mar. Biol.,* **40,** 197–205.
Chapman, A. R. O. & Lindley, J. E., 1980. *Mar. Biol.,* **57,** 1–5.
Choat, J. H., 1982. *Ann. Rev. Ecol. Syst.,* **13,** 423–449.
Choat, J. H. & Schiel, D. R., 1982. *J. exp. mar. Biol. Ecol.,* **60,** 129–162.
Clarke, T. A., 1970. *Ecol. Monogr.,* **40,** 189–212.

Connell, J. H. 1974. In, *Ecology: Field Experiments in Marine Ecology,* edited by R. N. Mariscal, Academic Press, New York, pp. 21–54.
Connell, J. H. & Sousa, W. P., 1983. *Am. Nat.,* **121,** 789–824.
Cook, R., 1980. In, *Demography and Evolution in Plant Populations,* edited by O. T. Solbrig, University of California Press, Riverside, pp. 107–129.
Cousens, R., 1982. *Botanica Mar.,* **25,** 191–195.
Cousens, R. & Hutchings, M. J., 1983. *Nature, Lond.,* **301,** 240–241.
Cowen, R. K., 1983. *Oecologia (Berl.),* **58,** 249–255.
Cowen, R. K., Agegian, C. R. & Foster, M. S., 1982. *J. exp. mar. Biol. Ecol.,* **64,** 189–201.
Coyer, J. A. & Zaugg-Haglund, A. C., 1982. *Phycologia,* **21,** 399–407.
Dawson, E. Y., 1951. *J. mar. Res.,* **10,** 39–58.
Dawson, E. Y., Neushul, M. & Wildman, R. I., 1960. *Pacif. Nat.,* **1,** 1–81.
Dayton, P. K., 1971. *Ecol. Monogr.,* **41,** 351–389.
Dayton, P. K., 1973. *Ecology,* **54,** 433–438.
Dayton, P. K., 1975. *Fish. Bull. NOAA,* **73,** 230–237.
Dayton, P. K., 1984. In, *Ecological Communities: Conceptual Issues and the Evidence,* edited by D. R. Strong *et al.*, Princeton University Press, Princeton, pp. 181–197.
Dayton, P. K., Currie, V., Gerrodette, T., Keller, B. D., Rosenthal, R. & Ven Tresca, D., 1984. *Ecol. Monogr.,* **54,** 253–289.
Dayton, P. K. & Tegner, M. J., 1984a. *Science,* **224,** 283–285.
Dayton, P. K. & Tegner, M. J. 1984b. In, *Novel Approaches to Interactive Systems,* edited by P. W. Prince *et al.*, J. Wiley & Sons, New York, pp. 457–481.
Dean, T. A., Schroeter, S. C. & Dixon, J. D., 1984. *Mar. Biol.,* **78,** 301–313.
DeVinny, J. S. & Kirkwood, P. D., 1974. *Botanica Mar.,* **17,** 100–106.
DeVinny, J. S. & Volse, L. A., 1978. *Mar. Biol.,* **48,** 343–346.
DeWreede, R. E., 1983. *Phycologia,* **22,** 153–160.
DeWreede, R. E., 1984. *Mar. Ecol. Prog. Ser.,* **19,** 93–100.
Deysher, L. E. & Norton, T. A., 1982. *J. exp. mar. Biol. Ecol.,* **56,** 179–195.
Deysher, L. E. & Dean, T. A., 1984. *J. Phycol.,* **20,** 520–524.
Dix, T. G., 1969. *Pacif. Sci.,* **23,** 123–124.
Dix, T. G., 1970. *N.Z. J. mar. freshwat. Res.,* **4,** 91–116.
Druehl, L. D., 1967a. *J. Fish. Res. Bd Canada,* **24,** 33–46.
Druehl, L. D., 1967b. *J. Phycol.,* **3,** 103–108.
Druehl, L. D., 1969. *Proc. Int. Seaweed Symp.,* **6,** 161–170.
Druehl, L. D., 1970. *Phycologia,* **9,** 237–247.
Druehl, L. D., 1981. In, *The Biology of Seaweeds,* edited by C. S. Lobban & W. J. Wynne, University of California Press, Berkeley, pp. 306–325.
Duggins, D. O., 1980. *Ecology,* **61,** 447–453.
Duggins, D. O., 1981a. *Oecologia (Berl.),* **48,** 157–163.
Duggins, D. O., 1981b. *Limnol. Oceanogr.,* **26,** 391–394.
Duggins, D. O., 1983. *Ecology,* **64,** 1610–1619.
Ebeling, A. W., Laur, D. R. & Rowley, R. J., 1985. *Mar. Biol.,* **84,** 287–294.
Edelstein, T., Craigie, J. S. & Mclachlan, J., 1969. *J. Fish. Res. Bd Can.,* **26,** 2703–2713.
Estes, J. A., Jameson, R. J. & Johnson, A. M., 1981. In, *The Worldwide Furbearer Conference Proceedings, Vol. 1,* edited by J. A. Chapman & D. Pursley, Worldwide Furbearer Conference, Inc., Frostburg, Maryland, pp. 606–641.
Estes, J. A., Jameson, R. J. & Rhode, E. B., 1982. *Am. Nat.,* **120,** 242–258.
Estes, J. A. & Palmisano, J. F., 1974. *Science,* **185,** 1058–1060.
Estes, J. A., Smith, N. S. & Palmisano, J. S., 1978. *Ecology,* **59,** 822–833.
Estes, J. A. & Van Blaricom, G. R., in press. In, *Conflicts Between Marine Mammals and Fisheries,* edited by R. A. Beverton *et al.*, Allen & Unwin, London.

Field, J. G., Griffiths, C. L., Griffiths, R. J., Jarman, N., Zoutendyk, P., Velimirov, B. & Barnes, A., 1980. *Trans. R. Soc. E. Afr.,* **44,** 145–192.
Fletcher, R. L., 1980. *Botanica Mar.,* **23,** 425–432.
Fletcher, R. L. & Fletcher, S. M., 1975. *Botanica Mar.,* **18,** 149–156.
Foreman, R. E., 1977. *Helgoländer wiss. Meeresunters.,* **30,** 468–484.
Foster, M. S., 1972. *Proc. Int. Seaweed Symp.,* **7,** 55–60.
Foster, M. S., 1975a. *Mar. Biol.,* **32,** 313–329.
Foster, M. S., 1975b. *Mar. Biol.,* **32,** 331–342.
Foster, M. S., 1982. In, *Synthetic and Degradative Processes in Marine Macrophytes,* edited by L. Srivastava, W. de Gruyter, Berlin, pp. 185–205.
Foster, M. S., Dean, T. A. & Deysher, L. E., 1985. In, *Handbook of Phycological Methods: Ecological Methods for Macroalgae,* edited by M. M. Littler & D. S. Littler, Cambridge University Press, Cambridge, pp. 199–231.
Foster, M. S. & Schiel, D. R., 1985. *The Ecology of Giant Kelp Forests in California: A Community Profile,* U. S. Fish & Wildlife Service Biol. Rep. 85(7.2), Washington, D. C., 153 pp.
Gagne, J. A., Mann, K. H. & Chapman, A. R. O., 1982. *Mar. Biol.,* **69,** 91–101.
Gaines, S. D. & Lubchenco, J., 1982. *Ann. Rev. Ecol. Syst.,* **13,** 111–138.
Garnick, E., 1978. *Oecologia (Berl.),* **37,** 77–84.
Gerard, V. A., 1982a. *J. exp. mar. Biol. Ecol.,* **62,** 211–224.
Gerard, V. A., 1982b. *Mar. Biol.,* **66,** 27–35.
Gerard, V. A., 1982c. *Mar. Biol.,* **69,** 51–54.
Gerard, V. A., 1984. *Mar. Biol.,* **84,** 189–195.
Gerard, V. A. & Kirkman, H., 1984. *Botanica Mar.,* **27,** 105–109.
Gerard, V. A. & Mann, K. H., 1979. *J. Phycol.,* **15,** 33–41.
Ghelardi, R. J., 1971. *Nova Hedwigia,* **32,** 381–420.
Grant, J. J., Wilson, K. C., Grover, A. & Togstad, H., 1982. *Mar. Fish. Rev.,* **44,** 53–60.
Harper, J. L., 1977. *Population Biology of Plants.* Academic Press, New York, 892 pp.
Harris, L. G., Ebeling, A. W., Laur, D. R. & Rowley, R. J., 1984. *Science,* **224,** 1336–1338.
Harrold, C. & Reed, D., 1985. *Ecology,* **66,** 1160–1169.
Himmelman, J. H. & Steele, D. H., 1971. *Mar. Biol.,* **9,** 315–322.
Himmelman, J. H., Lavergne, Y., Axelsen, F., Cardinal, A. & Bourget, E., 1983 *Can. J. Fish. Aquat. Sci.,* **40,** 474–486.
Hines, A. H. & Loughlin, T. R., 1980. *Fish. Bull. NOAA,* **78,** 159–163.
Hixon, M. A., 1980. *Ecology,* **61,** 918–931.
Hubbell, S. P., 1979. *Science,* **203,** 1299–1309.
Hurlbert, S. H., 1984. *Ecol. Monogr.,* **54,** 187–211.
Jackson, G. A., 1977. *Limnol. Oceanogr.,* **22,** 979–995.
Jones, N. S. & Kain, J. M., 1967. *Helgoländer wiss. Meeresunters.,* **15,** 460–466.
Jupp, B. P. & Drew, E. A., 1974. *J. exp. mar. Biol. Ecol.,* **15,** 185–196.
Kain, J. M., 1962. *J. mar. biol. Ass. U.K.,* **42,** 377–385.
Kain, J. M., 1963. *J. mar. biol. Ass. U.K.,* **43,** 129–151.
Kain, J. M., 1966. In, *Light as an Ecological Factor,* edited by R. Bainbridge *et al.* Blackwell, Oxford, pp. 319–334.
Kain, J. M., 1971. *J. mar. biol. Ass. U.K.,* **51,** 387–408.
Kain, J. M., 1975a. *J. mar. biol. Ass. U.K.,* **55,** 567–582.
Kain, J. M., 1975b. *J. Ecol.,* **63,** 739–765.
Kain, J. M., 1976. *J. mar. biol. Ass. U.K.,* **56,** 267–290.
Kain, J. M., 1977. *J. mar. biol. Ass. U.K.,* **57,** 587–607.
Kain, J. M., 1979. *Oceanogr. Mar. Biol. Ann. Rev.,* **17,** 101–161.
Kain, J. M., Drew, E. A. & Jupp, B. P., 1976. In, *Light as an Ecological Factor. II,* edited by G. C. Evans *et al.,* Blackwell, Oxford, pp. 63–92.

Kain, J. M. & Jones, N. S., 1966. *Proc. Int. Seaweed Symp.*, **5**, 139–141.
Kain, J. M. & Svendsen, P., 1969. *Sarsia*, **38**, 25–30.
Kastendiek, J., 1982. *J. exp. mar. Biol. Ecol.*, **62**, 201–210.
Keats, D. W., South, G. R. & Steele, D. H., 1982. *Phycologia*, **21**, 189–191.
Kennelly, S. J., 1983. *J. exp. mar. Biol. Ecol.*, **68**, 257–276.
Kennelly, S. J. & Larkum, A. W. D., 1983. *Aquat. Bot.*, **17**, 275–282.
Kenyon, K., 1969. *N. Amer. Fauna*, **68**, 1–352.
Kirkman, H. 1981. *J. exp. mar. Biol. Ecol.*, **55**, 243–254.
Kitching, J. A., 1941. *Biol. Bull. mar. biol. Lab., Woods Hole*, **80**, 324–337.
Kitching, J. A., Macan, T. T. & Gilson, H. C., 1934. *J. mar. biol. Ass. U.K.*, **19**, 677–705.
Koehl, M. A. R. & Wainwright, S. A., 1977. *Limnol. Oceanogr.*, **22**, 1067–1071.
Lang, C. & Mann, K. H., 1976. *Mar. Biol.*, **36**, 321–327.
Larkum, A. W. D., 1972., 1972. *J. mar. biol. Ass. U.K.*, **52**, 405–418.
Larson, B. R., Vadas, R. L. & Keser, M., 1980. *Mar. Biol.*, **59**, 49–62.
Larson, R. J., 1980. *Ecol. Monogr.*, **50**, 221–239.
Lawrence, J. M., 1975. *Oceanogr. Mar. Biol. Ann. Rev.*, **13**, 213–286.
Lawrence, J. M. & Sammarco, P. W., 1982. In, *Echinoderm Nutrition*, edited by M. Jangoux & J. M. Lawrence, A. A. Balkema, Rotterdam, pp. 409–519.
Lees, D. C. & Carter, G. A., 1972. *Ecology*, **53**, 1127–1133.
Leighton, D. L., 1971. *Nova Hedwigia*, **32**, 421–453.
Leighton, D. L., Jones, L. G. & North, W. J., 1966. *Proc. int. Seaweed Symp.*, **5**, 141–153.
Lindauer, V. W., Chapman, V. J. & Aiken, M., 1961. *The Marine Algae of New Zealand. II. Phaeophyceae.* Nova Hedwigia III, 350 pp.
Lissner, A. L., 1980. *J. exp. mar. Biol. Ecol.*, **48**, 185–193.
Lowry, L. F., McElroy, A. J. & Pearse, J. S., 1974. *Biol. Bull. mar. biol. Lab., Woods Hole*, **147**, 386–396.
Lowry, L. F. & Pearse, J. S., 1973. *Mar. Biol.*, **23**, 213–219.
Lubchenco, J., 1978. *Am. Nat.*, **112**, 23–39.
Lubchenco, J. & Gaines, S. D., 1981. *Ann. Rev. Ecol. Syst.*, **12**, 405–437.
Lüning, K., 1979. *Mar. Ecol. Prog. Ser.*, **1**, 195–207.
Lüning, K., 1980. *J. Phycol.*, **16**, 1–15.
Lüning, K., 1981. *Proc. int. Seaweed Symp.*, **10**, 35–55.
Lüning, K., 1982. In, *Synthetic and Degradative Processes in Marine Macrophytes*, edited by L. M. Srivastava, Walter de Gruyter, Berlin, pp. 47–67.
Lüning, K. & Neushul, M., 1978. *Mar. Biol.*, **45**, 297–309.
Mann, K. H., 1972. *Mar. Biol.*, **12**, 1–10.
Mann, K. H., 1973. *Science*, **182**, 975–981.
Mann, K. H., 1977. *Helgoländer wiss. Meeresunters.*, **30**, 1733–1738.
Mathieson, A. C. & Fralick, R. A., 1973. *Rhodora*, **75**, 52–64.
Mattison, J. E., Trent, J. D., Shanks, A. L., Akin, T. B. & Pearse, J. S., 1977. *Mar. Biol.*, **39**, 25–30.
McLean, J. H., 1962. *Biol. Bull. mar. biol. Lab., Woods Hole*, **122**, 95–114.
Michanek, G., 1979. *Botanica Mar.*, **22**, 375–391.
Miller, R. J., 1985. *Mar. Biol.*, **84**, 275–286.
Miller, R. J. & Colodey, A.G., 1983. *Mar. Biol.*, **73**, 263–267.
Moitoza, D. J. & Phillips, D. W., 1979. *Mar. Biol.*, **53**, 299–304.
Moore, L. B., 1943. *Trans. R. Soc. N.Z.*, **72**, 333–341.
Moreno, C. A. & Sutherland, J. P., 1982. *Oecologia (Berl.)*, **55**, 1–6.
Nelson, B. W. & Vance, R. R., 1979. *Mar. Biol.*, **51**, 251–258.
Neushul, M., 1965. In, *Proceedings of the Fifth Marine Biology Symposium*, Acta Universitatis, Göteberg, pp. 161–176.
Neushul, M., 1967. *Ecology*, **48**, 83–94.
Neushul, M., 1971a. *Nova Hedwigia*, **32**, 211–228.

Neushul, M., 1971b. *Nova Hedwigia,* **32,** 241–254.
Neushul, M., Foster, M. S., Coon, D. A., Woessner, J. W. & Harger, B. W., 1976. *J. Phycol.,* **12,** 397–408.
Neushul, M. & Harger, B. W., in press. *J. Solar Energy Engineer.*
Noro, T., Masaki, T. & Akioka, H., 1983. *Bull. Fac. Fish. Hokkaido Univ.,* **34,** 1–10.
North, W., Gerard, V. & Kuwabara, J., 1982. In, *Synthetic and Degradative Processes in Marine Macrophytes,* edited by L. Srivastava, W. de Gruyter, Berlin, pp. 247–264.
North, W. J., 1971a. (Editor). *The Biology of Giant Kelpbeds (*Macrocystis) *in California. Nova Hedwigia,* **32,** 1–600.
North, W. J., 1971b. *Nova Hedwigia,* **32,** 1–97.
North, W. J., 1971c. *Nova Hedwigia,* **32,** 123–168.
North, W. J., 1972. In, *Contribution to the Systematics of Benthic Marine Algae of the North Pacific,* edited by I. A. Abbot & M. Kuragi, Japanese Society of Phycology, Kobe, pp. 75–92.
North, W. J. & Pearse, J. S., 1970. *Science,* **167,** 209 only.
Norton, T. A., 1978. *J. mar. biol. Ass. U.K.,* **58,** 527–536.
Novaczek, I., 1981. *J. Brit. Phycol. Soc.,* **16,** 363–371.
Ostfield, R. S., 1982. *Oecologia (Berl.).,* **53,** 170–178.
Pace, D., 1981. *Proc. int. Seaweed Symp.,* **8,** 457–463.
Paine, R. T., 1979. *Science,* **205,** 685–687.
Paine, R. T. & Vadas, R. L., 1969a. *Limnol. Oceanogr.,* **14,** 710–719.
Paine, R. T. & Vadas, R. L., 1969b. *Mar. Biol.,* **4,** 79–86.
Parker, B. C. & Bleck, J., 1966. *Ann. Missouri Bot. Gard.,* **53,** 1–16.
Pearse, J. S., Costa, D. P., Yellin, M. B. & Agegian, C. R., 1977. *Fish Bull. NOAA,* **75,** 645–648.
Pearse, J. S. & Hines, A. H., 1979. *Mar. Biol.,* **51,** 83–91.
Prentice, S. A. & Kain, J. M., 1976. *Estuar. cstl mar. Sci.,* **4,** 65–70.
Pringle, J. D., Sharp, G. J. & Caddy, J. F., 1980. *Can. Tech. Rep. Fish. Aquat. Sci.,* **954,** 237–251.
Quast, J. C., 1971. *Nova Hedwigia,* **32,** 229–240.
Reed, D. C., & Foster, M. S., 1984. *Ecology,* **65,** 937–948.
Rosenthal, R. J., Clarke, W. D. & Dayton, P. K., 1974. *Fish. Bull. NOAA,* **72,** 670–684.
Russo, A. R., 1979. *J. Biogeog.,* **6,** 407–414.
Santelices, B. & Ojeda, F. P., 1984a. *Mar. Ecol. Prog. Ser.,* **14,** 165–173.
Santelices, B. & Ojeda, F. P., 1984b. *Mar. Ecol. Prog. Ser.,* **14,** 175–183.
Schiel, D. R., 1982. *Oecologia (Berl.),* **54,** 379–388.
Schiel, D. R., 1985a. *J. Phycol.,* **21,** 99–106.
Schiel, D. R., 1985b. *J. Ecol.,* **73,** 199–217.
Schiel, D. R., & Choat, J. H., 1980. *Nature, Lond.,* **285,** 324–326.
Schmitt, R. J., 1982. *Ecology,* **63,** 1588–1601.
Schonbeck, M. & Norton, T. A. 1978. *J. exp. mar. Biol. Ecol.,* **31,** 303–313.
Schonbeck, M. & Norton, T. A. 1980. *J. exp. mar. Biol. Ecol.,* **43,** 131–150.
Schroeter, S. C., Dixon, J. & Kastendiek, J., 1983. *Oecologia (Berl.),* **56,** 141–147.
Shepherd, S. A. & Womersley, H. B. S., 1970. *Trans. R. Soc. S. Aust.,* **94,** 105–138.
Shepherd, S. A. & Womersley, H. B. S., 1971. *Trans. R. Soc. S. Aust.,* **95,** 155–167.
Shepherd, S. A. & Womersley, H. B. S., 1976. *Trans. R. Soc. S. Aust.,* **100,** 177–191.
Shimek, S. J., 1977. *Calif. Fish & Game Fish. Bull.,* **63,** 120–122.
Simenstad, C.A., Estes, J. A. & Kenyon, K. W., 1978. *Science,* **200,** 403–411.
Sousa, W. P., 1984. *Ecology,* **65,** 1918–1935.
Sousa, W. P., Schroeter, S. C. & Gaines, S. D., 1981. *Oecologia (Berl.),* **48,** 297–307.

Steinberg, P. D., 1984. *Science,* **223,** 405–407.
Stephenson, M. D., 1977. *Calif. Fish & Game Fish. Bull.,* **63,** 117–120.
Stephenson, T. A. & Stephenson, A., 1949. *J. Ecol.,* **36,** 289–305.
Stephenson, T. A. & Stephenson, A., 1972. *Life Between Tidemarks on Rocky Shores,* Freeman, San Francisco, 425 pp.
Strong, D. R., Simberloff, D., Abele, L. G. & Thistle, A. B., 1984. (Editors). *Ecological Communities: Conceptual Issues and the Evidence.* Princeton University Press, Princeton, 613 pp.
Sundene, O., 1964. *Nytt Mag. Bot.,* **11,** 83–107.
Tegner, M. J. & Dayton, P. K., 1977. *Science,* **196,** 324–326.
Tegner, M. J. & Dayton, P. K., 1981. *Mar. Ecol. Prog. Ser.,* **5,** 255–268.
Tegner, M. J. & Levin, L. A., 1982. In, *International Echinoderms Conference, Tampa Bay,* edited by J. M. Lawrence, A. A. Balkema, Rotterdam, pp. 265–271.
Tegner, M. J. & Levin, L. A., 1983. *J. exp. mar. Biol. Ecol.,* **73,** 125–150.
Thompson, W. C., 1959. In, *Proceedings of the International Oceanographic Congress,* Am. Assoc. Adv. Sci., Washington, p. 568 only.
Underwood, A. J., 1979. *Adv. Mar. Biol.,* **16,** 111–210.
Underwood, A. J., 1981. *Oceanogr. Mar. Biol. Ann. Rev.,* **19,** 513–605.
Underwood, A. J., in press. In, *Community Ecology—Pattern and Process,* edited by D. J. Anderson & J. Kikkawa, Blackwell Scientific, Oxford.
Underwood, A. J. & Denley, E. J., 1984. In, *Ecological Communities: Conceptual Issues and the Evidence,* edited by D. R. Strong *et al.,* Princeton University Press, Princeton, pp. 151–180.
Underwood, A. J., Denley, E. J. & Moran, M. J., 1983. *Oecologia (Berl.),* **56,** 202–219.
Underwood, A. J. & Jernakoff, P., 1981. *Oecologia (Berl.),* **48,** 221–233.
Underwood, A. J. & Jernakoff, P., 1984. *J. exp. mar. Biol. Ecol.,* **75,** 71–96.
Vadas, R. L., 1977. *Ecol. Monogr.,* **47,** 337–371.
Vadas, R. L., Elner, R. W., Garwood, P. E. & Babb, I. G., 1986. *Mar. Biol.,* **90,** 433–448.
Van Blaricom, G. R. & Estes, J. A., in press. Editors. *The Community Ecology of Sea Otters.* Springer-Verlag, Berlin.
Vance, R. R., 1979. *Ecology,* **60,** 537–546.
Vance, R. R. & Schmitt, R. J., 1979. *Oecologia (Berl.).,* **44,** 21–25.
Van den Hoek, C., 1982. *Biol. J. Linn. Soc.,* **18,** 81–144.
Velimirov, B., Field, F. G., Griffiths, C. L. & Zoutendyk, P., 1977. *Helgoländer wiss. Meeresunters.,* **30,** 495–518.
Velimirov, B. & Griffiths, C. L., 1979. *Botanica Mar.,* **22,** 169–172.
Watanabe, J. M., 1983. *J. exp. mar. Biol. Ecol.,* **71,** 257–270.
Watanabe, J. M., 1984a. *Ecology,* **65,** 920–936.
Watanabe, J. M., 1984b. *Oecologia (Berl.),* **62,** 47–52.
Wharton, W. G. & Mann, K. H., 1981. *Can. J. Fish. Aquat. Sci.,* **38,** 1339–1349.
Wheeler, W. N., 1982. In, *Synthetic and Degradative Processes in Marine Macrophytes,* edited by L. M. Srivastava, W. de Gruyter, New Jersey, pp. 121–137.
Wilson, K. C. & North, W. J., 1983. *J. World Maricul. Soc.,* **14,** 347–359.
Winer, B. J., 1971. *Statistical Principles in Experimental Design.* McGraw-Hill Kogakusha, Tokyo, 907 pp.
Womersley, H. B. S., 1954. *Univ. Calif. Publs Bot.,* **27,** 109–132.
Zimmerman, R. C. & Kremer, J. N., 1984. *J. mar. Res.,* **42,** 591–604.
ZoBell, C. E., 1971. *Nova Hedwigia,* **32,** *269*–314.

Oceanogr. Mar. Biol. Ann. Rev., 1986, **24**, 309–377
Margaret Barnes, Ed.
Aberdeen University Press

VARIATION AND NATURAL SELECTION IN MARINE MACROALGAE

G. RUSSELL
Department of Botany, The University, Liverpool L69 3BX, England

INTRODUCTION

The Darwinian theory of evolution is based upon two propositions: first, that there is random variation within species and, secondly, that the forces of natural selection have the power to order the randomness by actively favouring the most advantageous variants (*i.e.*, the most fit). As a consequence of ecological, geographical and/or reproductive isolation, this process may lead to speciation and, ultimately, to the evolution of higher taxa. Evolution is, therefore, initiated at and below the rank of species, higher taxa being the products of more ancient evolutionary events.

Early advocates of evolution were taxonomists. Haeckel (1866) was perhaps first to publish a phylogenetic tree incorporating algae, depicting these as a monophyletic group comprising four major taxa. W. H. Harvey, a contemporary of Darwin, was the first phycologist to recognize the crucial importance of chloroplast pigmentation in the circumscription of higher algal taxa (Boney, 1978). In *Phycologia Britannica* (Harvey, 1846–51), and in other floristic works, he also revealed a profound and extensive knowledge of seaweed species in the field. Harvey was acutely aware of infraspecific variation and of the fact that certain algal forms could be associated with particular kinds of habitat. He was also familiar with the ideas of temporal change in organisms which had been advanced from geological evidence. Much of his major work preceded the publication of *The Origin of Species* (Darwin, 1859) but, in a late popular book (Harvey, 1857) he expressed the opinion that species have arisen progressively through time as a result of special creation. Later phycologists seem to have accepted, either explicitly or implicitly, the truth of Darwinian or neo-Darwinian theory, although they have chosen to interpret it in a variety of ways.

The terms phylogeny and evolution have sometimes been treated as synonyms, and ideally perhaps they ought to be synonymous. Indeed it might be considered odd to suggest that they describe different things, but phycological literature gives much support for making such a distinction. Phylogeny may be regarded as a historical reconstruction of taxonomic lineages. Although covering periods of time which are, in all probability, very long, algal phylogeny owes little to fossils. Fossil records are mostly restricted to those forms with mineralized cell walls and to those that cause mineral precipitations to accumulate around their cells. The former group includes siliceous algae (silicoflagellates and diatoms) and calcified algae (certain Codiaceae, Dasycladaceae, Charophyceae, Rhodophyceae, and the coccoliths of Prymnesiophyceae). The latter group includes certain Cyanophyceae. A good summary of fossil algae has been given by Round (1973) and a more detailed account of the calcified taxa of limestone formations by Johnson (1961). The lack of fossil evidence on algae with unmineralized thalli, and the difficulty of assigning known examples to even major taxa, has been commented upon by Dixon (1973) and Clayton (1984). Algal phylogeny is, therefore, the reconstruction of lineages from phenetic relationships of existing taxa, and its practitioners are almost always taxonomists. Phenetic relationships have been construed chiefly from morphological and anatomical attributes, although biochemical evidence has also been used (Klein & Cronquist, 1967).

Evolution, in a Darwinian sense, begins as a process of change which organisms undergo in response to changes in their external environment. If Darwinian theory is correct, it is a consequence of variation and selection, and it is concerned with the natural habitat and with fitness. Clayton (1984) has contended that " . . . to attempt to explain the adaptive value of the features characteristic of brown algal taxa requires leaps of the imagination and a high proportion of guesswork". This analysis is very probably correct for the taxonomic rank with which she was chiefly concerned (the order Fucales): taxa at this level of the hierarchy are likely to be ancient and the environmental background to their inception is, at best, known sketchily. To measure adaptive qualities of algae, however, need involve neither leaps nor luck if applied to living plants studied in their local populations or in culture. Attributes of these plants may be measured, hypotheses on their adaptive status made, and these hypotheses tested by appropriate experimentation.

The differences in emphasis between phylogenetic and evolutionary phycology are not, however, simply methodological, for they extend also to the characteristics of the plants being studied. Thus, evolutionary importance may be ascribed to traits such as thallus size, longevity, mechanical strength, and physiological tolerances; it may also be attributable to the frequency, abundance and age of onset of reproduction. Phylogeny, in contrast, appears to attach greater importance to attributes such as planes of cell division, patterns of flagellar organization and sporangial shapes and types.

These differences in emphasis have emerged against a background of changing concepts of taxa and their circumscription. The crudest definition of the species seems to be that of the practical species concept of Mayr (1942): " . . . A species is a systematic unit which is considered a species by a competent systematist." The morphological species concept (Mayr, 1942) and its synonym the nomenspecies of Ravin (1963) is based purely on the degree of morphological distinction shown by its members in certain key characteristics. The taxonomic type specimen serves as the morphological yard-stick against which individuals are judged. Most algal species have been established in this way. Another concept, that of the taxospecies, was advanced by Ravin (1963), apparently in response to the growing influence of numerical taxonomy in microbiology at that time, species being defined according to calculations of overall similarity or affinity. Ravin at the same time introduce the concept of the genospecies based upon experimental studies of genetic transfer and recombination. His proposals were warmly recommended by Forest (1968) as suitable for algae but they have yet to receive wider support. The biological species concept of Mayr (1942, 1970) although intended for zoological use is, nevertheless, applicable to plants. With its emphasis on the importance of the population as the primary unit and its recognition of variation as the key to ecological and evolutionary success, it has particular appeal to experimental botanists. For a professional plant taxonomist, however, it is a concept that raises too many practical difficulties. The explicit rejection of the type as the paradigm of the species is a source of particular taxonomic uncertainty, and the importance attached to sexual compatibility is irrelevant to asexual species. In practice, there is no generally accepted species concept for algae and it is evident that certain difficult groups may even need to be treated according to special criteria (Woelkerling, 1971).

Changing attitudes in algal taxonomy may be detected in Figure 1 which gives the numbers of species, genera and infraspecific taxa of brown algae reported for the British Isles over the past 135 years by Harvey (1849), Holmes & Batters (1890), Batters (1902), Newton (1931), Parke (1953), and Parke & Dixon (1976). The line of species numbers follows a trajectory that might be expected for a smallish territory subject to continuous botanical study, there being an early period of rapid increase followed by one of relative stability. During the latter phase new species have been added but others, for a variety or reasons, have been removed; the stability in numbers is, therefore, more apparent than real. Calculations of numbers of infraspecific taxa is rendered difficult by uncertainty whether authors had in mind only the described variant, or both the variant and the typical forms. The numbers of infraspecific taxa shown in Figure 1 are based on a consistent assumption that the latter alternative is true, and they may, therefore, be somewhat inflated. The marked rise and fall in numbers

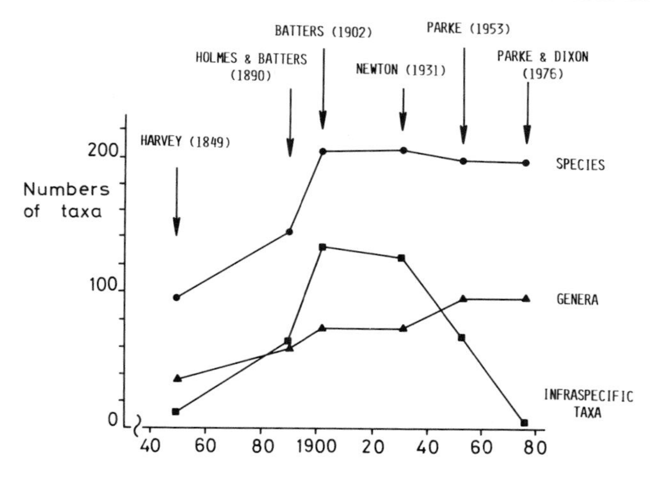

Fig. 1.—Changes in the numbers of species, genera and infraspecific taxa of brown algae recorded for the British Isles since the mid-nineteenth century.

TABLE I

Classificatory units of biosystematics and how they are defined: columns 1, 2, and 3 denote the respective terminologies of Turesson (1922), Danser (1929), and Gilmour & Heslop-Harrison (1954); reproduced from Stace (1980) with permission

<table>
<tr><th>Behaviour of individuals within the group</th><th>1</th><th>2</th><th>3</th><th>Behaviour of group towards other such groups</th></tr>
<tr><td>Individuals capable of hybriizing with one another</td><td rowspan="2">Coenospecies</td><td>Comparium</td><td>Syngamodeme</td><td>Group incapable of hybridizing with any other groups</td></tr>
<tr><td>Individuals capable of hybridizing with one another to give hybrids showing some fertility</td><td>Commiscuum</td><td>Coenogamodeme</td><td>Group capable of hybridizing with other groups but such hybrids are sterile</td></tr>
<tr><td>Individuals capable of hybridizing with one another to give fully fertile hybrids</td><td>Ecospecies</td><td rowspan="3">Convivium</td><td>Hologamodeme</td><td>Group capable of hybridizing with other groups to give hybrids showing some fertility</td></tr>
<tr><td>Individuals occupying a particular habitat and forming an interbreeding population which differs genotypically from other such populations</td><td>Ecotype</td><td>Genoecodeme</td><td>Group capable of hybridizing with other groups to give hybrids showing complete fertility</td></tr>
<tr><td>Individuals occupying a particular habitat and adapted to it phenotypically but not genotypically</td><td>Ecophene</td><td>Plastoecodeme</td><td>As previous category</td></tr>
</table>

would be evident, however, whatever method of calculation had been adopted. The third component of this taxochronology, the numbers of genera, has shown a sustained if unspectacular increase.

Many of these changes since 1931 have been brought about by the advent of numerical or experimental studies and, in particular, by widespread use of improved algal culture methods. The unstable nature of algal taxonomy has been discussed by Mathieson, Norton & Neushul (1981), but as experimental phycology is a rapidly expanding discipline, further changes must be regarded as likely.

Experimental innovations have been associated with increasing use by phycologists of the classificatory units of biosystematists. Biosystematics (Stace, 1980), or genecology (Langlet, 1971), or experimental taxonomy (Heslop-Harrison, 1953), is essentially a 20th century phenomenon although its origins have been traced by Langlet (1971) to the mid-18th century. Its scope and its relationships with orthodox taxonomy have been detailed by Davis & Heywood (1963) and, more recently, by Stace (1980) from whose work Table I has been reproduced. Table I gives the experimental categories proposed by Turesson (1922), Danser (1929), and Gilmour & Heslop-Harrison (1954) together with the criteria by which they are defined. The Turesson classification, particularly the terms ecotype and ecophene (or ecad), has been most commonly adopted by phycologists, although the -deme terminology of Gilmour & Heslop-Harrison has been recommended by Russell (1978).

None of the classifications in Table I corresponds very exactly with that of orthodox taxonomy, and it is usually held that the two should be retained as strictly separate systems (Gregor, 1948; Gilmour, 1961; Stace, 1980). It is also important that biosystematics terminologies are applied solely on the evidence of experimental investigations. The remark by Turrill (1952) that ". . . careful field observations often enable ecads to be distinguished from ecotypes with a high degree of probability . . ." would be regarded by most seaweed botanists as sanguine.

Turesson (1922) initially held to the view that his was essentially an ecological classification, but later (Turesson, 1926), in response to an allegation that his system was based upon Lamarckian ideas, he published a reply which was firmly expressed in Darwinian, or rather neo-Darwinian, terms.

The aim of this essay is to explore briefly some of the ways in which evolutionary theory may apply to seaweeds.

VARIATION WITHIN SPECIES

In the opening chapter of *The Origin of Species*, Darwin made the point that variation within certain species has been harnessed by mankind to produce, by means of artificial selection, numerous domesticated and cultivated varieties. Mariculture is, in technical terms, a more primitive activity than agriculture but there is now good evidence to show that the principle established by Darwin applies equally to algae.

Chapman (1977) has made a detailed study of morphological and anatomical variation within and between populations of the commercially important red seaweed *Gracilaria* on the northeastern coast of North America from New England to the Gulf of St Lawrence. At the same time, a *Gracilaria* species from these shores has been the subject of cultivation experiments by Edelstein, Bird & McLachlan (1976) and Edelstein (1977) which demonstrated the existence of interpopulation differences in productivity, and went on to show which culture conditions may promote greatest crop yield. Patwary & Meer (1983a,b) have greatly increased the variation available for selection by culturing natural and induced mutants of *G. tikvahiae*. These mutants could be distinguished morphologically as well as physiologically, and seven were chosen for comparison with four wild populations. Two of the mutants, coded as MP-40 and MP-44, proved particularly superior to the wild types in agar production and one of them (MP-40) produced agar of greatly superior gel strength to any of the wild types. This mutant was also characterized by excellent resistance to epiphytes. Large-scale cultivation of this strain, or its successors, may therefore lead to greatly increased commercial exploitation of *Gracilaria*.

Chondrus crispus is another red seaweed which has been harvested for its gel-forming compounds. Taylor & Chen (1973) and Chen & Taylor (1980) have made detailed observations on natural populations in Nova Scotia and have described the very considerable amount of morphological variation present. Chen (1980) has also shown that form variations may be stable, their characteristic features being retained throughout prolonged laboratory culture. The existence of a lot of natural variation within *C. crispus* has been confirmed by Cheney & Mathieson (1979) using isozyme electrophoresis. These authors demonstrated the existence of genetic variation on a local ecological scale which was as great as that between geographically separated populations.

Taylor (1971) has made field observations on *Chondrus* productivity at various sites and over extended periods of time. Such observations provide a good basis for the selection of potentially high-yielding cultivars. Neish & Fox (1971) began an important series of trials on the green-house cultivation of several *Chondrus* strains and they identified one, coded as T4 with particularly good growth characteristics. Later, Shacklock, Robson & Simpson (1975) described the culture conditions which gave the best yields from T4. The continuing success of T4 as a cultivar may be seen in the numerous papers which describe improvements and refinements in the methods for its cultivation: Neish, Shacklock, Fox & Simpson (1977), Simpson, Neish, Shacklock & Robson (1978), Simpson & Shacklock (1979), Bidwell, Lloyd & McLachlan (1984), Bidwell, McLachlan & Lloyd (1985).

T4 is a gametophyte and therefore produces only K-carrageenan. Cultivation of *Chondrus* tetrasphorophytes which produce λ-carrageenan were initially less successful because these plants tend to disintegrate when they reach reproductive maturity. There was, therefore, a need to find an "aposporogenous" strain as a potential source of λ-carrageenan. Chen, Morgan & Simpson (1982) collected 689 infertile tetrasporophytes from the field, their carrageenan type being the basis of their determination as tetrasporophytes. These were then isolated in culture and, as they attained reproductive maturity, fertile plants were discarded. Eventually 100 infertile thalli remained and further experiments on growth rates led to the selection of the best (coded ARL16) for investigation as a possible future cultivar.

Hansen, Packard & Doyle (1981), in their brief account of the mariculture of commercially important red algae, point to the opportunities for selective breeding programmes to produce the best strains. There is no reason why this process should not also apply to commercially important algae belonging to the other major groups. Meer (1981a) in his brief account of the domestication of seaweeds has stressed the importance of genetics as the key to these developments.

The pressures that operate upon a natural community are necessarily more complex than those acting on a clonal culture. It has been suggested by Mather (1961) and Berry (1977) that natural selection has three distinct rôles: these are stabilizing selection, disruptive selection, and directional selection. It seems probable that all three can influence seaweed populations in the field. Stabilizing selection favours phenotypes which are at or close to the mean value for the population. For example, maximum plant size in a dense population of macrophytes may be subject to stabilizing selection in that smaller-than-average individuals are liable to shading and taller-than-average individuals more susceptible to wave action. Disruptive selection, in contrast, operates against the average and favours two extreme phenotypes. Examples of this kind of selection seem to exist in those species with heteromorphic life-histories (see p. 348 *et seq.*). Directional selection favours one extreme phenotype, and its influence is most evident in natural environments that are extreme in one particular and readily identifiable factor. It has been in environments such as these that algal ecotypes in the sense of Turesson (see Table I) have most frequently been located and described. Patterns of genetic divergence in populations of marine algae have been discussed by Innes (1984).

A good example of directional selection may be found in the evolution of heavy-metal tolerance, first reported in algae by Russell & Morris (1970). These authors compared the growth characteristics in culture of the brown seaweed *Ectocarpus siliculosus* from a natural rocky shore and from the hulls of two ships which had been treated with copper-based antifouling paints. The ship-fouling isolates proved to be much more tolerant to copper-enriched culture media than the wild type. Later, Russell & Morris (1973) extended their study to a larger number of natural and artificial populations and identified a copper-tolerant strain growing on rock containing high concentrations of copper ore. At about the same time Stokes, Hutchinson & Krauter (1973) reported the existence of heavy-metal tolerance in the freshwater algae of metal-contaminated lakes near Sudbury, Ontario, Canada; the freshwater blue-green alga *Anacystis nidulans* has since been shown capable of evolving zinc tolerance by Shehata & Whitton (1982). Copper-zinc co-tolerance has recently been identified in a copper-tolerant population of *Ectocarpus siliculosus* by Hall (1980), and copper-tolerance has been reported in the green seaweed *Enteromorpha compressa* by Reed & Moffatt (1983). The tolerance mechanism in *Enteromorpha* is evidently one of internal detoxification as copper uptake and concentration by tolerant and non-tolerant strains seem to be similar. In *Ectocarpus*, on the other hand, a mechanism for copper-exclusion by the tolerant strains has been postulated by Hall, Fielding & Butler (1979) and Hall (1981). Copper exclusion has also been reported in copper-tolerant *Chlorella vulgaris* by Foster (1977).

The damage wrought by heavy metals to cell membranes may be equalled by that arising from violent changes in external salinity. The ways in which such changes affect cells of marine algae and the types of response they evoke have recently been discussed by Russell (in press). Salinity has also operated as an agent of directional selection leading to the evolution of ecotypes within species belonging to all the major algal taxa. The blue-green alga *Phormidium* is widespread in marine and fresh waters, the marine strains being inherently more halotolerant than the latter (Stam & Holleman, 1975). In the brown algae, estuarine populations of *Ectocarpus siliculosus* have been shown to possess greater tolerance to low salinities than plants from a strictly marine environment (Russell & Bolton, 1975). In a more detailed study of a particular estuary, and one of its plants, *Pilayella littoralis*, Bolton (1979) has demonstrated the existence of greater tolerance to low salinities in the estuarine samples than in those from the sea coast away from the estuary. Bolton's observations have been confirmed recently by Reed & Barron (1983) who worked on *Pilayella* from a different estuary. They reported also some morphological differences (cell size) between estuarine and marine plants, and have demonstrated differences in the mechanism of turgor regulation in these ecotypes.

Three different salinity ecotypes of the green alga *Enteromorpha intestinalis* have been described by Reed & Russell (1979) from four contrasting natural environments. These populations were located in the low eulittoral zone and in the high littoral fringe of a rocky shore, in freshwater ponds in the maritime zone close to the littoral fringe populations, and in an estuary. The eulittoral-zone, littoral-fringe and estuarine ecotypes each had its own characteristic salt-tolerance profile. The maritime zone population was identical in salt-tolerance with that from the littoral fringe and Reed & Russell concluded that it had been recruited from the latter. Gene flow, which is an important element in the evolutionary process, is also evident in the results of Bolton's (1979) work on *Pilayella* (see Russell, in press). The salt-tolerance spectra of the *Enteromorpha* population studied by Reed & Russell (1979) were possessed also by their progeny grown in culture indicating that salt-tolerance is a stable, heritable attribute. In the case of *Pilayella*, the stability of the observed salt tolerances was inferred from their retention by strains that had been grown in standard sea-water culture for an extended period prior to experimentation (Bolton, 1979).

In the red algae, the existence of populations that are tolerant to reduced salinities in habitats subject to naturally low salt concentrations has been reported for a number of taxa. Such ecotypes have been found in species of *Audouinella (Rhodochorton)* by West (1972), of *Caloglossa* and *Bostrychia* by Yarish, Edwards & Casey (1979) and Yarish & Edwards (1982), and of *Polysiphonia* by Reed (1983). Reed also described differences in the cell sizes of the marine and estuarine ecotypes of *P. lanosa* and differences in osmoticum content.

Differences both in salt tolerance and in morphology are present between many of the marine algae of the Baltic Sea and their North Atlantic counterparts. The reduced, depauperate thalli characteristic of many Baltic marine algae is evidently associated with tolerance to greatly reduced salinity. This combination of morphological and physiological traits has been reported in *Delesseria sanguinea* and *Fucus serratus* by Nellen (1966) and in *Chorda filum* by Russell (1985a). It has been postulated by Russell (1985b) that the progressive desaliniation of the Baltic Sea over the past 3000 years has been accompanied by extensive ecotype evolution among its marine algal species.

The process of ecotype evolution in response to saline selective pressure has not, however, been confined to marine algae. The discovery of salt-tolerant ecotypes of the freshwater green alga *Stigeoclonium* in brackish waters has been reported by Francke & Rhebergen (1982). Also of considerable interest is the observation by Wilkinson (1974) that estuarine plants of the marine green alga *Eugomontia sacculata* are not only more tolerant to reduced salinity than those from a sublittoral marine population but also more tolerant to high light intensity.

Ecotypes are known to have evolved in response to a number of other environmental factors. Nutrient ecotypes have been reported in *Stigeoclonium* (freshwater) by Francke & Cate (1980) and in *Laminaria longicruris* by Espinoza & Chapman (1983). Geographically distinct populations of the brown alga *Scytosiphon lomentaria* have been found to differ in the way daylength induces erect thallus formation. This discovery by Lüning (1980a) has been discussed together with other photoperiodic effects on marine algae by Dring (1984). Experiments by Meer & Simpson (1984) on the cryopreservation of their very considerable bank of culture strains of *Gracilaria tikvahiae* revealed these to differ also in their viability in response to treatment. Wave action ecotypes have been identified in *Sargassum cymosum* by Paula & Oliveira (1982) on the basis of field transplant experiments. These experiments elicited no changes in the morphological characteristics of the respective populations, which the authors therefore concluded to be inherent and stable. Stable morphological differences between populations of two Californian species of the red alga *Sorella* have been described by Stewart (1977), but in this case the stability was observed in plants maintained in controlled laboratory culture. Similar results have been obtained by McLachlan, Chen & Edelstein (1971) who cultured isolates belonging to four species of *Fucus* and found each to retain the morphological features of its parent population throughout the period in culture.

Field observations generally provide rather insubtantial indications of genetic variation although a few species generate quite good evidence. The morphological variation which exists between individual plants of the long-lived fucoid *Ascophyllum nodosum*, and the lack of variation between axes of the same plant, do point to a genetic basis for thallus form, as Baardseth (1970) contends. Also the marked differences in growth rate observed by Burrows (1955) in exposed- and sheltered-shore populations of *Fucus vesiculosus*, in the same geographical area and over the same time span, would appear to denote innate differences. However, in *Ascophyllum*, at least, a great deal by phenotypic plasticity is also involved (Cousens, 1985, 1986).

Morphological variability of the blue-green algae, seems certain to have at least in part a genetic base. Forest (1968) has concluded that almost every collection and culture of these organisms differs a little from the others and, in the particular case of *Anabaena*, Stulp & Stam (1984) have shown by means of enzyme electrophoresis and a variety of DNA techniques that morphological differences are associated with genetic differences.

One possible consequence of prolonged culture is that the procedure itself may generate a kind of selection pressure which gradually reduces and perhaps eventually eliminates genetic variation. Thus the repeated subculturing of a myrionemoid variant of the brown alga *Ectocarpus fasciculatus* by Baker & Evans (1971) gradually led to the loss of the macrothallus which was present in their original culture.

One of the contentious issues surrounding the use of the ecotype concept by flowering plant biosystematists has been concerned with environmental gradients. Turesson seems to have concluded that ecotypes are sharply distinguishable entities and that any transitional habitats will, therefore, be narrow zones containing mixtures of ecotypes from the two adjoining populations. Faegri (1937), on the other hand, took the view that ecotypes may have been identified from too few samples taken from too sharply contrasting environments, and concluded that transitional environments could be populated by transitional forms. Huxley (1938) considered the existence of genetic gradients, which he termed clines, an important fact with which taxonomists have to take serious account.

Marine algae provide some evidence to support the positions both of Turesson and Faegri. For example, a transition area between exposed-shore and sheltered-shore populations of *Fucus vesiculosus* on the Isle of Man was found by Russell (1978) to be populated by morphologically heterogeneous plants rather than those of uniformly intermediate character. Similarly the local form variation in *Agarum cribrosum* described by Yamada (1974) on the coasts of Japan seems to be discontinuous in character.

Continuous, clinal variation has also been reported, however. The geographical differences in the photoperiodic responses of *Scytosiphon* populations described by Lüning (1980a) are plainly clinal in character. Bolton (1983) has demonstrated the existence of a geographical cline in the temperature growth optima and temperature survival limits of *Ectocarpus siliculosus*. A geographical cline in the morphology of the red alga *Bangia* has been reported by Sheath & Cole (1984) while a combined salt-tolerance and morphological cline has been postulated by Russell (1985b) for several marine algae along the salinity gradient between the Baltic Sea and the North Atlantic ocean. Not all clinal variation is, however, on such vast scales. The variation in salt tolerance of estuarine and open-shore *Pilayella* described by Bolton (1979) is also clinal although much more locally distributed.

The lowest biosystematic category defined by Turesson is the ecophene or ecad (Table I), and its members are identified by phenotypic characters which are caused directly by environmental factors, that is they are plastic. Phenotypic plasticity has been demonstrated in marine algae from so many diverse taxonomic groups that it may safely be regarded as omnipresent, and many examples have been described by Norton, Mathieson & Neushul (1981). It is very probable that most algal taxonomists share the opinion of Dixon (1973) that phenotypic plasticity is a major problem, perhaps the major problem, in making effective circumscription of taxa. Huxley (1938) has stated that non-genetic characters, are valueless for taxonomic purposes although Turrill (1952) has argued an opposing viewpoint.

As with ecotypes, the occurrence of ecophenes of marine algae has been demonstrated (or inferred) from field observations or from experiments, which are either field or laboratory based. Thus, Kristiansen (1984) has shown that Danish populations of *Scytosiphon lomentaria* are capable of producing erect thalli under a variety of day-length conditions (*cf.* Lüning, 1980a). The initiation of the erect thallus in this locality is, therefore, more a general expression of plant growth than of physiologically distinct strains. Field observations made by Cousens (1982) on *Ascophyllum nodosum* in southeastern Canada led to the discovery of continuous variation in internode length, pigment concentration, and receptacle size and weight in relation to wave action. Continuous morphological variation in *Colpomenia* species (Clayton, 1975), in *Chordaria flagellitormis* (Munda, 1979), and in *Pterocladia pyramidale* (Stewart, 1968) has been associated with habitat factors and concluded to be environmentally determined. Differences in the reproductive phenology of Japanese and European population of *Bonnemaisonia hamifera* have also been associated with geographical differences in day-length and temperature (Lüning, 1981).

Laboratory culture experiments have greatly extended our knowledge of phenotypic plasticity in marine algae. Blue-green algae have proved extremely responsive to changed culture conditions, Prud'homme van Reine & Hoek (1966). Murray & Dixon (1973, 1975) have shown that by varying light intensity and day-length, it is possible to alter the rates of axial cell division and cell expansion in the red alga *Pleonosporium squarrulosum*. Dring (1984), however, has emphasized the importance of introducing a light-break into experimental design before reaching any firm conclusion on the existence of a photoperiodic response. Garbary (1979a) has by this means demonstrated the absence of a truly photoperiodic influence on the morphogenesis of four species of Ceramiaceae (red algae), although he has confirmed the Murray & Dixon observations that light does indeed affect morphology through induced variation in rates of cell division and expansion.

Reed & Russell (1978) have shown that the formation of branches in the green alga *Enteromorpha intestinalis*, leading to a characteristic "bottle brush" form is a plastic response effected by treatment in either very high or very low salinity. The same "bottle brush" form has, however, been obtained in culture by temperature manipulations by Moss & Marsland (1976).

The simpler brown algae have proved to be equally plastic in response to changes in culture conditions. Thallus morphology in *Feldmannia* has been found to be readily alterable by changing its culture condtions (Knoepffler-Peguy, 1977), and the introduction of blue light has been shown to bring about hair formation and the induction of two-dimensional growth in the brown alga *Scytosiphon lomentaria* (Dring & Lüning, 1975). Brown algal plasticity has also been demonstrated in response to the nutrient status of the culture media. In the case of *Cladosiphon zosterae*, nitrogen (as ammonium) seems to be an important morphogenetic factor (Lockhart, 1979) whereas in *Scytosiphon lomentaria* a plastic response to phosphate has been reported by Roberts & Ring (1972). Lockhart (1979) has also deduced that the presence of certain bacteria in her *Cladosiphon* cultures was important in determining algal morphology. The morphogenetic impact of bacteria on the morphology of *Ulva lactuca* has since been very elegantly demonstrated by Provasoli & Pintner (1980). Axenic cultures of the green alga *Monostroma oxyspermum* are unable to develop a normal thallus form, but will do so in the presence of certain strains of bacteria and other species of red and brown (but not green) algae (Tatewaki, Provasoli & Pintner, 1983).

Field experiments to demonstrate phenotypic plasticity have also been numerous. Russell (1967) subjected members of three species of ectocarpoid brown algae to life in an unattached state by maintaining them in a semi-enclosed apparatus located in the sea at the Isle of Man. On termination of the experiment it was found that the branching patterns had altered radically in all three species. Transplant experiments involving salt-marsh forms of the fucoid algae *Ascophyllum nodosum* and *Pelvetia canaliculata* by Brinkhuis & Jones (1976) and by Oliveira & Fletcher (1980), respectively, have demonstrated the readily alterable nature of these forms.

Transplant experiments with species of kelp have revealed the presence of phenotypic plasticity, especially with regard to the various dimensions of blade morphology and anatomy. The development of a broad *cucullata* blade in plants transferred to sheltered conditions has been reported in *Laminaria digitata* (Sundene, 1964), in *L. hyperborea* (Svendsen & Kain, 1971), and in *Saccorhiza polyschides* (Norton, 1969). By inference, the deeply and narrowly divided *stenophylla* forms are likely to be plastic responses to strong wave action or water currents (Sundene, 1964; Norton, 1969).

Of special interest is a recent report of plasticity in the kelp *Macrocystis integrifolia* on the Pacific coast of Canada by Druehl & Kemp (1982). Samples from morphologically and ecologically different populations were transferred to a common environment and observed over an extended period. The authors found that initial differences in blade morphology and in biomass growth rate gradually decreased as the morphologies of the plants from the various populations converged. The importance of this observation is that plasticity may not be immediately evident in an experimental study; consequently, some of the evidence of the existence of ecotypes may need to be re-examined in light of more critical experimental conditions. Another problem, which may lie concealed in the body of experimental evidence so far discussed, is related to the differences in susceptibility to environmental change shown by juvenile and adult developmental states. Mathieson (1982) has demonstrated clearly that juvenile and adult thalli of the brown alga *Phaeostrophion irregulare* differ considerably in their tolerances to a wide range of temperature, salinity, and desiccation conditions. The possibility that such differences may be responsible for conflicting accounts of salt tolerance in the brown alga *Chorda filum* has been discussed by Russell (1985a).

A further complication exists in that plastic and genetic attributes are not always readily distinguishable. For example, Geesink (1973) has described a series of experiments in which he gradually acclimated a freshwater population of the red alga *Bangia atropurpurea* to sea water. Likewise, he acclimated a marine strain to very low salinities. It, therefore, seemed very likely that the two populations were interchangeable as Geesink contended, in spite of the fact that some persistent differences were evident in his results. The very considerable saline plasticity of this species has been confirmed by Sheath & Cole (1980). More recently, however, experiments by Reed (1980) and Sheath & Cole (1984) have disclosed the presence of inherent differences in salt tolerance in this species. That phenotypic plasticity and genetically transmitted traits can be difficult to distinguish, seems also to be true of the *simplices* complex of *Laminaria* in eastern Canada. Chapman (1974, 1975) has made a study of these algae by means of reciprocal transplant experiments and by quantitative analyses of the morphology of sporophytes resulting from the crossing of gametophytes from different populations. Characters such as stipe morphology proved to be fairly stable indicating a relatively high genetic component of interpopulation differences. Mucilage canal production, on the other hand, seemed rather plastic having a relatively large environmental component of the population differences. One population, that from Nova Scotia, however, bred true for mucilage canal production suggesting a genetic contribution to the expression of this character. Another example of the interplay of plasticity and genetics may be seen in the complanate form of the brown alga *Scytosiphon lomentaria* found in southern Australia. Clayton (1976) has shown that the ralfsioid microthallus may exist in day-length sensitive and insensitive forms, a fact which may also have a bearing on the reported differences in responses to day-length shown by populations of this species from the northern hemisphere. *Scytosiphon* illustrates the real difficulty that may be experienced in distinguishing plastic and genetic chacteristics in marine algae. Those isolates which are sensitive to day-length are plastic in the sense that it is day-length which brings the observed morphological change into being, but the ability to respond to day-length is an expression of the genetic apparatus of the plant.

It is possible, as Bennett (1970) and Bradshaw (1984) have argued, that phenotypic plasticity is a mechanism by means of which a taxon may retain genetic variablity in the presence of powerful natural selection. Strongly directional selection might eventually lead to a loss of variability, but plasticity could provide a means of avoiding genetic uniformity.

Whatever evolutionary importance phenotypic plasticity may prove to have for marine algae, the evident difficulty of distinguishing plastic and stable characters does raise a more immediate question; are the ecotype and the ecophene as defined by Turesson units of any practical value? Some experimental observations would suggest that their value is small. The transplantation of *Macrocystis integrifolia* by Druehl & Kemp (1982) showed blade morphology to be plastic but also that rates of frond initiation were stable. Branching pattern in *Ectocarpus siliculosus* is plastic (Russell, 1967) but copper tolerance is not (Russell & Morris, 1970, 1973). Branch production in *Enteromorpha intestinalis* is plastic (Reed & Russell, 1978) but salt tolerance is not (Reed & Russell, 1979). Thus, the same populations of all three species could legitimately be regarded as ecotypes or ecophenes depending upon the attributes under consideration. It follows, therefore, that if these terms are to be employed in any meaningful way, it can only be with reference to the particular factor or factors involved. Perhaps the real importance of Turesson to the study of plant populations rests less with his system of units than with his approach to experimenting with plants.

It may be concluded then that there is an enormous amount of variation within species of marine algae. It is evident that some of this variation has been subject to natural selection and is adaptive in character, although Lewontin (1978) should be consulted for a discussion of the pitfalls of arguments about adaptation. Natural selection is, therefore, likely to play a major rôle in the evolution of infraspecific differences, as has the artificial selection of seaweed cultivars for commercial exploitation.

Given then that some seaweed species at least are very variable entities, there remains the problem of identifying the sources of this variation. Conventional wisdom holds to the view that mutation plays a small rôle in the evolutionary process. Stebbins (1974) has probably expressed the view of most evolutionists in his remark that mutations occur so rarely that they must be regarded as exceptional rather than the usual basis for evolutionary trends in flowering plants. Dobzhansky (1959) takes a somewhat different view in ascribing a relatively important evolutionary function to mutation. In practice, marine algae seem to be rather mutable organisms although the literature indicates also that mutation is common only in certain well-worked taxonomic groups. The green algal order Ulvales, for example, has proved a particularly rich source of mutants (Fjeld & Lovlie, 1976; Baca & Cos, 1979). Other orders of algae have yielded fewer examples or none at all. Most mutants seem to be morphological, although Kornmann (1970a) has described a life-history mutant of the green alga *Derbesia marina* which is haploid and asexual. Mutant heavy-metal tolerant strains have been reported in the marine brown alga *Ectocarpus siliculosus* by Russell & Morris (1973) and in the freshwater blue-green alga *Anacystis nidulans* by Shehata & Whitton (1982).

The most impressive corpus of knowledge on the incidence of mutation in marine algae and its evolutionary importance has, however, come about through the research of J. P. van der Meer and his co-workers, who have worked exclusively on red algae. Meer (1981c) has isolated nine spontaneous pigment mutants of *Chondrus crispus*. Six of these showed strictly maternal inheritance, while the remaining three gave classical Mendelian transmission ratios. The mutations were, therefore, inherited by both nuclear and non-nuclear mechanisms. Patwary & Meer (1982) have studied nine spontaneous and induced mutants of *Gracilaria tikvahiae*. These differed morphologically and all were transmitted as single, and usually recessive, traits. Patwary & Meer (1983b,d) have also compared *G. tikvahiae* mutants and wild types in culture. The former had faster growth rates than the latter at both high and low densities, they also removed nitrogen (ammonium) from sea water faster and had superior resistance to epiphytes. Many marine algae are known to reproduce only asexually; for such species, mutation may be the sole source of variation.

Another important and interesting source of variation in red algae has been identified by Meer & Todd (1977) in the form of mitotic recombination. This was detected in certain tetrasporophytes of *G. tikvahiae* which were heterozygous for two genes affecting plant colour. The spontaneous development of homozygous tissues in these thalli could therefore be detected by their colour differences. These homozygous patches then gave rise to functional but diploid male and female gametes from which polyploids were duly obtained. A variety of polyploid lineages of *G. tikvahiae* have since been described by Meer (1981b) and Patwary &

Meer (1984) who found triploids to be more robust then tetraploids which were, in turn, more robust than pentaploids, but the growth rates of even the most robust polyploids were found to be no greater than those of the normal diploid tetrasporophytes. It seems likely that mitotic recombination is the cause of the simultaneous occurrence of tetrasporangia and carposporophytes on the same thallus in a number of species of red algae as reported by Dixon (1961), Edelstein & McLachlan (1967), and Lawson & Russell (1967). By bringing about differences in ploidy within species, mitotic recombination must have been responsible for introducing a considerable amount of genetic variation to these taxa. Differences in ploidy may be associated with speciation, and indeed Bird, Meer & McLachlan (1982) seem to feel that English plants of *G. verrucosa* (n = 32) and those from Vancouver (n = 24) are correctly separate species in spite of the fact that they are virtually indistinguishable morphologically and anatomically. A rather different conclusion has been reached by Kapraun (1978b) who has retained at the rank of variety two morphologically and cytologically distinct strains of *Polysiphonia harveyi* (var. *arietina*, n = 32:var. *oneyi*, n = 28).

Genetic recombination is usually regarded as the principal source of variation within species and, therefore, the most important source of evolutionary change. Morphological differences that have been interpreted as plastic responses usually (and unsurprisingly) give no evidence of character inheritance from population crossing experiments: see Sundene (1958) on the interfertility of forms of *Laminaria digitata*, and Svendsen & Kain (1971) on *L. hyperborea*. The more sharply circumscribed local forms of *Agarum cribrosum* on the shores of Japan have also proved interfertile although giving some evidence of inherited physiological differences (Nakahara & Yamada, 1974). Four distinctive forms of the Australasian fucoid *Hormosira banksii* have also been found to be sexually compatible by Clarke & Womesley (1981) although the inheritance or otherwise of their morphological features remains unclear. The population crosses carried out by Chapman (1974, 1975) on members of the *simplices* complex of *Laminaria* on the Atlantic coast of Canada have demonstrated heritability of several phenetic traits (see p. 321). When an extended programme of crosses within this complex was, however, carried out involving populations from both sides of the Atlantic ocean, a much greater amount of inherited variation was revealed. Lüning, Chapman & Mann (1978) who carried out this investigation, crossed *Laminaria* plants from four populations, Halifax, Nova Scotia (HFX), the Isle of Man (IOM), Britanny (BRI), and Heligoland (HEL). HFX × HFX plants all possessed short but wide blades while those of HEL × HEL were long and narrow. Reciprocal crosses between HFX and HEL were both long and wide; and their respective dimensions also gave some evidence of sex linkage. Most of these population differences in morphology were quantitative but a qualitative character was noted in the wrinkling (bullation) of the blade. All phenotypes were bullate except for HEL × HEL which was smooth. The smooth surface was, therefore, controlled by a single recessive gene and expressed only in the homozygous state. Smoothness did not appear to have any particular selective value over the bullate state until it was found to be associated with greater tolerance to the high sea-water temperatures which may occur in the shallow water of Heligoland in summer. Blade smoothness may thus be an instance of pleiotropy (multiple effects of a single gene) in algae.

Another example of a morphological character controlled by a single gene is branch production in the red alga *Antithamnion plumula*. Two varieties of this species, var. *plumula* and var. *bebbii* have been investigated by Rueness & Rueness (1975). The varieties, which differ in their patterns of branching proved to be interfertile in culture. Heterozygous tetrasporophytes resulting from crosses were always of one form (var. *plumula*) but the tetraspores obtained from these yielded equal numbers of *plumula* and *bebbii* individuals. Thus a single gene was involved with *bebbii* as the recessive condition. It is an interesting fact that population hybrids in seaweeds do not seem to possess conspicuous hybrid vigour. This absence has also been noted by Patwary & Meer (1983c) in the course of their work on *Gracilaria tikvahiae*. Perhaps the nearest to hybrid vigour recorded in various experiments is in the HFX × HEL crosses of *Laminaria* described by Lüning *et al.* (1978).

Two additional possible sources of variation should be considered briefly. Tveter-Gallagher & Mathieson (1980) have reported the coalescence of haploid and diploid sporelings of the red alga *Chondrus crispus*, which was in some cases complete, giving rise to a thallus of mixed origin. The evolutionary consequences of this behaviour, if any, and its frequency in nature remain uncertain.

Another interesting aspect of phenetic variation was discovered by Bonneau (1977) in a haploid axenic culture of *Ulva lactuca*. This strain produced, in successive cultures, crops of plants which were distromatic, partially distromatic or tubular. The progeny from swarmers of any of these morphological types yielded a similar mixture of forms. The origin of this variation was thought unlikely to be mutation and Bonneau concluded that thallus morphogenesis comprised a series of developmental events each of which was under control of one or more "switches", the positions of these (off or on) being in turn controlled by environmental conditions. These conditions were thought to undergo change through time even within the confines of a small culture vessel. The existence of such a mechanism in other algae in culture or, more important, in nature is, however, uncertain.

Most experimental evidence from marine algae shows genetic variation to be quantitative rather than qualitative in expression. This may be true of biochemical attributes as well as of morphological characters, as shown for example by the agar constitutions of different *Gracilaria* genotypes by Craigie, Wen & Meer (1984). Evolution is therefore most likely to have occurred through selection acting upon quantitative variations, as Arthur (1984) has proposed.

TRANSSPECIFIC EVOLUTION

If we are to have any reasonably clear picture of speciation, or transspecific evolution (Stebbins, 1974), in algae then we need also to know the taxonomic criteria on which these species have been based. Mayr (1982), in the most recent exposition of his biological species concept, defines the species as " . . . a reproductive community of populations (reproductively isolated from others) that occupies a specific niche in nature''. The importance he attaches to sexual reproductive barriers has been accepted by many taxonomists, and it has been agreed also among biosystematists. The higher biosystematic categories given in Table I (p. 312) are essentially measures of sexual compatibility.

The biological species concept, built as it is upon these criteria, is clearly inapplicable to the numerous cases of algae which do not reproduce sexually. The red acrochaetioid algae studied by Garbary (1978, 1979b), the brown ectocarpoid taxa studied by Pedersen (1984), and the green chaetophoroid algae studied by Yarish (1985) and Nielsen (1978, 1980) exemplify this problem. All these algae are small and structurally simple, many being epiphytes or endophytes, and they generate relatively few morphological taxonomic characters. Many, too, are known only as asexual plants. These taxa, together with most (or all) marine blue-green algae, cannot be considered as biological species. Indeed, attempts have been made, in the case of the acrochaetioid forms, to develop a special concept (White & Boney, 1969; Garbary, 1979c). It would appear likely that many of these asexual taxa are really only microspecies, and some may be the results of isolation of ''fragments'' of more complex life-histories (Bernatowicz, 1958; Kornmann, 1970a).

Speciation in marine algae seems likely to have been associated with isolation processes of three principal types. These are reproductive isolation, ecological isolation, and geographical isolation.

Reproductive (sexual) isolation would seem at first to be easily verifiable technically and unequivocal in its implications. Unfortunately, it is often difficult to draw firm conclusions from hybridization experiments partly because they may present considerable technical problems but mainly because the criteria for sexual compatibility are themselves at variance. Many workers have assumed compatibility on the evidence of successful syngamy, others have gone beyond syngamy and observed the growth of a daughter thallus, others to the onset of mitotic spore production by the daughter, but the only really stringent test of crossability is completion of meiosis to yield viable progeny. In spite of these differences in approach, a growing body of literature exists on the sexual isolation of marine species; an excellent summary of this work has been given by Mathieson, Norton & Neushul (1981) who provide a very complete list of these experiments and their results. A chosen few will be discussed below.

In 1970, Wynne & Edwards described a new species of red algae, *Polysiphonia boldii* which Edwards (1970) attempted to hybridize with *P. denudata*. His experiment, the first of its kind to be completed in red algae, failed and the species status of *P. boldii* seemed secure. Rueness (1973), however, successfully crossed *P. boldii* with yet another *Polysiphonia* species, *P. hemisphaerica*, and followed development as far as production of the tetrasporophyte generation. It was only after meiosis, when very reduced tetraspore viability was observed, that the incomplete compatibility of the two taxa could be seen. *P. boldii* was, therefore, reduced in rank to a variety of *P. hemisphaerica*.

Bird & McLachlan (1982) have invoked certain attributes of the red algal genus *Gracilaria*, including type of spermatangia, chromosome numbers, and infertility, as a means of establishing species limits. This follows the discovery by Bird, Meer & McLachlan (1982) that the English and Vancouver populations of *G. verrucosa* have different chromosome numbers, although morphologically and anatomically indistinguishable. An earlier paper by McLachlan, Meer & Bird (1977) revealed that a Canadian population of *G. foliifera* failed to cross successfully with an English *Gracilaria* population although the two shared the same chromosome number. In this latter case syngamy was thought to have occurred but the cystocarps which were produced subsequently proved to be functionless. It was concluded that these Canadian and English populations of *Gracilaria* belonged to different species, a view which may be contrasted with that of Kapraun (1978b) who has retained only at varietal rank two populations of *Polysiphonia harveyi* which possess different chromosome numbers.

Rueness (1978) has made a very thorough review of hybridization in red algae and its taxonomic implications. He showed that a Baltic species *Ceramium tenuicorne*, formerly thought to be endemic, was conspecific with *C. strictum* because of successful crossing in culture although heritable morphological differences were observed. In cases of geographically isolated populations it is perhaps pertinent, however, to ask whether or not the crossings demonstrated in the laboratory ever in practice take place in nature. Masuda *et al.* (1984) found that *Gigartina pacifica* and *G. ochotensis* hybridized with 78% success. The two taxa were considered to be conspecific despite incomplete sexual compatibility.

Guiry & West (1983) have isolated into culture a large number of strains of *G. stellata* from many localities in the North Atlantic. They found two kinds of plant, those with an asexual, direct type of life-history and others in which the sexual cycle took place normally. A programme of crossing experiments involving 32 female and 27 male gametophytes has revealed the existence of two virtually non-interbreeding sexual populations with a high degree of geographical separation. The existence of intraspecific sterility barriers with an ecological or geographical basis is evident also in the case of the red alga *Antithamnion plumula* (Sundene, 1959, 1975).

Sexual isolation between northern and southern European populations of the green alga *Ulva lactuca* was first demonstrated by Föyn (1955). Closer study of these populations then revealed a wealth of phenetic differences and the populations were duly recognized as separate species. Two other members of the order Ulvales, *Enteromorpha intestinalis* and *E. compressa*, have been the subject of a study by de Silva & Burrows (1973). These authors found that the sterility barrier between the two species was incomplete and postulated that tidal levels may have been the isolating mechanism separating the populations. The outcome of this investigation was the conclusion that the two entities involved are conspecific and that they should be retained taxonomically only at the rank of subspecies. Larsen (1981), however, has found that these *Enteromorpha* species are totally isolated in reproductive terms, although his sampling sites differed from those of de Silva & Burrows.

The kind of incomplete sexual isolation observed by de Silva & Burrows *(1973) and Sundene (1975) has also been found in the brown alga Ectocarpus siliculosus*. Müller (1976b, 1979) has observed sexual attraction between gametes and their syngamy in a number of *Ectocarpus* isolates from 11 geographically separated areas in Europe, Atlantic, North America, and Australia. Müller's results are shown in Figure 2; filled circles

(●) denote the occurrence of syngamy, open circles (○) the occurrence of sexual attraction between gametes as revealed by their behaviour but without consummation, the points (·) denote the absence of sexual behaviour of any kind. It is evident that most isolates, including that from Australia, have no barriers to syngamy but also that some sexual isolation exists between these and the isolates from Port Aransas and, more particularly, that from Woods Hole.

Müller's work is of additional interest because of the information it contains on gametangium morphology. It is evident from Figure 2 that the two most similar gametangia are those of the Bergen and Woods Hole (5 and 6) isolates; yet these are quite intersterile. Thus, phenotypic similarity in *Ectocarpus* is not related to reproductive affinity. Finally, the facts that these gametangia are quantitatively rather than qualitatively different and that they are haploid, show that a number of genes are responsible for gametangium morphology in this species.

The *simplices* section of the genus *Laminaria* has been the subject of a number of experimental crosses. Populations from both sides of the Atlantic have proved interfertile (Lüning *et al*., 1978), suggesting that a single species *L. saccharina* can accommodate the entire complex of heritable phenetic differences observed within the group. If speciation is active in this complex, then isolation is still at an extremely early, possibly ecotypic, stage.

In the genus *Fucus*, the evidence for the presence of reproductive isolation is very contradictory. Burrows & Lodge (1951) have reported crossing successes of 91% and 99% in reciprocal matings of *F. vesiculosus* and *F. serratus* but Bolwell, Callow, Callow & Evans (1977) have found complete sexual isolation between these species. The discrepancy between these observations has been largely explained by Bolwell *et al*., who have shown that sterility barriers operate effectively in gametes released from freshly collected plants, but that these break down permitting interspecific fertilization if the parent plants have been stored in the laboratory for a time. Nevertheless, hybrids between species of *Fucus* species continue to be reported in nature; Lein (1984b) has recently described *F. distichus* × *F. serratus* hybrids in the inner Oslofjord. Perhaps the polluted waters in this locality also break down the natural sterility barriers, as they appear to do in the river Mersey (Burrows & Lodge, 1951).

The potential for successful hybridization in *Fucus* was demonstrated by Evans (1962) when he reported that all four common species of *Fucus* in Britain have the same chromosome number (n = 32). In the order Laminariales a more complicated situation seems to exist. North American species of *Alaria* have a haploid chromosome complement of 14 whereas the only British member of this genus, *A. esculenta* has n = 28 (Robinson & Cole, 1971; Evans, 1965). All four British species of *Laminaria* and *Saccorhiza polyschides* have, however, the same numbers (n = 31). The prospect of elucidating brown algal evolution from chromosome counts may be regarded as remote (see Cole, 1967, 1968).

Ecological isolation seems certain to have played an important rôle in the speciation of seaweeds, perhaps most important in the case of intertidal seaweeds. Schonbeck & Norton (1978) have concluded that the upper levels of fucoid species on British rocky shores are largely determined by their

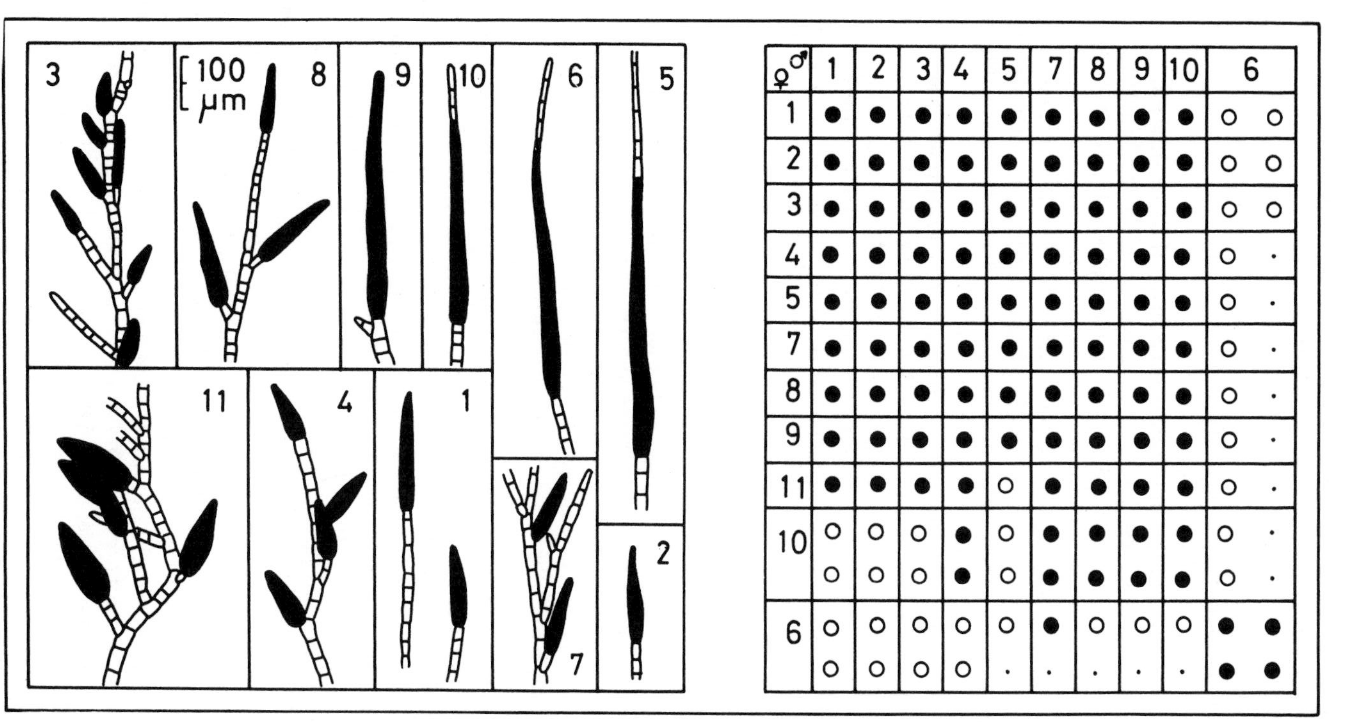

♀ \ ♂	1	2	3	4	5	7	8	9	10	6	
1	●	●	●	●	●	●	●	●	●	○	○
2	●	●	●	●	●	●	●	●	●	○	○
3	●	●	●	●	●	●	●	●	●	○	○
4	●	●	●	●	●	●	●	●	●	○	·
5	●	●	●	●	●	●	●	●	●	○	·
7	●	●	●	●	●	●	●	●	●	○	·
8	●	●	●	●	●	●	●	●	●	○	·
9	●	●	●	●	●	●	●	●	●	○	·
11	●	●	●	●	○	●	●	●	●	○	·
10	○	○	○	●	○	●	●	●	●	○	·
	○	○	○	●	○	●	●	●	●	○	·
6	○	○	○	○	○	●	○	○	○	●	●
	○	○	○	○	·	·	·	·	·	●	●

Fig. 2.—Plurilocular gametangia (left) of *Ectocarpus siliculosus* from 11 geographically different localities: note continuous variation in gametangial morphology. Table of sexual compatibilities (right) of these isolates as indicated by gametic behaviour: ●, denote successful syngamy; ○, indicate occurrence of sexual attraction, but without syngamy; ·, indicate total absence of sexual behaviour. Diagrams redrawn from Müller (1979); Key to localities: 1, Naples; 2, Villefranche-sur-Mer; 3, Roscoff; 4, Isle of Man; 5, Bergen; 6, Woods Hole; 7, Beaufort, North Carolina; 8, Wilmington, North Carolina; 9, Tampa, Florida; 10, Port Aransas, Texas; 11, Victoria, Australia.

respective sensitivies to the stresses of the physical environment at different shore heights. This has been confirmed by Dring & Brown (1982) who recorded the times required for recovery of pre-dehydration photosynthetic activity in thalli that had been subject to desiccation. The species studied differed significantly in their recovery times and in a sequence that echoed their zonal distributions in nature.

Dahl (1959) has concluded that the physical stresses of the intertidal region have played an important rôle in speciation, although he attached greatest significance to salinity as an agency of evolution. Observations by Biebl (1952, 1967) and Russell (in press) confirm Dahl's hypothesis in that upper-shore algae possess the widest salinity tolerances while the narrowest tolerances are possessed by species from the sublittoral zone. Desiccation and salinity effects are both manifestations of changes in external water potential, but other components of the intertidal physical environment have almost certainly played a part in algal speciation.

Estuaries also present algae with considerable saline problems which have been met in a variety of ways by populations within species (see p. 316) Most estuaries are, however, vegetated by euryhaline species with fairly wide ecological amplitudes, and endemic species are uncommon. *Fucus ceranoides* is a species which is restricted to waters with some freshwater input, and Burrows (1963) has shown that its fertilized eggs develop tolerance to freshwater very much faster than do those of the more stenohaline *F. serratus*. Khfaji & Norton (1979) have shown that thalli of *F. ceranoides* grow more successfully in diluted sea water than do those of the marine species *F. vesiculosus*. There are possibly some differences in tolerance of adult and juvenile thalli in that juveniles of *F. ceranoides* may be more amenable to acclimation (Lein, 1984a).

Seasonal climatic factors may be involved in the development of ecological isolation. For example, three species of *Sargassum* observed by McCourt (1984) in northern California seem to differ a little in their growth and reproductive behaviour over the year. A rather clearer picture of seasonal differences emerges from the observations by Lee & Kurogi (1971) on ten species of the order Rhodymeniales (red algae) on the coast of Hokkaido, Japan. These authors sampled mainly from the lower intertidal region, so there are no zonal differences to be confused with seasonal effects. Lee & Kurogi observed very considerable differences within, as well as between, species in growth and reproductive phenology. It is noteworthy that Burrows & Lodge (1951) have attributed the scarcity of *Fucus serratus* × *F. vesiculosus* hybrids in nature to the marked differences in reproductive phenology of these species.

A particularly elegant demonstration of the importance of ecological factors in algal speciation has been carried out by Lüning (1979). He compared the thallus growth characteristics of three European species of *Laminaria* which occur naturally at different water depths. Two of these species *L. hyperborea* and *L. saccharina*, which are able to inhabit deep waters, ceased blade growth in early summer and entered a period in which metabolites were accumulated as food reserves for the coming winter. The third species, *L. digitata*, is in nature restricted to the uppermost levels of the sublittoral and the lowermost parts of the intertidal regions. This species showed a much more extended period of blade growth which Lüning interpreted as being necessary for thallus renewal in a habitat where wave action and blade abrasion were likely. A consequence of this extended growth would, however, be smaller amounts of stored photosynthate, which in turn should make *L. digitata* a poor performer in deeper sublittoral habitats. The appropriate transplant experiment confirmed the inability of *L. digitata* to survive permanently at great depths. This investigation serves also to illustrate very clearly the existence of evolutionary constraints (see Lewontin, 1978), adaptation with respect to one particular selection pressure being accompanied by loss of adaptation to another.

Ecological isolation may also operate on a much smaller scale in the relation of epiphytes and their host plants. Novak (1984) has applied rigorous sampling and analytical methods to the distribution of microscopic epiphytes of the sea-grass, *Posidonia*, and described the process of epiphyte community development in some detail. He has coined the term "ultra-ecology" for his approach to epiphyte community ecology. Sea-grass epiphytes have also been studied by Jacobs, Hermelink & Geel (1983) who noted that the pioneer species were also the most frequent and characteristic members of the host's epiflora. Ballantine (1979) has concluded from a study of algal epiphytes on macrophytes at Puerto Rico that host species vary in suitability as substrata and that, on all host species, epiphyte distribution is a function of host tissue age. This does not, however, explain how particular host-epiphyte associations arise, rather it gives force to the opinion of Harlin (1975) that we need to look more to the ecological relations of the participants in the association.

The *Laminaria* blade is a frequent substratum for epiphytes, many of which are brown, ectocarpoid taxa. It is also a substratum which generates a certain amount of selection pressure in that it is continuously losing tissue from its distal end and producing new tissue from its primary meristem at the junction between stipe and blade. Thus, any epiphyte settling at the distal end of the blade is at risk of being shed together with the senescent tissues of its host. Settlement at a more proximal position will allow a longer period of time for epiphyte development but if the epiphyte is to attain reproductive maturity and give rise to progeny then its cycle of development must keep pace with the growth of its substratum. Experiments by Russell (1983a,b) suggest that the most frequent components of the *Laminaria* blade epifloras are successful because their reproductive cycles are in phase with the growth of the host. It may be postulated that selection pressure of this kind has led to the evolution of host dependency, as in the case of parasitic red algae (see Kraft, 1981; Brawley & Wetherbee, 1981).

Additional interest in host-epiphyte relationships has been generated by Filion-Myklebust & Norton (1981) and Moss (1982) who have described recurrent shedding of "skin", in the form of sheets or flakes of outer cell wall material, from a number of fucoid species. They interpreted this behaviour as having an antifouling function, an opinion shared by Russell & Veltkamp (1984) who discuss also the means by which certain species of epiphytes are able to circumvent the cleaning action of skin-shedding. Hosts and epiphytes evidently have a subtle ecological relationship which deserves further research. The various species of a kelp epifauna in southern California have been examined by Bernstein & Jung (1979) who concluded that co-evolution has occurred in the community. In algae at least, an element of pre-adaptation may well exist as evidenced by the fact that the arrival of the alien macrophyte *Sargassum muticum* on British shores was quickly followed by its acquisition of a varied epiflora (Withers *et al.*, 1975).

Geographical isolation has long been considered an important factor in marine algal speciation. Some accounts such as that of Womersley (1959) which deals with Australian coasts, of South (1978) with the Antarctic and Southern oceans, of Santelices (1980) with the temperate coast of Pacific South America, and of Druehl (1970) with the Laminariales of the north-eastern Pacific, are primarily floristic analyses with much attention given to matters of endemism and species migrations. Others, on the other hand, seem directed more towards establishing phytogeographical correlations with some environmental factor, usually temperature; the works of Abbott & North (1972), Thom (1980), and Murray & Littler (1981) on Californian algal phytogeography are of this character.

The importance of sea-water temperature as an agency of algal evolution has become clear as a result of the recent research by C. van den Hoek. An investigation of benthic algal species distributions of north Atlantic coasts led to a description of temperature-based phytogeographic provinces Hoek (1975) which were later examined in the light of the distributions of species of *Cladophora* in the same region (Hoek, 1979). The patterns of distribution of *Cladophora* species agreed closely with the previously described phytogeographical provinces. Hoek (1982) has made use of published information on the effects in culture of temperature on algal life-histories, and on geographical distribution, in a remarkable synthesis of algal phytogeography on a world-wide basis. Recent experimental evidence on the importance of temperature and geographical isolation in species delimitation has been reported by Yarish, Breeman & Hoek (1984). A new text-book by Lüning (1985) contains an excellent account of recent ideas on marine algal phytogeography.

The importance of sea-water temperature in determining algal geographical limits is certainly very great. Sundene (1962) established this fact in relation to the kelp *Alaria esculenta* by means of field transplant experiments, and subsequent research by other workers on other species has confirmed its importance. Bolton & Lüning (1982) have measured the temperature tolerances of nine isolates of four species of *Laminaria* collected from a variety of North Atlantic sites. In this instance, tolerance was measured in terms of sporophyte growth and gametophyte survival in laboratory culture. The authors concluded that there were no significant population differences in temperature tolerance in spite of the wide variations in temperatures between collecting sites, and that the success of *Laminaria* over such a wide temperature range was due more to plasticity than to the selection of temperature races or ecotypes. Nevertheless, their own data seem to indicate that the highest temperature tolerable to gametophytes of the Arctic *L. solidungula* was 5 °C lower than that of *L. saccharina*. More recently, McLachlan & Bird (1984) have shown that populations of the red alga *Gracilaria bursapastoris* from England and Hawaii have different temperature tolerances. It may, therefore, be premature to assume that marine algal species are immutable with respect to temperature. Indeed if species distributions are closely integrated with temperature, it seems illogical not to expect divergences to have occurred below the species level as a prelude of speciation. Moreover, it would be regrettable if temperature were to become generally accepted as the sole driving force of marine algal evolution. In the Baltic Sea, which is admittedly too young for speciation to have occurred, it seems likely that evolutionary divergence has been as much in response to salinity changes as to temperature (Russell, 1985b). Furthermore, evidence suggests that the importance of temperature in determining species limits may have been over-stated. Stewart (1984) has found the temperature optima for growth in culture of several geographically distinct species of Californian red algae to be virtually indistinguishable.

Floristic changes, it is now clear, are very liable to occur as a consequence of introductions by man. Recent introductions in British coastal waters have been reported by Jones (1974), in Australia by Skinner & Womersley (1983) and in New Zealand by Parsons (1985). The introduction of *Eucheuma* species to Fanning Island, Kiribati, for mariculture evaluation was accompanied by the appearance of four additional species, thought to have been epiphytes of *Eucheuma* (Russell, 1982). Floristic changes accomplished in this way will make phytogeographical interpretation increasingly difficult. Rates of natural floristic change seem, however, to vary widely. Irvine (1982) has concluded that the flora of the Faeroes has remained stable since the nineteenth century, but Hansen, Garbary, Oliveira & Scagel (1981) have recently described a number of new records and range extensions in the Alaskan flora.

Any consideration of changes in species distributions, or of the impact of environmental changes upon ecologically or geographically fixed populations, is liable to dissolve in the general uncertainty about what actually constitutes an algal species. In his revision of the European species of the order Ulvales, Bliding (1963) made extensive use of cross-fertilization experiments to delimit sexually reproducing species but Koeman & Hoek (1980, 1982) working with the same group of plants have defined species almost entirely in morphological terms. The order Ulvales is a taxonomically "difficult" group and the status of many of its species is questionable. A similar situation obtains in the brown algal order Ectocarpales, but in this case a major source of confusion arises from the common occurrence of reproductivity fertile thalli at early states in thallus ontogeny. Such developmental variation has been observed in culture by Russell (1966) and Ravanko (1970). The latter author has concluded that there are only a few true species in this order, but this is a view which probably would not be shared by more recent researchers such as Clayton (1974) and Pedersen (1984).

In an interesting recent paper, Amsler & Kapraun (1985) have stressed the importance of recording the ontogeny in culture of these structurally simple brown algae, but at the same time warning against the risks of morphological artifacts arising from faulty culture technique. These authors argue that evidence of morphological similarity in culture is by itself insufficient reason for merging forms that appear distinct in nature, and hence of regarding them as environmentally induced. Differences retained in culture are, however, to be regarded as strong support for the view that different species of life-history phases are involved. This recommendation is very similar to an argument advanced by Söderström (1965) in relation to species delimitation in *Cladophora*.

Strict application of this criterion would undoubtedly overturn a great deal of published work, but Amsler & Kapraun have made a point which any observer of algae in culture should bear in mind before abolishing taxa. Their proposal, unfortunately, gives no assistance on the problem of continuous variation in field samples. Abbott (1972) and Clayton (1975) have reported the presence of continuous variation in thallus morphologies of *Iridaea* (red algae) and *Colpomenia* (brown algae), respectively, and have emphasized the problem this raises in discrimination of species effectively. In ligulate and filiform species of the brown algal genus *Desmarestia* there would appear to be a great deal of continuous variation which has led Chapman (1972a,b), who quantified it, to combine many species.

A suggestion that use of ratios of morphological traits might assist taxonomic circumscription has been made by Norton, Mathieson & Neushul (1981), and ratios have been adopted by Goodband (1971) and Amsler & Kapraun (1985). It is not easy to see how ratios of attributes that vary continuously can do other than compound the problem and this has indeed proved to be the case, at least in *Fucus*. Marsden, Evans & Callow (1983a) have argued the importance of establishing the normality of distribution of morphometric traits before making analyses of data, and subsequently when they calculated ratios of such traits they found these not to be normally distributed (Marsden, Evans & Callow, 1983b). Moreover, ratios failed to distinguish even the readily separable *F. vesiculosus* and *F. serratus*.

The need to quantify the morphological attributes of algae, and to put species discrimination on as rational a basis as possible, has prompted a number of exercises in numerical taxonomy, one of which by Rice & Chapman (1985) on *F. distichus* also surveys the earlier literature. This investigation resulted in good evidence for the existence of two species of the *F. distichus* complex in the North Atlantic, *F. distichus* and *F. evanescens*. Dendrogams which result from numerical analyses such as that of Rice & Chapman are simply expressions of morphological affinity and they have no phylogenetic implications whatever. Some confusion over this point still, however, persists; the use of the term "cladogram" by Pedersen (1984) when referring to dendrograms implies a misinterpretation both of their construction and of their purpose.

Phylogenetic relations of groups of species as deduced from their morphology is still an area of very considerable uncertainty. The phylogenetic diagrams representing the relationships of *Cystophora* and *Fucus* species, for example (Womersley, 1964; Powell, 1963) are based upon relatively few selected attributes. Roberts (1978) in her account of active speciation in the genus *Cystoseira* makes the interesting point that non-adaptive features are the basis for species separation in this group. Ercegovic (1952), who worked on the same genus in the Adriatic Sea, has nevertheless reported a considerable ecological and geographical component to the species distributions. It is possible, therefore, that an adaptive element is present in the morphological evolution of *Cystoseira* species. The observations of Ercegovic give reason for agreeing with Neushul (1972) and Norton, Mathieson & Neushul (1982) in their plea for more investigation of the relationship between form and function in marine algae.

Rice & Chapman (1985) have stated that the biological species concept of Mayr (1942) is inappropriate for marine algae because of the incidence of hybrids between species. Their analysis of this concept is in that respect correct but sexual isolation is a relative phenomenon; there are many degrees of sexual compatibility (Chapman, 1978). A similar range of variation exists in ecological isolation; the amount of genetic variation detected in species of *Eucheuma* by Cheney & Babbel (1978) using enzyme electrophoresis was very considerable, and the more ecologically diverse species possessed the greater variability. Variation of this sort is evident also on a geographical scale, as revealed by the analyses of agars extracted from three strains of *Gracilaria tikvahiae* from north American and those from three other species of *Gracilaria* from China. Such quantitative differences are analogous to the morphological variation already mentioned.

Perhaps the principal reason for the absence of a clear and universally applicable species concept for algae is because there is no qualitative difference between micro-evolutionary events and transspecific evolution. All the reproductive morphological, ecological, and geographical differences between species have their counterparts below the level of species. The essential continuity of evolution that this implies has also been argued forcefully by Stebbins (1950) and Charlesworth (1982).

MACRO-EVOLUTION

The evolution of higher algal taxa is, in the words of Loeblich (1974) an "... interesting and highly speculative field". He concludes, mainly from fossil evidence, that the differentiation of the major algal groups was a series of events which took place during the Precambrian period, and which preceded the diversification of the animal kingdom. As already mentioned (see p. 309 *et seq.*), the algal fossil record has proved disappointingly incomplete, and macro-evolutionary speculation has been based principally upon comparative studies of existing taxa. These studies have been mostly observational in character in spite of the fact that experimental deductions are possible, as Raven (1984) has shown in his recent synthesis of physiological and anatomical evidence bearing on the evolution of vascular plants.

A recent, and very elegant, experimental examination of the theory of complementary chromatic adaptation has been carried out by Ramus & Meer (1983) and Ramus (1983). The idea tested was that the red seaweeds had evolved primarily for life in deep water, the absorbance characteristics of phycoerythrin, their red accessory pigment, being complementary to that of submarine light. Ramus & Meer compared the chlorophyll contents, photosynthetic performances and growth characteristics of two normally-pigmented wild-type culture isolates of *G. tikvahiae* with those of two phycoerythrin-deficient mutants derived from these isolates. Thus, genetic differences were, as far as possible, restricted to those genes responsible for pigmentation. The two sets of isolates were grown under controlled conditions and it was found that their responses were dependent solely upon irradiance, and that they were independent of the spectral distribution of the light. In all cases, the mutants performed as well as their wild-type progenitors. In a similar set of experiments, Ramus (1983) used seven species of red, brown, and green algae, and once again he found that photosynthetic performances were independent of spectral distribution and were, in most cases, dependent upon irradiance. Thus, accessory pigments do not appear to provide a functional base on which to construct an adapative radiation of the major algal groups. Nevertheless, there remains an environmental dimension to many of the higher algal taxa; groups like the red and brown algae are almost exclusively marine while others like the charophytes are almost all freshwater. Only a few, such as *Cladophora* break rank (Round, 1984).

Pigments, however, remain the principal attributes in the taxonomic discrimination of major algal taxa. Edwards (1976) has postulated the existence of seven plant kingdoms within two super-kingdoms (Procaryota and Eucaryota). The seven kingdoms are diagnosed by cytological and biochemical characters, the three eucaryotic algal kingdoms being distinguished according to their chlorophylls. They are the Erythrobionta (chlorophyll a or $a+d$), and Chlorobionta ($a+b$) and the Ochrobionta ($a+c$). Edward's algal kingdoms are thus similarly defined to those of Christensen (1964) who adopted the more familiar names Rhodophyta and Chlorophyta for the first two, but coined a new name, the Chromophyta for the chlorophylls (a + c) group. Leedale (1974) has posed the question, how many are the kingdoms of organism? It is an extremely pertinent question, because if a four-kingdom evolutionary scheme is postulated then each of the three eucaryotic kingdoms is, on the basis of existing knowledge, highly polyphyletic. If, however, we insist upon having monophyletic kingdoms then a multi-kingdom system must inevitably follow (Leedale suggests a total of 18). Whatever number of highest algal taxa may come to be generally accepted, it seems likely that unifying concepts at this level are going to be about rather fundamental aspects of plant cell metabolism or such matters as the endosymbiotic origin of plastids (Cavalier-Smith, 1982; Round, 1984), although Chadefaud (1982) has evidently concluded that in all classes of algae with flagellate cells, the flagellar organization may be derived from a prasinophycean "archetype".

The evolutionary importance of ontogeny has been expounded by Gould (1977). Phylogenetic relationships at lower levels in the algal taxonomic hierarchy have been deduced from ontogenic evidence as well as from comparisons of perfect thallus states. Perestenko (1972) has reconstructed the phylogeny of the brown algae on the basis of their patterns of early development. His conclusions seem to agree substantially with those of Fott (1974) and Jensen (1974) in tracing many of the parenchymatous and pseudoparenchymatous orders to an ectocarpoid origin. At odds with this rather traditional view is the suggestion by Pedersen (1984) that the primitive brown algal thallus evolved from a parenchymatous chrysophycean ancestor, and was itself parenchymatous, although of a fairly undifferentiated type. Pedersen does not argue his case from ontogenetic evidence, and this would indeed be difficult to do, because his own very extensive observations and culture experiments have shown the presence of filamentous thalli early in the development of the numerous algae studied. Pedersen has shown a filamentous organization to be normally present in the microthalli, but also to occur in the early development of the macrothalli, some of which may remain filamentous in part, even in the perfect state.

Ontogenetic observations have been deployed to make taxonomic decisions at the levels of genus and species in the brown algae by Russell (1966), Ravanko (1970), and Amsler & Kapraun (1985) and by Chamberlain (1984) and O'Kelly & Yarish (1980) on red and green algae, respectively.

Phylogenetic systematics (cladistics) has been hailed by some as illuminating and clarifying the phylogenetic process, and rejected by others as mere obfuscation. Cladistics has been expounded by Patterson (1982a,b) as follows. Phylogenetic relationships are defined by means of branching diagrams (cladograms) and are evinced only by shared derived characters (synapomorphies). Shared primitive characters (symplesiomorphies) are inherited from a more remote common ancestor and are irrelevant or misleading in the search for phylogenetic relationships. A third type of character, those unique to any single groups are referred to as autapomorphies. Three types of groups may be distinguished, monophyletic groups being those containing all and only the descendants of a common ancestor. Paraphyletic groups are characterized by shared primitive characters, and polyphyletic groups are those whose defining characters are thought not to have existed in their common ancestor, *i.e.* they are characterized by convergent characters. Only monophyletic groups are acceptable. The rules of cladogram construction also require the most parsimonious expression of the assembled data, *i.e.* that which neglects fewest apomorphies is to be adopted.

At risk of gross over-simplification, cladistics is first a means of character weighting (derived or primitive) and secondly a techinque for assembling the resulting information. The technical part is logical and efficient, but there is cause for uneasiness over the choosing and weighting of characters. Particular difficulty may arise in distinguishing homologous from analogous character states, a problem discussed by Bock (1963) and Cain (1982). According to Simpson (1959), analogous similarity may arise in four different ways. These are: (1) convergence in which formerly dissimilar features become similar; (2) parallelism in which features possessed by different groups undergo a similar change or changes; (3) reversal in which the character state changes from a derived condition to one that closely resembles an earlier condition (*i.e.*, secondarily primitive); and (4) independent origin, in which a similar or identical feature or condition appears independently in two or more groups. The existence of analogous phenetic characters in algae belonging to very different higher taxa has been known for a very long time (see Fristsch, 1935, 1945) and there is no evidence to show that their frequency has been exaggerated. Nevertheless, the would-be algal cladist has to decide whether a character state is derived or primitive. This question may be resolved by regarding as primitive those features which extend over diverse groups, derived character states being those which are more narrowly confined. This criterion is open to the danger of circular argument Round (1984). Furthermore, shared primitive characters, although irrelevant to phylogenetic reconstruction may be vitally important in a Darwinian sense.

A particularly elegant example of phylogenetic systematics has been executed recently by Prud'homme van Reine (1982) as part of a taxonomic revision of the European Sphacelariales (brown algae). The cladogram resulting from this research demonstrated the existence of species groupings which were similar to those previously obtained by means of numerical taxonomy. Figure 3 which is based on his results illustrates very clearly the way in which cladistics is highly productive of taxa at every level of the hierarchy. Pedersen (1984) has worked with rather simpler brown algae and has expressed the view that cladistics provides a satisfactory means of classification, although it is not at all clear how his cladograms have been constructed and he fails to give a clear inventory of the important character states and their respective primitive and advanced condtions. Pedersen also criticizes the numerical taxonomic approach to classification of the simpler brown algae and reprimands Russell & Garbary (1978) for adopting this method. His criticisms are, first that the character states chosen by these authors are mainly primitive, but numerical taxonomy conventionally ignores character weighting of this kind as both unnecessary and unjustifiable (Sokal & Sneath, 1963). Pedersen's second criticism is that certain taxa in the dendrograms of Russell & Garbary (1978) have unacceptably high

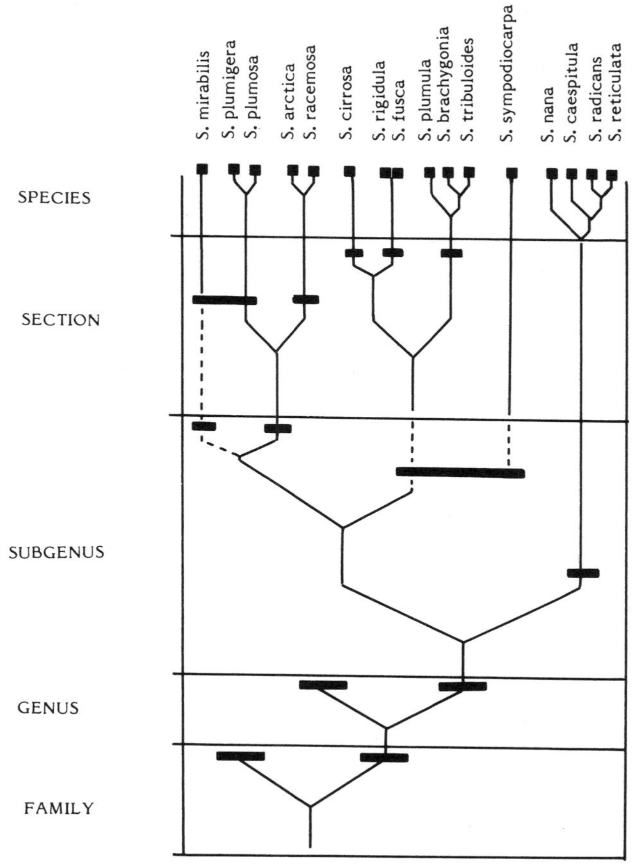

Fig. 3.—Cladogram showing possible phylogenetic relationships of European species of *Sphacelaria*: squares denote species and rectangles other unnamed higher taxa; note the effect of cladistics in generating a hierarchy of taxa; cladogram redrawn from Prud'homme van Reine (1982).

coefficients of similarity. This point too has little meaning as structurally simple plants generate relatively few characters states and high similarities are, therefore, likely to arise. In any case, it is the overall relative similarities within the group in the dendrogram that are important; to take the similarity of any two taxa, in isolation, as an absolute measure of their affinity is misleading.

It is possible to treat cladistics as purely an exercise in classification Patterson (1982a,b), but this seems more than a little disingenuous as the method remains firmly rooted in phylogenetic evaluations of character states. At the same time, it is hard to accept that organismic evolution really has proceeded according to the statutes of cladogram construction. Cain (1983), a critic of phylogenetic systematics, attaches particular importance to the high incidence in nature of convergent and parallel character states, but he also inveighs against the concept of parsimony of which, he argues, evolution knows nothing. McNeill (1982) also holds to the view that cladistics should not be equated with phylogenetic reconstruction. If cladistics is, however, to be regarded simply as a classificatory technique, then it is one which will generate a "special-purpose" classification (Gilmour, 1961). The virtues of special-purpose classifications are not always evident to, or of practical value to, the general users of floras or taxonomic revisions, as has proved to be the case with biosystematics classifications (see Table I, p. 312).

A very good example of the problems of reconstructing phylogenies of higher algal taxa is found in the order Fucales (brown algae). Fucalean evolution has been the subject of an excellent review by Clayton (1984), and on the basis of information from a variety of sources, together with her own observations, she has concluded that the first fucoid algae probably evolved on South Australasian shores during the Mesozoic period. From this region and at that time, the radiation of the fucoid algae subsequently occurred, with the assistance of plate tectonics.

The matter of establishing primitive and advanced character states in fucoid algae has been the subject of several investigations. Manton (1964) examined the fine structure of sperm from three genera of Cystoseiraceae and found them to be structurally less specialized than those of *Fucus*. Manton therefore concluded that *Fucus* was not a primitive fucoid, in spite of its relatively simple thallus construction. Indeed, Manton made the interesting suggestion that the *Fucus* thallus structure is related to its intertidal habitat, which she considered not to be a primitive environment. Jensen (1974) has pointed to the difficulty in establishing phylogenetic series in the Fucales but has suggested that the presence of a single egg per oogonium is a more advanced attribute than the presence of several, a view shared by Clayton (1984). According to this criterion, members of the Cystoseiraceae could be regarded as advanced. Clayton has suggested that reduction in egg numbers, from eight per oogonium as in *Fucus*, may have occurred independently more than once during the evolution of the various families of Fucales. This too seems very possible: in *Pelvetia canaliculata* there are two eggs per oogonium but these are shed and fertilized while retained within a layer of oogonial wall (mesochiton). Moss (1974) and Russell (in press) have discussed the rôle of mesochiton as a protective device against the physical stresses of the upper-shore levels occupied by this plant. The two-egg oogonium may, therefore, be an advanced, or a secondarily advanced character.

Clayton (1984) has postulated that the hermaphrodite condition is primitive in Fucales, and this character is present in the Cystoseiraceae. Its presence also in upper-shore fucoids such as *P. canaliculata* and *Fucus spiralis*, however, could be interpreted as an advanced mechanism to maximize fertilization success in a demanding environment. The resulting in-breeding could also be interpreted as an advanced mechanism to reduce variation among well-adapted populations.

The vegetative thallus of the Cystoseiraceae and other related families is quite elaborate with a radial or helical arrangement of branches. This radial pattern has also been detected in the distribution of conceptacles in the reproductive tissues (receptacles) of these plants (Jensen, 1974; Clayton, 1984). This radial symmetry is, however, sometimes visible in receptacles of *Fucus*, although absent from the mature vegetative thallus. Figure 4 (prepared with the assistance of C. J. Veltkamp) illustrates spirality in young receptacles of *F. vesiculosus* (top right and left). The bifid receptacle (bottom left) shows a less marked radial arrangement, although it does reveal some synchrony in conceptable initiation in the two receptacle halves. The last illustration (bottom right) shows very little sign of any helical pattern and, in the population studied this nearly-random arrangement was possessed by about 90% of the receptacles examined. If we assume that spirality is a primitive character, then randomness is presumably derived; but one may reasonably ask whether it is a true evolutionary novelty or simply the consequence of the gradual loss of spirality from the species?

A final and possibly optimistic note on which to end these comments on fucalean phylogeny has recently come to light in the discovery by R. H. Reed (pers. comm.) of the presence of altritol in the Australasian fucoid *Hormosira*. Altritol, an unusual algal polyol, was identified by Chudek, Foster, Davison & Reed (1984) in *Himanthalia*, in which it plays an important rôle in turgor regulation. Recently, also, Moestrup (1982) has proposed a close phylogenetic relationship between *Himanthalia* and *Hormosira* on the basis of similarity in their sperm flagellar. Thus, two very different attributes now link two geographically remote taxa, although the question of the degree of primitiveness of altritol remains at present uncertain.

Below the rank of order, phylogenetic speculation is made particularly difficult by widespread uncertainty about the status of the taxa in question. In part this can be a consequence of inconsistency of taxonomic criteria. The subgenera of the red alga *Porphyra* found on the coasts of Japan have been distinguished on features of thallus anatomy and chloroplast morphology (Kurogi, 1972). Criteria such as these are widely used by brown algal taxonomists to diagnose full genera. By changing the criteria. taxonomists may also produce re-alignments of taxa, as O'Kelly & Yarish (1981) have shown for the green algal *Entocladia*. Similarly, Garbary (1978) has drawn a variety of phylogenetic diagrams involving the family Acrochaetiaceae (red algae) showing different relationships resulting from the application of different criteria. By giving greater emphasis to ultrastructural attributes, Garbary, Hansen & Scagel (1980) have materially altered the taxonomy of the Bangiophyceae (red algae).

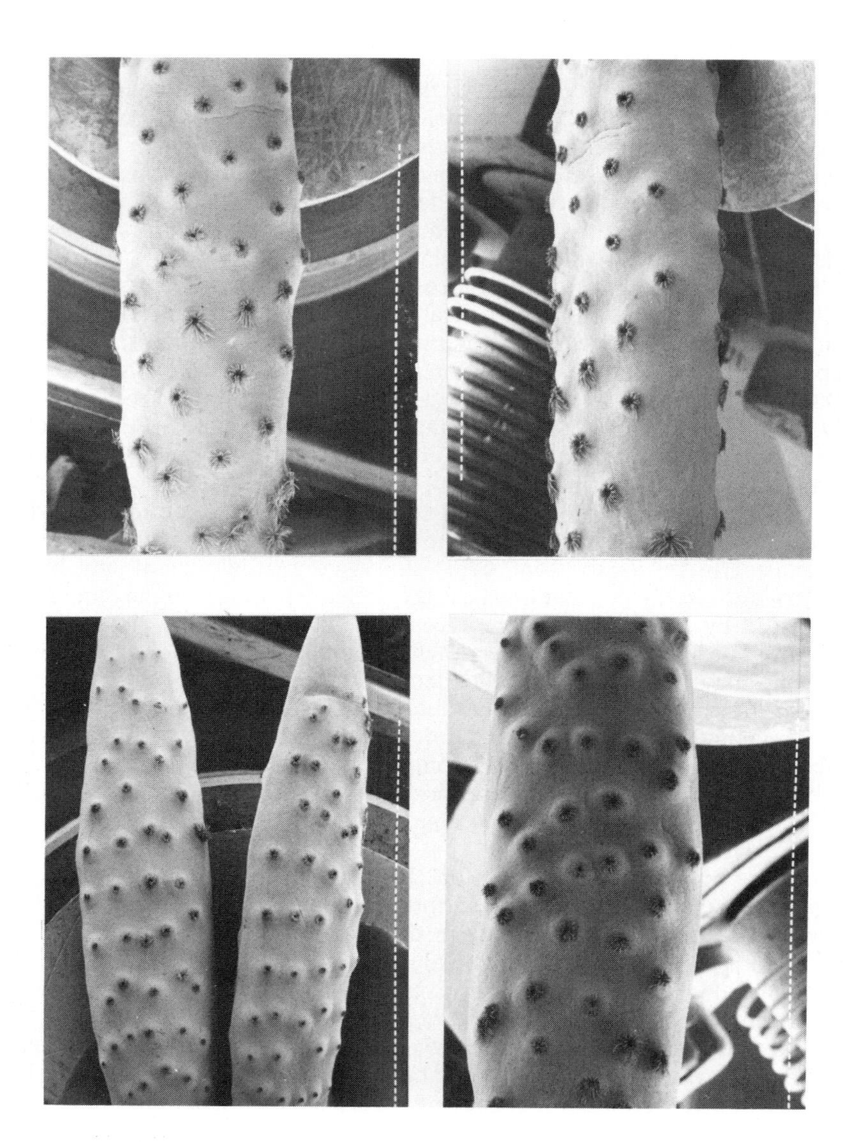

Fig. 4.—Scanning electron microscope preparatons of young receptacles of *Fucus vesiculosus* from Scarlett Point, Isle of Man: top left, face view of receptacle showing helical arrangement of conceptacle ostioles (scale bar, 91 μm); to right, side view of receptacle showing helical pattern (scale bar, 83 μm); bottom left, bifid receptacle showing synchrony of conceptable initiation (scale bar, 100 μm); bottom right, receptacle with more random arrangement of conceptacles (scale bar, 67 μm).

Such a change in emphasis has resulted in the description of a new class of green algae, the Codiolophyceae, by Kornmann (1973). This taxon was defined by the occurrence of unicellular sporophytes (*Codiolum* phases) in the life histories of its members. This new class brought together genera such as *Ulothrix*, *Monostroma*, *Acrosiphonia*, and *Spongomorpha* which had previously been distributed among different higher taxa.

The question of generic rank has been particularly troublesome among those algae which are structurally simple. Applications of numerical taxonomic methods to the Acrochaetiaceae (red algae) by Garbary (1979b), to the Ectocarpaceae (brown algae) by Russell & Garbary (1978), and to the genus *Chlorodesmis* (green algae) by Ducker, Williams & Lance (1965) have led all these authors to the conclusion that a smaller number of genera should be recognized than was at the time of publication generally accepted. Such conclusions, it could be argued, are artifacts of numerical taxonomy but, from the evidence provided by the brown algal flora of the British Isles at least, there is reason for suspecting that taxonomists have been over-enthusiastic in describing genera. It has been shown by Williams (1964) that the numbers of genera containing 1, 2, 3, *etc.* species have a frequency distribution which usually follows the line of a logarithmic series. Such series are commonplace in the natural world, *e.g.* in numbers of individual organisms obtained in samples. When the species composition of the genera of British brown algae are set out in this way (Fig. 5) and the best-fitting logarithmic series superimposed, then it is clear that there is an excess of genera with single species. It is equally evident that there is a general and serious short-fall in multi-species genera. Not all of the single-species genera on the British list are truly monotypic, but a sufficiently large number of them are so to justify some disquiet over this very peculiar generic organization. As Mathieson, Norton & Neushul (1981) have stated, all is not well with algal taxonomy.

In the red algae, the existence of morphologically intermediate taxa, or those with certain important attributes, has been used to illuminate relationships within and between higher taxa (see Womersley, 1965; Kraft, 1975, 1977; Sears & Brawley, 1982). In the brown algal order Laminariales, it has been shown by Sanbonsuga & Neushul (1978) that *Macrocystis angustifolia* and *Pelagophycus porra* could be successfully crossed giving rise to viable hybrid sporopytes. These authors were also able to hybridize *Macrocystis angustifolia* with *Nereocystis leutkeana*, again resulting in viable sporophytes. Thus, comparative-morphological and sexual-compatibility studies both lead to the same suspicion that macro-evolution is not fundamentally different in character from speciation or from micro-evolution, as has been argued also by Greenwood (1979) and Charlesworth (1982).

LIFE HISTORIES

Marine algal life histories have been described in an excellent general text by Hoek (1978), and more detailed reviews are available; see West & Hommersand (1981) for the red algae, Wynne & Loiseaux (1976) and Pedersen (1981) for the brown algae, and Tanner (1981) for the green algae. The sequences of morphological phases and reproductive events which these and

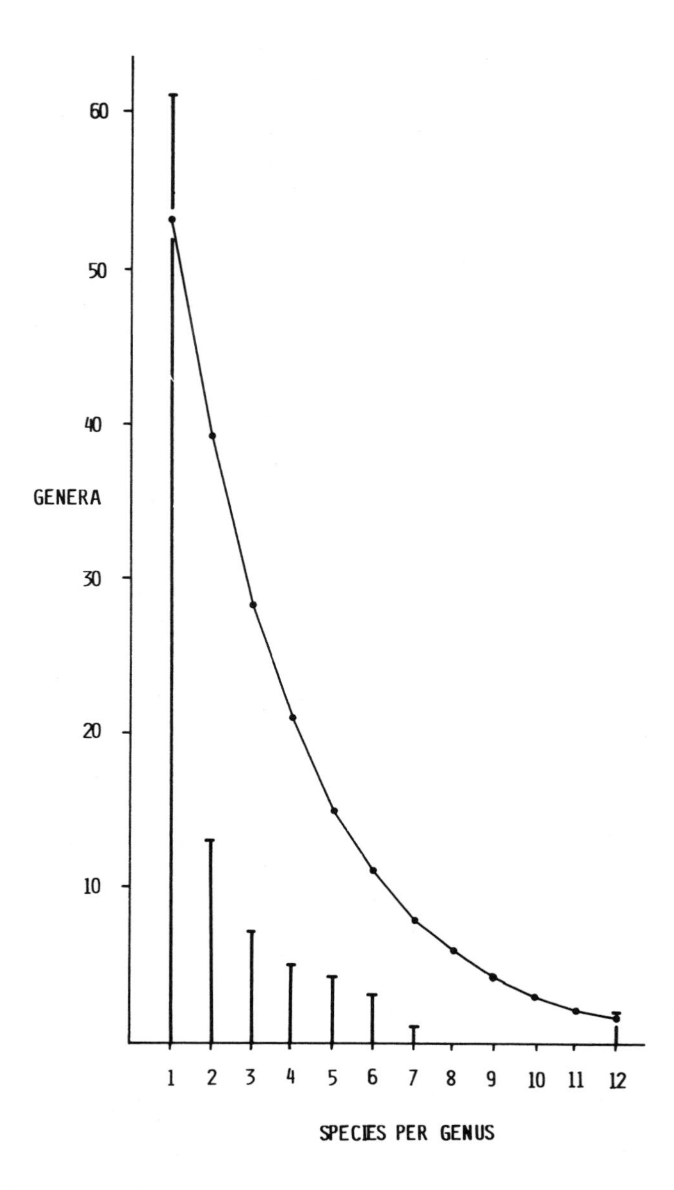

Fig. 5.—Bar diagram numbers of genera of brown algae containing 1, 2, 3, . . . *etc*. species, as reported for the British Isles by Parke & Dixon (1976): superimposed on bar diagram is line of best logarithmic fit calculated according to Williams (1964); note excess of genera containing only one species and short-fall in multispecies genera.

other authors describe, are usually summarized in the form of flow diagrams. Emphasis is given to the number of different thallus forms and to their description rather than to a consideration of their respective functional characteristics. Emphasis is also given to the different kinds of reproductive bodies rather than to their rôle in the maintenance of the species or population. Such preoccupations, together with the enormous variety of recorded life histories, give these accounts a somewhat abstract quality. It is not flow diagrams which evolve or which experience selection pressure but particular phenotypes. Nevertheless, features of algal histories have long been regarded by a number of authors as important in reconstructing phylogenies (Feldmann, 1952): see also Kylin (1937) and Wynne & Loiseaux (1976) for life-history phylogenies of the brown algae, and Magne (1967a, 1972, 1982) for the red algae.

Hoek, Cortel-Breeman, Rietama & Wanders (1972) have commented that authors' perceptions of life-history events are often influenced by pre-existent models. Human factors of this kind are probably unavoidable but they have not halted the flow of new life histories reported for all major taxa. For example, the life history of the freshwater red alga *Lemanea*, in which the site of meiosis is the thallus apical cell (Magne, 1967b) defied all pre-existent models; a similar phenomenon has since been reported for another freshwater red alga, *Batrachospermum* by Stosch & Theil (1979). The life history of *Palmaria palmata* (red alga) described by Meer & Todd (1980) was also novel in that female gametophytes were observed to be minute and to attain sexual maturity when a few days old, whereas the macroscopic male gametophytes required a 9–12 month period of growth before reaching maturity. This life history was also unusual, although not unprecedented, in that the diploid tetrasporophyte generation developed directly from the fertilized female gametophyte. Hsiao & Druehl (1971) have similarly observed male gametophytes of *Laminaria saccharina* to attain sexual maturity 6–9 days earlier than females. Perhaps the most important evolutionary aspect here is the difference in maturation rates of the respective sexes, which will undoubtedly serve to minimize in-breeding. The incidence of in-breeding might also be reduced if sexual thalli differ in longevity, as seems to be the case in the red alga *Constantinea rosa-marina* (Lindstrom, 1981).

The life histories of kelp species were probably regarded as uniform and stereotype, involving only diploid sporophytes bearing unilocular meiosporangia and haploid microscopic gametophytes, until Nakahara & Nakamura (1973) published their observations on *Alaria crassifolia*. In this species, unfertilized eggs and gameotophytes proved capable of apogamous development into haploid sporophytes. Vegetative cells of diploid sporophytes could, by apospory, develop into diploid gametophytes which were monoecious. Sexual reproduction by these resulted in the formation of tetraploid sporophytes. Thus, in the case of *A. crassifolia*, the pre-existent model of the kelp life history failed to accommodate the range of events observed by Nakahara & Nakamura.

Environmental control over the events in life histories is a long-established fact and the very great importance of day-length and temperature has been discussed by Lüning (1980a). In certain species, photoperiodic effects are evidently paramount and these have been very comprehensively reviewed by

Dring (1984). In other species temperature appears to have a greater power to modify the life history, although day-length may not be entirely ineffective. The brown alga *Sphacelaria furcigera* appears to be in this category, the formation of unilocular sporangia (the site of meiosis), being triggered by low temperatures and (possibly) short days (Hoek & Flinterman, 1968; Coliijn & Hoek, 1971). These observations resemble those made by Müller (1962) on *Ectocarpus siliculosus*, another brown alga. Müller reported exclusive production of unilocular sporangia at 13 °C and of plurilocular sporangia at 19 °C while at 16 °C both types were formed. Müller also noted that at 13 ° and 19 °C the type of sporangium produced was unaffected by other environmental factors. At 16 °C, however, the relative numbers of sporangia could be altered by manipulating irradiance, day-length and salinity. Subsequently, Müller (1972) has demonstrated the occurrence of polyploidy in this species, that sex is determined genotypically and that syngamy is accomplished by the production of a pheromone by the female gamete. Earlier reports that zoospores from unilocular sporangia could function as gametes were shown to be unfounded by Müller (1975). The very considerable importance of temperature as a life-history controlling factor in a third brown alga, *Desmotrichum undulatum* has also been demonstrated by Rietema & Hoek (1981).

The three common European species of *Laminaria* are ecologically somewhat isolated (see p. 329). Their gametophytes have recently been shown by Lüning (1980b) to differ in tolerance to high temperatures, to white and to UV irradiance. The most tolerant gametophytes proved to be those of *L. saccharina* which may sometimes occur at high-shore levels. It is perhaps reasonable to point out that any ecological advantage gained from tolerances such as these is, however, liable to be obviated by effects of other factors *e.g.* desiccation). In the red alga, *Rhodochorton purpureum*, the formation of tetrasporangia and hence the occurrence of meiosis, is photoperiodically controlled (Dring & West, 1983), but there is also considerable variation in the critical day-length for sporangium initiation. This variation is linked with geographical differences, the critical day-length being usually greater in culture strains obtained from higher latitudes. Geographical variation in response to light and temperature has been reported also from the red alga *Polysiphonia urceolata* by Kapraun (1978a) who has identified these variant strains as ecotypes or, possibly, "sibling species".

In some polyploid red algae, escape from sterility has been achieved by the evolution of special means of asexual reproduction. In *Plumaria elegans*, triploid plants have been shown to reproduce over several generations in culture by means of paraspores (Rueness, 1968). Rueness's view that these triploids have a wider and more northerly distribution than haploid and diploid plants has been confirmed by Whittick (1977). The inevitable reproductive isolation of parasporic individuals such as these has been found also to be associated with ecological isolation in a population of *Callithamnion hookeri* from the Swedish west coast (Rueness & Rueness, 1978).

Variation in life history below the level of species is now known to be widespread and, in many cases, the existence of a geographical or ecological component to the variation is evident. A very good example of this is to be found among the voluminous literature on the brown alga *Scytosiphon*. Kristiansen & Pedersen (1979) investigated 48 strains of *S. lomentaria* from the shores of Denmark and, while they found these to have a life history involving crustose or other microthalli, the alternation between these and the macrothallus was always asexual. Pedersen (1980) has studied both the complanate and tubular forms of *Scytosiphon* from Greenland and likewise reported these to have only asexual reproduction. Because of the stable morphological differences, Pedersen (1980) proposed raising the complanate form to species rank.

Clayton (1980, 1981a,b) has worked on Australian populations of complanate *Scytosiphon* and has rejected Pedersen's decision to raise this form to the rank of species. More important from an evolutionary angle, she has demonstrated the existence of sexual reproduction in the Australian material. She examined 13 culture strains and found these to possess five different life-history patterns. She also observed a great deal of asexual reproduction by gametophytes, indeed sexual reproduction was rather rare in the species. Clayton, however, concluded that copious asexual reproduction served to maintain a large standing crop of gametophytes and hence to improve the chances of fertilization taking place. It is likely that the amount of genetic variation will greatly increase when there is even a little sexual reproduction. Variation observed by Kristiansen & Pedersen (1979) and Pedersen (1980) in their northern European populations may have its origins in mutation. The existence of rare sexual reproduction in *Colpomenia peregrina* from Australia has also been demonstrated by Clayton (1979).

Clayton's observation of differences in life-history pattern within a single species has its parallels in the green algae. Tatewaki & Iima (1984) have described seven different life-history patterns in *Blidingia minima* in populations from Hokkaido, Japan, which they group into four main types. This species, formerly regarded as completely asexual, was found to have a sexual capability and so to have the potential for greatly increased variation arising from genetic recombination. It is not clear what, if any, isolating factors separate these populations but since all were sampled from the same region of Japan, they are presumably ecological rather than geographical.

A very marked geographical factor distinguishes life histories in the green alga *Bryopsis*, as observations by Bartlett & South (1973), Diaz-Piferrer & Burrows (1974), Rietema (1975), and Jonsson (1980) on cultured isolates from a variety of north Atlantic localities have shown (see Tanner, 1981). *Derbesia marina*, another siphonaceous green alga, has been reported by Sears & Wilce (1970) to exist in two ecologically and reproductively isolated populations on the New England coast. The deep-water population, which has a completely asexual life history is considered by Sears & Wilce to be in the process of transspecific evolution.

The notoriously small size of marine algal chromosomes continues to present problems in the elucidation of life histories and of relationships. For example, the brown algal epiphyte *Elachista fucicola* has been observed in culture by Blackler & Katpitia (1963) and by Koeman & Cortel-Breeman (1976). The sequence of phases and the observed patterns of reproduction were very similar, both accounts finding only asexual reproduction. The chromosome number reported by the former authors is, however, 9-10 while 20-24 are reported by the latter. It is impossible to know with any certainty whether these differences represent evolutionary differences in ploidy, or are simply evidence of technical difficulties. A similar problem exists in the green alga *Acrosiphonia arcta*. According to Jonsson (1964) syngamy in this species need not be followed by caryogamy. Only true diploid zygotes give rise to unicellular sporophytes; the other cells resulting from syngamy retain their parental nuclei as separate entities, and the thallus that develops from these is, in effect, a heterokaryon This observation has been rejected by Kornmann (1970b) and re-affirmed by Jonsson (1971). These differences in opinion have evolutionary implications because if syngamy is not followed by nuclear fusion, then meiosis will not take place and neither, in turn, will genetic recombination. Kornmann (1970b) also describes interspecific differences in life history in *Acrosiphonia*. These involve transition from heteromorphic to isomorphic and eventually to entirely asexual species.

Members of the brown algal order Fucales have relatively simple life histories, and most of these have been known for many years. Interpretation of the events in the fucoid life history also remained uncontroversial for many years until an old argument was revived by Caplin (1968). Caplin contended that the fucoid thallus, being diploid, was essentially a sporophyte and that the cells in which meiosis took place, as a prelude to gametogenesis, were homologous with unilocular sporangia. It followed, therefore, that the gametangia represent reduced gametophytes. Caplin's interpretation has been accepted by Jensen (1974) and by Clayton (1984) although rejected by Chadefaud (1980) who takes the more traditional line that the fucoid thallus is simply a diploid gametophyte thallus. Cytological evidence in support of the Caplin position has been obtained by Setzer (1980) who describes a cellular organization of the female structures ("oogonia") in *Cystoseira*, an observation which is consistent with the presence of a reduced gametophyte. This argument, although plainly far removed from questions of selection pressure and other agencies of evolutionary change, is important in phylogenetic reconstruction because if two generations are held to be present, then the fucoid life history has greater similarity to that of other brown algae, especially the kelps, than if only one generation is involved. This argument also points to the inherent weaknesses of the various systems for life-history classification. For example, according to Drew (1955), the fucoid life history is monomorphic and diplontic but, if the Caplin interpretation is adopted then the same sequence should be described as dimorphic and diplohaplontic.

Fletcher (1980) and Norton (1981) have reported the interesting fact that in *Sargassum muticum*, fertilized eggs remain attached to the parent thallus for a brief period of gestation before being released. In this species, zygole (or germling) release follows a lunar pattern with maximum release taking place shortly after the largest tide of each set of spring tides. Lunar periodicity in reproduction has also been reported in *Dictyota* by Müller (1962) and in the green alga *Enteromorpha intestinalis* by Christie & Evans (1962).

The life history of *Fucus* although simple and cyclical in character is nevertheless expressed differently in populations living under different conditions. For example, receptacle initiation in a Baltic population of

F. vesiculosus has been reported by Russell (1985b) to occur under a different set of temperature and day-length conditions from populations on British shores. Receptacle growth in both populations was, however, strongly influenced by temperature. In *F. distichus*, receptacle initiation seems to be a short-day response and is unaffected by either total irradiance or temperature (Bird & McLachlan, 1976).

Life-history flow diagrams seldom indicate the occurrence of vegetative reproduction, except perhaps for those cases such as the brown alga *Sphacelaria*, in which special organs of vegetative reproduction (propagules) have evolved. Dixon (1965) has pointed out that reproduction by vegetative means can be responsible for very considerable differences between the "theoretical" life history and the sequence of events that actually obtains. Cheney & Mathieson (1978) similarly attach considerable ecological and evolutionary significance to vegetative reproduction in seaweeds.

Vegetative reproduction may be accomplished in diverse ways. Bonneau (1978) has shown that vegetative cells sloughed from senescent thalli of *Ulva lactuca* behave like settled swarmers and can easily give rise to new thalli. This behaviour he considered to be of adaptive value to *Ulva* living in stressed environments. In the brown alga *Elachista stellaris*, filaments may function as stolons, with new thalli being initiated at the tips (Wanders, Hoek & Schillern-van nes, 1972). This method is, however, not likely to match the efficiency with which tropical green algae such as *Caulerpa* can colonize new substrata by rhizomatous growth. In practice, detached pieces of algal thallus can have very considerable powers of secondary attachment, as Fletcher (1977) has demonstrated; and, in the case of *Laminaria sinclairii*, reproduction by means of detached portions of hapteron seems to be particularly important. Markham (1968) observed that *L. sinclairii* sporophytes are able to regenerate new fronds by proliferation from the holdfast and that dispersal of the species can be by means of detached pieces of hapteron. Markham concludes that this method of vegetative reproduction has largely replaced sexual reproduction in a population subject to scouring and burial by sand. Members of the order Fucales have frequently been found as detached and vegetatively-reproducing populations (see Fritsch, 1945), but McLachlan & Chen (1972) have shown this propensity for vegetative proliferation to be possessed even by very early stages of development.

Two recent cases of ecologically isolated and vegetatively reproducing populations may be mentioned. In North Carolina, the brown alga *Acinetospora crinita* does not produce sporangia, and cultured isolates remain sterile under a range of conditions (Amsler, 1984). This plant, therefore, depends entirely upon vegetative growth of detached fragments for its continued survival, and its sterility appears to have become a stable inherent characteristic. A similar absence of sporangia has been reported in the unattached population of the brown alga *Pilayella littoralis* in Nahant Bay, Massachusetts, by Wilce, Schneider, Quinlan & Bosch (1982). This population has a characteristic ball form and it reproduces by fragmentation. Although known to reproduce solely by this means, the Nahant Bay *Pilayella* is extremely productive and may occur in very great abundance. The efficiency of vegetative reproduction in this population is due mainly to the presence of a fungal parasite, *Eurychasma dicksonii* which kills *Pila-*

yella cells and thus causes the pieces of filament on either side of the infection to separate. An element of coevolution in this curious partnership would seem possible.

The green alga *Prasiola stipitata*, which lives normally at high levels on rocky shores, reproduces commonly and perhaps mainly by means of walled cells (akinetes). These structures are likely to tolerate the very considerable fluctuations in physical conditions that obtain at such levels. *P. stipitata*, however, also reproduces sexually, and the sexual thalli are found at lower-shore levels than the vegetatively-reproducing plants Friedmann (1959). The existence of sexual reproduction within an otherwise asexual population ought to permit greater genetic variability but an isolated and uniform population may nevertheless remain fairly pure through inbreeding. Enzyme electrophoresis provides a means of revealing the extent of variability of populations and it has recently been used by Innes (1984) and Innes & Yarish (1984) for that purpose. Innes & Yarish sampled over 1000 *Enteromorpha linza* plants from four localities in Long Island Sound and recorded the presence of a variety of enzyme phenotypes. The distribution of variation departed significantly from that predicted by the Hardy–Weinberg equilibrium, and in a way which led the authors to conclude that the population reproduced asexually. This hypothesis was then tested by means of culture experiments and found to be true. It should be borne in mind, however, that the Hardy–Weinberg equilibrium is a rather abstract concept because the conditions necessary for it to be upheld include the occurrence of random mating, which may not always apply in *Enteromorpha* (see de Silva & Burrows, 1973, and p. 326), and the absence of natural selection, which scarcely ever applies. Enzyme analyses such as this are nevertheless of enormous value in demonstrating the scale of variation available to the forces of selection.

Evidence that these forces have a powerful effect on heteromorphic algal life histories has come about through a series of careful field experiments conducted on various rocky shores of the United States of America. The first of these were carried out by Lubchenco & Cubit (1980) who observed that the appearance of the erect macrothallus of several species (including *Scytosiphon)* in nature corresponded with the period of maximum inactivity among herbivores (winter and early spring). The crustose alternative phase, on the other hand, persisted throughout summer when grazing pressure was intense. By the simple experimental procedure of excluding herbivores, Lubchenco & Cubit were able to demonstrate the growth of erect macrothalli in summer, the "wrong" season. They concluded that the upright and crustose thalli represent mutually exclusive adaptations to fluctuations in grazing pressure, the upright phase being adapted for high rates of growth and reproduction, and the crusts for high grazing pressure. In environments where grazing pressure stays constant, Lubchenco & Cubit predicted that isomorphic species would predominate; a prediction that has yet to be tested.

The publication of Lubchenco & Cubit's observations coincided with that of another set of very similar results and conclusions. Slocum (1980) investigated the red alga *Gigartina papillata*, the life history of which involves alternation between an upright blade form and a crustose thallus. These phenotypes were observed by Slocum to occupy similar shore levels but to differ significantly with respect to grazing pressure. Herbivore exclusion experiments resulted in a decline in the number of crustose plants, due to their being overgrown by other algae. In contrast, the numbers of blade forms increased. Thus, the crust is persistent in the presence of grazing but is susceptible to overgrowth, while the more productive blade is more susceptible to grazing.

Scytosiphon and *Gigartina* are, respectively, brown and red algae; they are therefore likely to possess few homologous attributes, and those possessed are certain to be primitive. That they have evolved the same, or similar, solutions to the problem of herbivory also provides us with another example of analogous similarity. The production of two different phenotypes in each of these species also identifies the selection pressure as disruptive selection (see p. 315).

Scytosiphon has been the subject of further experiments. Dethier (1981) noted the presence of upright and crustose phases in the same rock-pools and the occurrence of some overlap in their respective seasonal distributions. Experimental manipulation of herbivores demonstrated that they do not graze the crustose phase, and are therefore not responsible for its seasonal fluctuations, but that they are crucial to its persistence because they eliminate potential competitors. Littler & Littler (1983a) also measured the effects of herbivory (by sea urchins) on *Scytosiphon*. After a 48-h period of grazing, the loss of standing crop of complanate *Scytosiphon* amounted to 82·7%, that of the cylindrical form was 81·4% while the crustose phenotype lost only 16·2%. Littler & Littler concluded that opposing (*i.e.* disruptive) selective factors have resulted in the evolutionary divergences observed in alga with heteromorphic life histories. They also argued that other workers may have over-emphasized the selective rôle of grazing because the crustose form is equally well adapted to withstand physical stresses.

The effects of disruptive selection may also be identified in a single algal thallus. The calcified red alga *Corallina officinalis* has a branched upright frond borne from an expanded crustose base. Littler & Kauker (1984) have recently demonstrated that the frond is more productive than the crust, but less heavily calcified and more susceptible to sea-urchin grazing. The crust therefore maintains possession of primary space (substratum) and, through growth, is the means of the plant acquiring additional space. Thus, the two thallus components, although united in the same individual, manifest the same kind of divergent effects of disruptive selection as seen in heteromorphic species.

An analogous example of structural variation within a thallus has been described for the brown alga *Durvillea potatorum* by Cheshire & Hallam (1985). In this species the alginates of the lamina and stipe are rich in mannuronic acid whereas those of the holdfast are rich in guluronic acid. The authors propose that this variation in alginate composition is related to the functions of the tissues in providing flexibility (lamina and stipe) and rigidity (holdfast).

REPRODUCTION

The ecological and evolutionary functions of sexual and asexual reproduction are somewhat different (Clayton, 1982). Asexual reproduction is an economic means of population increase but one which offers little scope for genetic diversification; sexual reproduction has the potential for greatly enhanced variability, but is liable to combine increased expenditure of resources with a greater risk of reproductive failure. In practice, seaweeds usually indulge in both methods. Many species of green and brown algae whose means of sexual reproduction are motile isogametes may be able to effect these methods simultaneously in that unmated gametes may be able to develop independently into functional thalli. This auxiliary asexual capability takes an unusual form in Australian populations of the brown alga *Scytosiphon*. Gametophytes of this taxon reach reproductive maturity at an age of 8–10 weeks but the initial zoids are all asexual; sexual competence does not develop until the plants are about 14 weeks old (Clayton, 1981a).

Structurally simple brown algae such as *Ectocarpus* were once thought to possess relative sexuality, a condition characterized by certain sexually ambiguous gametes which were held to function either as male or female when in the presence of other gametes with more pronounced mating qualities. Relative sexuality, in *Ectocarpus* at least, has now been found by Müller (1976a) not to exist. The sexuality of *Ectocarpus* gametes is genetically determined and unambiguously bipolar.

In brown algae, the risk of mating failure may have been reduced by the evolution of sexual hormones. These pheromones are complex olefinic hydrocarbons, and are volatile with a characteristic scent. They are produced by female gametophytes and they stimulate discharge of male gametes as well as their attraction to the female cells. The rôle of these pheromones has been very thoroughly investigated by Müller (1981) who has written a good review of their various characteristics and functions, Pheromones seem to be present in simpler ectocarpoid algae with isogamous reproduction (Müller, 1980) as well as in the more complex oogamous taxa such as *Desmarestia* (Müller & Lüthe, 1981).

Of particular interest is the fact that these pheromones are evidently not species specific in their rôle as male attractants. The pheromones of *E. siliculosus* and *E. draparnaldioides* appear to be identical chemically and to bring about mutual attraction of gametes, but syngamy does not follow (Müller, 1980). A similarly broad taxonomic expression of pheromone activity has been demonstrated by Lüning & Müller (1978) who cultured gametophytes of six kelp species belonging to the genera *Alaria*, *Laminaria*, *Macrocystis*, and *Pterygophora*. The sea water in which the female gametophytes of each species had matured caused the discharge and attraction of sperm in all 36 combinations of male and female gametophytes. This demonstration of mutual attraction among geographically isolated members of the same order is of undoubted phylogenetic interest but the immediate evolutionary advantage of a non-specific pheromone is not altogether clear. Perhaps the pheromone serves to attract a large body of male gametes of different taxa but syngamy occurs only when chemical recognition by members of the same species takes place. In this way, the probability of mating success may be increased but the non-specificity of the pheromone action must do little to reduce the expenditure of resources in the form of 'wasted' gametes.

Biochemical non-specificity has been reported in several other algal characteristics. Spencer, Yu, West & Glazer (1981) have shown that the phycoerythrins of Norwegian and North American isolates of *Callithamnion byssoides* are chemically indistinguishable, in spite of the fact that the isolates are sexually incompatible. It has been concluded that phycoerythrins are of only limited value in discriminating genera and species (Glazer, West & Chan, 1982). In similar vein, Marsden, Callow & Evans (1984) have experienced difficulties in drawing clear taxonomic distinctions between brown algal taxa using enzyme electrophoretic methods. Refinement of analytical techniques may, however, bring about improved discrimination of taxa.

The development of Darwin's theory of evolution was greatly influenced by the ideas of Malthus on human population growth. It can scarcely be doubted that Darwin would have been greatly impressed had he been aware of the huge disparity between zoospore production and sporophyte numbers in a stand of *Laminaria*. It has been estimated that, over a three-month reproductive period, *L. saccharina* will produce 64 million zoospores from 1 cm^2 of blade surface (Parke, 1948). In a forest of *L. hyberborea* such as is obtained at Port Erin, Isle of Man, 3·3 million zoopores are produced for every mm^2 of rock surface (Kain, 1975b).

The fate of *Laminaria* zoospores has been the subject of a detailed study by Chapman (1984), who worked on *L. longicruris* and *L. digitata* on the southwestern coast of Nova Scotia. Sporophytes of *L. longicruris* occurred at a density of 1·2 plants$\cdot m^{-2}$ and from these, zoospore production was $8 \cdot 9 \times 10^9 \cdot m^{-2} \cdot yr^{-1}$. The number of microbenthic stages was reduced to $8 \cdot 89 \times 10^6 \cdot m^{-2} \cdot yr^{-1}$ but new macrothalli numbered only $1 \cdot m^{-2} \cdot yr^{-1}$. The density of *L. digitata* sporophytes was 3·2 plants$\cdot m^{-2}$ and zoospore production $20 \cdot 02 \times 10^9 \cdot m^{-2} \cdot yr^{-1}$. Microbenthic stages numbered $0 \cdot 98 \times 10^6 \cdot m^{-2} \cdot yr^{-1}$ and these were reduced to 2 macrothalli$\cdot m^{-2} \cdot yr^{-1}$. Chapman's observations thus confirm those of Kain (1975b) and Parke (1948) that maintenance of a *Laminaria* forest involves vast losses of zoospores and microscopic stages.

Chapman (1984) observed spore settlement on artificial substrata, a technique also used by Koetzner & Wood (1970) to demonstrate seasonal variation in algal colonization on the shore of Long Island, New York. There is scope for further application of this method to elucidate the very uncertain period between spore production and the advent of the next cohort of macrothalli. Equal or possibly greater uncertainty surrounds the behaviour of released spores, and the factors which control their movement and attachment. Study of spore behaviour by Suto (1950) led him to define seven ecological spore-groups based on criteria such as positive and negative phototaxis, but the practicality of his classification is not at all clear.

It is now evident that considerable variation in seaweed reproductive behaviour may be present within species. The crustose red alga *Lithophyllum incrustans* is common on rocky shores throughout the British Isles and its reproductive effort has been measured by Ford, Hardy & Edyvean (1983) and Edyvean & Ford (1984a). They have observed that the proportion of gametophytes is very small in all populations, but that gametophytes commit rather more of their thalli to reproduction than do the asexual thalli. The incidence of sexual plants seems to vary with latitude—16% of all plants examined on the southern coast of England were gametophytes, but this proportion fell to 4% on the Northumberland coast, while in northern Scotland, the entire population proved to be asexual. The overall preponderance of asexual plants led Ford *et al.* (1983) to conclude that most sporangia were apomeiotic. *L. incrustans* thalli may be aged by counting the numbers of layers of conceptacles present, and by this means it is possible to quantify the age structure of a population. Examination of the age structures of populations revealed some to have greatest numbers of individuals in the first year class and fewer in each succeeding year class (Ford *et al.*, 1983; Edyvean & Ford, 1984). An age distribution of this type is consistent with the foregoing observations by Chapman (1984) on *Laminaria*. Variation in age structure was also found to occur both on geographical and local scales. Unstable age structures were common in rock-pools of low volume and high surface area, as were plants with low reproductive effort (Edyvean & Ford, 1984b). Such variation is more likely to be due to extrinsic environmental factors than to inherent differences. In perennial algae such as *Lithophyllum*, reproductive failure in a single season need not have serious long-term repercussions if the thallus remains viable, as seems to be the case with the brown alga *Bifurcaria* near its northern geographic limit (de Valera, 1962).

Differences between populations in reproductive effort may also be observed in *Fucus*. Russell (1979) has reported that populations of *F. vesiculosus* which experience particularly heavy losses of individuals through the effects of intense grazing and/or wave action allocate more thallus biomass to reproductive tissue than do plants from sheltered and relatively ungrazed habitats. Sheltered-shore *F. vesiculosus* is, on British coasts, liable to experience strong competition for light and primary space from the larger fucoid *Ascophyllum nodosum*. In these circumstances, selection will operated in favour of *Fucus* plants with a heavy commitment to vegetative growth. Differences in reproductive effort among *Ascophyllum* populations are, however, largely plastic (Cousens, 1986).

The reproduction of fucoid algae on British shores has also been the subject of an investigation by Vernet & Harper (1980). These authors observed that in *F. spiralis* (a hermaphrodite and upper-shore species) the biomass allocated to sperm is extremely low in relation to that of eggs. Also, *F. spiralis* sperm proved to be smaller than those of dioecious *Fucus* species. Vernet & Harper point out that in higher plants it is common for inbreeders to allocate less resource to male reproductive activities, and they conclude that inbreeding is common among hermaphrodite upper-shore fucoids. Inbreeding in *F. spiralis* has since been confirmed by Müller (1985). The outcome of this inbreeding is likely to be "congealed" genotypes whose evolution has been dominated by repeated exposure to the same physical demands of life at high-shore levels. Lower shores are more stable physically but are also floristically and faunistically richer. Dioecy in fucoids at this level is considered by Vernet & Harper to reflect the advantages of "uncongealed" genotypes in environments dominated by "evolving biological interaction". Not all reproductive traits, however, proved as amenable to ecological or evolutionary interpretation; egg sizes were quite unrelated to zonal levels, for example.

Sampling of dioecious *Fucus* species by Vernet & Harper (1980) sometimes revealed departures from the expected 1:1 ratio of male and female plants. Ecological differentiation of this kind may be present in other species. In the red alga *Chondrus crispus* on the shores of Nova Scotia, tetrasporophytes were found by Craigie & Pringle (1978) to be consistently more abundant at the seaward ends of transects. Similarly, Norall, *et al.*(1981) report higher frequencies of tetrasporophytes of *Ptilota serrata* in deeper waters, and of cystocarpic plants in shallow populations. An excess of tetrasporophytes may occur in certain populations of red algae and while this may be a consequence of failure in meiosis (see Ford *et al.* 1983) it may also arise if tetrasporophytes have superior properties of perennation or if carpospores have superior survival characteristics to tetraspores (Kain, 1982). The former explanation has been found to be correct for the red alga *Gracilaria* in Nova Scotia where tetrasporophytes have greater longevity than gametophytes (Bird *et al.*, 1977). Asexual and sexual reproductive effort in six sympatric tropical *Gracilaria* species is adaptive to seasonal conditions (Hay & Norris, 1984).

The ecological and evolutionary importance of reproductive phenology may also be seen in *Laminaria hyperborea*. Kain (1975b) has shown that this species on the Isle of Man produces most zoospores in early Janaury. Kain points out the advantage to the species in reproducing at a time when the area of denuded rock surface available for colonization is at its greatest and recolonization by competing algal species is low. The existence of both spatial and temporal niche differences in sporophytes and gametophytes has been detected by Liddle (1971) in the brown alga *Padina sanctaecrucis*.

Changes through time in the numbers of plants present in a particular seaweed population are the products of the rates of reproduction (births) and of death. They are also affected by rates of emigration from the population and of immigration from other populations. Demographic studies of seaweed populations are still fairly rare although some very thorough investigations have been made. The brown alga *Pelvetia fastigiata* forms small and relatively stable communities on the shores of southern California. Gunnill (1980a) has shown that the survival of recruits to these populations is similar regardless of season or year of recruitment. Their mortality rates decrease with age so that the mean life expectancy increases throughout the first year. Losses of juveniles were found to be high and only 9% survived to reproductive maturity ($\approx 1 \cdot 5$ yr). Gunnill also noted that individuals within aggregations of large plants lived longer than very dispersed individuals, plant aggregations remaining intact throughout the period of study. *Cystoseira osmundacea*, another Californian fucoid, has been the subject of demographic study by Schiel (1985b). In this species too, losses among juveniles is extremely heavy but older thalli have much greater life expectancy.

Leathesia difformis is a very different kind of brown alga in that it is ephemeral, the life span of plants on the coast of Nova Scotia being only about 12 weeks from mid-June until early September. Nevertheless, it may occur in considerable abundance and observations on its demography during this period have been carried out by Chapman & Goudey (1983). These authors observed that the rate of mortality of *Leathesia* increased as crowding increased, the plants detaching one another, in effect by mutual jostling. Experimental thinning greatly reduced mortality although wave action was an additional source of plant losses.

The effect of a dense aggregation of mature plants upon recruitment of juveniles is a matter of some controversy. Most reports indicate that the presence of an adult canopy has a negative effect upon recruitment. This is certainly the case with *Durvillea antarctica* (Santelices, Castilla, Cancino & Schmiede, 1980), with *Sargassum muticum* (Deysher & Norton, 1982), with *Ecklonia radiata* (Kirkman, 1982), and with *Lessonia nigrescens* (Santelices & Ojeda, 1984a; Santelices *et al.*, 1980). Removal of adult sporophytes of *Laminaria longicruris* and *L. digitata* was, however, found by Chapman (1984) to have no effect on recruitment of new macrothalli. Chapman did, however, observe a ten-fold increase in *Laminaria* recruitment when he removed the turf of red understory plants. The importance of competing species such as these in preventing macrophyte recruitment has been noticed also by Deysher & Norton (1982) and by Santelices & Ojeda (1984a,b).

Changes in recruitment, and in stand sizes, of one species of green algae and five of brown algae have been recorded by Gunnill (1980b) over a four-year period. The observations, made near La Jolla, California, were based upon large (25 m^2) permanent quadrats. Gunnill concluded that the reproductive seasonal variation of all species was "adaptive" in relation to seasonal climatic and hydrographic conditions for the region; he also concluded that fluctuations in population size were due to local variation in these environmental factors.

In higher plants, if a population is sampled through time then the mathematical relationships between the logarithm of plant weight and that of plant density is a line of $-1 \cdot 5$ slope, the so-called $-3/2$ power law. The basis of this law is that production and mortality are density dependent. Schiel & Choat (1980) have measured the total yield, plant lengths, plant dry weights, and the dry weights of reproductive tissues in monospecific stands of several algal macrophytic species in New Zealand. These stands were of different densities, and their observations led Schiel & Choat to the conclusion that all species fared better at high densities, *i.e.* that the $-3/2$ power law did not apply. This research was criticized first by Brawley & Adey (1981a). These authors pointed out that supplies of water and nutrients are obviously density-dependent in terrestrial plant populations, and that these considerations are inapplicable to stands of marine vegetation. While accepting that terrestrial models may not hold true for marine plant communities, Brawley & Adey expressed the view that the effects of density should await additional data obtained from other seaweed stands. This information was duly produced by Cousens & Hutchings (1983). Their observations on a number of different species of algal macrophytes showed that the $-3/2$ law did indeed hold and, consequently, that the Schiel & Choat interpretation was 'incorrect'. Nevertheless, Schiel (1985a) has confirmed that in *Sargassum sinclairii* and *Carophyllum maschalocarpum*, individual plants grew faster and attained larger size in high density stands. Clarification of this point calls for an experimental programme involving thinning to different densities, as indicated by Cousens & Hutchings (1983).

In a recent important paper, Dayton *et al.* (1984) have investigated the structural stability of kelp patches in coastal waters of southern California, combining measures of recruitment with strategic characteristics (see p. 355) of the plants.

STRATEGIES

The terms "adaptation" and "strategy" have been roundly criticized by Harper (1982) for the teleological thinking which seems to govern their use. His strictures are likely to prove ineffectual. Both terms, for all their teleology and vagueness are probably now ineradicable from the language of biology, and certainly both have come into widespread use with respect to marine organisms. As is often the case with related terms, frequency of use has led to blurring of distinctions between the two. "Adaptation" is, however, usually employed fairly narrowly to describe an organism's ability to cope with a particular component of its environment. "Strategy" seems to be used altogether more broadly with the aim of defining the particular ecological niche for which the organism has evolved. It is unfortunate that biology has purloined this particular military expression because, teleology aside, war history shows that for every successful strategic design there has been another of notable fallibility.

Species zonation on rocky shores was, until recently, largely a matter of pattern analysis and descriptive terminology. The entry of strategic considerations into zonation studies may be seen in the view expressed by Connell (1972) that the upper limits of intertidal species are set by their tolerance to physical factors and their lower limits by biological interactions. The importance of physical factors such as desiccation, high temperatures and extreme salinities in regulating the upper limits of benthic algae has since been demonstrated by Schonbeck & Norton (1978), Dring & Brown (1982), and Russell (in press). The second part of Connell's argument has also been confirmed by Schonbeck & Norton (1980) who cleared small areas of intertidal rock and so permitted downward extension of the range of the species from a higher zone. This technique was also adopted by Lubchenco (1980) on New England shores. By selective removal of *Chondrus crispus*, she was able to demonstrate the downward extension of *Fucus vesiculosus*. Thus, competition from *Chondrus* set the lower limit of *Fucus*, the upper limit of *Chondrus* being ascribed to desiccation. Similar methods and similar conclusions have been reached by Santelices, Montalva & Oliger (1981) and Santelices & Ojeda (1984a,b).

Confirmation of this 'explanation' of zonation does not altogether eradicate the suspicion that it may be a little too simple. Any frontier between two populations of seaweeds is likely to involve mutual interactions rather than a unilateral effect. The recent demonstration by Hawkins & Hartnoll (1985) that selective removal of algae from intertidal populations can result in upward movement of species from lower zones is important in that respect. Upper limits may also be influenced by interspecific competition.

Competition between seaweeds for available resources has been observed in mixed cultures by Russell & Fielding (1974). Their experiments involved a green, a brown, and a red alga grown together in pairs and starting with different amount of inocula. Their method was thus a variant on the familiar De Wit replacement series technique. Inhibition of growth of a crustose brown alga in culture by another has been demonstrated by Fletcher (1975), who attributed the antagonistic effect of toxin release. Manipulation of mixed cultures of seaweeds is both difficult and time-consuming and the technique is probably applicable only to the smaller and more amenable thallus forms. Perhaps for these reasons, we have more information on the rôle of competition between seaweeds as a result of field experiments based upon the selective removal of a hypothetical competitive dominant.

This technique was adopted by Dayton (1975) who has devised a classification of seaweeds in which three principal strategies are recognized. These are : (1) canopy species (tall species which monopolize the available light and hence attain competitive dominance), (2) fugitive species (small species with rapid growth and early reproduction which readily exploit any disturbance and which are common as early-succession species), and (3) obligate understory species (those species which are associated with the canopy species and which either die completely or die back to the base when the canopy is removed). There is a circularity in the reasoning behind these definitions which is unsatisfactory: obligate understory species are those which die back after canopy removal, these species have died back, therefore they are obligate understory species.

In practice, however, this kind of distinction would appear to have some merits. The hypothetical dominant investigated by Dayton (1975) was the kelp *Hedophyllum sessile*, and its removal from a number of sites elicited the predicted response. Obligate understory species consistently declined and were replaced by fugitive species. These were in turn eliminated by the return of the competitively superior *Hedophyllum* which also brought about the return of the obligate understory species.

The Dayton classification has since been validated by Ojeda & Santelices (1984) and by Santelices & Ojeda (1984b), who have investigated different kelp communities on the coast of Chile. These authors did, however, observe a rather more complex pattern of species interaction which seems to be related to the presence of a third stratum of vegetation located between the canopy and ground layers.

Foster (1975b) has shown that the presence of a canopy may reduce algal diversity and cover underneath. Conversely, Harkin (1981), by removal of blades of *Laminaria hyperborea*, caused a great increase in biomass of red and brown algae on the remaining stipes. It is not certain if these algae were entirely fugitive species, as would be expected from the Dayton model, or whether some obligate understory species were involved. Both accounts demonstrate the importance of an algal canopy in determining the structure of the plant community.

Marine benthic communities differ from those in terrestrial habitats in that competition for substratum space may occur between plants and sedentary animals. Dayton (1973) demonstrated competition of this kind between the annual kelp *Postelsia palmaeformis* and the mussel *Mytilus californianus*. Dayton showed that the effective spore dispersal distance for *Postelsia* was only about 3 m from an existing clump of plants. He also concluded that *Postelsia* was able to persist in the presence of the competitively superior mussels because they settled on mussel shells and, as they grew in size, caused the animal to be removed by wave action. This particular association has also been studied by Paine (1979) who demonstrated the importance of wave action in maintaining the presence of *Postelsia*. Wave action on exposed sites caused removal of mussels and so released space for *Postelsia* colonization. Paine made the point of distinguishing catastrophes (infrequent and extreme events which destroy large portions of a population) and disasters (disruptive influences of a more local scale and greater relative frequency). Thus, *Polstelsia* survives only on exposed shores because of the high incidence of disasters among the mussel beds.

The relationship between mussels and algae has also been investigated on New England rocky shores by Lubchenco & Menge (1978). In this locality, the red alga *Chondrus crispus* is common only on sheltered shores because abundant mussel predators reduce the numbers of *Mytilus*, the superior competitor. On exposed shores, however, predators are scarce and unable to control the growth of mussels which successfully outcompete *Chondrus*.

Mussels are not only space competitors with algae. As filter-feeders, they may also remove algal propagules from the surrounding water, although the rôle of such feeding mechanisms in determining the structure of benthic vegetation is still not very certain. Herbivores which consume benthic algae, have a more clearly understood and probably more decisive influence on vegetation structure. This influence is exercised in two ways, the removal of algal biomass and the selective consumption of the most palatable taxa. An investigation into the effects of grazing on seaweed zonation was carried out by May, Bennett & Thompson (1970) who systematically removed herbivores from a permanent transect. Herbivore exclusion was found not to cause any species of alga to appear outside the zone in which it normally occurred, although differences in species abundance did arise in response to treatment. Their transect was rather narrow (1 m), and herbivores were removed fortnightly, so their conclusion may not have been completely warranted. It has recently been contradicted by the results of some experiments carried out on the Chilean coast by Moreno & Jaramillo (1983). In this case, herbivore exclusion resulted in serious disruption of the pre-existing zonation pattern.

Kitching & Ebling (1961) removed large numbers of sea urchins from a shallow sublittoral area of Lough Ine, Ireland, and observed a rapid growth of *Enteromorpha* vegetation. Subsequent re-introduction of urchins resulted in depletion of the *Enteromorpha*. A similar preference for *Enteromorpha* and other ephemeral green algae has been detected in the snail *Littorina littorea* from field observations by Lein (1980). This animal and its feeding preferences were studied in the laboratory by Lubchenco (1978). Her experiments demonstrated grazing selectivity by *L. littorea*, the animals preferring ephemeral algae such as *Enteromorpha*, *Ulva*, *Ectocarpus*, *Scytosiphon*, *Ceramium*, and *Porphyra*, and avoiding tougher more cartilagenous algae such as *Fucus*, *Laminaria*, and *Chondrus*. Lubchenco (1978) then applied her knowledge of the feeding habits of *Littorina littorea* to a study of New England rock-pool communities. She observed that high *Littorina* density was associated with seaweed vegetation dominated by *Chondrus*; low snail density was accompanied by a rich growth of *Enteromorpha*. By experimentally manipulating the density of *Littorina littorea* she was able to bring about the predicted changes in vegetation.

Dietary selectivity has been demonstrated in herbivorous fish by Horn, Murray & Edwards (1982) and Sammarco (1983), and in amphipods by Brawley & Adey (1981b). Their findings, based on laboratory experiments, analyses of gut contents and field observations, show that for these animals too, delicate ephemeral algae are eaten in preference to tougher perennial algal forms. It is also evident that a simple hierarchy of algae in terms of dietary preference could, however, be misleading. Some herbivores are evidently specialist feeders, while other may graze selectively on certain mixtures of algae (Kitting, 1980).

The rôle of herbivory as an important agency of natural selection in algae is becoming increasingly clear. Wanders (1977), working on the shallow-reef communities at Curaçao, showed that reef-building coralline algae died when herbivores had been excluded, probably because they had become covered with a dense vegetation of filamentous and fleshy algae. Crustose corallines were only able to settle and grow on substrata which were subject to grazing.

The crustose brown alga *Ralfsia* and its relationship with the territorial limpet *Patella longicosta* have been described by Branch (1981). These limpets maintain algal gardens within which the *Ralfsia* is more productive than in areas outside. Branch attributes this greater productivity to the removal of epiphytes from the *Ralfsia* crust by the grazing activity of the animal. The snail *Littorina littorea* has been found also to promote the development of crustose algae (*Ralfsia* and the uncalcified red alga *Hildenbrandia*) by selectively removing erect algae (Bertness, Yund & Brown, 1983). Observations such as these suggest that crustose algae may not simply be resistant to grazing but, as Slocum (1980) has suggested, dependent on grazing. A recent investigation by Steneck (1982) gives strong support to that point of view. Steneck studied the association between the limpet *Acmaea testudinalis* and the crustose coralline alga *Clathromorphum circumscriptum* and found that their abundances were positively correlated. *Acmaea* was shown to graze *Clathromorphum*, which was an abundant and dependable source of food; it also provided a good attachment surface for the animal. *Clathromorphum* was found to require grazing as the means of removing potentially lethal epiphytes. Its growth zones were protected from grazing damage by the presence of a thick layer of tissue, and its reproductive tissues were protected by being submerged in the thallus and by being produced in winter. Steneck concluded that this association was, therefore, one of mutual dependency or co-evolution. Damage to crustose algae by herbivores may be considerable (Padilla, 1985).

It is evident, therefore, that herbivory may act selectively against ephemeral algae and in favour of hard crustose forms. It may also operate against canopy species. Dayton (1975) observed that high densities of the sea urchin *Strongylocentrotus* were associated with very poor development of the kelp *Hedophylum*, but experimental exclosures of urchin eventually became vegetated by *Hedophyllum*.

In coastal waters of Nova Scotia, sea urchins are the natural prey of lobsters and over-fishing of the latter resulted in a dramatic increase in the former. This increase led to the destruction of the kelp bed and hence to the removal of protective cover for small lobsters. It was concluded that this ecosystem had reached a new stable state (Mann 1977). An outbreak of an unidentified disease, however, caused mass mortality among sea urchins in this region, an event which was followed by an increase in algal biomass and a recovery of the kelp community (Miller, 1985). Thus, the consequence of intense grazing for the strategically different canopy species and fugitive species can be very similar.

Dietary selectivity seems likely to be expressed also by small herbivores. A preference for algal epiphytes to the tissues of the host species (*Sargassum muticum*) has been found among some members of the associated fauna by Norton & Benson (1983). Selective grazing of epiphytes on *Fucus vesiculosus* by the snail *Littorina* has similarly been reported by Lubchenco (1983). Laboratory experiments with small crustaceans (*Gammarus* and *Idotea* species) by Shacklock & Doyle (1983) showed that these animals could maintain cultures of the red algae *Gracilaria* and *Gymnogongrus* free from epiphytic *Ectocarpus* and *Enteromorpha*. The relationships between seaweed vegetation and the associated fauna seem certain to be complicated. Even in a relatively simple community dominated by the small fucoid alga *Pelvetia fastigiata* there is a delicate balance between the numbers of different animals and the sizes and dispersion patterns of the plants Gunnill, 1982).

An additional source of complication is the possibility that certain algae may pass through the digestive trace of the herbivore and emerge in a viable state. Santelices, Correa & Avila (1983) have cultured the gut contents of the Chilean sea urchin *Tetrapygus niger* and classified 84·6% of the surviving algae as opportunists and the remaining 15·4% as late-succession forms. These authors suggest that the ability to survive digestion may be an important survival mechanism for these ephemeral algae but more needs to be known of the initial grazing behaviour of the animals before a firm conclusion can be reached.

Observations on the processes of plant succession on new or cleared substrata have also been subject to strategic interpretation. Doty (1967a,b) recorded species succession on new lava shores in Hawaii and observed the colonization process to begin with ephemeral algae; the large perennial algae were characteristic of late succession stages. Doty concluded that a state of community equilibrium (climax) could be attained in six to ten years. He also noted that the randomness characteristic of algal distribution during the pioneer ephemeral phase was gradually replaced by a more structured distribution pattern as succession proceeded. The association of ephemeral species of algae with early stages in plant succession has been well established in other investigations. On intertidal substrata these are commonly green algae such as *Enteromorpha* (Burrows & Lodge, 1950); but in sublittoral habitats the pioneers are more usually brown and red algae (Jones & Kain, 1967; Foster, 1975a; Kain, 1975a). It would be wrong, however, to conclude that plant succession proceeds in an entirely predictable manner. Hawkins (1981) established a series of clearances on a rocky shore on the Isle of Man and observed no fixed sequence of algae and, in one case, the ephemeral stage was omitted altogether. Murray & Littler (1978), in contrast, found the vegetation of a sewage polluted locality in California to be vegetated permanently by early succession species.

Plant competition and grazing may both be involved in succession, as Lubchenco (1982) has demonstrated. Fucoid algae are absent from New England rock-pools, although abundant on adjacent rock, and Lubchenco could only bring about the establishment of these plants in pools by removal of herbivores and algal competitors. Removal of only one of these two components was ineffective in promoting fucoid establishment.

Plant succession although liable to vary considerably, is not an altogether random process. For this reason, several attempts have been made to construct a theory of the succession process. Connell & Slatyer (1977) have described three possible models. The first of these, the facilitation model, proposed that only certain species are able to colonize a new substratum, and these alter the environment making it possible for others to settle. In this way an orderly succession of competitively more competent species is obtained. The second, or tolerance model, states that any species capable of survival as an adult can be the colonizing species and that its presence has little or no effect on the settlement of other species. Succession is, therefore, a product of differences in competitive ability. The third, or inhibition model, also states that any species capable of survival as an adult can function as a colonizing species. It also states, however, that the colonizer will exclude all other species unless death or disturbance opens up new space for settlement.

The observation by Hawkins (1981) that different species of algae may function as colonizers would appear to argue against the facilitation model. Lubchenco (1983) observed that a stand of the early succession alga, *Enteromorpha*, excluded the late succession, *Fucus vesiculosus*, unless grazing by *Littorina* created free space, thus providing evidence against the tolerance model. Sousa (1979) noted that new substratum on a southern Californian shore was normally colonized by *Ulva* which prevented further succession unless removed by herbivores. Likewise algae of each succession stage would slow down or prevent the appearance of other species. These assorted pieces of evidence can be accommodated satisfactorily only in the framework of the inhibition model; but that is not to prove it flawless on every occasion.

Some authors have chosen to interpret the growth characteristics of algae in strategic terms. Thus, Mann (1973) has referred to the seasonal growth cycle of north Atlantic kelp species and to their storage of photosynthate, which supports growth during winter, as a growth strategy. The same term has been adopted by Rosenberg & Ramus (1981) to describe differences in growth characteristics of *Gracilaria foliifera* and *Ulva* observed in culture. At low light intensities the growth rate of *Gracilaria* could exceed that of *Ulva* but under natural levels of irradiance the reverse was true.

The rates of production of numerous species of north American algae have been measured by Littler (1980) and by Littler & Arnold (1982). Their results led them to devise a strategic classification, based mainly on the gross morphology of the thallus. Thus, the highest rates of production were recorded by members of the sheet-group, followed in turn by the filamentous-group, the coarsely branched-group, the thick leathery group, the jointed calcareous-group and finally by the crustose-group. The rate of production was observed to undergo a two-fold reduction (approximately) between successive groups. The strategic implications of this apparently rather crude analysis are evident from the observations on rates of nutrient uptake made on Baltic macroalgae by Wallentinus (1984). She recorded highest uptake rates in sheet and filamentous thalli such as those of *Enteromorpha* and *Cladophora* species while the lowest rates were measured in coarse species with low surface area:volume ratios such as *Fucus* and *Furcellaria*. The form groups defined by Littler have also been linked closely, and in the same sequence, with resistance to grazing by sea urchins and fish. The most resistant to grazing were the members of the crustose-group and the least resistant, the sheet-group (Littler, Taylor & Littler, 1983). The classification of algal thallus forms based on gross features of external morphology has been attempted before (see Feldmann, 1951). It is interesting to see from the observations of Littler and his co-workers that these classifications have a functional basis.

Phenotypic plasticity has also been the subject of strategic interpretation by Taylor & Hay (1984). These authors investigated the red algae *Corallina*, *Gelidium*, and *Rhodoglossum* all of which exist in two forms: an intricate turf-form which was present on open rock, and a solitary form which occurred in rock-pools. Turfs of each species developed the solitary morphology after being transplanted into rock-pools. The reciprocal transplant resulted in the development of turf-forms, although rather less successfully. The turf-form proved to be more resistant to desiccation than the solitary form but also more expensive energetically, its productivity being 23–48% less than that of the solitary forms. Taylor & Hay concluded that phenotypic plasticity allows seaweeds to adopt morphological features that maximize fitness in a wide variety of habitats without being "developmentally committed".

Another ecological function of the seaweed thallus has been reported as the "whiplash effect" (Dayton, 1971). This is simply the sweeping action of the frond over the rock surface in response to wave action. Velimirov & Griffiths (1979) have described how sweeping by *Laminaria* fronds maintains a belt of bare rock around each clump of plants, separating it from herbivores. They concluded that sweeping provided protection against grazing as well as producing clear substratum for further expansion of the clump. Similar whiplash effects have been reported by Santelices & Ojeda (1984a).

Sand scouring and recurrent burial by sand is a form of physical disturbance which may also act selectively on algae, as has been shown recently by Littler, Martz & Littler (1983). Such environments tend to be vegetated by perennial forms which tolerate periods of disturbance and ephemeral forms which are able to exploit temporary periods of stability. The contrasting selection pressures generated on stable and unstable substrata have also been discussed by Littler & Littler (1984).

It is evident that recent literature contains numerous accounts, and diverse interpretations, of algal fitness. The ecological implications of these have been discussed by Chapman (1986), and the particular effects of herbivory on vegetation structure have been examined by Hawkins & Hartnoll (1983), both excellent reviews. These works simply underline the fact that the volume of published information which links characteristics of algal phenotypes with the selection pressures of their natural environments is immense. A need has, therefore, been perceived by some authors to bring this array of material to order by means of a simple classification of strategies.

The notion of classifying plants along strategic lines is an old one, the origins of which may be traced back to the concept of life-forms (Russell & Fielding, 1981). Life-form categories were recognized mainly on details of gross morphology and life history, and although they were intended primarily for comparison of geographical or regional floras, they have also been employed on a more local scale in analyses of marine algal community structure (see Russell & Fielding, 1981). Life-forms are nevertheless based firmly upon floristics whereas strategic classifications put much greater emphasis on identification of the principal forces of natural selection and their evolutionary effects.

Under favourable conditions the numbers of plants in a population will tend to increase exponentially at a rate (*r*) characteristic of the particular taxon. Sooner or later, resources will become limiting and competition for these will increase. Population increase will then cease and the numbers of individuals present should remain roughly stable at a level determined by the carrying capacity (*K*) of the environment. The values of *r* and *K* for any given population vary according to species; some species reproduce copiously and early in life while others allocate more resources to maintaining a numerically stable and competitive population structure. These two kinds of evolutionary response are referred to as r- and K-strategies and the forces which have brought them into being as r- and K-selection (Berry, 1977).

Vadas (1977) has adopted the idea of r- and K-selection with respect to marine algae. He identifies two mechanisms by which r-strategists may survive, that is to be short-lived (escape in time) and to be able to exploit available free space (escape in space). He similarly identifies two means of K-strategist survival, the structural defence (having a hard, inedible or indigestible thallus) and the chemical defence (having an unpalatable or toxic thallus). An example of the latter condition has been described by Sieburth & Conover (1965) and others are cited by Hawkins & Hartnoll (1983).

A different approach to the classification of higher-plant strategies has been advocated by Grime (1977). Grime accepts that K-selection has led to the evolution of large, competitive and relatively long-lived plants. He suggests, however, that r-selection, as defined above, conceals two quite different evolutionary forces. One of these he refers to as stress, which he defines as those conditions that restrict plant productivity (*e.g.* shortages of light and nutrients, suboptimal temperatures). The other force he refers to as disturbance and this he defines as conditions that destroy plant biomass by such means as herbivory, erosion, and pathogens. Use of the term "stress" in biology has been deplored by Harper (1982) for the good reason that it is seldom clear whether stimuli or their effects are being considered. Nevertheless, the term is now in such wide and frequent use that its abandonment must be considered unlikely. Hawkins & Hartnoll (1983) have discussed the interactive effects of stress, grazing, and algal competitive abilities in determining the structure of benthic marine communities, although they do not refer to the Grime system. Dring (1982) has stated that marine algal strategies may indeed be amenable to the Grime approach and suggests that the three types of strategist recognized may be exemplified in the thallus types defined by Dayton (1975). Dring does not, however, go much beyond registering approval of the Grime approach.

The only way of testing the efficacy of this system is, therefore, to try it and this Figure 6 attempts to do using the triangular configuration recommended by Grime. At the bottom left hand corner, which represents highly stressed conditions, have been located maritime lichens, mat-forming cyanobacteria and other such organisms that can survive unproductive

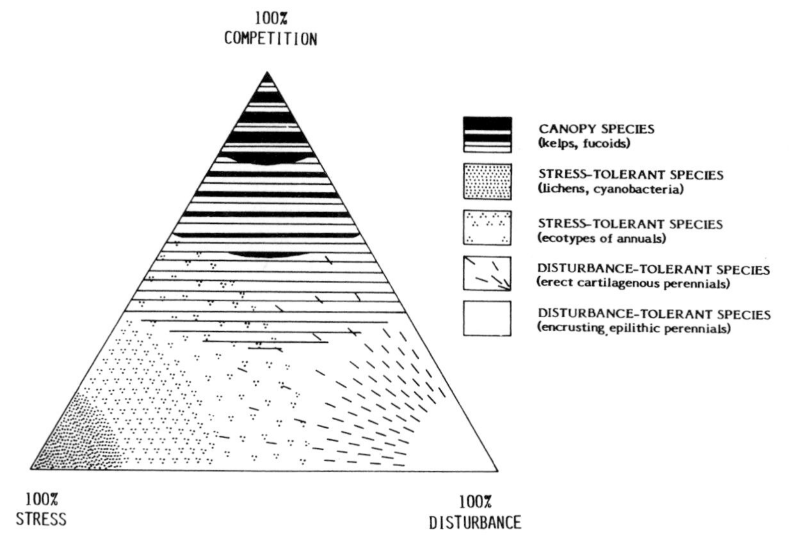

Fig. 6.—Ordination of benthic marine algae according to the strategic classificatory system proposed by Grime (1977), and making use of the three principle criteria (competition, stress and disturbance) recognized by him: the reasons for locating certain algae in particular sectors of the triangle are given in the text; it is sufficient to record here that the system is not wholly satisfactory for use with marine algae.

environments such as those experienced on the upper levels of a rocky shore. Still stress-tolerant, but less markedly than these, are annual species which have proved tolerant to extreme salinities, heavy metals *etc.*, usually by the evolution of specific ecotypes. At the bottom right hand corner have been located encrusting, epilithic perennial algae. Many of these have proved to be very resistant to grazing and some may even have evolved dependence on grazing. Further to the left have been located those algae of tough cartilagenous habit which have been shown to be low in herbivore preferences; taxa such as *Chondrus* and *Gigartina* are included here. Under benign environmental conditions, selection should operate in favour of competitive ability and, therefore, the canopy-forming kelps and fucoids have been placed in the upper section of the triangle. Their location here is based upon the results of canopy-removal experiments such as those described on previous pages.

A cursory inspection of Figure 6 might give some confidence in the Grime system but its deficiencies become evident with closer examinaton. Dring (1982) has suggested that *Ascophyllum* has the longevity and unpalatability of many stress-tolerant plants whereas *Fucus* is a faster-growing and more competitive plant. Nevertheless, *Ascophyllum* has been shown to possess superior competitive ability to *Fucus* by Hawkins & Hartnoll (1985), on the basis of *Ascophyllum* removal experiments.

Enteromorpha, on all available evidence, should be regarded as an r-strategist, and one which is stress-tolerant rather than grazing (disturbance)-tolerant. It has, however, also been shown that *Enteromorpha* can effectively exclude canopy species such as *Fucus* (Lubchenco, 1983) and grazing-resistant species such as *Chondrus* (Lubchenco, 1978). In similar vein, *Chondrus* has proved to be competitively superior to the canopy-forming *Fucus* (Lubchenco, 1980).

Littler & Littler (1983a) have suggested that previous work may have over-emphasized the selective rôle of grazing in the evolution of the crustose thallus form and they point out that such forms are also resistant to physical stresses. As an example of these physical stresses they cite sand scouring, an unfortunate choice perhaps as this could be interpreted as another manifestation of disturbance. The underlying point, that stress and disturbance tolerances may, however, be combined in the same individual taxon is an astute one. Crustose, coralline algae have proved to be resistant to disturbance in the form of grazing but they also occur naturally at very considerable depths in the ocean (Clokie, Scoffin & Boney, 1981). Indeed the deepest known plant life is a crustose coralline found by Littler, Littler, Blair & Norris (1985) at the astonishing depth of 268 m off San Salvadore, Bahamas. An environment rendered unproductive by such extremely low levels of irradiance could only be interpreted as stressed. It follows, therefore, that species which possess co-tolerance to two of the three principal selection forces recognized by Grime cannot easily be located within the axes defined by these forces. The additional fact that many species of marine algae have proved capable of evolving ecotypes with different tolerances to specific environmental conditions (see p. 315 *et seq.*) again makes it difficult to locate a species within a triangle, although it may be possible to locate a particular population.

It seems very possible that the Grime classification of strategies is fundamentally unsuitable as a means of expressing the complexities of selection pressures in the sea, and the evolutionary divergences which they promote. Lubchenco & Gaines (1981) have concluded that single-factor explanations of distribution and abundance patterns of algae are liable to be over-simplified. Except in cases of extremely directional selection, this analysis is probably correct. It is perhaps pertinent also to indicate the hazards of over-simplification from Lubchenco's own research. Most of her conclusions on the status of *Chondrus* in relation to herbivory have been arrived at from a study of the snail *Littorina littorea*. Other species of herbivore have been shown by Shacklock & Croft (1981) to consume *Chondrus* more actively than *Littorina*, and this knowledge puts the grazing tolerance of this plant in a different perspective. Natural selection has many components and its products are unlikely ever to be accommodated easily in simple classificatory schemes such as these.

CONCLUSIONS

1. There is a great deal of phenotypic variation within species of marine algae. This variation may be expressed in terms of physiology, biochemistry, morphology, and/or reproduction. It may be continuous or discontinuous in distribution and it may occur in ecological, geographical or temporal isolation.

2. Phenotypic traits may be stable or plastic in character. Some stable traits have been shown to be, and others probably are, heritable. Some plastic attributes (*e.g.* resonses to day-length) have a stable, heritable basis. The plasticity of every trait is, in any case, subject to limitations imposed by the genotype.
3. Any individual phenotype is the simultaneous expression of a variety of stable and plastic traits, and no individual can therefore be judged as simply stable or plastic. This makes it impossible to identify a plant or population unequivocally, as a member of one of the lower units of biosystematics.
4. Sources of phenotypic variation include genetic recombination and mutation, but our knowledge of gene expression, transmission, linkage, dominance *etc.* is still too narrowly confined to a few groups of related taxa. Recent research has shown that when experiments are designed in compliance with the methods of geneticists, then extremely interesting patterns of character heritability can be revealed.
5. Phenotypic variation tends to be quantitative rather than qualitative in character.
6. The process of transspecific evolution and that of evoluton of higher algal taxa does not seem to differ in any fundamental way from the events in microevolution. The morphological differences, and the patterns of spatial, temporal and/or reproductive isolation, that separate species all have their counterparts at the population level. Even in terms of magnitude of difference, species are not always clearly discernible from higher taxa or from infraspecific taxa.
7. There is no generally accepted species concept for marine algae. Even sexual compatibility among taxa can in practice be quantitative rather than qualitative. As a criterion of species status, sexual compatibility is, in any case, inapplicable to the numerous asexual taxa. Perhaps we should look rather to the ways attributes have responded to selection pressure. Traits that are labile and responsive, and which are active in the evolution of particular local populations, may be less important in species delimitation than those which are relatively constant in expression. It seems probable, however, that the rates of evolutionary change are themselves liable to vary through time, and constancy may therefore be only a transient condition. The serious limitations now evident in the classificatory schemes of biosystematists (see Table I, p. 312) makes it necessary to appreciate the species as an evolutionary unit, and one which has the natural plant population as its fundamental subunit.
8. The deduction of ancient lineages is fraught with difficulties, three of which are expressed in the following quotation from Klein & Cronquist (1967): " . . . Few of the papers examined were based on comparative studies utilizing a sufficiently broad range of selected organisms. A second problem has been the tendency for specialists in any given field to assume that the item of their speciality is, by itself, almost sufficient to serve as the primary base for erecting a phylogeny. The inadequacy of this point of view is apparent. Another problem has been the evident lack of appreciation of what might be primitive or advanced in terms of evolutionary specialization". This last

problem is possibly insoluble. Convergence, parallelism, and independent origin of traits are widespread in algae; so too may be the incidence of secondarily advanced or primitive character states. It therefore becomes very difficult to distinguish homologous and analogous similarities.

9. The problem of identifying the forces of natural selection and of measuring their evolutionary effects is also a difficult one. Cain (1982) has remarked that ". . . even the most apparently trivial and variable characters in wild populations when adequately investigated show the influence of considerable natural selection". On the other hand, as Clutton-Brock & Harvey (1979) have pointed out, ". . . the logical structure of many adaptive arguments is weak: a difference in behaviour or morphology is demonstrated, an explanation is constructed to fit the facts, and subsequent citations quote the initial evidence as proof of the hypothesis that is derived from it". It is evident from a large, and growing, body of literature that much of the variation between and within species of marine algae is functionally advantageous with respect to particular conditions of their natural environments. The selective effect of these conditions is often clear. It is, however, equally evident that some algal traits cannot be explained so readily in terms of natural selection. Some of these may be simply evolutionary hang-overs, products of some ancient evolutionary event. Others may even be disadvantageous, but are linked with certain positive responses to selection pressure. Evolutionary constraints are likely to be as real for marine algae as for higher plants and animals.
10. It is often possible to distinguish adaptive and non-adaptive traits by appropriate experiment; performances of plants from chosen populations or species being assessed under conditions that are deemed strongly selective. Another approach, advocated by Clutton-Brock & Harvey (1979), is to investigate the functions of those features which have arisen independently in phylogenetically different groups of species (*i.e.* analogous traits). The various strategies of marine algae, and the functional aspects of thallus morphology as expressed, for example, by crustose and canopy-forming types, are in line with this approach. Cases of analogous phenotypic plasticity should also be considered; the functional properties of turf and solitary forms of thalli in several very different species of red algae are evidently adaptive.
11. Finally, and to paraphrase a remark by Bradshaw (1984), it seems certain that future progress in understanding evolution in marine algae depends upon a plurality of approaches and a community of effort between ecologists, physiologists, geneticists, and systematists.

ACKNOWLEDGEMENT

I am grateful to C. J. Veltkamp for his assistance with scanning electron microscopy.

REFERENCES

Abbott, I. A., 1972, In, *Contributions to the Systematics of Benthic Marine Algae of the North Pacific,* edited by I. A. Abbot & M. Kurogi, Japanese Society of Phycology, Kobe, pp. 253–264.

Abbott, I. A. & North, W. J., 1972. *Proc. Int. Seaweed Symp.,* **7,** 72–79.

Amsler, C. D., 1984. *Phycologia,* **23,** 377–382.

Amsler, C. D. & Kapraun, D. F., 1985. *J. Phycol.,* **21,** 94–99.

Arthur, W., 1984. *Mechanisms of Morphological Evolution.* J. Wiley, Chichester, 275 pp.

Baardseth, E., 1970. *F.A.O. Fisheries Synopsis,* No. 38, pag. var.

Baca, B. J. & Cos, E. R., 1979. *Phycologia,* **18,** 369–377.

Baker, J. R. J. & Evans, L. V., 1971. *Br. phycol. J.,* **6,** 73–80.

Ballantine, D. L., 1979. *Botanica mar.,* **23,** 107–111.

Bartlett, R. B. & South, G. R., 1973. *Acta bot. neerl.,* **23,** 1–5.

Batters, E. A. L., 1902. *J. Bot. Lond.,* **40** (suppl.), 1–107.

Bennett, E., 1970. In, *Genetic Resources in Plants—Their Exploration and Conservation,* edited by O. H. Frankel & E. Bennett, I. B. P. Handbook No. 11, Blackwell, Oxford, pp. 115–129.

Bernatowicz, A. J., 1958. *Biol. Bull. Mar. biol. Lab., Woods Hole,* **115,** 232 only.

Bernstein, B. B. & Jung, N., 1979. *Ecol. Monogr.,* **49,** 335–355.

Berry, R. J., 1977. *Inheritance and Natural History.* Collins, London, 350 pp.

Bertness, M. D., Yund, P. O. & Brown, A. F., 1983. *J. exp. mar. Biol. Ecol.,* **71,** 147–164.

Bidwell, R. G. S., Lloyd, N. D. H. & McLachlan, J., 1984, *Hydrobiologia,* **167/117,** 292–294.

Bidwell, R. G. S., McLachlan, J. & Lloyd, N. D. H., 1985, *Botanica mar.,* **28,** 87–94.

Biebl, R., 1952. *J. mar. biol. Ass. U.K.,* **31,** 307–315.

Biebl, R., 1967. *Naturwiss. Rundsch.,* **20,** 248–252.

Bird, C. J., Edelstein, T. & McLachlan, J., 1977. *Naturaliste Can.,* **104,** 257–266.

Bird, C. J. & McLachlan, J., 1982. *Botanica mar.,* **25,** 557–562.

Bird, C. J., Meer, J. P. van der & McLachlan, J., 1982. *J. mar. biol. Ass. U.K.,* **62,** 453–459.

Bird, N. L. & McLachlan, J., 1976. *Phycologia,* **15,** 79–84.

Blackler, H. & Katpitia, A., 1963. *Trans. Proc. bot. Soc. Edinb.,* **39,** 392–395.

Bliding, C., 1963. *Opera Bot.,* **8**(3), 1–160.

Bock, W. J., 1963. *Am. Nat.,* **97,** 265–285.

Bolton, J. J., 1979. *Estuar. cstl mar. Sci.,* **9,** 273–280.

Bolton, J. J., 1983. *Mar. Biol.,* **73,** 131–138.

Bolton, J. J. & Lüning, K., 1982. *Mar. Biol.,* **66,** 89–94.

Bolwell, G. P., Callow, J. A., Callow, M. E. & Evans, L. V., 1977. *Nature, Lond.,* **268,** 626–627.

Boney, A. D. 1978. In, *Modern Approaches to the Taxonomy of Red and Brown Algae,* edited by D. E. G. Irvine & J. H. Price, Academic Press, London, pp. 1–19.

Bonneau, E. R., 1977. *J. Phycol.,* **13,** 133–140.

Bonneau, E. R., 1978. *Botanica mar.,* **21,** 117–121.

Bradshaw, A. D., 1984. In, *Evolutionary Ecology,* edited by B. Shorrocks, Blackwell Sci. Publs, Oxford, pp. 1–25.

Branch, G. M., 1981. *Oceanogr. mar. Biol. Ann. Rev.,* **19,** 235–280.

Brawley, S. H. & Adey, W. H., 1981a. *Nature Lond.,* **292,** 177 only.

Brawley, S. H. & Adey, W. H., 1981b. *Mar. Biol.,* **61,** 167–177.

Brawley, S. H. & Wetherbee, R., 1981. In, *The Biology of Seaweeds,* edited by C. S. Lobban & M. J. Wynne, Blackwell Sci. Publs, Oxford, pp. 248–299.

Brinkhuis, B. H. & Jones, R. F., 1976. *Mar. Biol.,* **34,** 339–348.

Burrows, E. M., 1955. *Proc. Int. Seaweed Symp.*, **2**, 163–170.
Burrows, E. M., 1963. *Proc. Int. Seaweed Symp.*, **4**, 166–170.
Burrows, E. M. & Lodge, S. M., 1950. *Rep. mar. biol. Stn Port Erin*, No. 62, 30–34.
Burrows, E. M. & Lodge, S. M., 1951. *J. mar. biol. Ass. U.K.*, **30**, 161–176.
Cain, A. J., 1982. In, *Problems of Phylogenetic Reconstruction*, edited by K. A. Joysey & A. E. Friday, Academic Press, London, pp. 1–19.
Cain, A. J., 1983. In, *Protein Polymorphism: Adaptive and Taxonomic Significance*, edited by G. S. Oxford & D. Rollison, Academic Press, London, pp. 391–397.
Caplin, S. M., 1968. *Bioscience*, **18**, 193 only.
Cavalier-Smith, T., 1982. *Biol. J. Linn. Soc.*, **17**, 289–306.
Chadefaud, M., 1980. *Cryptogamie: Algologie*, **1**, 213–217.
Chadefaud, M., 1982. *Cryptogamie: Algologie*, **3**, 241–256.
Chamberlain, Y. M., 1984. *Phycologia*, **23**, 433–442.
Chapman, A. R. O., 1972a. *Syesis*, **5**, 1–20.
Chapman, A. R. O., 1972b. *Phycologia*, **11**, 225–231.
Chapman, A. R. O., 1974. *Mar. Biol.*, **24**, 85–91.
Chapman, A. R. O., 1975. *Br. phycol. J.*, **10**, 219–223.
Chapman, A. R. O., 1977. *Botanica mar.*, **20**, 149–153.
Chapman, A. R. O., 1978. In, *Modern Approaches to the Taxonomy of Red and Brown Algae*, edited by D. E. G. Irving & J. H. Price, Academic Press, London, pp. 423–432.
Chapman, A. R. O., 1984. *J. exp. mar. Biol. Ecol.*, **78**, 99–109.
Chapman, A. R. O., 1986. *Adv. mar. Biol.*, **23**, 1–161.
Chapman, A. R. O. & Goudey, C. L., 1983, *Can. J. Bot.*, **61**, 319–323.
Charlesworth, B., 1982. In, *Darwin Up to Date*, edited by J. Cherfas, I. P. C. Magazines Ltd, London, pp. 23–26.
Chen, L. C.-M., 1980. *Botanica mar.*, **23**, 441–448.
Chen, L. C.-M., Morgan, K. & Simpson, F. J., 1982. *Botanica mar.*, **25**, 35–36.
Chen, L. C.-M. & Taylor, A. R. A., 1980. *Botanica mar.*, **23**, 435–440.
Cheney, D. P. & Babbel, G. R., 1978. *Mar. Biol.*, **47**, 251–264.
Cheney, D. P. & Mathieson, A. C., 1978. *J. Phycol.*, **14** (suppl.), 27 only.
Cheney, D. P. & Mathieson, A. C., 1979. *Isozyme Bull.*, **12**, 57 only.
Cheshire, A. C. & Hallam, N. D., 1985. *Phycologia*, **24**, 147–153.
Christensen, T., 1964. In, *Algae and Man*, edited by D. F. Jackson, Plenum Press, New York, pp. 59–64.
Christie, A. O. & Evans, L. V., 1962. *Nature Lond.*, **193**, 193–194.
Chudek, J. A., Foster, R., Davison, I. R. & Reed, R. H., 1984. *Phytochem.*, **23**, 1081–1082.
Clarke, S. M. & Womersley, H. B. S., 1981. *Aust. J. Bot.*, **29**, 497–505.
Clayton, M. N., 1974. *Aust. J. Bot.*, 743–813.
Clayton, M. N., 1975. *Phycologia*, **14**, 187–195.
Clayton, M. N., 1976. *Mar. Biol.*, **38**, 201–208.
Clayton, M. N., 1979. *Br. phycol. J.*, **14**, 1–10.
Clayton, M. N., 1980. *Br. phycol. J.*, **15**, 105–118.
Clayton, M. N., 1981a. *Proc. R. Soc. Vict.*, **92**, 113–118.
Clayton, M. N., 1981b. *Phycologia*, **20**, 358–364.
Clayton, M. N., 1982. *Botanica mar.*, **25**, 111–116.
Clayton, M. N., 1984. In, *Progress in Phycological Research, Vol. 3*, edited by F. E. Round & D. J. Chapman, Biopress, Bristol, pp. 11–46.
Clokie, J. J. P., Scoffin, T. P. & Boney, A. D., 1981. *Mar. Ecol. Prog. Ser.*, **4**, 131–133.
Clutton-Brock, T. H. & Harvey, P. H., 1979. *Proc. R. Soc. Ser. B*, **205**, 547–565.
Cole, K., 1967. *Can. J. Genet. Cytol.*, **9**, 519–530.
Cole, K., 1968. *Can. J. Genet. Cytol.*, **10**, 670–672.
Coliijn, F. & Hoek, C. van den, 1971. *Nova Hedwigia*, **21**, 899–922.

Connell, J. H., 1972. *Ann. Rev. Ecol. Syst.*, **3**, 169–192.
Connell, J. H. & Slatyer, R. O., 1977. *Am. Nat.*, **111**, 1119–1144.
Cousens, R., 1982. *Botanica mar.*, **25**, 191–195.
Cousens, R., 1986 *J. exp. mar. Biol. Ecol.*, **92**, 231–249.
Cousens, R., 1986. *Estuar. cstl shelf Sci.*, **22**, 495–507.
Cousens, R. & Hutchings, M. J., 1983. *Nature Lond.*, **301**, 240–241.
Craigie, J. S. & Pringle, J. D., 1978. *Can. J. Bot.*, **56**, 2910–2914.
Craigie, J. S., Wen, Z. C. & Meer, J. P. van der, 1984. *Botanica mar.*, **27**, 55–61.
Dahl, E., 1959. *Archo. Oceanogr. Limnol.*, **11** (Suppl.), 227–236.
Danser, B. H., 1929. *Genetica*, **11**, 399–450.
Darwin, C., 1859. *On the Origin of Species by Means of Natural Selection, or the Observation of Favoured Races in the Struggle for Life.* J. Murray, London (re-issued 1982 by Penguin Books, Harmondsworth, 477 pp.).
Dayton, P. K., 1971. *Ecol. Monogr.*, **41**, 351–389.
Dayton, P. K., 1973. *Ecology*, **54**, 433–438.
Dayton, P. K., 1975. *Ecol. Monogr.*, **45**, 137–159.
Dayton, P. K., Currie, V., Gerrodette, T., Keller, B. D., Rosenthal, R. & ven Tresca, D., 1984. *Ecol. Monogr.*, **54**, 253–289.
Davis, P. H. & Heywood, V. H., 1963. *Principles of Angiosperm Taxonomy.* Oliver & Boyd, Edinburgh, 556 pp.
De Silva, M. W. R. N. & Burrows, E. M., 1973. *J. mar. biol. Ass. U.K.*, **53**, 895–904.
De Valera, M., 1962. *Proc. R. Ir. Acad.*, **62B**, 77–79.
Dethier, M. N., 1981. *Oecologia (Berl.)*, **4**, 333–339.
Deysher, L. & Norton, T. A., 1982. *J. exp. mar. Biol. Ecol.*, **56**, 179–195.
Diaz-Piferrer, M. & Burrows, E. M., 1974. *J. mar. biol. Ass. U.K.*, **54**, 529–538.
Dixon, P. S., 1961. *Can. J. Bot.*, **39**, 541–543.
Dixon, P. S., 1965. *Bot. gothoburg.*, **3**, 67–74.
Dixon, P. S., 1973. *Biology of the Rhodophyta.* Oliver & Boyd, Edinburgh, 285 pp.
Dobzhansky, T., 1959. *Proc. Am. phil. Soc.*, **103**, 252–263.
Doty, M. S., 1967a. *Blumea*, **15**, 95–105.
Doty, M. S., 1967b. *Bull. Sth. Calif. Acad. Sci.*, **66**, 175–194.
Drew, K. M., 1955. *Biol. Rev.*, **30**, 343–390.
Dring, M. J., 1982. *The Biology of Marine Plants.* E. Arnold, London, 199 pp.
Dring, M. J., 1984. In, *Progress in Phycological Research, Vol. 3,* edited by F. E. Round & D. J. Chapman, Biopress, Bristol, pp. 159–192.
Dring, M. J. & Brown, F. A., 1982. *Mar. Ecol. Prog. Ser.*, **8**, 301–308.
Dring, M. J. & Lüning, K., 1975. *Z. PflPhysiol.*, **75**, 107–117.
Dring, M. J. & West, J. A., 1983. *Planta*, **159**, 143–150.
Druehl, L. D., 1970. *Phycologia*, **9**, 237–247.
Druehl, L. D. & Kemp, L., 1982. *Can. J. Bot.*, **60**, 1409–1413.
Ducker, S. C., Williams, W. T. & Lance, G. N., 1965. *Aust. J. Bot.*, **13**, 489–499.
Edelstein, T., 1977. *J. exp. mar. Biol. Ecol.*, **30**, 249–259.
Edelstein, T., Bird, C. J. & McLachlan, J., 1976. *Can. J. Bot.*, **54**, 2275–2290.
Edelstein, T. & McLachlan, J., 1967. *Br. phycol. Bull.*, **3**, 185–187.
Edwards, P., 1970. *Nature, Lond.*, **226**, 467–468.
Edwards, P., 1976. *Taxon*, **25**, 529–542.
Edyvean, R. G. J. & Ford, H., 1984a. *Biol. J. Linn. Soc.*, **23**, 353–363.
Edyvean, R. G. J. & Ford, H., 1984b. *Biol. J. Linn. Soc.*, **23**, 365–374.
Ercegovic, A., 1952. *Fauna Flora Adriatica*, **2**, 1–112.
Espinoza, J. & Chapman, A. R. O., 1983. *Mar. Biol.*, **74**, 213–218.
Evans, L. V., 1962. *Ann. Bot.*, **26**, 345–360.
Evans, L. V., 1965. *Ann. Bot.*, **29**, 541–562.
Faegri, K., 1937. *Bot. Rev.*, **3**, 400–423.
Feldmann, J., 1951. In, *Manual of Phycology—An Introduction to the Algae and their Biology,* edited by G. M. Smith, Waltham, Mass., pp. 313–334.

Feldmann, J., 1952. *Revue Cytol. Biol. veg.*, **13**, 1–49.
Filion-Myklebust, C. C. & Norton, T. A., 1981. *Mar. Biol. Lett.*, **2**, 45–51.
Fjeld, A. & Lovlie, A., 1976. In, *The Genetics of Algae,* edited by R. A. Lewin, Blackwell Sci. Publs, Oxford, pp. 219–235.
Fletcher, R. L., 1975. *Nature Lond.*, **253**, 534–535.
Fletcher, R. L., 1977. *Proc. Int. Congr. mar. Corrosion Fouling,* **4**, 169–177.
Fletcher, R. L., 1980. *Botanica mar.*, **23**, 425–432.
Ford, H., Hardy, F. G. & Edyvean, R. G. J., 1983. *Biol. J. Linn. Soc.*, **19**, 211–220.
Forest, H., 1968. In, *Algae, Man and the Environment,* edited by D. F. Jackson, Syracuse University Press, Syracuse, New York, pp. 185–199.
Foster, M. S., 1975a. *Mar. Biol.*, **32**, 313–329.
Foster, M. S., 1975b. *Mar. Biol.*, **32**, 331–342.
Foster, P. L., 1977. *Nature Lond.*, **269**, 322–323.
Fott, B., 1974. *Taxon*, **23**, 446–461.
Föyn, B., 1955. *Pubbl. Staz. zool Napoli,* **27**, 261–270.
Francke, J. A. & Cate, H. J. ten, 1980. *Br. phycol. J.*, **15**, 343–355.
Francke, J. A. & Rhebergen, L. J., 1982. *Br. phycol. J.*, **17**, 135–145.
Friedmann, I., 1959. *Ann. Bot.*, **23**, 571–594.
Fritsch, F. E., 1935. *The Structure and Reproduction of the Algae. Vol. 1.* Cambridge University Press, Cambridge, 791 pp.
Fritsch, F. E., 1945. *Structure and Reproduction of the Algae. Vol. 2.* Cambridge University Press, Cambridge, 939 pp.
Garbary, D., 1978. *Br. phycol. J.*, **13**, 247–254.
Garbary, D., 1979a. *Helgoländer wiss. Meeresunters.*, **32**, 213–227.
Garbary, D., 1979b. *Botanica mar.*, **22**, 477–492.
Garbary, D., 1979c. *Bot. Notiser,* **132**, 451–455.
Garbary, D. J., Hansen, G. I. & Scagel, R. F., 1980. *Nova Hedwigia,* **33**, 145–166.
Geesink, R., 1973. *J. exp. mar. Biol. Ecol.*, **11**, 239–247.
Gilmour, J. S. L., 1961. In, *Contemporary Botanical Thought,* edited by A. M. MacLeod & L. S. Cobley, Oliver & Boyd, Edinburgh, pp. 27–45.
Gilmour, J. S. L. & Heslop-Harrison, J., 1954. *Genetica,* **27**, 147–161.
Glazer, A. N., West, J. A. & Chan, C., 1982. *Biochem. Syst. Ecol.*, **10**, 203–215.
Goodband, S. J., 1971. *Ann. Bot.*, **35**, 957–980.
Gould, S. J., 1977. *Ontogeny and Phylogeny.* Harvard University Press, Cambridge, Mass., 501 pp.
Greenwood, P. H., 1979. *Biol. J. Linn. Soc.*, **12**, 293–304.
Gregor, J. W., 1948. *Trans. Proc. bot. Soc. Edinb.*, **34**, 378–391.
Grime, J. P., 1977. *Am. Nat.*, **111**, 1169–1194.
Guiry, M. D. & West, J. A., 1983. *J. Phycol.*, **19**, 474–494.
Gunnill, F. C., 1980a. *Mar. Biol.*, **59**, 169–179.
Gunnill, F. C., 1980b. *Mar. Ecol. Prog. Ser.*, **3**, 231–243.
Gunnill, F. C., 1982. *Mar. Biol.*, **69**, 263–280.
Haeckel, E., 1866. *Generelle Morphologie der Organismen.* Reimer, Berlin.
Hall, A., 1980. *New Phytol.*, **85**, 73–78.
Hall, A., 1981. *Botanica mar.*, **24**, 223–228.
Hall, A., Fielding, A. H. & Butler, M., 1979. *Mar. Biol.*, **54**, 195–206.
Hansen, G. I., Garbary, D. J., Oliveira, J. C. & Scagel, R. F., 1981. *Syesis,* **14**, 115–123.
Hansen, J. E., Packard, J. E. & Doyle, W. T., 1981. *Mariculture of Red Seaweeds.* Calif. Sea Grant, Coll. Program Rep. T-CSGCP-002, Univ. Calif., La Jolla,42 pp.
Harkin, E., 1981. *Proc. Int. Seaweed Symp.*, **10**, 303–308.
Harlin, M. M., 1975. *Aquatic Bot.*, **1**, 125–131.
Harper, J. L., 1982. In, *The Plant Community as a Working Mechanism,* edited by E. I. Newman, Blackwell Sci. Publs, Oxford, pp. 11–25.

Harvey, W. H., 1846–1851. *Phycologia Britannica, Vols. 1–4,* London.
Harvey, W. H., 1849. *A Manual of the British Marine Algae.* Van Voorst, London 2nd edition, 252 pp.
Harvey, W. H., 1857. *The Sea-Side Book.* Van Voorst, London, 324 pp.
Hawkins, S. J., 1981. *J. mar. biol. Ass. U.K.,* **61,** 1–15.
Hawkins, S. J. & Hartnoll, R. G., 1983. *Oceanogr. mar. Biol. Ann. Rev.,* **21,** 195–282.
Hawkins, S. J. & Hartnoll, R. G., 1985. *Mar. Ecol. Prog. Ser.,* **20,** 265–271.
Hay, M. E. & Norris, J. N., 1984. *Hydrobiologia,* **116/117,** 63–94.
Heslop-Harrison, J., 1953. *New Concepts in Flowering-Plant Taxonomy.* W. Heinemann, London, 134 pp.
Hoek, C. van den, 1975. *Phycologia,* **14,** 314–330.
Hoek, C. van den, 1978. *Algen.* Thieme, Stuttgart, 418 pp.
Hoek, C. van den, 1979. *Helgoländer wiss. Meeresunters.,* **32,** 374–393.
Hoek, C. van den, 1982. *Biol. J. Linn. Soc.,* **18,** 81–144.
Hoek, C. van den, Cortel-Breeman, A. M., Rietema, H. & Wanders, J. B. W., 1972. *Soc. bot. Fr. Mem.,* 1972, 45–66.
Hoek, C. van den & Flinterman, A., 1968. *Blumea,* **16,** 193–242.
Holmes, E. M. & Batters, E. A. L., 1890. *Ann. Bot.,* **5,** 63–107.
Horn, M. H., Murray, S. N. & Edwards, T. W., 1982. *Mar. Biol.,* **67,** 237–246.
Hsiao, S. I. C. & Druehl, L., 1971. *Can. J. Bot.,* **49,** 1503–1508.
Huxley, J., 1938. *Nature Lond.,* **142,** 219–220.
Innes, D. J., 1984. *Helgoländer Meeresunters.,* **38,** 401–407.
Innes, D. J. & Yarish, C., 1984. *Phycologia,* **23,** 311–320.
Irvine, D. E. G., 1982. *Bull. Br. Mus. nat. Hist. (Bot.),* **10,** 109–131.
Jacobs, R. P. W. M., Hermelink, P. M. & Geel, G. van, 1983 *Aquatic Bot.,* **15,** 157–173.
Jensen, J. B., 1974. *Univ. Calif. Publs Bot.,* **68,** 1–61.
Jones, N. S. & Kain, J. M., 1967. *Helgoländer wiss. Meeresunters.,* **15,** 460–466.
Jones, W. E., 1974. In, *The Changing Flora and Fauna of Britain,* edited by D. L. Hawksworth, Academic Press, London, pp. 97–113.
Johnson, J. H., 1961. *Limestone-Building Algae and Algal Limestones.* Colorado School of Mines, Golden, 297 pp.
Jonsson, S., 1964. *C.r. Acad. Sci., Paris,* **258,** 2145–2148.
Jonsson, S., 1971. *Helgoländer wiss. Meeresunters.,* **22,** 281–294.
Jonsson, S., 1980. *Cryptogamie: Algologie,* **1,** 51–60.
Kain, J. M., 1975a. *J. Ecol.,* **63,** 739–765.
Kain, J. M., 1975b. *J. mar. biol. Ass. U.K.,* **55,** 567–582.
Kain, J. M., 1982. *Br. phycol. J.,* **17,** 321–331.
Kapraun, D. F., 1978a. *J. Phycol.,* **14** (suppl.), 25 only.
Kapraun, D. F., 1978b. *Phycologia,* **17,** 152–156.
Khfaji, A. K. & Norton, T. A., 1979. *Estuar. cstl mar. Sci.,* **8,** 433–439.
Kirkman, H., 1982. *J. exp. mar. Biol. Ecol.,* **55,** 243–254.
Kitching, J. A., & Ebling, F. J., 1961. *J. anim. Ecol.,* **30,** 373–384.
Kitting, C. L., 1980. *Ecol. Monogr.,* **50,** 527–550.
Klein, R. M. & Cronquist, A., 1967. *Q. Rev. Biol.,* **42,** 105–296.
Knoepffler-Peguy, M., 1977. *Revue Algol.,* **12,** 111–128.
Koeman, R. P. T. & Cortel-Breeman, A. M., 1976. *Phycologia,* **15,** 107–117.
Koeman, R. P. T. & Hoek, C. van den, 1980. *Br. phycol. J.,* **16,** 9–53.
Koeman, R. P. T. & Hoek, C. van den, 1982. *Arch. Hydrobiol. Suppl.* No. 63, 279–330.
Koetzner, K. & Wood., R. D., 1970. *Bull. Torrey bot. Club,* **97,** 120–121.
Kornmann, P., 1970a. *Helgoländer wiss. Meeresunters.,* **21,** 1–8.
Kornmann, P., 1970b. *Helgoländer wiss. Meeresunters.,* **21,** 292–304.
Kornmann, P., 1973. *Helgoländer wiss. Meeresunters.,* **25,** 1–13.
Kraft, G. T., 1975. *Br. phycol. J.,* **10,** 279–290.

Kraft, G. T., 1977. *Phycologia,* **16,** 43–51.
Kraft, G. T., 1981. In, *The Biology of Seaweeds,* edited by C. S. Lobban & M. J. Wynne, Blackwell Sci. Publs, Oxford, pp. 6–51.
Kristiansen, A., 1984. *Nord. J. Bot.,* **4,** 719–724.
Kristiansen, A. & Pedersen, P. M., 1979. *Bot. Tidsskr.,* **74,** 31–56.
Kurogi, M., 1972. In, *Contributions to the Systematics of Benthic Marine Algae of the North Pacific,* edited by I. A. Abbott & M. Kurogi, Japanese Society of Phycology, Kobe, pp. 167–192.
Kylin, H., 1937. *Lund. Univ. årsskr.,* **33,** 1–34.
Langlet, O., 1971. *Taxon,* **20,** 653–722.
Larsen, J., 1981. *Nord. J. Bot.,* **1,** 128–136.
Lawson, R. P. & Russell, G., 1967. *Br. phycol. Bull.,* **3,** 249–250.
Lee, I. K. & Kurogi, M., 1971. *Proc. Int. Seaweed Symp.,* **7,** 131–134.
Leedale, G. F., 1974. *Taxon,* **23,** 261–270.
Lein, T. E., 1980. *Sarsia,* **65,** 87–92.
Lein, T. E., 1984a. *Sarsia,* **69,** 75–81.
Lein, T. E., 1984b. *Blyttia,* **42,** 71–75.
Lewontin, R. C., 1978. *Scient. Am.,* **239,** 156–169.
Liddle, L. B., 1971. *J. Phycol.,* **7** (suppl.), 4 only.
Lindstrom, S., 1981. *Jap. J. Phycol.,* **29,** 251–257.
Littler, M. M., 1980. *Botanica mar.,* **22,** 161–165.
Littler, M. M. & Arnold, K. E., 1982. *J. Phycol.,* **18,** 307–311.
Littler, M. M. & Kauker, B. J., 1984. *Botanica mar.,* **27,** 37–44.
Littler, M. M. & Littler, D. S., 1983a. *J. Phycol.,* **19,** 425–431.
Littler, M. M. & Littler, D. S., 1983b. *J. exp. mar. Biol. Ecol.,* **74,** 13–34.
Littler, M. M., Littler, D. S., Blair, S. M. & Norris, J. N., 1985. *Science,* **227,** 57–59.
Littler, M. M., Martz, D. R. & Littler, D. S., 1983. *Mar. Ecol. Prog. Ser.,* **11,** 129–139.
Littler, M. M., Taylor, P. R. & Littler, D. S., 1983. *Coral Reefs,* **2,** 111–118.
Lockhart, J. C., 1979. *Am. J. Bot.,* **66,** 836–844.
Loeblich, A. R., 1974. *Taxon,* **23,** 277–290.
Lubchenco, J., 1978. *Am. Nat.,* **112,** 23–39.
Lubchenco, J., 1980. *Ecology,* **61,** 333–344.
Lubchenco, J., 1982. *J. Phycol.,* **18,** 544–550.
Lubchenco, J., 1983. *Ecology,* **64,** 1116–1123.
Lubchenco, J. & Cubit, J., 1980. *Ecology,* **61,** 676–687.
Lubchenco, J. & Gaines, S. D., 1981. *Ann. Rev. Ecol. Syst.,* **12,** 405–437.
Lubchenco, J. & Menge, B. A., 1978. *Ecol. Monogr.,* **48,** 67–94.
Lüning, K., 1979. *Mar. Ecol. Prog. Ser.,* **1,** 195–207.
Lüning, K., 1980a. In, *The Shore Environment, Vol. 2,* edited by J. H. Price, D. E. G. Irvine & W. F. Franham, Academic Press, London, pp. 915–945.
Lüning, K., 1980b. *J. Phycol.,* **16,** 1–15.
Lüning, K., 1981. *Ber. dt. bot. Ges.,* **94,** 401–417.
Lüning, K., 1985. *Meeresbotanik.* Georg Thieme, Stuttgart, 375 pp.
Lüning, K., Chapman, A. R. O. & Mann, K. H., 1978. *Phycologia,* **17,** 291–296.
Lüning, K. & Müller, D. G., 1978. *Z. PflPhysiol.,* **89,** 333–341.
Magne, F., 1967a. *Botaniste,* **50,** 297–308.
Magne, F., 1967b. *C.r. Acad. Sci., Paris,* **264,** 2632–2633.
Magne, F., 1972. *Soc. bot. Fr. Mem.,* 1972, 247–268.
Magne, F., 1982. *Cryptogamie: Algologie,* **3,** 265–271.
Mann, K. H., 1973. *Science,* **182,** 975–981.
Mann, K. H., 1977. *Helgoländer wiss. Meeresunters.,* **30,** 455–467.
Manton, I., 1964. *New Phytol.,* **63,** 244–254.
Markham, J. W., 1968. *Syesis,* **1,** 125–131.
Marsden, W. J. N., Callow, J. A. & Evans, L. V., 1984. *Botanica mar.,* **27,** 521–526.

Marsden, W. J. N., Evans, L. V. & Callow, J. A., 1983a. *Botanica mar.*, **26**, 383–392.
Marsden, W. J. N., Evans, L. V. & Callow, J. A., 1983b. *Botanica mar.*, **26**, 527–531.
Masuda, M., West, J. A., Ohni, Y. & Kurogi, M., 1984. *Bot. Mag. Tokyo*, **97**, 107–125.
Mather, K, 1961. In, *Contemporary Botanical Thought*, edited by A. M. MacLeod & L. S. Cobley, Oliver & Boyd, Edinburgh, pp. 47–94.
Mathieson, A. C., 1982. *Botanica mar.*, **25**, 93–99.
Mathieson, A. C., Norton, T. A. & Neushul, M., 1981. *Bot. Rev.*, **47**, 313–347.
May, V., Bennett, I. & Thompson, T. E., 1970. *Oecologia (Berl.)*, **6**, 1–14.
Mayr, E., 1942. *Systematics and the Origin of Species*, Columbia University Press, New York, 334 pp.
Mayr, E., 1970. *Populations, Species and Evolution*. Harvard University Press, Cambridge, Mass., 453 pp.
Mayr, E. 1982. *The Growth of Biological Thought. Diversity, Evolution and Inheritance*. Harvard University Press, Cambridge, Mass., 974 pp.
McCourt, R. M., 1984. *J. exp. mar. Biol. Ecol.*, **74**, 141–156.
McLachlan, J. & Bird, C. J., 1984. *Helgoländer Meeresunters.*, **38**, 319–334.
McLachlan, J. & Chen, L. C-M., 1972. *Can. J. Bot.*, **50**, 1841–1844.
McLachlan, J., Chen, L. C-M. & Edelstein, T., 1971. *Can. J. Bot.*, **49**, 1463–1469.
McLachlan, J., Meer, J. P. van der & Bird, N. L., 1977. *J. mar. biol. Ass. U.K.*, **57**, 1137–1141.
McNeill, J., 1982. *Zool. J. Linn. Soc.*, **74**, 337–334.
Meer, J. P. van der, 1981a. *Bioscience*, **33**, 172–176.
Meer, J. P. van der, 1981b. *Can. J. Bot.*, **59**, 787–792.
Meer, J. P. van der, 1981c. *Proc. N.S. Inst. Sci.*, **31**, 187–192.
Meer, J. P. van der & Simpson, F. J., 1984. *Phycologia*, **23**, 195–202.
Meer, J. P. van der & Todd, E. R., 1977. *Can. J. Bot.*, **55**, 2810–2817.
Meer, J. P. van der & Todd, E. R., 1980. *Can. J. Bot.*, **58**, 1250–1256.
Miller, R. J., 1985. *Mar. Biol.*, **84**, 275–286.
Moestrup, O., 1982. *Phycologia*, **21**, 427–528.
Moreno, C. A. & Jaramillo, E., 1983. *Oikos*, **41**, 73–76.
Moss, B. L., 1974. *Phycologia*, **13**, 317–322.
Moss, B. L., 1982. *Phycologia*, **21**, 185–188.
Moss, B. L. & Marsland, A., 1976. *Br. phycol. J.*, **11**, 309–313.
Müller, D. G., 1962. *Botanica mar.*, **14**, 140–155.
Müller, D. G., 1972. *Soc. bot. Fr. Mem.*, 1972, 87–98.
Müller, D. G., 1975. *Br. phycol. J.*, **10**, 315–321.
Müller, D. G., 1976a. *Arch. Microbiol.*, **109**, 89–94.
Müller, D. G., 1976b. *J. Phycol.*, **12**, 252–254.
Müller, D. G., 1979. *Phycologia*, **18**, 312–318.
Müller, D. G., 1980. *Naturwissenschaften*, **67**, 462 only.
Müller, D. G., 1981. *Proc. Int. Seaweed Symp.*, **10**, 57–70.
Müller, D. G., 1985. *Abstr. 2nd. Int. Phyc. Congr.*, Copenhagen, p. 113 only.
Müller, D. G. & Lüthe, N. M., 1981. *Br. phycol. J.*, **16**, 351–356.
Munda, I., 1979. *Nova Hedwigia*, **31**, 567–591.
Murray, S. N. & Dixon, P. S., 1973. *J. exp. mar. Biol. Ecol.*, **13**, 15–27.
Murray, S. N. & Dixon, P. S., 1975. *J. exp. mar. Biol. Ecol.*, **19**, 165–176.
Murray, S. N. & Littler, M. M., 1978. *J. Phycol.*, **14**, 506–512.
Murray, S. N. & Littler, M. M., 1981. *J. Biogeogr.*, **8**, 339–351.
Nakahara, H. & Nakamura, Y., 1973. *Mar. Biol.*, **18**, 327–332.
Nakahara, H. & Yamada, I., 1974. *J. Fac. Sci. Hokkaido Univ., ser. 5 (Bot.)*, **10**, 49–54.
Neish, A. C. & Fox, C. H., 1971. *Tech. Rep. Atlantic Regional Lab., Halifax, N.S. Can.*, No. 12, 35 pp.

Neish, A. C., Shacklock, P. F., Fox, C. H. & Simpson, F. J., 1977. *Can. J. Bot.*, **55,** 2263–2271.
Nellen, U. R., 1966. *Helgoländer wiss. Meeresunters.*, **13,** 288–313.
Neushul, M., 1972. In, *Contributions to the Systematics of Benthic Marine Algae of the North Pacific,* edited by I. A. Abbott & M. Kurogi, Japanese Society of Phycology, Kobe, pp. 47–74.
Newton, L., 1931. *A Handbook of the British Seaweeds.* British Museum (Nat. Hist.), London, 478 pp.
Nielsen, R., 1978. *J. Phycol.*, **14,** 127–131.
Nielsen, R., 1980. *Br. phycol. J.*, **15,** 131–138.
Norall, T. L., Mathieson, A. C. & Kilar, J. A., 1981. *J. exp. mar. Biol. Ecol.*, **54,** 119–136.
Norton, T. A., 1969. *J. mar. biol. Ass. U.K.*, **49,** 1025–1045.
Norton, T. A., 1981. *Botanica mar.*, **24,** 465–470.
Norton, T. A. & Benson, M. R., 1983. *Mar. Biol.*, **75,** 169–177.
Norton, T. A., Mathieson, A. C. & Neushul, M., 1981. In, *The Biology of Seaweeds,* edited by C. S. Lobban & M. J. Wynne, Blackwell Sci. Publs, Oxford, pp. 421–451.
Norton, T. A., Mathieson, A. C. & Neushul, M., 1982. *Botanica mar.*, **25,** 501–510.
Novak, R., 1984. *P.S.Z.N.I. Mar. Ecol.*, **5,** 143–190.
Ojeda, F. P. & Santelices, B., 1984. *Mar. Ecol. Prog. Ser.*, **19,** 83–91.
O'Kelly, C. J. & Yarish, C., 1980. *J. Phycol.*, **16,** 549–558.
O'Kelly, C. J. & Yarish, C., 1981. *Phycologia,* **20,** 32–45.
Oliveira, E. C. de & Fletcher, A., 1980. *Botanica Mar.*, **23,** 409–417.
Padilla, D. K., 1985. *Mar. Biol.*, **90,** 103–109.
Paine, R. T., 1979. *Science,* **205,** 685–687.
Parke, M., 1948. *J. mar. biol. Ass. U.K.*, **27,** 651–709.
Parke, M., 1953. *J. mar. biol. Ass. U.K.*, **32,** 497–520.
Parke, M. & Dixon, P. S., 1976. *J. mar. biol. Ass. U.K.*, **56,** 527–594.
Parsons, M. J., 1985. *New Zealand J. mar freshw. Res.*, **19,** 131–138.
Patterson, C., 1982a. In, *Evolution Now. A Century after Darwin,* edited by J. M. Smith, Nature/Macmillan, London, pp. 110–120.
Patterson, C. 1982b. In, *Darwin Up To Date,* edited by J. Cherfas, I.P.C. Magazines Ltd, London, pp. 35–39.
Patwary, M. U. & Meer, J. P van der, 1982. *Can J. Bot.*, **60,** 2556–2564.
Patwary, M. U. & Meer, J. P. van der, 1983a. *Botanica mar.*, **26,** 295–299.
Patwary, M. U. & Meer, J. P. van der, 1983b. *Aquaculture,* **33,** 207–214.
Patwary, M. U. & Meer, J. P. van der, 1983c. *Proc. N.S. Inst. Sci.*, **33,** 95–99.
Patwary, M.U. & Meer, J. P. van der, 1983d. *Can. J. Bot.*, **61,** 1654–1659.
Patwary, M. U. & Meer, J. P. van der, 1984. *Phycologia,* **23,** 21–27.
Paula, E. J. de & Oliveira, E. C. de, 1982. *Phycologia,* **21,** 145–153.
Pedersen, P. M., 1980. *Br. phycol. J.*, **15,** 391–398.
Pedersen, P. M., 1981. In, *The Biology of Seaweeds,* edited by C. S. Lobban & M. J. Wynne, Blackwell Scientific Publs, Oxford, pp. 194–217.
Pedersen, P. M., 1984. *Opera Bot.*, **74,** 1–76.
Perestenko, L. P., 1972. *Botan. Zhurnal,* **57,** 750–764.
Powell, H. T., 1963. In, *Speciation in the Sea,* edited by J. P. Harding & N. Tebble, Systematics Association Publ. No. 5, London, pp. 63–77.
Provasoli, L. & Pintner, I. J., 1980. *J. Phycol.*, **16,** 196–201.
Prud'homme van Reine, W. F., 1982. *A Taxonomic Revision of the European Sphacelariaceae (Sphacelariales, Phalophyceae).* Leiden University Press, Leiden, 288 pp.
Prud'homme van Reine, W. F. & Hoek, C. van den, 1966. *Blumea,* **14,** 227–283.
Ramus, J., 1983. *J. Phycol.*, **19,** 173–178.
Ramus, J. & Meer, J. P. van der, 1983. *J. Phycol.*, **19,** 86–91.

Ravanko, O., 1970. *Nova Hedwigia,* **20**, 179–252.
Raven, J. A., 1984. *Bot. J. Linn. Soc.,* **88,** 105–126.
Ravin, A. W., 1963. *Am. Nat.,* **97,** 307–318.
Reed, R. H., 1980. *Br. phycol. J.,* **15,** 411–416.
Reed, R. H., 1983. *J. exp. mar. Biol. Ecol.,* **68,** 169–193.
Reed, R. H. & Barron, J. A., 1983. *Botanica mar.,* **26,** 409–416.
Reed, R. H. & Moffat, L., 1983. *J. exp. mar. Biol. Ecol.,* **69,** 85–103.
Reed, R. H. & Russell, G., 1978. *Br. phycol. J.,* **13,** 149–153.
Reed, R. H. & Russell, G., 1979. *Estuar. cstl mar. Sci.,* **8,** 251–258.
Rice, E. L. & Chapman, A. R. O., 1985. *J. mar. biol. Ass. U.K.,* **65,** 433–459.
Rietema, H., 1975. Ph.D. thesis, University of Groningen, Netherlands, 130 pp.
Rietema, H. & Hoek, C. van den, 1981. P.S.Z.N.I. *Mar. Ecol.,* **4,** 321–335.
Roberts, M., 1978. In, *Modern Approaches to the Taxonomy of Red and Brown Algae,* edited by D. E. G. Irvine & J. H. Price, Academic Press, London, pp. 399–422.
Roberts, M. & Ring, F. M., 1972. *Soc. Bot. Fr. Mem.,* 1972, 117–128.
Robinson, G. G. C. & Cole, K., 1971. *Botanica mar.,* **14,** 59–62.
Rosenberg, G. & Ramus, J., 1981. *Botanica mar.,* **24,** 583–589.
Round, F. E., 1973. *The Biology of the Algae,* E. Arnold, London, 2nd edition, 278 pp.
Round, F. E., 1984. In, *Systematics of the Green Algae,* edited by D. E. G. Irvine & D. M. John, Academic Press, London, pp. 1–27.
Rueness, J., 1968. *Nytt Mag. Bot.,* **15,** 220–224.
Rueness, J., 1973. *Phycologia,* **12,** 107–109.
Rueness, J., 1978. In, *Modern Approaches to the Taxonomy of Red and Brown Algae,* edited by D. E. G. Irvine & J. H. Price, Academic Press, London, pp. 247–262.
Rueness, J. & Rueness, M., 1975. *Phycologia,* **14,** 81–85.
Rueness, J. & Rueness, M., 1978. Norw. J. Bot., **25,** 201–205.
Russell, D. J., 1982. *Micronesia,* **18,** 35–44.
Russell, G., 1966. *J. mar. biol. Ass. U.K.,* **46,** 267–294.
Russell, G., 1967. *Helgoländer wiss. Meeresunters.,* **15,** 155–162.
Russell, G., 1978. In, *Modern Approaches to the Taxonomy of Red and Brown Algae,* edited by D. E. G. Irvine & J. H. Price, Academic Press, London, pp. 339–369.
Russell, G., 1979. *Estuar. cstl mar. Sci.,* **9,** 659–661.
Russell, G., 1983a. *Mar. Ecol. Prog. Ser.,* **11,** 181–187.
Russell, G, 1983b. *Mar. Ecol. Prog. Ser.,* **13,** 303–304.
Russell, G., 1985a. *J. mar. biol. Ass. U.K.,* **65,** 343–349.
Russell, G., 1985b. *Br. phycol. J.,* **20,** 87–104.
Russell, G., in press. In, *Physiological Ecology of Amphibious and Intertidal Plants,* edited by R. M. M. Crawford, British Ecological Society Symposium, London, in press.
Russell, G. & Bolton, J. J., 1975. *Estuar. cstl mar. Sci.,* **3,** 91–94.
Russell, G. & Fielding, A. H., 1974. *J. Ecol.,* **62,** 689–698.
Russell, G. & Fielding, A. H., 1981. In, *The Biology of Seaweeds,* edited by C. S. Lobban & M. J. Wynne, Blackwell Sci. Publs, Oxford, pp. 393–420.
Russell, G. & Garbary, D., 1978. *J. mar. biol. Ass. U.K.,* **58,** 517–525.
Russell, G. & Morris, O. P., 1970. *Nature Lond.,* **228,** 288–289.
Russell, G. & Morris, O. P., 1973. *Proc. Int. Congr. mar. Corrosion Fouling,* **3,** 719–730.
Russell, G. & Veltkamp, C. J., 1984. *Mar. Ecol. Prog. Ser.,* **18,** 149–153.
Sammarco, P. W., 1983. *Mar. Ecol. Prog. Ser.,* **13,** 1–14.
Sanbonsuga, Y. & Neushul, M., 1978. *J. Phycol.,* **14,** 214–224.
Santelices, B., 1980. *Phycologia,* **19,** 1–12.

Santelices, B., Castilla, J. C., Cancino, J. & Schmiede, P., 1980. *Mar. Biol.,* **59,** 119–132.
Santelices, B., Correa, J. & Avila, M., 1983. *J. exp. mar. Biol. Ecol.,* **70,** 263–269.
Santelices, B., Montalva, S. & Oliger, P., 1981. *Mar. Ecol. Prog. Ser.,* **6,** 267–276.
Santelices, B. & Ojeda, F. P., 1984a. *Mar. Ecol. Prog. Ser.,* **19,** 73–82.
Santelices, B. & Ojeda, F. P., 1948b. *Mar. Ecol. Prog. Ser.,* **14,** 165–173.
Schiel, D. R., 1985a. *J. Ecol.,* **73,** 199–217.
Schiel, D. R., 1985b. *J. Phycol.,* **21,** 99–106.
Schiel, D. R. & Choat, J. H., 1980. *Nature Lond.,* **285,** 324–326.
Schonbeck, M. W. & Norton, T. A., 1978. *J. exp. mar. Biol. Ecol.,* **31,** 303–313.
Schonbeck, M. W. & Norton, T. A., 1980. *J. exp. mar. Biol. Ecol.,* **43,** 131–150.
Sears, J. R. & Brawley, S. H., 1982. *Am. J. Bot.,* **69,** 1450–1461.
Sears, J. R. & Wilce, R. T., 1970. *J. Phycol.,* **6,** 381–392.
Setzer, R. B., 1980. *Am. Zool.,* **20,** 849.
Shacklock, P. F. & Croft, G. B., 1981, *Aquaculture,* **22,** 331–342.
Shacklock, P. F. & Doyle, R. W., 1983. *Aquaculture,* **31,** 141–151.
Shacklock, P. F., Robson, D. R. & Simpson, F. J., 1975. *Tech. Rep. Atlantic Regional Lab. Halifax, N.S., Canada,* No. 21, 27 pp.
Sheath, R. G. & Cole, K. M., 1980. *J. Phycol.,* **16,** 412–420.
Sheath, R. G. & Cole, K. M., 1984. *Phycologia,* **23,** 383–396.
Shehata, F. H. A. & Whitton, B. A., 1982. *Br. phycol. J.,* **17,** 5–12.
Sieburth, J. McN. & Conover, J. T., 1965. *Nature Lond.,* **208,** 52–53.
Simpson, F. J., Neish, A. C. Shacklock, P. F. & Robson, D. R., 1978. *Botanica mar.,* **21,** 229–235.
Simpson, F. J. & Shacklock, P. F., 1979. *Botanica mar.,* **22,** 295–298.
Simpson, G. G., 1959. *Proc. Am. phil. Soc.,* **103,** 286–306.
Skinner, S. & Womersley, H. B. S., 1983. *Trans. R. Soc. S. Aust.,* **107,** 59–68.
Slocum, C. J., 1980. *J. exp. mar. Biol. Ecol.,* **46,** 99–110.
Söderström, J., 1965. *Botanica mar.,* **8,** 169–182.
Sokal, R. R. & Sneath, P. H. A., 1963. *Principles of Numerical Taxonomy.* W. H. Freeman, San Francisco, 350 pp.
Sousa, W. P., 1979. *Ecol. Monogr.,* **49,** 227–254.
South, G. R., 1978. In, *Proc Int. Symp. on Biogeogr. and Evolu. in the Sth. Hemiphere,* Auckland, N.Z., pp. 85–108.
Spencer, K. G., Yu, M. H., West, J. A. & Glazer, A. N., 1981. *Br. phycol. J.,* **16,** 331–343.
Stace, C. A., 1980. *Plant Taxonomy and Biosystematics.* E. Arnold, London, 279 pp.
Stam, W. T. & Holleman, H. C., 1975. *Acta. bot. neerl.,* **24,** 379–390.
Stebbins, G. L., 1950. *Variation and Evolution in Plants.* Columbia University Press, New York, 643 pp.
Stebbins, G. L., 1974. *Flowering Plants: Evolution Above the Species Level.* Harvard University Press, Cambridge, Mass., 399 pp.
Steneck, R. S., 1982. *Ecology,* **63,** 507–522.
Stewart, J. G., 1968. *J. Phycol.,* **4,** 76–84.
Stewart, J. G., 1977. *Bull. Sth. Calif. Acad. Sci.,* **76,** 5–15.
Stewart, J. G., 1984. *Bull. Sth. Calif. Acad. Sci.,* **83,** 57–68.
Stokes, P. M., Hutchinson, T. C. & Krauter, K., 1973. *Can. J. Bot.,* **51,** 2155–2168.
Stulp, B. K. & Stam, W. T., 1984. *Br. phycol. J.,* **19,** 287–301.
Stosch, H. A. von & Theil, G., 1979. *Am. J. Bot.,* **66,** 105–107.
Sundene, O., 1958. *Nytt Mag. Bot.,* **6,** 121–128.
Sundene, O., 1959. *Nytt Mag. Bot.,* **7,** 181–187.
Sundene, O., 1962. *Nytt Mag. Bot.,* **9,** 155–174.
Sundene, O., 1964. *Nytt Mag. Bot.,* **11,** 83–107.
Sundene, O., 1975. *Norw. J. Bot.,* **22,** 35–42.

Suto, S., 1950. *Bull. Jap. Soc. Sci. Fish.,* **16,** 1–9.
Svendsen, P. & Kain, J. M., 1971. *Sarsia,* **46,** 1–22.
Tanner, C. E., 1981. In, *Biology of Seaweeds,* edited by C. S. Lobban & M. J. Wynne, Blackwell Sci. Publs, Oxford, pp. 218–247.
Tatewaki, M. & Iima, M., 1984. *J. Phycol.,* **20,** 368–376.
Tatewaki, M., Provasoli, L. & Pintner, I. J., 1983. *J. Phycol.,* **19,** 409–416.
Taylor, A. R. A., 1971. *Proc. Int. Seaweed Symp.,* **7,** 263–267.
Taylor, A. R. A. & Chen, L. C.-M., 1973. *Proc. N.S. Inst. Sci.,* **27** (Suppl.), 1–21.
Taylor, P. R. & Hay, M. E., 1984. *Mar. Ecol. Prog. Ser.,* **18,** 295–302.
Thom, R. M., 1980. *J. Phycol.,* **16,** 102–108.
Turesson, G., 1922. *Hereditas,* **3,** 211–350.
Turesson, G., 1926. *Hereditas,* **8,** 157–160.
Turill, W. B., 1952. *Nature Lond.,* **169,** 388 only.
Tveter-Gallagher, E. & Mathieson, A. C., 1980. *Scanning Electron Microscopy,* **3,** 571–580.
Vadas, R. L., 1977. *Ecol. Monogr.,* **47,** 337–371.
Velimirov, B. & Griffiths, C. L., 1979. *Botanica mar.,* **22,** 169–172.
Vernet, P. & Harper, J. L., 1980. *Biol. J. Linn. Soc.,* **13,** 129–138.
Wallentinus, I., 1984. *Mar. Biol.,* **80,** 215–225.
Wanders, J. B. W., 1977. *Aquatic Bot.,* **3,** 357–390.
Wanders, J. B. W., Hoek, C. van den & Schillern-van nes, E. N., 1972. *Neth. J. sea Res.,* **5,** 458–491.
West, J. A., 1972. In, *Contribution to the Systematics of Benthic Marine Algae of the North Pacific,* edited by I. A. Abbott & M. Kurogi, Japanese Society of Phycology, Kobe, pp. 213–230.
West, J. A. & Hommersand, M. H., 1981. In, *The Biology of Seaweeds,* edited by C. S. Lobban & M. J. Wynne, Blackwell Sci. Publs, Oxford, pp. 133–193.
White, E. B. & Boney, A. D., 1969. *J. exp. mar. Biol. Ecol.,* **3,** 246–274.
Whittick, A., 1977. *J. exp. mar. Biol. Ecol.,* **29,** 223–230.
Wilce, R. T., Schneider, C. W., Quinland, A. V. & Bosch, K van den, 1982. *Phycologia,* **21,** 336–354.
Wilkinson, M., 1974. *J. exp. mar. Biol. Ecol.,* **16,** 19–27.
Williams, C. B., 1964. *Patterns in the Balance of Nature.* Academic Press, London, 324 pp.
Withers, R. G., Farnham, W. F., Lewey, S., Jephson, N. A., Haythorn, J. M. & Gray, P. W. G., 1975. *Mar. Biol.,* **31,** 79–86.
Woelkerling, W. J., 1971. *Aust. J. Bot.,* Suppl. 1, 1–91.
Womersley, H. B. S., 1959. *Bot. Rev.,* **25,** 545–614.
Womersley, H. B. S., 1964. *Aust. J. Bot.,* **12,** 53–110.
Womersley, H. B. S., 1965. *Aust. J. Bot.,* **13,** 435–450.
Wynne, M. J. & Edwards, P., 1970. *Phycologia,* **9,** 11–16.
Wynne, M. J. & Loiseaux, S., 1976. *Phycologia,* **15,** 435–452.
Yamada, I., 1974. *J. Fac. Sci. Hokkaido Univ., Ser. 5 (Bot.),* **10,** 32–47.
Yarish, C., 1975. *Nova Hedwigia,* **26,** 385–420.
Yarish, C., Breeman, A. M. & Hoek, C. van den, 1984. *Helgoländer Meeresunters.,* **38,** 273–304.
Yarish, C. & Edwards, P., 1982. *Phycologia,* **21,** 112–124.
Yarish, C., Edwards, P. & Casey, S., 1979. *J. Phycol.,* **15,** 341–346.

Oceanogr. Mar. Biol. Ann. Rev., 1986, **24**, 379–480
Margaret Barnes, Ed.
Aberdeen University Press

THE *ACANTHASTER* PHENOMENON*

P. J. MORAN
Australian Institute of Marine Science, P.M.B. No. 3, Townsville, M.C. 4810, Queensland, Australia

INTRODUCTION

The crown-of-thorns starfish (*Acanthaster planci* Linnaeus 1758) has become one of the most well-known animals in coral reef ecosystems. This notoriety has developed not because of its beauty or its commercial value but because it forms large aggregations or outbreaks, which can lead to the destruction of extensive areas of coral. Over the last 20 years numerous observations and opinions have been recorded about this starfish. These have ranged from scientific papers and reviews on various aspects of the biology and ecology of this animal to discussions of its effects on the tourist industry. Most debate on this topic has addressed two main questions: first, what causes outbreaks and secondly, are they influenced by man's activities?

In the light of such debate this paper has several aims. First, to define the bounds of our current knowledge of *A. planci* by focusing on those aspects of the phenomenon that are best known. Secondly, to indicate areas of conflict and debate among scientists and to highlight anomalies in the available data. Thirdly, and of equal importance it is the aim of this review to define those aspects that are least known, but are important to our understanding of the phenomenon. Fourthly, it will highlight the inadequacies of the hypotheses at present put forward to explain the origin of outbreaks. Finally, the paper identifies the various problems confronting scientists in their attempts to understand a phenomenon that is large in scale, enormously complex, and exhibits interesting inconsistencies and synchronies.

Several small reviews exist reporting current research or particular aspects of the phenomenon (*e.g.* Talbot & Talbot, 1971; Caso, 1972; O'Gower, McMichael & Sale, 1973; Sale, Potts & Frankel, 1976; Rowe & Vail, 1984a,b) yet few major reviews (*e.g.* Endean, 1973b, 1976) have been undertaken on this topic. The most recent (Potts, 1981) covered all research conducted until 1978. Since then a further series of outbreaks has occurred in various parts of the Indo-Pacific region leading to a resurgence in research and the development of several new ideas concerning outbreaks and their possible causes. These events have provided a justification for presenting this review as well as the desire to report this new information in

*Contribution Number 325 from the Australian Institute of Marine Science

the context of past results and hypotheses. Wherever possible the terminology in this paper follows that used by Potts (1981) who made a conscious attempt to avoid emotive terms, such as plague and infestation, because of their association with events that are somehow considered to be unpleasant, disastrous, and often unnatural. Consequently, the less emotional term, outbreak, has been used to describe large aggregated populations of starfish in this paper.

In essence, the *Acanthaster* phenomenon is a predator–prey interaction where the predator, *A. planci*, feeds on its prey, the corals. The two are intimately linked and should not be studied in isolation. Both must be investigated in order to comprehend fully the phenomenon, as the abundance of one changes in response to the abundance of the other (Bradbury, Hammond, Moran & Reichelt, 1985). Such dynamics are most readily seen in the interactions observed in various terrestrial ecosystems such as that between the lynx and snowshoe hare in Canada (Tanner, 1975). From a scientific viewpoint the occurrence of outbreaks of *A. planci* are unusual as this animal is a carnivore. Most references to outbreaks in the literature commonly involve herbivores such as locusts and other pest species (*e.g.* Ricklefs, 1979) and even sea urchins (North & Pearse, 1970). It is rare that a carnivore outbreaks on its own and that such outbreaks are not linked to increases in the abundance of its prey. There are even fewer reports of starfish outbreaking in the field. A notable exception is *Asterias forbesi* which has been recorded to outbreak in oyster grounds in the United States (Kenny, 1969).

Scientifically, outbreaks of *Acanthaster planci* are interesting as they provide an excellent opportunity first, to address certain key questions relating to the regulation of populations and secondly, to understand more about the dynamics of coral reef systems. Potts (1981) lamented the fact thatecologists had failed to use this natural experiment to their advantage. Perturbations on this scale offer scientists the opportunity to gain a deeper insight into the processes involved in structuring reefal systems as the systems themselves alter in response to the disturbance. While the results gained from studies of *A. planci* may have broad ramifications in several areas of marine ecology they may have importance in other scientific fields. For example, to date it has been used for testing neuropharmacological drugs (Buznikov, Malchenko, Turpaev & Tien, 1982), for synthesizing corticosteroids (Sheikh & Djerassi, 1973) and for investigating the physiological properties of echinoderm tissues (Motokawa, 1982).

THE *ACANTHASTER* DEBATE

The debate surrounding the *Acanthaster* phenomenon has developed into a very complex and emotional issue as the outbreaks themselves have involved people from many different parts of society and have affected the livelihoods of many people (*e.g.* those associated with the tourist industry). As a result, debates concerning the cause of outbreaks have involved a mass of opinions ranging from emotional calls for action to be undertaken (an understandable feeling if the effects of a large outbreak have been observed at first hand), through to informed and un-informed viewpoints from politicians and the general public. Enmeshed within these views and often swamped by them are those of the scientists. Like the public, their opinions also have varied since they have not been immune from the emotional aspects of the debate (*e.g.* Dwyer, 1971; Endean, 1971b; Hazell, 1971; Talbot, 1971; O'Gower, Bennett & McMichael, 1972; O'Gower, McMichael & Sale, 1973; Bradbury, 1976; Bradbury, Done *et al*, 1985; Rowe & Vail, 1985). Kenchington (1978) has given a thorough account of the various forces (*e.g.* scientific, historical, sociological, political, and economic) which were responsible for the controversy that surrounded the occurrence of outbreaks in Australia during the 1960s and 1970s. It would appear that in some instances the outbreaks were all but forgotten in the rush to enter the debate.

In some ways, scientists have only themselves to blame for the turmoil which has developed from this issue. When outbreaks were first reported (at a time when very little was known about them) numerous dire predictions were made by members of the scientific community. Some warned that outbreaks of starfish might lead to the mass erosion of reefs in the Indo-Pacific region which in turn might expose previously protected coastlines to erosional forces (Chesher, 1969a; Weber, 1970; Antonius, 1971). Others predicted that outbreaks would result in the destruction of the fishing industry and the loss of tourism (Chesher, 1969a; Vine, 1972). In addition, it was suggested that they may cause an increase in ciguatera poisoning (an algal-derived toxin in edible fish) since outbreaks produced large areas of substratum dominated by algae (Barnes, 1966) (see p. 441). To date, none of these predictions has been confirmed. This has led to a certain ambivalence on the public's part, towards the opinions and views of scientists (Raymond, 1984).

Since the early 1970s a number of committees of inquiry have been established to investigate the *Acanthaster* phenomenon and many of these have taken place in Australia. The first two committees formed (by the Federal and Queensland Governments) in this country (Walsh *et al.*, 1970, 1971) reported on what was known about *A. planci* at that time and sought to ascertain whether the starfish constituted a threat to the Great Barrier Reef. They also determined whether control measures should and could be implemented. Both committees recommended that extensive research be carried out on the phenomenon although the second committee concluded, on the basis of it findings, that *A. planci* did not "constitute a threat to the Great Barrier Reef as a whole" (Walsh *et al.*, 1971: p. 6). The conclusions of that committee, particularly the one just mentioned, were challenged and debated (*e.g.* Dwyer, 1971; Endean, 1971b; Talbot, 1971; O'Gower *et al.*, 1972; James, 1976). During 1971 an advisory committee was established to implement the policies of the second committee of inquiry and to coordinate future research. Over the ensuing years many aspects of the biology of *A. planci* were studied. The progress of these studies was reported in a document prepared by the advisory committee (Walsh, Harvey, Maxwell & Thomson, 1976) and in it further research was recommended particularly on the ecology and population dynamics of the starfish and its coral prey. With the decline of starfish outbreaks during the latter half of the 1970s research on *A. planci* waned. A further committee was established (by the Great Barrier Reef Marine Park Authority) not long after a second outbreak was reported at Green Island at the end of 1979. Once again the results of previous research were reviewed and the significance of those outbreaks was assessed to determine whether further research was warranted. That committee considered the situation serious enough to recommend that several types of research be undertaken, addressing a number of broad aspects of the phenomenon (Advisory Committee on the Crown of Thorns Starfish, 1980). Some of this research was implemented although field studies on the ecology of the starfish were largely neglected. In view of the seriousness of the current series of outbreaks on the Great Barrier Reef another committee was formed with similar aims to those preceding it (Crown of Thorns Starfish Advisory Committee, 1985). In constrast to the findings of the second committee of inquiry this committee concluded that "the destruction of hard coral by aggregations of *A. planci* poses a serious threat to the organisation and functional relationships within some reef communities within the Great Barrier Reef, at least in the short term" (*loc. cit.*, p. 1). It also recognized that outbreaks of starfish posed a "major management problem in some areas of the Great Barrier Reef" (*loc. cit.*, p. 1). On the basis of its findings the committee recommended that a co-ordinated programme of research be conducted over five years at an estimated cost of $A 3 million. Despite the reviews of these committees and the impetus that they gave to research they have not managed to quell the questions and debates, in Australia at least, concerning the phenomenon.

Frequently the debate surrounding the occurrence of outbreaks has been reduced to whether they are seen to be a problem, or threat to the reef and, ultimately, whether they are natural or man-induced events. Logic would have it that if they are natural then nothing is required except to adopt sensible management in areas of commercial interest. If unnatural then action may be required. Reducing the debate to this simplistic level at this time is trivial, as our knowledge of the phenomenon is inadequate to make rational decisions even in regard to these questions. All opinions, even those of scientists intimately associated with the phenomenon, are based to varying degrees on inadequate information. In conclusion, it is more realistic to suggest that outbreaks are a problem not because they may be natural or unnatural but because so little is known about them.

GENERAL BIOLOGY OF *A. PLANCI*

INTRODUCTION

Perhaps more is known about the general biology of *A. planci* than any other aspect of this animal. Research since the late 1960s has tended to concentrate on biological aspects; first, in order to gain a better understanding of the animal and secondly, as a means of establishing a store of knowledge upon which future experimentation may be based. Much of this research has been carried out in the laboratory (Potts, 1981) and has involved studies in the following five general areas: morphology, systematics, life history, growth, feeding and movement of *A. planci*. While these studies have increased our knowledge of *A. planci* they also have caused further controversy as some laboratory results have been found to be inconsistent with those obtained from field studies (these inconsistencies will be discussed in the following sections). The validity of results from laboratory studies has been questioned as they are derived under conditions which may be more artificial and simplistic than those found in the field. Even though this criticism may be justified care also should be undertaken when interpreting the significance of results from field studies as very little is known about the ecology and dynamics of *A. planci* populations. It should also be borne in mind that the results from field studies may not reflect the effects of the variable being tested, but a complex of variables which are poorly understood.

MORPHOLOGY

A. planci (Fig. 1) is a carnivorous starfish found on reefs throughout the Indo-Pacific region. A detailed description of the external features of this animal has been given by Caso (1970). It is a large asteroid which may grow to more than 700 mm in diameter (from arm tip to arm tip) in the wild (see Lucas, 1984). Measurements conducted throughout the world have shown that adults normally range in size from 250–350 mm (Campbell & Ormond, 1970; Nishihira & Yamazato, 1972; Cheney, 1974; Ormond & Campbell, 1974; Kenchington, 1977).

A. planci is multi-coloured and individuals have been reported to range from purplish blue with red-tipped spines (Clark, 1921) to green with yellow-tipped spines (Branham, 1973). The general colour of an individual, which depends on the degree of extension of the dermal papulae (Clark, 1921), may vary through time (Barnes & Endean, 1964; Barham, Gowdy &

Fig. 1.—The crown-of-thorns starfish (*Acanthaster planci*) seen on recently dead (white) coral.

Wolfson, 1973). This variation is thought to be related to the effects of diet (Branham, 1973).

Adults generally possess from 8–21 arms or rays, although this figure has been found to vary from place to place (Table I). A number of small rounded plates known as madreporites are situated on the aboral surface of the oral disc. Their relative position has been used, in conjunction with other variables, to identify individual starfish (p. 421). Adult *A. planci* may have between 3 and 16 madreporites (Hyman, 1955; Caso, 1970; Barham *et al.*, 1973; Glynn, 1982b). Adults have also been found to possess from 1–6 anuses (Glynn, 1982b).

TABLE I

Variation in the number of arms or rays of adult starfish from different areas of the Indo-Pacific region

Area	No. of arms	Reference
Great Barrier Reef	14–17	Endean, 1969
Guam	14–18	Cheney, 1974
Gulf of California	12–15	Barham *et al.*, 1973
Gulf of Thailand	8–17	Piyakarnchana, 1982
Indonesia	10–18	Aziz & Sukarno, 1977
Okinawa	11–21	Nishihira & Yamazato, 1972
Red Sea	13–15	Ormond & Campbell, 1971

The exterior of *A. planci* is covered by numerous spines up to 40–50 mm in length (Endean, 1973b) which may grow at a rate of 1·3 mm per month (Pearson & Endean, 1969). Caso (1970) identified six types of spines on the aboral and oral surfaces of *A. planci* (lateral, marginal, ventral, adambulacral, ambulacral, and buccal). Recently, Walbran (1984) compiled an atlas of the most common skeletal components of this starfish. This included a discussion on the morphology, micro-structure, and architecture of preserved fragments as well as those found in sediments. A comparison was also made between these skeletal components and those from other starfish commonly occurring on the Great Barrier Reef. Walbran (1984) concluded that the skeletal components of *A. planci* (even those found in sediments) could be differentiated readily from those of other starfish on the basis of morphology, colour, and micro-structure.

TOXICITY

Apart from being abundant and structurally diverse the spines of *A. planci* can inflict a toxic reaction. As well as inflicting a painful wound they may cause several other symptoms including nausea, vomiting, and swelling in humans (Barnes & Endean, 1964; Pope, 1964; Weber, 1969; Odom & Fischermann, 1972; Williamson, 1985). No evidence has been found to indicate that a venom is actively injected into the wound created by a spine (Fleming, Howden & Salathe, 1972). Toxic compounds have been isolated from the spines of *A. planci* by Croft, Fleming & Howden (1971) and Taira, Tanahara & Funatsu (1975). The substance isolated by Croft *et al.* (1971)

was found to be a saponin which was present in the tissue overlying the spines. It was thought that this compound was present in insufficient quantities to cause the toxic reactions normally associated with this starfish (Croft *et al.*, 1971). At present it is not known what causes these reactions which are sometimes severe. Crude extracts of material isolated from the surface of spines have been found to have a haemolytic effect on human red blood cells (Everitt & Jurevics, 1973). Biochemical studies by Heiskanen, Jurevics & Everitt (1973) have indicated that inflammation around the wound may be mediated by the activities of histamine-like compounds whereas the pain associated with being pierced by a spine may be due to another cause.

Because they are abundant, large, and toxic the spines of *A. planci* are thought to represent a specialized adaptation which serves to protect the animal from predation (Cameron, 1977; Moore, 1978). This may be true but it is not known to what extent they prevent predation nor is it known how toxic they are to other marine animals. Indeed, very little is known about the quantitative aspects of predation of this starfish (see pp. 414–418).

HABITAT

Studies of the distribution of *A. planci* on reefs have shown that it prefers to live in sheltered environments such as lagoons and also in deeper water on the windward slopes of reefs (Chesher, 1969a; Pearson & Endean, 1969; Ormond & Campbell, 1974; Moran, Brabury & Reichelt, 1985). In general, this starfish avoids shallow or exposed locations where it is susceptible to wave action. Aggregations of starfish have been recorded to depths of 30 m (Branham, Reed, Bailey & Caperon, 1971) while individual starfish have been observed at approximately 40 m (Devaney & Randall, 1973). It is likely that they inhabit greater depths as an *A. planci* was dredged from almost 64 m near Euston Reef in the Great Barrier Reef (Great Barrier Reef Marine Park Authority, unpubl. data). An unconfirmed report exists of a starfish being found off Hawaii at a depth of 100 m (Chesher, 1969a).

SYSTEMATICS

A. planci has been known for many years. It was first described by Rumphius in 1705 and later by Plancus and Gualtieri in 1743 (Vine, 1973) and named in 1758 by Linnaeus. An historical account of the early description and classification of *A. planci* has been given by several authors (Weber, 1969; Branham, 1973; Vine 1973). There continues to be confusion as to the number of valid species referable to the genus *Acanthaster*. Madsen (1955), in reviewing the genus recognized two distinct species, *A. planci* and *A. ellisii*, the latter being found only in the eastern Pacific region. A third species, *A. brevispinus*, which at that time was known only from the Philippines was proposed. Its status, however, as a separate species was thought to be doubtful. Caso (1961) considered *A. ellisii* to be a valid species and divided it into two subspecies, *A. e. ellisii* and *A. e. pseudoplanci*. Barham *et al.* (1973) also argued for the separation of *A.*

ellisii from *A. planci* on the basis of its different behavioural characteristics; tending not to be cryptic during daylight hours, and appearing not to aggregate or migrate. They also pointed out that the disc diameter/arm length ratio for this species is different from that of *A. planci*. These distinguishing features have, however, been regarded by Glynn (1974, 1976) as being minor and reflecting the normal range of variability present in *A. planci*.

Attempts were made by Lucas & Jones (1976) to evaluate the status of *A. planci* and *A. brevispinus* by crossing individuals from an area of neighbouring sympatry on the Great Barrier Reef. Although both species were shown to have a high degree of genetic compatibility they were regarded as sibling species. Lucas & Jones (1976) argued that hybrids did not occur naturally on the Great Barrier Reef as ecological barriers prevented the exchange of genetic information between the two species. Unlike *A. planci*, on the Great Barrier Reef *A. brevispinus* is not found on reefs but occurs in deep water between reefs. Also it does not feed on corals but is thought to be an omnivore, preferring a more general diet. The results from recent studies have substantiated the claim that there is no exchange of genes between these species. Lucas, Nash & Nishida (1985) have demonstrated that larvae from F2 and hybrid x parental crosses are of low viability and suffer a high rate of developmental abnormalities. Furthermore, while the two species share common alleles for most gene loci they are homozygous for different alleles at one presumptive locus. From this biochemical genetic evidence they concluded that *A. planci* had evolved recently from a more generalist ancestor similar to *A. brevispinus*.

While there is good evidence to support the separation of *A. planci* and *A. brevispinus*, the taxonomic status of *A. ellisii* remains uncertain. A biochemical genetic study has indicated that this species is very similar to *A. planci* (Lucas *et al.*, 1985). At present gene frequencies for *A. ellisii* and *A. brevispinus* exist only for a single population of each ($n = 53$ and 11, respectively) and nothing is known of their variation between populations. The only information on this topic for *A. planci* comes from studies conducted by Nash (1983) who analysed seven populations within a large area, from Lizard Island to One Tree Island (Capricorn-Bunker Group: see Fig. 7, p. 431), on the Great Barrier Reef. He found that the genetic composition of starfish populations over this region was generally homogeneous. A population at Green Island was found to be genetically different from the others but the reasons for this were unable to be established.

Taxonomic uncertainty has occurred at the family level as well as the species level. The two species *A. planci* and *A. brevispinus* form part of the monogeneric family Acanthasteridae which was recently aligned with the family Oreasteridae on the basis of certain skeletal characteristics (Blake, 1979). In doing so the family was assigned from the order Spinulosida to the order Valvatida. This alteration is at variance with the findings of Mochizuki & Hori (1980) who suggested on the basis of immunological and morphological evidence, that a close affinity existed between the families Acanthasteridae (*A. planci*), Solasteridae and Asterinidae, all of which occur in the order Spinulosida. A close affinity was also proposed between these families and the Ophidiasteridae in the order Valvatida.

There is very little fossil evidence to support theorized phylogenies within the genus *Acanthaster*. A possible relative of this genus may extend back to the Cenozoic period, but this conclusion is based on incomplete fossil evidence (Blake, 1979).

REPRODUCTION AND LIFE CYCLE

A. planci is a gonochoristic (dioecious) species which reproduces sexually. Unlike some starfish (*e.g. Linckia* spp.) it is not known to reproduce asexually by arm autotomy or somatic fission (Yamaguchi, 1975b). Studies in the field have found that the ratio of males to females is almost one to one (Pearson & Endean, 1969; Nishihira & Yamazato, 1974). Like many other invertebrates, planktonic larvae of *Acanthaster planci* are produced by external fertilization. Estimates have been made of the number of eggs that may be spawned by a single female during one season. Pearson & Endean (1969) calculated that females may contain from 12–24 million eggs. Recently, Conand (1983) suggested that large individuals (400 mm diameter) may produce as many as 60 million eggs during one season.

There is still some uncertainty regarding the timing and duration of the spawning season of *A. planci*. On the Great Barrier Reef spawning has been reported between December and January when the water temperature is above 28 °C (Pearson & Endean, 1969; Lucas, 1973). Reports from other areas in the Indo-Pacific region, indicate that it may vary according to location (Table II). In addition, while spawning in some areas is relatively restricted, occurring over a few months of the year (*e.g.* Great Barrier Reef), there are places where it seems to be more prolonged occurring intermittently over a number of months (*e.g.* Gulf of California). Indeed, in some areas fertile eggs have been collected from starfish throughout the year (Branham *et al.*, 1971; Yamazato & Kiyan, 1973) suggesting that there is an almost year-round potential for spawning. This potential has been

TABLE II

Spawning period of A. planci reported *for different locations in the Indo-Pacific region*

Location	Spawning period	Reference
Fiji	Dec.–Feb.	Owens, 1971
Great Barrier Reef	Dec.–Jan.	Pearson & Endean, 1969; Lucas, 1973
Guam	Nov–Dec.	Chesher, 1969a
	Sept.–Oct.	Cheney, 1974
Gulf of California	Apr.–	Dana & Wolfson, 1970
Hawaii	Apr.–May	Branham *et al.*, 1971
Java	Apr.–	Mortensen, 1931
New Caledonia	Nov.–Feb.	Conand, 1983
Okinawa	June–July	Yamazato & Kiyan, 1973
Panama	Jan.–	Glynn, 1974
Red Sea	July–Aug.	Roads & Ormond, 1971
Western Australia	Nov.–Jan.	Wilson & Marsh, 1975
Western Samoa	Dec.–Jan.	Garlovsky & Bergquist, 1970

demonstrated also for the Great Barrier Reef but there is no evidence of this actually occurring (Lucas, 1973). While there is the potential for prolonged spawning it would appear that this is not a significant occurrence and that spawning normally takes place within a well-defined period of a few months. The data presented in Table II indicate that spawning is concentrated between May and August in the Northern Hemisphere and November to February in the Southern Hemisphere. A similar breeding season for areas north of the equator was proposed by Birkeland (1982). Yamazato & Kiyan (1973) have argued that the spawning period for *A. planci* is more extended in the tropics than in the higher latitudes as those areas experience longer periods of high water temperature.

Apart from geographical variations in the timing and duration of the spawning period of *A. planci* it has been reported also that spawning in some areas is variable from year to year (Wilson & Marsh, 1975). These variations may reflect local changes in environmental factors such as temperature (Cheney, 1974) which is important in influencing spawning (Cheney, 1972a; Lucas, 1984). In addition, they may reflect the different methods (gonad index, gonad dissection, gonad histology) used to determine the reproductive state of starfish (Lucas, 1972). Additional variability in these determinations may occur as gonad size and state has been found to vary widely in isolated *A. planci* yet remain uniform in aggregated individuals (Cheney, 1974).

There have been several reports of *A. planci* spawning in the field. Owens (1971) and Branham *et al.* (1971) observed spawning starfish in Fiji and Hawaii, respectively. Perhaps the best account of this phenomenon is that given by Pearson & Endean (1969) who described a spawning event on the Great Barrier Reef. In particular, they noted the behaviour of starfish before, and during the release of gametes which lasted for approximately 30 min in both sexes (Fig. 2). Although most were males, one female was seen to spawn in their vicinity. Spawning has been reported in the laboratory by Branham *et al.* (1971) and Misaki (1974, 1979). Lucas (1984) also reported that a group of hybrids spawned in his aquaria during winter. Spawning can be induced artificially by injecting ripe adults with a prepared solution of 1-methyladenine (Yamaguchi, 1973b) provided the starfish are in the final stages of gametogenesis.

A substance has been isolated from the gonads of both male and female starfish which is thought to synchronize the release of gametes by starfish (Beach, Hanscomb & Ormond, 1975). Similar amounts of this compound were found in both sexes and neither showed a contrasting sensitivity to it. Experiments conducted in the laboratory have demonstrated that this substance is released during spawning and that once released it induces nearby starfish to spawn. In addition, its release stimulated the movement of starfish in the direction of the spawning individual. Apart from the limited observations of Pearson & Endean (1969) there are very few eyewitness accounts of synchronized spawning. The results of Beach *et al.* (1975), however, suggest that this pheromone-like compound may be an important factor in determining the numbers of larvae produced during a spawning period since it has the potential to induce starfish to aggregate and spawn synchronously. No experiments have been conducted in the field to determine whether the degree of fertilization of eggs is positively correlated with adult density. Lucas (1975) considered that normal, non-aggregated,

Fig. 2.—Typical behaviour of a spawning adult (photograph taken by J. Davidson).

populations of *A. planci* were likely to produce few larvae as such adult densities would not lead to a high rate of fertilization.

Information on the life cycle of *A. planci* has come mainly from studies conducted in the laboratory (Potts, 1981). Six main stages have been identified and they are summarized in Table III. The last four stages were described by Lucas (1984) from laboratory studies. The general life cycle of this starfish is presented diagrammatically in Figure 3. Once fertilized, the egg of *A. planci* develops from an embryo into a larva which feeds on phytoplankton (Henderson & Lucas, 1971). During its planktonic life, which may be up to a month in duration (Yamaguchi, 1972a, 1973b), it passes through several developmental phases after which time it settles and metamorphoses into a five-armed juvenile. The last process takes about two days (Henderson & Lucas, 1971). Initially, this juvenile starfish is thought to eat mostly encrusting and epiphytic algae and its growth rate is relatively slow. After approximately six months it has the morphology of an adult and changes its diet to corals. Although the growth rate of this starfish is high during this period it is not capable of reproducing and hence it is termed a coral-feeding juvenile. The general term juvenile (*e.g.* Laxton, 1974) is often applied to starfish belonging to either of the first two post-metamorphic categories delineated by Lucas (1984). According to laboratory growth studies *A. planci* begins reproducing towards the end of its second year (Lucas & Jones, 1976) and is then referred to as a coral-feeding adult. Lucas (1984) found that the growth rate of starfish during this stage

TABLE III

Stages in the life cycle of A. planci *(after Lucas, 1984): *age from metamorphosis*

No.	Stage	Age	Size (mm)	Growth rate	Food	Feeding rate	Fecundity
i	Laval	0–28 days	0·1–1·2	—	Unicellular algae	—	—
ii	Settlement/Meta-morphosis	0–2 days	0·3–0·5	—	—	—	—
iii	Algal feeding juvenile	0–6 months*	1–10	Low (exponential)	Encrusting and epiphytic algae	—	No
iv	Coral feeding juvenile	6 months–2 yr*	10–200	High (von Bertalanffy)	Coral	High	No
v	Coral feeding adult	2–3 yr*	200–350	Decreasing	Coral	Decreasing	High
vi	Senile adult	>3 yr*	350	Low	Coral	Low	Low

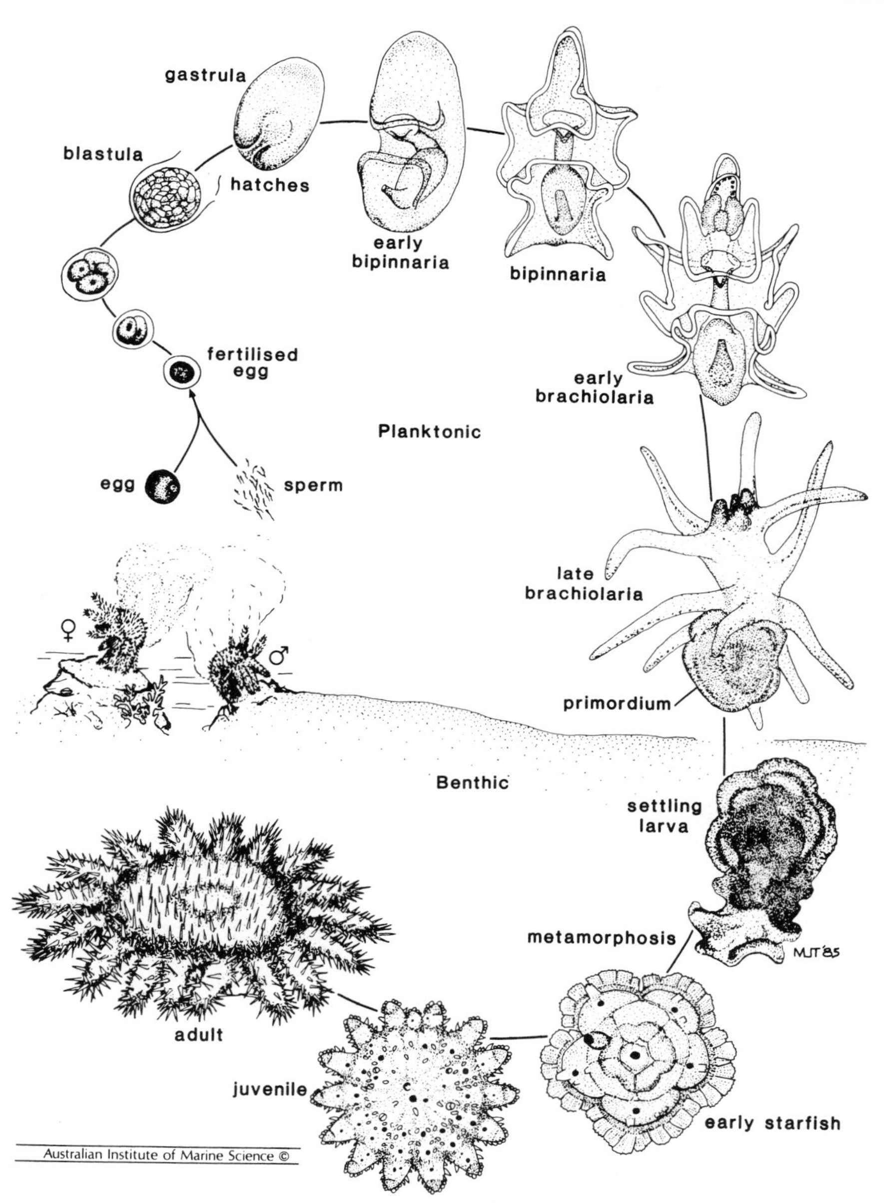

Fig. 3.—Diagrammatic representation of the life cycle of *Acanthaster planci* developed from ideas by J. Lucas, R. Olson and the author: the two juvenile stages redrawn from Yamaguchi (1973b).

decreases and that their reproductive output is high. At the end of three years, laboratory reared adults have been found to enter a "senile" phase in which they experience little or no growth. Gamete production also decreases sharply during this time. This senile period may last for a further two years after which time the starfish are likely to die (Lucas, 1984). Several inconsistencies in the biological data presented for the phases identified by Lucas (1984) will be discussed in a later section (see pp. 393–397).

LARVAL BIOLOGY

Lucas (1982) recognized seven distinct stages in the development of larvae some of which are shown in Figure 3. This process began with the hatching of the embryo as a gastrula larva and proceeded through the following six stages; early bipinnaria, advanced bipinnaria, early brachiolaria, mid brachiolaria, late brachiolaria, settlement and metamorphosis. Prior to the commencement of this process, the eggs when shed by the female are approximately 0·2 mm in size, light yellow in colour (Henderson & Lucas, 1971; Yamaguchi, 1973b, 1977) and are negatively buoyant (R. Olson, pers. comm.). The sperm are much smaller and possess a spherical head (0·002 mm in diameter), middle section and long (0·04–0·05 mm) flagellum (Henderson, 1969). After fertilization, the embryo develops to the blastula stage (which has a wrinkled exterior) within seven hours (Hayashi, Komatsu & Oruro, 1973) and hatches as a free-swimming gastrula after about 30 hours (Henderson, 1969). It then develops into a bipinnaria larva and begins feeding on unicellular algae. Development to the bipinnaria stage may take from two to four days (Yamaguchi, 1977; Henderson, 1969).

Mortensen (1931) was the first to rear larvae to the brachiolaria stage in the laboratory. Other early larval studies were unsuccessful. Henderson (1969), Branham *et al.* (1971), and Henderson & Lucas (1971) reared larvae in the laboratory to the juvenile starfish stage. These studies demonstrated that the rate of development of larvae to the brachiolaria stage may be affected greatly by small changes in temperature. The time taken for larvae to develop to the brachiolaria stage has been variously reported to take from 9 days (Lucas, 1982) to 12 days (using normal sea water that was high in phytoplankton) (Yamaguchi, 1977) at 28 °C, 16 days at 27 °C (Mortensen, 1931), and 23 days at temperatures ranging between 24 and 29 °C (Henderson & Lucas, 1971). Larvae exposed to continual temperatures of 24–25 °C did not advance past the early bracholaria stage. It would seem that the development of larvae is completed only within the temperature range of 25–32 °C (Lucas, 1973) and that maximum survival and development is achieved between 28 and 32 °C.

Salinity changes have been noted to affect the development of larval *A. planci*. Bipinnaria larvae were found to tolerate a wide salinity range (21–33‰) while later stages were less tolerant (Henderson & Lucas, 1971). Lucas (1973) reported that larvae completed their development in salinities as low as 26‰. He found, however, that survival of larvae was enhanced threefold in a salinity of 30‰ (Lucas, 1973, 1975).

Observations from laboratory studies suggest that while in the plankton, larvae exhibit negative geotaxis and are photopositive actively swimming towards the water surface, although it is possible that this movement may be disrupted by wave motion and water currents (Yamaguchi, 1973b). Very little is known about the dispersal of larvae in the field (Lucas, 1975) or the effects of water currents on larval dispersal and recruitment. Plankton trawls were undertaken by Pearson & Endean (1969) in a bid to study larval dispersal but they were largely unsuccessful.

Towards the end of the brachiolaria stage when the larvae are about 1–1·2 mm in size they begin to drift downward and explore substrata to find a suitable surface on which to settle. It has been suggested that they settle mainly on dead corals and under boulders (Ormond *et al.*, 1973). Yamaguchi (1973a) observed in the laboratory that some larvae settled on dead coral covered with coralline (*Porolithon*) and other epiphytic algae. This was also noted by Henderson & Lucas (1971) although they found that the larvae did not settle on other substrata. They suggested that larvae may not settle if a suitable substratum is not found. Experiments by Lucas (1975) provided evidence to indicate that larvae did not require a particular surface but only one that possessed a biological film. Apart from not knowing what type of surface the larvae settle on in the field it is not known in which areas of the reef they settle. If they remain in the upper layers of the water column, it might be expected that they would settle in shallow areas on reefs (Ormond & Campbell, 1974). However, towards the end of the brachiolaria stage when a primordium is beginning to develop the larvae become negatively buoyant and tend to sink (Olson, 1985). This behaviour may result in larvae settling in areas of deeper water. This aspect will be discussed in more detail in a later section (see pp. 413–414).

GROWTH AND LONGEVITY

Studies of the growth of *A. planci* essentially have addressed the following four questions each of which will be discussed in turn in this section.

(1) Does *A. planci* grow at the same rate throughout its life?
(2) Does it grow continually throughout its life; *i.e.* is growth determinate or indeterminate?
(3) Is it possible to determine the age of a starfish from its size?
(4) How long does *A. planci* live?

One of the first studies of the growth rate of *A. planci* was undertaken by Pearson & Endean (1969) who obtained growth data from individual starfish kept in the field as well as the laboratory. From these studies they found that adults grew at a rate of between 9–14 mm a month while over the same time juveniles increased their diameter by 11 mm. They thought that growth after metamorphosis was rapid as they had found individuals up to 33·8 mm in size only two months after the spawning period. As a result they estimated that starfish could attain a size of 140 mm in almost 12 months assuming growth occurred at a linear rate of approximately 10 mm a month.

Studies have since demonstrated that starfish do not grow at a linear rate and that the initial growth of juveniles prior to transformation is slow. The newly metamorphosed juvenile starfish is between 0·3 and 0·5 mm in size (Henderson & Lucas, 1971; Yamaguchi, 1973b). Over the next four to five months the starfish, which feeds on algae grows to 8–10 mm in diameter (Yamaguchi, 1972a,b). The growth rate at this stage is exponential and the starfish may develop new arms at the rate of one every 9–10 days (Yamaguchi, 1975b). After about six months the juvenile starfish possesses all the external features of an adult, with about 16–18 arms. At this point it begins to feed on corals (Yamaguchi, 1973b, 1974; Lucas, 1975). Once this transformation has been completed, the growth rate increases and starfish may reach a size of 60–70 mm within 1 year and 200 mm after 1·5–2 years (Yamaguchi, 1974b; Lucas & Jones, 1976). This phase of Von Bertalanffy-type growth (Lucas, 1984) continues at least until the starfish reaches sexual maturity. Most starfish become sexually mature late in their second year of life (Lucas & Jones, 1976).

Until starfish attain sexual maturity their growth is sigmoidal and may best be described by a logistic growth curve (Yamaguchi, 1975a). Studies by Yamaguchi (1974b) and Lucas (1984) demonstrated that the growth rate of starfish declines greatly once sexual maturity is attained. In other words, the growth of starfish tapers off approximately 20 months after metamorphosis. From his laboratory studies Lucas (1984) recognized a phase of non-growth in starfish at three years of age (about 350 mm). During this "senile" period the size of some individuals was found to decrease and gametogenesis also began to decline. Lucas (1984) indicated that this phase may last for several years after which time the starfish may die.

Laboratory studies by Yamaguchi (1973b, 1974b, 1975b), Lucas & Jones (1976), and Lucas (1984) indicate not only that the growth rate of *A. planci* is variable throughout its lifetime but also that its growth is determinate. Whether the growth of *A. planci* in the field is determinate has not been resolved. The studies mentioned previously have been criticized on the grounds that the results may be an artifact of laboratory conditions and that they may represent the effects of such factors as disease and infection (Kenchington, 1977). This was also suggested by Lucas (1984) who stated that the senile phase he observed in the laboratory may have been a consequence of a number of factors including; the size and volume of the aquaria, the absence of predation and the lack of environmental variability. He considered, however, that there was circumstantial evidence of senility occurring amongst starfish in the field. He referred to the studies of Branham *et al.* (1971) and Kenchington (1977) where they had measured the growth of isolated starfish populations over a year. The individuals in these populations grew very little during these studies and this was interpreted by Lucas (1984) as evidence of senility. Kenchington (1977), on the other hand, regarded that this lack of growth in both studies was due to local conditions which reduced the availability of food.

The debate surrounding the mode of growth of *A. planci* has continued because there are few data on the growth of individuals in the field. Kenchington (1977) stated that the growth of this starfish is indeterminate since individuals up to 700 mm have been found on reefs. This is not an isolated occurrence as there have been several reports of starfish greater than 500 mm in size on reefs (Chesher, 1969b; Laxton, 1974; Stanley, 1983; Moran, Bradbury & Reichelt, 1985). These findings would appear to conflict with those of Yamaguchi (1974b, 1975a) and Lucas (1984) who proposed that starfish ceased growing at a diameter of approximately 350–400 mm. Kenchington (1977) attempted to obtain information on growth from the field by analysing size frequency data which had been collected at different times from a number of reefs in the Great Barrier

Reef. By analysis of modes within size frequency distributions he identified a number of size classes within each sample group which were thought to be related to the age of starfish. A total of six year classes were determined for the entire number of samples ($n=7143$). Kenchington (1977) derived a growth curve from this information which gave values similar to those obtained in the field by Pearson & Endean (1969), although they did not reflect the early rapid growth phase (coral-feeding juvenile phase) described by Yamaguchi (1974b) and Lucas (1984).

The model presented by Kenchington (1977) demonstrated that *A. planci* grew initially at an exponential rate followed by an arithmetic increase in size. The growth curve derived from the size frequency data suggested that the growth of *A. planci* was indeterminate and that it did not stop after about three years of age as was proposed by Yamaguchi (1974b) and Lucas (1984). Kenchington (1977) argued that large starfish (>350 mm) found in the field may have undergone longer periods of exponential, or faster, growth. He concluded that if this were the case then large sexually immature animals should be found in populations on reefs. As this type of starfish has not been found in the field Kenchington (1977) proposed that the growth of *A. planci* was indeterminate. Lucas (1984), on the other hand, claimed that the occurrence of these large animals in the field would not be evidence of this type of growth but that they may have arisen due to genotypic and environmental variations, although he did not elaborate on these suspected causal factors. This debate will remain unresolved until intensive growth studies of individual starfish are carried out in the field over long periods.

The results of Kenchington's study have received further criticism. Ebert (1983) has stated that the data analysed by Kenchington (1977) were unsuitable as there had been little continuous sampling of the same sites through time. He pointed out as a consequence that it was not possible to define whether or not successful recruitment had occurred during the same year over the entire Great Barrier Reef.

While undertaking this study Kenchington (1977) made three assumptions; first, that the spawning period of *A. planci* was restricted to late December or January, secondly, that the growth of starfish was the same over all areas of the Great Barrier Reef and finally, that modes in a size frequency distribution corresponded to age classes that were separated by one year. This study has been criticized mainly on the grounds that there is little relationship between the size of a starfish and its age (Lucas, 1984). Other studies have been conducted which have ascertained the age of starfish on reefs from an analysis of the size frequency distribution of the population (Ormond & Campbell, 1971, 1974; Nishihira & Yamazato, 1972; Endean & Stablum, 1973b; Laxton, 1974) and they too have been criticized for the same reasons.

In the laboratory, Lucas (1984) demonstrated that the growth and size of starfish are governed by diet. He showed that the diameter of starfish fed on coral may be twenty times that of starfish fed on coralline algae for the same period of time. On the basis of this information he suggested that it was erroneous to assume that there was a correlation between the size of an individual and its age. It follows that the growth of starfish in the field will depend on the types of food available (Ormond & Campbell, 1971). Should larvae recruit to an area which has a high coverage of encrusting algae and little coral, then their growth and size may be severely restricted. If they are unable to find coral, juvenile starfish may continue to feed on algae and their size may be much less than those of larvae that fortuitously settle in an area of high coral cover and that were able to change their diet quickly once they had attained adult morphology. Recently, studies have been conducted in Fiji where the growth of a large number of juvenile starfish was followed for over a year (L. Zann, *pers. comm.*). Initially they were similar in size but as the study progressed the size range of the starfish increased considerably. This gap was found to widen as some starfish began to feed on corals while others continued to feed on algae.

Observations in the field indicate that the growth and thus size of starfish may also be altered by several factors other than diet. Branham *et al.* (1971) reported that the diameter of starfish increased and decreased both before and after spawning. They also suggested that the mean size of individuals may be determined by population density. In Hawaii they found that the mean size of aggregated starfish was smaller (240 mm) than individuals that were sparsely aggregated (350 mm). The size of starfish may also be affected by handling. Yamaguchi (1974b) found that handling of *A. planci* may cause them to reduce their size by up to 20%. These findings highlight the need for care when interpreting the results of size frequency distributions (Feder & Christensen, 1966). It would seem that the usefulness of such a practice may be confined to making general statements about population structure (*e.g.* defining the occurrence of juvenile and adult sub-populations) rather than attempting to describe more detailed characteristics such as the age of various subgroups within a population.

Ebert (1983) considered that Kenchington's model indicated that *A. planci* was a relatively short-lived species since the populations declined to low levels several years after they appeared. While this model does indicate that the majority of starfish disappear from reefs (die?) it does not preclude the possibility that a small number of starfish may remain and live for many years. This raises the question as to what happens to the large numbers of starfish which seemingly vanish from reefs at the end of an outbreak. Do they move off into deeper water or to another reef or do they die? Studies by Glynn (1984b) have indicated that it may take at least 4 days for starfish to decompose in the field. Why then are newly-dead or decomposing starfish not sometimes observed on reefs, given the densities of individuals which may be present in outbreaks? The life expectancy of starfish in the field is unknown. Cameron & Endean (1982) hypothesized that *A. planci* must be a long lived species because it has specialized defensive structures (long venomous spines), few parasites and has certain specialized feeding adaptations (see p. 400). Chesher (1969b) suggested that *A. planci* may live for up to eight years but he gave no evidence to support this statement. In the light of studies by Lucas (1984) this figure may be realistic as he managed to keep some starfish in aquaria for almost this length of time despite the fact that others had died earlier from disease. Ebert (1973) applied a growth model to data from Hawaii (Branham *et al.*, 1971) and predicted that it would take almost 30 years for starfish to reach full size. In the light of current knowledge of the biology of *A. planci* this model would seem to be unrealistic. Accurate information on the longevity of *A. planci* in the field may not be forthcoming until a true field study of the population dynamics of this starfish is undertaken.

FEEDING BIOLOGY

Experiments conducted in the laboratory by Lucas (1982) using diets of single species of unicellular algae have shown that there is an inverse relationship between the filtration rate of bipinnaria and brachiolaria larvae and food concentration. The maximum rates of filtration for these larval stages were recorded to be from 1·3–6·6 μl per min. While this relationship is common to a number of echinoderms it was noted to be complex for certain of the larval stages of *A. planci*. Although the filtration rate declined as food concentration increased it was generally insufficient to cause a reduction in the rate at which food was ingested. Thus, there was a positive relationship between ingestion rate and food concentration. The highest rate of development and survival was achieved with food concentrations from $5–10 \times 10^3$ cells per ml (Lucas, 1982). During these studies seven species of algae were tested for their effects on larval development and survival. Of these, *Dunaliella primolecta* and *Phaeodactylum tricornutum* supported the most rapid larval development and highest survival. After comparing the results of these experiments with data (phytoplankton abundance and chlorophyll *a* concentrations) available from the Great Barrier Reef, Lucas (1982) concluded that the levels of phytoplankton normally found in the field were insufficient for the development of larvae of *Acanthaster planci*. Consequently, he postulated that "food is a major environmental influence on survival and development of *A. planci* larvae in these waters" (Lucas, 1982: p. 173).

This statement addresses the much debated issue of the relationship between larval abundance and phytoplankton concentration which Thorson (1950) considered an important problem in larval ecology. Despite being recognized as an important issue for many years, very little is known about whether larval starvation occurs in the field. It is generally thought that the two major causes of mortality of invertebrate larvae are predation and starvation (Vance, 1974). Starvation may affect the survival of larvae directly, by causing the death of the organism. It may also affect survival indirectly, by lengthening the larval phase (Lucas, 1982), thereby reducing the 'vitality' of larvae and increasing the potential for predation.

Lucas (1982) used chlorophyll *a* as a measure of phytoplankton biomass and compared his results (where larvae were fed on a single algal species) with concentrations in the field. While phytoplankton productivity in coral reef areas is generally considered to be low (Kinsey, 1983), it is not clear whether these conditions cause mass larval starvation. Rather, their effects on larval survival may be compensated for by the presence of a diverse range of phytoplankton species. Several studies have shown that a mixed diet of phytoplankton is beneficial to the survival of invertebrate larvae (Bayne, 1965; Gaudy, 1974). As yet, nothing is known about the likely benefits such a diet would have on the survival of larvae of *A. planci* in the field.

Bacteria and dissolved organic matter are two other possible sources of nutrition for larvae. Very little is known about their abundance in coral reef waters or their nutritional importance to the larvae of *A. planci*. Lucas (1982) considered that bacteria may not be an important nutritional source as the larvae of *A. planci* may not be able to feed on them efficiently. He admitted, however, that there were no data on bacterial numbers in Great Barrier Reef waters. Dissolved organic matter has been shown to be used as a source of nutrition by echinoderm embryos (Strathmann, 1975). While Lucas (1982) agreed that organic molecules may be absorbed by echinoderm larvae, he stated that it was unlikely to be a major source of nutrition for the larvae of *A. planci*. Recent studies by Manahan, Davis & Stephens (1983) indicate, however, that 79% of the energy requirements of larvae of the echinoid *Stronglyocentrotus purpuratus* could be supplied by amino acids which exist in a number of forms in sea water. Thus the rôle of dissolved organic matter in the nutrition of larvae of *Acanthaster planci* may prove to be more important than first thought.

Clearly, a great deal more research needs to be undertaken in this area. It is particularly important to determine the concentrations of phytoplankton, bacteria, and dissolved organic matter that are normally found in coral reef waters. In addition, experiments concerned with the effects of multiple species diets and alternative food sources such as bacteria and dissolved organic matter on the larval survival of *A. planci* need to be conducted. A more direct way of approaching the question of whether the larvae of *A. planci* normally starve in the field is to attempt to observe their development *in situ*. Olson (1985) has demonstrated that this type of approach is feasible. Using specially developed culturing systems he reared larvae of *A. planci in situ*, under nutrient conditions (at 5 and 15 m depths) which were thought to approximate natural food levels. Although survival was low, Olson succeeded in rearing larvae to the mid-brachiolaria stage. Larvae need to be reared to the late brachiolaria stage before conclusions can be drawn as to whether starvation is an important factor in their mortality. Olson (1985) pointed out that if starvation is important then it probably occurs during this later stage.

In reviewing the diet of *A. planci* Jangoux (1982a) considered that it was essentially a carnivore on corals (corallivore) and that it rarely fed on other animals. This statement, while correct, does not apply to the first six months of this starfish's life when it feeds on coralline and epiphytic algae (Yamaguchi, 1973b, 1974b, 1975b; Lucas, 1975). Even when it changes its diet, corals are not the only food that this starfish is capable of eating. There are numerous references in the literature to *A. planci* feeding on other types of food (Table IV). These range from anemones to soft corals and encrusting organisms. Most of the information given in Table IV has come from observations in the field. In captivity *A. planci* may be fed on fish, squid, and scallop meat as well as beef and echinoids (Branham, 1973; Yamaguchi, 1975b; Lucas, 1984). Cannibalism has also been observed under these conditions (Barnes, 1966; pers. obs.). It is likely that these foods are only eaten in captivity and would not be common food sources in the field. Sloan (1980) has discussed the effects that captivity may have on asteroid feeding.

From field observations it would appear that adult *A. planci* commonly feeds on corals and that it only feeds on other sources of food when there is very little coral available (Chesher, 1969b). Sloan (1980) has suggested that

TABLE IV

Alternative foods of A. planci: **field observation*

Food	Reference
Other *A. planci* (cannibalism)	Barnes, 1966; Branham, 1973
Algae	
Coralline*	Barham *et al.*, 1973
Other*	Dana & Wolfson, 1970; Vine, 1972
Clams*	Pearson & Endean, 1969
Echinoids	Yamaguchi, 1975b
Encrusting organisms*	Branham, 1973
Fish, squid, scallop meat	Branham, 1973; Cannon, 1975; Lucas, 1984
Gastropods*	Clark, 1950
Gorgonians*	Dana & Wolfson, 1970; Barham *et al.*, 1973
Hydrozoan corals*	Chesher, 1969b; Barnes *et al.*, 1970
Sea anemones*	Verwey, 1930
Soft corals*	Pearson & Endean, 1969; Chesher, 1969a Laxton, 1974

Fig. 4.—The ventral surface of a starfish (12 cm diameter) showing the mouth, and inside, part of the stomach (centre) (photograph taken by L. Brady).

A. planci is a specialist coral-feeder. Consideration of its feeding biology confirms this, as *A. planci* is an extraoral feeder (Jangoux, 1982b). When feeding it everts its stomach through its mouth and spreads this membraneous structure over an area of the coral surface equal to that of the oral disc (Goreau, 1964) (Fig. 4). The tube feet are used to position the stomach around the irregularities of the coral (Brauer, Jordan & Barnes, 1970). Once this has been accomplished the stomach secretes an enzyme which digests the coral tissue and the products are then absorbed (Goreau, 1964; Endean, 1973a). The feeding process may take from 4–6 h (Brauer *et al.*, 1970). The enzyme which is secreted is thought to have a proteolytic action (*i.e.* it hydrolyses proteins) as collagenase has been isolated from the stomach of *A. planci* (Yomo & Egawa, 1978). An additional enzyme (N-acetyl-B-D-glucosaminidase) has also been found in the pyloric caecum by Yomo & Tokumoto (1981). Proteolytic action in the stomach was greatest at pH 8·4 (Shou-Hwa, 1973); this may be an adaptation for extracellular uptake of food in sea water. Optimal proteolytic activity for the pyloric caecum was at pH 7·6.

In a recent review Jangoux (1982a) commented that the digestive system of *A. planci* was similar to that of starfish in the family Solasteridae although its stomach was much larger. The anatomy of the stomach as well as the pyloric and rectal caeca have been described by Hayashi (1939). *A. planci* is considered to be a specialist coral-predator (Cameron & Endean, 1982) partly because it has the unique ability to hydrolyse cetyl palmitate which is a major wax energy reserve in corals (Bensen, Patton & Field, 1975). A common feature of asteroids is that they can live for long periods (sometimes years) without feeding (Sloan, 1980). Observations of adults in captivity indicate that *A. planci* also has this ability and it may survive for up to six months without food (J. S. Lucas, pers. comm.; pers. obs.). Pearson & Endean (1969) starved three caged adults in the field for four months; at the end of that time they were alive and apparently healthy.

FEEDING BEHAVIOUR

There is much conflicting evidence concerning whether *A. planci* feeds nocturnally or diurnally. Several studies have indicated that the feeding behaviour of this starfish is related to population density. Chesher (1969b) stated that it was a nocturnal feeder when in low population densities. This was also confirmed by Pearson & Endean (1969) and Endean (1974) although they suggested that up to 90% of individuals in aggregations fed during the day on the Great Barrier Reef. This type of feeding behaviour was also noted in Hawaii (Branham *et al.*, 1971) and Micronesia (Cheney, 1974). These results conflict with those found elsewhere. In the Red Sea *A. planci* was primarily a nocturnal feeder even when in dense aggregations (Roads & Ormond, 1971). Similarly, Ormond & Campbell (1974) found that only 12% of starfish actively fed diurnally, irrespective of whether or not they were in aggregations. In contrast, Dana & Wolfson (1970) observed, in the Gulf of California, that starfish (*A. ellisii*?) fed during the day even though they were not aggregated. This was also recorded in Panama (*A. planci*) by Glynn (1972). In Western Australia, 30–50% of starfish were reported to feed during the day whether aggregated or not (Wilson & Marsh, 1974, 1975). From these findings it would seem that the feeding behaviour of *A. planci* is varied and shows little relationship to population density

(Kenchington, 1975a; Kenchington & Morton, 1976). Experiments conducted in the laboratory have demonstrated that feeding behaviour may be dependent on the time of day and the physiological state of the animal. Brauer *et al.* (1970) found that a high proportion of starfish in aquaria fed during the night. During the following day the starfish showed a marked decline in their desire to feed when they were presented with coral extracts. In the field this behaviour is likely to be far more complex and variable as it may be affected by a number of factors including: location (*i.e.* type and density of coral) (Potts, 1981); environmental conditions such as temperature (Yamaguchi, 1973c, 1974a), exposure (Endean, 1973b); age of starfish (Goreau, Lang, Graham & Goreau, 1972; Laxton, 1974); time of year (*i.e.* during spawning season) (Beach, Hanscomb & Ormond, 1975); light levels (Rosenberg, 1972).

On encountering a live coral or extracts of coral *A. planci* has been observed to rear its arms and retract its tube feet (Barnes, Brauer & Jordan, 1970). This aversive response was shown to be due to the nematocysts released by the coral and also the chemicals derived from coral tissue (Moore & Huxley, 1976). The intensity of this response was found to depend on the nutritional state of the starfish (Barnes *et al.*, 1970). As it was initiated before contact with the coral this withdrawal response was thought to aid in protecting the tube feet of individuals. Starfish may overcome the effects of nematocysts in corals when attempting to feed by moving on their arms and spines (Barnes *et al.*, 1970). It would appear that these effects are not as pronounced during feeding as the stomach is less sensitive than the tube feet (Barnes *et al.*, 1970).

A number of studies have attempted to determine the factors responsible for inducing feeding in *A. planci*. Observations in the field by Ormond *et al.* (1973) indicated that *A. planci* preferentially attacked damaged corals or those already being eaten. Using Y-shaped aquaria they demonstrated that a chemical attractant was released when starfish fed which stimulated others to move towards the corals being eaten. Beach *et al.* (1975) found that movement could be induced by presenting the starfish with extracts of live coral. Earlier, Brauer *et al.* (1970) showed that feeding (stomach eversion) in *A. planci* also could be induced using these extracts. In a series of biochemical experiments conducted in the laboratory Collins (1974) was able to produce two sorts of responses from starfish using extracts of coral tissue. He was able to invoke a settlement (*i.e.* mounting and positioning of starfish on coral colony) and stomach eversion response and an arm retraction or avoidance response. High and low molecular weight fractions, which were separated and isolated from live coral tissue, were found to cause the settlement and stomach eversion of starfish. Collins (1975a) discovered that the low molecular weight fraction comprised amino acids and small peptides. The other fraction was macromolecular and was thought to be a glycoprotein. The entire coral extract was found to cause the withdrawal of arms and sometimes the retraction of tube feet. Collins (1975a) identified the substance which was primarily responsible for this avoidance response. It was chemically similar to the amino acid proline. During further experiments Collins (1975b) found that the intensity of the avoidance response could be altered by using extracts from different types of corals. Later experiments by Hanscomb, Bennett & Harper (1976) showed that high molecular weight mucoproteins from coral mucus produced a feeding response in *A. planci*.

Sloan & Campbell (1982) have thoroughly discussed the evidence for the chemical perception of corals by *A. planci*. They pointed out that under certain conditions asteroids may be "pursuers" rather than "searchers" of prey. That is, they have the ability to perceive their prey, at short distances, and hunt them down. *A. planci* would also appear to have this ability although it may be affected by local environmental conditions (Sloan & Campbell, 1982).

There are some data on the feeding rate of *A. planci* and this has been derived from studies in the field and in the laboratory. Pearson & Endean (1969) determined the feeding rate of individual adult starfish of average size which were kept in cages on a reef. They found that these starfish consumed between 116 and 187 cm^2 of coral tissue per day. This represents a feeding rate of about 5·8 m^2 of coral tissue per year (Potts, 1981). Feeding rates of between 5 and 6 m^2 per year have been reported in the field from studies conducted in Panama (Dana & Wolfson, 1970; Glynn, 1973). Chesher (1969b) reported that starfish in Guam fed on 378 cm^2 of coral tissue per day or approximately 12 m^2 per year. This rate is twice that recorded in other parts of the world and must be treated with some scepticism as Chesher (1969b) gave very little information as to how this figure was derived. In the laboratory Yamaguchi (1974b) found that a juvenile of average size may kill around 200 g of *Pocillopora damicornis* in a day (based on the amount of dry coral skeleton mass killed). This may increase to about 300 g of coral per day or 100 kg per year for adult specimens. In general the feeding rate of *Acanthaster planci* will depend upon the same factors affecting its feeding behaviour.

FEEDING PREFERENCES

Experiments conducted in the laboratory have shown that *A. planci* prefers to feed upon certain types of corals (Brauer *et al.*, 1970). Coral extracts from *Acropora* and *Pocillopora* were found to produce stomach eversion whereas those from *Porites* mainly caused withdrawal responses. Collins (1975b) demonstrated that the type of coral consumed by a starfish may depend on its previous dietary experience. In the laboratory he demonstrated that *Acanthaster planci* learnt to differentiate between corals it had eaten previously and those that it had not encountered before. In general, *Acropora* spp. were found to be acceptable as food irrespective of the previous diet of the starfish. Ormond, Hanscomb & Beach (1976) also reported this type of learnt behaviour. They found that *Acanthaster planci* would feed more readily on corals that it had experienced before and that over a given time it reduced its feeding responses to coral extracts. Exploring this learnt behaviour further Collins (1975b) showed that starfish could be conditioned to eat species of coral (*e.g. Fungia*) which they may initially refuse. This was also reported by Huxley (1976) and Ormond *et al.* (1976) who stated that this type of learnt behaviour may persist for some time. Huxley (1976) commented that starfish learnt in time to determine the difference between coral extracts and live coral. He proposed that they may be able to detect the lack of some important dietary requirements as the coral extracts aged over the period of the experiments. Therefore, the acceptability of a particular type of coral may well depend on its nutritional value (Ormond *et al.*, 1976).

From the results of these laboratory studies, *Acanthaster planci* would appear to favour feeding on commonly occurring corals such as *Acropora*. This is probably why *Acanthaster planci* has been observed to prefer this coral in the Great Barrier Reef (Pearson & Endean, 1969), where it tends to be the most common genus (J. E. N. Veron, pers. comm.), and not in some other parts of the Indo-Pacific region (*e.g.* Red Sea) where it may be less common (Ormond *et al.*, 1973). Potts (1981), however, has pointed out that while this starfish may feed predominantly on more abundant corals they may not be the most preferred species. Apart from learnt behaviour, a variety of factors are likely to influence the feeding preferences of starfish in the field. Those factors (some already have been discussed in previous sections) which may be responsible for determining the types of corals which are consumed by *A. planci* are as follows.

(1) Nutritional state of starfish (Brauer, Jordan & Barnes, 1970).
(2) Release of substances (*e.g.* nematocysts, mesenteric filaments) by corals (Barnes, Brauer & Jordan, 1970; Goreau *et al.*, 1972).
(3) Release of chemical attractants by corals (Ormond *et al.*, 1973).
(4) Learnt behaviour of starfish (Collins, 1975b; Huxley, 1976; Ormond *et al.*, 1976).
(5) Abundance and distribution of corals (Ormond *et al.*, 1973).
(6) Accessibility of corals (Barnes *et al.*, 1970).
(7) Enviromental conditions (Endean, 1973b; Ormond *et al.*, 1973).
(8) Morphology of corals (Chesher, 1969a; Ormond & Campbell, 1974; Menge, 1982).
(9) Commensal organisms in corals (Glynn, 1976, 1977, 1980, 1982a).
(10) Nutritional value of corals (Ormond *et al.*, 1976).

Most of the information on feeding preferences in the field has come from qualitative studies as noted by Potts (1981). Of the numerous reports available on this topic there are really only three studies where *A. planci* has been demonstrated to show a preference for a particular type of coral or corals. In two of these studies starfish were reported to feed on corals which were considered to be less abundant. Branham *et al.* (1971) noted in Hawaii that 80–90% of *A. planci* fed on *Montipora verrucosa* despite the fact that this coral made up only 5% of the total coral cover. Similarly, Glynn (1974, 1976) showed that almost 50% of the diet of starfish in Panama was comprised of species that were comparatively rare (*i.e.* comprising only 7·2% of total coral cover). Laboratory and field experiments showed that *Acanthaster planci* tended to avoid the most common coral (*i.e. Pocillopora*) because it contained symbionts (the shrimp *Alpheus lottina* and the crab *Trapezia* spp.) which used chemical cues to detect and subsequently attack it when feeding. These animals were 31% effective in preventing *Acanthaster planci* from mounting and feeding on this coral (Glynn, 1976, 1980). In contrast to these results, Ormond *et al.* (1976) stated that in the Red Sea *A. planci* preferred the most abundant corals (*e.g. Pocillopora* and *Acropora*). This preference was, however, not well defined as the information they presented was somewhat conflicting. For a more detailed account of the results of these studies refer to Potts (1981).

While all three of the studies described above provide the best information to date on feeding preferences in the field they were inadequate for either of two reasons. First, they relied fully or partly on qualitative assessments of the amount of coral eaten and the abundance of each coral genus (*e.g.* Branham *et al.*, 1971). Secondly, they attempted to demonstrate feeding preference by comparing the proportion of a particular type of coral eaten with its proportion at a community or reef level. As the distribution of corals may be patchy over different scales of the system (Reichelt & Bradbury, 1984; Bradbury, Hammond *et al.*, 1985) this comparison may have little meaning. Indeed, feeding preference may also vary in conjunction with these changes in coral distribution. Perhaps a better method of assessment would involve a comparison at the coral colony level rather than at the community level.

Apart from these studies there has been reference to the feeding preference of *Acanthaster planci* in a number of areas in the Indo-Pacific. Goreau (1964) noted that this starfish appeared to favour no one particular coral species in the Red Sea. In the Gulf of California *A. planci* was considered an "obligate feeder" as feeding preference depended on the distribution of corals (Barham, Gowdy & Wolfson, 1973). Coral genera such as *Pocillopora* (Glynn, 1976), *Porites*, *Galaxea* (Barnes *et al.*, 1970), and *Diploastrea* (Endean & Stablum, 1973a,b) have been reported to be not eaten by starfish in the field. This is not consistent, however, as in other parts of the world some of these corals (*e.g. Porites*, *Pocillopora*) have been observed to be eaten by *Acanthaster planci* (Dana & Wolfson, 1970; Goreau *et al.*, 1972; Nishihira & Yamazato, 1972, 1973; Nishihira *et al.*, 1974; Aziz & Sukarno, 1977). Of all corals *Acropora* (particularly tabular and branching forms) appears to be one of the most preferred (Chesher, 1969a,b; Pearson & Endean, 1969; Roads, 1969; Garlovsky & Bergquist, 1970; Nishihira & Yamazato, 1972; Aziz & Sukarno, 1977). There are several other studies which have reported additional information concerning the feeding preferences of *Acanthaster planci* (Campbell & Ormond, 1970; Weber & Woodhead, 1970; Ormond & Campbell, 1974). It is difficult to determine from the information given above whether *A. planci* shows definite feeding preferences as most of the evidence is qualitative. A series of intensive quantitative field studies involving densities of starfish and different types of coral communities is needed to enable a more accurate understanding of feeding preferences in *A. planci*.

MOVEMENT

Some information is available on the rate of movement of adult starfish in the field (Table V). Pearson & Endean (1969) reported that adults were capable of moving at a rate of about 20 m per hour over sand. The maximum rates of movement in other parts of the Indo-Pacific (Gulf of California, Red Sea, and Indonesia) were found to be almost half this figure. Data on the movement of juvenile starfish have come from aquarium studies. Yamaguchi (1973b) found that juveniles of 1 mm in diameter (2 weeks old) moved at a rate of about 1·0 mm per min (0·06 m per h).

Larger juveniles (19–70 mm) were recorded to move at speeds of between 1·4–4·0 m per hour (Pearson & Endean, 1969) (Table V). In addition, it has been reported that certain arms may lead during periods of movement indicating that *A. planci* may have a posterior-anterior axis (Rosenberg, 1972).

While *A. planci* is capable of relatively fast movement over coral reef substrata it is not known how long this rate can be maintained. More long-term studies of starfish movement have indicated that they may move up to 580 m in a week (Roads & Ormond, 1971). In Guam, individuals travelled a distance of approximately 250 m over the same period of time (Chesher, 1969a).

Other studies have shown that the movement of starfish is non-random over the scale of metres. Uni-directional movement, of several hours duration, has been observed in transplanted starfish (Branham *et al.*, 1971) and using time-lapse photography on the Great Barrier Reef (P. W. Sammarco, pers. comm.). In American Samoa, Beulig, Beach & Martindale (1982) studied the movement of groups of starfish of three different types of densities. Over 24 hours each group moved consistently in a different direction.

It is likely that several factors determine the rate and direction of movement of starfish. Barham *et al.* (1973) has suggested that the rate of movement is dependent on the density of coral. They found that *A. planci* moved at 0·6 m per hour in areas with low coral cover and at 0·25 m per hour at sites where the corals were more dense. Ormond & Campbell (1974) also proposed that starfish movement may be affected by environmental factors, particularly wave action, exposure, and perhaps light. Apart from coral density and various environmental factors it is possible that other variables are important in influencing the movement of starfish. These include: age, condition and nutritional state of the starfish; time of day; and type of substratum.

It is suspected that starfish move in large populations from one reef to another once the supply of food is exhausted (Endean, 1969; Talbot & Talbot, 1971). There are two main reasons for proposing this and they are based on circumstantial evidence. First, it has been reported that starfish first appear in deep water and then move up the reef slope consuming corals as they go (Moran, Bradbury & Reichelt, 1985). Secondly, the starfish comprising these initial stages of the outbreaks are not usually juveniles but

TABLE V

Rate of movement of juvenile and adult A. planci: **field observation*

Starfish	Rate ($m \cdot h^{-1}$)	Reference
Juvenile	0·06	Yamaguchi, 1973b
Juvenile	1·4–4·0	Pearson & Endean, 1969
Adult*	20·0	Pearson & Endean, 1969
Adult*	10·0	Barham *et al.*, 1973
Adult*	5·0–10·0	Goreau, 1964
Adult*	0·3–8·0	Aziz & Sukarno, 1977

tend to be 2–3 years old (Endean, 1973b). While these observations may provide evidence to support the notion that starfish move between reefs, the sudden appearance of large starfish in deep water may be explained equally well if they originated from larvae which settled in deep water at the base of a reef.

There is indirect evidence to suggest that starfish are capable of moving large distances between reefs. The information presented above indicates that they can move rapidly over various types of reef terrain. They are able to go for long periods of time without feeding. This information, however, was based on animals in captivity and not on ones that were highly active. Starfish have been observed to cross large expanses of sand between patch reefs (Pearson & Endean, 1969). They have also been dredged from deep water (64 m) between reefs (Great Barrier Reef Marine Park Authority, unpubl. data). Unfortunately there is no conclusive proof that starfish move in large numbers between reefs.

PHYSIOLOGY

Few studies have been reported on the physiology of *A. planci*. Those by Yamaguchi (1973c, 1974a) represent essentially the only attempts to investigate this particular facet of the biology of this animal. From these experiments it was demonstrated that *A. planci* is a "metabolic conformer" as its rate of oxygen consumption is determined by changes in environmental temperature and possibly other variables. Maintenance of a normal metabolism and behaviour occurred up to a temperature of 31 °C. Increases in temperature to 33 °C were observed to cause abnormal behaviour, the cessation of feeding, and disruption to the metabolic activity of individual starfish. Prolonged exposure (about 1 week) to this temperature regime caused the eventual death of starfish. This led Yamaguchi (1974a) to postulate that adult *A. planci* may avoid reef flat environments, where high temperatures may occur, as they may not be able to maintain a constant oxygen metabolism.

METABOLISM OF STEROIDS

Numerous studies have been carried out which have sought to isolate sterols and other steroid-related compounds from *A. planci*. Experiments of this sort have been conducted on a variety of echinoderms (Voogt, 1982). They are important from a theoretical perspective as echinoderms are thought to be closely related to vertebrates. Many of the metabolic processes identified in echinoderms parallel those found in vertebrates. Isolation of sterols and steroid-related substances from echinoderms may provide a better understanding of the metabolic processes of vertebrates and how they evolved.

In vertebrates, sterols are important structural components of cell membranes and are the antecedents of steroids and cholic acids (Voogt, 1982). A number of sterols were identified in *A. planci* by Gupta & Scheuer (1968). The chemical structure of one of those (acansterol) was isolated and described in detail by Sheikh, Djerassi & Tursch (1971). They argued that this sterol was a derivative of gorgosterol which occurs in coelenterates such as corals. The existence of this pathway was verified by Kanazawa,

Teshima, Ando & Tomita (1976) who succeeded in isolating an intermediate compound called gorgostanol.

The occurrence of gorgosterol indicated that the composition of sterols in *A. planci* may be a function of diet. Experiments by Sato, Ikekawa, Kanazawa & Ando (1980) identified the chemical structures of the various sterols present in *A. planci*. The most dominant group of sterols in this starfish were Δ^7 sterols. Kanazawa, Teshima, Tomita & Ando (1974) showed that this starfish contained sterols which were similar to those found in other coral reef organisms thus indicating that they may have been derived through the food chain. Experiments by Teshima, Kanazawa, Hyodo & Ando (1979) also demonstrated that the sterols in *A. planci* may be transferred to a known predator, *Charonia tritonis* (giant triton), as the sterol composition of both animals was found to be similar.

Another group of compounds called saponins have been isolated from *Acanthaster planci*. These substances are derived from steroids and are toxic to various marine animals (Voogt, 1982). Numerous studies by Croft, Fleming & Howden (1971), Sheikh, Tursch & Djerassi (1972a,b), Shimizu (1971, 1972), Sheikh & Djerassi (1973), Sheikh, Kaisin & Djerassi (1973), Howden, Lucas, McDuff & Salathe (1975), and Fleming, Salathe, Wyllie & Howden (1976) have resulted in the identification of at least four different saponins from adult *A. planci*. Similar compounds have been found in comparable amounts in the eggs, ovaries, and larvae of this starfish (Howden *et al.*, 1975; Lucas, Hart, Howden & Salathe, 1979). Further characterization of the chemical sub-units of the saponins isolated from *A. planci* have been conducted by Kitagawa, Kobayashi, Sugawara & Yosioka (1975), Kitagawa & Kobayashi (1977, 1978), Kitagawa, Kobayashi & Sugawara (1978) and Komori *et al.* (1980, 1983a,b). The potential antipredator rôle of saponins in larvae is discussed later (see p. 415).

ECOLOGY OF *A. PLANCI* POPULATIONS

INTRODUCTION

Many studies conducted in the field have involved surveys which have attempted to determine the distribution and abundance of starfish and/or corals on reefs. Numerous surveys have been undertaken on the Great Barrier Reef (see p. 431) and on reefs in Micronesia (Chesher, 1969a; Marsh & Tsuda, 1973) over the last 20 years. From the information given in the previous section it becomes apparent that very few attempts have been made to investigate more detailed aspects of the ecology of *A. planci* populations. This general lack of research has occurred at all stages of the life history of this animal. Studies of the larval ecology of *A. planci* have included preliminary, *in situ*, rearing experiments (Olson, 1985) and extensive plankton sampling programmes on the Great Barrier Reef (Pearson & Endean, 1969; Walsh *et al.*, 1976). Despite intensive efforts these latter studies were unsuccessful as no *A. planci* larvae were identified. Only one intensive field study of juveniles has so far been reported (see p. 396). Similarly, there has also been a lack of studies on adults in the field. Those that have been conducted have concentrated on investigating the movement and behaviour of individuals or aggregations over relatively short time intervals, measuring the size of individual starfish, recording the density of starfish over well-defined but small areas and determining the feeding preferences of starfish. To date, little information is available about the population dynamics of *A. planci*. Field data on growth, longevity, mortality, and to a lesser extent, movement and feeding rate are inadequate. These data are essential in order to achieve a basic understanding of the dynamics of *A. planci* populations.

The general lack of field studies on all aspects of the life history of *A. planci* has arisen mainly because experimentation on larvae, juveniles or adults has proved logistically difficult. The larvae of *A. planci* are difficult to study since they are very small and may often be dispersed by ocean currents. While juveniles are much bigger than larvae they are none the less difficult to find in the field because they are extremely cryptic and capable of inhabiting very small crevices and holes in the reef substrata. In contrast to larvae and juveniles, adults may be easily found on reefs from time to time especially during outbreaks but it is difficult to study these types of starfish in the field as they are not amenable to tagging and hence individuals cannot be recognized or followed over long periods. Most field studies of adults were designed to obtain data about entire populations (*e.g.* size frequency data) and have not presented long term information on individual starfish. It is obvious that future research must be concentrated in these areas if a greater understanding of the *Acanthaster* phenomenon is to be achieved.

DISTRIBUTION AND ABUNDANCE

Copious data exist on the distribution and abundance of starfish following numerous surveys conducted throughout the Indo-Pacific region over the last 20 years. The data are not definitive assessments of the distribution and abundance but represent only broad estimates of starfish populations. There are several reasons for this.

(1) Reefs may be such large and complex structures that they cannot be surveyed accurately using current techniques.
(2) The starfish are often cryptic and their abundances are difficult to estimate; this may become even more arduous when outbreaks occur, as often their abundances are so high that they cannot be counted effectively.
(3) Animals may be distributed unevenly over the reef surface and the results obtained for one area may not reflect those on the reef as a whole; thus extrapolating the results for small areas to indicate the likely abundance of starfish over entire reefs requires care.

Starfish abundance has little meaning if it is not compared with some type of standard measure. There have been several attempts at standardization when assessing starfish populations. For example, Pearson & Endean (1969) observed 405 starfish in 5 min of searching at the Frankland Islands. They also reported finding 1150 individuals in 20 min at Green Island. Different figures have been reported from other parts of the Indo-Pacific. Glynn

(1974) recorded a maximum density of 1 starfish per 50 m^2 on reefs in Panama while Branham *et al.* (1971) found 158 starfish in a circular area, 10 m in radius. In Okinawa, Nishihira & Yamazato (1974) reported finding an average of 241·5 starfish for every 10 min of searching while in the Fijian Islands Randall (1972) recorded 510 starfish in 100 min. There are many more references in the literature to this type of information (see Potts, 1981). Perhaps the best estimates of the maximum number of starfish which may occur during outbreaks come from the numbers of individuals killed during control programmes (see Table XIV, p. 450). Those conducted in Hawaii destroyed two groups, each of about 10 000 individuals, within a two-year period (Branham *et al.*, (1971). Approximately 44000 starfish were removed from a small area on Green Island over about 18 months while over a similar length of time almost 490 000 individuals were destroyed in American Samoa (Birkeland & Randall, 1979). The results obtained from control programmes indicate that outbreaks may consist of hundreds of thousands, perhaps even millions of starfish (Yamaguchi, in press). Similar levels of abundance were reported for the outbreak that occurred at Green Island during 1979–1981 (Endean, 1982). Taken as a whole, data from starfish surveys and control programmes serve to highlight the extreme variability that can occur in the abundance of starfish on reefs.

Not only is the abundance of starfish on reefs highly variable but so also is their distribution. Numerous surveys have demonstrated that starfish do not occur evenly over the surface of reefs but tend to form localized concentrations or aggregations (Fig. 5) (Ormond & Campbell, 1974; Endean & Stablum, 1975; Birkeland, 1979). These are thought to be the result of several factors whose effects may be cumulative over a given period (Sloan,

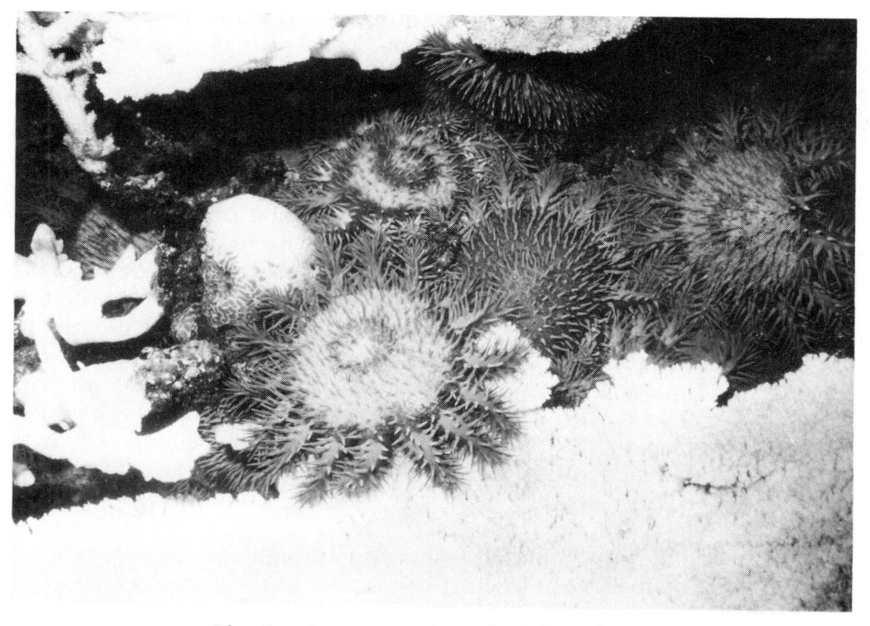

Fig. 5.—An aggregation of adult starfish.

1980). Factors which may be important in causing the formation of such aggregations are: presence of spawning attractants and coral extracts, age of starfish, distribution and abundance of coral, type of coral, stage of outbreak, and environmental preferences (*e.g.* depth, type of substratum, light and type of exposure) (Endean, 1974; Sloan, 1980). While the spatial distribution of starfish is uneven on reefs it also may vary temporally as the aggregations themselves may move. Endean (1969) suggested that this occurs once the food supply has been exhausted on a section of reef. Aggregations have been reported to move at a rate of approximately 100 m per month (Ormond *et al.*, 1973; Ormond & Campbell, 1974) and persist for up to 2 years. In Guam, Chesher (1969a) reported that aggregations travelled approximately 3 km in a month although feeding during this time was probably reduced as the movement occurred over dead or poorly developed reef.

The distribution and abundance of starfish on reefs varies over both temporal and spatial scales (Moran, Reichelt & Bradbury, 1985). While most surveys have managed to demonstrate that starfish abundances vary spatially few, if any, have described the pattern of change in the distribution and abundance of starfish over a complete outbreak cycle (*i.e.* before, during, and after an outbreak). This is because most surveys have not been carried out repeatedly over the same areas through time. The few attempted (*e.g.* Kenchington & Morton, 1976) were not conducted at short enough intervals of time nor were they undertaken over long enough periods. Recently Moran, Bradbury & Reichelt (1985) reported the results of surveys where changes in the abundance and distribution of starfish were followed before, during and after an outbreak over 2 years on John Brewer Reef. Prior to the outbreak, starfish were rare; only four individuals over 10 km of the reef perimeter were observed. Within 12 months the population had, however, increased dramatically with an average of up to 100 starfish being recorded for each two-min manta tow. After a further three months starfish numbers had declined again to relatively low levels (<7 per two-min tow). An interesting result emerged from these surveys. At the start of the oubreak starfish were concentrated on the fore-reef slopes but over a 9-month period they became more abundant in sheltered back-reef areas. This pattern of change in the distribution of starfish has been reported from other reefs in this region. For example, Laxton (1974) reported that this took at least two years to occur on Lodestone reef. In another study Kenchington (1976) found that the change in the distribution of starfish from front- to back-reef areas occurred within 12 months on several reefs. The tendency for adult starfish to seek sheltered back-reef areas may be because their powers of adhesion appear to decline with age (Ormond & Campbell, 1971; Goreau *et al.*, 1972). The rate at which this change occurs may depend on the size and structure of the reef, the distribution and abundance of live coral, the age and physiological state of the starfish, and environmental conditions (Moran, Bradbury & Reichelt, 1985).

While surveying the population repeatedly through time on John Brewer Reef, Moran, Bradbury & Reichelt (1985) measured the diameter (from arm tip to arm tip) of some 1200 starfish. Their mean size, 346·0 mm, indicated that the outbreak consisted primarily of adults. This is a feature common to most outbreaks of starfish. Measurements in other areas have demonstrated that outbreaks generally consist of adults ranging in size from 250–350 mm (Chesher, 1969a; Pearson & Endean, 1969; Branham *et al.*, 1971; Ormond & Campbell, 1971; Owens, 1971; Cheney, 1974). Investigations of the size frequency distributions of these outbreaking populations have often shown them to be unimodal, comprising essentially one size class. Dana, Newman & Fager (1972) concluded that populations of starfish in Saipan, Kapingamarangi, and the Gulf of California were characteristically unimodal as were also populations in Fiji and Panama (Owens, 1971; Glynn, 1973). While some populations have been reported to be unimodal others have been considered to be polymodal, consisting of two or more size classes. Such populations have been reported on the Great Barrier Reef (Endean, 1973b), Japan (Suzuki, 1975, Moyer, 1978; Fukuda & Miyawaki, 1982; Matsusita & Misaki, 1983), and the Red Sea (Ormond & Campbell, 1971). Whether or not a population is represented by a polymodal or unimodal size frequency distribution may well depend on the time at which the measurements were undertaken. Moran, Bradbury & Reichelt (1985) found that initially the starfish population on John Brewer Reef was unimodal and dominated by one size class (at about 350 mm). Following additional measurements of the population on this reef after six months two distinct size classes were identified; one mode at approximately 300 mm and the other at 100 mm. They postulated that this latter mode represented recruitment possibly from the previous spawning period. Juveniles as small as 30 mm were recorded in this outbreaking population. This shift in the modal structure of outbreaks, caused by the influx of juveniles into adult populations has been mentioned by other authors (*e.g.* Endean, 1973b). Such events have been reported on rare occasions on several reefs in the Indo-Pacific (Pearson & Endean, 1969; Birkeland, 1982; L. Zann, pers. comm.).

At best, single, one-off surveys of reefs give only a snapshot view of outbreaking populations and do not indicate any temporal changes that may be occurring. Repeated intensive surveys of reefs are required to understand more about the dynamics of the behaviour of outbreaks, particularly whether they are declining or changing their position and whether additional recruitment to the population has occurred. This information is particularly relevant when attempting to undertake effective control programmes (see pp. 448–454).

RECRUITMENT

A feature common to many tropical marine species is that adults are conspicuous in the field while juveniles are rarely seen (Yamaguchi, 1973b). This is particularly noticeable in the case of *A. planci* where outbreaks of adults are commonly observed yet those of juveniles are not. As a consequence while there is some information on the ecology of adult *A. planci* there is practically no information on the processes that occur in the field between the time an egg is fertilized to its first appearance as an adult (about 250–350 mm in diameter). This is a grey area in the ecology of *A. planci* and is perhaps the main reason why the *Acanthaster* phenomenon is so poorly understood. One way of overcoming this situation is to obtain information on the recruitment of this starfish. In a recent review Ebert (1983) defined recruitment as the "addition of new individuals to a population" (p. 169) and stated that this may be a result of immigration or reproduction. An investigation of recruitment is extremely important because it may lead to a greater appreciation of the reasons for fluctuations in the distribution and abundance of adult populations. The recruitment of many coral reef species is highly variable (Ebert, 1983) since it has been shown to be sporadic, varying in time and place (Frank, 1969; Sale, 1980). Thorson (1961) suggested that this type of recruitment was indicative of animals that were highly fecund and whose larvae were widely dispersed.

It has been postulated that large fluctuations in the abundance of *A. planci* are the result of differential survival of larvae (Birkeland, 1982) rather than any other stage. Yamaguchi (1937b) has pointed out that the survival of larvae and early juvenile stages may be variable but emphasized that there was hardly any information to verify this supposition. Lucas (1975) suggested the following factors which may be important in affecting the survival of these stages: degree of fertilization, abundance of food, temperature, salinity, extent of predation, dispersal and availability of suitable substrata for settlement. The last two are considered below (see earlier sections for a discussion of the other factors).

One of the most difficult things to determine about larvae is their likely dispersal before settlement and metamorphosis. Cheney (1974) has suggested that the increased recruitment of *A. planci* in Micronesia, specifically Guam, may well be a result of eddy systems which capture larvae and prevent them from being transported into deep oceanic water where they would most probably die. He found evidence which indicated that often oubreaks of *A. planci* were found on reefs where these eddy systems were prevalent. Such self-seeding of reefs may be important in Micronesia as many of the reefs are separated by large expanses of deep water. Rowe & Vail (1984b) have argued in a similar vein, suggesting that eddies and gyres may be responsible for the retention of *A. planci* larvae on some reefs in the Great Barrier Reef. They postulated that these current patterns may lead to recurrent outbreaks of starfish on the same reefs. More recently, Williams, Wolanski & Andrews (1984) have developed a model of the current patterns in the central section of the Great Barrier Reef. Using this model they showed that there was a tremendous potential for larvae to be dispersed over large distances in this region. During summer, the currents in shallow water (*i.e.*$<$40 m) were found to move in a net southerly direction at a rate of up to 300 mm per s. In deeper water the currents moved in the same direction but at about one third the speed. Given the relatively long larval life of *A. planci*, it is possible that a cloud of larvae released from mid-shelf reefs off Cairns may, after three weeks, be located adjacent to reefs near Townsville, a distance of some 300 km (see Fig. 7, p. 431). The model of Williams *et al.* (1984) lends support to that put forward by Kenchington (1977) which was based on analyses of the size frequency distributions of starfish from a number of different reefs. The model proposed by Kenchington suggested that recruitment of starfish in areas south of Cairns occurred in a series of three major waves (*i.e.* reefs off Innisfail–1964/66: reefs off Townsville–1967/69: reefs south of Townsville–1970/72) moving southwards (Fig. 7). The actual pattern of outbreaks observed in this region broadly agrees with the models put forward by Williams *et al.* (1984) and Kenchington (1977) (see p. 434). When combined with the larval recruitment hypothesis (Pearson, 1975b) or the terrestrial run-off hypothesis (Birkeland, 1982) (see p. 462), these models provide an extremely plausible mechanism for the propagation of outbreaks on the Great Barrier Reef. Some questions are, however, still unresolved regarding this mechanism. First, it relies on the fact that a concentrated patch of larvae (larval cloud) is produced in areas where there are outbreaking populations. Despite intensive efforts to locate larvae near such populations none have ever been identified in plankton trawls (Pearson & Endean, 1969). Secondly, in view of the findings of Lucas (1982) (who suggested that natural food levels were insufficient for the survival of larvae), the larval cloud presumably would need to travel with, or pass through nutrient-rich patches or regions of food that would keep the larvae viable for the length of time they were in the water column. This is quite likely to be many days as the model by Kenchington (1977) indicated that the larval cloud may travel up to 100 km. Finally, each successive wave of larvae would need to be synchronized with the occurrence of these phytoplankton blooms. To date, there is no evidence to suggest that such conditions ever occur on the Great Barrier Reef (see p. 397).

Yamaguchi (in press) has proposed that larvae may be dispersed over great distances in accounting for the occurrence of outbreaks on the mainland of Japan and at Miyake Island (Fig. 6, p. 422). He postulated that outbreaks in Japan were a result of larvae that were transported by the warm Kuroshio Current from the Ryukyu Islands. Outbreaks at Miyake Island were thought to have originated when this current changed its course and left the main coast of Japan. As with the models proposed for the Great Barrier Reef, that postulated by Yamaguchi does not indicate whether the larvae are able to survive these long periods of travel. Larvae released as a cloud in the Ryukyu Islands would need to travel approximately 700–800 km to reach Miyake Island. More detailed oceanographic and planktonic studies are required in order to determine whether this is possible.

The survival of larvae depends not only on dispersion but also on whether there are suitable surfaces available upon which the larvae can settle. The extent to which this takes place can be gauged by estimating the numbers of small juveniles present on reefs. As mentioned previously, this is a difficult task as they are extremely cryptic and hard to locate. Yamaguchi (1973b) found few juveniles on the reefs at Guam despite searching intensively. This task is made even harder since it is not definitely known where larvae settle in the field. As the juvenile stages of *A. planci* feed on coralline algae it is often presumed that they settle on substrata where this food is available. Indeed, Yamaguchi (1973b) observed them to settle on these types of substrata in the laboratory, although Lucas (1975) considered that there was some evidence to indicate that all they required was a substratum that possessed a biological film. Because coral colonies killed by *A. planci* are quickly covered by epiphytic and coralline algae it has been suggested that this starfish provides an attractive substratum for the settlement of its own progeny (Chesher, 1969a; Ormond *et al.*, 1973). As yet no studies have been conducted to determine whether the larvae of *A. planci* prefer specific types of coralline algae on which to settle. Indeed, little is known about the factors which are important in governing the settlement of these larvae. Research on abalone larvae has indicated that the settlement and metamorphosis of invertebrate larvae on coralline algae may be induced by a peptide, similar to the neurotransmitter gamma-aminobutyric acid (Morse, Hooker, Duncan & Jensen, 1979; Trapido-Rosenthal & Morse, in press). Further studies by Baloun & Morse (1984) have demonstrated that this may be inhibited or enhanced by altering the external concentrations of potassium ions.

While the mechanisms responsible for the settlement of *A. planci* larvae are poorly understood it is generally assumed that they settle in shallow water on reefs. This is because the few juveniles that have been found in the field have mostly been reported in these areas. Pearson & Endean (1969) discovered 46 juveniles (11–69 mm) in sheltered water (2–6 m depth) at Green Island on the Great Barrier Reef. They also found another 142 individuals (15–79 mm) at Fitzroy Island in a similar location. There are other reports of juveniles being found in the field; they, however, relate mainly to starfish which are bigger than 70 mm and could be small adults. More recently, numerous small juveniles (<50 mm) have been recorded by Moran, Bradbury & Reichelt (1985) on the Great Barrier Reef and by Zann (pers. comm.) in Fiji. One feature common to all reports is that the juveniles were located not only in shallow water but also a few years after an outbreak of adult starfish. Endean (1973b) considered that they may be the progeny of these adults and had been retained on the same reef as a result of water current patterns. If this is true then their occurrence in shallow water may be determined by the distribution and abundance of corals left after the initial outbreak of adults. Corals in shallower locations on reefs are commonly left by starfish (see p. 438). Perhaps the progeny of adults are distributed over a wide area of the reef, including shallow and deep water, but only those which settle in areas of high coral cover manage to change their diet, from algae to coral tissue, and survive. This does not, however, indicate where the starfish of initial oubreaks on reefs settle. If they normally settle in shallow water, it is strange that they are not seen until they are adults. Once they switch their diet to corals their presence on reefs becomes progressively more obvious with the increase in the size and number of feeding scars. One might presume that large numbers of smaller sized starfish (70–120 mm) would be reported more often if they settled initially in shallow water. If larvae become negatively buoyant prior to settlement as suggested earlier, then it is possible that the larvae responsible for the initial outbreaks on reefs may settle in deeper water at the base of reef slopes. More intensive searches of cryptic habitats in these areas on reefs prior to outbreaks may resolve this question.

PREDATORS

Twelve species of animals have been observed to feed on apparently healthy *A. planci*. These data have come from the field and the laboratory and are listed in Table VI. Predation of all four of the major stages in the cycle of *A. planci* (*i.e.* gametes, larvae, juveniles, and adults) have been reported.

Pearson & Endean (1969) have provided the only account of a damselfish (*Abudefduf curacao*) eating the eggs of a spawning starfish in the field.

TABLE VI

Animals that have been observed to feed on A. planci

Predator Type	Predator Name	*A. planci* stage predated	Reference
Anemone	*Stoichactis* sp.	Adults	Chesher, 1969a
Coral	*Pocillopora damicornis*	Larvae	Yamaguchi, 1973b; Ormond *et al.*, 1973
	Pocillopora damicornis	Juveniles	Yamaguchi, 1974b
Crab	*Promidiopsis dormia*	Adults	Alcala, 1974
	Xanthid	Juveniles	Lucas, 1975
Fish	*Abudefduf curacao*	Eggs	Pearson & Endean, 1969
	Chromis dimidiatus	Larvae	Lucas, 1975
	Arothron hispidus	Juveniles/Adults	Ormond & Campbell, 1974
	Balistoides viridescens	Juveniles/Adults	Ormond *et al.*, 1973
	Pseudobalistes flavimarginatus	Juveniles/Adults	Ormond & Campbell, 1974; Owens, 1971
Gastropod	*Bursa rubeta*	Juveniles/Adults	Alcala, 1974
	Charonia tritonis	Juveniles/Adults	Endean, 1973b
Shrimp	*Hymenocera picta*	Juveniles/Adults	Wickler & Seibt, 1970; Wickler, 1973; Rainbow, 1974; Glynn, 1982a, 1984b
	Neaxius glyptocercus	Juveniles/Adults	Brown, 1970
Worm	*Pherecardia striata*	Juveniles/Adults	Glynn, 1982a, 1984b

Several studies have been conducted in the laboratory to investigate the predation of the eggs and larvae of *Acanthaster planci*. Yamaguchi (1973a) and Ormond *et al.* (1973) reported that larvae were eaten by corals. In addition, Yamaguchi (1974b, 1975) found that certain asteroid larvae and eggs, including those of *A. planci*, were either avoided or actively expelled by some species of fish. Experiments by Lucas (1975) demonstrated that while the larvae of *A. planci* were consumed by fish (Pomacentridae) they were not preferred and were discriminated against when there was a choice of larval species. These observations indicated that the larvae and perhaps eggs of *A. planci* contained substances that may repel predators.

This proved correct as Howden *et al.* (1975) managed to isolate toxic chemical compounds (saponins) from the eggs, ovaries, and body of *A. planci*. In a series of experiments Lucas *et al.* (1979) were able to show that these substances were partly responsible for the observed rejection of the eggs and larvae of *A. planci* by some species of fish. Those authors also observed that the fish varied in their discrimination and demonstrated that this may depend on the tastiness (*e.g.* whether the larvae and eggs are yolky or non-yolky) of the prey and also on the degree of hunger of the predator. Dana *et al.* (1972) have postulated that it is highly probable that predation of larvae in the field is extensive as the reef is composed of a vast array of plankton-feeders such as corals. Despite this claim they acknowledged that it was not known whether this type of predation was extensive.

Small juveniles of *A. planci* have been reported to be preyed upon in the laboratory by xanthid crabs (Lucas, 1975; pers. obs.). Yamaguchi (1974b) observed that they were badly damaged by the mesenteric filaments of corals once they had attained adult morphology and changed their diet from algae. Several individuals were so severely damaged that they lost arm tips or complete arms. Most often these lost parts were regenerated within a few months. This type of damage was not recorded once the starfish had reached a sufficient size to avoid attack. These results highlight the fact that the mortality of starfish may be particularly high during the early stages just after metamorphosis. Indeed, it is possible that predation of young juveniles may be important in limiting the number of adult starfish on reefs.

A variety of other animals have been reported to feed on juvenile and adult *A. planci*. Endean (1969, 1977, 1982) proposed that the giant triton, *Charonia tritonis*, was a major predator of large juvenile and small adult starfish and was capable of altering their abundances in the field. This gastropod was reported to feed on *Acanthaster planci* by Pearson & Endean (1969). Using caged individuals they demonstrated that *Charonia tritonis* preferred starfish other than *Acanthaster planci* if given a choice. Indirect evidence from the field tended to support these findings. Of 28 tritons collected during two years of research on the Great Barrier Reef only seven regurgitated material associated with *A. planci*. An additional 12 tritons regurgitated parts of starfish, 11 of which were *Linckia* sp. and one was from *Culcita* sp. (Pearson & Endean, 1969). While appearing to prefer other starfish *Charonia tritonis* was also found to consume *Acanthaster planci* at a relatively slow rate. Pearson & Endean (1969) recorded that it ate only 0·7 starfish per week over a period of three months. Observations from Micronesia also suggested that attacks on *A. planci* by tritons were not always fatal and the animal was often able to escape and regenerate any damaged parts (Chesher, 1969a).

Another animal which has been proposed as a major predator of *A. planci* is the painted shrimp *Hymenocera picta* (Wickler, 1970; Wickler & Seibt, 1970). Experiments in aquaria indicated that this animal seeks out starfish using its antennules as chemoreceptors (Rainbow, 1974). The shrimp was observed to turn over small starfish and feed on their gonads and soft tissues (Wickler, 1970; Wickler & Seibt, 1970). One study indicated that this occurred only when *H. picta* was very hungry (Wickler, 1973). While these attacks caused the death of some individuals, Rainbow (1974) suggested that *H. picta* would not seriously injure adult starfish which were more than three times larger than the shrimp, but may affect juveniles. Therefore, he concluded that this shrimp was unlikely to control the abundance of adult starfish in the field. More recently, Glynn (1977) estimated the abundance of *H. picta* on lower fore-reef slopes in Panama and found that densities ranged from 1–118 individuals per hectare. From the results of field and laboratory studies he hypothesized that this shrimp was able to limit the abundance of *Acanthaster planci* as it was compelled to prey on it because of a lack of other more preferred starfish species (*e.g. Linckia* spp., *Nardoa* spp.). This produced a decrease in the rate of coral mortality in this area (see p. 417).

Four species of fish are known to feed on *A. planci*. This information comes from direct observations of predation or finding parts of *A. planci* in the stomachs of animals. Ormond *et al.* (1973) and Ormond & Campbell (1974) observed three species, *Arothron hispidus*, *Balistoides viridescens* and *Pseudobalistes flavimarginatus*, to feed on starfish in the Red Sea. By recording the frequency with which the remains of starfish (these were considered to be unique for fish attacks) were sighted they estimated that approximately 200–800 adults were killed each year by these fish predators. This was thought to account for the gradual decline in starfish numbers which had been recorded over a two-year period. Predation by these species has not been reported to any great extent in other parts of the Indo-Pacific (*e.g.* Glynn, 1982a), although Wilson, Marsh & Hutching (1974) found spines and skeletal ossicles in the gut contents of a specimen of *Arothron hispidus* from the waters of Western Australia. Endean (1977) considered it unlikely that these species would be responsible for controlling starfish populations on the Great Barrier Reef since they were not common in this area. On the other hand, he maintained that the groper *Promicrops lanceolatus* was an important predator of large juvenile starfish on the Great Barrier Reef (Endean, 1982). He gave no real quantitative evidence to support this statement but he did report finding parts of juvenile starfish in the stomach of several specimens of this species (Endean, 1974, 1977). It is not, however, known whether the starfish were alive or dead when eaten. Indeed, Glynn (1984b) found that a variety of different animals including polychaetes, echinoids, crustaceans and fish fed on starfish which were either mutilated or dead. This activity was thought to hasten the rate of decomposition of these starfish.

So far, the only study to provide quantitative evidence of predation was that undertaken by Glynn (1982a, 1984b) in Panama. He found that starfish were often killed as a result of attacks by the shrimp *Hymenocera picta* and the annelid *Pherecardia striata*. Using a combination of laboratory and field experiments he demonstrated that 5–6% of starfish at any time were being preyed upon by *Hymenocera picta* and that 0·6% of individuals were being attacked by both predators. He used mortality and immigration rates to predict the abundance of starfish, which approximated that observed for this are over three years. From these results Glynn (1982a, 1984b) concluded that these two predators appeared to be responsible for preventing an increase in starfish numbers in the area studied.

There have been several other reports of animals preying on *Acanthaster planci* in the field (Chesher, 1969a; Brown, 1970; Alcala, 1974) (Table VI). It is unlikely that any of these species would be important predators of this starfish given their biological characteristics. A further group of animals has been suggested as possible predators of *A. planci* but there is little or no evidence to support these assertions. They are: *Cassis cornuta* (Endean, 1969), *Cheilinus undulatus* (Endean, 1982), sharks (Dixon, 1969), *Murex* sp. (Chesher, 1969a), *Dardanus* sp. and *Cymatorium lotorium* (Ormond & Campbell, 1974).

At present there are little direct, quantitative data to suggest that predation plays an important rôle in limiting the numbers of starfish on reefs. There is some indirect evidence to suggest that juveniles and adults suffer extensive predation in the field. This comes from surveys, conducted in several parts of the Indo-Pacific, which have looked at the proportion of starfish with missing or regenerating arms and tissues. From these surveys it was found that from 17–60% of individuals in populations had suffered recent damage. The results of these surveys are given in Table VII. While the study of Glynn (1982a) demonstrated that the predation of juveniles and adults may be relatively high in the field there is little evidence to suggest that

TABLE VII

Proportion of starfish with missing or regenerating arms

Location	Proportion	Reference
Great Barrier Reef	26–60%	Pearson & Endean, 1969
Guam	43%	Glynn, 1982b
Hawaii	60%	Branham, 1973
Panama	17%	Glynn, 1982b
Papua New Guinea	50%	Pyne, 1970
Red Sea	30%	Ormond & Campbell, 1971
Western Australia	38%	Wilson *et al.*, 1974

it is important during the planktonic phase of *Acanthaster planci*. Results from laboratory studies have indicated that the eggs and larvae of this starfish may not be extensively preyed upon since they contain toxic saponins. Unfortunately, there are few data on the predation of these stages in the field. Until information is obtained many questions relating to the occurrence and propagation of outbreaks will remain unanswered.

ORGANISMS ASSOCIATED WITH *A. PLANCI*

Cannon (1972) listed a total of 34 organisms which were considered to be associated with *A. planci*. About five of these may have been duplicate records resulting from taxonomic errors. Another nine organisms were regarded as predators of *A. planci* while the association exhibited by many of the remaining organisms was uncertain. Eldredge (1972) presented a list of 15 organisms that were possibly associated with *A. planci*; at least six of these animals were known to be predators of this starfish.

A list has been prepared of those organisms which are known to be symbiotically associated with *A. planci*, are not predators of this animal and which have been identified taxonomically. This information is given in Table VIII. Besides the turbellarian, *Pterastericola* sp. all the other animals listed in Table VIII are regarded as commensal associates of *Acanthaster planci*. Little is known about the interrelationship between each of these animals and this starfish, although Cannon (1975) stated that the association between *Pterastericola* sp. and *Acanthaster planci* was a host–parasite one. As a result of his investigations into these organisms Cannon (1975) concluded that parasites and diseases were not significant determinants of starfish numbers as none could be found.

TABLE VIII

Animals found in association with A. planci

Copepod	*Onochopygus impavidus*	Humes & Cressey, 1958
	Stellicola acanthasteris	Humes, 1970
Fish	*Siphamia fuscolineata*	Allen, 1972; Eldredge, 1972
	Carapus mourlani	Cheney, 1973a
	Encheliophis gracilis	Cheney, 1973a
Polychaete	*Hololepidella nigropunctata*	Eldredge, 1972
Shrimp	*Periclimenes soror*	Hayashi, 1973
Turbellarian	*Pterastericola* sp.	Cannon, 1972, 1975

Recently, Lucas (1984) reported that starfish were subject to or affected by a disease while undertaking a series of laboratory experiments using a recirculating sea-water system. The spread of this disease could be checked with antibiotics, although sometimes individuals died after contracting this infection. Lucas (1984) described the early, advanced, and severe symptoms of this disease. The severe symptoms produced ulcerations and necrotic tissue and often led to the death of individuals several days after they had been observed. Coelomic fluid taken from infected starfish prior to the occurrence of necrosis was found to contain large numbers of bacteria. The disease was transmitted throughout the entire aquarium system.

The occurrence of this pathogen indicates a possible cause for the rapid disappearance of large aggregations which has been observed in the field (Moran, Bradbury & Reichelt, 1985). Experiments are at present in progress to isolate any pathogens which may possibly cause this (J. S. Lucas, pers comm.).

POPULATION DYNAMICS AND TAGGING

In ecology a group of individuals of the same type or species is referred to as a population. One thing that can be said in all certainty about populations is that they will fluctuate in size (Pielou, 1977). The study of the decrease and the increase of populations (population dynamics) has received great attention from both biologists and mathematicians. In essence, a population is thought of as a single entity which may be defined by a certain set of parameters. These parameters, which are similar for most populations, include; density, birth and death rates (*i.e.* natality and mortality), immigration and emigration rates, age distribution, growth rate of the population, dispersion and movement, longevity, size of individuals and sex ratio. A study of them can lead to a greater understanding of the ecology of a species, its relationship to the ecosystem and the reasons for its increase or decrease (Krebs, 1978).

There are few field data on the population dynamics of *A. planci*. One reason for this is that it is extremely difficult to recognize individuals in the field and follow them for long periods of time. O'Gower, McMichael & Sale (1973) stated that it has not been possible to undertake long-term field studies on *A. planci* due to the difficulties involved in tagging or marking starfish. Consequently, information relating to population parameters such as growth, longevity, mortality and movement is lacking. Up to the present time several tagging methods have been employed in a number of studies but they have proved largely unsuccessful. This problem does not relate solely to *A. planci* but is a problem common to echinoderms in general.

In the first studies which attempted to address this problem tags were attached through the body of starfish (Pearson & Endean, 1969; Branham *et al.*, 1971; Ormond & Campbell, 1974; Wilson & Marsh, 1975). This technique proved unsuccessful for a number of reasons. First, the starfish was able to release the tag by creating an opening in its body wall. Secondly, in some instances the starfish autotomized that part of the body (normally an arm) to which the tag was attached. Thirdly, some starfish became diseased and died. To overcome these responses, tags were tied around an arm or part of the oral disc using monofilament nylon or stainless steel wire (Pearson & Endean, 1969). This method also was unsuccessful as the starfish were able to extricate themselves from their harnesses. This technique has been tested on other types of starfish with little success (Kvalvagnaes, 1972).

Another method, reported by Roads & Ormond (1971), involved attaching coloured bands to the spines of starfish. This was carried out in a bid to follow the movement of starfish over a 24-hour period. While the method proved successful over this short time, experiments by Pearson & Endean (1969) demonstrated that the tagged spines would be shed within a few days.

Instead of attaching a marker to the body of a starfish several attempts have been made to recognize individuals by altering their external appearance. This has mainly involved clipping spines or removing arms. Cheney (1972a) and Ormond & Campbell (1974) clipped the spines of starfish as a means of following individuals in the field. Success was limited since the spines regenerate within a few months (Glynn, 1982b) and they are lost naturally from starfish (by way of predation) which may cause some confusion in identifying marked individuals. Consequently, this method of tagging is only useful for following a small population of starfish over a relatively short time (Vine, 1972).

Another method of marking individuals involved removing an arm or ray. Owens (1971) found that it took 116 days for a new arm to grow 10 mm and postulated that a medium sized individual may be recognizable for at least two years. Again this technique is of limited application and would be useful only for following a small number of starfish. As *A. planci* has a relatively small number of arms the number of starfish that are able to be individually marked is similarly small. Also, some uncertainty may arise when distinguishing between marked individuals in the field as a significant proportion have been observed to have missing or regenerating arms (see Table VII).

Aziz & Sukarno (1977) used natural external features (*e.g.* colour patterns, size, number of arms, position of broken spines) to identify starfish in the field. As the density of starfish was low, plastic markers were placed beside each animal to avoid mis-identifications. It is unlikely that this method would be suitable for studying starfish in outbreaking populations as only a small number of features are used and they may not vary sufficiently between individuals to enable a large number of starfish to be identified (Glynn, 1982b).

Perhaps the most extensive series of experiments on tagging *A. planci* were those undertaken by Glynn (1982b) who tested several different methods including; branding with hot steel rods, applying dyes (Nile-blue sulphate solution, Neutral Red dye) to the arm and aboral surface, staining the aboral disc with saturated solutions of silver nitrate and iodine, inserting T-bar fasteners into the aboral disc, looping cable ties around arms, inserting stainless steel wire and monofilament nylon under the dermis, injecting India ink subcutaneously, inserting insulated wire through the disc and clipping spines. All these methods proved unsatisactory with some producing death in animals. The staining, branding and dye techniques did not produce permanent, recognizable marks and the spines regenerated within four to five months. All tags which were attached to starfish were shed within one to two weeks.

Glynn (1982b) developed a technique for recognizing individual starfish in the field which consisted of using a series of natural characters which were more variable than those utilized by Aziz & Sukarno (1977). These characters included the natural arrangement and numbers of arms, madreporites, anuses, spines, and scars. Data relating particularly to the madreporites and anuses were selected to calculate a madreporite/arm code. Use of this code in conjunction with the other characters was shown in most instances to produce an individual descriptor for each starfish. As the coding process is relatively time-consuming this method would be unsuitable for following large numbers of animals (Glynn, 1982b). Given a small population of intransient starfish then this method, however, is useful and has the added advantages that the starfish are not handled extensively or subjected to injury which may result in changes in their biology, behaviour, and longevity.

DISTRIBUTION OF *A. PLANCI*

A. planci has been recorded throughout the Indo-Pacific region from reefs off the eastern coast of Africa to those in the Gulf of California and Panama. Whilst *A. planci* is known to be associated with coral reefs, it has not been observed on reefs in the Atlantic (Vine, 1973). The reason for this is not known. Predictions of catastrophe have been made should a sea level passage be constructed through central America joining the Pacific and the Atlantic. It has been suggested that this may lead to outbreaks of starfish in the Carribean as a result of larval input from the Pacific (Johannes, 1971). There is no evidence to support this allegation.

The locations where *A. planci* has been observed in the Indo-Pacific are listed in Table IX and shown in Figure 6. Where this starfish has occurred at various locations within the same general area or territory they are listed under the one region (*e.g.* Great Barrier Reef, Mariana Islands). More specific information on the locations of starfish has been given where the reports are for isolated areas (*e.g.* Phuket). The reports themselves have been taken as much as possible from the scientific literature and the references from which these reports were obtained are given in the Table. This list of references is not exhaustive as those presented in the Table represent the major sources of information on starfish for that particular location. Areas where *A. planci* has been reported as "abundant" or "common" have been identified in the Table. For most locations an attempt has been made to define the period during which the starfish were observed. In some instances they were reported over several years at the one location (*e.g.* Ryukyu Islands) and in certain areas a second series of population increases have been observed (*e.g.* Great Barrier Reef) and these also have been noted.

Several conclusions can be made regarding the distribution of outbreaks of *A. planci* from the information contained in Table IX.

(1) Not only have there been outbreaks on a wide variety of reefs thoughout the Indo-Pacific region but they have occurred in isolated areas separated from other reefs by large distances of deep water; examples are the Hawaiian and Cocos-Keeling Islands, Wake Island, and Elizabeth and Middleton reefs.

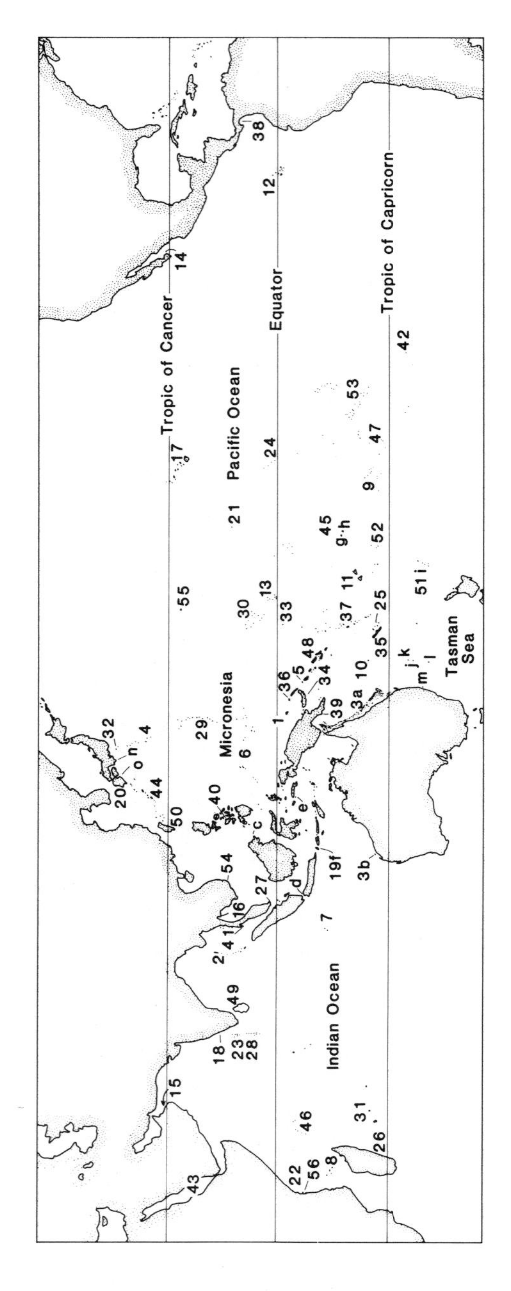

Fig. 6.—Distribution of *A. planci* in the Indo-Pacific region.

TABLE IX

Distribution of A. planci *in the Indo-Pacific region: the map numbers given refer to Figure 6: *starfish abundant or common; †second population increase*

Map No.	Location	Date	References
1	Admiralty Islands	1969*	Pyne, 1970
2	Andaman Islands	1953	Madsen, 1955
3	Australia:		
	a. Great Barrier Reef	1962–1977*, 1979†	Potts, 1981*; Kenchington & Pearson, 1982†
	b. Western Australia	1971–1974*	Wilson, 1972; Wilson & Marsh, 1974, 1975
4	Bonin Islands	—	Yamaguchi, 1977
5	Buka	1968*	Endean & Chesher, 1973
6	Caroline Islands	1969–1972*	Chesher, 1969a; Cheney, 1973b
7	Cocos-Keeling Islands	1949, 1976*	Clark, 1950; Colin, 1977*
8	Comoro Islands	1973*	Polunin, 1974
9	Cook Islands	1969–1970*	Devaney & Randall, 1973
10	Coral Sea:		
	Chesterfield Reef	1970	Endean & Chesher, 1973
11	Fiji	1969*, 1979†	Owens, 1969*, Robinson, 1971*; Zann, pers. comm.†
12	Galapagos Islands	1889	Sladen, 1889
13	Gilbert Islands	1969*	Weber & Woodhead, 1970
14	Gulf of California	1970*	Dana & Wolfson, 1970
15	Gulf of Oman	1982	Stanley, 1983
16	Gulf of Thailand	1973*	Piyakarnchana, 1982
17	Hawaiian Islands	1969*	Branham *et al.*, 1971
18	India:		
	Goa	1743	Vine, 1972
19	Indonesia:		
	c. Sabah	1967*	Yonge, 1968; Morris, 1977
	d. Pulau Pari Islands	1975*	Aziz & Sukarno, 1977
	e. Ambon Island	1973*	Soegiarto, 1973
	f. Bali	1982	Kenchington, pers. comm.
20	Japan:		
	n. Kushimoto	1973*	Hayashi & Tatsuki, 1975; Hayashi, 1975; Yamaguchi, in press
	o. Ashizuri-Uwakai	1972–1983*	Tada, 1983; Ito, 1984; Yamaguchi, in press
21	Johnston Islands	1969*	Chesher, 1969a
22	Kenya	1972	Polunin, 1974
23	Laccadive Islands	1976–1979	Sivadas, 1977; Murty *et al.*, 1979
24	Line Islands	1933	Edmondson, 1933
25	Loyalty Islands	1983	Conand, 1983
26	Madagascar	1958	Humes & Cressey, 1958
27	Malaysia	1968*	Chesher, 1969a
28	Maldive Islands	1963	Clark & Davies, 1965
29	Mariana Islands	1967–1972*, 1979†	Chesher, 1969a*; Marsh & Tsuda, 1973*, Birkeland, 1982†
30	Marshall Islands	1969–1970*	Chesher, 1969a; Branham, 1971
31	Mauritius	1972*	Endean & Chesher, 1973; Fagoonee, 1985a
32	Miyake Island	1977–1980*	Moyer, 1978

TABLE IX (*continued*)

Map No.	Location	Date	References
33	Nauru	1971	Randall, 1972
34	New Britain	1968*	Endean, 1969
35	New Caledonia	1969* 1982	Chesher, 1969a; Conand, 1983
36	New Hanover	1968*	Pyne, 1970
37	New Hebrides	1970*	Endean & Chesher, 1973
38	Panama	1970*	Glynn, 1973, 1974
39	Papua New Guinea	1968–1970*	Pyne, 1970
40	Philippines	1972*	Beran, 1972
41	Phuket	1969*	Chesher, 1969a
42	Pitcairn Group	1970	Devaney & Randall, 1973
43	Red Sea	1968–1970*	Roads & Ormond, 1971
44	Ryukyu Islands	1957–1958*	Yamazato, 1969; Nishihira & Yamazato, 1972, 1973
		1969–1985*	Fukuda, 1976; Fukuda & Okamoto, 1976; Fukuda & Miyawaki, 1982; Matsusita & Misaki, 1983; Ui, 1985; Yamaguchi, in press
45	Samoa:		
	g. Western	1969–1970*	Garlovsky & Bergquist, 1970
	h. American	1977–1979*	Birkeland & Randall, 1979
46	Seychelles	1972	Endean & Chesher, 1973
47	Society Islands	1969–1971*	Chesher, 1969a; Devaney & Randall, 1973
48	Solomon Islands	1969–1971*	Garner, 1971
49	Sri Lanka	1971*	Vine, 1972; De Bruin, 1972; De Silva, 1985
50	Taiwan	1971*	Randall, 1972; Endean & Chesher, 1973
51	Tasman Sea:		
	i. Kermadec Islands	1978	McKnight, 1978
	j. Elizabeth Reef	1979, 1981*	McKnight, 1979; Veron, pers. comm.*
	k. Middleton Reef	1981*	Done, pers. comm.
	l. Lord Howe Island	—	Rowe & Vail, 1984a
	m. Solitary Islands	—	Rowe & Vail, 1984b
52	Tonga	1969*, 1976	Weber & Woodhead, 1970*; Francis, 1981
53	Tuamotu Archipelago	1970	Devaney & Randall, 1973
54	Vietnam	1981	Buznikov *et al.*, 1982
55	Wake Island	1969*	Randall, 1972
56	Zanzibar	1921	Caso, 1970

(2) Some outbreaks have been recorded in areas of relatively high latitude; for example, Ashizuri-Uwakai, Kushimoto (Japan), Miyake Island (all between 33–34° N), and Elizabeth and Middleton reefs (approximately 30° S).

(3) Most outbreaks have been over the same general period throughout the world. Major outbreaks were reported in many areas during the 1960s and 1970s (*e.g.* Great Barrier Reef, Ryukyu Islands and Micronesia). This synchrony of outbreaks was also apparent in the late 1970s with renewed population increases in several areas, notably the Great Barrier Reef, Guam and Fiji.

Major outbreaks involving large numbers of starfish and large scale coral destruction have occurred essentially in three areas in the Indo-Pacific region; the Great Barrier Reef, Micronesia, and the Ryukyu Islands. The first recorded outbreak of *A. planci* in the world was at Miyako Island in the Ryukyu Islands in 1957 (Nishihira & Yamazato, 1972). This was soon followed by reports of outbreaks in 1962 at Green Island on the Great Barrier Reef (Barnes & Endean, 1964), in 1967 at Guam in Micronesia (Chesher, 1969a) and in 1969 on the west coast of Okinawa approximately 320 km to the north of Miyako Island (Nishihira & Yamazato, 1972). Outbreaks have occurred continually in the Ryukyu Islands over the last 15 years (Yamaguchi, in press) leading to large scale control efforts. Soon after they were reported in Micronesia extensive surveys were undertaken to determine the extent of the starfish populations and the coral damage caused by them (Chesher, 1969a). Large populations of *A. planci* have also been recorded in the Red Sea, Fiji, Panama, Samoa, and the Cook Islands. The Hawaiian Islands also experienced outbreaks of *A. planci* towards the end of the 1960s; they appeared, however, to have little effect on the coral communities in this area (Branham *et al.*, 1971).

OUTBREAKS OF *A. PLANCI*

DEFINITION OF OUTBREAKS

It has come to be realized that outbreaks are not all the same but are highly variable phenomena. Despite this variability, attempts have been made to define what is meant by outbreaking and normal populations of *A. planci*. These definitions are important when trying to compare the populations on different reefs and summarize the extent of the phenomenon. As the definitions were derived using a variety of survey techniques, in general they cannot be directly compared and consequently a standardized definition for the two population states has not been formulated. The various definitions proposed are given in Table X. All of those listed define outbreaking and

TABLE X

Definitions of an outbreaking and normal reef

Definition	Reference
Outbreaking	
14 starfish per 1000 m^2	Endean & Stablum, 1975b
40 starfish per 20 min swim	Pearson & Endean, 1969
100 starfish per 20 min. swim or manta tow	Chesher, 1969a
10 starfish per 1 min spot check	Pearson & Garrett, 1976
Normal	
About 1 starfish per 100 m^2 of reef	Dana *et al.*, 1972
About 6 starfish per km^2 of reef	Endean, 1974
Between 4–5 starfish per km of reef	Chesher, 1969a
Between 5–20 starfish per km of reef	Ormond *et al.*, 1973
Less than 14 starfish per 1000 m^2	Endean & Stablum, 1975b
Less than 10 starfish per 20 min swim	Pearson & Endean, 1969
Less than 20 starfish per 20 min swim	Chesher, 1969a

normal reefs in terms of the number of starfish observed over some unit of the survey. As discussed earlier, this is a difficult task since starfish may not be distributed evenly over reefs. Furthermore, their cryptic behaviour and colour make them difficult to observe in the field, particularly when the reef structure is patchy (Kenchington & Morton, 1976). Chesher (1969a) defined several types of populations which he considered to be "normal". In analysing this data, Dana, Newman & Fager (1972) concluded that very few populations met Chesher's criteria and that the definitions for outbreaking and normal populations were inadequate. Dana and his colleagues based their conclusions on the following information obtained from the surveys conducted in Micronesia.

(1) Outbreaks were not evenly distributed on reefs.
(2) The populations varied temporally.
(3) There were different types of outbreaks each with continuously varying densities of starfish.
(4) Some large outbreaks caused little coral mortality.

Kenchington & Morton (1976) considered that it was not possible to define a normal population since little was known about the rôle of *A. planci* in the ecology of the reef. It would appear from these opinions that there is no real solution to this problem and that the terms "outbreak" and "normal" will continue to be defined in such imprecise terms until more is known about the ecology of *A. planci* and a standard method of survey is adopted. Such a survey should include not only data on the abundance of the predator but also that of the coral prey as the two are inextricably linked. More accurate descriptions of starfish populations may also be obtained if the extent of these abundances was presented in some standardized manner. Without a doubt Potts (1981) was correct when he stated that "outbreaks cannot be recognized by any single qualitative or quantitative character" (p. 66).

PRIMARY AND SECONDARY OUTBREAKS

Often outbreaks are classified into two types, primary and secondary (Potts, 1981). In essence, primary outbreaks involve increases in starfish abundance that are associated with the changes in certain local factors in and around reefs and have not arisen from nearby populations (Endean, 1973b; Potts, 1981). On the other hand, secondary outbreaks have been defined as those which have resulted from nearby outbreaks either due to larval input from areas of primary outbreaks or by adult migration (Endean, 1973b). This distinction is relatively clear cut but it is not a simple task to classify outbreaks on this basis since it requires some knowledge about the processes which have lead to their existence. Often it is not possible to determine whether an outbreak is primary or secondary as little quantitative data are available concerning these processes. Primary outbreaks have been demonstrated by implication rather than by direct evidence. Their existence has been inferred particularly in areas isolated by large distances of deep water, such as some of the reefs in Micronesia. It is an extremely remote possibility that outbreaks could have originated in these areas due to input of larvae or adults from other areas and it has been assumed that they arose in response to changes in local conditions. While primary outbreaks may have occurred in these essentially simple reefal systems it is much more difficult to demonstrate their existence on reefs which are large and heterogeneous in structure, such as the Great Barrier Reef. In this instance, primary outbreaks are difficult to infer since the reef is not one single structure but is composed of a multitude of individual reefs which are separated by relatively short distances of shallow water (about 60 m deep). Outbreaks on them may arise both as a result of changes in local factors or due to larval input or adult migration. Unfortunately, there do not appear to be any differences in the manner in which these outbreaks occur which would enable them to be readily identified. One possible way of determining where primary outbreaks have taken place is to ascertain where outbreaks are likely to have begun. In hindsight, this might be accomplished by obtaining information on the pattern and extent of outbreaks over these large complex reefal systems. From this type of information Kenchington (1977) proposed that primary outbreaks were present on the Great Barrier Reef in an area just to the north of Green Island during the late 1950s (see p. 434). These outbreaks were thought to have triggered a wave of secondary outbreaks which moved increasingly southwards. While this model parallels observations on the Great Barrier Reef for that period, unfortunately there are few data to indicate where primary outbreaks originated, despite numerous surveys. Ebert (1983) has suggested an alternative explanation for the pattern of outbreaks recorded. He proposed that the apparent southward movement in the centre of outbreaks may be the result of differential growth of starfish in areas of varying latitude or temperature. Ebert (1983) further postulated that this movement may stem from one major primary outbreak or a series of simultaneous primary outbreaks. While this alternative model would seem plausible, there is insufficient evidence to indicate that variations in the growth rates of starfish in different latitudes would be large enough to account for outbreaks occurring at least a decade apart.

Potts (1981) suggested that the Great Barrier Reef was probably the only area in the world where extensive secondary outbreaks have occurred. Information given by Yamaguchi (in press) now suggests that outbreaks of similar magnitude and type have also taken place in the Ryukyu Islands. Primary outbreaks were thought to have occurred in this region during 1953–1957 (Potts, 1981). A further series of extensive outbreaks (which are still occurring) took place after that time throughout this region and were first reported in Okinawa in 1969. It is not clear whether they arose from new primary outbreaks or whether the outbreaks originated from those which were present in the late 1950s. Those recently reported on mainland Japan and at Miyake Island were considered to represent secondary outbreaks (Yamaguchi, in press).

Outbreaks of *A. planci* have occurred in a similar manner on the Great Barrier Reef and in the Ryukyu Islands. Both regions have experienced extensive secondary outbreaks that have been extremely prolonged. They have been occurring intermittently in these regions for the last 25 years at least. This may be partly related to their structure. As mentioned earlier both areas are composed of many reefs separated by relatively short distances of water, which are often shallow. As the reefs in these areas are close to one another then the chance of large-scale recruitment to some reefs is likely to be high. This is based on the assumption that larvae coming from a nearby reef upstream would be less diluted than those from reefs separated by large distances of water. In some instances where reefs are separated by narrow, shallow channels of water there is also the potential for outbreaks to be perpetuated by adults immigrating from nearby areas (Endean, 1973b). These mechanisms may lead to a higher proportion of reefs being affected by outbreaks and they may also result in the development of chronic outbreaks on some reefs. Thus, there may be a greater potential for reefal complexes such as the Great Barrier Reef and the Ryukyu Islands to suffer protracted outbreaks than reefs isolated by deep water and long distances. This may also depend on a variety of other factors (*e.g.* water currents) which are poorly understood, but it is clear that care must be taken when deciding whether an outbreak (primary) has arisen *de novo* (Potts, 1981) or as a consequence of other mechanisms (secondary). Presumably, this distinction will become clearer when more is known about the causes of outbreaks.

POPULATION MODELS

Until recently, very few mathematical models had been developed as a means of identifying some of the major processes underlying the *Acanthaster* phenomenon. Perhaps this deficiency reflects the lack of suitable information for such modelling procedures. To date, two types of models have been developed and both focus on the interaction between the starfish predator and its coral prey. This was first attempted by Antonelli & Kazarinoff (1984) who considered the interaction representative of that between a herbivore and vegetation. The aggregative behaviour of *A. planci* was incorporated into this model by employing a quadratic co-operative term. This mathematical term was responsible for producing stable limit cycles. The interaction that was modelled was that between two types of corals and one starfish. Two important aspects of the model were the aggregative behaviour of the starfish and the preference shown by the starfish towards its prey. The stability of this interaction was analysed using Hopf bifurcation theory. If no preference was demonstrated by the starfish then the model was found to be "neutrally" stable. The model, however, exhibited stable limit cycle behaviour as the starfish began to prefer one coral over another. The stability of these limit cycles was found to strengthen as coral preference became more asymmetric. As they had demonstrated that natural mechanisms could be responsible for cyclic fluctuations in populations of *A. planci*, Antonelli & Kazarinoff (1984) hypothesized that outbreaks may be natural phenomena akin to those observed in other herbivore–plant interactions.

Bradbury, Hammond *et al.* (1985a) questioned whether the asymmetry used by Antonelli & Kazarinoff was appropriate. They constructed a model from qualitative data on the abundance of starfish and corals from a number of reefs on the Great Barrier Reef. When combined, these data produced a composite view of the interaction. By considering the topological properties of this interaction they demonstrated the existence of cycles that were argued to be the qualitative analogues of stable limit cycles.

Four distinct phases were identified from this qualitative interaction between *A. planci* and corals; (1) coral phase (where coral cover is at a maximum and few starfish are present); (2) outbreaking phase (where the corals are diminishing in abundance and the starfish are rapidly increasing in number); (3) Crown-of-thorns phase (where coral abundance is at a minimum and starfish numbers are at a maximum); and (4) recovery phase (where coral cover is once again increasing and the abundance of crown-of-thorns has declined). These phases reflect the sorts of changes in the abundance of starfish and corals which have been observed in the field. The stability of this cycle was inferred rather than analytically derived since it was demonstrated on reefs whose coral communities were structurally dissimilar. Bradbury *et al.* (1985a) considered that the qualitatively stable cycles may be driven by endogenous factors (forces operating from within the interaction), such as delays in the interaction which may occur due to the structure of the reef. They suggested that this was a more important asymmetry in their model than that employed by Antonelli & Kazarinoff (1984). They postulated that the cycles may also be driven by exogenous factors (forces acting on the interaction from outside) (*e.g.* terrestrial run-off, predation) which may prevent the cycle from heading towards a stable point or level.

Bradbury, Hammond, Moran & Reichelt (1985b) extended this model to include data on starfish and coral abundances that had been collected on each of two occasions in one year at each of almost 100 reefs on the Great Barrier Reef. They once again utilized the qualitative aspects of the data to observe the underlying processes in the interaction. In doing so, they employed the principles of graph theory to plot values for each reef as discrete points in a lattice of two dimensions. The axes of this lattice corresponded to predator and prey abundance categories. This technique revealed three types of dynamic behaviour in the interaction; stable points, stable cycles, and chaos. Their existence had been defined in earlier models of predator-prey interactions (May, 1975). Unlike the earlier studies, the results of Bradbury *et al.* (1985b) indicated that these three states may occur at the same time within the one interaction. Consequently, they argued that the interaction may be a result of endogenous forces that stem from differences in the life history of the predator and its prey.

While the results of these studies are of interest they create a simplistic representation of the phenomenon. At present these models can be considered in their infancy and no doubt they will become increasingly sophisticated as more accurate biological and ecological informaton becomes available.

ACANTHASTER OUTBREAKS ON THE GREAT BARRIER REEF

INTRODUCTION

The Great Barrier Reef is the largest reefal system in the world, comprising of some 2500 individual reefs and extending along almost the entire Queensland coast of Australia for a distance of about 2000 km (Great Barrier Reef

Marine Park Authority, 1981). This section which deals specifically with outbreaks on the Great Barrier Reef has been included for the following reasons.

(1) Outbreaks which have occurred in this region are probably the most extensive in the world having been reported in an area from the Swain reefs to those near Princess Charlotte Bay (see Fig. 7), a distance of about 1200 km.
(2) With the possible exception of the Ryukyu Islands, the Great Barrier Reef is thought to be the only place in the world where secondary outbreaks have occurred (Potts, 1981).
(3) Two series of extensive outbreaks, the second occurring at present, have taken place on the Great Barrier Reef and they are probably the most well documented of the outbreaks that have been reported in the Indo-Pacific region. Extensive scientific surveys have been undertaken on the Great Barrier Reef for almost 20 years providing the most accurate account of starfish outbreaks to date. The only other place where extensive surveys have been undertaken is in Micronesia and these were conducted in the late 1960s and early 1970s. No major surveys have been undertaken in the Ryukyu Islands apart from that reported by Nishihira & Yamazato (1972) at Okinawa. Most efforts in this area have been directed towards undertaking control programmes.

The information presented in this section provides an excellent example of the way in which outbreaks may develop in large reef systems. This information will also be used to discuss the pattern of outbreaks and the problems associated with attempts to determine the extent of these phenomena.

Outbreaks of the crown-of-thorns starfish were first recorded on the Great Barrier Reef in 1962 at Green Island. It has been reported that large numbers of starfish were observed before this time during 1954 on Lodestone reef and 1957 in the Swain reefs but such observations remain unsubstantiated (Vine, 1970; D. Tarca, pers. comm.). Outbreaks continued until 1977 and then for a period of over two years no reports were received. At the end of 1979 new outbreaks were once again reported at Green Island. It is not known whether outbreaks were present on the Great Barrier Reef between 1977 and 1979 or whether this reflects the fact that no scientific surveys were undertaken during this period. Despite this, for convenience, the following account of outbreaks on the Great Barrier Reef has been divided into the two periods; those that occurred between 1962 and 1977 and those reported from 1979 to the present. Information about outbreaks in most instances has been drawn from the results of both published and to a lesser degree, unpublished scientific surveys. The extent and duration of these surveys are given in Figure 7.

OUTBREAKS: 1962–1977

Large numbers of *A. planci* were recorded at Green Island in 1962 (Barnes & Endean, 1964). Over the next two years they increased to plague proportions which resulted in the loss of almost 80% of the live coral on

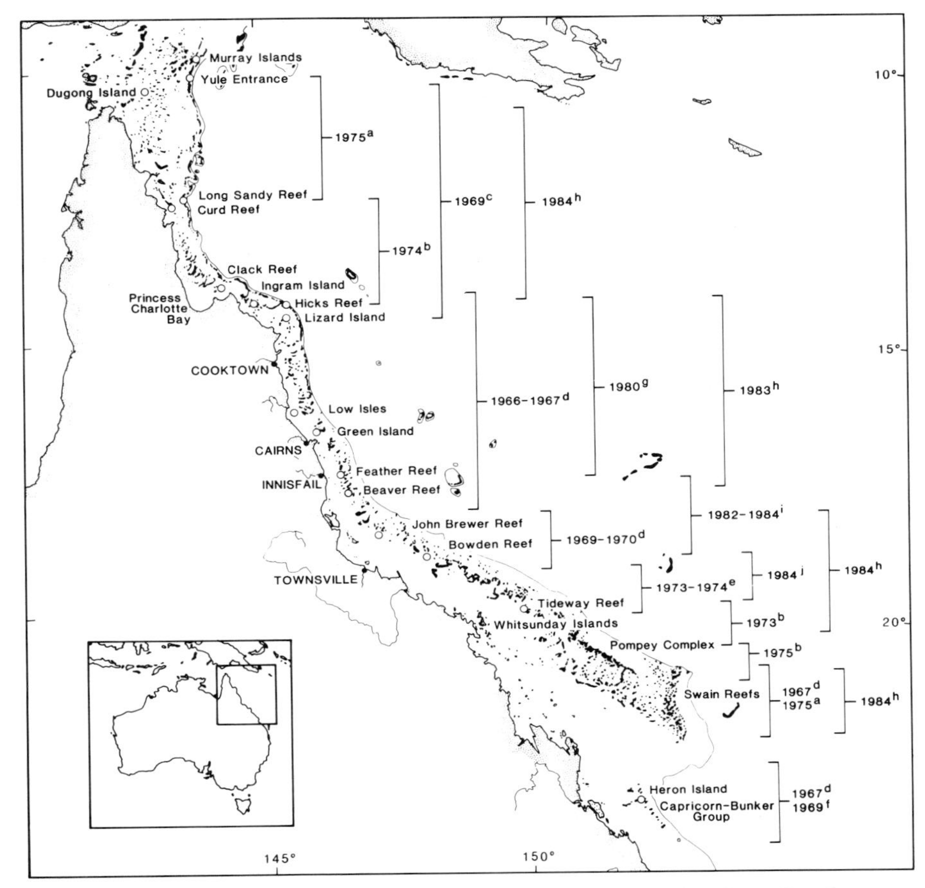

Fig. 7.—Major crown-of-thorns surveys carried out on the Great Barrier Reef: a, Pearson & Garrett (1978); b, Pearson & Garrett (1976); c, Vine (1970); d, Pearson & Endean (1969); e, Morton (1975), Endean & Stablum (1975a), Pearson & Garrett (1975), Kenchington (1975a,b; 1976), Kenchington & Morton (1976); f, Pearson (1972b); g, Nash & Zell (1982); h, Great Barrier Reef Marine Park Authority (unpubl. data); i, Hegerl (1984b); j, Great Barrier Reef Marine Park Authority (1985).

that reef (Pearson & Endean, 1969). The population on Green Island persisted until about 1967 as few starfish were observed after this time (Endean, 1974). Barnes (1966) has given a detailed description of the movement of this population and its effects on the coral communities on the reef. Since this outbreak had such a catastrophic effect on the corals at Green Island evidence of large populations was sought in other areas of the Great Barrier Reef. Throughout the next decade a number of surveys were carried out to determine the extent of these outbreaks (Fig. 7).

By 1966 many of the inner platform reefs between Michaelmas Reef (near Green Island) and Beaver Reef were found to carry large populations of starfish (Pearson & Endean, 1969; Endean, 1974), while there were very few starfish on reefs to the south (Endean & Stablum, 1973b). In 1966 to 1967 outbreaks were beginning to appear on the reefs off Innisfail (Pearson, 1974) and some reefs even further to the south (*e.g.* Otter and Rib Reefs) (Endean & Stablum, 1975b). Coral destruction on many of these reefs, particularly Feather and Peart, was estimated to have exceeded that at Green Island (Pearson, 1974). By 1969 their populations had diminished to low levels (Pearson, 1981).

Surveys conducted from 1966 to 1968 showed that reefs as far north as Lark Reef (just north of Cooktown) had large numbers of *A. planci* (Pearson & Endean, 1969; Endean & Stablum, 1973b). For example, the reefs around Low Isles were noted to have many *A. planci* on them in 1966 and 1967 (Pearson & Endean, 1969). In 1968, none were observed in a survey of 21 reefs in the northern section of the Great Barrier Reef, from Lizard Island to Thursday Island (Vine, 1970). Surveys of Lizard Island and nearby Carter Reef in 1973 revealed few starfish and little coral mortality (Endean, 1974).

Many of the reefs off Townsville (*e.g.* Slashers, Britomart, John Brewer, Lodestone, and Trunk) were found to possess large populations of *A. planci* (Kenchington, 1975a) by 1970. Prior to this, starfish had been very rare on these reefs (Kenchington, 1975a, 1976). Several reefs, particularly John Brewer, were extensively damaged during this period (Pearson, 1981). The starfish populations on reefs in this region had begun to decline substantially by 1971 (Endean & Stablum, 1973b). Starfish outbreaks were recorded on several reefs further to the south (*e.g.* Bowden, Mid, Prawn, Shrimp and Shell) during 1972 to 1973 (Endean, 1974) and 1973 to 1974 (Kenchington, 1975a,b, 1976). These same reefs did not have starfish aggregations on them in 1970 (Endean & Stablum, 1973b; Pearson & Garrett, 1976). At the same time some reefs even further south had experienced outbreaks (*e.g.* Hope, Gould, Rafter, and Line) while others in between had not (*e.g.* Stanley, Old, and the Darley complex) (Endean & Stablum, 1973b). By 1973 the southernmost extent of these outbreaks was thought to be at Tideway Reef (Pearson & Garrett, 1975).

Increasing numbers of starfish were reported on several reefs in the Swain complex during 1973 to 1974 (Pearson & Garrett, 1976). Previously only one starfish had been recorded in this region during a survey of several reefs in 1967 (Endean, 1969). Surveys conducted during 1975, however, found outbreaks of starfish on a number of reefs in the northeastern sector of the Swain complex (Pearson & Garrett, 1976). Extensive coral mortality was also observed on these reefs.

While starfish outbreaks had been recorded during this time on reefs in the Swain complex they had not been reported on reefs immediately to the north in the Pompey complex. Surveys undertaken in 1975 failed to locate any evidence of starfish outbreaks in this area (Pearson & Garrett, 1976). During the 1960s and 1970s no major outbreaks were recorded on reefs to the south in the Capricorn-Bunker Group. Small populations of adult starfish were recorded in the lagoons on some reefs (*e.g.* Llwellyn Reef, Lady Musgrave Island) during 1967 and 1969 although they appeared to be causing minimal coral mortality (Pearson, 1972b). These were thought to be "resident" populations in equilibrium with the surrounding coral communities.

Surveys in the northern sections of the Great Barrier Reef in 1974 found evidence of outbreaks of *A. planci* at Clack Reef and Ingram Island near Princess Charlotte Bay. No evidence of starfish outbreaks was, however, recorded north of this area (to Curd Reef) during this time (Pearson & Garrett, 1976). Towards the end of 1975, abnormal amounts of coral mortality were found on reefs between Long Sandy Reef and Dugong Island (north of Cape York). This damage was attributed to the crown-of-thorns starfish despite the fact that very few were observed and the damage was not recent (Pearson & Garrett, 1978). There is no direct evidence to suggest that major outbreaks have occurred north of Princess Charlotte Bay, although two localized aggregations of starfish were reported in the Torres Strait at the Murray Islands in 1975 (Hegerl, 1984a) and Yule Entrance in 1974 (Pearson & Garrett, 1976) (Fig. 7). By 1977, the only known large populations of *A. planci* were restricted to the eastern section of the Swain Reefs (Kenchington & Pearson, 1982).

OUTBREAKS: 1979–1985

No outbreaks were recorded after 1977 on the Great Barrier Reef for almost two years, perhaps because major scientific surveys were not conducted during this period. From late 1979 to early 1980 another large population of starfish was, however, observed at Green Island (Kenchington & Pearson, 1982). By December 1979, it was estimated that approximately 60% of hard corals had been killed and the starfish population comprised between 350 000 and 2 000 000 individuals (Endean, 1982). Some two months later almost 90% of the live hard coral cover on Green Island had been killed and by the end of the year the starfish population had declined dramatically (Endean, 1982). As a consequence, surveys were renewed in a bid to locate further large populations of *A. planci* (Fig. 7).

Four reefs between Hicks Reef and Ellison Reef (near Beaver Reef; Fig. 7) were found to carry large starfish populations on them in early 1980, while several others exhibited recent coral damage (Nash & Zell, 1982). By 1983 *A. planci* was observed on 23 reefs in this region, although only two reefs were considered to have had large populations of this starfish (Great Barrier Reef Marine Park Authority, unpubl. data). An additional five reefs were considered to have low coral cover which was presumed to be due to *Acanthaster* predation. Towards the end of 1984, 24 mid-shelf reefs in this region were found to have extensive areas of dead coral which was attributed to *Acanthaster* predation (Great Barrier Reef Marine Park Authority, 1985). Although few starfish were seen on these reefs large numbers were recorded on several of the ribbon reefs east of Lizard Island (Hegerl, 1984b). Just prior to this, surveys were conducted on reefs to the north, from Princess Charlotte Bay to Whyborn Reef near the tip of Cape York. No evidence of recent outbreaks was found (Hegerl, 1984b).

During 1983 and 1984 (approximately four years after the start of the outbreak on Green Island) outbreaks were observed on a number of reefs near Townsville (Bradbury, Done *et al.* 1985; Bradbury *et al.*, 1985a). Nineteen of 42 reefs in the central section of the Great Barrier reef were reported to have *Acanthaster* on them during surveys conducted in late 1984. Of these 19, only 12 had large numbers of starfish on them and nearly all were mid-shelf reefs located near Townsville (Great Barrier Reef Marine Park Authority, unpubl. data). Surveys conducted to the south of this region during 1983 and 1984 on reefs east of the Whitsunday Islands, in the Pompey and Swain complexes and the Capricorn-Bunker group failed to find any evidence of outbreaks (Great Barrier Reef Marine Park Authority, unpubl. data, 1985). It would appear that by the middle of 1985, the southern and northern limits of this second series of outbreaks were to be found at reefs near Townsville and Lizard Island, respectively.

PATTERN OF OUTBREAKS

It has been suggested on numerous occasions that outbreaks moved in a southerly direction during the 1960s and 1970s (Talbot & Talbot, 1971; Pearson, 1972b; Endean, 1974). In addition, Kenchington (1977) suggested that this pattern was initiated by primary outbreaks that had occurred on reefs to the north of Green Island in the mid 1950s. While there is evidence to support the notion that outbreaks tended to be in more southerly latitudes with time there are a number of inconsistencies in this model. First, there is no direct evidence that primary outbreaks were present on reefs north of Green Island in the 1950s. In fact, outbreaks were observed on many reefs in this region during 1966–1968 (Pearson & Endean, 1969). Secondly, a consistent southward trend in the pattern of outbreaks is not evident. For example, Rib Reef (located just north of Townsville) had a large population of starfish on it in 1966 several years before the majority of reefs in this areas and at a time when reefs further north off Innisfail were only just beginning to experience them. Similarly reefs such as Hope, Gould, Rafter and Line were observed to have outbreaks in 1972 and 1973 at the same time as those much further to the north (*e.g.* Bowden, Prawn, and Shrimp). Those at the southern end of the Great Barrier Reef in the Swain region also were experiencing outbreaks during this period despite the fact that a vast area of reefs further to the north (Pompey complex) were not (Birtles *et al.*, 1976).

While some of these anomalies may be due to inadequate data it is clear that the southward movement model proposed by several authors provides only a general description of the pattern of outbreaks in the 1960s and 1970s. Indeed, it is possible that this model is derived in part by the fact that the surveys themselves moved in a southerly direction with time (Fig. 7). Most important, however, the model cannot be applied to the entire Great Barrier Reef since it relates only to those reefs in the southern half of it (*i.e.* reefs south of Green Island).

The results of surveys completed since 1979 tend to support the notion of a general southward movement of outbreaks as reefs off Innisfail experienced them in 1981 and 1982 (some two years after those at Green Island) and those off Townsville in 1983 and 1984. If this pattern continues then reefs between Townville and the Whitsunday Islands will outbreak over the next few years. During this time surveys need to be undertaken repeatedly on reefs to provide a more accurate description of the movements of outbreaks in this region.

Only one attempt has been made to analyse the large volume of information collected during surveys conducted over the last 20 years. In this study starfish abundances, recorded on reefs throughout the Great Barrier Reef between 1979 and 1984, showed a strong temporal component in the pattern of outbreaks rather than a spatial one (Bradbury, Done *et al.*, 1985). From the analysis it was concluded that this indicates "some sort of long-term cyclicity at the whole GBR scale" (Bradbury, Done *et al.*, 1985, p. 108). While this may be true it does not invalidate the southward movement model which relates to the pattern of outbreaks occurring over a completely different scale (*i.e.* reefs south of Green Island). More analyses of this type are needed if realistic models of the *Acanthaster* phenomenon are to be achieved. In order to do this a more homogeneous data set is, however, needed (Bradbury, Done *et al.*, 1985).

No general pattern in the occurrence of outbreaks can be readily discerned for reefs to the north of Green Island. This is partly because the region is more remote and was surveyed less intensively than reefs in the southern half of the Great Barrier Reef. Large numbers of starfish were recorded on several reefs as far north as Lizard Island during 1966 and 1967. By 1974 they were observed further north near Princess Charlotte Bay and extensive areas of dead coral were reported on reefs near Cape York in 1975. It is not known whether these observations reflect either a northward movement in the outbreaks or indeed the surveys, or whether they suggest the occurrence of earlier primary outbreaks. Repeated surveys of reefs in this region during the next few years may provide a more accurate picture of the pattern of spread of these outbreaks.

There are two other interesting features relating to the pattern of outbreaks on the Great Barrier Reef that should be mentioned.

(1) No outbreaks have been observed in the Capricorn-Bunker Group at the far southern end of the Reef. Surveys during the late 1960s did not find any evidence of outbreaks although small resident populations of starfish were reported in sheltered locations on some reefs (Pearson, 1972b). Since that time no evidence of outbreaks has been reported on these reefs (Done, Kenchington & Zell, 1982; Great Barrier Reef Marine Park Authority, 1985). Lucas (1973, 1975) has suggested that outbreaks may not occur in this region as the temperature regime is less favourable for the survival of large numbers of larvae. This would seem unlikely, however, as outbreaks occurred on reefs nearby in the Swain region and were also observed on Elizabeth and Middleton Reefs (Table IX, see p. 424) which are located approximately 850 km to the southeast.

(2) Certain reefs appear to be more susceptible to outbreaks than others. Recent information has indicated that 16 of 21 reefs that had large numbers of *A. planci* in 1983 also had outbreaks on them during 1966 to 1970 (Great Barrier Reef Marine Park Authority, 1984a). Of these reefs several mid-shelf reefs between Cairns and Townsville have experienced catastrophic outbreaks on both occasions which involved large numbers of starfish and resulted in extensive coral mortality. Outbreaks of this sort were experienced on Green Island (Pearson & Endean, 1969), Feather (Pearson, 1974), Rib (Pearson & Endean, 1969), and John Brewer (Pearson, 1981) reefs from 1960 to 1970. These same reefs suffered outbreaks of a similar magnitude between 1979 and 1985 (Endean, 1982; Hegerl, 1984b).

In contrast some reefs, particularly those on the outer edge of the continental shelf, do not seem to be susceptible to outbreaks. For example, only one outer barrier reef was recorded to have large numbers of starfish on it during surveys conducted between 1966 and 1969 (Pearson & Endean, 1969; Pearson, 1970). In addition, some outer shelf reefs (*e.g.* Myrmidon Reef located off Townsville) have never been reported to have suffered an outbreak (Endean & Stablum, 1975b; Great Barrier Reef Marine Park Authority, unpubl. data). On the other hand, mid-shelf reefs appear to have a higher incidence of outbreaks. Almost all those surveyed off Townsville during 1984 were found to have large numbers of starfish (Hegerl, 1984b; Great Barrier Reef Marine Park Authority, unpubl. data). There are inconsistencies, however, as some reefs in this region (*e.g.* Wheeler and Davies) had few starfish on them during the 1970s despite being situated close to reefs (John Brewer, Lodestone, and Keeper) that had large outbreaks (Endean & Stablum, 1975b). Why some reefs should be more likely to experience an outbreak than others is not understood. Perhaps factors such as the morphology and position of reefs, water currents, temperature, and salinity are important in determining the "outbreak behaviour" of individual reefs. These factors may operate on both the adult and larval stages of the life cycle of *A. planci.*

EXTENT OF OUTBREAKS

Information on outbreaks of *A. planci* over the last 20 years has been compiled by the Great Barrier Reef Marine Park Authority. This information is based on reports of the presence or absence of starfish not only from scientific surveys but also other reef users (*e.g.* sport divers, tourist operators). Up until 1983 reports dating from 1957 had been compiled for 516 reefs or approximately 20% of the total number of reefs comprising the Great Barrier Reef system (Great Barrier Reef Marine Park Authority, 1984a; Kenchington, 1985). A summary of the information received for 1984 indicated that *A. planci* was not observed on 57% of the 178 reefs for which reports were received. It was deemed to be uncommon (<10 starfish observed) on 18% of reefs and common (10–39 starfish observed) on a further 9% of reefs. Aggregations of 40 or more starfish were reported on the remaining 16% of reefs (Great Barrier Reef Marine Park Authority, 1984b).

Despite the large amount of information relating to the abundance of *A. planci* on reefs at no time over the last 20 years has it been of sufficient detail to provide an accurate assessment of the extent of outbreaks on the Great Barrier Reef. As a consequence, great controversy has surrounded this question and it has involved both the public and the scientific community (Kenchington, 1978). Even recently little agreement has been reached among scientists as to the extent of the Great Barrier Reef affected during the second series of outbreaks since 1979 (Crown of Thorns Starfish Advisory Committee, 1985; see also Endean & Cameron, 1985). Accurate information which will enable definitive statements to be made regarding the extent of outbreaks has been difficult to obtain for the following reasons.

(1) It is impossible to survey entirely the Great Barrier Reef since it is so large and heterogeneous a structure. Such an undertaking would

require unlimited resources as well as personnel and time. During 1985 surveys of approximately 10% of the total number of reefs in the Great Barrier Reef were conducted (as part of an employment programme) at a cost of A$ 1 million (Bradbury, Done *et al.*, 1985).

(2) Outbreaks are not uniform phenomena, but vary substantially in population size and the extent of reef that they encompass (Moran, Bradbury & Reichelt, 1985). Recent studies on John Brewer Reef have demonstrated that major temporal and spatial changes in the distribution and abundance of starfish may be quite rapid occurring in the order of months rather than years (Moran, Reichelt & Bradbury, 1985).

(3) Estimates of the abundance of starfish and corals have been conducted using different methods making it difficult to compare and analyse the data collected.

Given the problems listed above it is clear that the degree of information required to provide an 'error-free' assessment of the extent of outbreaks on the Great Barrier Reef or any other large reefal system will never be attained. For this reason it should be recognized that debates focusing on this issue may never be fully resolved. Despite the fact that a definitive answer is not likely to be forthcoming, information may be obtained which will allow reliable predictions to be made regarding the extent of outbreaks. Information of this type can be gained by repeatedly surveying a smaller proportion of reefs situated uniformly throughout the reefal system. This will generate a homogeneous information base which will be amenable to mathematical analysis.

EFFECTS OF OUTBREAKS

EXTENT OF CORAL MORTALITY

Pearson (1981) in reviewing the information available on the recovery and recolonization of coral communities stated that outbreaks of *A. planci* caused coral mortality that was more extensive and dramatic than any other natural or man-made disturbance (Fig. 8a,b,c). Various estimates have been given of the extent of coral mortality which can be inflicted by outbreaks of starfish. Chesher (1969a) reported that outbreaks in Guam were destroying corals at an average rate of 1 km per month. On the Great Barrier Reef outbreaks were indicated to have killed approximately 80% of all corals down to a depth of 40 m at Green Island (Pearson & Endean, 1969). Higher figures of coral mortality were given for Fitzroy Island (Pearson & Endean, 1969) and recently for Green Island (Endean, 1982). These reports indicate that outbreaks of *A. planci* are capable of killing large areas of coral, but not all outbreaks produce such destruction. For example, the outbreak in Hawaii was found to have had little effect on the coral populations by the time it had dispersed (Branham *et al.*, 1971). In addition, Glynn (1973, 1974) considered that although starfish were common on reefs in Panama their level of predation was not enough to alter coral community structure as they preferred to feed on less abundant corals. These findings opposed those of Porter (1972, 1974) who suggested that *A. planci* fed preferentially on competitively dominant species (*Pocillopora damicornis*) and thus was responsible for creating a more diverse coral assemblage composed of less preferred species. Glynn (1976) argued that the surveys undertaken by Porter were inadequate as a large proportion of them were carried out in shallow water and not in areas where most *Acanthaster planci* were found. He also showed that *Pocillopora damicornis* is not a preferred food source due to the occurrence of symbionts (*Alpheus* sp. and *Trapezia* sp.) which live in the coral and prevent the starfish from feeding (Glynn, 1976, 1980). Glynn (1976) suggested several other factors which could account for the community patterns identified by Porter (1972) and these have been summarized by Menge (1982).

The information given above suggests that outbreaks of *Acanthaster planci* may not kill all the coral in an area (Rowe & Vail, 1984b). Outbreaks themselves are variable phenomena, both spatially and temporally and the amount of coral damage on a reef is not always evenly apportioned. For example, it has been reported on numerous occasions that corals in shallow water tend to survive starfish outbreaks because of the turbulent conditions (Endean & Stablum, 1973a; Colgan, 1982; Moran, Bradbury & Reichelt, 1985). Also, certain species of corals, particularly massive forms, may be left after outbreaks because generally they are not a preferred food (*e.g. Pocillopora damicornis*, *Porites* spp., *Diploastrea* sp.) (Glynn, 1976; Endean & Stablum, 1973a; Pearson, 1974). Recently, Done (1985) suggested that the mortality of some of these corals may be a function of the size of the colony. In a series of extensive surveys on John Brewer Reef they found that massive colonies of *Porites* spp. greater than 500–600 mm were less susceptible to predation by *Acanthaster planci*. Even large massive *Porites* that had suffered predation were most often not killed entirely and the live surfaces were observed to regrow over the dead surfaces forming knob-like protrusions (Woodhead, 1971; Done, 1985). Done also demonstrated that areas which may have suffered almost 100% loss of coral cover may contain large numbers (up to 59 per m^2) of small, remnant colonies (10–100 mm) that had escaped starfish predation. Thus the term "devastation" must be used with caution when describing reefs that have suffered extensive coral mortality.

Since coral mortality is not uniform on reefs and some corals survive starfish predation better than others, it becomes very difficult to determine the extent of coral damage caused by outbreaks. This is particularly true when trying to assess the amount of mortality over an entire reef surface since survey methods which may be suitable for recording coral mortality in small areas (*e.g.* line transects and quadrats) may not give accurate information over this much larger area. More broad-scale survey techniques (*e.g.* manta towing and spot checks) may be needed in order to obtain information on coral mortality at the whole reef scale (Kenchington & Morton, 1976; Pearson & Garrett, 1975, 1976, 1978). Care must, however, be taken when conducting such surveys as areas of dead white coral (termed feeding scars when caused by *Acanthaster planci*) (Fig. 8b) may also be caused by other means. A variety of different animals have been reported to feed on coral although most of them are unlikely to produce extensive areas of mortality (Endean, 1971a; Glynn, 1985). While this is generally true, several animals are capable of causing large areas of dead white coral which may be mistaken for the recent predatory activities of *A. planci*. For example, the gastropods *Drupella fragum* and *D. rugosa* have been demonstrated to be responsible for causing extensive coral mortality (up to 35% of coral cover destroyed) in Japan and the Philippines, respectively (Moyer, Emerson & Rose, 1982). Similarly, the starfish *Culcita novaeguineae* may be a significant cause of dead white coral in certain parts of the Indo-Pacific region (Goreau *et al.*, 1972). The gastropod *Jenneria pustula* was recorded by Glynn, Stewart & McCosker (1972) to occur in large populations (up to 18 000 individuals) on reefs in Panama. At those densities it was found that this animal could destroy 5·26 metric tons of *Pocillopora damicornis* per hectare per year. This rate of destruction was estimated to be equivalent to that generated by a population of *Acanthaster planci* at a density of 30 individuals per hectare. Other animals which have been reported to produce significant amounts of coral mortality are hermit crabs, puffer fish (Glynn, 1974) and the starfish *Pharia pyramidata* (Dana & Wolfson, 1970). In addition, not all dead white coral may result from the feeding activities of animals alone. Coral bleaching which has been reported recently in Panama (Glynn, 1983, 1984a) and on the Great Barrier Reef (Harriott, 1985; Fisk & Done, 1985; Oliver, 1985) may affect up to 50–80% of coral cover on some reefs. It follows that recently dead coral on reefs may not indicate the presence of large numbers of *Acanthaster planci*, but may represent the effects of other biotic and abiotic factors. Thus care must be taken when using the abundance of feeding scars as a measure of the extent of the activities of this starfish.

Recently, Cameron & Endean (1985) have suggested that the severity of

Fig. 8(a)

Fig. 8(b)

Fig. 8c

Fig. 8.—Before (a), during (b) (note patches of white coral), and after (c) an outbreak of starfish.

an outbreak should be judged not only according to how much coral mortality it produces but also on the types of coral species killed. They argued cogently that long-lived corals (*e.g. Porites* spp.), rather than transient species, were the main architects of coral community structure. Based on this argument they used the term "ecocatastrophe" to describe the second outbreak on the Great Barrier Reef as they claimed that many massive corals had beeb killed (Endean & Cameron, 1985). They used this emotive term since they considered that this outbreak, unlike the first, had succeeded in removing those corals which made up the very fabric of the Reef. They did not, however, present much quantitative evidence to bolster these assertions.

CORAL RECOVERY

One of the first changes which has been observed after an outbreak has occurred is the recolonization of the dead surfaces by algae. It has been commonly reported that once a coral has been killed the bare white surface is quickly colonized by these organisms (Pearson & Endean, 1969; Endean & Stablum, 1973a; Nakasone *et al.*, 1974). This lead Cameron & Endean (1982) to suggest that after outbreaks the reefs are dominated initially by algae. A number of studies have investigated this process in more detail. In general, it would seem that the rate and pattern of recolonization vary according to location. In the Red Sea, Biggs & Eminson (1977) found that corals predated by *Acanthaster planci* were rapidly covered by algae which reached their maximum growth after two weeks. They suggested that feeding scars could only be recognized in the field within ten days of the death of the coral. On the basis of these findings they advised that feeding scars may not be reliable indices of the extent of predation on reefs. Price (1972, 1975) found algal colonization of dead corals to be much slower on the Great Barrier Reef. Algae did not become apparent until almost two weeks after the death of the coral. Recording the recolonization of algae over a period of 77 weeks he reported that turf algae and blue green algae were important during the early stages of the development of the algal community. The turf algal coverage declined after about one year and encrusting algal forms then dominated the community. As these forms were thought to consolidate the corals, Price (1975) considered it unlikely that they would be eroded as suggested by Fishelson (1973). Belk & Belk (1975) studied the processes involved in the recolonization of algae on recently killed *Acropora aspera* colonies in Guam. They discovered that dead surfaces were covered by three species of blue-green algae and two species of red algae within 24 hours. Most algal species recorded during this study had settled within nine days. Two species of blue-green algae dominated the substrata within the first 25 days after which time they were dominated by the brown alga, *Giffordia indica*.

The processes involved in the recolonization and development of hard corals have been investigated in several studies. So far, however, no long term study has been conducted where coral community structure has been surveyed before and after an outbreak of starfish at the same site. Surveys of community structure prior to an outbreak may provide information which can be used to assess accurately the extent of recovery. Studies of this sort have only been recently initiated (Done, 1985; Moran, Bradbury & Reichelt, 1985).

During the series of outbreaks that occurred on the Great Barrier Reef in the 1960s and 1970s a broad series of surveys of coral recolonization were conducted by Endean & Stablum (1973a). Recovery was assessed visually as well as by photographic techniques which had been developed earlier by Laxton & Stablum (1974). Pearson (1981) has given a thorough account of this latter method and has raised certain doubts about its accuracy. Endean & Stablum (1973a) found little if any recolonization on reefs affected by outbreaks, but as Pearson (1981) pointed out it is not possible to determine how long these processes had been underway on these reefs. The major findings of the studies undertaken by Endean & Stablum (1973a) were as follows.

(1) Recolonization was most rapid in shallow areas on seaward slopes where many corals had survived.
(2) Recolonization in deeper water and in sheltered locations (back reefs and lagoons) was found to be slow and often dominated by soft corals and algae.
(3) Encrusted skeletons of dead coral may remain *in situ* for several years although there was an indication that the skeletons of some colonies (*e.g. Acropora hyacinthus*) tend to collapse due perhaps to the activities of boring animals.
(4) In the early stages of recolonization soft coral cover was noted to have increased on many reefs.

(5) The most common recolonizing hard coral species were: *Pocillopora damicornis*, *Seriatopora hystrix*, *Stylophora pistillata*, *Acropora hyacinthus*, *A. humilis*, *A. variabilis*, *A. formosa*, *A. cuneata*, *A. echinata*, *Porites* spp., and *Turbinaria* spp.

Endean & Stablum (1973a) and Endean (1974, 1976, 1977) considered that recovery of corals from outbreaks of *Acanthaster planci* may take from 20–40 years although they indicated that it may be slow or even retarded in some areas particularly if the skeletons of dead corals were eroded. There have been several other qualitative or visual reports of the recovery and recolonization of coral communities. For example, Branham (1973) recorded substantial recolonization of corals in Hawaii which had suffered extensive damage as a result of an outbreak which had occurred only three years previously. In Japan, Nishihira & Yamazato (1974) noted that recolonization was variable in different areas being dominated not only by hard corals but also by soft corals and algae. Only two quantitative long-term studies on the recovery of coral communities have been conducted after outbreaks of starfish, one on the Great Barrier Reef (Pearson, 1972a, 1973, 1974, 1975a, 1977, 1981) and the other at Guam (Randall, 1973a,b,c,d; Colgan, 1981, 1982).

Pearson (1974, 1981) investigated the recovery of coral communities using permanent 10 m × 1 m study plots on several reefs (Feather, Ellison, and John Brewer) and 1 m × 1 m quadrats spaced evenly along transects laid down the seaward slopes of 18 reefs between Innisfail and Townsville (Fig. 7, see p. 431). In this series of studies all permanent plots and quadrats were established a few years after the outbreaks had disappeared. In order to assess the extent of recovery the results from these surveys were compared with those obtained from nearby reefs which had not been affected by the outbreaks. The results from the surveys of quadrats and permanent plots are given in Tables XI and XII, respectively. The major findings were as follows.

(1) The pattern of recovery was variable between reefs and within specific locations on reefs (Pearson, 1981).
(2) There is potential for the rapid recovery of reefs since recruitment (202 new colonies were found in seven months in one plot) and growth of some species was rapid (*Acropora* spp. were found to reach 200 mm in diameter three years after settlement on artificial substrata) (Pearson, 1974, 1975a).
(3) The most common recolonizers were *Acropora* spp. and *Porites* spp. and the rapid increase in hard coral cover was due mainly to the growth of tabular colonies (some had reached 500–1000 mm in diameter in approximately ten years) (Pearson, 1975a, 1977, 1981).
(4) Coral recovery was slower in more unfavourable environments such as reef flats than in deeper locations on seaward slopes (Pearson, 1981).
(5) Coral cover and density (of colonies and species) reached levels similar to those on nearby undamaged reefs within 10–15 years.

In discussing the results of these surveys Pearson (1981) suggested that a number of factors may influence the type and speed of recovery including;

TABLE XI

Comparison of coral communities on reefs recovering from starfish outbreaks (in 1966–1967) with those on reefs unaffected by them: this information is from quadrat surveys conducted off Innisfail by Pearson (1974, 1981)

Reef			Hard corals			Number of transects
State	Number	Year	Cover(%)	Species (m^{-2})	Colonies (m^{-2})	
Undamaged	7	1971	28·2	7·2	10·5	10
Recovering	10	1971	8·2	5·6	5·1	15
Recovering	1	1975	16·0	7·7	7·0	2
Recovering	10	1977	21·4	11·4	10·8	15

TABLE XII

*Information on the recovery of coral communities in permanent study areas (2–3 m depth) on three reefs in the Great Barrier Reef: this information is taken from Pearson (1981); *from Done, 1985*

Reef	Year	Coral cover (%)	Colony size (cm) ($\bar{x}$)	Common recolonizers Genus	(% proportion)
Feather	1972	12·3	5·1	*Acropora*	(39·6)
				Porites	(22·3)
				Galaxea	(10·9)
Feather	1974	63·0	—	—	—
Feather	1975	60·0	7·3	*Acropora*	(43·7)
				Galaxea	(7·2)
				Porites	(6·7)
				Pocillopora	(5.5)
Ellison	1972	2·6	4·4	*Acropora*	(59·0)
				Favia complex	(10·0)
				Porites	(9·3)
Ellison	1975	60·0	11·6	—	—
Ellison	1978	80·6	—	*Acropora*	(71·2)
				Seriatopora	(5·1)
				Porites	(5·1)
John Brewer	1974	6·0	3·6	—	—
John Brewer	1978	45·2	6·9	*Acropora*	(30·0)
				Favia complex	(6·7)
				Fungia	(4·7)
	1982*	78·0	—	—	—

type of subtratum (whether or not it is algal-covered), sedimentation, growth rates of species, predation, further disturbances, environmental variables (*e.g.* light intensity, water circulation), location, recruitment, and settlement. He also stated that there was no evidence to support the notion that soft corals overgrew areas of hard corals killed by *Acanthaster planci*.

The studies instigated by Randall (1973a,b,c,d) at Guam were not all undertaken at the same locations. Coral recovery was followed in a number of reef zones at Tanguisson Bay from 1970 to 1974. The recovering communities were compared with those at Tumon Bay (about 10 km south) that had been surveyed prior to the outbreaks (during 1968). Unfortunately, at this site systematic surveys were not undertaken in the submarine terrace (6–18 m) or seaward slope (18–35 m) zones where coral mortality was greatest. This study was recently extended by Colgan (1981, 1982) who reported on the recovery of the coral communities up to 1980 in these two zones, as well as the reef front zone (0–6 m). The data obtained from all these studies are given in Table XIII. It is difficult to compare them directly as three different survey methods (line transect, quadrats, point-quarter) were utilized during the period of the entire study. Despite this Colgan (1982) identified five stages in the recovery of the communities at Tanguisson Bay.

(1) Dominance of crustose and filamentous algae.
(2) Recruitment of planulae.
(3) Differentiation of growth forms (from encrusting to massive and corymbose).

TABLE XIII

*Information on coral communities obtained before (during 1968 at Tumon Bay) and after (from 1970–1981 at Tanguisson Bay) an outbreak of starfish, for three reef zones at Guam: these data were taken from Randall (1973a,b,c,d) and Colgan (1981, 1982); *total number of species recorded over all reef zones for each year*

	Reef front					Submarine terrace					Seaward slope				
	1968	1970	1971	1974	1981	1968	1970	1971	1974	1981	1968	1970	1971	1974	1981
Coral cover (%)	49·0	20·9	21·9	24·8	43·7	59·1	0·9	4·0	12·0	65·9	50·1	0·5	2·1	6·3	36·2
No. colonies (m^{-2})	—	16·7	20·8	24·2	24·7	—	8·0	15·8	24·7	31·1	—	4·0	13·1	21·5	43·8
No. species (m^{-2})	—	5·4	6·7	7·5	7·5	—	4·8	6·4	8·0	10·8	—	2·1	6·0	8·0	13·1
Species richness	98	70	68	—	—	73	47	70	—	—	57	32	61	—	—
	(139)	(91)	(105)	(131)	(137)*										
Growth form (% frequency)															
Massive	25·4	20·3	15·5	32·4	33·3	—	12·8	7·0	32·1	22·6	—	8·8	10·8	46·4	36·9
Encrusting	24·4	59·5	62·2	26·4	25·8	—	78·1	84·3	58·2	61·0	—	80·8	80·0	39·1	41·0
Branching	43·6	8·5	21·4	39·2	38·3	—	1·6	5·2	8·8	15·0	—	5·0	4·6	9·9	14·8
Other	6·6	11·7	0·9	2·0	2·6	—	7·5	3·5	0·9	1·4	—	5·4	4·6	4·6	7·3

(4) Expansion of colonies (this lead to a reduction in the number of coral colonies).
(5) Competition between corals.

Like Pearson (1981), Colgan (1982) found that coral cover had regenerated to levels recorded before the outbreak in about 11 years. He also noted that the species richness and size frequency distribution of the communities had recovered in this time. Initially, recovery in the areas at Tanguisson Bay was slow until adults became re-established and this was thought to occur as a result of regrowth from small remnants and recruitment from nearby surviving adults. It was evident that numerous small colonies survived, since after the outbreak the zones were found to have relatively high species diversity and species richness although coral cover was only 1%. In 1970 (about two years after the outbreak), 87% of the corals at Tanguisson Bay were less than 100 mm (Randall, 1973b). As noted earlier, Done (1985) reported high densities of small remnants (10–100 mm) in areas of John Brewer Reef just after they had experienced extensive starfish outbreaks. Colgan (1981) considered that these survivors enabled rapid recovery of the coral communities at the local scale.

The studies of Pearson and Colgan have demonstrated that some of the variables which characterize coral communities may return to their original levels within 10–15 years. Despite this they have not shown whether the structure of these communities may recover in such a period. Colgan (1981) stated that the species diversity of the communities at Tanguisson Bay (gauged using diversity indices) had "approached or exceeded" the values recorded at Tumon Bay prior to the outbreak, although no values were given for this latter area. Pearson (1981) felt that it may take several decades for coral communities to recover completely. More long-term, before and after studies are needed in order to resolve this question.

OTHER COMMUNITIES

The severity of starfish outbreaks has been gauged most often according to the extent of their effects on the hard coral communities which are a major component of the physical structure of reefs. Perturbations of this scale are likely to influence the distribution and abundance of other organisms which may interact with or depend on this complex assemblage of corals. Few studies have been conducted to ascertain whether the creation of large areas of dead coral has 'downstream' effects on other communities. References to such effects have mainly come from incidental field observations and have not resulted from quantitative studies. For example, Laxton (1974) suggested that the distribution of the blue starfish *Linckia* sp. had been extended on some reefs as outbreaks had caused an increase in the cover of coralline algae. Similarly, Garlovsky & Bergquist (1970) noted that the annelid *Palola siciliensis* had declined in abundance in Western Samoa in conjunction with the increase in numbers of *Acanthaster planci*. In this instance outbreaks were thought to have caused the destruction of much of the habitat of this animal. Outbreaks have also been thought to lead to the dominance of soft corals on certain reefs (Endean, 1971a), presumably because they are generally not eaten by starfish and have the opportunity to grow into those areas of coral destroyed by *A. planci*. While Garrett (1975) found that some soft corals are competitively dominant over smaller hard corals, evidence from studies of coral recovery have indicated that they are not a particularly important component of this process (Pearson, 1981).

Most references to 'downstream' effects on other communities have been in relation to fish communities. Observations by Chesher (1969a), Cheney (1972c), and Endean & Stablum (1973b) indicated that algal feeding fish such as acanthurids and scarids were more common after outbreaks but that chaetodontids and serranids gradually disappeared sometime after these events. In general, these conclusions have been supported by the results of two studies which have investigated the effects of outbreaks on coral reef fish communities. The first by Sano, Shimizu & Nose (1984) involved manipulative experiments on a small number of coral colonies. The results from these experiments were then compared with observations of fish communities near corals that had been killed by *A. planci*. From these studies they predicted that coral feeding species (*e.g.* chaetodontids and serranids) may decrease significantly in abundance after outbreaks. They postulated that this was due to the lack of available food. They also predicted a decrease in the species richness of resident fish that used the corals as a habitat (*e.g.* pomacentrids and apogonids) and a decline in the overall diversity of fish species. Not all results in this study were clear cut. For example, they found that the quantitative increase in algal cover had little or no effect on the species richness and abundance of herbivorous and omnivorous fish. The study by Sano *et al.* (1984) utilized information on the change of fish communities in single coral colonies to predict the likely changes which may occur over large areas of reef. This procedure assumes that interactions at the coral colony level will also be manifested at the coral community level.

Studies by Williams (1986) attempted to determine whether fish communities were affected over large areas by extensive outbreaks of *A. planci*. During these studies detailed visual surveys of fish were undertaken at several locations on reefs before and after outbreaks of starfish. The outbreaks were found to have caused a significant reduction in the abundance of chaetodontids. This was the only major change that could be readily attributed to the effects of the outbreaks. Williams suggested that they may have more long term effects by indirectly altering the growth rates and fecundity of fish, as well as making them more susceptible to predation. Changes in the type of substratum by outbreaks were thought possibly to lead to variation in the recruitment of fish. Further studies of these fish communities may indicate whether these expected long term changes in fish community structure will eventuate.

CONTROL OF *A. PLANCI*

Since the first outbreaks of *A. planci* were recorded during the late 1950s a number of control programmes have been conducted in various parts of the Indo-Pacific region. In general, control programmes were conducted in a bid to protect coral communities from widespread destruction although some (*e.g.* the programme undertaken at Nauru) were carried out for little apparent reason and yielded few starfish. A list of these programmes is given in Table XIV.

It is clear from the scientific literature that an enormous number of starfish (approximately 14·6 million) have been killed or removed from reefs throughout the Indo-Pacific since the late 1950s. Relatively few control programmes have been undertaken on the Great Barrier Reef despite it being the largest reef system in the world and they have been mainly centred in areas used for tourism (*e.g.* Green Island, John Brewer Reef). By far the largest control programme has been that undertaken in the Ryukyu Islands. Yamaguchi (1985, in press) recently stated that almost 13 million starfish were removed from the reefs in this area over the period from 1970 to 1983 at an estimated cost of 600 million yen (almost $A 2·6 million). Other large control programmes have been conducted in Micronesia and Samoa (Table XIV).

A variety of different techniques were employed in these programmes to reduce starfish numbers. Initially at Green Island starfish were killed by chopping them up. This was abandoned as at the time it was thought that the starfish could regenerate from its severed parts (Gouldthorpe, 1968). Although other species of starfish are capable of doing this there is little evidence to suggest that *A. planci* has this ability. Owens (1971) cut a specimen of *A. planci* in half and found that within one week the two halves had rejoined. By re-separating them he was able to generate two new individuals which appeared after one month to be healthy and capable of feeding. An additional experiment by Owens (1971) failed to reproduce this result. Similar tests by Pearson & Endean (1969) also resulted in the death of starfish. For fear of increasing the population, later control programmes at Green Island involved collecting starfish and burying them on land. This method has been used in the majority of programmes conducted throughout the Indo-Pacific. Many of those killed in Micronesia and the Ryukyu Islands were destroyed in this fashion (Cheney, 1973; Yamaguchi, in press).

Collection by hand is both time consuming and labour intensive (Endean, 1969) and consequently several other methods have been tested as a means of efficiently reducing starfish numbers. Most involve the injection of various substances such as: 100% formalin, 10% acetic acid and 90% formalin (Owens, 1971), 18% ammonium hydroxide (Nishihira & Yamazato, 1972) and household ammonia (Branham *et al.*, 1971). While each method was found to be time efficient not all of the starfish injected were killed. Kenchington & Pearson (1982) reported the results of a test comparing three methods of control; collection by hand, collection with compressed air, and injection with copper sulphate. Killing starfish by copper sulphate injection was found to be the most efficient (132 starfish killed per hour per diver) of the three methods tested. Apart from these methods, Endean (1969) has reported that application of quicklime to the surfaces of starfish may be a useful way of controlling starfish numbers. This method, which was found to kill *A. planci* within approximately 24–48 hours, has been used to control outbreaks of *Asterias forbesi* in the United States (Loosanoff & Engle, 1938, 1942).

All of the above methods involve controlling starfish numbers at the level of the individual. During the 1960s and early 1970s an attempt was made to develop a mass control method which could be used to exclude starfish from

TABLE XIV

Acanthaster planci control programmes in the Indo-Pacific region: *denotes total for region

Region	Year	Number reported killed	References
Cook Islands	1973	80 974	Birkeland, 1982
Fiji	1970	9 860	Owens, 1971
Great Barrier Reef		93 312*	
Green Island	1966–1968	44 000	Barnes, 1966, Harding, 1968
Green Island	1980	25 850	Kenchington & Pearson, 1982
John Brewer Reef	1983–1984	3 000	Tarca, pers. comm.
Beaver Reef	1966–1968	11 000	*Sunday Sun* 16 Oct. 1983
Beaver Reef	1983–1984	9 000	*Sunday Sun* 16 Oct. 1983
Comorant Pass	1984	462	GBRMPA, unpubl. data
Hawaii		20 000*	
Molokai	1970	10 000	Branham, 1971
Molokai	1972	10 000	Branham, 1973
Japan		13 220 000*	
Ashizuri-Uwakai	1973–1983	87 000	Yamaguchi, in press
Kushimoto	1973–1979	1 400	Yamaguchi, in press
Miyake Island	1980	3 500	Crown of Thorns Starfish Advisory Committee, 1985
Ryukyu Islands	1957	220 000	Yamazato, 1969
	1970–1983	12 908 100	Yamaguchi, in press
Malaya	1971	350	Endean & Chesher, 1973
Micronesia		658 830*	
Mariana Islands	1968–1972	77 000	Tsuda, 1971, 1972; Cheney, 1972b, 1973b; Ikehara, 1972
Caroline Islands	1969–1972	226 750	Tsuda, 1971, 1972; Aisek, 1972; Marsh, 1972a,b; Nakamura, 1972; Wass, 1972; Cheney, 1973b
	1977	354 470	Birkeland, 1982
Marshall Islands	1971–1972	610	Sablan, 1972
Nauru	1971	150	Endean & Chesher, 1973
Philippines	1974–1976	1 000	Alcala, 1976
Samoa:			
Western	1969–1970	13 847	Garlovsky & Bergquist, 1970
American	1977–1978	486 933	Birkeland & Randall, 1979
Solomon Islands	1971	1 000	Randall, 1972
Tahiti	1971	10 000	Devaney & Randall, 1973
Wake Island	1969	300	Randall, 1972

large areas of reef. The system consisted of perforated nylex tubing which contained copper sulphate gel (Walsh *et al.*, 1971). Once submerged the copper sulphate was slowly released through the holes in the nylex tubing. This method was tested on the Great Barrier Reef but was found to be unsuccessful (Walsh *et al.*, 1976).

While a variety of toxic substances has been used to kill large numbers of *Acanthaster planci* no studies have been conducted to determine whether other reef organisms would be affected by these methods if they were undertaken on a large scale. It is possible that substances such as copper sulphate and formalin, which are known to be highly toxic to marine organisms may leach into the water after the starfish has decomposed. In addition, predators may be affected should they feed on starfish that have been injected with these substances.

During the course of outbreaks several other control methods have been suggested which include the use of electric barriers, electric guns, and suction dredging (Vine, 1970). As an alternative to employing methods which rely on direct control by man, Endean (1969) proposed that outbreaks of *A. planci* may be biologically controlled using a known predator, the giant triton shell, *Charonia tritonis*. There are several reasons why this method should be avoided. First, experience in other ecosystems has shown that methods involving biological control frequently fail, often producing many additional problems (Krebs, 1978). Secondly, very little is known about the population dynamics of the target species. This information is needed in order to implement an effective biological control programme (Krebs, 1978). Thirdly, information from several studies suggests that *C. tritonis* is not the sole predator of *Acanthaster planci* (see Table VI, p. 415) nor is *A. planci* the only prey of *Charonia tritonis*. In reality there are practically no quantitative data concerning the interaction between *C. tritonis* and its prey. Fourthly, it is not known what long-term effects this method would have on *Acanthaster planci* or the reefal communities with which it interacts. Finally, as *Charonia tritonis* is generally present in low densities on reefs (Endean, 1974) there may be an insufficient number of predators available for use as biological controls.

With the onset of outbreaks and the extensive death of corals there was considerable debate as to whether control programmes should be implemented to limit the numbers of starfish on reefs. Chesher (1970) argued that outbreaks of *Acanthaster planci* were not normal and that they may cause the permanent destruction of reefs if allowed to continue. As a result of this 'everything to gain, nothing to lose' approach Chesher proposed that control methods be implemented to limit starfish numbers. This view was also supported by Endean (1971a) and O'Gower, Bennett & McMichael (1972) who suggested that control measures will have done little harm even if outbreaks subsequently were shown to be a natural phenomenon. Newman (1970) opposed these views contending that outbreaks were probably natural events and that *A. planci* was an integral part of the ecology of reefs. He emphasized that there was insufficient evidence to indicate that coral reefs would be permanently destroyed if control methods were not implemented. On the basis of this 'everything to lose, nothing to gain' approach he advocated that it would be unwise to undertake such drastic measures.

Despite views to the contrary, control programmes were initiated throughout the Indo-Pacific. As many of these programmes were conducted at a time (late 1960s–early 1970s) when starfish populations were generally declining throughout this region it is difficult to determine whether they were successful. In general, the control programmes conducted in Micronesia were considered successful although they did not eradicate all starfish from the reefs in this region. Cheney (1973b) considered that they had caused a marked decline in starfish numbers thereby "reducing the potential for the destruction of coral cover" (p. 179). Both Tsuda (1971) and Wass (1973) also felt that these programmes had been successful although the latter author stated that the starfish populations had declined only in those areas where control measures had been introduced. It would appear that they had little effect outside these areas. Marsh & Tsuda (1973) were more cautious in evaluating the success of the control programmes in Micronesia. They were reluctant to attribute the population decline entirely to the implementation of control programmes as the starfish population on one island (Aguijan) declined substantially even though control methods were not undertaken.

Control programmes conducted in other parts of the Indo-Pacific appear to have been less successful than those in Micronesia. For example, programmes undertaken in Hawaii were only partly successful in attempting to eradicate three large aggregations of starfish. While one of the populations was eradicated another survived for a year despite being reduced to half its original size (Branham, 1973). In Australia, attempts to protect a small coral viewing area at Green Island were unsuccessful even though a diver was permanently stationed in this area to collect starfish. During the course of two years 44 000 starfish were removed from the area. Unfortunately, these intensive control measures were unable to prevent the starfish from causing considerable damage to the coral communities in this area (Barnes, 1966; Harding, 1968). Flanigan & Lamberts (1981) reported that control measures did not significantly alter the starfish populations in American Samoa despite the fact that almost 500 000 individuals were killed in this region.

It must be seriously questioned whether control programmes are of value in limiting starfish numbers and preventing widespread coral mortality given the results of programmes conducted in the Ryukyu Islands. Yamaguchi (1985, in press) concluded that they had been largely ineffective in preventing the destruction of considerable areas of reef even though an enormous number of starfish were killed. He stated that there were two main reasons why these costly programmes had failed to achieve their objectives. First, for fiscal reasons they were slow in commencing which meant that some outbreaks had been established on reefs for up to a year before control measures were implemented. This lag ensured that the outbreaks often were left undisturbed for at least one spawning period allowing them the opportunity to propagate and thus increase their size and distribution. Secondly, as control measures were conducted on the basis of collecting efficiency relatively large numbers of starfish were left after the programmes had finished. Outbreaks have occurred on many reefs in the Ryukyu Islands over the last 15 years. The persistence of these populations is thought to have caused a wave of outbreaks at Ashizuri-Uwakai,

Kushimoto (Japan) and at Miyake Island (Yamaguchi, in press) (see Fig. 6). Control measures were also implemented at these three locations. The control programme at Miyake Island was the only one considered to have been successful in eradicating the starfish (Yamaguchi, in press).

These results aside, the usefulness of control programmes must be further doubted with the re-occurrence of outbreaks of *A. planci* in areas where control measures had been undertaken up to a decade before. Examples of such areas are Green Island (Kenchington & Pearson, 1982), Guam and Palau (Birkeland, 1982). In hindsight it would seem that control programmes may represent only a short-term solution to the problem of widespread outbreaks of *A. planci* in the Indo-Pacific region. While control measures may change the local abundance of *A. planci* on a reef they may have little, if any effect on the occurrence of outbreaks in the future. This conclusion was reached by Bradbury *et al.* (1985a) who constructed a qualitative model of outbreaks based on data from the Great Barrier Reef. Using this model they provided evidence that outbreaks in this reefal system displayed a stable cyclicity whose trajectory was unlikely to be altered by attempts to change the abundance of *A. planci*.

With these conclusions in mind reef managers are faced with three possible choices. The first is to continue conducting extensive control programmes in the hope that they are in some way helping to eradicate *A. planci* and so saving our reefs from imminent destruction. While this perhaps could be likened to someone trying to hold back the tide it is none the less alluring, particularly if the outbreaks are widespread and are causing considerable coral mortality. Of course, the effects of outbreaks may be magnified if they occur in conjunction with man-related activities such as dredging, blasting, fishing, and pollution (Fagoonee, 1985a,b; De Silva, 1985; Muzik, 1985). The combination of these processes may lead to gross economic, management, and conservational problems which in turn may generate tremendous pressure to eradicate *A. planci*. The second option is to concentrate control efforts on a much smaller scale in areas which have some importance (*e.g.* tourist areas). Besides being relatively less expensive these types of control programmes may be more successful in the long-term, in protecting small areas of coral than those programmes undertaken on a much larger scale. Results so far suggest that control programmes carried out in small isolated areas (hundreds of kilometres away from other populations) have the greatest likelihood of success (*e.g.* Miyake Island). The third and final option is to do nothing. As control programmes of any sort are becoming increasingly more expensive and thus harder to justify this approach also has much to recommend it.

In the final analysis, the option chosen must depend on a host of variables which will differ according to the situation. These variables relate to the following.

(1) Size and areal extent of the population to be contained (*e.g.* is the population too large or widespread to enable control methods to be effective?)
(2) Position of the population (*i.e.* is it located in a remote area or an area which is easily accessible?)

(3) Importance of the area affected (*i.e.* what is the use of the area—is the area used for tourism or some other commercial venture?)
(4) Distance of the population from other reproductively active populations (*i.e.* is there the likelihood that recruitment may occur after the control methods have been conducted?)
(5) Funds available (*i.e.* are sufficient funds available to complete the programme?)
(6) Time at which funds become available (*i.e.* is the time at which funds become available the most appropriate time for initiating an effective control programme?)

The effectiveness of a control programme will depend on each of these factors being addressed in the correct manner. It may depend as much on the time when funds become available as the numbers of starfish involved in the outbreak. Recent studies on the Great Barrier Reef have demonstrated that the distribution and abundance of starfish may change quite rapidly on reefs proceeding from high densities to relatively low population levels within the space of six months (Moran, Bradbury & Reichelt, 1985). Lags in funding may cause a control programme to be ineffective by ensuring that it was implemented after the outbreak had reached its zenith. It becomes apparent, therefore, that the timing of control programmes (relative to the state of the outbreak) has a major bearing on their success.

Bearing in mind that outbreaks may move over reefs with great speed then the success of a programme will also depend on the proportion of starfish in an outbreak which are killed. Yamaguchi (in press) attributed the ineffectiveness of the control programme in the Ryukyu Islands to be partly due to the fact that only starfish which could be collected quickly (*e.g.* those in shallow waters) were killed, leaving behind a large number of individuals. By the same token removing every starfish from a particular area and disregarding those in surrounding territories may have little success. This was found at Green Island. In some areas an 'all or nothing' response may need to be considered if a control programme is going to be successful.

On a cost-benefit basis it would appear that the undertaking of concentrated control programmes in discrete locations offers the best opportunity for success using the methods presently available. The general pessimism regarding control programmes has, however, reached the stage that even this alternative is considered doubtful by some (Kenchington & Pearson, 1982). Apart from the results of the programmes at Green Island it is not clear whether such a plan can be used as a general purpose control by reef managers. It is, however, clear that it is time to re-assess the rationale governing the use of control measures and to undertake quantitative studies with the aim of developing a coherent and effective management policy in relation to the control of outbreaks of *A. planci*.

UNANSWERED QUESTIONS

While a great deal of information has been presented about various facets of the *Acanthaster* phenomenon there is still much that is not known. From that presented so far it is clear that the amount of information available on each facet is very different and some are more well-defined than others. For example, much more is known about the biology of *A. planci* than its ecology and population dynamics. Often this lack of information arises because of experimental difficulties (*e.g.* evaluating the dispersal of larvae).

It can be argued from an inspection of the scientific literature and various media reports that the intense debate and controversy surrounding the *Acanthaster* phenomenon have been exacerbated because many views and hypotheses have arisen from an ignorance of what is known as well as what is not known about the subject. Previous sections in this paper have defined what is known about the phenomenon. It is also equally important to define what is not known about this subject for several reasons. First, it earmarks those facets of the subject which are most poorly understood and hence need to be intensively studied. Secondly, it stimulates the development of research questions and the subsequent delineation of research priorities. Finally, it provides a sound basis for the formulation and refutation of new theories and hypotheses.

The aim of this section is to present, in question form, those facets which are most important in understanding the phenomenon and about which there is little if any information. The following questions, which are divided into three groups (1) larvae and juveniles; (2) adults; and (3) effects on communities and processes—remain unanswered. Comprehensive information on each will provide an understanding of the following.

(1) Why outbreaks occur and whether they are natural or unnatural phenomena.
(2) Whether they play an important part in reefal processes and the development of reef structure.
(3) Why some reefs are more susceptible to outbreaks than others.
(4) Why some outbreaks cause extensive coral mortality while others do not.
(5) How outbreaks are propagated over large distances.
(6) Whether special management policies need to be formulated in order to prepare for the occurrence of future outbreaks.

LARVAE AND JUVENILES

(1) Are high nutrient conditions needed for the enhanced survival of larvae in the field?
(2) Do these types of conditions occur frequently in the field? If so, do they coincide with observed spawning periods and how long do they occur?
(3) Can larvae develop and settle under 'non-bloom' nutrient conditions in the field. If so, can high densities of larvae be sustained under these conditions?
(4) How important is diet in influencing the survival of larvae? Is survival more dependent on the diversity rather than density of food species? What other factors influence the survival of larvae?
(5) Do certain physical conditions occur in the field that cause the increased survival of larvae? Do these conditions act in conjunction with any other factors?

(6) How long do larvae spend in the plankton before settling? What is the maximum period of time they can spend in this phase and yet still be able to settle?
(7) How far can larvae be dispersed in the field?
(8) What factors are important in causing their dispersal?
(9) Is there a positive correlation between larval density, recruitment density, and adult density?
(10) Where do larvae occur in the water column? Does their position vary throughout their planktonic period? What factors are responsible for determining their position?
(11) Where do larvae settle in the field? Is it in shallow or deep water on reefs? Do they settle in high densities?
(12) Do larvae tend to settle on a particular type of surface? What factors are important in determining the type of surface chosen by larvae for settlement?
(13) Are there particular areas on reefs which are more suitable for settlement than others?
(14) Do larvae tend to settle on those reefs from which they were propagated or do they generally recruit to reefs other than the parent reef?
(15) Do juveniles tend to be in shallow or deep water on reefs? Does this location vary depending on whether or not the reef has recently suffered an outbreak of adults?
(16) What are the mortality rates of larvae and juveniles in the field?
(17) Is predation important in determining the density of larvae and juveniles? What are the main predators of each stage?
(18) Apart from predation what other factors are important in causing the mortality of juveniles (*e.g.* disease, lack of nutrients)?
(19) What type of food do juveniles eat in the field?
(20) How fast do juveniles grow in the field? Is it similar to that recorded in the laboratory? How important is diet in determining the growth rate of juveniles?
(21) How far do juveniles move in the field? Do they show any feeding preferences?

ADULTS

(1) Are adults capable of moving between reefs?
(2) How rapidly do they grow in the field? Is their rate of growth similar to that recorded in the laboratory?
(3) Can the age of a starfish be determined from its size?
(4) How long do adults survive in the field?
(5) What are the rates of mortality for adults in the field?
(6) What is the rate of predation on adults on reefs? What are the main predators of adult starfish? Are these predators sufficient to limit adult population levels? Do the densities of these predators fluctuate markedly through time?
(7) Are there any other factors which are important in causing the mortality of adult starfish (*e.g.* disease)?

(8) Do adult starfish enter a senile phase in the field where their growth declines greatly and they become infertile?
(9) What causes the rapid disappearance of adult starfish which has been observed in the field? Is it related to density dependent factors (*e.g.* crowding causing loss of condition)? What happens to the majority of starfish? Do they die (*e.g.* from disease) or do they move to another reef?
(10) Do the skeletal components of starfish accumulate in the sediments after times of outbreaks? Do more spines tend to accumulate during outbreaks than during times when starfish densities are low?
(11) Do adults show a distinct preference for certain types of coral?

EFFECTS ON COMMUNITIES AND PROCESSES

(1) Do coral communities recover from outbreaks of starfish? How long does this take? Is the pattern of recovery similar for most types of reefs and for different scales of disturbance?
(2) What effect do outbreaks have on other communities (*e.g.* fish, soft corals)? Is this effect permanent or do these communities recover from such a disturbance?
(3) What effect do outbreaks have on reef processes such as calcification, primary production, and reef growth?

REASONS FOR OUTBREAKS

HYPOTHESES

A number of hypotheses have been formulated to account for the occurrence of outbreaks in the Indo-Pacific region over the last 25 years. The hypotheses that have been developed fall into one of two categories; those based on the premise that outbreaks of *A. planci* are natural phenomena and those that assume them to be unnatural. Hypotheses which emphasize that outbreaks are natural phenomena are based on the view that the variability observed in the population structure of *A. planci* over both temporal and spatial scales is normal (Dana, 1970; Newman, 1970; Vine, 1970). Such a view assumes that the wide population fluctuations of *A. planci* are representative of the normal variability which can occur within animal populations on coral reefs. Moore (1978) has argued from a theoretical approach that large population fluctuations of *A. planci* can occur naturally without man's intervention. He considered that the life history characteristics of *A. planci* were indicative of a binomic behaviour which conformed broadly to the exploitation of habitats. Since organisms with this type of ability (denoted *r*-strategists) exhibit large scale population fluctuations Moore (1978) regarded outbreaks of *A. planci* to be normal and "inherent of this mode of living" (p. 57). These conclusions must be questioned as they were based on limited information. Although very little is known about the movement of starfish Moore surmised that migration was an important factor in the rate of mortality of starfish. Similarly, he assumed that *A. planci* reproduced repeatedly throughout its life cycle although this may not be true in the light of results presented by Lucas (1984) who showed that adults may enter a senile phase.

Besides arguing from a theoretical perspective, attempts have been made to verify the belief that outbreaks are natural by demonstrating that they have occurred in the past. From information of this sort it is inferred that the present starfish outbreaks are periodic or cyclic phenomena which are natural in origin. Several authors have used historical records to show that outbreaks have occurred previously (Dana, 1970; Newman, 1970). Vine (1973) suggested that *A. planci* was abundant and had a wide distribution in much earlier times. As the records he used did not give accurate data on starfish numbers and were largely anecdotal it is, however, difficult to assess the value of this information. This criticism has been levelled at all studies which have adopted this approach (Endean, 1973b). In fact, several authors have insisted that there is no historical evidence to indicate that outbreaks of *A. planci* have taken place prior to the 1950s (Chesher, 1969a; Randall, 1972; Endean, 1977, 1982; Cameron & Endean, 1982). Also, Branham (1973) stated that anecdotal references to large concentrations of starfish may in fact refer to normal aggregations of individuals during spawning.

In addition to the above argument it has been proposed that outbreaks occurred in the past but that they went largely unnoticed and it only has been with the advent of SCUBA equipment and the increased use of coral reef environments (for tourism and research) that they have been recorded recently (Newman, 1970; Weber & Woodhead, 1970). It is no doubt true that these factors have been responsible for our greater awareness of the distribution and abundance of *A. planci*; it is, however, sheer speculation to suggest anything more than this. Randall (1972) considered that it would be unlikely for such dramatic events to be overlooked, particularly in areas which were close to human settlements and which had been used over many years for diving and fishing.

Birkeland & Randall (1979) provided evidence, after interviewing a number of old fishermen, that outbreaks of *A. planci* may have occurred at the beginning of this century in Samoa. In some instances the information they collected was, however, conflicting. Some Samoan fisherman reported that *A. planci* (termed "Alamea") had been abundant in 1916 but had been scarce since then, while others suggested that large numbers of this starfish also had been present during 1932. Again, as no records of the numbers of starfish were given it is difficult to ascertain whether they refer to normal populations of *A. planci* or outbreaking populations.

Flanigan & Lamberts (1981) proposed that *A. planci* had been well known in Samoa for many years as information on this animal could be found in records of the verbal history, linguistics, and proverbs of this country's culture. Birkeland (1981) used a similar approach to show that outbreaks were a natural and recurring phenomenon in Micronesia. He maintained that this could be implied since several high cultures in this region were familiar with this starfish, each having their own particular name for this species and advice on how to cure its sting (by applying the stomach of the starfish to the wound). Birkeland (1981) suggested that *A. planci* must have been present perhaps abundantly, for many years for this type of information to have been incorporated into these cultures.

It is not surprising that *A. planci* has been known to these cultures for many years as the earliest records of this starfish date back to 1705 when it was first described by Rumphius. The occurrence of past outbreaks in Micronesia and Samoa cannot, however, be inferred from the information given by Flanigan & Lamberts (1981) and Birkeland (1981) as the importance of *A. planci* in these cultures may result from other factors (*e.g.* its appearance, toxic nature) rather than a recognition of its having occurred in large numbers at some stage in the past.

There have been very few studies that have tried to provide direct evidence of the occurrence of outbreaks in the past. During the early 1970s Maxwell (1971) reported finding skeletal debris in sediment samples from various reefs in the Great Barrier Reef, which suggested that there had been an increase in the number of echinoderms about 300, 800, and 1500 years ago. He also found that the percentage of echinoderm fragments in sediment samples increased from south to north along the Reef. As the remains of *A. planci* could not be differentiated from those of other echinoderms these results were considered to have little relevance to the debate concerning previous outbreaks of *A. planci* (Endean, 1971b; Talbot, 1971). A more extensive geological study was undertaken on the Great Barrier Reef a few years later by Frankel (1975a,b, 1977, 1978). In this study he sought to demonstrate, by searching for the presence of skeletal remains of *A. planci* in surface and sub-surface sediments, that outbreaks were a recurring phenomenon. He obtained 54 sub-surface samples from 27 reefs between Lizard Island and Gould Reef and found skeletal remains of *A. planci* in horizons from 16 of these reefs. The age of these remains were determined by dating the sediment surrounding them. From this work Frankel concluded that outbreaks of *A. planci* had occurred up to 3355 years B.P. and that they were "natural phenomena".

The conclusions of Frankel have been both accepted and criticized. Endean (1977, 1982) stated that the studies conducted by Frankel did not provide evidence of previous aggregations for three main reasons. First, *A. planci* has probably been a component of reefal ecosystems for a number of years and it is natural that skeletal remains from this animal would be deposited in the sediments. Secondly, no mass mortalities of starfish have been recorded on reefs; this is important as it is presumed that more skeletal fragments accumulate in sediments during outbreaks as a result of the mass mortality of starfish. Finally, Endean (1977, 1982) argued that it was difficult to determine the significance of skeletal debris since it was not known how many skeletal fragments were needed in a sediment horizon to constitute a past outbreak.

With the occurrence of another, recent series of outbreaks in various parts of the Indo-Pacific (Birkeland, 1982; Endean, 1982) the debate concerning the occurrence of past outbreaks has re-emerged. Several authors have considered that the data presented by Frankel (1977, 1978) provide direct evidence that large aggregations of *A. planci* occurred in the geological past (Bennett, 1981; Potts, 1981; Birkeland, 1982; Rowe & Vail, 1984b). Recently, this data was statistically re-evaluated by Moran, Reichelt & Bradbury (1986). Using the results from starfish surveys (Fig. 7) they were able to establish whether reefs had experienced a recent outbreak prior to being sampled. This information was combined with Frankel's own results for reefs where recent skeletal material was found. The combined data were arrayed in a contingency table and analysed using a Fisher Exact Probability Test. The analyses indicated that the occurrence of skeletal remains in recent sediments was independent of whether or not the reef had suffered a recent outbreak. Moran, Reichelt & Bradbury (1986) proposed from these results that it was erroneous to infer the occurrence of outbreaks in the past from similar debris in much older sediments. They concluded that while Frankel's data may demonstrate that *A. planci* had existed for a long time it did not prove that outbreaks of this animal had occurred in the geological past.

Randall (1972) has criticized the idea that outbreaks of starfish are natural phenomena which may occur regularly or periodically. He argued that the levels of coral community structure on the Great Barrier Reef and other reefs in the Indo-Pacific such as Guam could never have been attained if starfish outbreaks occurred regularly. Chesher (1969a) also adopted this line of reasoning suggesting that it was improbable that outbreaks had taken place on Guam over the last 200 years. Randall (1972) further criticized the notion that outbreaks are cyclical or periodic events on the basis that if they had occurred repeatedly over the years then reefs should be composed primarily of species that are least preferred by *A. planci* (*e.g.* *Porites* spp.). He felt that this was not the case on many reefs. The arguments raised by Randall (1972) are themselves open to question as they assume that outbreaks in the past were similar in duration and intensity to those recorded recently. At present there is no evidence to indicate that this is correct. Also, the latter criticism by Randall is based on the assumption that past outbreaks have occurred at relatively short intervals. Recent information has indicated that coral cover may regenerate to original levels within 10–15 years and that preferred corals such as *Acropora* spp. may tend to dominate these developing communities (see p. 447).

From surveys carried out in the South Pacific, Weber & Woodhead (1970) stated that *Acanthaster planci* was more common on reefs than is generally believed. This notion has been used by Dana & Newman (1972) and Dana, Newman & Fager (1972) to form the basis of the adult aggregation hypothesis (Potts, 1981). These authors considered that primary outbreaks originated when adult starfish are forced to aggregate after catastrophic events such as severe storms. The reasoning is that *A. planci* is normally common (but not necessarily obvious, visually) on reefs and under these conditions food is not a limiting factor. When large areas of coral are destroyed during tropical storms individuals aggregate in areas which have not been destroyed and where a large source of food is available. Mass mortality of corals during these conditions was thought to be a result of mechanical damage, sedimentation and freshwater input. Dana *et al.* (1972) used data from surveys carried out in Micronesia (Chesher, 1969a) to substantiate this hypothesis. They classified this information according to habitat type and starfish abundance. In doing so it was discovered, despite variability in the data, that the largest numbers of *A. planci* occurred on the leeward side of exposed reefs. Dana *et al.* (1972) postulated that these protected areas would be most susceptible to the formation of starfish aggregations as they often had an abundance of corals and supported a relatively large number of scattered starfish. They also were able to show from the survey data that the abundance of starfish in these areas (from 0·5–1·0 starfish per 100 m^2) would be sufficient to cause the largest aggregations seen in Guam. In conjunction with these analyses Dana *et al.* (1972) showed that there was a positive correlation between the occurrence of typhoons and cyclones and the formation of outbreaks in Guam and the Great Barrier Reef.

This hypothesis is appealing since it can account for the fact that outbreaks initially appear to be composed of adults. Despite this however, the hypothesis has received little attention in the scientific literature. Potts (1981) suggested that it was one of the simplest hypotheses that had been put forward to account for the occurrence of outbreaks. While certain features recommend it, it also suffers several shortcomings. Although Dana & Newman (1972) and Dana *et al.* (1972) maintained that *A. planci* is generally common on reefs others have suggested that it is normally a rare animal (Pearson, 1975b; Endean, 1977). Potts (1981) considered that these authors may have under-estimated the true abundance of starfish under 'normal' conditions as they were derived from surveys that were carried out in relatively shallow water. He referred to data which showed that *A. planci* may also be found in deep water and thus considered that the surveys of Pearson (1975b) and Endean (1977) may have only sampled part of the habitat of this starfish. This debate is unresolved as there is little information on the abundance of starfish in deep water beyond the slope of reefs. A logical inconsistency in the aggregation hypothesis is that if an appreciable proportion of starfish is located in deeper water it is not exactly clear why these animals should aggregate as it is unlikely that these habitats would be as greatly affected by the types of disturbances mentioned by Dana *et al.* (1972). A further problem is encountered when trying to ascertain whether the abundances of starfish observed in primary outbreaks could have arisen from a dispersed normal population which has been forced to aggregate. Dana *et al.* (1972) maintained that the outbreak at Guam, which was estimated to comprise approximately 38 000 starfish, could have developed this way. On the other hand, it is much more difficult to believe that the supposed primary outbreak at Miyako Island in 1957 (Yamazato & Kiyan, 1973), which contained at least 220 000 starfish, could have arisen as a result of the aggregation of a normal population of dispersed individuals.

Apart from this, the adult aggregation hypothesis has been questioned by Pearson (1975b) who argued that these disturbances need not necessarily cause the mass mortality of corals and that large areas of coral may survive. This argument was based on observations he had made at a reef off Townsville which had recently suffered the effects of a major cyclone. Potts (1981) suggested that these observations did not invalidate Dana *et al.*'s (1972) hypothesis as it required only that intense coral mortality be confined to a localized area. This argument is somewhat pedantic as the term "localized area" may be defined in several different ways depending on the size of the reef. For an aggregation of starfish to take place in the manner suggested by Dana *et al.* (1972) the mortality of corals would have to occur over a large area. Newman & Dana (1974) have suggested that this hypothesis could be tested empirically by limiting the amount of food available (either by removing coral or increasing starfish numbers) and observing whether the starfish move into areas with abundant coral.

Two other hypotheses have been proposed that suggest that outbreaks are a consequence of natural processes. One of them proposes that the recruitment of larvae of *A. planci* is enhanced during times of favourable environmental conditions and can be termed the larval recruitment hypothesis. This hypothesis was based on the results from laboratory experiments, which showed that the survival of larvae is improved under conditions of lowered salinity (about 30‰) and higher temperature (around 28 °C) (Lucas, 1973, 1975). Lucas (1972) proposed that the survival rate of larvae may be increased if these conditions occurred in the field. From this he hypothesized that a slight alteration in the survival rate of larvae could lead to large increases in the number of individuals that settle and this may result in population outbreaks of *A. planci* in later years. Pearson (1975b) demonstrated that these sorts of conditions may occur within 50 km of the North Australian coast (between Ingham and Mossman). Like Dana *et al.* (1972) and Nishihira & Yamazato (1974), he considered that these conditions may be associated with periods of heavy run-off as a large proportion of rivers were located in this region. He proposed that there would be a greater chance of outbreaks occurring if there was a higher survival of larvae. Pearson (1975b) suggested that high larval survival may not always occur after times of flood as the right rainfall conditions would need to be combined with periods of light wind (which would prevent the water layers from being mixed) and the availability of large areas of suitable substratum.

In this hypothesis natural processes are seen to be the primary cause of starfish outbreaks. It also allows for the fact that the frequency of occurrence of these processes, and thus outbreaks, may have been increased indirectly by man's activities (Dana, 1970). For example, the development of land may have increased the amount of run-off into the sea thus leading to more frequent starfish outbreaks. With this in mind it has been pointed out that nearly all of the outbreaks which have occurred in the Indo-Pacific region have occurred on reefs near high islands or mainland continents (Tsuda, 1971; Pearson, 1975b).

Another hypothesis also explains the occurrence of outbreaks in terms of natural processes. This hypothesis, developed by Birkeland (1982), has several features in common with that of the larval recruitment hypothesis. It also emphasizes the importance of run-off in creating outbreaks of starfish and, therefore, can be referred to as the terrestrial run-off hypothesis. While Pearson (1975b) stressed that run-off from landmasses created environmental conditions (decreased salinity and increased temperature) which enhanced larval survival, the terrestrial run-off hypothesis suggested that the nutrients in run-off from high islands and continental land masses caused phytoplankton blooms which acted as a food source for larvae, thus promoting their survival. This is also based on results which have emanated from studies conducted in the laboratory. The findings of Lucas (1982) suggested that food availability was important in determining the survival of larvae (see p. 397). Birkeland (1982) has adopted this view and made it the central theme of his hypothesis. Implicit in it is the belief that under normal conditions the survival of larvae is low due to a lack of food. The high larval mortality under these conditions may be the result of starvation or predation (see p. 397). During times when there is sufficient food, such as when phytoplankton blooms occur, the survival of larvae is enhanced. Like Lucas (1972), Birkeland (1982) suggested that a small percentage increase in the survival of larvae could lead to a great increase in the number of adults on reefs.

By correlating rainfall data with information on outbreaks he showed that outbreaks of *A. planci* follow some three years after periods of heavy rainfall (*i.e.*>100 cm in three months or 30 cm in 24 h) which themselves have followed times of drought (*i.e.*<25 cm in four months). From these analyses he found that outbreaks did not occur after "dry" typhoons (which produce little rain) but only followed from "wet" typhoons. He also showed that they tended to take place around high islands but not coral atolls. On the basis of this information he successfully predicted an outbreak of starfish at Saipan in 1981.

Birkeland (1982) pointed out that one of the main advantages of his hypothesis was that it could account for the sudden appearance of large numbers of starfish which he considered characteristic of outbreaks. He stated that this feature indicated that outbreaks arise from periods of successful recruitment and not from a decrease in predator pressure, which he considered would result in the gradual build-up of individuals over a number of years. In addition, he maintained that since outbreaks occurred at so many localities and were composed of such large numbers of starfish it could be implied that the increased survival of larvae was the main factor involved.

While the larval recruitment and terrestrial run-off hypotheses have much to commend them they fail to address several points. For example, it is assumed for outbreaks to arise that spawning must have been successful in terms of the proportion of eggs fertilized. The hypotheses do not explain how large numbers of larvae are produced from a population which under normal conditions would be dispersed. It is not known what percentage of eggs are fertilized in the field when adults are dispersed. Presumably, the extent to which fertilization occurs depends on adult density although perhaps a threshold level of individuals is needed before large numbers of larvae are produced. It is possible that adults in a normal population aggregate during spawning due to biochemical means; few such aggregations have, however, been observed in the field.

The validity of these two hypotheses has also been questioned on the grounds that they require the synchronization of a number of different and highly variable processes (Potts, 1981). They require that the spawning of adults occurs within a short time after the onset of heavy run-off. Birkeland (1982) maintained that spawning occurred at the start of the wet season on either side of the equator at a time when phytoplankton blooms are most likely to arise. The synchronization of these two processes presupposes that the bloom conditions remain intact and undispersed for at least several weeks. Similarly, prolonged conditions of reduced salinity are required under the larval recruitment hypothesis. In addition, for outbreaks to occur on isolated reefs (such as in Micronesia) during periods of bloom conditions or optimal physical conditions then a favourable hydrodynamic regime must prevail so that larvae are not dispersed away from these areas.

It can be seen that synchronization of a number of variable events is assumed for both hypotheses. Potts (1981) stressed that there was no direct evidence to support the larval recruitment hypothesis and that no outbreak of juveniles has been recorded in the field. This criticism may also be levelled at the terrestrial run-off hypothesis as both hypotheses suggest that primary outbreaks will arise from the settlement of high densities of larvae. While it is true that no such outbreaks (apart from those mentioned earlier) of juveniles have ever been observed this may not be a valid criticism as little is known about where larvae settle on reefs. If they settled in deep water off, or at the base of reef slopes then it is possible that high densities of juveniles may go unnoticed until they become adults and capable of moving and feeding over large distances.

Cameron (1977) and Cameron & Endean (1982) have also raised doubts about the validity of the terrestrial run-off hypothesis in explaining the occurrence of outbreaks in various parts of the Indo-Pacific region. They argued that the life history of *A. planci* was not unique among other asteroids. For example, they compared its life history with that of *Culcita* sp. and found that both starfish were carnivores, had similar distributions and larval biologies, and were large in size. In view of this similarity they stated that the terrestrial run-off hypothesis failed to explain why animals with similar life histories to *Acanthaster planci* did not outbreak. This question will remain unanswered until more detailed data are obtained on the larval ecology of these starfish.

While the hypothesis proposed by Birkeland (1982) may hold for isolated areas in Micronesia and the south Pacific where primary outbreaks can be presumed with some degree of certainty, there are no data as yet to indicate that it can be applied to outbreaks on the Great Barrier Reef. The data presented by Birkeland (1982) for this area are at variance with the pattern of outbreaks recorded for that particular time. These data indicated that outbreaks of *A. planci* occurred in 1962 on reefs along the Queensland coast between Townsville and Bowen. This is incorrect as outbreaks were not reported in this area until the early 1970s. By 1962 they had only just been reported at Green Island (Pearson & Endean, 1969) (see p. 432). Another inconsistency is that the outbreaks which occurred in this area were correlated with an intense cyclone (the third most severe on record) which crossed the coast in 1959. According to the terrestrial run-off hypothesis outbreaks take place after a period of high rainfall which itself has been preceded by a period of dry weather. The date (1959) given by Birkeland (1982) is inconsistent with this explanation since according to Dana *et al.* (1972) a severe cyclone had also affected the area in question in 1958. Dana *et al.* also mentioned that increased cyclonic activity was experienced along much of the Queensland coast during the period from 1958–1961. Clearly, more accurate data on past and present weather conditions and outbreaks are needed before it can be determined whether the terrestrial run-off hypothesis can be applied to the Great Barrier Reef system.

The idea that outbreaks are unnatural phenomena is mainly based on the premise that they have not occurred in the past. Another underlying assumption is that coral reefs are complex systems that are biologically stable and predictable. It is postulated that the inertia of these systems prevents species or groups of species from undergoing marked changes in their population structures. The homeostatic mechanisms responsible for this reside in a system that is highly diverse and dominated by co-evolved relationships among species.

Cameron (1977) and Cameron & Endean (1982) considered that *A. planci* was rare, large in size, relatively long-lived, morphologically and chemically specialized for feeding and defence, and had few parasites. Like Moore (1978), they based their conclusions on the life history and ecology of the animal. They, however, regarded it as a rare and specialized carnivore and not an opportunistic species. From a theoretical point of view they stated that outbreaks of this starfish were a unique event within complex systems such as coral reefs and, therefore, were indicative of a novel sort of perturbation. Although the ideas of Moore (1978) and Cameron & Endean (1982) are divergent they are important as they represent the first attempts to link present concepts in theoretical ecology with information on the *Acanthaster* phenomenon. They both suffer from inadequate data. For example, Cameron (1977) and Cameron & Endean (1982) presumed that *A. planci* was long-lived although there are no data on the longevity of starfish in the field. Also, they stated that *A. planci* had very few parasites although recently it has been suggested that they may suffer from a bacterial infection (see p. 419).

Cameron & Endean (1982) have emphasized that outbreaks of *A. planci* represent a novel event in complex tropical reef systems. Birkeland (1982, 1983) disagreed with this view and has proposed that certain animal populations apart from *A. planci* may fluctuate widely in their abundance. He gave the examples of *Diadema setosum* and *Echinothrix diadema* which were recorded in large numbers at Guam in 1977. He further stated that there were many species of planktonic larvae and tropical invertebrates (*e.g.* insects) whose populations were characterized by large fluctuations. He stated that there was no empirical basis for suggesting that coral reefs were predictable, stable systems. This view was also supported by Sale (1980) who concluded that the available evidence showed that coral reef fish communities were predominantly unstable and suffered from large fluctuations in recruitment.

In general, hypotheses which have emphasized that outbreaks are unique or unnatural phenomena have explained their occurrence in terms of man-induced perturbations. It has been pointed out that all major outbreaks have occurred near centres of human populations (Chesher, 1969a). Chesher (1969a) proposed that increases in blasting and dredging in Micronesia may have been responsible for creating large areas of clear space which would favour the settlement of larvae, thus increasing their survival. He suggested that larval mortality was normally high as a result of predation by benthic organisms such as corals. The destruction of large areas of reef by these activities was thought to enhance the survival of larvae by reducing predation and to provide an abundance of suitable substrata for settlement. Chesher (1969a) presumed that this would perhaps allow more starfish to settle and that these centres of settlement would in turn develop into "seed" populations. In support of this hypothesis he gave examples of several areas in Micronesia, particularly Guam, where outbreaks of starfish had occurred after dredging and blasting had been undertaken. This hypothesis has been criticized for a number of reasons. First, Endean (1977) maintained that there was insufficient evidence to support this hypothesis. Secondly, Branham (1973) stated that it did not account for why outbreaks occurred during the same period throughout the Indo-Pacific. Thirdly, it has been pointed out by Randall (1972) that outbreaks of starfish have occurred on reefs where such activities had never been reported. Finally, the hypothesis does not explain why outbreaks of starfish were not recorded during or immediately after the Second World War on many reefs in Micronesia which experienced extensive blasting, dredging and bombing (Endean, 1977).

Nishihira & Yamazato (1974) found, on the island of Okinawa, that starfish outbreaks appeared to occur more intensively on reefs which were affected by human activities. They did not, however, suggest a reason for this. Hypotheses have been put forward by Fischer (1969) and Randall (1972) which link man's activities with the occurrence of outbreaks. They stem from the observation that all major starfish aggregations have occurred near populated areas. In both hypotheses it is proposed that the increased input of chemical pollutants into the sea by man has been responsible for reducing the predators of larval and adult crown-of-thorns starfish. This in turn has allowed far greater numbers of starfish, particularly larvae, to survive. There is very little evidence to support this hypothesis or indeed the notion that coral reefs are being polluted by chemicals such as pesticides. The results of a study by Tranter (1971) showed that the tissues of three animals (*Acanthaster*, *Linckia*, and *Tridacna*), collected from sites near human populations on the Great Barrier Reef, contained only very low amounts of chlorinated hydrocarbons. He concluded from this that they were not greatly polluted by pesticides. A study reported by Haysom (1972) also indicated that there was little evidence to indicate that chemical pollutants were in abnormally high concentrations in waters of the Great Barrier Reef. No significant difference was found in the pesticide levels of oysters from several different locations. Further studies by McCloskey & Deubert (1972) found no correlation between starfish abundance and organochlorine concentrations in the gonads of starfish from areas within the Great Barrier Reef, Micronesia, and Hawaii. They also discovered that not all the highest levels of these pesticides came from starfish in areas close to human populations.

Of all the hypotheses which focus on man-induced causes that proposed by Endean (1969) has received the greatest attention in the scientific literature. This hypothesis emphasizes that outbreaks of *Acanthaster planci* are unique events which arise because man has removed the predators of this starfish. Thus it can be termed the predator removal hypothesis (Potts, 1981). Initially, the major predator controlling starfish numbers on the reef was thought to be the giant triton (*Charonia tritonis*) (Endean, 1969). Endean (1973a) stated that this animal was a predator of large juvenile and small adult starfish, a fact which had been well documented in the scientific literature. Large adults were not thought to experience heavy predation because of their greater defensive capabilities (*i.e.* size, toxicity of spines, behaviour) and because they had been observed to escape from attack by *C. tritonis* and regenerate any damaged tissue (Chesher, 1969a). This is indicated in the field by the high percentage of starfish which have been found to have missing or regenerating arms (see Table VII, p. 418). Small starfish presumably would not suffer high levels of predation because of their size and ability to inhabit small crevices and spaces which could not be invaded by *C. tritonis* (Chesher, 1969a). Endean (1973a) claimed that collection of *C. tritonis* by man had occurred increasingly since the end of the Second World War and had resulted in an increase of starfish on some reefs. It was proposed that this build-up of adult starfish to a threshold level culminated in the production of large numbers of larvae which drifted to other reefs causing primary outbreaks. Further support for this hypothesis comes from Fagoonee (1985a) who reported that *Acanthaster planci* had greatly increased in numbers at Mauritius, at a time when the abundance of *Charonia tritonis* had decreased due to its collection by man.

Birkeland (1982) has disagreed with this hypothesis on the grounds that the mechanism proposed by Endean (1973a) would lead to a gradual increase in starfish numbers over several years whereas observations in the field indicate that outbreaks build-up very suddenly. The validity of the predator removal hypothesis also has been questioned by Chesher (1969a) and Vine (1970) who claimed that *C. tritonis* is normally rare on reefs and, therefore, could not be responsible for controlling the abundance of juvenile and adult starfish. This view has received additional support since experiments with caged starfish showed that *C. tritonis* may eat less than one starfish per week and that it prefers to consume other species (*e.g. Linckia* sp.) if given a choice (Chesher, 1969a; Pearson & Endean, 1969). Potts (1981) has provided a detailed summary of the results of these studies. To date, no experiments have been conducted to determine the extent to which *C. tritonis* preys on juvenile and adult starfish in the field.

In more recent years Endean (1977, 1982) has extended this hypothesis to include the effects of fish predators such as the groper *Promicrops lanceolatus*. Other species that have been observed to feed on *Acanthaster planci* (*e.g. Balistoides viridescens*, *Pseudobalistes flavimarginatus*, and *Arothron hispidis*) (Ormond & Campbell, 1974) were not included in the hypothesis as it was doubted whether they were important predators on the Great Barrier Reef. This extended version of the predator removal hypothesis stressed that *Charonia tritonis* was a major predator of large juvenile and small adult starfish whereas *Promicrops lanceolatus* preyed on juvenile *Acanthaster planci*. Endean (1969, 1974, 1977) considered it unlikely that starfish abundance would be controlled by the predation of eggs and larvae for two reasons. First, this type of predation would not be specific to *A. planci* and, therefore, should lead to population increases in other similar animals. The fish *Abudefduf curacao* was observed to feed on eggs of *Acanthaster planci* (Pearson & Endean, 1969), although Endean (1974) pointed out that there was no evidence to indicate that there had been a decline in the predation of starfish eggs in recent years. Secondly, experiments by Howden *et al.* (1975) and Lucas (1975) have shown that the eggs and larvae may not be eaten by fish because they contain toxic saponins. Endean (1982) claimed that the collection of triton shells and overfishing of some reefs may have been responsible for recent starfish outbreaks. He maintained that this hypothesis correlated well with the history of starfish outbreaks on the Great Barrier Reef.

Rowe & Vail (1984a), in reviewing current knowledge of the *Acanthaster* phenomenon stated that the predator removal hypothesis was no longer accepted by most scientists. This comment may be true but it fails to point out that scientists in general are not in a position to be able to make an objective decision concerning its validity on the grounds that very little is known about predation in general. This is a feature of each of the hypotheses presented in this section. Clearly, an extensive series of field experiments is needed in order to generate a more informed debate of this issue. Recently, it has been recommended that modelling studies be undertaken to test hypotheses which incorporate man-induced triggers (Crown of Thorns Starfish Advisory Committee, 1985) but even these studies require more empirical information on factors influencing the distribution and abundance of *A. planci* than is at present available.

The predator removal hypothesis, like all hypotheses has weaknesses. Potts (1981) regarded it as "the least satisfactory model" on the grounds that it relied on some invalid assumptions and there was little before and after information on triton numbers which would allow the hypothesis to be tested. He disputed the notion that *Charonia tritonis* was a "specialist" predator of *Acanthaster planci* and that it was capable of controlling the numbers of juvenile and adult starfish. Chesher (1969a) has raised some further doubts concerning this hypothesis particularly in relation to its application to outbreaks in other parts of the Indo-Pacific. He considered that it was possible that the collection of tritons had lead to outbreaks of starfish on some reefs in Micronesia. He, however, made the point that outbreaks occurred on some isolated reefs in this region (*e.g.* Ponape, Tinian, Ant, and Truk) where fishing and shell collecting were unlikely to have been carried out. Conversely, he suggested that outbreaks of starfish were not recorded on several reefs (*e.g.* Ifalik, Woleai, Kapingamarangi) where these activities were intensively conducted.

Two other features of the predator removal hypothesis require explanation. First, Endean (1974) stated that the collection of triton shells had occurred until 1969 when it became a protected species. Almost 16 years have elapsed since then and it is presumed that the triton populations have begun to recover. What is not readily apparent is why outbreaks are at present occurring on the Great Barrier Reef when this animal has been protected for so many years? Secondly, it is not known whether the progressive removal of tritons over a number of years would lead to a gradual build-up in starfish numbers, as suggested by Birkeland (1982), or whether it would cause the rapid appearance of outbreaks. If it produced a gradual increase in starfish abundance then most likely this would have been manifested on a number of reefs on the Great Barrier Reef as shell collecting and fishing have been carried out over a large part of this region. This facet of the hypothesis should be testable using population models of the phenomenon.

This completes a discussion of the main hypotheses which have been raised to account for the occurrence of starfish outbreaks. Other mechanisms have been postulated, such as genetic mutations of *A. planci* (Antonius, 1971), but there is no evidence to support them.

CAUSE OR CAUSES?

It should be recognized that the hypotheses discussed above have some basis in fact or offer apparently plausible reasons for the occurrence of outbreaks. As there are, however, inconsistencies within each, no one hypothesis fully explains the occurrence of outbreaks of *A. planci*. This is for several reasons. First, many are based on a correlative approach and consequently they do not demonstrate true cause and effect. Kendall & Stuart (1979) have discussed the problems involved in establishing causation by studying the interdependence of two variables. Secondly, because so little is known about *A. planci* in the field all the hypotheses suffer, to varying extents, from a lack of supporting evidence. Thirdly, in some instances they are based on evidence which is equivocal and can be interpreted in a number of different ways. Finally, some stem from information which has been derived from outbreaks in specific areas and, therefore, inconsistencies emerge when they are extrapolated to account for the global pattern of outbreaks. Perhaps a criticism which may be levelled at all hypotheses is that they tend to be overly simplistic and seek to explain the occurrence of outbreaks in terms of a single (global) process (*e.g.* predation, terrestrial run-off, pollution). Randall (1972) and Endean (1977) suggested that the probability of outbreaks occurring concurrently in different reefal areas, separated by large distances, was low and that this was indicative of a single controlling factor. No doubt there is some truth to this statement as there are a number of similarities among the outbreaks that have occurred throughout the Indo-Pacific. These are given below.

(1) All major outbreaks in the world have occurred near landmasses (*e.g.* Great Barrier Reef, Ryukyu Islands, Micronesia, Fiji, Samoa, Hawaii, and Tahiti). Exceptions to this are the outbreaks that have been recorded on Elizabeth and Middleton reefs in the Tasman Sea. These may, however, have resulted from an influx of larvae from the Great Barrier Reef.
(2) Most outbreaks appear to be synchronized, having occurred over the same general period (*i.e.* 1960s–1970s). Of late, outbreaks have once again arisen in several areas at about the same time (*i.e.* late 1970s–1980s). Examples of these are the Great Barrier Reef, Guam, Palau, Saipan, and Fiji.
(3) Several reefs appear to have suffered extensive outbreaks on both occasions (*e.g.* Great Barrier Reef, Fiji, Guam and Palau).

While there are certain similarities among outbreaks that have occurred in the Indo-Pacific there is no reason to suppose that this is because they each originated as the result of the same single process (Weber & Woodhead, 1970). Indeed the fact that no one hypothesis can account fully for their occurrence suggests that a number of processes may be involved. The information presented earlier in this paper indicates that outbreaks may be caused by a complex interaction of factors which are poorly understood. Perhaps a more accurate explanation of the global occurrence of outbreaks may be achieved by considering the effects of a number of processes which may vary in their importance and their relationship with each other in different areas. To date, no hypothesis has incorporated this type of approach probably because there is so much that is not known about this starfish.

It is worth pondering whether our understanding of the *Acanthaster* phenomenon is hamstrung because there is a tendency to rely on hypotheses which may provide simplistic answers to what may be a far more complex question?

Perhaps the real answer may lie in a collage of the main hypotheses proposed earlier. A similar suggestion was also made by Potts (1981). It is possible that adults may aggregate under natural conditions as proposed by Dana *et al.* (1972). If the spawning of these adults coincided with times of heavy run-off, high food abundances (Birkeland, 1982) and optimal physical conditions (Pearson, 1975b), then this may lead to the increased survival of larvae. The settlement of large numbers of larvae and the establishment of dense aggregations of juveniles may occur provided predation is not extensive (Endean, 1982). This hypothetical example still allows for the possibility that outbreaks may be man-induced or that their frequency of occurrence has been increased by man. Answers to this question may involve a much more intensive study of each process and the relationships between them in order to determine the critical pathways in the system.

Of course, this explanation may also be inadequate, but one must be alert to the possibility that this phenomenon may not be explained easily and that to trust to one hypothesis is akin to putting on blinkers. In the future as more is known about *A. planci*, particularly its ecology, there must be a willingness to modify and extend hypotheses. Otherwise our knowledge of this phenomenon may stagnate and will revolve around a debate of the same ideas and issues; this has happened already to a certain extent. Only in this way may we be able to appreciate more fully the *Acanthaster* phenomenon. Obviously with the difficulties faced by scientists in undertaking studies in the field several aspects of this animal's biology may never be fully comprehended. Hence Bradbury, Done *et al.*'s (1985) warning that more research may not lead to a complete understanding of the phenomenon. The success of future research may well depend on addressing the right research questions at the correct time. For this to happen it is imperative in dealing with this episodic animal that the availability of funds, the formulation of research questions, and the occurrence of outbreaks be synchronized. Unfortunately this has not happened to date, despite the large number of committees and governmental bodies that have been formed to look into this problem (*e.g.* in Australia alone—Walsh *et al.*, 1970, 1971, 1976; Advisory Committee on the Crown of Thorns Starfish, 1980; Crown of Thorns Starfish Advisory Committee, 1985; Milton, 1985).

In conclusion, future research on the *Acanthaster* phenomenon is important for several reasons. First, it will extend our knowledge of outbreaks of invertebrate populations. Secondly, and just as important, it offers scientists the unique opportunity to obtain a greater understanding of coral reefs and the processes that are important in structuring them. Finally, it will help scientists to decide whether the phenomenon is a problem, in the sense that it may be causing irreparable and unnatural changes to many of the world's coral reefs.[1]

[1]At the end of 1985 the Australian Government allocated $971,000 to the Great Barrier Reef Marine Park Authority to initiate what is anticipated will be a four-year research programme on the crown-of-thorns starfish.

ACKNOWLEDGEMENTS

This paper has benefited enormously from discussions I have had with numerous people over the last few years. Accordingly, I wish to thank: R. Bradbury, A. Dartnall, T. Done, R. Kenchington, J. Lucas, R. Olson, R. Pearson, R. Reichelt, J. Stoddart, and L. Zann for their thoughts, criticisms, and helpful advice. I am particularly indebted to R. Bradbury, A. Dartnall, R. Reichelt, R. Olson, and J. Stoddart for making many valuable comments on earlier drafts of this paper. This work was carried out while I was a Research Associate at the Australian Institute of Marine Science and was funded by a grant from the Queen's Fellowship and Marine Research Allocations Advisory Committee. I thank the Great Barrier Reef Marine Park Authority for access to unpublished reports and survey data. I greatly appreciated the help of Debbie Bass in painstakingly checking all the references. The manuscript was typed with great care by Frances Hickey, Liz Howlett, and Kim Truscott and the figures were prepared by Marty Thyssen. Most important I thank my wife for her patience and love while this task was being undertaken.

REFERENCES

Advisory Committee on the Crown of Thorns Starfish, 1980. *Report to the Great Barrier Reef Marine Park Authority*, 13 pp.

Aisek, K., 1972. In, *Proc. University of Guam—Trust Territory* Acanthaster planci *(crown-of-thorns starfish) Workshop*, edited by R. T. Tsuda, *Univ. of Guam, Mar. Lab. Tech. Rep.*, No. 3, 27–28.

Alcala, A. C., 1974. *Silliman J.*, **21**, 174 only.

Alcala, A. C., 1976. *Silliman J.*, **23**, 279–285.

Allen, G. R., 1972. *Copeia*, 1972, No. 3, 595–597.

Antonelli, P. L. & Kazarinoff, N. D., 1984. *J. theor. Biol.*, **107**, 667–684.

Antonius, A., 1971. *Int. Rev. ges. Hydrobiol.*, **56**, 283–319.

Aziz, A. & Sukarno, 1977. *Mar. Res. Indonesia*, **17**, 121–132.

Baloun, A. J. & Morse, D. E., 1984. *Biol. Bull. mar. biol. Lab., Woods Hole*, **167**, 124–138.

Barham, E. G., Gowdy, R. W. & Wolfson, F. H., 1973. *Fish. Bull. NOAA*, **71**, 927–942.

Barnes, D. J., Brauer, R. W. & Jordan, M. R., 1970. *Nature, Lond.*, **228**, 342–344.

Barnes, J. & Endean, R., 1964. *Med. J. Aust.*, **1**, 592–593.

Barnes, J. H., 1966. *Aust. nat. Hist.*, **15**, 257–261.

Bayne, B. L., 1965. In, *Marine Mussels: Their Ecology and Physiology*, edited by B. L. Bayne, Cambridge University Press, London, pp. 81–120.

Beach, D. H., Hanscomb, N. J. & Ormond, R. F. G., 1975. *Nature, Lond.*, **254**, 135–136.

Belk, M. S. & Belk, D., 1975. *Hydrobiologia*, **46**, 29–32.

Bennett, I., 1981. *The Great Barrier Reef*. Lansdowne Press, Sydney, 184 pp.

Benson, A. A., Patton, J. S. & Field, C. E., 1975. *Comp. Biochem. Physiol.*, **52B**, 339–340.

Beran, B. D., 1972. *Silliman J.*, **19**, 381–386.

Beulig, A., Beach, D. & Martindale, M. Q. 1982. In, *Proc. Fourth Int. Coral Reef Symp.*, **2**, 755 only.

Biggs, P. & Eminson, D. F., 1977. *Biol. Conserv.*, **11**, 41–47.

Birkeland, C., 1979. *Report to Office of Marine Resources*, Marine Laboratory, University of Guam, 30 pp.
Birkeland, C., 1981. *Atoll Res. Bull.,* No. 255, 55–58.
Birkeland, C., 1982. *Mar. Biol.,* **69,** 175–185.
Birkeland, C., 1983. *The Siren,* **22,** 13–17.
Birkeland, C. & Randall, R. H., 1979. *Report to Office of Marine Resources,* Marine Laboratory, University of Guam, 53 pp.
Birtles, R. A., Collins, J. D., Kenchington, R. A., Price, I. R., Lucas, J. S., Williams, W. T., Barnes, D. J., Crossland, C. J., Vernon, J. E. N., Frankel, E. & Pearson, R., 1976. *The Bulletin* (*Aust.*), 1 May, 30 only.
Blake, D. B., 1979. *J. nat. Hist.,* **13,** 303–314.
Bradbury, R. H., 1976. *Search,* **7,** 461–462.
Bradbury, R. H., Done, T. J., English, S. A., Fisk, D. A., Moran, P. J., Reichelt, R. E. & Williams, D. McB., 1985. *Search,* **16,** 106–109.
Bradbury, R. H., Hammond, L. S., Moran, P. J. & Reichelt, R. E., 1985a. *J. theor. Biol.,* **113,** 69–80.
Bradbury, R. H., Hammond, L. S., Moran, P. J. & Reichelt, R. E., 1985b. In, *Proc. Fifth Int. Coral Reef Congress,* **5,** 303–309.
Branham, J. M., 1971. *Am. Zool.,* **11,** 694 only.
Branham, J. M., 1973. *BioScience,* **23,** 219–226.
Branham, J. M., Reed, S. A., Bailey, J. H. & Caperon, J., 1971. *Science,* **172,** 1155–1157.
Brauer, R. W., Jordan, M. R. & Barnes, D. J., 1970. *Nature, Lond.,* **228,** 344–346.
Brown, T. W., 1970. *Hemisphere,* **14,** 31–36.
Buznikov, G. A., Malchenko, L. A., Turpaev, T. M. & Tien, V. D., 1982. *Biol. Morya,* **6,** 24–29 (in Russian, English abstract).
Cameron, A. M., 1977. In, *Proc. Third Int. Coral Reef Symp.,* **1,** 193–199.
Cameron, A. M. & Endean, R., 1982. In, *Proc. Fourth Int. Coral Reef Symp.,* **2,** 593–596.
Cameron, A. M. & Endean, R., 1985. In, *Proc. Fifth Int. Coral Reef Congress,* **6,** 211–215.
Campbell, A. C. & Ormond, R. F. G., 1970. *Biol. Conserv.,* **2,** 246–251.
Cannon, L. R. G., 1972. In, *Proc. Crown-of-thorns Starfish Sem.,* 25 Aug. 1972, AGPS, Canberra, pp. 10–18.
Cannon, L. R. G., 1975. In, *Proc. Crown-of-thorns Starfish Sem.,* 6 Sept. 1974, AGPS, Canberra, pp. 39–54.
Caso, M. E., 1961. *An. Inst. Biol. Univ. Nal. Auton. Mexico,* **32,** 313–331 (in Spanish).
Caso, M. E., 1970. *An. Inst. Biol. Univ. Nal. Auton. Mexico,* **41,** 63–78 (in Spanish, English abstract).
Caso, M. E., 1972. *Rev. Soc. Mex. Hist. nat.,* **33,** 51–84 (in Spanish).
Cheney, D. P., 1972a. In, *Proc. University of Guam—Trust Territory* Acanthaster planci (*crown-of-thorns starfish*) *Workshop,* edited by R. T. Tsuda, *Univ. of Guam, Mar. Lab. Tech. Rep.,* No. 3, 11 only.
Cheney, D. P., 1972b. In, *Proc. University of Guam—Trust Territory* Acanthaster planci (*crown-of-thorns starfish*) *Workshop,* edited by R. T. Tsuda, *Univ. of Guam, Mar. Lab. Tech. Rep.,* No. 3, 19 only.
Cheney, D. P., 1972c. *Guam Recorder,* **1,** 74–80.
Cheney, D. P., 1973a. *Micronesica,* **9,** 159 only.
Cheney, D. P., 1973b. *Micronesica,* **9,** 171–180.
Cheney, D. P., 1974. In, *Proc. Second Int. Coral Reef Symp.,* **1,** 591–594.
Chesher, R. H., 1969a. *Report to U.S. Dept. Interior No. PB187631,* Westinghouse Electric Co., 151 pp.
Chesher, R. H., 1969b. *Science,* **165,** 280–283.
Chesher, R. H., 1970. *Science,* **167,** 1275 only.

Clark, A. H., 1921. *Publs Carnegie Instn Wash.*, 214, *Dept. Mar. Biol.*, **10**, 224 pp.
Clark, A. H., 1950. *Bull. Raffles Mus.*, No. 22, 53–67.
Clark, A. M. & Davies, P. S., 1965. *Ann. Mag. nat. Hist.*, **8**, 597–612.
Colgan, M. W., 1981. *Univ. of Guam, Mar. Lab. Tech. Rep.*, No. 76, 69 pp.
Colgan, M. W., 1982. In, *Proc. Fourth Int. Coral Reef. Symp.*, **2**, 333–338.
Colin, P. L., 1977. In, *Proc. Third Int. Coral Reef Symp.*, **1**, 63–68.
Collins, A. R. S., 1974. *J. exp. mar. Biol. Ecol.*, **15**, 173–184.
Collins, A. R. S., 1975a. *J. exp. mar. Biol. Ecol.*, **17**, 69–86.
Collins, A. R. S., 1975b. *J. exp. mar. Biol. Ecol.*, **17**, 87–94.
Conand, C., 1983. *Office de la Recherche Scientifique et Technique Outre-mer. Centre de Noumes Océanographic*, 43 pp. (in French, English abstract).
Croft, J. A., Fleming, W. J. & Howden, M. E. H., 1971. *Contributions to the International Marine Science Symposium of the Institute of Marine Science*, University of New South Wales, Sydney, pp. 80–90.
Crown of Thorns Starfish Advisory Committee, 1985. *Report to Great Barrier Reef Marine Park Authority*, 49 pp.
Dana, T. & Wolfson, A., 1970. *Trans. San Diego Soc. nat Hist.*, **16**, 83–90.
Dana, T. F., 1970. *Science*, **169**, 894 only.
Dana, T. F. & Newman, W. A., 1972. *Micronesica*, **9**, 205 only.
Dana, T. F., Newman, W. A. & Fager, E. W., 1972. *Pacif. Sci.*, **26**, 355–372.
De Bruin, G. H. P., 1972. *Bull. Fish. Res. Stn Sri Lanka (Ceylon)*, **23**, 37–41.
De Silva, M. W. R. N., 1985. In, *Proc. Fifth Int. Coral Reef Congress*, **6**, 515–518.
Devaney, D. M. & Randall, J. E., 1973. *Atoll Res. Bull.*, No. 169, 23 pp.
Dixon, B., 1969. *New Scientist*, **44**, 226–227.
Done, T. J. 1985. In, *Proc. Fifth Int. Coral Reef Congress*, Tahiti. **5**, 315–320.
Done, T. J., Kenchington, R. A. & Zell, L. D., 1982. In, *Proc. Fourth Int. Coral Reef Symp.*, **1**, 299–308.
Dwyer, P. D., 1971. *Search*, **2**, 262–263.
Ebert, T. A., 1973. *Oecologia* (*Berl.*), **11**, 281–298.
Ebert, T. A., 1983. In, *Echinoderm Studies*: *Vol. 1*, edited by M. Jangoux & J. M. Lawrence, Balkema, Rotterdam, pp. 169–202.
Edmondson, C. H., 1933. *Spec. Publ. Bishop Mus.*, **22**, 67 only.
Eldredge, L. G., 1972. In, *Proc. University of Guam—Trust Territory* Acanthaster planci (*crown-of-thorns starfish*) *Workshop*, edited by R. T. Tsuda, *Univ. of Guam, Mar. Lab. Tech. Rep.*, No. 3, 15–17.
Endean, R., 1969. *Qld. Dept of Primary Ind. Fisheries Branch*, Brisbane, 30 pp.
Endean, R., 1971a. *J. mar. biol. Ass. India*, **13**, 1–13.
Endean, R., 1971b. *Operculum*, **2**, 36–48.
Endean, R., 1973a. *Koolewong*, **2**, 6–9.
Endean, R., 1973b. In, *Biology and Geology of Coral Reefs, Vol. II, Biology 1*, edited by O. A. Jones & R. Endean, Academic Press, New York, pp. 389–438.
Endean, R., 1974. In, *Proc. Second Int. Coral Reef Symp.*, **1**, 563–576.
Endean, R., 1976. In, *Biology and Geology of Coral Reefs. Vol. III, Biology 2*, edited by O. A. Jones & R. Endean, Academic Press, New York, pp. 215–254.
Endean, R., 1977. In, *Proc. Third Int. Coral Reef Symp.*, **1**, 185–191.
Endean, R., 1982. *Endeavour*, **6**, 10–14.
Endean, R. & Cameron, A. M., 1985. In, *Proc. Fifth Int. Coral Reef Congress*, **5**, 309–314.
Endean, R. & Chesher, R. H., 1973. *Biol. Conserv.*, **5**, 87–95.
Endean, R. & Stablum, W., 1973a. *Atoll Res. Bull.*, No. 168, 26 pp.
Endean, R. & Stablum, W., 1973b. *Atoll Res. Bull.*, No. 167, 62 pp.
Endean, R. & Stablum, W., 1975a. In, *Proc. Crown-of-thorns Starfish Sem.*, 6 Sept. 1974, AGPS, Canberra, pp. 21–37.
Endean, R. & Stablum, W., 1975b. *Environ. Conserv.*, **2**, 247–256.
Everitt, B. J. & Jurevics, H. A., 1973. *Proc. Aust. Physiol. Pharmacol. Soc.*, **4**, 46–47.

Fagoonee, I., 1985a. In, *Proc. Fifth Int. Coral Reef Congress,* **2,** 127 only.
Fagoonee, I., 1985b. In, *Proc. Fifth Int. Coral Reef Congress,* **2,** 128 only.
Feder, H. M. & Christensen, A. M., 1966. In, *Physiology of Echinoderms,* edited by R. A. Boolootian, Interscience Publishers, New York, pp. 87–127.
Fischer, J. L., 1969. *Science,* **165,** 645 only.
Fishelson, L., 1973. *Mar. Biol.,* **19,** 183–196.
Fisk, D. A. & Done, T. J., 1985. In, *Proc. Fifth Int. Coral Reef Congress,* **6,** 149–154.
Flanigan, J. M. & Lamberts, A. E., 1981. *Atoll Res. Bull.,* No. 255, 59–62.
Fleming, W. J., Howden, M. E. H. & Salathe, R., 1972. In, *Proc. Crown-of-thorns Starfish Sem.,* 25 Aug. 1972, AGPS, Canberra, pp. 83–95.
Fleming, W. J., Salathe, R., Wyllie, S. G. & Howden, M. E. H., 1976. *Comp. Biochem. Physiol.,* **53B,** 267–272.
Francis, M. P., 1981. *Atoll Res. Bull.,* No. 255, 63–68.
Frank, P. W. 1969. *Oecologia* (*Berl.*), **2,** 232–350.
Frankel, E., 1975a. In, *Proc. Crown-of-thorns Starfish Sem.,* 6 Sept. 1974, AGPS, Canberra, pp. 159–165.
Frankel, E., 1975b. In, *Proc. Pacif. Sci. Congr.,* **13,** 125 only.
Frankel, E., 1977. In, *Proc. Third Int. Coral Reef Symp.,* **1,** 201–208.
Frankel, E., 1978. *Atoll Res. Bull.,* No. 220, 75–93.
Fukuda, T., 1976. *Mar. Parks J.,* **38,** 7–10 (in Japanese).
Fukuda, T. & Miyawaki, I., 1982. *Mar. Parks J.,* **56,** 10–13 (in Japanese).
Fukuda, T. & Okamoto, K., 1976. *Sesoko mar. Sci. Lab. Tech. Rep.,* No. 4, 7–17.
Garlovsky, D. F. & Bergquist, A., 1970. *South Pacif. Bull.,* **20,** 47–49.
Garner, D., 1971. In, *Regional Symposium on Conservation of Nature—Reef and Lagoons,* South Pacific Commission, Noumea, pp. 195–201.
Garrett, R. N., 1975. In, *Proc. Crown-of-thorns Starfish Sem.,* 6 Sept. 1974, AGPS, Canberra, pp. 135–147.
Gaudy, R., 1974. *Mar. Biol.,* **25,** 125–141.
Glynn, P. W., 1972. In, *The Panamic biota: Some observations prior to a sea-level canal,* edited by M. L. Jones, *Bull. Biol. Soc. Wash.,* **2,** 13–30.
Glynn, P. W., 1973. *Science,* **180,** 504–506.
Glynn, P. W., 1974. *Environ. Conserv.,* **1,** 295–304.
Glynn, P. W., 1976. *Ecol. Monogr.,* **46,** 431–456.
Glynn, P. W., 1977. In, *Proc. Third Int. Coral Reef Symp.,* **1,** 209–215.
Glynn, P. W., 1980. *Oecologia* (*Berl.*), **47,** 287–290.
Glynn, P. W., 1982a. In, *Proc. Fourth Int. Coral Reef Symp.,* **2,** 607–612.
Glynn, P. W., 1982b. *Coral Reefs,* **1,** 89–94.
Glynn, P. W., 1983. *Environ. Conserv.,* **10,** 149–154.
Glynn, P. W., 1984a. *Environ. Conserv.,* **11,** 133–146.
Glynn, R. W., 1984b. *Bull. mar. Sci.,* **35,** 54–71.
Glynn, P. W., 1985. In, *Proc. Fifth Int. Coral Reef Congress,* **4,** 183–188.
Glynn, P. W., Stewart, R. H. & McCosker, J. E., 1972. *Geol. Rundsch.,* **61,** 483 only.
Goreau, T. F., 1964. *Sea Fish. Res. Stn Haifa Bull.,* No. 35, 23–26.
Goreau, T. F., Lang, J. C., Graham, E. A. & Goreau, P.D., 1972. *Bull. Mar. Sci.,* **22,** 113–152.
Gouldthorpe, K., 1968. *Life* (*Aust.*), 24 June, pp. 55–60.
Great Barrier Reef Marine Park Authority, 1981. *Nomination of the Great Barrier Reef for inclusion in the World Heritage List.* UNESCO, 37 pp.
Great Barrier Reef Marine Park Authority, 1984a. *Reeflections,* **14,** 3–4.
Great Barrier Reef Marine Park Authority, 1984b. *Information Kit on the Crown of Thorns starfish* Acanthaster planci. 22 pp.
Great Barrier Reef Marine Park Authority, 1985. *Reeflections,* **15,** 6–7.
Gupta, K. C. & Scheuer, P. J., 1968. *Tetrahedron,* **24,** 5831–5837.

Hanscomb, N. J., Bennett, J. P. & Harper, G., 1976. *J. exp. mar. Biol. Ecol.,* **22,** 193–197.
Harding, J., 1968. *Sea Frontiers,* **14,** 258–261.
Harriott, V. J., 1985. *Mar. Ecol. Prog. Ser.,* **21,** 81–88.
Hayashi, K., 1973. *Proc. Jap. Soc. Syst. Zool.,* **9,** 29–35.
Hayashi, K., 1975. *Bull. Mar. Park Res. Stn,* **1,** 1–9.
Hayashi, K., Komatsu, M. & Oruro, C., 1973. *Proc. Jpn Soc. Syst. Zool.,* **9,** 59–61.
Hayashi, K. & Tatsuki, T., 1975. *Bull. Mar. Park Res. Stn,* **1,** 11–17.
Hayashi, R., 1939. *Stud. Palao Trop. Biol. Stn,* **3,** 417–447.
Haysom, N. M., 1972. In, *Proc. Crown-of-thorns Starfish Sem.,* 25 Aug. 1972, AGPS, Canberra, pp. 59–64.
Hazell, M. J., 1971. *Mar. Pollut. Bull.,* **2,** 166–169.
Hegerl, E., 1984a. *Bull. Aust. Litt. Soc.,* **7,** 1–2.
Hegerl, E., 1984b. *Bull. Aust. Litt. Soc.,* **2,** 2–4.
Heiskanen, L. P., Jurevics, H. A. & Everitt, B. J., 1973. *Proc. Aust. physiol. pharmacol. Soc.,* **4,** 57 only.
Henderson, J. A., 1969. *Queensland Dept of Harbours and Marine, Fisheries Notes,* **3,** 69–75.
Henderson, J. A. & Lucas, J. S., 1971. *Nature, Lond.,* **232,** 655–657.
Howden, M. E. H., Lucas, J. S., McDuff, M. & Salathe, R., 1975. In, *Proc. Crown-of-thorns Starfish Sem.,* 6 Sept. 1974, AGPS, Canberra, pp. 67–79.
Humes, A. G., 1970. *Publs Seto mar. biol. Lab.,* **17,** 329–338.
Humes, A. G. & Cressey, R. F. 1958. *J. Parasitol.,* **44,** 395–408.
Huxley, C. J., 1976. *J. exp. mar. Biol. Ecol.,* **22,** 199–206.
Hyman, L. H., 1955. *The Invertebrates, Vol. 4.* McGraw-Hill, New York, 763 pp.
Ikehara, I. I., 1972. In, *Proc. University of Guam—Trust Territory* Acanthaster planci (*crown-of-thorns starfish*) *Workshop,* edited by R. T. Tsuda, *Univ. of Guam, Mar. Lab. Tech. Rep.,* No. 3, 7–8.
Ito, T. 1984. *Animals and Nature,* **14,** 19–24 (in Japanese).
James, P., 1976. *Requiem for the Reef.* Foundation Press, Brisbane, 84 pp.
Jangoux, M., 1982a. In, *Echinoderm Nutrition,* edited by M. Jangoux & J. M. Lawrence, Balkema, Rotterdam, pp. 235–272.
Jangoux, M., 1982b. In, *Echinoderm Nutrition,* edited by M. Jangoux & J. M. Lawrence, Balkema, Rotterdam, pp. 117–159.
Johannes, R. E., 1971. *Mar. Pollut. Bull.,* **2,** 9–10.
Kanazawa, A., Teshima, S., Ando, T. & Tomita, S., 1976. *Mar. Biol.,* **34,** 53–57;
Kanazawa, A., Teshima, S., Tomita, S. & Ando, T., 1974. *Bull. Jap. Soc. scient. Fish.,* **40,** 1077 only.
Kenchington, R. A., 1975a. In, *Proc. Crown-of-thorns Starfish Sem.,* 6 Sept. 1974, AGPS, Canberra, pp. 1–7.
Kenchington, R. A., 1975b. In, *Proc. Pacif. Sci. Congr.,* **13,** 128 only.
Kenchington, R. A., 1976. *Biol. Conserv.,* **9,** 165–179.
Kenchington, R. A., 1977. *Biol. Conserv.,* **11,** 103–118.
Kenchington, R. A., 1978. *Environ. Conserv.,* **5,** 11–20.
Kenchington, R. A., 1985. In, *Proc. Fifth Int. Coral Reef Congress,* **2,** 201 only.
Kenchington, R. A. & Morton, B., 1976. *Report of the Steering Committee for the crown of thorns survey, March 1976.* AGPS, Canberra, 186 pp.
Kenchington, R. A. & Pearson, R., 1982. In, *Proc. Fourth Int. Coral Reef Symp.,* **2,** 597–600.
Kendall, M. & Stuart, A., 1979. *The Advanced Theory of Statistics, Vol. 2.* Charles Griffin, London, 748 pp.
Kenny, J. E., 1969. *Ocean Industry,* **4,** 20–22.
Kinsey, D. W., 1983. In, *Perspectives on Coral Reefs,* edited by D. J. Barnes, Brian Clouston, Canberra, pp. 209–220.

Kitagawa, I. & Kobayashi, M., 1977. *Tetrahedron Lett.,* No. 10, 859–862.
Kitagawa, I. & Kobayashi, M., 1978. *Chem. Pharm. Bull.,* **26,** 1864–1873.
Kitagawa, I., Kobayashi, M. & Sugawara, T., 1978. *Chem. Pharm. Bull.,* **26,** 1852–1863.
Kitagawa, I., Kobayashi, M. Sugawara, T. & Yosioka, I., 1975. *Tetrahedron Lett.,* **11,** 967–970.
Komori, T., Matsuo, J., Itakura, T., Sakamoto, K., Ito, Y., Taguchi, S. & Kawasaki, T., 1983a. *Liebigs Ann. Chem.,* **1,** 24–36.
Komori, T., Nanri, H., Itakura, T., Sakamoto, K., Taguchi, S., Higuchi, R., Kawasaki, T. & Higuchi, T., 1983b. *Liebigs Ann. Chem.,* **1,** 37–55.
Komori, T., Sanechika, Y., Ito, Y., Matsuo, J., Nohara, T., Kawasaki, T. & Schulten, H. R., 1980. *Liebigs Ann. Chem.,* **5,** 653–668 (in German, English abstract).
Krebs, C. J., 1978. *Ecology. The Experimental Analysis of Distribution and Abundance.* Harper and Row, New York, 678 pp.
Kvalvagnaes, K., 1972. *Sarsia,* **49,** 81–88.
Laxton, J. H., 1974. *Biol. J. Linn. Soc.,* **6,** 19–45.
Laxton, J. H. & Stablum, W. J., 1974. *Biol. J. Linn. Soc.,* **6,** 1–18.
Loosanoff, V. L. & Engle, J. B., 1938. *Science,* **88,** 107–108.
Loosanoff, V. L. & Engle, J. B., 1942. *U.S. Bureau of Sports Fisheries and Wildlife,* Res. Rep., No. 2, 1–29.
Lucas, J. S., 1972. In, *Proc. Crown-of-thorns Starfish Sem.,* 25 Aug. 1972, AGPS, Canberra, pp. 25–36.
Lucas, J. S., 1973. *Micronesica,* **9,** 197–203.
Lucas, J. S., 1975. In, *Proc. Crown-of-thorns Starfish Sem.,* 6 Sept. 1974, AGPS, Canberra, pp. 103–121.
Lucas, J. S., 1982. *J. exp. mar. Biol. Ecol.,* **65,** 173–194.
Lucas, J. S., 1984. *J. exp. mar. Biol. Ecol.,* **79,** 129–147.
Lucas, J. S., Hart, R. J., Howden, M. E. & Salathe, R., 1979. *J. exp. mar. Biol. Ecol.,* **40,** 155–165.
Lucas, J. S. & Jones, M. M., 1976. *Nature, Lond.,* **263,** 409–411.
Lucas, J. S., Nash, W. J. & Nishida, M., 1985. In, *Proc. Fifth Int. Coral Reef Congress,* **5,** 327–332.
Madsen, F. J., 1955. *Vidensk. Meddr dansk naturh. Foren.,* **117,** 179–192.
Manahan, D. T., Davis, J. P. & Stephens, G. C., 1983. *Science,* **220,** 204–206.
Marsh, J. A., 1972a. In, *Proc. University of Guam—Trust Territory* Acanthaster planci *(crown-of-thorns starfish) Workshop,* edited by R. T. Tsuda, *Univ. of Guam, Mar. Lab. Tech. Rep.,* No. 3, 24 only.
Marsh, J. A., 1972b. In, *Proc. University of Guam—Trust Territory* Acanthaster planci *(crown-of-thorns starfish) Workshop,* edited by R. T. Tsuda, *Univ. of Guam, Mar. Lab. Tech. Rep.,* No. 3, 20 only.
Marsh, J. A. & Tsuda, R. T., 1973. *Atoll Res. Bull.,* No. 170, 16 pp.
Matsusita, K. & Misaki, H., 1983. *Mar. Parks J.,* **59,** 14–16 (in Japanese).
Maxwell, W. G. H., 1971. In, *Report of the Committee Appointed by the Commonwealth and Queensland Governments on the Problem of the Crown-of-Thorns Starfish* (Acanthaster planci), Appendix G, pp. 43–45.
May, R. M., 1975. *J. theor. Biol.,* **49,** 511–524.
McCloskey, L. S. & Deubert, K. H., 1972. *Bull. Env. Contam. Toxicol.,* **8,** 251–256.
McKnight, D. G., 1978. *N.Z. Oceanog. Inst. Rec.,* **4,** 17–19.
McKnight, D. G., 1979. *N.Z. Oceanog. Inst. Rec.,* **4,** 21–23.
Menge, B. A., 1982. In, *Echinoderm Nutrition,* edited by M. Jangoux & J. M. Lawrence, Balkema, Rotterdam, pp. 521–551.
Milton, P., 1985. In, *Report, House of Representatives Standing Committee on Environment and Conservation,* AGPS, Canberra, 39 pp.
Misaki, H., 1974. *Mar. Pavilion,* **3,** 52 only (in Japanese).

Misaki, H., 1979. *Mar. Pavilion,* **8,** 59 only (in Japanese).
Mochizuki, Y. & Hori, S. H., 1980. *Comp. Biochem. Physiol.,* **65B,** 119–126.
Moore, R. E., & Huxley, C. J., 1976. *Nature, Lond.,* **263,** 407–409.
Moore, R. J., 1978. *Nature, Lond.,* **271,** 56–57.
Moran, P. J., Bradbury, R. H. & Reichelt, R. E., 1985. In, *Proc. Fifth Int. Coral Reef Congress,* **5,** 321–326.
Moran, P. J., Reichelt, R. E. & Bradbury, R. H., 1986. *Coral Reefs.* **4,** 235–238.
Morris, P. G., 1977. *Malay. Nat. J.,* **30,** 79–85.
Morse, D. E., Hooker, N., Duncan, H. & Jensen, L., 1979. *Science,* **204,** 407–410.
Mortensen, T., 1931. *K. Danske. Vidensk. Selsk. Skr. Naturvid. Math. Afd.,* **9,** 1–39.
Morton, B. R., 1975. In, *Proc. Crown-of-Thorns Starfish Sem.,* 6 Sept. 1974, AGPS, Canberra, pp. 11–17.
Motokawa, T., 1982. *Comp. Biochem. Physiol.,* **73C,** 223–229.
Moyer, J. T., 1978. *Mar. Parks J.,* **44,** 17 only (in Japanese).
Moyer, J. T., Emerson, W. K. & Rose, M., 1982. *Nautilus,* **96,** 69–82.
Murty, A. V. S., Subbaraju, G., Pillai, C. S. G., Josanto, V., Livingston, P. & Vasantha Kumar, R., 1979. *Marine Fisheries Information Service, Technical and Extention Series,* Cochin, India, **13,** 10–12.
Muzik, K., 1985. In, *Proc. Fifth Int. Coral Reef Congress,* **6,** 483–490.
Nakamura, K., 1972. In, *Proc. University of Guam—Trust Territory* Acanthaster planci *(crown-of-thorns starfish) Workshop,* edited by R. T. Tsuda, *Univ. of Guam, Mar. Lab. Tech. Rep.,* No. 3, 23 only.
Nakasone, Y., Yamazato, K., Nishihira, M., Kamura, S. & Aramoto, Y., 1974. *Ecol. Stud. Nat. Cons. Ryukyu Is.,* **1,** 213–236 (in Japanese, English abstract).
Nash, W. & Zell, L. D., 1982. In, *Proc. Fourth Int. Coral Reef Symp.,* **2,** 601–605.
Nash, W. J., 1983. M.Sc. thesis, James Cook University of North Queensland, Australia, 163 pp.
Newman, W. A., 1970. *Science,* **167,** 1274–1275.
Newman, W. A. & Dana, T. F., 1974. *Science,* **183,** 103 only.
Nishihira, M. & Yamazato, K., 1972. *Sesoko mar. Sci. Lab. Tech. Rep.,* No. 1, 1–20.
Nishihira, M. & Yamazato, K., 1973. *Sesoko mar. Sci. Lab. Tech. Rep.,* No. 2, 17–33.
Nishihira, M. & Yamazato, K., 1974. In, *Proc. Second Int. Coral Reef Symp.,* **1,** 577–590.
Nishihira, M., Yamazato, K., Nakasone, Y., Kamura, S. & Aramoto, Y., 1974. *Ecol. Stud. Nat. Cons. Ryukyu Is.,* **1,** 237–254 (in Japanese, English abstract).
North, W. J. & Pearse, J. S., 1970. *Science,* **167,** 209 only.
Odom, C. B. & Fischermann, E. A., 1972. *Hawaii Med. J.,* **31,** 99–100.
O'Gower, A. K., Bennett, I. & McMichael, D. F., 1972. *Search,* **3,** 202–204.
O'Gower, A. K., McMichael, D. F. & Sale, P. F., 1973. *Search,* **4,** 368–369.
Oliver, J., 1985. In, *Proc. Fifth Int. Coral Reef Congress,* **4,** 201–206.
Olson, R. R. 1985. *Mar. Ecol. Prog. Ser.,* **25,** 207–210.
Ormond, R. F. G. & Campbell, A. C., 1971. *Symp. zool. Soc. Lond.,* **28,** 433–454.
Ormond, R. F. G. & Campbell, A. C., 1974. In, *Proc. Second Int. Coral Reef Symp.,* **1,** 595–619.
Ormond, R. F. G., Campbell, A. C., Head, S. H., Moore, R. J., Rainbow, P. R. & Saunders, A. P., 1973. *Nature, Lond.,* **246,** 167–169.
Ormond, R. F. G., Hanscomb, N. J. & Beach, D. H., 1976. *Mar. Behav. Physiol.,* **4,** 93–105.
Owens, D., 1969. *Dept Fish. Suva, Fiji,* **419,** 1–6.
Owens, D., 1971. *Fiji agric. J.,* **33,** 15–23.
Pearson, R. G., 1970. *Newsl. Qld Litt. Soc.,* Jan.-Feb. 1970, pp. 1–10.

Pearson, R. G., 1972a. In, *Proc. Crown-of-thorns Starfish Sem.,* 25 Aug. 1972, AGPS, Canberra, pp. 66–71.
Pearson, R. G., 1972b. *Nature, Lond.,* **237,** 175–176.
Pearson, R. G., 1973. *Micronesica,* **9,** 223 only.
Pearson, R. G., 1974. In, *Proc. Second Int. Coral Reef Symp.,* **2,** 207–215.
Pearson, R. G., 1975a. In, *Proc. Crown-of-thorns Starfish Sem.,* 6 Sept. 1974, AGPS, Canberra, pp. 127–129.
Pearson, R. G., 1975b. In, *Proc. Crown-of-thorns Starfish Sem.,* 6 Sept. 1974, AGPS, Canberra, pp. 131–134.
Pearson, R. G., 1977. *Mar. Res. Indonesia,* **17,** 119 only.
Pearson, R. G., 1981. *Mar. Ecol. Prog. Ser.,* **4,** 105–122.
Pearson, R. G. & Endean, R., 1969. *Queensland Dept of Harbours and Marine, Fisheries Notes,* **3,** 27–55.
Pearson, R. G. & Garrett, R. N., 1975. In, *Proc. Crown-of-thorns Starfish Sem.,* 6 Sept. 1974, AGPS, Canberra, pp. 123–126.
Pearson, R. G. & Garrett, R. N., 1976. *Biol. Conserv.,* **9,** 157–164.
Pearson, R. G. & Garrett, R. N., 1978. *Micronesica,* **14,** 259–272.
Pielou, E. C., 1977. *Mathematical Ecology.* Wiley, New York, 383 pp.
Piyakarnchana, T., 1982. In, *Proc. Fourth Int. Coral Reef Symp.,* **2,** 613–617.
Polunin, N. V. C., 1974. *Nature, Lond.,* **249,** 589–590.
Pope, E. C., 1964. *Aust. Nat. Hist.,* **14,** 350 only.
Porter, J. W., 1972. *Am. Nat.,* **106,** 487–492.
Porter, J. W., 1974. *Science,* **186,** 543–545.
Potts, D. C., 1981. In, *The Ecology of Pests,* edited by R. L. Kitching & R. E. Jones, CSIRO, Melbourne, pp. 55–86.
Price, J. R., 1972. In, *Proc. Crown-on-Thorns Starfish Sem.,* 25 Aug. 1972, AGPS, Canberra, pp. 37–46.
Price, J. R., 1975. In, *Proc. Crown-of-thorns Starfish Sem.,* 6 Sept. 1974, AGPS, Canberra, pp. 181–191.
Pyne, R. R., 1970. *Papua New Guinea Agric. J.,* **21,** 128–138.
Rainbow, P. S., 1974. *Afr. J. trop. Hydrobiol. Fish.,* **3,** 183–191.
Randall, J. E., 1972. *Biotropica,* **4,** 132–144.
Randall, R. H., 1973a. *Publs Seto mar. biol. Lab.,* **20,** 469–489.
Randall, R. H., 1973b. *Micronesica,* **9,** 119–158.
Randall, R. H., 1973c. *Micronesica,* **9,** 213–222.
Randall, R. H., 1973d. *Univ. de Guam, Mar. Lab. Tech. Rep.,* No. 7, 94–145.
Raymond, R., 1984. *The Bulletin* (*Aust.*), **106,** 68–77.
Reichelt, R. E. & Bradbury, R. H., 1984. *Mar. Ecol. Prog. Ser.,* **17,** 251–257.
Ricklefs, R. E., 1979. *Ecology.* Nelson, Sunbury-on-Thames, 966 pp.
Roads, C. H., 1969. *Geogr. Mag.* (*Lond.*), **41,** 524–529.
Roads, C. H. & Ormond, R. F. G. 1971. Editors. *Report of the Third Cambridge Red Sea Expedition 1970,* Cambridge Coral Starfish Research Group, Cambridge, 124 pp.
Robinson, D. E., 1971. *J. R. Soc. N.Z.,* **1,** 99–112.
Rosenberg, D. L., 1972. In, *Proc. University of Guam—Trust Territory* Acanthaster planci (*crown-of-thorns starfish*) *Workshop,* edited by R. T. Tsuda, *Univ. of Guam, Mar, Lab. Tech. Rep.,* No. 3, 12 only.
Rowe, F. W. E. & Vail, L., 1984a. *Aust. nat. Hist.,* **21,** 195–196.
Rowe, F. W. E. & Vail, L., 1984b. *Search,* **15,** 211–213.
Rowe, F. W. E. & Vail, L., 1985. *Search, 1,* 109 only.
Sablan, B., 1972. In, *Proc. University of Guam—Trust Territory* Acanthaster planci (*crown-of-thorns starfish*) *Workshop,* edited by R. T. Tsuda, *Univ. of Guam, Mar. Lab. Tech. Rep.,* No. 3, 21–22.
Sale, P. F., 1980. *Oceanogr. Mar. Biol. Ann. Rev.,* **18,** 367–421.
Sale, P. F., Potts, D. C. & Frankel, E., 1976. *Search,* **7,** 334–338.

Sano, M., Shimizu, M. & Nose, M., 1984. *Pacif. Sci.,* **38,** 51–79.
Sato, S., Ikekawa, N., Kanazawa, A. & Ando, T., 1980. *Steroids,* **36,** 65–72.
Sheikh, Y. M. & Djerassi, C., 1973. *Tetrahedron Lett.,* No. 31, 2927–2930.
Sheikh, Y. M. Djerassi, C. & Tursch, B. M., 1971. *J. Am. chem. Soc.,* **5(D),** 217–218.
Sheikh, Y. M., Kaisin, M. & Djerassi, C., 1973. *Steroids,* **22,** 835–850.
Sheikh, Y. M., Tursch, B. M. & Djerassi, C., 1972a. *J. Am. chem. Soc.,* **94,** 3278–3280.
Sheikh, Y. M., Tursch, B. M. & Djerassi, C., 1972b. *Tetrahedron Lett.,* No. 35, 3721–3724.
Shimizu, Y., 1971. *Experientia,* **27,** 1188–1189.
Shimizu, Y., 1972. *J. Am. chem. Soc.,* **94,** 4051–4052.
Shou-Hwa, C., 1973. *J. Singapore Nat. Acad. Sci.,* **3,** 1–4.
Sivadas, P., 1977. *Mahasagar,* **10,** 179–180.
Sladen, W. P., 1889. *Rep. sci. Res. H.M.S. Challenger, Zool.,* **30,** 535–538.
Sloan, N. A., 1980. *Oceanogr. Mar. Biol. Ann. Rev.,* **18,** 57–124.
Sloan, N. A. & Campbell, A. C., 1982. In, *Echinoderm Nutrition,* edited by M. Jangoux & J. M. Lawrence, Balkema, Rotterdam, pp. 3–23.
Soegiarto, A., 1973. *Micronesica,* **9,** 181 only.
Stanley, S. O., 1983. *Interim Report No. 4, Natural Environment Research Council,* 59 pp.
Strathmann, R., 1975. *Am. Zool.,* **15,** 717–730.
Suzuki, T., 1975. *Mar. Pavilion,* **4,** 2 only.
Tada, M., 1983. *The Drama Beneath the Sea,* Hoikusha, Tokyo, 133 pp. (in Japanese).
Taira, E., Tanahara, N. & Funatsu, M., 1975. *Sci. Bull. Coll. Agr. Univ. Ryukyus,* **22,** 203–213.
Talbot, F. H., 1971. *Search,* **2,** 192–193.
Talbot, F. H. & Talbot, M. S., 1971. *Endeavour,* **30,** 38–42.
Tanner, J. T., 1975. *Ecology,* **56,** 855–867.
Teshima, S., Kanazawa, A., Hyodo, S. & Ando, T., 1979. *Comp. Biochem. Physiol.,* **64B,** 225–228.
Thorson, G., 1950. *Biol. Rev.,* **25,** 1–45.
Thorson, G., 1961. In, *Oceanography,* edited by M. Sears, A.A.A.S., No. 67, 455–474.
Tranter, D. J., 1971. In, *Report of the Committee Appointed by the Commonwealth and Queensland Governments on the Problem of the Crown-of-Thorns Starfish* (Acanthaster planci), Appendix F, pp. 41–42.
Trapido-Rosenthal, H. & Morse, D. E., in press. *Bull. mar. Sci.*
Tsuda, R. T., 1971. Editor. *Univ. of Guam, mar. Lab. Tech. Rep., No. 2,* 127 pp.
Tsuda, R. T., 1972. In, *Proc. University of Guam—Trust Territory* Acanthaster planci *(crown-of-thorns starfish) Workshop,* edited by R. T. Tsuda, *Univ. of Guam, Mar. Lab. Tech. Rep.,* No. 3, 4–5.
Ui, S., 1985. *Mar. Parks J.,* **64,** 13–17 (in Japanese).
Vance, R. R., 1974. *Am. Nat.,* **107,** 353–361.
Verwey, J., 1930. *Treubia,* **12,** 305–366.
Vine, P. J., 1970. *Nature, Lond.,* **228,** 341–342.
Vine, P. J., 1972. *Underwat. J.,* **4,** 64–73.
Vine, P. J., 1973. *Atoll Res. Bull.,* No. 166, 10 pp.
Voogt, P. A., 1982. In, *Echinoderm Nutrition,* edited by M. Jangoux & J. M. Lawrence, Balkema, Rotterdam, pp. 417–436.
Walbran, P. D., 1984. *Report to Great Barrier Reef Marine Park Authority,* 15 pp.
Walsh, R. J., Day, M. F., Emments, C. W., Hill, D., Martyn, D. F., Rogers, W. P. & Waterhouse, D. F., 1970. *Rep. Aust. Acad. Sci.,* No. 11, 20 pp.
Walsh, R. J., Harris, C. L., Harvey, J. M., Maxwell, W. G. H., Thomson, J. M. & Tranter, D. J., 1971 *Report of the Committee Appointed by the Common-*

wealth and Queensland Governments on the Problem of the Crown-of-Thorns Starfish (Acanthaster planci). CSIRO, Melbourne, 45 pp.
Walsh, R. J., Harvey, J. M., Maxwell, W. G. H. & Thomson, J. M., 1976. *Report on Research Sponsored by the Advisory Committee on Research into the Crown of Thorns Starfish*. AGPS, Canberra, 35 pp.
Wass, R. C., 1972. In, *Proc. University of Guam—Trust Territory* Acanthaster planci *(crown-of-thorns starfish) Workshop,* edited by R. T. Tsuda, *Univ. of Guam, Mar. Lab. Tech. Rep.,* No. 3, 25 only.
Wass, R. C., 1973. *Micronesica,* **9,** 167–170.
Weber, J. N., 1969. *Earth and Mineral Sciences,* **38,** 37–41.
Weber, J. N., 1970. *Geol. Soc. Am. Abs., 2,* 39–40.
Weber, J. N. & Woodhead, P. M. J., 1970. *Mar. Biol.,* **6,** 12–17.
Wickler, W., 1970. *Naturw. Rdsch.,* **23,** 368–369.
Wickler, W., 1973. *Micronesica,* **9,** 225–230.
Wickler, W. & Seibt, U., 1970. *Z. Tierpsychol.,* **27,** 352–368.
Williams, D. McB., 1986. *Mar. Ecol. Prog. Ser.* **28,** 157–164.
Williams, D. McB., Wolanski, E. & Andrews, J. C., 1984. *Coral Reefs,* **3,** 229–236.
Williamson, J., 1986. *The Marine Stinger Book*. Surf Life Saving Association of Australia, Brisbane, 3rd edition, 85 pp.
Wilson, B. R., 1972. In, *Proc. Crown-of-thorns Starfish Sem.,* 25 Aug. 1972, AGPS, Canberra, pp. 47–58.
Wilson, B. R. & Marsh, L. M., 1974. In, *Proc. Second Int. Coral Reef Symp.,* **1,** 621–630.
Wilson, B. R. & Marsh, L. M., 1975. In, *Proc. Crown-of-thorns Starfish Sem.,* 6 Sept. 1974, AGPS, Canberra, pp. 167–179.
Wilson, B. R., Marsh, L. M. & Hutching, B., 1974. *Search,* **5,** 601–602.
Woodhead, P. M. J., 1971. In, *Report of the Committee Appointed by the Commonwealth and Queensland Governments on the Problem of the Crown-of-Thorns Starfish* (Acanthaster planci), Appendix E, pp. 34–40.
Yamaguchi, M., 1972a. In, *Proc. University of Guam—Trust Territory* Acanthaster planci *(crown-of-thorns starfish) Workshop,* edited by R. T. Tsuda, *Univ. of Guam, Mar. Lab. Tech. Rep.,* No. 3, 10 only.
Yamaguchi, M., 1972b. *Guam Rail.,* **6,** 4–5.
Yamaguchi, M., 1973a. *Micronesica,* **9,** 207–212.
Yamaguchi, M., 1973b. In, *Biology and Geology of Coral Reefs, Vol. 11, Biology 1,* edited by O. A. Jones & R. Endean, Academic Press, New York, pp. 369–387.
Yamaguchi, M., 1973c. *Univ. of Guam, Mar. Lab. Tech. Rep.,* No. 7, 147–158.
Yamaguchi, M., 1974a. *Pacif. Sci.,* **28,** 139–146.
Yamaguchi, M., 1974b. *Pacif. Sci.,* **28,** 123–138.
Yamaguchi, M., 1975a. *Oecologia* (*Berl.*), **20,** 321–332.
Yamaguchi, M., 1975b. *Biotropica,* **7,** 12–23.
Yamaguchi, M., 1977. *Micronesica,* **13,** 283–296.
Yamaguchi, M., 1985. In, *Proc. Fifth In Coral Reefs Congress,* **2,** 415 only.
Yamaguchi, M., in press. *Coral Reefs*.
Yamazato, K., 1969. *The Ryukyus Today,* **13,** 7–9 (in Japanese).
Yamazato, K. & Kiyan, T., 1973. *Micronesica,* **9,** 185–195.
Yomo, H. & Egawa, Y., 1978. *Agric. biol. Chem.,* **42,** 1591–1592.
Yomo, H. & Tokumoto, M., 1981. *J. agric. chem. Soc Jpn,* **55,** 15–21 (in Japanese, English abstract).
Yonge, C. M., 1968. *Proc. R. Soc. Ser. B,* **169,** 329–344.

Oceanogr. Mar. Biol. Ann. Rev., 1986, **24**, 481–520
Margaret Barnes, Ed.
Aberdeen University Press

THE EFFECTS OF TEMPERATURE AND SALINITY ON THE TOXICITY OF HEAVY METALS TO MARINE AND ESTUARINE INVERTEBRATES

DONALD S. McLUSKY, VICTORIA BRYANT
and RUTH CAMPBELL
Department of Biological Science, The University, Stirling FK9 4LA, Scotland

INTRODUCTION

Despite the considerable effort expended on the toxicity testing of potentially harmful substances to aquatic organisms by many research organizations, relatively few studies have examined the effect of the temperature and salinity on toxicity (Laws, 1981). The majority of studies have examined the toxicity of a substance at one temperature and salinity regime. For marine species this has meant studies in sea water (often, however, with the salinity of that water unstated), and for freshwater species this has meant the locally available tap or pond water (often, however, with the water hardness or pH unstated), with measurements at the local ambient temperature. There have been fewer studies of the toxicity of substances to estuarine animals, and even fewer cases where a range of temperatures and salinities have been included in the experimental design.

"Although rather extensive bibliographies give the impression that there is a vast amount of literature on the effects of temperature on aquatic organisms, when one tries to apply this information to specific interactions such as the effect of temperature changes on chemical toxicity to aquatic organisms, often very little of the evidence is applicable." Thus wrote Cairns, Heath & Parker (1975) who continued, "Even this body of literature is not adequate to make any scientifically justifiable generalisations!"

The measurement of acute toxicity either as median lethal concentration (LC_{50}) at an agreed time limit (usually 96 h) or as median lethal time (LT_{50}) at a specified concentration may rightly be criticized as being an extreme measurement of the toxicity of a compound. Instead it might be desirable to seek sub-lethal, chronic or incipient criteria as a better measure of the toxicity of a substance. Sub-acute criteria, however, may be difficult to specify in a manner that can be applicable to a wide range of species and phyla. There is also no good reason to suppose that there is a constant relationship for different pollutants of different species between the dose needed to kill and that needed to impair an organism (Moriarty, 1983). Therefore, as Moriarty concludes, given the difficulties of studying an ecosystem, the most effective way to predict biological effects is likely to be by discerning the least exposure that produces a deleterious response in individual organisms and then examining the extent to which different environmental conditions alter this minimum exposure. The present review examines the effect of temperature and salinity, which are the principal environmental factors which may affect the inhabitants of estuaries and inshore marine waters, on toxicity of heavy metals. The effect of these pollutants on the estuarine and coastal ecosystem may thus hopefully be determined subsequently. The acute toxicity criteria of LC_{50} and LT_{50} have remained as the most common criteria used by legislative authorities (Lloyd, 1979; Sprague, 1969, 1970, 1971) and this review is predominantly concerned with acute toxicity measurements.

This review will examine the effects of temperature and salinity on the toxicity of heavy metals (arsenic, cadmium, chromium, copper, lead, mercury, nickel, and zinc) to marine and estuarine invertebrates. It is not directly concerned with the uptake or accumulation of the metals, unless such studies have also measured toxicity. Bryan (1971, 1976, 1980, 1984), Coombs (1979, 1980), Moore (1981), and Murphy & Spiegel (1982, 1983) have reviewed many papers on metal uptake in marine and estuarine invertebrates and the development of metal-tolerance in some species. Reish *et al.* (1983, 1984) have also reviewed the effects of many pollutants, including heavy metals, on salt-water organisms. The present review excludes the vertebrates (fish, *etc*) and is solely concerned with the invertebrate phyla, predominantly Mollusca, Crustacea and Annelida.

Because mercury and cadmium are included on the "black list" or "List I" of toxic substances, most work has been done with these metals. The other metals are usually considered as members of the "grey list" or "List II" of pollutants and have been less studied (O'Donnell & Mance, 1984). Throughout the review it should be noted that 1 ppb (part per billion) = $1\ \mu g \cdot l^{-1}$, that 1 ppm (part per million) = $1\ mg \cdot l^{-1}$ and that 1 ppt (part per thousand) = $1 g \cdot l^{-1}$ also expressed as ‰ for salinity (NaCl).

Several papers examining the combined effect of salinity, temperature, and heavy metals on estuarine and marine animals have used response surface analysis techniques to present their results. Schnute & McKinnell (1984) have shown that quantitative and qualitative errors may have been incurred in these analyses, and have defined new parameters for the quadratic model used in such analyses. Accordingly papers which were written prior to Schnute & McKinnell's approach should be treated with appropriate caution.

ARSENIC

Arsenic is probably the least studied of the heavy metals with regard to its toxicity, and only the recent study of Bryant, Newbery, McLusky & Campbell (1985a) has reported on the effect of temperature and salinity on arsenic toxicity. Using three estuarine species (*Corophium volutator*, *Macoma balthica*, and *Tubifex costatus*) at three temperatures (5, 10, 15 °C) and a range of salinities (5–35‰) with time intervals of up to 384 h, they found that toxicity increased as temperature and concentration of arsenic increased, but that salinity had no significant effect (Figs 1 and 2). Bryant *et al.* (1985a) used pentavalent arsenic (as sodium arsenate). Pentavalent arsenic is known to compete with phosphate ions in the process of oxidative phosphorylation and, as this process is temperature dependent, this may explain the temperature effect observed. For the other metals covered in this review and where a distinct salinity effect was observed, the metal was presented as a cation and in this form competes with calcium and magnesium at uptake sites for osmoregulation (Phillips, 1980). The arsenic, however, was presented as the anion arsenate and in this form may thus not interfere directly in osmoregulatory processes.

Unlu & Fowler (1979) designed radioactive trace experiments to study the effects of temperature, salinity, arsenic concentration and mussel size on arsenic accumulaion and elimination processes in *Mytilus galloprovincalis*. Arsenic (as arsenate) uptake increased with increasing arsenic concentration in the water but the response was not proportional, indicating that accumulation was partially suppressed at higher external arsenic concentra-

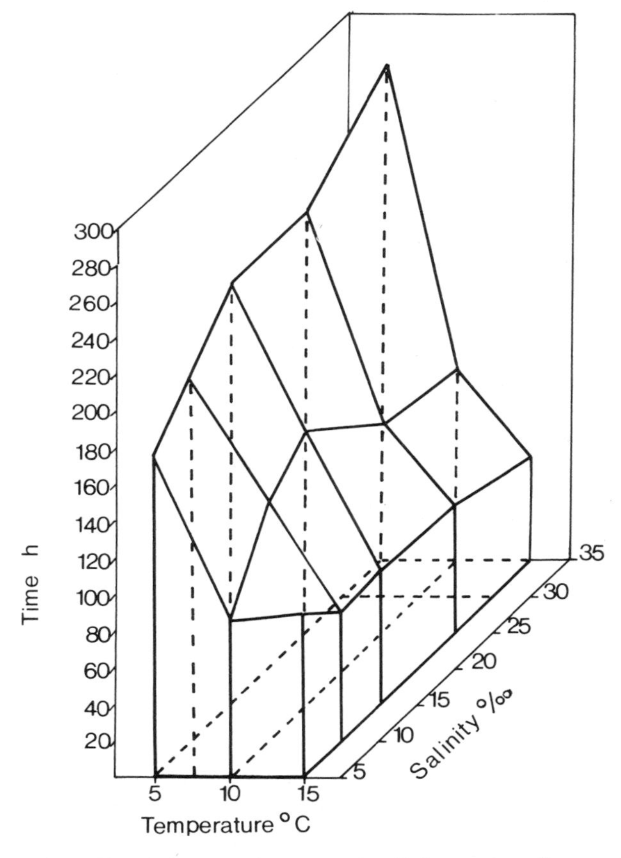

Fig. 1.—The effect of temperature and salinity on median periods of survival, in hours, of *Corophium volutator* at an arsenic concentration of 8 $mg \cdot l^{-1}$ (from Bryant, McLusky & Campbell, 1984).

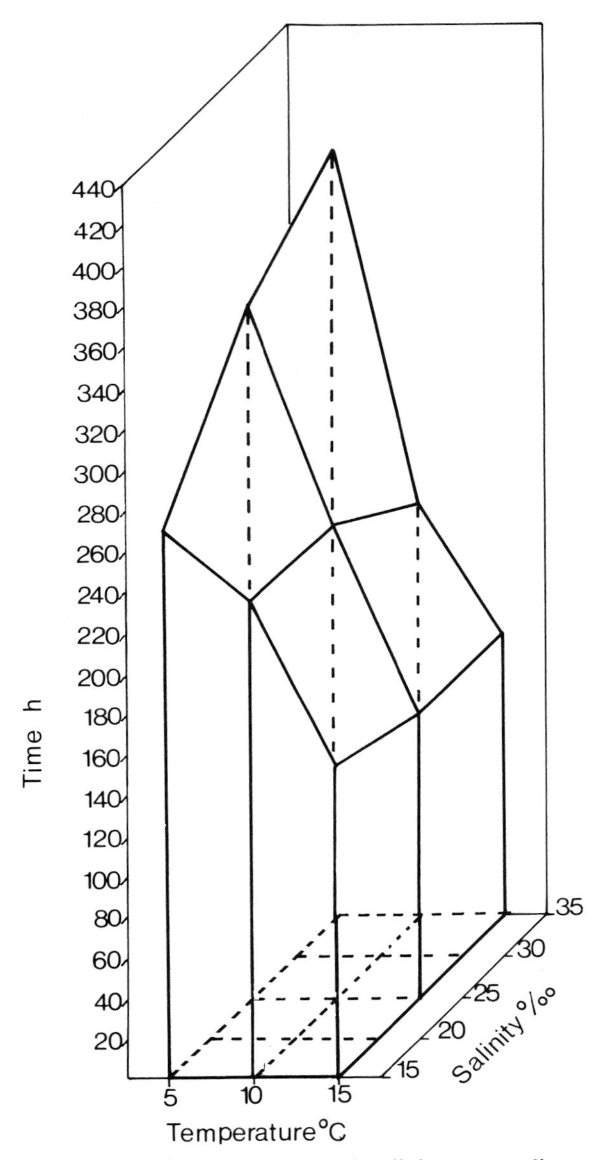

Fig. 2.—The effect of temperature and salinity on median periods of survival, in hours, of *Macoma balthica*, at an arsenic concentration of 30 mg·l^{-1} (from Bryant, McLusky & Campbell, 1984).

tions. Increased temperature enhanced both arsenic uptake and loss. Mussels in sea water at 19‰ salinity accumulated approximately three times more than those held at 38‰ salinity. Arsenic loss was much less affected by salinity, with only a tendency for greater arsenic retention noted at lower salinities.

Zaroogian & Hoffman (1982) studied arsenic uptake and loss at ambient sea-water salinity and temperature in the American oyster, *Crassostrea virginica*, with a view to using this organism as an indicator of arsenic pollution. They were, however, unable to establish a relationship between tissue arsenic concentration and sea-water arsenic concentration, and thus *Crassostrea* would be unsuitable for this purpose.

There are few studies on the acute toxicity of arsenic to marine invertebrates. Nelson *et al.* (1976) report a 96 h LC_{50} of 3·49 mg·l^{-1} arsenic as As^{III} to *Argopecton irradians* at 20 °C and 25‰ salinity and noted that arsenic was less toxic than mercury, silver or cadmium. Calabrese, Collier, Nelson & MacInnes (1973) reported a 48 h LC_{50} value of 7·5 mg·l^{-1} arsenic as As^{III} to embryos of *Crassostrea virginica* at 26 °C, 25‰ salinity. Curtis, Copeland & Ward (1979) report a 96 h LC_{50} of 24·7 mg·l^{-1} As^{III} for the prawn, *Penaeus setiferous*.

CADMIUM

GENERAL REVIEW

As a "black list" substance, cadmium has been extensively studied, and indeed Taylor (1981a) listed over 200 papers dealing with the acute and sub-lethal effects of cadmium on aquatic animals. Even that list, however, is not fully comprehensive in that bio-accumulation was not considered. Obviously, most of the reports covered by Taylor refer to exclusively marine or freshwater animals, and only a minority of the references are to estuarine invertebrates or temperature-salinity effects. In general, the 96 h LC_{50} values in Taylor are in the range 0·1–10 mg·l^{-1}, with most values between 1 and 10 mg·l^{-1}. Taylor (1981b) developed his earlier review further, with 250 references, including 126 freshwater species and 95 marine species. He concluded that, in general, the lethal effects of cadmium in its soluble ionic form appear at concentrations greater than 2 μg·l^{-1} in fresh waters and 100 μg·l^{-1} in marine waters. The corresponding concentrations for sub-lethal effects are 1 μg·l^{-1} and 7 μg·l^{-1}. For freshwater animals he concluded that 95% of all reported "effect concentrations" (*i.e.* 5 percentile value) are >1·3 μg·l^{-1}, also noting that freshwater algae were relatively insensitive, whilst fish were the most sensitive. Marine animals are considerably less sensitive to cadmium than freshwater animals, with the 5 percentile value for marine species being 95 μg·l^{-1}. The toxicity of cadmium in the aquatic environment is influenced by a wide range of factors, the most important being the chemical species present. Larval and embryonic stages are usually more sensitive than adults, and freshwater fish appear to be the most sensitive of all organisms yet studied. Increased salinity, or water hardness, and decreased temperature all tend to decrease the toxicity of cadmium. Taylor (1984) has subsequently argued that cadmium should no longer be regarded as a "black list" substance.

Theede (1980) has reviewed the responses of estuarine animals to cadmium pollution and shows that acute cadmium toxicity is strongly modified by abiotic factors, mainly temperature and salinity. In his own studies on marine hydrozoans he found polyps most tolerant to cadmium at low temperatures (7·5 °C) and high salinities (25‰), with maximum toxicity at 17·5 ° and 10‰ *S*. Taking his own results as well as those in the literature, he concluded that marine organisms living near their distributional limits in estuaries and brackish waters tend to suffer at pollution levels considerably lower than those living under optimal environmental conditions.

A number of both synergistic and antagonistic responses to environmental factors in association with cadmium have been reported. Engel & Fowler (1979) found that the toxicity of dissolved cadmium to a variety of marine organisms was related to salinity with decreased toxicity observed at high salinities. The toxicity of cadmium to shrimps and fish is reported to be a function of the free cadmium ion concentration, which is controlled by the chloride content of the water. As the chloride ion concentration (*i.e.* salinity) increases, so the concentration of free cadmium ion decreases relative to the total metal concentration, due to its complexation with chloride ions. The toxicity of cadmium to biota, and accumulation of the metal by biota, are both significantly affected by the ambient salinity (Phillips, 1980). In almost all the examples reported to date, a decrease in salinity levels leads to an increase in toxicity, or net uptake, of the metal. The observed salinity effects are not considered to be based on a simple physiological effect such as, for example, changes in the organism's filtration rate, rather the salinity effect must be based on a specific change in the metal uptake processes, or excretion. Phillips, following George & Coombs (1977), supports an interaction with calcium ion, such as competition between cadmium and calcium at uptake sites, as explaining much of the salinity dependence observed in studies of the toxic effects of cadmium.

Phillips (1980) noted that, in general, increased temperature increases the toxicity and/or accumulation of cadmium in marine biota. He favoured specific effects as the basis for the effects of temperature, operative at uptake sites, in a similar manner to salinity effects. In a general study of cadmium in food chains, Amiard-Triquet *et al.* (1983) found higher accumulation of metal in invertebrates than in their fish predators, and higher concentrations in estuarine than marine species.

The relationship between the chemical speciation and toxicity of cadmium to *Palaemonetes pugio*, and salinity (4–29‰), was determined by Sunda, Engel & Thiotte (1978). They clearly showed that shrimp mortality decreased with increasing salinity, and that the protective effect of high salinity was attributable to complexation of cadmium with chloride. The mortality is determined by the free cadmium in concentration; with 50% mortality at a free cadmium ion concentration of 4×10^{-7} M.

Calabrese, Thurberg & Gould (1977) reviewed their own series of earlier papers, and showed that early life stages of marine animals appeared to be more sensitive to mercury and silver than cadmium and that mercury and silver were taken up more readily than cadmium. In adults, cadmium was, however, found to be more toxic than mercury and silver despite the lower rate of cadmium uptake. Eisler & Hennekey (1977) studied six species of estuarine animals, *viz. Asterias forbesi* (starfish), *Nereis virens* (sandworm), *Pagurus longicarpus* (hermit crab), *Mya arenaria* (clam), *Nassarius obsoletus* (snail), and *Fundulus heteroclitus* (fish) with five metals, at one salinity (20‰) and one temperature (20 °C), and found a rank order of toxicity: Hg^{2+}>Cd^{2+}>Zn^{2+}>Cr^{6+}>Ni^{2+}. Cadmium toxicity decreased markedly between 96 h and 168 h, and it was clear that 168 h was insufficient time to assess the toxicological effects of cadmium hence, based on the available literature, they recommended up to 17 weeks of study were needed for subsequent work. Comparing their results with those of other workers shows that increasing toxicity was related both to decreasing salinity and increasing temperature.

The chemistry of cadmium in estuarine waters and its complexation with humic material has been described by Mantoura, Dickson & Riley (1978).

MOLLUSCS

Phillips (1980) reviewed the extensive literature on cadmium toxicity and accumulation in estuarine and marine biota, but had to conclude that although the effects of temperature and salinity on the uptake of cadmium by molluscs had been well studied, no data were available concerning the effect of temperature and salinity on cadmium toxicity in molluscs. This position has since been only partially remedied.

Axiak & Schembri (1982) studied the effect of temperature (15–30 °C) on the gastropod *Monodonta turbinata*, at one salinity, and found that mortality increased with increasing temperature. The mortality caused by increasing temperature and cadmium concentration together was greater than the sum of the mortalities caused by increasing the two stresses individually.

Lehnberg & Theede (1979) studied the combined effects of temperature. (10–30 °C) and salinity (5–40‰) and cadmium on the development, growth and mortality of *Mytilus edulis* larvae from the western Baltic Sea. The results were analysed by surface response techniques, and revealed that maximal survival (>90%) without cadmium was at 16 °C and 33‰. At a cadmium concentration of 150 $\mu g \cdot l^{-1}$, maximal survival was at 10 °C and 25‰, and the point of minimum mortality shifted with increasing concentration of cadmium. The mortality rate rose first in supraoptimal salinity, or suboptimal temperature, and thereafter in suboptimal salinity and supraoptimal temperature. It was clear from their results that cadmium disturbed the relationship between temperature, salinity and larval survival. Sunila (1981) studied the toxicity of cadmium to *Mytilus edulis* from the Gulf of Finland (*S* 6‰), and found that small individuals were more resistant than large ones. His LC_{50} of 4 $mg \cdot l^{-1}$ was higher than values in other studies made at higher salinities, which is the opposite of the conventional view that low salinity increases toxicity. He attributed this difference to genetic factors and adaptation to high metal levels in the mussel's natural habitat.

Eisler (1977a) subjected adults of *Mya arenaria* continously to a flowing sea-water mixture containing a mixture of salts of manganese, zinc, lead, nickel, copper, and cadmium (at 1$\mu g \cdot l^{-1}$) in a series of experiments in summer (16–22 °C) and winter (0–10 °C). He found that accumulations and death rates were accelerated at the higher temperatures.

Many papers on the uptake of cadmium by molluscs and crustaceans have been excluded from this review of toxicity. In general it should be noted that cadmium uptake is increased by a reduction in salinity, or an increase in temperature.

ANNELIDS

Phillips (1980) has shown that the 96 h LC_{50} values for annelids are generally similar to those for molluscs, being $1-10\ mg \cdot l^{-1}$, except for *Nereis diversicolor*, the commonest British estuarine annelid, which has a value of up to 100 $mg \cdot l^{-1}$ at 17‰ *S*. He was, however, unable to report any data on the effect of natural variables such as temperature and salinity on the toxicity of the metal.

Roed (1979) studied the effect of interacting salinity, cadmium and mercury on 48 h LC_{50} of *Dinophilus gyrociliatus*, and found a significant interaction for both 2 and 3 factor interactions. Cadmium and mercury together and separately had an increased effect as salinity decreased. Roed (1980) has subsequently studied the annelid *Ophryotrocha labronica*, cultured in the laboratory, with sub-lethal levels of cadmium and salinities of 20, 25 and 30‰. Low salinity levels, and the presence of cadmium resulted in reduced growth rates, prolonged times to reach sexual maturity, and reduced size at maturity. Cadmium and salinity were fatal in the combination of 0·4 $mg \cdot l^{-1}$ and 20‰ for 30 days, and 4 $mg \cdot l^{-1}$ at all salinities if a 48-h LC_{50} was measured instead.

Chapman, Farrell & Brinkhurst (1982a) investigated the tolerances of 12 oligochaete species to five pollutants (cadmium, mercury, pentachlorophenol, pulpmill effluent, and sewage sludge) and four environmental factors (pH, temperature, salinity, and anoxia) both with and without sediment. Salt-water oligochaetes were found to be among the most tolerant species to cadmium, and the presence of sediment resulted in increased tolerance for all species. Chapman, Farrell & Brinkhurst (1982b) also investigated the toxicity (as 96 h LC_{50} values) for five aquatic oligochaete species (three freshwater, two marine-estuarine), in relation to four pollutants (cadmium, mercury, pentachlorophenol, and "black liquor") with different combinations of pH, temperature and salinity. All oligochaetes tested were more tolerant to cadmium at lower than at higher temperatures (range 1–20 °C). All species, even the freshwater species, were more tolerant to cadmium under saline conditions (20‰) than in fresh water. These results show that increased salinity may protect against various environmental effects even in predominantly freshwater species. Increased tolerance to cadmium due to increased salinity was, however, too great to be explained merely in terms of physiological adjustments. Rather there is evidence that cadmium is rendered less toxic, perhaps by complexation with choloride, when salinity is increased.

CRUSTACEANS

Alone of the major groups of marine and estuarine invertebrates, there has been considerable study of the effect of temperature and salinity on the toxicity of cadmium to crustaceans. Crustaceans are generally regarded as being more sensitive to cadmium than other groups such as fish or molluscs (Phillips, 1980). Results for crustaceans are often complicated by problems due to the effects of moulting and cannibalism, and the exact effect of these factors is little understood.

O'Hara (1973) reported for the fiddler crab (*Uca pugilator*) that at any given time and salinity, increases in water temperature increased the cadmium toxicity; and at any given time and temperature, increases in salinity decreased the toxicity of cadmium. The crabs were most susceptible (240 h LC_{50}, 2·9 $mg \cdot l^{-1}$) at 30 °C and 10‰ , and least susceptible (240 h LC_{50}, 47·0 $mg \cdot l^{-}$) at 10 °C and 30‰ . Sullivan (1977) found similar effects for *Paragrapsus gairmardii*.

Vernberg, De Coursey & O'Hara (1974) kept zoeae of the fiddler crab *Uca pugilator* in 13 salinity-temperature combinations, with and without 1 μg $Cd \cdot l^{-1}$. Even at this low level, survival was affected by the cadmium, with salinity tolerance narrowed and temperature tolerance shifted towards lower temperatures. They found that adult crabs were less sensitive to cadmium poisoning than larvae, but there were no differences between males and females. Cadmium is most toxic at high temperatures and low salinities, and both the distribution and total body burden of cadmium is dependent upon environmental conditions. Rosenberg & Costlow (1976) studied the effect of salinity and cycling or constant temperature on the toxicity of cadmium to larvae of *Callinectes sapidus* (blue crab) and *Rhithropanopeus harrisii* (mud crab). 150 μg $Cd \cdot l^{-1}$ was lethal to both species at the lowest salinities (10‰) in which the control animals could live.

The influence of temperature, salinity (hypersaline), dissolved oxygen concentration and population density on the toxicity of cadmium to the copepod *Tisbe holothuriae* has been studied by Verriopoulos & Moraitou-Apostolopoulou (1981). The results show that the toxicity of cadmium was significantly affected by the dissolved oxygen concentration, the population density and the interaction between temperature and salinity. At 11 °C, survival was significantly affected by the elevation of salinity (from 33 to 46‰). At temperatures higher than 11 °C (up to 26 °C) the combined effects of temperature and cadmium cause a heavy stress to *Tisbe* so that the salinity effect is masked.

In an extensive study of six species of isopod, Jones (1975a) examined mortality at two temperatures (5 and 10 °C), two concentrations (10 and 20 $mg \cdot l^{-1}$), and seven salinities (0·34–34‰). Maximum mortality was at low salinities and high temperatures, although estuarine species were more resistant than marine species (Figs 3 and 4). Reviewing both his own and other workers' reports, Jones (1975b) stated that marine species living at their normal salinities are less susceptible to heavy metal pollution than are estuarine species living in salinities near the lower limit of their normal salinity range. Clearly, the action of heavy metals when combined with stressful conditions of salinity results in increased toxicity of metals and, therefore, represents a severe threat to estuarine species which are subjected daily to dilute sea water.

Denton & Burdon-Jones (1982) studied the influence of temperature and salinity on the acute toxicity of heavy metals to the banana prawn, *Penaeus merguiensis*. The toxicity of all metals, including cadmium, increased with increasing temperature. Cadmium appeared to be more toxic at lower salinities but the differences were not statistically significant.

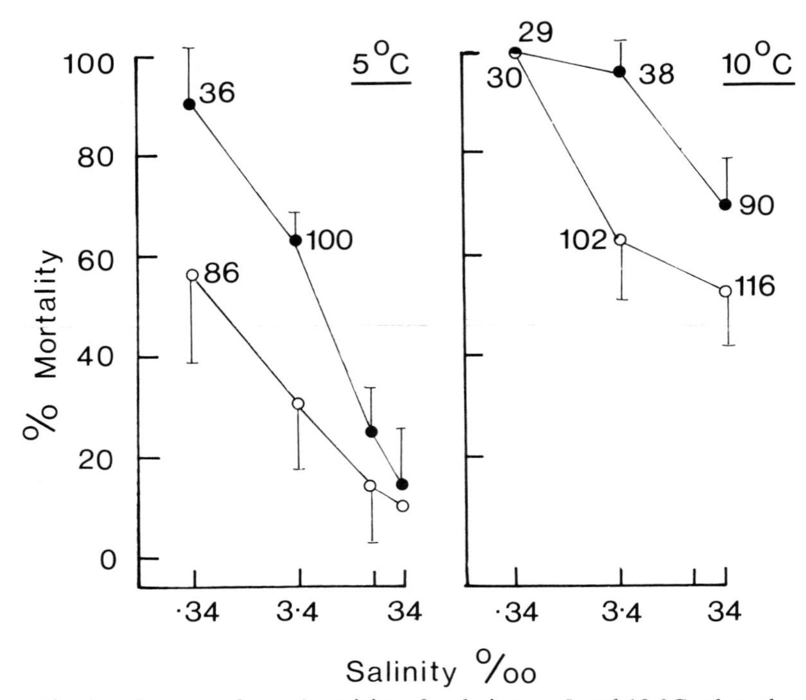

Fig. 3.—*Jaera nordmanni*: toxicity of cadmium at 5 and 10 °C, plotted as mean per cent mortality (after 120 h) of three experiments, together with standard deviations; lack of bar lines indicates zero deviation; LT_{50} times (in h) are also shown; ○,10 $mg \cdot l^{-1}$ cadmium; ●,20 $mg \cdot l^{-1}$ cadmium; after Jones, 1975a.

In contrast to most other studies, Bahner & Nimmo (1975) found no apparent toxic effect of lowered salinity after exposures of estuarine the shrimp, *Penaeus duorarum*, to various cadmium-pesticide mixtures or components. Ahsanullah *et al.* (1981) concluded that 14 days was insufficient time to assess the lethal effect of cadmium on the shrimp, *Callianassa australiensis*, and later Negilski, Ahsanullah & Mobley (1981) showed that when paired with zinc or copper, cadmium showed an interactive response. The blue crab, *Callinectes sapidus*, shows increased mortality to cadmium when in combination with low salinity, over the range 1–35‰ (Frank & Robertson, 1979).

Sub-lethal effects of cadmium have also been well studied in crustaceans (Phillips, 1980), and have been shown to interfere with osmoregulation, oxygen consumption, gill structure, moulting, limb regeneration, and larval development. Weis (1978) showed that cadmium affects limb regeneration in fiddler crabs in normal sea water, and that the effects are greatly intensified in lower salinities, so that the growth of limb buds is extremely slow, if it occurs at all. The synergistic effects of cadmium and salinity on larval development of estuarine crabs were demonstrated by Rosenberg & Costlow (1976) who also showed that fluctuating temperatures had a stimulating effect on survival compared with a stable temperature.

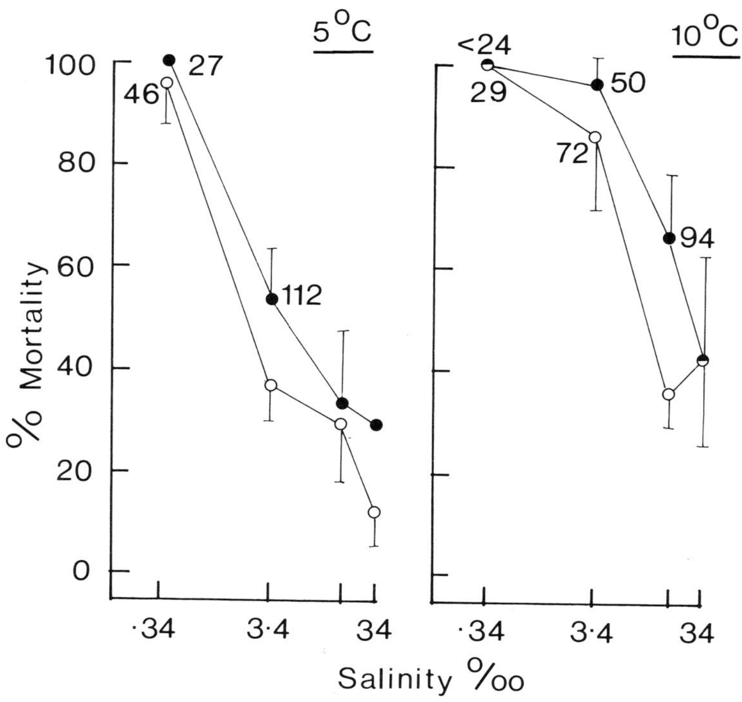

Fig. 4.—*Jaera albifrons*: toxicity of cadmium at 5 and 10 °C, plotted as mean per cent mortality (after 120 h) of three experiments, together with standard deviations; lack of bar lines indicates zero deviation; LT_{50} times (in h) are also shown; ○,10 mg·l⁻¹ cadmium; ●,20 mg·l⁻¹ cadmium; after Jones, 1975a.

The effect of exposure of developing *Palaemonetes pugio* embryos to cadmium on subsequent larval sensitivity to combined cadmium and salinity was investigated by Middaugh & Floyd (1978), but found to have no additive effect on the sensitivity of the larvae to cadmium exposure and salinity stress for 14 days after hatching. Davis *et al.* (1981) conclude that the dominant uptake route of cadmium to crabs is through their diet, and in several studies not reviewed here it has been shown that cadmium uptake is increased in elevated temperatures and decreased salinities (*e.g.* Hutcheson, 1972; Wright, 1977).

Cadmium has been found to cause osmotic elevation of crab serum, above its normal hyperosmotic state and to reduce the rate of oxygen consumption in *Carcinus maenas* and *Cancer irroratus* at salinities from 17–32‰ (Thurberg, Dawson & Collier, 1973).

Thorp, Giesy & Wineriter (1979) subjected crayfish, *Cambarus latimarius*, to low levels of cadmium in a continuous flow system, over five months at ambient temperatures. At the end of the 5-month period they were subjected to daily elevations of temperature. They found that mortality increased with higher cadmium concentrations (up to 10 $\mu g \cdot l^{-1}$), but neither growth nor thermal tolerance to elevated temperatures was related to cadmium concentration in the tissues. Examining the toxicity of cadmium to the estuarine crab, *Varuna littorata*, Kulkarni (1983) showed that mortality was greater at high temperatures, but that high temperature and low salinity together made cadmium even more toxic.

OTHER GROUPS

The acute toxicity of cadmium to the hydrozoan, *Laomedea loveni*, is strongly modified by abiotic factors (Theede, Scholz & Fischer, 1979). At low temperature (7·5 °C) and high salinity (25‰) it is more tolerant to cadmium contamination than at high temperature (17·5 °C) and low salinities (10‰) (Fig. 5).

No other reports of the effects of temperature and salinity on cadmium toxicity for other groups have been traced.

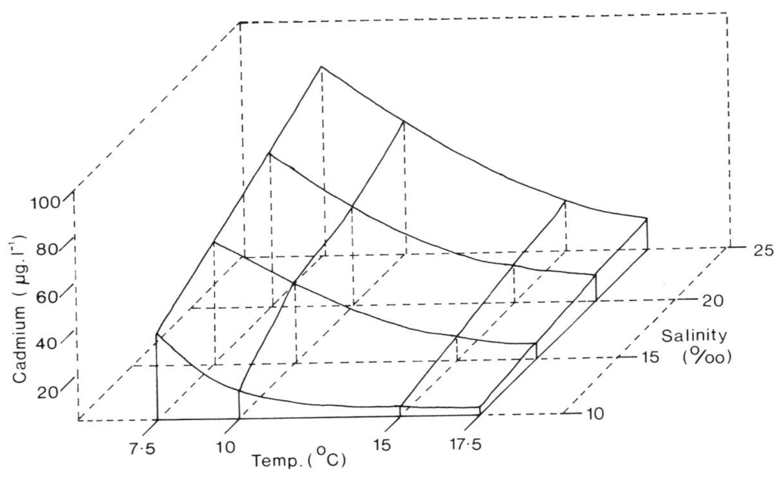

Fig. 5.—*Laomedea loveni*: 50% retraction of polyps due to sea water contaminated with cadmium at different combinations of temperature and salinity; time of exposure, 7 days; after Theede *et al.*, 1979.

CHROMIUM

The limited data available on the acute toxicity of chromium to vertebrates and invertebrates in both marine and freshwater environments have been summarized by Taylor (1981c) and Mance (1984). Apart from one paper by the present authors, data on the effects of variations in salinity and temperature which are comparable with those encountered in estuaries are, however, not available.

Bryant, McLusky, Roddie & Newbery (1984) studied the acute toxicity of hexavalent chromium (as potassium dichromate) to three estuarine animals (*Corophium volutator*, *Macoma balthica*, and *Nereis diversicolor*) at three temperatures (5, 10, 15 °C) and a range of salinities (5 to 40‰ in 5‰ increments) at time intervals of up to 384 h. They found that for all species toxicity increased as temperature increased and as salinity decreased. Typical results are shown in Figures 6, 7, and 8. Significant interactions between temperature, salinity, and chromium concentration were found from an analysis of variance. For *Corophium*, the combined interactions of chromium concentration × salinity, temperature × salinity, and chromium concentration × temperature all significantly decreased median survival times. For *Macoma* only the interactive effect of salinity × chromium concentration significantly decreased median survival times. Frank & Robertson (1979) reported on the effect of chromium on juvenile blue crabs, *Callinectes sapidus*, at three salinities and one temperature, and

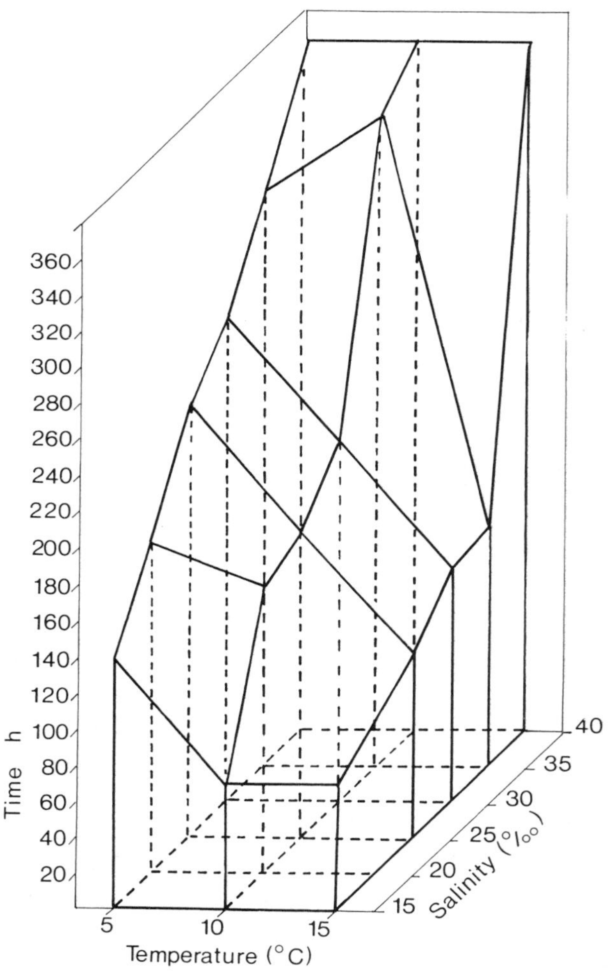

Fig. 6.—The effect of temperature and salinity on median survival time (h) of *Macoma balthica* at a chromium concentration of 64 mg·l^{-1} (after Bryant, McLusky, Roddie & Newbery, 1984).

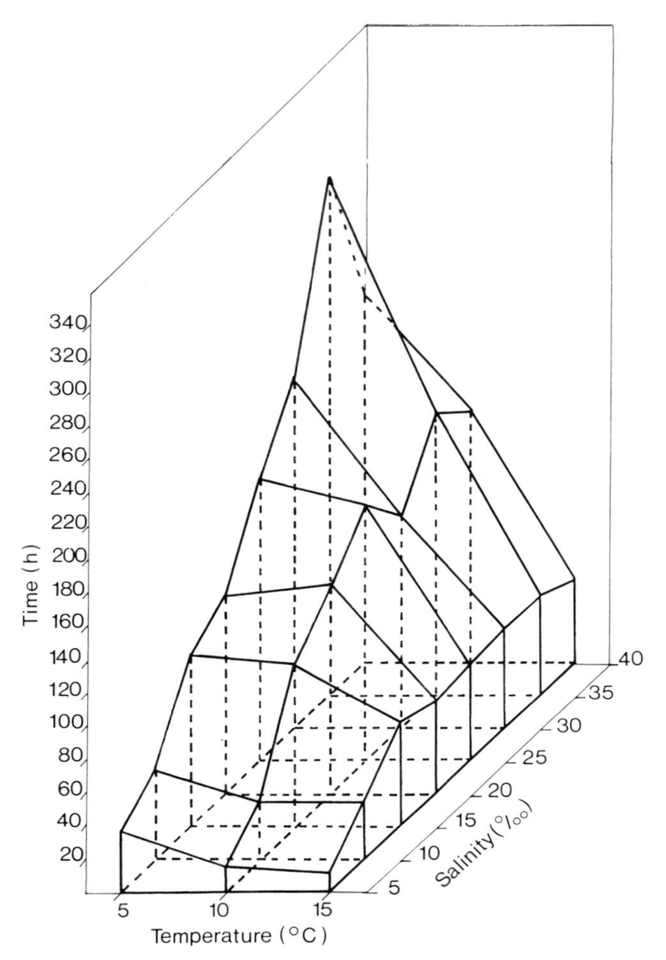

Fig. 7.—The effect of temperature and salinity on median survival time (h) of *Corophium volutator* at a chromium concentration of 16 mg·l⁻¹ (after Bryant, McLusky, Roddie & Newbery, 1984).

found that toxicity decreased with increasing salinity (48 h LC_{50}:34 $mg \cdot l^{-1}$ at 1‰ to 98 $mg \cdot l^{-1}$ at 35‰). Eisler & Hennekey (1977) compared the toxicity of several estuarine invertebrates at one temperature and salinity and found the rank order of toxicity to be: $Hg^{2+} > Cd^{2+} > Zn^{2+} > Cr^{6+} > Ni^{2+}$ with the taxonomic ranking of toxicity to chromium being annelids>crustaceans>molluscs.

It should be emphasized that the range of toxicity values for a single estuarine species exposed to different salinity and temperature combinations by Bryant, McLusky, Roddie & Newbery (1984) (*Nereis* 0·7–80 $mg \cdot l^{-1}$; *Corophium* 3–120 $mg \cdot l^{-1}$; *Macoma* 6–640 $mg \cdot l^{-1}$) all exceed the total range of toxicity values for all species in marine and freshwater conditions (2–100 $mg \cdot l^{-1}$) reported by Taylor (1981c) and Mance (1984). The fact that the range of toxicity values for chromium for a single species under different temperature and salinity conditions can

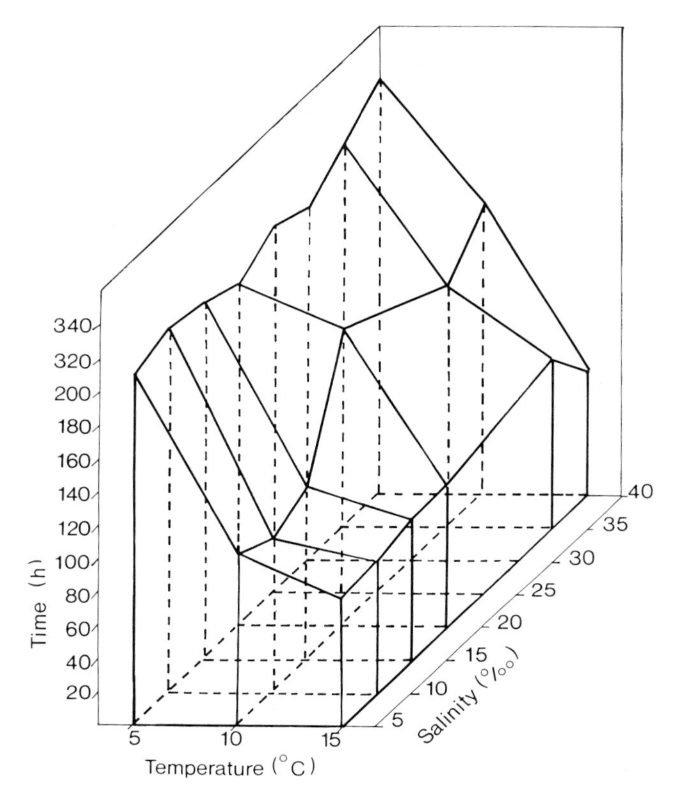

Fig. 8.—The effect of temperature and salinity on median survival time (h) of *Nereis diversicolor* at a chromium concentration of 16 $mg \cdot l^{-1}$ (after Bryant, McLusky, Roddie & Newbery, 1984).

exceed the reported range for several species emphasizes that single values for the toxicity of chromium, and perhaps other metals, should be treated with extreme caution.

COPPER

GENERAL REVIEW

Copper toxicity has been shown to be influenced by both temperature and salinity; the shrimp, *Pandalus danae*, was found to be more sensitive to copper at 20 °C than at 10 °C (Gibson, Thatcher & Apts, 1976), while groups of soft-shell clams exhibited greater reduced survival during immersion in copper salts at summer temperatures (22 °C) than at autumn or winter temperatures (17 and 4 °C).

It is possible that increased toxicity at reduced salinities may be linked with osmoregulatory impairment (Jones, 1975b). Thurberg *et al.* (1973) reported that a characteristic feature of copper is the depression of the blood osmotic pressure in decapod crustaceans; this has also been reported for fish (Lewis & Lewis, 1971). Increased toxicity at reduced salinities may also be related to changes in chemical speciation (Sunda & Guillard, 1976) and competitive interactions with major cations for sensitive sites (Von Westernhagen, Rosenthal & Sperling, 1974). According to Fletcher (1970), in the case of estuarine organisms, when the negative potential difference of the inner body wall increases with decreased salinity, ion transport into organisms consequently increases. In addition, in the low salinity or freshwater environment there is no competition between calcium and magnesium as cation exchangers. Thus, copper may replace either calcium or magnesium in ion transport at low salinities. The case for a synergistic effect of salinity and temperature on copper toxicity has been proposed by some authors, *e.g.* Jones (1975b) who suggests that concentrations of copper which are sub-lethal at optimum conditions become increasingly toxic as environmental stress is increased.

MOLLUSCS

Davenport (1977) studied the survival and behaviour of *Mytilus edulis* exposed to continuous and discontinuous copper combined with fluctuating salinity. He found that while continuous copper caused mortalities within 1–2 days, discontinuous copper in a 6 h on: 6 h off cycle caused no damage after 5 days. This was due to the ability of *Mytilus* to detect copper in the environment and close its shell valves. In conditions of fluctuating salinity and copper delivery, survival at low salinities was enhanced due to the interaction of closure responses to copper and to low salinities. In real, polluted estuarine conditions it is often the case that the pollutant is borne by fresh water, and hence both influences are closely linked. The shell closure response to either pollutants or fresh water will give some measure of protection. Sunila (1981) investigated the toxicity of copper to *Mytilus edulis* at a salinity of 6‰ and found small individuals to be more sensitive to copper than large ones. Copper induced valve-closing in 3 min, then mortality increased after a lag of 5–6 days. An LC_{50} at 6 °C of 400 $\mu g \cdot l^{-1}$ was found for copper, higher than most recorded values for higher salinities. He suggested that this may be due to genetic factors and adaptation of the trial animals to high levels of metal in the natural environment.

MacInnes & Calabrese (1977) studied the acute toxicity of a range of metals to *Crassostrea virginica* at three temperatures. They found the order of metal toxicity to be Hg>Cu>Ag>Zn. Copper toxicity was not significantly influenced by temperature.

Coglianese (1982) monitored embryonic development of *Crassostrea gigas* under combinations of copper, silver, and salinity. Between 22·7 and 33‰ salinity with 0–10 $\mu g \cdot l^{-1}$ copper and 0–16 $\mu g \cdot l^{-1}$ silver administered there was no effect on development. Below 22·7‰ salinity the effects of copper were, however, highly deleterious. Response surface analysis indicated a significant interaction between salinity and metal concentration in causing embryo mortality. A shift from salinity to a more pronounced concentration effect occurred as the copper concentration was increased. Optimum conditions occurred at 23–33‰ salinity with less than 2 $\mu g \cdot l^{-1}$

copper. Thus, low salinities may represent a significant threat to oyster embryos. Low salinities appear to act synergistically with low metal concentration causing a reduction in normal embryonic development, so while a low level of metal alone may not significantly affect normal development, alteration in salinity could intensify the toxicity of low metal levels through increased metal ion uptake at low salinities.

MacInnes & Calabrese (1979) looked at the combined effects of salinity, temperature, and copper on two life stages of the American oyster, *Crassostrea virginica*. Embryos and larvae were exposed to copper at 20, 25 and 30 °C, and 17·5, 22·5 and 27·5 ‰ salinity. For embryos a range of 0–20 $\mu g \cdot l^{-1}$ copper was used, and for larvae 0–90 $\mu g \cdot l^{-1}$. The greatest salinity effect on embryos was at 0, 5 and 10 $\mu g \cdot l^{-1}$ copper, but at the highest copper concentration the temperature effect was equivalent to that of salinity. The apparent loss of ability of the embryos to adapt was confirmed by a shifting response centre. For the embryos the greatest temperature effects were at 30, 60 and 90 $\mu g \cdot l^{-1}$ copper with a significant temperature-salinity interaction only occurring at higher levels of copper. Veliger larvae were considerably more tolerant to temperature and salinity changes than were developing embryos, thus low levels of copper may stress embryos during periods of persistently low salinity and high temperature, possibly producing intolerable stress upon recruitment of oyster embryos.

Eisler (1977b) investigated the effects of copper and temperature on the soft-shell clam, *Mya arenaria*. Soft-shell clams at 30‰ salinity were found to be more resistant at lower temperatures displaying increasing survival with decreasing temperature. For copper 336 h LC_{50} values at 17 and 4 °C were 0·086 and above 3·0 $mg \cdot l^{-1}$, respectively. He also found that copper was accumulated more rapidly by *Mya* at summer than at winter temperatures.

ANNELIDS

Jones, Jones & Radlett (1976) investigated copper toxicity to *Nereis diversicolor* from two sites in the Humber estuary, one of low, the other of high salinity. The worms were exposed to three copper concentrations up to 0·8 $mg \cdot l^{-1}$ and 5 to 34‰ salinity. 96 h LC_{50} values were calculated, as were tissue concentrations of copper after death. Worms from the high salinity area appeared to be more tolerant to high salinity plus copper than worms from the low salinity area, and also less tolerant of low salinity plus copper (Fig. 9). Copper toxicity was greatest for all worms at 5‰ salinity. Worms kept at intermediate salinities accumulated greater amounts of copper yet survived for a longer period, suggesting that the final copper content of the body is not a critical factor in causing mortality. McLusky & Phillips (1975) suggested that the rate of uptake of copper may be the lethal factor for *Phyllodose maculata* rather than the amount of copper accumulated.

CRUSTACEANS

Olson & Harrel (1973) determined "median tolerance limits" for copper for *Rangia cuneata* from the Neches River of southeastern Texas. Three salini-

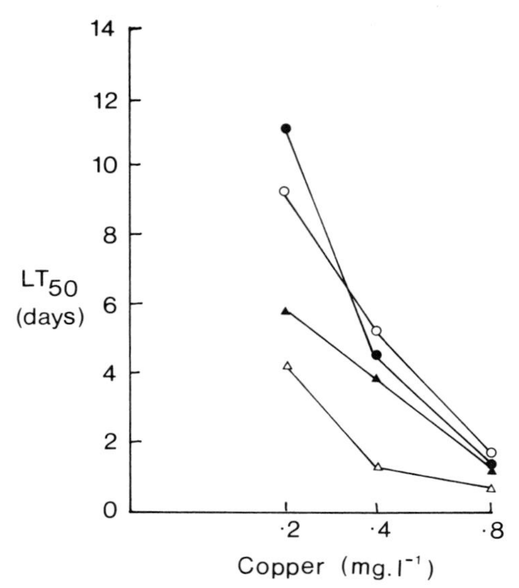

Fig. 9.—The effect of salinity on the LT_{50} (in days) of *Nereis diversicolor* exposed to 0·2, 0·4 and 0·8 mg·l^{-1} of copper at salinities of 34‰ (▲), 17·5‰ (○), 10‰ (●), and 5‰ (△): controls survived for more than 24 days; after Jones *et al.*, 1976.

ties were used: less than 1, 5·5 and 22‰ to represent the normal range of the river. Copper was found to be most toxic in fresh water, with less than 1 mg·l^{-1} required for a 48 h LT_{50}, whereas more than 10 mg·l^{-1} was required for the same effect at 22‰.

Jones (1975a,b) investigated the effect of copper on the survival and osmoregulation of several marine and brackish water isopods. He found that 10 mg·l^{-1} was highly toxic at all trial salinities (100 to 1% sea water) and that LT_{50} values decreased with decreasing salinity *i.e.* copper toxicity increased. 1 mg·l^{-1} copper caused a significant lowering of haemolymph osmotic pressure, particularly in dilute salt solutions. Jones suggests that copper acts synergistically with salinity such that concentrations of copper sub-lethal at optimal concentrations become increasingly toxic as environmental stress increases (Fig. 10).

Thurberg, Dawson & Collier (1973) investigated the effects of copper on osmoregulation and oxygen consumption of two crabs, *Carcinus maenas* and *Cancer irroratus*, at five salinities. Under normal circumstances these species regulate blood solutes so that they are hyperosmotic to the surrounding medium at lower salinities, so a change in the normal habitat caused by heavy metal pollution may induce alteration of osmoregulation. This was in fact the case. Crabs exposed to a variety of copper concentrations from 2·5 to 40 mg·l^{-1} at a range of five salinities exhibited a loss of osmoregulatory function. The greatest copper-induced disruption was at lower salinities, and mortality occurred above 5 mg·l^{-1} copper in *Cancer irroratus*, and above 40 mg·l^{-1} in *Carcinus maenas*.

Lang, Forward, Miller & Marcy (1980) examined behavioural effects (*i.e.* swimming speed and phototaxis) in nauplii of the barnacle *Balanus*

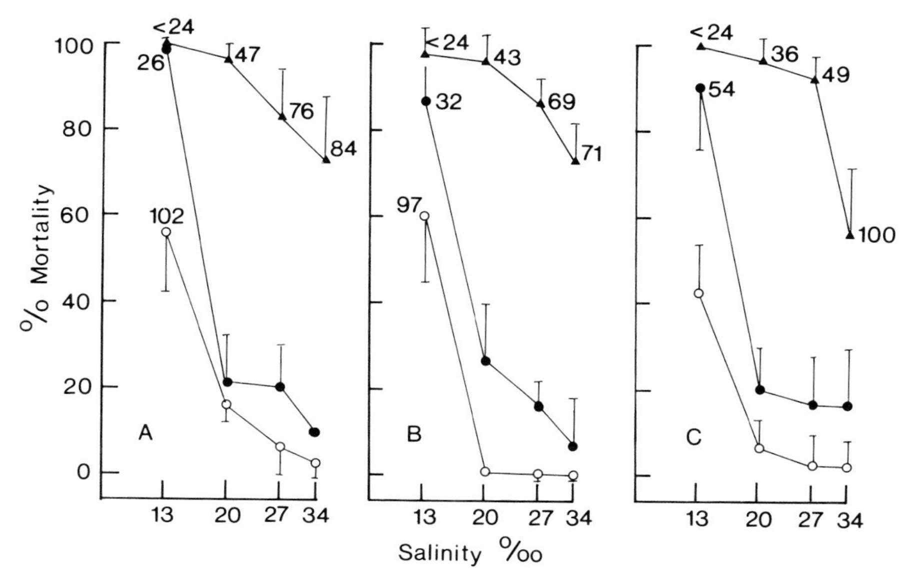

Fig. 10.—Cumulative per cent mortality (after 120 h) in salinities of 13–34‰ for *Idotea emarginata* (A), *I. neglecta* (B), and *I. baltica* (C) plotted as the mean of three experiments, together with the standard deviations: LT_{50} values (in h) also given; ○, control; ●, 1 $mg \cdot l^{-1}$ copper; ▲, 10 $mg \cdot l^{-1}$ copper; after Jones, 1975b.

improvisus, following sub-lethal copper exposure and 24 h LC_{50}. The 24 h LC_{50} at 15‰ salinity was 88 $\mu g \cdot l^{-1}$ while for 30‰ it was greater than 200 $\mu g \cdot l^{-1}$. A similar pattern of behavioural effects occurred with disruption to swimming speed and photo-behaviour caused by adding 150 $\mu g \cdot l^{-1}$ copper at 30‰ equivalent to that caused by adding 50 $\mu g \cdot l^{-1}$ copper at 15‰.

McCleese (1974) studied the toxicity of copper to the American lobster, *Homarus americanus* at two temperatures and three salinities. LT_{50} was determined at 5 and 13 °C, at 20, 25 and 30‰ salinity. At a specific concentration the LT_{50} was longer at 5 than at 13 °C, but was not affected by differences in salinity. For lobsters exposed to a particular lethal concentration of copper McCleese found that the LT_{50} increased by a factor of about two with a temperature change from 13 to 5 °C.

Denton & Burdon-Jones (1981) investigated the effects of temperature and salinity on acute toxicity of metals including copper to juvenile banana prawns, *Penaeus merguiensis*. These animals experience dramatic changes in temperature and salinity during the summer monsoon season and it was during this time of year that animals were collected. Bioassays were conducted at 35, 30, and 20 °C, at 36‰ and 20‰ over 96 h. Metal toxicity was in the order Hg>(Cu, Cd, Zn)>Ni>Pb. Copper toxicity was found to increase with increased temperature, most notably in high salinity. The effect of salinity was not as marked as the effect of temperature and, in fact, was only significant at 20 °C. At this temperature the animals were markedly more susceptible to copper at low salinities.

LEAD

GENERAL REVIEW

Lead as a pollutant has assumed particular importance due to its relative toxicity and increased environmental contamination via car exhaust and highway run-off. Effects of lead in the estuarine environment with the combined effects of fluctuating salinity and temperature, however, have not been widely studied and relevant literature is scarce. This may be partly due to certain experimental difficulties with lead such as the gross precipitation which occurs in solutions which led Denton & Burdon-Jones (1982) to suggest that LC_{50} values should be based on soluble concentration alone as opposed to total lead, although the gill clogging action of the precipitate may affect gaseous exchange.

Freedman, Cunningham, Schindler & Zimmerman (1980) looked at the effect of lead speciation on toxicity to the amphipod, *Hyallela azteca*. Using different lead and phosphate concentrations at a number of pH levels they found the highest mortality rates to be associated with the highest free lead concentrations. Gray (1974) found that for a marine ciliate population under optimum conditions lead at 0·3 $mg \cdot l^{-1}$ reduced the growth rate by 11·7%. Anderson (1978) found that increased lead concentration caused a decrease in the ability of the crayfish, *Orconectes virilis*, to take up oxygen, although it was able to acclimate in the short term by increasing its ventilation volume thereby compensating for decreased gill efficiency.

MOLLUSCS

Hrs-Brenko, Claus & Bubic (1977) investigated the effects of lead, salinity, and temperature on *Mytilus edulis* embryos using a $4 \times 6 \times 3$ factorial experimental design, with 100 to 1000 $\mu g \cdot l^{-1}$ Pb^{2+} and 25 to 37·5‰ salinity at 15, 17·5 and 20 °C. They found that salinity had more effect on development than temperature. Optimum conditions existed at 34·8‰ and 15·6 °C, under which conditions lead had a minimal effect. The deleterious effect of lead was greatest at maximum temperature, lead causing a delay or inhibition of development with a large number of abnormal larvae. The results were subjected to regression analysis, which showed that development was most affected by salinity changes, less so by temperature. At all temperatures the effect of lead was minimal in optimum salinity, but lead concentration was more determinant for the developmental rate than salinity. An increase in temperature from 15 to 20 °C had a pronounced effect on development. At 20 °C the development reached was below 80% at all lead concentrations but at 17·5 °C 80% was observed in concentrations below 100 $\mu g \cdot l^{-1}$and at 15 °C 80% was observed below 200 $\mu g \cdot l^{-1}$. Thus, the species prefers lower temperatures in its early stages, and is also more sensitive to salinity changes. These narrow ecological preferences interfere with the toxic effect of lead ions on mussel larvae. The effect of lead is minimal under optimum conditions, and greater than that of salinity at all temperatures. So the effects of moderate pollution (up to 250 $\mu g \cdot l^{-1}$) on the life cycle of the mussel in the zone of 35–37‰ salinity are relatively small. In the shore zone springs of fresh water, however, lower the normal salinity, increasing the possible deleterious effects of lead.

Phillips (1976) investigated the effects of environmental variables on lead uptake in *Mytilus edulis*. Environmental variables included position in the water column, season, salinity, and temperature. Lead levels in the mussel varied with the depth of the mussel and distance from the source of the pollutant. Mussels were exposed to a range of conditions: 15 and 35‰, each at 10 and 18 °C for 21 days. Low salinity decreased the lead uptake while low temperature had no effect.

CRUSTACEANS

Jones (1975a) studied the effect of lead on the mortality of marine and estuarine isopod species under varying conditions of salinity and temperature. The estuarine species were exposed to 100, 50 and 10 and 1% sea water combined with 10 or 20 mg·l^{-1} lead at 5 and 10 °C, while marine species were exposed to the same lead concentrations and temperatures with 100, 80, 60 and 40% sea water. No difference was found in mortalities after

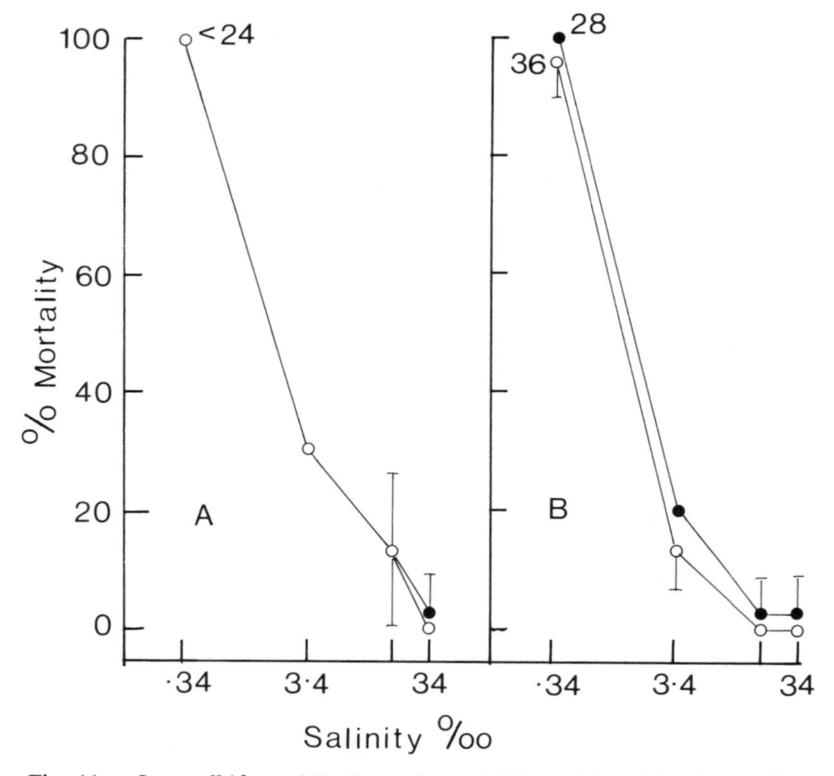

Fig. 11.—*Jaera albifrons* (A), *J. nordmanni* (B): toxicity of lead at 5 °C, plotted as mean per cent mortality (after 120 h) of three experiments, together with standard deviations: lack of bar lines indicates zero deviation; LT_{50} times (in h) are also shown; ○,10 mg·l^{-1} lead; ●,20 mg·l^{-1} lead; after Jones, 1975a.

120 h with 10 and 20 $mg \cdot l^{-1}$ lead. While there was less than 40% mortality in 100, 50 and 10% sea water, 100% occurred in 1% sea water (Fig. 11). Lead is, therefore, toxic and its toxicity is enhanced by stressful salinity conditions. Temperature had no significant effect.

Denton & Burdon-Jones (1982) studied the effects of temperature and salinity on the acute toxicity of lead and six other metals of juvenile banana prawns, *Penaeus merguiensis*. In this experiment a tropical species was chosen due to typical low salinity and high temperature combinations in the habitat, and juvenile banana prawns were specifically chosen as they are subjected to dramatic changes in salinity during the summer months when they are flushed from rivers and creeks into the coastal belt. Tests were conducted at 35, 30 and 20 °C at salinities of 36 and 20‰ over 96 h. Lead was the least toxic metal of the seven. Lead toxicity increased with increasing temperature, most noticeably in high salinity. Salinity only significantly affected metal toxicity at 20 °C; at this temperature prawns were markedly more susceptible to metals in low salinity sea water. The maximum 96 h LC_{50} recorded was 195 $mg \cdot l^{-1}$ at 20 °C and 36‰ while the minimum was 30 $mg \cdot l^{-1}$ at 35 °C and 20‰.

MERCURY

GENERAL REVIEW

Mercury is considered a non-essential but highly toxic element for living organisms, consequently mercury and its compounds are included in the "black list" of all international conventions. The literature on the sub-lethal and acute effects of inorganic and organic mercury is extensive and has been reviewed by Taylor (1979), however, most of the reports refer to marine and freshwater species and only a minority refer to estuarine species. Where direct comparisons were possible organic mercury compounds were found to be substantially more toxic than inorganic compounds.

Although research has shown that the toxicity of heavy metals generally increases with decreasing salinity and increasing temperature, this may not always be the case with mercury; changes in temperature and salinity from the optimum conditions required by a particular animal are, however, likely to lead to an increase in sensitivity to mercury. The chemistry of mercury in estuarine waters and its complexation with humic material has been described by Mantoura, Dickson & Riley (1978).

MOLLUSCS

MacInnes & Calabrese (1977) studied the acute toxicity of copper, mercury, silver, and zinc each singly and in combination to embryos of the American oyster, *Crassostrea virginica*, in natural sea water at 20, 25 and 30 °C. Mercury, whether added individually or in combination with other metals, was less toxic at 25 °C than at either 20 or 30 °C, although the differences were not significant. Less than additive effects were observed at 20 and 25 °C with mercury and silver in combination. Simple additive effects were noted at 30 °C for the mercury-silver mixture.

The toxicity of five heavy metals including mercury to the larvae of *Crassostrea virginica* and the hard clam, *Mercenaria mercenaria*, was reported by Calabrese, MacInnes, Nelson & Miller (1977). Mercury was the most toxic metal tested. Comparison with the work of Calabrese *et al.* (1973) and Calabrese & Nelson (1974) indicated that the larvae were less sensitive to mercury than the embryos of these species.

Axiak & Schembri (1982) studying the effects of temperature on the toxicity of mercury to the littoral gastropod, *Monodonta turbinata*, found that survival was significantly affected by mercury concentration and temperature (Fig. 12). Mortality caused by increasing the temperature together with mercury concentration was greater than the sum of the mortalities caused by increasing the two stresses individually. Thus, temperature acted synergistically with mercury concentration to decrease survival, this effect being more pronounced at high temperatures. It would, thus, be expected that a given level of heavy metal pollution would exert the greatest effect on this species during summer months.

Nelson, Calabrese & MacInnes (1977) undertook a study to determine whether salinity and temperature stress enhanced mercury toxicity to juvenile bay scallops, *Argopecten irradians* (Fig. 13). A $3 \times 3 \times 3$ factorial experimental design was used with three salinities, three mercury concentrations, and three temperatures. The single most important variable affecting scallop survival was mercury concentration; salinity was the next most important. The interaction between temperature and salinity and between mercury and temperature also significantly affected survival. Low salinity alone enhanced the toxicity of low concentrations of mercury, as did high temperature. High temperature and low salinity acted synergistically with

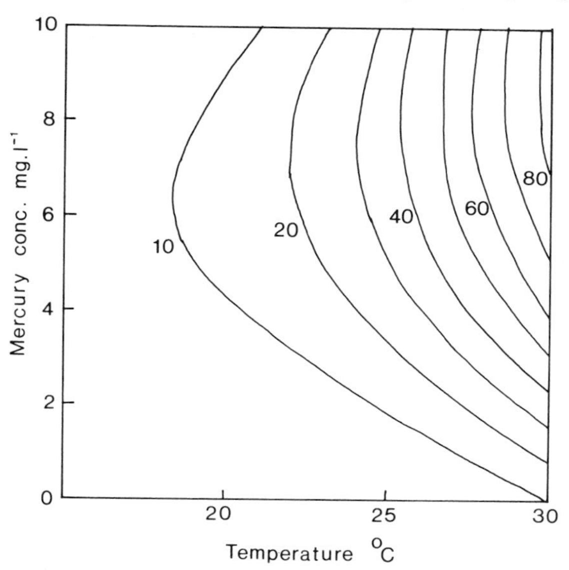

Fig. 12.—The response surface representing the effect of concentration of mercury and temperature on the mortality of *Monodonta turbinata* after 24 h exposure: contour lines drawn at every 10% mortality; after Axiak & Schembri, 1982.

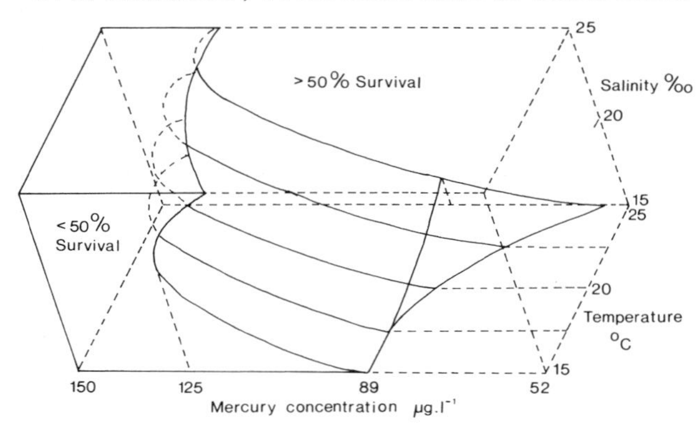

Fig. 13.—*Argopecten irradians*: polygon showing LT_{50} values for juveniles exposed for 96 h to mercury at various salinity and temperature combinations; after Nelson *et al.*, 1977.

mercury to increase mortality. At the highest mercury concentrations, mercury masked the effects of salinity and temperature and was the only significant cause of mortality.

PROTOZOANS

The only study of the effect of mercury toxicity on protozoans found in the literature is that of Gray (1974) who studied the interacting effects of mercury, zinc, and lead on the growth rates of a marine ciliate protozoan *Cristigera* sp. After establishing the effect of temperature and salinity on growth rate, he examined the effect of mercury ions on the exponential growth phase. Using response surface methodology, the optimum conditions for growth were found to be at 27·6‰ and 15·94 °C, however further experiments on the effects of mercury were done at 34·4‰ and 16 °C. Under these conditions concentrations of mercury as low as 2·5 $\mu g \cdot l^{-1}$ reduced growth rate by 9·5% and 5 $\mu g \cdot l^{-1}$ reduced it by 12·8%. He found that all three metals combined acted in a more than additive way, *i.e.* synergistically, on growth rate reduction at some metal levels, whereas at some concentrations interactions were antagonistic, *i.e.* less than additive. In general, interactions were antagonistic at low concentrations whereas at higher concentrations interactions were synergistic.

ANNELIDS

Roed (1979) used multivariate analysis to study the effects of interacting salinity, cadmium, and mercury on the growth of an archiannelid, *Dinophilus gyrociliatus*. A 48 h LC_{50} of 0·2 $mg \cdot l^{-1}$ mercury at 20 °C and 30‰ *S* was established and the long-term effects of 5% and 10% of this value together with varying salinity on population growth of *D. gyrociliatus* were investigated. Both population growth rate and time for first generation to start reproduction were studied. Significant effects were found for both 2 and 3 factor interactions. Using response surface methods cadmium and mercury in combination were shown to have an increased effect as salinity decreased and the two factor interactions were shown to change from antagonism at 30‰ to synergism at 25‰ *S*.

Eisler & Hennekey (1977) conducted acute bioassays with five metals on six species of estuarine animals including *Nereis virens* at 20‰ salinity and 20 °C, and found a rank order of toxicity Hg^{2+}>Cd^{2+}>Cr^{6+}>Ni^{2+}. There was a marked and progressive decline in LC_{50} values between 24 and 168 h.

Gray (1976) studied the effects of salinity, temperature and mercury on the mortality of trochophore larvae of *Serpula vermicularis*. Using response surface methods, 4-day old larvae were shown to be more euryplastic with respect to salinity and temperature than were gastrulae and 1-day-old larvae. Significant interaction was found between the effect of temperature and reduced salinity on mortality of larvae at each age tested. 4-day trochophore larvae were more resistant to reduced salinity at low temperatures than were gastrulae and 1-day-old larvae. There were significant interactions between salinity and temperature at each mercury concentration. As mercury concentration increased there was a linear additive effect between reduced salinity and temperature. The larvae were more tolerant at low temperatures and high salinity than at high temperatures and low salinity.

Reish, Martin, Pilz & Wood (1976) looked at the effect of several heavy metals on laboratory populations of juvenile and adult *Neanthes arenaceodentata* and *Capitella capitata*. Mercury was found to be most toxic to both species regardless of life stage, and adults were more tolerant than juveniles.

The relative tolerances of selected aquatic oligochaetes to combinations of pollutants and environmental factors have been studied by Chapman, Farrell & Brinkhurst (1982a,b). The two salt-water species tested were *Monopylephorus cuticulatus* and *Limnodrilus verrucosus*. Four pollutants were used, one of which was mercury, with different combinations of pH, temperature, and salinity. Acute 96 h bioassays were conducted with 24 h solution replacement. Both species showed similar tolerance to mercury at 1 and 10 °C. At 20 °C both species were significantly more tolerant than at 10°C. At pH 6 and 8, *Monopylephorus cuticulatus* was significantly less tolerant than at pH 7. At 10‰ salinity *M. cuticulatus* was significantly more tolerant to mercury than at 20‰. That is, increased salinities did not result in increased tolerances. Mercury toxicity to oligochaetes is highly variable, making it difficult to predict its effect in the natural environment.

CRUSTACEANS

The effect of salinity on the acute toxicity of three heavy metals, including mercury, for *Rangia cuneata* has been studied by Olson & Harrel (1973). The 96 h LC_{50} was determined for mercury at 1·0, 5·5, 22‰ salinity at 24 °C. Toxicity was greatest in fresh water, and *R. cuneata* was most resistant to mercury at the intermediate salinity of 5·5‰.

Roesijadi *et al.* (1974) studied the survival and chloride ion regulation of the porcelain crab, *Petrolisthes armatus*, exposed to mercury. As chloride ions account for approximately 90% of the total anionic concentration of crustacean blood, it was thought that the regulation of blood chloride levels might be a sensitive area in which a sub-lethal effect of mercury could be assessed. The experimental variables used were 1, 25, 50, 75 $\mu g \cdot l^{-1}$ Hg, and 7, 14, 21, 28, 35‰ salinity at 24 °C. Acute toxicity bioassays demonstrated that low mercury concentrations were lethal to *P. armatus*. 96 h LC_{50} values varied from 50–64 $\mu g \cdot l^{-1}$ depending on test salinities. There was a trend of decreased survival at lower salinities; lower salinities decreased LC_{50} values. Differences in survival after 96 h due to salinity were not significantly different. Exposure to 50 $\mu g \cdot l^{-1}$ Hg did not alter chloride ion regulation of either acclimated crabs or crabs adjusting to new salinities. Since exposure to mercury concentrations approaching lethal levels did not affect the chloride regulatory ability of *P. armatus*, it is apparent that the toxicity of mercury to these crabs is not related to a disruption of their chloride ion regulation.

Vernberg & Vernberg (1972) studied the effect of interactions between temperature, salinity, and mercury on the survival and metabolism of the adult fiddler crab, *Uca pugilator*. Under optimal conditions, 25 °C and 30‰ *S*, crabs survived a concentration of 0·18 $mg \cdot l^{-1}$ Hg for two months with only slight mortality. At this concentration of mercury fewer crabs survived at low temperature (5 °C) and low salinity (5‰) than at a combination of high temperature (35 °C) and low salinity. Males were far more sensitive to the stress than females. Metabolic rates of male and female crabs were affected by prolonged exposure to mercury both under optimum environmental conditions and under temperature and salinity stress. Metabolic rates of males were more adversely affected than those of females.

De Coursey & Vernberg (1972) reported that the toxicity of mercury to larval *U. pugilator* increased with larval age. Vernberg, De Coursey & Padgett (1973) reported on the synergistic effects of environmental variables on larvae of *U. pugilator*. The effects of sublethal concentrations of mercury in combination with stressful temperature–salinity regimes were considered for larval development. In all environmental regimes sub-optimal conditions caused increased mortality and high temperatures were especially stressful. Under high temperature (30 °C) and optimum salinity there was no difference between controls and zoeae exposed to mercury. Larvae exposed to mercury and maintained at high temperature and low salinity showed, however, a 27% increase in mortality compared with the larvae not exposed to mercury. At low temperature there was a marked increased in mortality at both optimal and low salinity with the addition of mercury. The following factors were significant at 5% level: temperature (T), salinity (S), Hg, T × Hg, S × Hg. The interaction of T × S was not significant.

Sub-optimal temperature-salinity conditions also generally depressed metabolic rate in larval *U. pugilator* and the effect of added mercury stress was temperature-dependent. At 25 and 30 °C mercury depressed metabolic rates; at 20 °C oxygen uptake was enhanced. Sub-optimal temperatures modified phototactic response of larvae, but salinity was not a critical factor. The addition of mercury further modified the response.

Vernberg, De Coursey & O'Hara (1974) summarized their own work on *U. pugilator* as follows. The effect of mercury on *U. pugilator* depends on a number of factors including stage of the life cycle, sex, thermal history, and environmental regime. Larvae are several orders of magnitude more sensitive to mercury than adults, and adult males are more sensitive to mercury stress than are females. Mercury is most toxic at low temperatures and low salinity. Warm-acclimated animals (summer animals) are less tolerant of mercury at low temperatures than cold-acclimated winter ones, and concentrations of mercury that are sub-lethal under optimum conditions of temperature and salinity become lethal when temperature-salinity regimes become stressful. Distribution of mercury in the tissues of the crab is dependent on the environmental regime, but the total body burden is not. They suggested that the cause of death at low temperatures may be because crabs seem unable to transport mercury from the gill tissue to the hepatopancreas, thus leaving high mercury residues in the gills.

Portmann (1968) conducted routine testing at 15 °C on the toxicity of mercury, copper, zinc, phenol and nickel to pink shrimps (*Pandalus*), brown shrimps (*Crangon*), shore crabs (*Carcinus*), and cockles (*Cardium*). *Pandalus* was the most susceptible genus; the most toxic chemical was mercury, followed by Cu>Zn>phenol>Ni. This sequence is interesting because it follows the relative solubilities of the four metals in sea water. Reduction in temperature from 22 to 5 °C increased the tolerance of brown shrimps to mercury by a factor of five. With cockles the effect was more pronounced and LC_{50} increased by a factor of 130.

Jones (1973) has reported the influence of salinity and temperature on the toxicity of mercury to marine and brackish water isopods. Under optimum conditions of 34‰ salinity in 10 °C all species tested tolerated 0·1 and 1 $mg \cdot l^{-1}$ Hg for 5 days without achieving 50% mortality. A decrease in salinity and an increase in temperature caused a dramatic increase in toxicity and reduced the LT_{50} values. An interspecific comparison indicates that species adapted to a fluctuating estuarine environment are more influenced by the extra stresses of heavy metal pollution than marine forms for which environmental variables are relatively more stable.

Jones (1975a,b) made a further study of the synergistic effects of salinity, temperature and four heavy metals on mortality and osmoregulation in marine and estuarine isopods. Only osmoregulation was studied in relation to mercury exposure at 10 °C for the estuarine species *Jaera albifrons sensu stricto*. 0·1 $mg \cdot l^{-1}$ Hg did not disrupt osmoregulatory ability in 30, 50 or 100% sea water, but significantly increased the blood osmotic pressure in 5% sea water. 1 $mg \cdot l^{-1}$ Hg only increased blood osmotic pressure in 100% sea water. Neither of the concentrations, however, affected osmoregulatory ability in 50% and 30% sea water.

Weis (1978) looked at interactions of methyl mercury, cadmium, and salinity on regeneration in the fiddler crabs *Uca pugilator*, *U. pugnax*, and *U. minax*. After multiple autonomy, both methyl mercury and cadmium retarded limb regeneration and ecdysis. When crabs in sea water are exposed to a mixture of both metals, the effect is increased, indicating that the two are interacting in an additive way. In 50% sea water (15‰ salinity) the effects of cadmium are greatly intensified so that growth of limb buds is extremely slow, if it occurs at all. When methyl mercury is present in 15‰ salinity at the same time, the severe effects of cadmium are somewhat ameliorated, indicating an antagonistic interaction of the two metals under these conditions.

The influence of temperature and salinity on the acute toxicity of heavy metals to the juvenile banana prawn *Penaeus merguiensis* has been studied by Denton & Burdon-Jones (1982). Ninety-six hour LC_{50} bioassays were conducted at 20, 30 and 35 °C at 20 and 36‰ salinity. At all salinity-temperate combinations, mercury was the most toxic metal. The toxicity of all metals increased with temperature, particularly in the high salinity experiments. The effect of salinity upon metal toxicity was not so marked as the effect of temperature. Although there was a trend of increased toxicity in lower salinities, the differences were not statistically significant.

McKenney & Costlow (1982) studied the effects of mercury on developing larvae of *Rhithropanopeus harrisii*. Developing larvae were most resistant to mercury under optimal temperature and salinity conditions (25 °C, 15–20‰), while mercury interacted significantly with both temperature and salinity to alter survival capacity of developing larvae. Both zoeae and megalopae exhibited reduced resistance to mercury at lower temperatures and, in turn, optimal temperatures for larval survival were higher with increasing concentrations of mercury. Tolerance to mercury during total larval development decreased at lower salinities, while the optimal salinity for complete larval development shifted upwards with increasing mercury concentrations. Mercury concentrations from 5–20 $\mu g \cdot l^{-1}$ reduced the survival capacity for developing larvae of *R. harrisii*. Exposure to such concentrations prolonged complete developmental duration by 3 to 4 days with zoeal developmental rates retarded more than megalopal rates. Developmental rates of megalopae were more reduced by mercury at higher salinities, and both zoeal and megalopal developmental rates were more retarded by mercury at lower temperatures.

NICKEL

GENERAL REVIEW

While some authors have studied acute toxicity and uptake of nickel, few have considered the effects of environmental variables on toxicity. Ahsanullah (1982) found the 96 h LC_{50} value for nickel to the amphipod *Allorchetes compressa* to be 34·68 $mg \cdot l^{-1}$, making it the second least toxic of a series including mercury, chromium, nickel, and molybdenum. Portmann (1968) investigated the toxicity of five metals to a variety of marine animals, chiefly pink shrimps, brown shrimps, shore crabs, and cockles. In order of relative toxicity, mercury was most toxic followed by copper, zinc, phenol, and nickel. Babich & Stotsky (1983) found nickel toxicity to microbes in marine systems was reduced by increasing salinity and decreasing temperature.

ANNELIDS

Petrich & Reish (1979) studied the effect of nickel on survival and reproduction in three species of polychaetes. 96 h LC_{50} values for *Capitella capitata*, *Ctenodrillus serratus*, and *Neanthes arenaceodentata* were 50, 17 and 49 $mg \cdot l^{-1}$, respectively. Suppression of reproduction with nickel was one or two orders of magnitude less than the 96 h LC_{50} value. The authors indicate that municipal waste water effluents in Southern California contain a similar level of nickel to that at which suppression of reproduction occurred (*i.e.* 0·14–3·6 $mg \cdot l^{-1}$) so this metal may already be at a critical level in nearshore waters adjacent to metropolitan areas.

CRUSTACEANS

Denton & Burdon-Jones (1982) conducted bioassays to determine the effects of temperature and salinity on the acute toxicity of six heavy metals including nickel to juvenile banana prawns *Penaeus merguiensis*. Tests were conducted at all combinations of 35, 30 and 20 °C, with 36 and 20‰ salinity for 96 h. The general order of metal toxicity was Hg>(Cu, Cd, Zn)>Ni>Pb. The toxicity of all metals increased with temperature, particularly in high salinity. The effect of salinity on metal toxicity was not as marked as the effect of temperature. For most metals toxicity, however, increased with decreased salinity particularly at 20 and 30 °C. In contrast to the above, the toxicity of nickel at 30 and 35 °C appeared to be greater at high salinity, but the differences were not statistically significant. Bryant, Newbery, McLusky & Campbell (1985b) found that the toxicity of nickel to the amphipod, *Corophium volutator*, decreased as salinity increased. The toxicity also decreased as temperature decreased, although the variation was not so great as that caused by variations in salinity (Fig. 14).

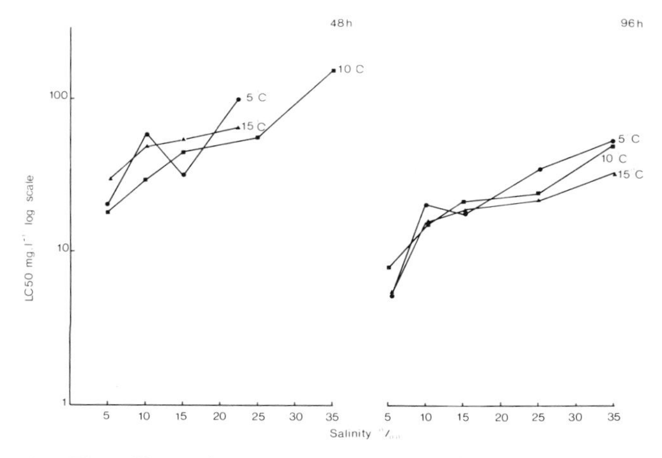

Fig. 14.—The effect of temperature and salinity on the median lethal concentration (LC_{50}) of nickel to *Corophium volutator* (after Bryant, McLusky & Campbell, 1984).

MOLLUSCS

Calabrese *et al.* (1973) looked at the acute toxicity of eleven heavy metals to *Crassostrea virginica* larvae. Nickel was the least toxic with a 48 h LC_{50} of 1·18 $mg \cdot l^{-1}$ at 25‰ salinity and 26 °C. Calabrese, MacInnes, Nelson &

Miller (1977) also studied the toxicity of nickel and other metals to larvae of *Crassostrea virginica* and *Mercenaria mercenaria*; LC_{50} values were calculated over 12 days. Effects on growth were also studied of a series of metals including mercury, silver, copper, zinc, and nickel. Nickel was the least toxic, but had the greatest effect on the larval growth rate and also caused abnormalities such as tissue extrusion from the shell. For *Crassostrea virginica* larvae the LC_{50} was 1·2 mg·l^{-1} nickel and for *Mercenaria mercenaria* it was 5·7 mg·l^{-1}. Eisler (1977b) found the 168 h LC_{50} for nickel to adults of the soft-shell clam, *Mya arenaria*, to be 50 mg·l^{-1}.

In another paper Eisler (1977a) subjected *M. arenaria* to a flowing sea-water solution containing a mixture of salts including Mn, Zn, Pb, Ni, and Cu, with Ni at 50 μg·l^{-1} to approximate highest levels of each found in surficial interstitial sediment waters from mid-Narragansett Bay, Rhode Island. Two tests were done: in winter for 112 days at 0–10 °C, and in summer for 16 days at 16–22 °C. In winter all were dead within 4–10 wk, and experimental animals took up three times more nickel than controls. In summer all died within 6–14 days and also accumulated three times more than controls. The animals were thought to have died due to the copper and zinc components of the mixture. Bryant *et al.* (1985b) found that the toxicity of nickel to the bivalve *Macoma balthica* increased with decreasing salinity over the range 35–15‰, but was not significantly affected by temperature over the range 5–15 °C (Fig. 15).

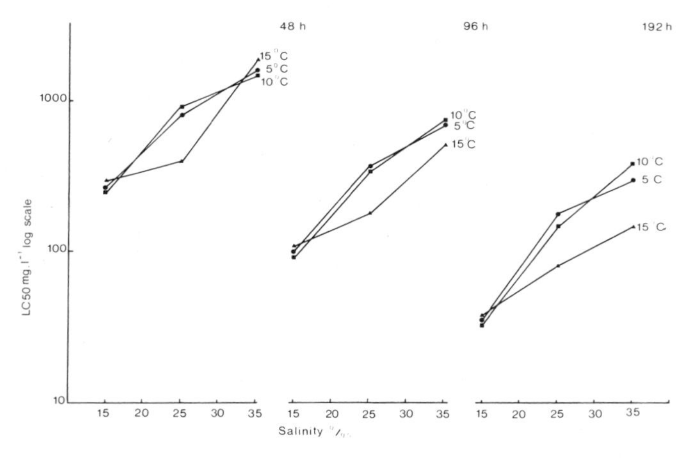

Fig. 15.—The effect of temperature and salinity on the median lethal concentration (LC_{50}) of nickel to *Macoma balthica* (after Bryant, McLusky & Campbell, 1984).

ZINC

GENERAL REVIEW

The combined effects of salinity and zinc have been investigated by several authors. Bryan & Hummerstone (1973) considered zinc uptake by *Nereis diversicolor* and found by using labelling techniques that zinc was more quickly absorbed in low salinity. Comparing animals from two different areas, the animals from the less polluted area were found to absorb zinc more quickly, but final body levels of zinc were not proportional to levels in the water, suggesting that it can be regulated. The authors suggest that increased resistance to heavy metals depends on a complexing system which detoxifies the metal and stores it in the epidermis and nephridia. This is comparable with certain zinc-tolerant land plants where zinc is bound strongly to cell walls protecting other parts of the cell.

Cairns, Heath & Parker (1975) reviewed the literature on the effects of temperature upon toxicity of chemicals to aquatic organisms. Most of the available literature regarding zinc concerned fish with one reference to freshwater pond snails. In general the effects of temperature on zinc toxicity varied greatly between species and no clear trends were apparent. It was clear, however, that a changing thermal environment can amplify the effects of trace metal toxicity far more than a constant high temperature.

MOLLUSCS

Calabrese, MacInnes, Nelson & Miller (1977) established the order of toxicity Hg>Cu>Ag>Zn>Ni for oyster larvae. MacInnes & Calabrese (1977) measured the acute toxicity of zinc for *Crassostrea virginica* and found that toxicity was not affected by temperature.

Eisler (1977a) exposed *Mya arenaria* to a mixture of heavy metals which included zinc at $2{\cdot}5\ mg{\cdot}l^{-1}$ which was equivalent to the highest measured levels in surficial interstitial sediment waters from mid-Narragansett Bay, Rhode Island. He used both a $2{\cdot}5$ and a $0{\cdot}5\ mg{\cdot}l^{-1}$ solution and conducted a winter and a summer study. In winter where the animals were kept in the $2{\cdot}5\ mg{\cdot}l^{-1}$ solution for 112 days at 0–10 °C all died within 4–10 wk, while in the $0{\cdot}5\ mg{\cdot}l^{-1}$ solution all survived. In summer animals were kept for 16 days at 16–22 °C and all were dead within 6–14 days. In winter the survivors had fifteen times more zinc in soft body parts than controls, while in summer survivors had eleven times more zinc than controls. Although a mixture of elements was used, Eisler proposed that the deaths were probably caused by copper and zinc.

Mya arenaria was considerably more resistant to zinc at 4 °C than at 17 °C; 336 h LC_{50} values at 4 and 17 °C were $25{\cdot}0$ and $2{\cdot}65\ mg{\cdot}l^{-1}$, respectively (Eisler, 1977b).

Cotter, Phillips & Ahsanullah (1982) used *Mytilus edulis* to investigate the possible effects of a heated effluent discharge. Following 14 days exposure to zinc at $0{\cdot}8\ mg{\cdot}l^{-1}$, a 24-h period at 29–31 °C resulted in greatly increased mortality. A second experiment tested the toxicity of zinc to *M. edulis* which had been first acclimated to a range of different temperatures and salinities. Following 14 days at 10, 16 or 22 °C and 22 or 35‰ salinity, animals were exposed to $0{\cdot}3$ or $1{\cdot}0\ mg{\cdot}l^{-1}$ zinc. It was found that zinc accumulation was independent of temperature and salinity. Zinc caused greater mortality at 22 °C and 35‰ than at low temperatures and salinities. The authors suggest that heated effluent will accelerate any toxic effects of zinc or low salinity which occur, posing a hazard in winter as well as in summer.

In two papers Phillips (1976, 1977) also investigated zinc uptake by *M.*

edulis. He found uptake was unaffected by low temperature or low salinity. Mussels collected from two areas, one of naturally fluctuating salinity and the other stable, were exposed to zinc. There was no significant difference in zinc uptake in mussels from either population.

Bryant *et al.* (1985b) found that the toxicity of zinc to the estuarine bivalve, *Macoma balthica*, increased as temperature increased and as salinity decreased (Fig. 16) over the range 5–15 °C, and 5–35‰, at time intervals of up to 384 h, although the effect of temperature was much less than the effect of salinity.

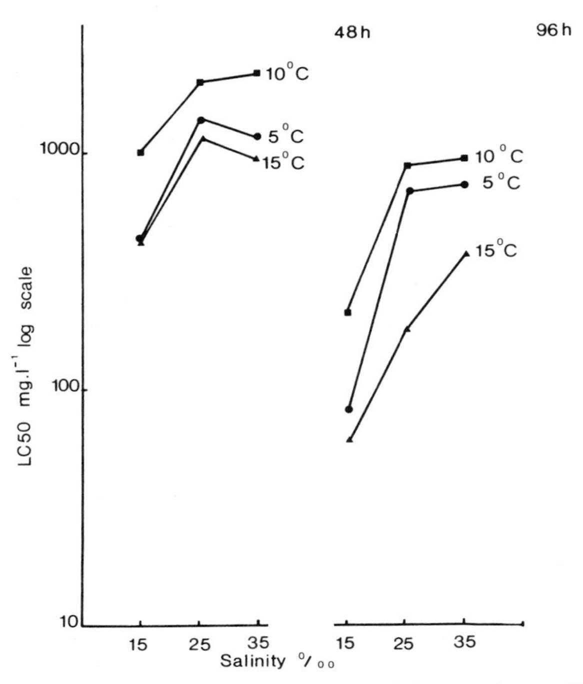

Fig. 16.—The effect of temperature and salinity on the median lethal concentration (LC_{50}) of zinc to *Macoma balthica* (after Bryant, McLusky & Campbell, 1984).

ANNELIDS

Fernandez (1983) calculated 96 h LC_{50} values for *Hediste* (=*Nereis*) *diversicolor* at four salinities 5, 10, 17·5 and 30‰ at 12 °C. The 96 h LC_{50} at 17·5‰ and 12 °C was 38 mg·l^{-1} while at 5‰ and 10‰ it was 7 and 19 mg l^{-1}, respectively, *i.e.* toxicity decreased with increasing salinity. When salinity was kept constant at 17·5% but temperature varied, the following 96 h LC_{50} values were recorded: 40 mg·l^{-1} at 6 °C, 32 mg·l^{-1} at 12 °C, and 9·1 mg·l^{-1} at 20 °C *i.e.* toxicity increased with increasing temperature. Uptake of zinc was measured and was highest at 20 °C and lowest at 6 °C. Similarly there was greater uptake at lower salinities. In another experiment the presence of sediment was found to reduce toxicity and body levels of zinc.

CRUSTACEANS

Denton & Burdon-Jones (1982) investigated the effects of salinity and temperature on the toxicity of zinc to the banana prawn, *Penaeus merguiensis*, over 96 h. Combinations of 35, 30 or 20 °C with 36 and 20‰ salinity were used. Toxicity was found to increase with increasing temperature, most noticeably at high salinity. Salinity itself did not, however, significantly influence metal toxicity. A combination of 35 °C and 36‰ gave a 96 h LC_{50} of 0·37 $mg \cdot l^{-1}$; 20 °C and 20‰ gave a 96 h LC_{50} of 4·8 $mg \cdot l^{-1}$.

Jones (1975a,b) looked at the effect of zinc on the mortality and osmoregulation of six species of isopods, four marine, and two estuarine. The marine species were *Idotea baltica*, *I. neglecta*, *I. emarginata*, and *Eurydice pulchra*. The estuarine species were *Jaera albifrons sensu stricto* and *J. nordmanni*. For all animals mortality rates in 100% sea water with 10 and 20 $mg \cdot l^{-1}$ zinc added were low, but as salinity decreased, toxicity increased resulting in lower LT_{50} values. Mortalities at 10 °C were greater than at 5 °C. Zinc did not lower the haemolymph osmotic concentration of *Idotea neglecta* at 100 or 80% sea water.

An interspecific comparison of the results by Jones (1975a,b) revealed a gradation in tolerance to heavy metals related to the ability of the animal to survive in stressful conditions. Thus, the marine species suffer over 40% mortality with heavy metals even in optimum salinity, and a decrease in salinity can result in a rapid increase in mortality. In 40% sea water there was a synergism between heavy metals and salinity which results in 100% mortality for marine species. In salinities from 100 to 10% sea water, mortality for estuarine species was below 40% at optimum temperature, but at 1% sea water which they normally tolerate well, very high mortalities resulted from an addition of heavy metals. *Jaera nordmanni* survived better than *J. albifrons* in these stressful conditions and this was reflected in the field by the greater penetration of *J. nordmanni* into freshwater conditions. Jones concludes that marine species living at their normal salinities are less susceptible to heavy metal pollution than are estuarine species living in salinities near the lower limit of their normal salinity range. Synergistic action of these heavy metals combined with stressful salinity conditions represents a serious threat to estuarine species subjected daily to dilute sea water.

McKenney & Neff (1979) used surface-response methodology to investigate the combined effects of salinity, temperature and zinc on larvae of the grass shrimp, *Palaemonetes pugio*. Salinities of 3–31‰ were combined with temperatures from 20 to 35 °C and zinc concentrations from 0 to 1·00 $mg \cdot l^{-1}$ to produce 80 combinations in which larvae were raised. Optimal conditions were found to exist at 17–27% salinity and 20–27 °C. Survival and adaptation to temperature were reduced by zinc concentrations from 0·25 to 1 $mg \cdot l^{-1}$ while developmental rates were decreased outside optimal conditions. There was significant interaction between zinc and temperature such that an increase in zinc concentration reduced survival and developmental rates more at sub-optimal temperatures. Resistance to zinc was least at supra-optimal salinities indicating a significant interaction between zinc and salinity. Zinc thus limits the euryplastic nature of the larvae limiting and broad areas of salinity and temperature within which the unexposed larvae can successfully develop. Zinc induced prolongation of the pelagic phase of the larval life cycle by four days, possibly increasing predation of the species (Thorson, 1950; Mileikovsky, 1971). Increased predation decreases the number available for recruitment into the parental benthic population or for dispersion and establishment of new populations in less severe environments.

Bryant *et al.* (1985b) found that the toxicity of zinc to the estuarine amphipod, *Corophium volutator*, increased as temperature increased and as salinity decreased over the range 5–15 °C, and 5–35‰ (Fig. 17). The effect of temperature was much less than the effect of salinity.

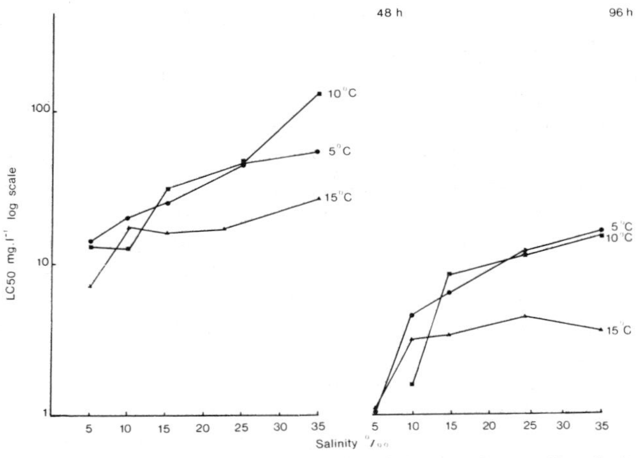

Fig. 17.—The effect of temperature and salinity on the median lethal concentration of (LC_{50}) of zinc to *Corophium volutator* (after Bryant, McLusky & Campbell, 1984).

OTHER SPECIES

Gray (1974) measured the effects of zinc sulphate on the growth rates of a population of marine ciliate protozoan (*Cristigera* spp.). Optimum growth conditions without added zinc were found at 27·64‰ and 15·94 °C. Further experimental work was, however, continued at one temperature and salinity only *i.e.* 34·4‰ and 16 °C. Under these conditions zinc sulphate at 0·25 mg·l^{-1} reduced the growth rate of the population by 13·7% compared with the control.

Castagna, Sinatra, Scalia & Capodicasa (1981) found that zinc interfered with sperm mobility and egg development in the echinoderm *Arbacia lixula*.

GENERAL DISCUSSION

The present review has revealed a considerable variety in the reports on the effect of temperature and salinity on the toxicity of different heavy metals to marine and estuarine animals.

In general, it is well known that estuarine animals are subject to highly variable temperature and salinity environments. Temperature variations may affect their survival, growth, and metabolic physiology. Salinity variations may affect these as well as imposing potential osmotic problems. Animals living in estuaries may have colonized the estuary from the freshwater or the seaward end; the majority are of marine origin, and the extent to which an individual species can enter the estuary depends on several factors, of which its ability to withstand lowered salinities is usually regarded as the most important (McLusky, 1971, 1981).

Most authors would regard animals living near to the limit of their tolerance range as being more susceptible to any stress, and in the present context this may be interpreted as animals living near to their salinity or temperature tolerance limits as being more susceptible to heavy metal stress. A contrary view has been propounded by Gray (1974) that since estuarine species have a wide tolerance to salinity, temperature, and oxygen levels, this tolerance pre-adapts the organisms to tolerate pollution stress, and may make them unsuitable as test species. Some support for this view was found by Howell (1984), who showed that euryhaline species of nematodes were less susceptible to metals than stenohaline species. We would certainly agree that many estuarine species are able to tolerate wider variations in temperature and salinity than their marine counterparts, but would argue strongly that estuarine species are the only suitable test species for assessing the impact of pollutants on the estuarine ecosystem. The principal conclusion that emerges from this review is that toxicity values determined under fixed (or single) temperature and salinity regimes are inappropriate for evaluating the effect of environmental factors in modifying the toxicity of metals to estuarine and marine invertebrates.

The range of toxicity values for a single species exposed to different temperatures and salinities may exceed the entire range of toxicity values previously published for a variety of marine or freshwater animals, as was seen for chromium for example (Bryant, McLusky, Roddie & Newbery, 1984).

The view that animals near to their salinity or temperature limits are more susceptible to heavy metal stress is also apparently confirmed by the present review. The common theme that is apparent from almost every metal reviewed is that toxicity increases as salinity decreases, and that toxicity increases as temperature increases. This effect is especially clear with cadmium and is in fact so marked that several authors claim that the distinct salinity with toxicity effect is due to an interaction between calcium and cadmium as they compete at uptake sites. Cadmium is one of the best studied metals, especially for crustaceans, and temperature and salinity effects upon toxicity have been clearly seen. Copper is also well studied and shows a clear temperature and salinity effect with again toxicity increasing as salinity decreases and temperature increases. Lead has been little studied, mainly because of the problems of precipitation of lead in sea water and the inability of the investigators to produce appropriate test solutions. Mercury, like cadmium, has been well studied with considerable effects of temperature and salinity being seen, although in several species no effects were seen and in several others the mercury concentration alone was found to be the dominant toxic factor. Nickel has been little studied, with temperature effects being seen, but a lesser effect of salinity. Zinc and chromium have also been little studied, but clear effects of temperature and salinity have been reported. Arsenic has been little studied and appears to show only a temperature effect on toxicity.

Two possible physiological mechanisms are involved in the effects of salinity and temperature on marine and estuarine invertebrates. First, osmoregulation is known to be affected by heavy metals (Bouquegneau & Gilles, 1979) with the metals competing with calcium and magnesium as cations at uptake sites. From the results presented in this review it would seem that such effects are widespread and all the metals reviewed, except arsenic, have been implicated in such effects. Secondly, temperature related effects have been observed, although precisely how temperature affects metal toxicity is not known. Denton & Burdon-Jones (1981, 1982) suggest that since permeability characteristics of membranes and penetration rates are important factors influencing metal toxicity, increased toxicity at elevated temperatures may be the result of increased metabolic activity which in turn may facilitate and hasten the rate of uptake. Collier, Miller, Dawson & Thurberg (1973) and Thurberg, Dawson & Collier (1973) found that heavy metals may inhibit oxygen transfer across crab gill membranes. This is of particular significance since oxygen demands are likely to increase with increased temperature. In the case of arsenic, it is known that this metal may inhibit oxidative phosphorylation and thus influence respiration. Respiration is well known to be temperature sensitive, and it may be suggested that the temperature effects noted for arsenic, and also other metals, indicate interference with respiratory enzymes by the metal (Cairns *et al.*, 1975).

The present review has confined itself to acute toxic effects. Because of the distinct effects of temperature and salinity which have, however, been observed in the acute effects of heavy metals, sub-lethal studies must also take account of environmental variables.

A rank-ordering of toxicity may be proposed from those studies which have examined several metals. This order would be mercury (most toxic)> cadmium>copper>zinc>chromium>nickel>lead and arsenic (least toxic). A taxonomic order may also be noted of: Annelida>Crustacea>Mollusca; with Annelida the most sensitive. Thus, in general, for estuarine animals heavy metal toxicity increases as salinity decreases and as temperature increases. Toxicity values determined under fixed (or single) temperature or salinity regimes have been shown to be generally inappropriate for evaluating the effect of heavy metals on estuarine or inshore marine ecosystems which may regularly expect to experience variations in temperature and/or salinity.

REFERENCES

Ahsanullah, M., 1982. *Aust. J. mar. freshwat. Res.,* **33,** 465–474.

Ahsanullah, M., Negilski, D. S. & Mobley, M. C., 1981. *Mar. Biol.,* **64,** 229–304.

Amiard-Triquet, C., Amiard, J. C., Robert, J. M., Métayer, C., Marchand, J. & Martin, J. L., 1983. *Cah. Biol. mar.,* **24,** 105–118.

Anderson, R. V., 1978. *Bull. environ. Contam. Toxicol.,* **20,** 394–400.

Axiak, V. & Schembri, J. L. 1982. *Mar. Pollut. Bull.,* **13,** 383–386.

Babich, H. & Stotsky, G., 1983. *Aquatic Toxicol.*, **3**, 195–208.
Bahner, L. H. & Nimmo, D. R., 1975. In, *Trace Substances in Environmental Health—IX*, edited by D. D. Hemphill, University of Missouri, Columbia Missouri, pp. 169–177.
Bouquegneau, J. M. & Gilles, R., 1979. In, *Mechanisms of Osmoregulation*, edited by R. Gilles, J. Wiley, New York, pp. 563–580.
Bryan, G. W., 1971. *Proc. R. Soc. Ser. B*, **177**, 389–410.
Bryan, G. W., 1976. In, *Effects of Pollutants on Marine Organisms, Vol. 2*, edited by A. P. M. Lockwood, Cambridge University Press, Cambridge, pp. 7–34.
Bryan, G. W., 1980. *Helgol. Meeresunters.*, **33**, 6–25.
Bryan, G. W., 1984. In, *Marine Ecology, Vol 5, Part 3*, edited by O. Kinne, J. Wiley, New York, pp. 1289–1431.
Bryan, G. W. & Hummerstone, L. G., 1973. *J. mar. biol. Ass. U.K.*, **53**, 859–872.
Bryant, V., McLusky, D. S. & Campbell, R., 1984. *Toxicity of Arsenic, Nickel and Zinc to Estuarine Invertebrates*. Report to Water Research Centre, Marlow, U.K. (MPC 4162), 60 pp.
Bryant, V., McLusky, D. S., Roddie, K. & Newbery, D. M., 1984. *Mar. Ecol. Prog. Ser.*, **20**, 137–149.
Bryant, V., Newbery, D. M., McLusky, D. S. & Campbell, R. 1985a. *Mar. Ecol. Prog. Ser.*, **24**, 129–137.
Byrant, V., Newbery, D. M., McLusky, D. S. & Campbell, R., 1985b. *Mar. Ecol. Prog. Ser.*, **24**, 139–153.
Cairns, J., Heath, A. G. & Parker, B. C., 1975. *Hydrobiologia*, **47**, 135–171.
Calabrese, A., Collier, R. S., Nelson, D. A. & MacInnes, J. R., 1973. *Mar. Biol.*, **18**, 162–166.
Calabrese, A., MacInnes, J. R., Nelson, D. A. & Miller, J. E., 1977. *Mar. Biol.*, **41**, 179–184.
Calabrese, A. & Nelson, D. A., 1974. *Bull. environ. Contam. Toxicol.*, **11**, 92–97.
Calabrese, A., Thurberg, F. P. & Gould, L., 1977. *Mar. Fish Rev.*, **39**, 5–17.
Castagna, A., Sinatra, F., Scalia, M., & Capodicasa, V., 1981. *Mar. Biol.*, **64**, 285–289.
Chapman, P. M., Farrell, M. A. & Brinkhurst, R. O., 1982a. *Aquatic Toxicol.*, **2**, 47–67.
Chapman, P. M., Farrell, M. A. & Brinkhurst, R. O., 1982b. *Aquatic Toxicol.*, **2**, 69–78.
Coglianese, M. P., 1982. *Arch. environ. Contam. Toxicol.*, **11**, 297–303.
Collier, R. S., Miller, J. E., Dawson, M. A. & Thurberg, F. P., 1973. *Bull. environ. Contam. Toxicol.*, **10**, 378–382.
Coombs, T. L., 1979. *Top. envir. Hlth*, **2**, 94–139.
Coombs, T. L., 1980. In, *Animals and Environmental Fitness*, edited by R. Gilles, Pergamon Press, Oxford, pp. 283–302.
Cotter, A. J. R., Phillips, D. J. H. & Ahsanullah, M., 1982. *Mar. Biol.*, **68**, 135–141.
Curtis, M. W., Copeland, T. L. & Ward, C. H., 1979. *Water Res.*, **13**, 137–141.
Davenport, J., 1977. *J. mar. biol. Ass. U.K.*, **57**, 63–74.
Davis, I. M., Topping, G., Graham, W. C., Falconer, C. R., McIntosh, A. D. & Saward, D., 1981. *Mar. Biol.*, **64**, 291–297.
De Coursey, P. J. & Vernberg, W. B., 1972. *Oikos*, **23**, 241–247.
Denton, G. R. W. & Burdon-Jones, C., 1981. *Mar. Biol.*, **64**, 312–317.
Denton, G. R. W. & Burdon-Jones, C., 1982. *Chem. Ecol.*, **1**, 131–143.
Eisler, R., 1977a. *Mar. Biol.*, **43**, 265–276.
Eisler, R., 1977b. *Bull. environ. Contam. Toxicol.*, **17**, 137–145.
Eisler, R. & Hennekey, R. J., 1977. *Arch. environ. Contam. Toxicol.*, **6**, 315–323.
Engel, D. W. & Fowler, B. A., 1979. *Environ. Health Perspect.*, **28**, 81–88.
Fernandez, T. V., 1983. Ph.D. thesis, University of Hull, Hull, 160 pp.

Fletcher, C. R., 1970. *J. exp. Biol.,* **53,** 425–443.
Frank, P. M. & Robertson, P. B., 1979. *Bull. environ. Contam. Toxicol.,* **21,** 74–78.
Freedman, M. L., Cunningham, P. M., Schindler, J. E. & Zimmerman, M. J., 1980. *Bull. environ. Contam. Toxicol.,* **25,** 389–393.
George, S. G. & Coombs, T. L., 1977. *Mar. Biol.,* **39,** 261–268.
Gibson, C., Thatcher, T. O. & Apts, C. W., 1976. In, *Thermal Ecology, Vol. 11,* edited by G. W. Esch & R. W. McFarlane, U.S. Energy Research and Development Administration, pp. 88–92.
Gray, J. S., 1974. In, *Pollution and Physiology of Marine Organisms,* edited by W. Vernberg & F. Vernberg, Academic Press, New York, pp. 465–485.
Gray, J. S., 1976. *J. exp. mar. Biol. Ecol.,* **23,** 127–134.
Howell, R., 1984. *Mar. environ. Res.,* **11,** 153–161.
Hrs-Brenko, M., Claus, C. & Bubic, S., 1977. *Mar. Biol.,* **44,** 109–115.
Hutcheson, M. S., 1972. M.Sc. thesis, University of South Carolina, Columbia, U.S.A., 65 pp.
Jones, L. H., Jones, N. V. & Radlett, A. J., 1976. *Estuar. cstl mar. Sci.,* **4,** 107–111.
Jones, M. B., 1973. *Estuar. cstl mar. Sci.,* **1,** 425–431.
Jones, M. B., 1975a. *Mar. Biol.,* **30,** 13–20.
Jones, M. B., 1975b. In, *Proc. 9th Europ. Mar. Biol. Symp.,* edited by H. Barnes, Aberdeen University Press, Aberdeen, pp. 419–431.
Kulkarni, K. M., 1983. *Environ. Ecol.,* **1,** 193–195.
Lang, W. H., Forward Jr, R. B., Miller, D. C. & Marcy, M., 1980. *Mar. Biol.,* **38,** 139–145.
Laws, X., 1981. *Aquatic Pollution.* J. Wiley & Son, Chichester, 482 pp.
Lehnberg, W. & Theede, H., 1979. *Helgoländer wiss. Meersesunters.,* **32,** 179–199.
Lewis, S. D. & Lewis, W. M., 1971. *Trans. Am. Fish Soc.,* **100,** 639–643.
Lloyd R., 1979. Toxicity tests with Aquatic Organisms. Lecture presented at the Sixth FAO/SIDA Workshop on Aquatic Pollution in relation to protection of living resources. Rome, FAO, TF-RAD 112 (SWE)(Suppl. 1), pp. 165–178.
MacInnes, J. R. & Calabrese, A., 1977. In, *Physiology and Behaviour of Marine Organisms, Proc. 12th Europ. Mar. Biol. Symp.,* edited by D. S. McLusky & A. J. Berry, Pergamon Press, Oxford, pp. 195–202.
MacInnes, J. R. & Calabrese, A. 1979. *Arch. environ. Contam. Toxicol.,* **8,** 553–562.
Mance, G., 1984. Proposed environmental quality standards for chromium in waters and associated materials. Water Research Centre, Stevenage, Tech. Report, No. 268-M, 59 pp.
Mantoura, R. F. C., Dickson, A. & Riley, J. P., 1978. *Estuar. cstl mar. Sci.,* **6,** 387–408.
McCleese, D. W., 1974. *J. Fish. Res. Bd Can.,* **31,** 1949–1952.
McKenney Jr, C. L. & Costlow Jr, J. D., 1982. *Estuar. cstl Shelf Sci.,* **14,** 192–213.
McKenney, C. L. & Neff, J. M., 1979. *Mar. Biol.,* **52,** 177–188.
McLusky, D. S., 1971. *Ecology of Estuaries.* Heinemann, London, 144 pp.
McLusky, D. S., 1981. *The Estuarine Ecosystem.* Blackie, Glasgow, 150 pp.
McLusky, D. S. & Phillips, C. N. K., 1975. *Estuar. cstl mar. Sci.,* **3,** 103–108.
Middaugh, D. P. & Floyd, D., 1978. *Estuaries,* **13,** 123–125.
Mileikovsky, S. A., 1971. *Mar. Biol.,* **10,** 193–213.
Moore, M. N., 1981. In, *Analysis of Marine Ecosystems,* edited by A. R. Longhurst, Academic Press, London, pp. 535–569.
Moriarty, F., 1983. *Ecotoxicology.* Academic Press, London, 233 pp.
Murphy, C. B. & Spiegel, S. J., 1982. *J. Water Pollut. Control Fed.,* **54,** 849–855.
Murphy, C. B. & Spiegel, S. J., 1983. *J. Water Pollut. Control Fed.,* **55,** 816–822.
Negilski, D. S., Ahsanullah, M. & Mobley, M. C., 1981. *Mar. Biol.,* **64,** 305–309.
Nelson, D. A., Calabrese, A. & MacInnes, J. R., 1977. *Mar. Biol.,* **43,** 293–297.

Nelson, D. A., Calabrese, A., Nelson, B. A., MacInnes, J. R. & Wenzloff, D. R., 1976. *Bull. environ. Contam. Toxicol.*, **16**, 275–282.
O'Donnell, A. R. & Mance, G., 1984. *Mar. Pollut. Bull.*, **15**, 284–288.
O'Hara, J., 1973. *Fishery Bull. NOAA*, **71**, 147–153.
Olson, K. R. & Harrel, R. C., 1973. *Contrib. mar. Sci.*, **17**, 9–13.
Petrich, S. M. & Reish, D. J., 1979. *Bull. environ. Contam. Toxicol.*, **23**, 698–702.
Phillips, D. J. H., 1976. *Mar. Biol*, **38**, 56–59.
Phillips, D. J. H., 1977. *Mar. Biol.*, **41**, 79–88.
Phillips, D. J. H., 1980. In, *Cadmium in the Environment, Part 1: Ecological Cycling*, edited by J. O. Nriagu, Wiley-Interscience, New York, pp 450–483.
Portmann, J. E., 1968. *Helgoländer wiss. Meersesunters.*, **17**, 247–256.
Reish, D. J., Geesey, G. G., Wilkes, F. G., Oshida, P. S., Mearns, A. J., Rossi, S. S. & Ginn, T. C., 1984. *J. Water Pollut. Control Fed.*, **56**, 758–774.
Reish, D. J., Martin, J. M., Pilz, F. M. & Wood, J. Q., 1976. *Water Res.*, **10**, 299–302.
Reish, D. J. Oshida, P. S., Wilkes, F. G., Mearns, A. J., Ginn, T. C. & Carr, R. S., 1983. *J. Water Pollut. Control Fed.*, **55**, 767–787.
Roed, K. H., 1979. *Sarsia*, **64**, 245–252.
Roed, K. H., 1980. *Helgoländer Meeresunters.*, **33**, 47–48.
Roesijadi, E., Petrocelle, R. R., Anderson, J. W., Presley, B. J. & Sims, R., 1974. *Mar. Biol.*, **27**, 213–217.
Rosenberg, R. & Costlow Jr, J. D., 1976. *Mar. Biol.*, **38**, 291–303.
Schnute, J. & McKinnell, S., 1984. *J. Fish aquat. Sci.*, **41**, 936–953.
Sprague, J. B., 1969. *Water Res.*, **3**, 793–821.
Sprague, J. B., 1970. *Water Res.*, **4**, 3–32.
Sprague, J. B., 1971. *Water Res.*, **5**, 245–266.
Sullivan, J., 1977. *Aust. J. mar. freshw. Res.*, **28**, 739–743.
Sunda, W. & Guillard, R. R. L., 1976. *J. mar. Res.*, **34**, 511–529.
Sunda, W. G., Engel, D. W. & Thiotte, R. M., 1978. *Environ. Sci. Tech.*, **12**, 409–413.
Sunila, I., 1981. *Ann. Zool. Fennici*, **18**, 213–223.
Taylor, D., 1979. *Residue Review*, **72**, 33–69.
Taylor, D., 1981a. A Review of the Lethal and Sub-lethal Effects of Cadmium on Aquatic Life. Lecture presented at the *3rd International Cadmium Conference*, Miami, U.S.A.
Taylor, D., 1981b. *A Summary of the Data on the Toxicity of Various Heavy Metals to Aquatic Life. Cadmium*, ICI, Brixham, Devon, 32 pp.
Taylor, D., 1981c. *A summary of the Data on the Toxicity of Various Heavy Metals to Aquatic Life. Chromium*, ICI, Brixham, Devon, 16 pp.
Taylor, D., 1984. *Mar. Pollut. Bull.*, **15**, 168–170.
Theede, H. 1980. *Helgoländer Meerseunters.*, **33**, 26–35.
Theede, H., Scholz, N. & Fischer, H., 1979. *Mar. Ecol. Prog. Ser.*, **1**, 13–19.
Thorp, J. H., Giesy, J. P. & Wineriter, S. A., 1979. *Arch. environ. Contam. Toxicol.*, **8**, 449–456.
Thorson, G., 1950. *Biol. Rev.*, **25**, 1–45.
Thurberg, F. P., Dawson, M. A. & Collier, R. S., 1973. *Mar. Biol.*, **23**, 171–175.
Unlu, M. Y. & Fowler, S. W., 1979. *Mar. Biol.*, **51**, 209–219.
Vernberg, W. B., De Coursey, P. J. & O'Hara, J., 1974. In, *Pollution and Physiology of Marine Organisms*, edited by F. J. Vernberg & W. B. Vernberg, Academic Press, New York, pp. 381–425.
Vernberg, W. B., De Coursey, P. J. & Padgett, W. J., 1973. *Mar. Biol.*, **22**, 307–312.
Vernberg, W. B. & Vernberg, F. J., 1972. *Fishery Bull. NOAA*, **70**, 415–420.

Verriopoulos, G. & Moraitou-Apostopoulou, M., 1981. *Arch. Hydrobiol.*, **91**, 287–293.
Von Westernhagen, H., Rosenthal, H. Sperling, K. R., 1974. *Helgoländer wiss. Meeresunters.*, **26**, 416–433.
Weis, J. S., 1978. *Mar. Biol.*, **49**, 119–124.
Wright, D. A., 1977. *J. exp. Biol.*, **67**, 137–146.
Zaroogian, G. S. & Hoffman, G. L., 1982. *Environ. Monitor. Assess.*, **19**, 345–358.

Oceanogr. Mar. Biol. Ann. Rev., 1986, **24**, 521–623
Margaret Barnes, Ed.
Aberdeen University Press

THE USE AND NUTRITIONAL VALUE OF *ARTEMIA* AS A FOOD SOURCE*

P. LÉGER
Artemia *Reference Center, Faculty of Agriculture, State University of Ghent, Rozier 44, B-9000 Ghent, Belgium*

D. A. BENGTSON
United States Environmental Protection Agency, Environmental Research Laboratory, South Ferry Road, Narragansett, Rhode Island 02882, U.S.A.

K. L. SIMPSON
Department of Food Science and Technology, Nutrition and Dietetics, University of Rhode Island, Kingston, Rhode Island 02881, U.S.A.

and

P. SORGELOOS
Artemia *Reference Center, Faculty of Agriculture, State University of Ghent, Rozier 44, B-9000 Ghent, Belgium*

INTRODUCTION

Successful rearing of larval stages of aquatic organisms is a challenge for aquarists, an aim and tool for aquatic ecologists and ecotoxicologists, and the determinant for the commercial success of the aquaculturist.

The primary problem in larval culturing is that of food (May, 1970; Houde, 1973; Barnabé, 1976; Girin & Person-Le Ruyet, 1977; Goodwin & Hanson, 1977). Ideally, one would feed fish and crustacean larvae with their natural diet characterized by a wide diversity of live organisms. Collecting and feeding natural plankton from rivers, lakes and seas may appear evident but already at the beginning of this century this method was designated as hardly dependable beyond aquarium scale (Fabre-Domergue & Bietrix, 1905). On a larger and industrial scale, similarly to intensive cattle and poultry farming where a reliably high culture performance is the objective, a readily available diet has to be selected which is easily accepted and digested and having a reproducibly high nutritional quality. An extensive list of potential organisms may meet the requirements of acceptability, digestibility, and (reproducibly high) nutritional quality. When it comes to availability, however, only a few organisms are left as possible candidates. The provision of adequate numbers of food organisms has been called a "sine qua non" for any rearing attempt (May, 1970) and "the main obstacle" (Barnabé, 1976) or "limiting factor" (Girin & Person-Le Ruyet, 1977) for a successful aquaculture. The provision of adequate numbers of food organisms appropriate to larval rearing has, moreover,

*Contribution No. 2339, Rhode Island Agricultural Experiment Station.

been quoted as the "only criterion for the success of a larval production system" (Paulsen, 1980).

The property of the small branchiopod crustacean *Artemia** (Fig. 1) of forming dormant eggs, so-called "cysts", may be the reason why it has, to a great extent, been designated a convenient, suitable and excellent larval food source. These cysts are available year-round in large quantities along the shorelines of hypersaline lakes, coastal lagoons, and solar saltworks scattered over the five continents (Persoone & Sorgeloos, 1980; Vanhaecke, 1983; Vanhaecke, Tackaert & Sorgeloos, 1985). After harvesting and processing the cysts are available as storable 'off the shelf' 'on demand' life food. Indeed, upon some 24-hours incubation in sea water the cysts release free-swimming nauplii that can be given directly as a nutritious, live source of food to the larvae of a variety of aquatic organisms.

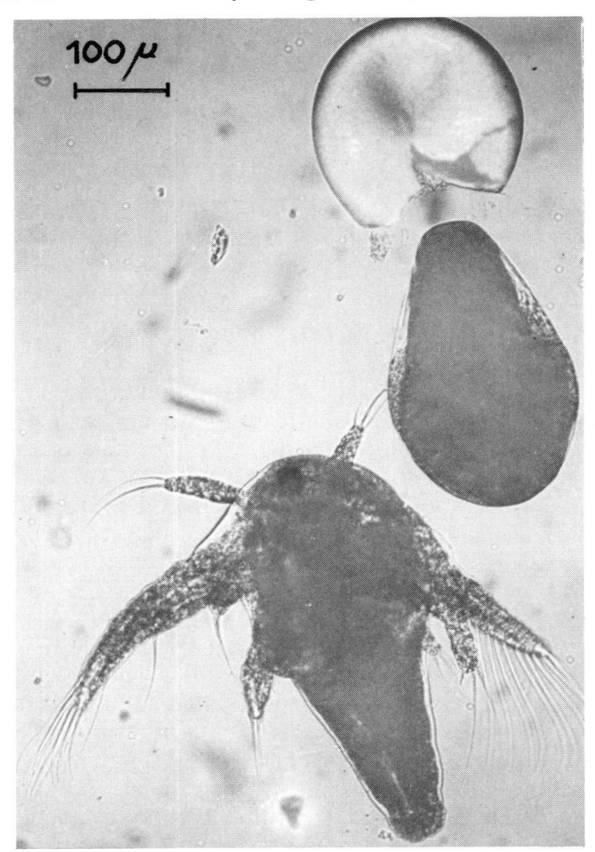

Fig. 1.—*Artemia* prenauplius shortly after breaking of a cyst and a freshly hatched instar I nauplius.

**Artemia* was first described by Schlösser in 1755 and later by Linnaeus in 1758 (Kuenen & Baas-Becking, 1938) under the binomen *Artemia salina*. Because crossing experiments of different *Artemia* populations revealed reproductive isolation of several groups of populations, it is suggested that until speciation in brine shrimp is more clearly understood, only the genus designation *Artemia* should be used (Persoone, Sorgeloos, Roels & Jaspers, 1980).

It is not the intention of the present article to compile all existing records of experiments using *Artemia* as a food source for this and that organism. We will rather go through the different applications of *Artemia*, the different forms of *Artemia* that are being used, the factors determining its nutritional value, its biochemical and chemical composition and, not least, the problems and constraints related to its use as a source of food. A better understanding of the nutritional value and constraints of *Artemia* as a food will, in the first place, lead to an optimized and more dependable culture performance and may ultimately constitute a more comprehensive basis for making it redundant through the formulation of artificial diets of equal merit.

ARTEMIA NAUPLII AND METANAUPLII

ARTEMIA NAUPLII AS A LIVE FOOD SOURCE

Artemia was described in the 18th century and has been extensively studied in the most diverse fundamental disciplines of biological sciences since the 19th century (Sorgeloos, 1980a). Its value as a suitable food organism was discovered only recently. Since Seale (1933), Gross (1937), and Rollefsen (1939) found that freshly hatched *Artemia* nauplii constituted an excellent food source for newborn fish larvae, its application in larval culture has been rampant.

The most diversified groups of organisms of the animal kingdom, *e.g.* foraminifers, coelenterates, flatworms, polychaetes, cnidarians, squids, insects, chaetognaths, fish, and crustaceans have been offered *Artemia* nauplii as a suitable food source (May, 1970; Kinne, 1977; Sorgeloos, 1980c). Kinne (1977) indeed stated that more than 85% of the marine animals cultivated so far have been offered *Artemia* as food source—either together with other foods or, more often, as a sole diet.

The ease with which *Artemia* nauplii are obtained from dry storable cysts has convinced most people involved with larval rearing, *i.e.* aquarists, aquatic ecologists and ecotoxicologists, and aquaculturists. In a digest for aquarists, Rakowicz (1972) stated that all aquarium fishes eat the slow-swimming baby brine shrimp and that those fishes show vigorous growth, excellent survival and best resistance to diseases. When comparing with alternative organisms, including those collected from wild sources or cultured at home, he concluded that brine shrimp nauplii emerge as one of the best of all live foods for most aquarium fishes.

In the cultivation of laboratory animals for scientific and applied purposes nearly all rearing attempts have employed *Artemia* nauplii (May, 1971). This is further confirmed by Kinne (1977), who noted that most investigators engaged in laboratory fish cultivation use *Artemia* nauplii, which in numerous instances proved to be a good food. Most workers culturing decapod larvae have also fed *Artemia* nauplii as a standard laboratory diet (Forster & Wickins, 1967; Provenzano, 1967; Roberts, 1972, 1974; Mootz & Epifanio, 1974; Provenzano & Goy, 1976). These authors cite the following advantages of using *Artemia*: its availability regardless of season, its suitable size for many decapod larvae and the fact that it allows complete development of the juvenile stage or beyond with reasonably consistent survival, intermoult duration and morphogenetic sequence.

Its success as a larval diet for laboratory animals was soon recognized widely among aquaculturists. Carlberg & Van Olst (1976) indeed designate *Artemia* nauplii among the most suitable food items for the controlled culture of larval stages of many commercial fish and shellfish. Girin & Person-Le Ruyet (1977) furthermore remark that 40 years after the first trials with *Artemia* as a food for fish larvae, its freshly hatched nauplii have now become an indispensable link in the larval rearing of most fish and marine crustacean species. More recently, Corbin, Fujimoto & Iwai (1983) agree that in aquaculture production around the world, *Artemia* nauplii are the principal food during the first weeks of larval rearing. Since Hudinaga in 1958 for the first time successfully reared *Penaeus japonicus* using *Artemia* nauplii during mysis and postlarval stages (Liao, Su & Lin, 1983), all commercial cultivation of penaeid shrimp species is at present using this practice (see comprehensive articles by Heinen, 1976; Hanson & Goodwin, 1977; Liao *et al.,* 1983). The culture of the freshwater prawn *Macrobrachium* sp. also heavily depends on the use of *Artemia* nauplii; the nauplii are used as the most successful diet throughout the larval rearing period, after one week mostly in combination with prepared diets (White & Stickney, 1973; Dugan, Hagood & Frakes, 1975; Aquacop, 1977; Hanson & Goodwin, 1977; Murai & Andrews, 1978; Corbin *et al.*, 1983).

Although it is common practice to feed adult *Artemia* to lobster larvae, Castell (1977) noticed better survival, colouration, activity and slightly better growth in *Homarus americanus* larvae raised with *Artemia* nauplii. Other decapod species with aquaculture potential such as spiny lobster (Dexter, 1972; Robertson, in Bardach, Ryther & McLarney, 1972; Roberts, 1974; Tholasilingam & Rangarajan, 1980) and *Palaemonetes* spp. (Broad, 1957; Forster & Wickins, 1967; Reeve, 1969a,b; Campillo, 1975; Sandifer & Williams, 1980; Anonymous, 1984) are also successfully cultured using *Artemia* nauplii.

Intensive larval rearing of commercial non-salmonid fish relies almost completely on the use of living food organisms despite considerable effort to develop artificial diets (Bryant & Matty, 1980; Paulsen, 1980). Nauplii of *Artemia* have most often been used as a convenient food for the larvae of cyprinids (Meske, 1973; Huisman, 1974; Bryant & Matty, 1980; Stroband & Dabrowski, 1981; Dabrowski, 1982), milkfish (Juario & Duray, 1981), flatfishes (Riley, 1966; Shelbourne, 1968; Girin, 1974a,b, 1979; Spectorova & Doroshev, 1976; Bromley, 1977; Gatesoupe, Girin & Luquet, 1977; Kingwell, Duggan & Dye, 1977; Dye, 1980; Fuchs, 1981/1982; Gatesoupe & Luquet 1981/1982; Bromley & Howell, 1983, Olesen & Minck, 1983), bass (Girin, Barahona-Fernandes & Le Roux, 1975; Barnabé, 1976, 1980; Barahona-Fernandes & Girin, 1977; Anonymous, 1978b), bream (Kittaka, 1977; Person-Le Ruyet & Verillaud, 1980), whitefish (Günkel, 1979; Flüchter, 1980, 1982), catfish (Hogendoorn, 1980), rabbitfish (Juario *et al.*, 1985), and sturgeons (Gun'ko, 1962; Gunk'ko & Pleskachevskaya, 1962; Azari Takami, 1976, 1985; Oleinikova & Pleskachevskaya, 1979; Binkowski & Czeskleba, 1980).

THE USE OF PREPARED FORMS OF *ARTEMIA* NAUPLII

In most cases live freshly hatched nauplii are used as a food for immediate use. Several authors, however, report experiments with live cold stored, killed, and other prepared forms of *Artemia* nauplii.

Live cold-stored Artemia *nauplii*

Mock, Fontaine & Revera (1980a) and Mock, Revera & Fontaine (1980b) recommend the use of chilled or frozen nauplii as a back-up to safeguard against a batch of cysts that are inferior in hatching quality. They note that freshly hatched *Artemia* nauplii can be concentrated and stored at 11 °C for several days, although careful monitoring is required to prevent mortality and decomposition. In order to minimize this risk they aerate the suspension of nauplii with an airstone and change the water every day. Léger, Vanhaecke & Sorgeloos (1983) described a technique for high density cold storage of *Artemia* nauplii. They showed that, except for the strains from Chaplin Lake (Canada) and Buenos Aires (Argentina), *Artemia* nauplii viability remains over 90% after 48 hours storage at 4 °C. Subsequent transfer to culture tank conditions (25 °C) did not affect *Artemia survival. Léger et al.* (1983) furthermore demonstrated that cold stored nauplii remained in the instar I stage (Hentschel, 1968) and that energetic losses were minimal (see also p. 587). Decreases in nutritional value of cold stored nauplii used as food for *Mysidopsis bahia* and *Cyprinus carpio* larvae are insignificant after 24-hours cold storage and minimal only for carp after 48 hours. This technique provides opportunities for automation in food distribution (Léger & Sorgeloos, 1982) and offers the possibility of frequent feedings without manual mediation over a two-day period (Fig. 2). Because the labour involved in feeding, especially in large-scale operations, is cumbersome and expensive (Fujimura & Okamoto, 1970; Goodwin & Hanson, 1977), this technique looks worth imitating, be it only to store left-overs of freshly hatched nauplii for later feeding.

Another advantage of using cold-stored nauplii is their initially slower movement from which the predator can benefit. Kahan (1979) indeed noticed that first-feeding mullet (*Mugil capito*) larvae were able to handle the slow-moving refrigerated nauplii, while other authors reported that mullet larvae could not handle *Artemia* nauplii prior to the 7th (Nash, Kuo & McConnel, 1974) or the 16th day (Liao, Lu, Huang & Lin, 1971). Sleet & Brendel (1983) have described a system for flow-through hatching and cold storage of the nauplii. They confirm that during cold storage the nauplii remain in their first larval stage, that viability is not affected even after transfer of the stored nauplii to 25 °C and that naupliar length after 48-hours cold storage only increased by 5·4% compared with 80% in the control (25 °C). It may be noticed that while Sleet & Brendel obtained good results with Canadian (Chaplin Lake) *Artemia*, Léger *et al.* (1983) reported poor storage performance for this strain as compared with others.

Frozen and freeze-dried nauplii

The use of killed forms of *Artemia* nauplii eliminates the drawback that the *Artemia* may compete for food with the predator larvae. Mock *et al.*

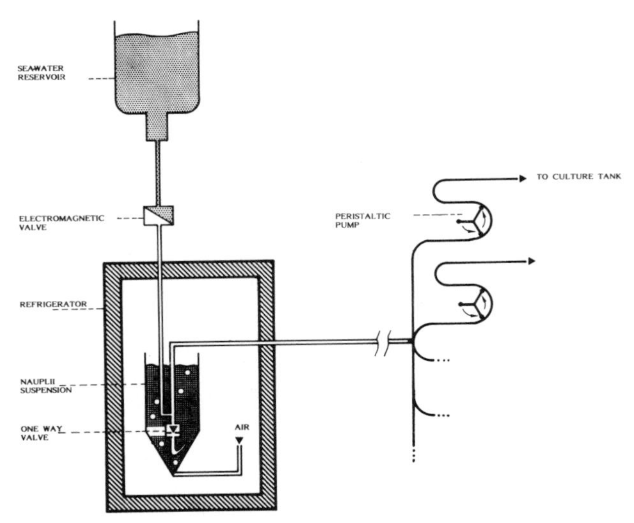

Fig. 2.—Schematic diagram of automatic distribution system for *Artemia* nauplii (modified from Léger & Sorgeloos, 1982).

(1980a,b) observed that *Artemia* nauplii very rapidly consume the algae which are still being fed to the penaeid shrimp larvae when the *Artemia* are first added. This usually results in the on-growing of the *Artemia* to such an extent that, because of their size and swimming speed, they are no longer ingestible by the shrimp larvae which, after all, are not very efficient hunters. To avoid this, Mock *et al.* fed frozen *Artemia* nauplii to zoeal shrimp larvae, *i.e.* a determined amount of *Artemia* was hatched, concentrated and stored after freezing. The frozen block could then either be thawed in sea water before feeding, or the frozen block could be placed directly in the culture tank. According to Mock *et al.* (1980a,b) penaeid shrimp larvae accept frozen nauplii equally well as live *Artemia*. The use of frozen *Artemia* provides, as Mock *et al.* state, a lot of advantages, *e.g.* it ensures a constant food supply, daily food requirements of the predator can be met with higher precision, no more fear that the *Artemia* grow into an unwanted food competitor.

In larval fish rearing frozen *Artemia* nauplii are being used, in the transition of live to artificial diets, aiming to facilitate the acceptance of non-living food. This practice has been described for seabass (*Dicentrarchus labrax*) (Anonymous, 1978b), and sole (*Solea* spp.) larvae (Girin, 1979; Metailler, Menu & Morinière, 1981; Cadena Roa, Huelvan, Le Borgne & Metailler, 1982a; Cadena Roa, Menu, Metailler & Person-Le Ruyet, 1982b; Gatesoupe & Luquet, 1981/1982). Gatesoupe & Luquet also used frozen nauplii as an attractant in re-hydratable extruded pellets.

In his experiments with whitefish (*Coregonus fera*) Günkel (1979) observed that the fry accepted dead nauplii, equally well as live *Artemia*, resulting in similar survival and growth. From these results he assumed that fry could be reared with dry diets. This appeared to be true if they were first fed *Artemia* nauplii and if proper weaning was allowed. Hogendoorn (1980) reported good results in rearing catfish (*Clarias lazera*) larvae using live or frozen *Artemia* nauplii in combination with a trout starter compared with other diets without *Artemia*. He, nevertheless, noticed significantly better growth and survival in the treatment including live nauplii. Fuchs (1981/1982), aiming to simplify the rearing methods for larval sole of Girin (1978), also compared live *versus* frozen *Artemia* nauplii as a food source. Fuchs also concluded that better survival, growth, and food conversion are obtained with live nauplii (Fig. 3). Similarly, Schauer, Richardson & Simpson (1979) and Seidel, Schauer, Katayama & Simpson (1980a) found largely better results feeding juvenile Atlantic silverside (*Menidia menidia*) with live instead of freeze-dried *Artemia* metanauplii. It was postulated by the last authors that something in the *Artemia* was lost or destroyed during the freeze-drying process.

Kentouri (1980) observed that seabass larvae, offered frozen prey which has been thawed for different times, only ingest the most freshly thawed product. He supposed that possible denaturation of vitamins and proteins, or lipid oxidation eventually aggravated by thawing procedures and especially thawing duration may explain inferior results obtained with a diet of frozen food organisms. Following Flüchter (1980) whitefish larvae metamorphose equally well whether they are fed live or shock-frozen (−196 °C) *Artemia* nauplii, but not when fed slow-frozen nauplii. The fish

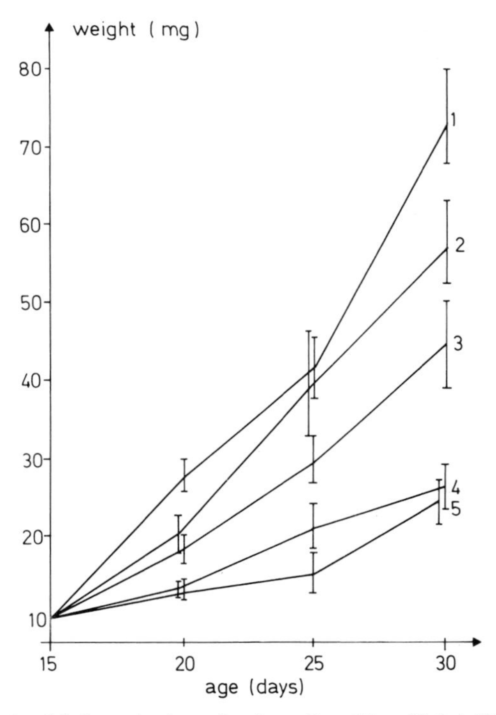

Fig. 3.—Growth of *Solea solea* juveniles from Day 15 to 30 fed different *Artemia* preparations: 1, live nauplii; 2, live plus frozen nauplii; 3, frozen nauplii (4 feeds); 4, frozen nauplii (distributed in 24 h); 5, frozen nauplii (distributed in 15 h); after Fuchs, 1981/1982.

larvae, however, eagerly took the slow-frozen *Artemia* from the bottom of the aquaria and even preferred them to live copepods abundantly present in the aquaria. Flüchter ascribed this feeding response to a strong smell or taste released by dead *Artemia* nauplii and concluded that a substance essential for whitefish larvae is lost during slow-freezing and not during shock-freezing. He assumed this substance to be largely insoluble in water, since during shock-freezing the expansion of the water in the body tissue causes the nauplii to burst. Furthermore, Flüchter postulated that this substance must be connected to the intermediate metabolism and absorbed through enzymatic action which does not stop immediately during slow freezing. Grabner, Wieser & Lakner (1981/1982) indeed proved that activities of proteases as well as enzymes of the intermediate metabolism in food organisms (including *Artemia*) are not diminished by freezing, freeze-drying and by storage at − 18 °C even for very long times. He noticed also that during the process of freezing or freeze-drying tissue cells of food organisms experience large scale damage explaining extensive leaching upon thawing, *i.e.* after 10 min at 9 °C about 70–75% of the activities of proteases and of LDH, and an even larger percentage of the free amino acids have disappeared from the food material and can be recovered in soluble form in the water. Following Grabner *et al.* (1981/1982), losses of essential nutrients during thawing are probably the most important reason why frozen food organisms have proved to be unsuitable for rearing the larvae of several fish species.

Other forms of non-living Artemia *nauplii*

In order to prevent food competition with algae, deterioration of water quality as when using frozen *Artemia*, and metabolism of the energy reserves as in live *Artemia*, Wilkenfeld, Lawrence & Kuban (1984) fed *Penaeus setiferus* larvae with UV-killed *Artemia* nauplii as an inactive food. UV-killed nauplii were obtained by exposing freshly hatched *Artemia* nauplii to four 30W germicidal tubes at 10 $mW \cdot cm^{-1} \cdot s^{-1}$ for one hour. Although they noted clumping of UV-killed *Artemia* and algae, they suggest their potential use as a food source during larval stages of penaeid shrimp. Further experimentation, however, is required to confirm their nutritional stability and possible effects on water quality.

When live *Artemia* nauplii were compared with preserved *Artemia* (dried, stored in brine or as a paste) as food for young sturgeons (*Acipenser stellatus*), the superiority of live *Artemia* was striking (Gun'ko & Pleskachevskaya, 1962; Pleskachevskaya, 1963, in Oleinikova & Pleskachevskaya, 1979), *e.g.* final sturgeon weight was 1141% of initial weight after 35 days when fed on live *Artemia* and only 75% when fed on dried *Artemia*; the weight increase was 764·8% and 53·5%, respectively. It was only 28·1% in larvae fed brined- and 22·5% in larvae fed pasted-*Artemia*.

FACTORS AFFECTING THE SUITABILITY AND NUTRITIONAL EFFECTIVENESS OF *ARTEMIA* NAUPLII

Although *Artemia* nauplii have been and are being used as a suitable food in the culture of numerous aquatic species, problems and constraints related to the use of *Artemia* have been reported by several authors. Besides an undesirable variation in hatching quality (Vanhaecke & Sorgeloos, 1983a) which will not be treated in this article, problems related to unreliable supply and high price, and especially the evidence of a varying nutritional quality have generated intensive research in looking for alternatives for *Artemia*. In this section we shall review and comment on factors affecting the suitability and the nutritional effectiveness of *Artemia* nauplii as a food source; *e.g.* the presence of cyst shells, microbial contamination, nauplius size, effect of feeding starved nauplii, differences in nutritional value of nauplii from different geographical origins.

The presence of cyst shells

Artemia nauplii harvested from the hatching suspension are often contaminated with empty cyst shells (for details on separation problems we refer to Sorgeloos *et al.*, 1983). Although these shells are undigestible (Stults, 1974; Bruggeman, Sorgeloos & Vanhaecke, 1980; MacDonald, 1980), they may be harmful when ingested by larvae. Herald & Rakowicz (1951) indeed observed young seahorses dying through obstruction of their gut by cyst shells. Morris (1956) noticed starvation effects in fish larvae which ingest shells as readily as nauplii and recommended that the nauplii be separated. Shrimp larvae apparently are not affected by the cyst shells as they are often introduced along with the nauplii in some outdoor operations (Heinen, 1976) or as cysts are sometimes incubated for hatching in the culture tank (Mock, pers. comm.). Even when no direct biological effect is seen, this practice is not advised for reasons of water quality. Dissolved hatching products, *e.g.* glycerol (Clegg, 1964) and contaminants carried by the cysts (see below) may indeed affect tank hygiene (MacFarlane, 1969). Several apparatus have been described for separating freshly hatched nauplii from their cyst-shells (Shelbourne, Riley & Thacker, 1963; Riley, 1966; Lenhoff & Brown, 1970; Jones, 1972; Persoone & Sorgeloos, 1972; Nash, 1973; Boyd, 1974; Ward, 1974; Smith *et al.*, 1978). Dissolved wastes and bacteria may be removed by simple washing (Austin & Allen, 1981/1982). The technique of decapsulation of *Artemia* cysts (Sorgeloos *et al.*, 1977, 1983; Bruggeman, Baeza-Mesa, Bossuyt & Sorgeloos, 1979; Bruggeman *et al.*, 1980) makes separation redundant and sterilizes the embryos at the same time.

Microbial contamination

Rakowicz (1972) preferred *Artemia* to natural plankton because the former are free from contagious diseases and parasites. Flüchter (1980) reported a reduced danger for disease introduction by feeding *Artemia* instead of natural zooplankton for coregonid and sturgeon larvae. So far no direct evidence for *Artemia*-borne infections in fish and crustacean larvae has been reported. Nonetheless *Artemia* cyst-shells are known to be contaminated with bacterial and fungal spores (Fig. 4; Wheeler, Yudin & Clark, 1979) and fish or shrimp might be infected *via* introductions with the *Artemia* hatching medium. Heavy bacterial loads have indeed been

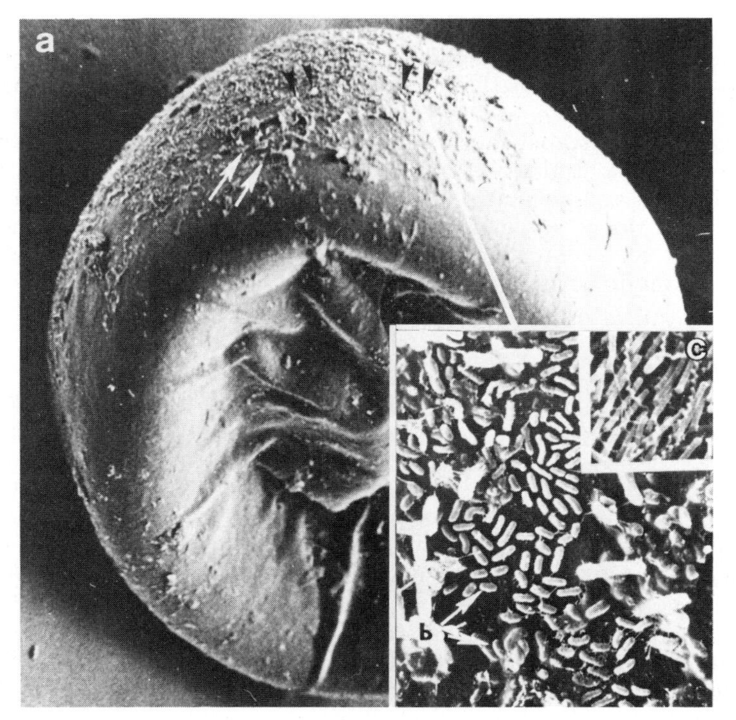

Fig. 4.—Dehydrated *Artemia* cyst covered with microbial material (arrows) a, ×412; b, ×2281; c, ×2500; after Wheeler, Yudin & Clark, 1979.

determined in canned *Artemia* cysts, *i.e.* after 20 to 48 h incubation in sterile sea water from 10^6 to 10^8 colony-forming units·ml^{-1} hatching medium have been counted by Gilmour, McCallum & Allan (1975), Coleman, Nakagawa, Nakamura & Chang (1980), and Austin & Allen (1981/1982). Austin & Allen, however, found no evidence of intimate bacterial colonization of the nauplii themselves and showed that bacteria surrounding *Artemia* nauplii may easily be removed by simple washing procedures. These authors reported the presence of *Bacillus*, *Erevinia*, *Micrococcus*, *Staphylococcus*, and *Vibrio* spp. In this regard several authors prefer to disinfect the *Artemia* cysts prior to their use. Lenhoff & Brown (1970), apprehending bacterial and fungal infections, decontaminate *Artemia* cysts using an 'Antiformin' solution (5·68 g NaOH and 3·2 g Na_2CO_3 in 100 ml of a 5·25% NaClO solution). These authors found the nauplii to be toxic when hatched from cysts disinfected with thiomersal as described by Provasoli & Shiraishi (1959). Sleet & Brendel (1983) sterilize *Artemia* cysts in sequential soakings of 1% sodium hypochlorite, 5% urea, and 13% benzalkonium chloride. After sterilization they resuspend the cysts in sterilized artificial sea water containing 10 μg·ml^{-1} gentamycin sulphate. Disinfection of cysts by hypochlorite treatment is also reported by Corbin *et al.* (1983) and by Artemia Systems (1985). An extreme form of disinfection is obained by decapsulation of the cysts, *i.e.* complete dissolution of the shell in a hypochlorite solution (Sorgeloos *et al.*, 1977,

1983). Coleman *et al.* (1980), in an attempt to increase hatchability, were successful in suppressing bacterial growth during hatching incubation using either 40 $mg \cdot l^{-1}$ veterinary grade chloramphenicol or 50 $mg \cdot l^{-1}$ research grade penicillin-streptomycin. They emphasized, however, the use of antibiotics for experimental testing only, not wishing to propagate their broad application at a production level. Using antibiotics may indeed induce selection and propagation of resistant bacteria and will increase operation costs. For use of *Artemia* on a large scale Coleman *et al.* (1980) suggest other means of suppressing bacterial growth *e.g.* UV-light, chlorination or washing. Oleinikova & Pleskachevskaya (1979) reported the development of moulds *e.g. Penicillium* spp. and *Aspergillus* spp. in unprocessed wet-stored cysts. Because the infested cysts loose their viability and infect the whole lot, the last two authors recommend the removal of mould-infested cysts (application of calcium hypochlorite or burning) and treatment of the rest with a 2% formalin solution before drying.

Nauplius size

The nutritional effectiveness of a food organism is in the first place determined by its ingestibility, and as a consequence by its size and configuration. This was clearly demonstrated by Sulkin & Epifanio (1975) who evaluated rotifers (*Brachionus plicatilis*, 45–180 μm), urchin gastrulae (*Lytechinus variegatus*, 110 μm) and *Artemia* nauplii (250 μm) as food sources for blue crab (*Callinectes sapidus*) larvae. Survival rates averaged 50, 5 and 0%, respectively, the last result being similar to that for the unfed control. They concluded that 110 μm was the maximum prey size for early larvae of the blue crab and suggested feeding rotifers during the first two zoea stages prior to a switch to *Artemia* nauplii (see also Sulkin, 1978). This confirms the observation of Roberts (1972) that *Callinectes sapidus* larvae (stages I, II and III) cannot capture nor ingest *Artemia* nauplii. The same author notes that some decapod species are indeed too small to handle *Artemia* nauplii or have mouth parts that are better suited for handling smaller food organisms. Roberts (1972) cites the example of hermit crab (*Pagurus longicarpus*) larvae which are able to capture *Artemia* nauplii but are often only removing and ingesting its appendages, leaving the body of the nauplius behind. The same observation was made for early zoea stages of *Penaeus marginatus* (Gopalakrishnan, 1976). With the further exception of all *Penaeus* spp. larvae which initially are phytoplankton filter-feeders, most decapod larvae can be reared on *Artemia* nauplii for their complete development (Rice & Williamson, 1970; Provenzano & Goy, 1976). On the contrary, most marine fish larvae cannot be fed *Artemia* nauplii at first-feeding. Morris (1956) indeed stated that the size of *Artemia* nauplii is a serious restriction to their use as food for marine fish larvae, and according to Houde (1973) most fish larvae, including those with relatively large mouths, begin feeding on organisms in the 50–100 μm range (size range of *Artemia* nauplii: 428–517 μm, Vanhaecke, 1983).

In his experiments with lemon sole (*Microstomus kitt*), Howell (1971) found that the fish larvae will first select small mussel trochophores and thereafter rotifers prior to the start of feeding on *Artemia* nauplii. In addition, Hirano & Oshima (1963) observed differences between fish species in the age at which they start to feed on *Artemia*. May (1970) relates this difference to varying morphometry and mouth size. He does not, however, exclude the fact of size differences between strains of *Artemia*. This was effectively demonstrated by Smith (1976) in his feeding tests with bluegill (*Lepomis macrochirus*) larvae. He indeed attributed early larval mortality using freshly hatched Great Salt Lake and older San Francisco Bay *Artemia* nauplii to the size of the *Artemia* nauplii. He observed starvation effects in the larvae fed Great Salt Lake nauplii. These bluegill larvae, however, resumed feeding when they were subsequently fed small freshly-hatched San Francisco Bay nauplii. This and other experiments with both *Artemia* strains allowed Smith to conclude that San Francisco Bay nauplii are smaller than Great Salt Lake nauplii, both varieties are smaller 4 h after hatching than they are when 2 days old, and within any of these groupings there is a substantial range in size.

Size differences between different *Artemia* strains have been reported by D'Agostino (1965), Claus, Benijts & Sorgeloos (1977) and Claus, Benijts, Vandeputte & Gardner (1979) and have been studied extensively by Vanhaecke & Sorgeloos (1980). Beck, Bengtson & Howell (1980) compared the biological effectiveness of freshly hatched nauplii from five geographical strains for the larvae of the Atlantic silverside (*Menidia menidia*). They observed an increasing mortality during the first three days, parelleling the results in the starved control, in the series fed the largest *Artemia* (Margherita di Savoia, Italy). After this critical period further mortalities did not differ from the ones observed in the treatments fed the smaller nauplii. From later culturing tests with the same species, offered eight different *Artemia* strains ranging in size from about 440 to 520 μm, Beck & Bengtson (1982) extrapolated a high correlation between early larval mortality and length of *Artemia* nauplius (Fig. 5). They calculated that the use of *Artemia* nauplii bigger than 480 μm could be expected to result in over 20% mortality in *Menidia menidia* larvae.

When size of freshly hatched *Artemia* nauplii is not normally limiting for ingestion by the predator, it may become so when no adequate feeding regimes are applied (see p. 533). Because prey catching, handling, and ingestion (*e.g.* swallowing compared with biting into species) differ from

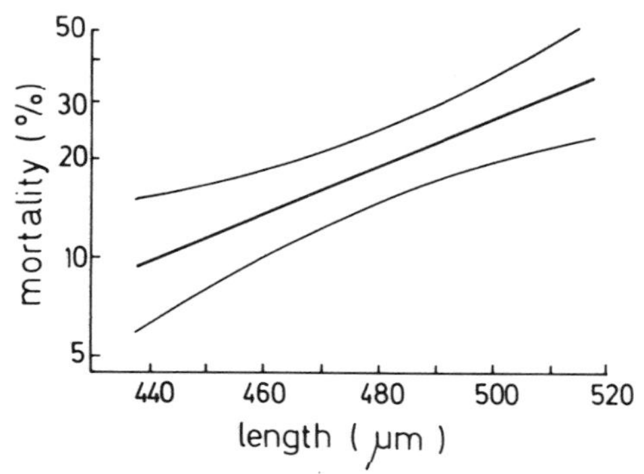

Fig. 5.—Correlation of mortality rate of *Menidia menidia* larvae and naupliar length of *Artemia* fed to the larvae: $\ln \text{mortality} = 15 \cdot 103 + 0 \cdot 0168 \times \text{length}$, or $\text{mortality} = 0 \cdot 006 \times e^{0 \cdot 0168} \times \text{length}$, $r^2 = 0 \cdot 792$; after Beck & Bengtson, 1982.

species to species, size in terms of length may not be the only criterion for morphometrical differences. Body volume of *Artemia* nauplii was considered important by Vanhaecke (1983) who noted very significant differences between strains, *e.g.* the largest difference as found between San Francisco Bay and Italian nauplii was as high as 80%.

Finally an advantage of *Artemia*, when trying to feed optimal sized prey, is that it can be reared to a larger size according to the requirements of the older predator larvae, which for energetical reasons need a larger prey (Sick & Beaty, 1974, 1975; Bryan & Madraisau, 1977). For this the use of ongrown *Artemia* looks most convenient (San Feliu, 1973; Kelly, Haseltine & Ebert, 1977; Girin, 1979; Paulsen, 1980). It was indeed found by Sick & Beaty (1974) that energy intake in *Macrobrachium rosenbergii* stage VIII is directly proportional not only to *Artemia* concentration but also to *Artemia* size. They demonstrated that, in the given experimental conditions, *Macrobrachium rosenbergii* stage VIII attained a maximum energy ingestion of 0·0066 cal·mg animal dry $wt^{-1} \cdot h^{-1}$ when fed 0·7-mm *Artemia* metanauplii, 0·062 when fed 1·5-mm *Artemia* larvae, and 1·014 when fed 5·5-mm *Artemia* juveniles.

Feeding regime

Various aspects related to feeding or 'food addition' *s.l.* appear to play an important rôle in successful shrimp- and fish-farming. The *Artemia* concentrations that are being applied will affect feeding rate, energy uptake and consequently growth, and survival of the predator. Besides, over-feeding may result in fouling stress and under-feeding in cannibalism (Gopalakrishnan, 1976) (Fig. 6). Sick & Beaty (1974) showed that *Macrobrachium rosenbergii* stage VIII larvae did not ingest *Artemia* metanauplii when fed at a concentration of $0 \cdot 1 \cdot ml^{-1}$. Increasing this up to $2 \cdot ml^{-1}$ gradually

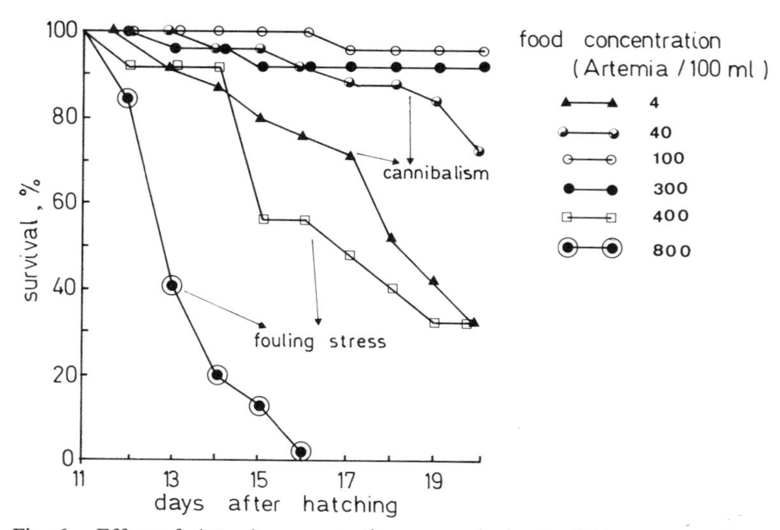

Fig. 6.—Effect of *Artemia* concentration on survival rate of *Penaeus marginatus* (after Gopalakrishnan, 1976).

improved ingestion rate and consequently energy uptake. Other authors (Reeve, 1969a,b; Mootz, 1973; Mootz & Epifanio, 1974; Vijayakumaran & Radhakrishnan, 1980) stress the importance of *Artemia* concentration on developmental rate in decapods. In this regard, Welch & Sulkin (1975) used an *Artemia* concentration of 40 nauplii $\cdot ml^{-1}$ and showed that lower levels increased developmental time; feeding 2 nauplii $\cdot ml^{-1}$ resulted in a significant delay in developmental rate.

Riley (1966) also showed that growth and survival of plaice larvae are markedly affected by the amount of nauplii available. High feeding levels are recommended for first-feeding fish larvae because of their low efficiency in prey catching (Flüchter, 1965; Rosenthal, 1969). Barahona-Fernandes & Girin (1977) agree with the low predatory efficiency in first-feeding fish larvae but advise strict limitation of daily rations of *Artemia* nauplii to match the intake capacity of the fish larvae. They observed that fish larvae eat more when more food is available, but do not grow faster; *i.e.* food conversion ratios appear to be about twice as good at the lowest feeding level as at the highest. Feeding excess food not only results in a lower feeding efficiency, it is a wasteful practice because of the cost of *Artemia* and may even be more dangerous, as a result of the accumulation of metabolites (Houde, 1975), than useful. Riley (1966) also cautioned that although higher feeding rates may increase survival in plaice larvae, excess food is detrimental due to fouling of the culture tanks. Similar observations have been reported in the culture of *Penaeus mondon* larvae (Gopalakrishnan, 1976) and of *Siganus lineatus* larvae (Bryan & Madraisau, 1977). High feeding levels were found to increase consumption in *Penaeus aztecus* mysis but this resulted in poorer survival in postlarval stages (Cook & Murphy, 1969). Roberts (1972) recommended high feeding levels (20 nauplii $\cdot ml^{-1}$) for crab larvae, but added that excessive amounts (80 $\cdot ml^{-1}$) may lead to oxygen depletion in static systems.

Another aspect in feeding practices is the progressive adjustment of the food concentration to the changing requirements of the developing larvae. It is logical to assume that the predator as it grows and develops will require more food. In this regard, Bryant & Matty (1980) have determined optimal *Artemia* rations for developing carp larvae, *i.e.* carp larvae were fed on quantified numbers of *Artemia* nauplii and growth rate was monitored for a 10-day period (Fig. 7). For optimal growth and food conversion, carp larvae were found to require 200–250% of their body weight of nauplii per day during the first 5 days of feeding and only 100–120% per day for the following 5 days. They claim that adjusting food concentrations according to changing requirements with age not only results in a faster growth of the larvae but also in considerable savings of *Artemia* cysts.

Food consumption rates also increase with progressive larval development in decapod larvae (Mootz & Epifanio, 1974), for several species of which daily consumption rates have been determined (*e.g.* Cook & Murphy, 1969; Reeve, 1969a; Omori, 1971; Uno, 1971; Zimmerman, 1973; Rodriguez, 1975; San Feliu, 1973; Shigueno, 1975; Gopalakrishnan, 1976; Heinen, 1976; Emmerson, 1977, 1980, 1984; Vijayakumaran & Rhadakrishnan, 1980; Yufera, Rodriguez & Lúbian, 1984). Differences found by these authors may reflect species specificity, experimental variability, as well as the use of different stages or strains of *Artemia* (*e.g.* varying size,

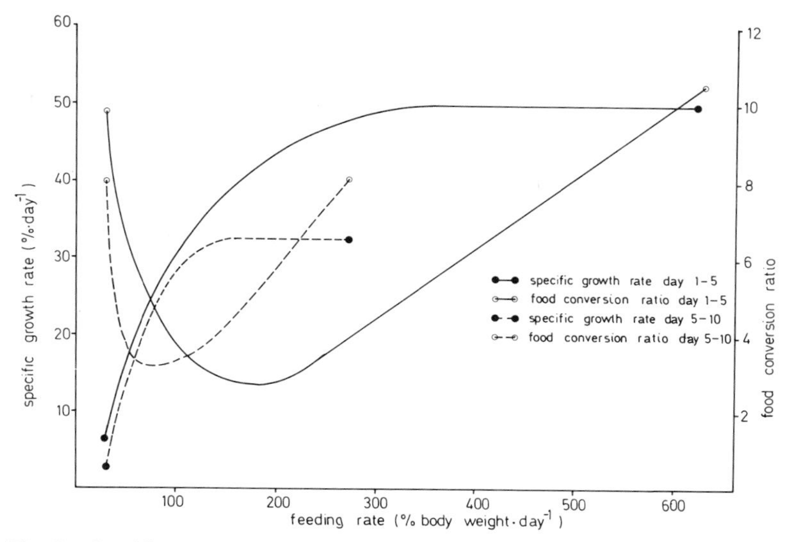

Fig. 7.—Specific growth rate and food conversion ratio of *Cyprinus carpio* larvae fed measured numbers of *Artemia* nauplii during two consecutive periods of five days each (after Bryant & Matty, 1980).

weight, energetic content, and possibly biochemical composition). Optimal feeding levels as established in laboratory studies cannot always be extrapolated to large scale cultures; *e.g.* in most experimental cases *Artemia* nauplii remaining from the previous feeding are removed daily or every other day; a practice which is inconceivable in production situations. Non-ingested *Artemia* nauplii, when not removed before moulting into the second instar stage, will start growing even when no food is available (D'Agostino, 1965; Hentschel, 1968; Sorgeloos, 1975; Smith, 1976; Claus *et al.*, 1979), swim faster (Miller *et al.*, 1979), and may reach a size which is no longer acceptable for the predator (Smith, 1976; Rollefsen in Morris, 1956). Even when acceptable, starved *Artemia* are not as nutritious as freshly hatched ones (see later). Furthermore, when food is available in the culture tank (*e.g.* algae) *Artemia* will not only grow but might also compete with the predator larvae for food and pollute the culture tank with its metabolites. This problem of the on-growing of *Artemia* is classical in penaeid shrimp farming and is aggravated when *Artemia* nauplii are fed during early protozoea stages. These stages eat little and are not very effective in catching and handling prey (Gopalakrishnan, 1976). Feeding protozoea II stage penaeids with *Artemia*, as suggested by Wilkenfeld *et al.* (1984), may indeed give better culture results on a laboratory scale; its application on a commercial scale, however, looks hardly feasible. A convenient solution to that may be the early administration of killed nauplii or decapsulated *Artemia* cysts as suggested by Mock *et al.* (1980a) and Wilkenfeld *et al.* (1984).

Instar-stage

In many cases the retention time in the culture tank of at least a part of the *Artemia* nauplii may exceed 24 h before they are ingested by the predator. This is particularly so when feeding is done *ad libitum* or when inappropriate feeding regimes are applied (see above). As a result part of the *Artemia* are in the second or third larval stage. Several scientists report storage of the freshly hatched nauplii for one or more days prior to feeding them to the predator (*e.g.* Jones, 1972; Tabb, Yang, Hirono & Heinen, 1972; Meske, 1973; Salser & Mock, 1974; L'Herroux, Metailler & Pilvin, 1977; Bengtson, Beck & Poston, 1978; Schauer *et al.*, 1979; Seidel *et al.*, 1980a; Duray & Bagarinao, 1984). Although this practice may be applied unintentionally some authors explicitly state that starvation of the *Artemia* for a few days enhances their nutritional value at least for some predators. Hauenschild (1954, 1956) indeed noticed that polyp stages of the hydrozoan *Hydractina echinata* did not do very well on a diet of freshly hatched *Artemia* but that metanauplii starved for 2 days constituted a better food for the polyps. He attributed this nutritional enhancement of the *Artemia* to a depletion of their fat reserves as a result of starvation. Werner (1968) also allowed *Artemia* nauplii to use up part of their energy-rich reserves prior to feeding them to hydrozoans.

Contrary to these observations with hydrozoans several authors have demonstrated that starved nauplii are nutritionally inferior to freshly hatched nauplii. In his experiments, Morris (1956) observed that when fish larvae were fed only older nauplii they did not grow well, although their guts were properly filled. He attributed this poorer nutritional performance of starved nauplii as food source to their reduced yolk reserves which were exhausted within 2 or 3 days. He noticed that the disappearance of the orange-red yolk was conspicuous in the nauplii even before transition to the second instar stage. Comparing newly hatched and starved nauplii he found the latter empty and chitinous and concluded that one of the primary attributes of the early nauplius, for at least some marine fishes, appears to be its yolk content. Similarly, Wickins (1976) postulated that when *Artemia* nauplii are starved, a depletion of their yolk reserves may result in qualitative or quantitative changes in their normally adequate amino-acid profile which may lead to a chronic nutritional deficiency in *Macrobrachium* larvae.

Dye (1980) and Paulsen (1980) also recommended the use of newly hatched nauplii rich in yolk reserves as food for fish larvae. Devrieze (1984) compared 24-h starved *Artemia* nauplii with newly hatched nauplii of the same strain (Macau, Brazil) as a food source for newborn carp (*Cyprinus carpio*) larvae. At the end of the first week only a slight reduction in growth was noticed in the series fed with starved mentanauplii but the difference became significant in the second week, *i.e.* 37% reduction in individual carp weight after 14 days as compared with the series fed with newly hatched nauplii. In order to satisfy their caloric requirements, the carp larvae apparently have to spend more energy in capturing enough metanauplii which in 24-h old Macau *Artemia* (25 °C) contain 32% less energy when compared with freshly hatched nauplii (Vanhaecke, 1983). This assumption confirms the earlier observations of Radhakrishnan & Vijayakumaran

(1980) that the ingestion rate of *Panulirus homarus* phyllosomae increases when fed with 2-day old instead of 1-day old *Artemia*, *i.e.* 19·3 and 15·1 nauplii ·day^{-1}, respectively. They further found that phyllosomae fed on 2-day old *Artemia* moulted to the fifth stage in 34 days while it took 31·2 days in the other case. Ablett & Richards (1980) also compared 1-day and 2-day starved *Artemia* nauplii for Dover sole (*Solea solea*) larvae. After 40 days mean length increase in fish was 10·4% higher in the 1-day old *Artemia* treatment and after 85 days this difference had grown to 16·3%. They also attributed this difference to the reduced carbohydrate and lipid levels in starved nauplii, *i.e.* even when fed *ad libitum* a greater feeding effort is required to maintain the same level of nutrition. The major reason for the reduced nutritional value of starved *Artemia* metanauplii is indeed the drastic reduction of their individual dry weight and consequently of their energy content during starvation (Paffenhöfer, 1967; Benijts, Vanvoorden & Sorgeloos, 1976; Oppenheimer & Moreira, 1980; Royan, 1980; Vanhaecke, Lavens & Sorgeloos, 1983). Von Hentig (1971) stated that from the onset of embryonic metabolism, the organic content in *Artemia* decreases until food uptake starts in the second instar stage (Hentschel, 1968; Benesch, 1969). Benijts *et al.* (1976) detected a drop in individual dry weight, organic content, energy content, total lipid and fatty acid content of, respectively, 20, 24, 27, 28 and 26% and an increase in ash content of 88% in San Francisco Bay nauplii which had moulted from the first into the second and third instar stage. Similarly, Oppenheimer & Moreira (1980) found a decrease in individual dry weight of approximately 18% in San Francisco Bay nauplii. Vanhaecke (1983) and Vanhaecke *et al.* (1983) studied decreases in individual dry weight and energy content from instar I to instar II and from II to III metanauplii in 15 different *Artemia* strains and measured differences from 16% (Shark Bay, Australia) to 34% (Buenos Aires, Argentina) in the first case and from 22% (Bahia Salinas, Puerto Rico) to 39% (Buenos Aires, Argentina) in the second. Vanhaecke *et al.* (1983) also noticed that for these various strains the dry weight and energy content of instar II-III metanauplii do not follow the ranking for the same characteristics in instar I nauplii; this allowed them to conclude that the rate of dry weight and energy consumption differs from strain to strain, eventually related to differences in swimming behaviour (Miller *et al.*, 1979). The data for dry weight decrease during nauplius starvation as reported by Paffenhöfer (1967) and Royan (1980) do not correspond well with those from the previous authors. Paffenhöfer noted a weight decrease of only 4% after 24 h and Royan reports a 50% decrease from instar I to instar III. It is to be noted, however, that Paffenhöfer did his experiment at 20 °C while Benijts *et al.* (1976) used 28 °C and Vanhaecke *et al.* (1983) 25 °C. Due to this lower temperature it is not impossible that only instar II metanauplii have been measured while Benijts *et al.* (1976) and Vanhaecke *et al.* (1983) analysed a mixed population of instar II-III. Royan does not report the temperature he used but his value applies to metanauplii which were all at instar III.

Oppenheimer & Moreira (1980) observed a 50% decrease in carbon and approximately 12% in nitrogen as *Artemia* moults from the instar I into instar II stage. They ascribe these changes to a period of ''self-absorption'' in *Artemia* during development of the rudimentary mandibles and of a feeding mechanism.

Claus *et al.* (1979) also studied starved compared with freshly hatched nauplii and reported an increase in protein and ash content and a decrease in carbohydrate and lipid content. Furthermore, they observed a change in fatty-acid profile; some fatty acids increased while others decreased. This was also noticed by Watanabe *et al.* (1978c), while Benijts *et al.* (1976) found that the relative proportions of the fatty acids were almost unchanged. The changes in the fatty-acid profile cannot be of great significance in explaining the lower nutritional value of starved metanauplii for marine larvae *i.e.,* the essential fatty acid 20:5ω3 even increases during starvation (Watanabe *et al.*, 1978c; Léger *et al.*, 1983). Claus *et al.* (1979), furthermore, found that the amino-acid profile changed little, but the essential amino acid methionine appeared absent in starved nauplii. Dabrowski & Rusiecki (1983) also analysed amino-acid profiles and contents in starved nauplii, and found some free amino acids to remain constant upon starvation while others decreased 4- to 2-fold; contrary to Claus *et al.* (1979), Dabrowski & Rusiecki measured some increase in methionine content in starved nauplii.

These observations do not minimize the first assumption that, provided their increased size does not interfere with ingestion problems, the reduced nutritional value of starved *Artemia* metanauplii is primarily determined by their reduced energy content. Proper attention has to be paid to the observation of Miller *et al.* (1979) that older nauplii swim faster than freshly hatched *Artemia*. This may indeed constitute an additional increase in energy demand and consumption for prey catching. Similarly important is the suggestion of Dendrinos, Dewan & Thorpe (1984) that loss of orange colour thus reducing the visibility of starved nauplii may to some extent explain their poorer nutritional effectiveness.

The assumption that viability of nauplii may be affected as a result of starvation (Forster & Wickins, 1972) has been rejected by Vanhaecke *et al.* (1983), who found starved *Artemia* nauplii to be very resistant; *i.e.* depending on the strain tested, median lethal time (LT_{50}) values ranged from 73 h to 177 h ($\bar{x}=118$ h) for animals submitted to starvation conditions at 20 °C, and from 42 h to 70 h ($\bar{x}=62$ h) at 30 °C. Even when starved in fresh water, Vanhaecke (1983) recorded LT_{50} values between 16 and 38 h ($x=29$ h).

From all these data it nevertheless looks evident that, perhaps with the exception for some Hydrozoa, freshly hatched instar I nauplii should be fed as a more nutritious food source than starved metanauplii. In order to achieve this prerequisite, application of standard hatching and harvest conditions, as well as proper knowledge of the hatching rate and hatching synchrony of the *Artemia* cysts used is essential. In this context application of the earlier mentioned techniques of cold storage and automated distribution for freshly hatched nauplii is very relevant (Léger & Sorgeloos, 1982; Léger *et al.*, 1983; Sleet & Brendel, 1983; see above). When size is not limiting, the use of fed or enriched metanauplii may be prefered because it solves the problem of nutritional deficiencies (see later).

Strain differences

Table I summarizes the results of culture tests evaluating different strains of *Artemia* for different predators; not all experiments treated in this table are discussed here.

Kuenen (1939) pointed out that the differences which he had observed among different geographical sources of *Artemia* were a potential source of significant variability in experiments in which *Artemia* were used as a food source. His prediction was eventually borne out by Shelbourne (1968) who had to switch from San Francisco Bay *Artemia*, because of their unavailability in early 1966, to Great Salt Lake nauplii for feeding his flatfish (*Pleuronectes platessa* and *Solea solea*) larvae; heavy larval mortality occurred 3 weeks after introducing Great Salt Lake nauplii in the culture tanks. In the same year, Slobodkin (1968) confirmed the poor nutritional value of Great Salt Lake *Artemia* for plaice larvae. He suggested that their "toxicity" could be related to bioaccumulation of residual insecticides from the lake area. Not only flatfish seemed to suffer from a Great Salt Lake *Artemia* diet. *Palaemon serratus* larvae during the first days of their life did equally well on Great Salt Lake as on San Francisco Bay *Artemia*, until metamorphosis, when heavy mortalities occurred in the former (Forster & Wickins, 1967). Forster & Wickens also demonstrated that the food value of Great Salt Lake *Artemia* could be improved in various ways, *e.g.* by mixing with San Francisco Bay nauplii, adding *Isochrysis* in the culture tanks or by feeding the nauplii for 4 days on this alga. They also noticed that no deleterious effects were encountered when Great Salt Lake nauplii were offered during the first 12 days only, followed by a diet of San Francisco Bay nauplii. Reeve (1969a) confirmed these findings with *Palaemon serratus* larvae which became lethargic on a Great Salt Lake diet and died during metamorphosis. Little (1969) and Reed (1969) described similar observations for other decapod larvae (*P. macrodactylus* and *Cancer magister*). In addition, Bookhout & Costlow (1970) reported that four crab species survived better on San Francisco Bay nauplii than on a Great Salt Lake diet; they ascribed the difference to the 3-fold higher concentration of DDT in the Great Salt Lake nauplii.

Wickins (1972) reviewed the available information, on the deleterious effects of Great Salt Lake nauplii as food for marine larvae; in general, negative effects (*e.g.* lethargy, lack of co-ordination, abnormal development, mortality) were manifested around the time of metamorphosis of the predator species. Wickins' (1972) own experiments with *Palaemon serratus* showed that newly hatched or starved Great Salt Lake nauplii were an inadequate food, but the same nauplii could be acceptable when fed on *Isochrysis*. His comparison of the chemical composition of newly hatched nauplii from Great Salt Lake and San Francisco Bay, in terms of pesticides, heavy metals, carotenoids, sterols, and fatty acids, yielded no differences that "could be confidently labelled as the cause of the poor food value of the Utah (Great Salt Lake) *Artemia* nauplii". In any case, the fact that feeding nauplii on *Isochrysis* improved their food value was an indication that the Great Salt Lake *Artemia* problem might be one of nutritional deficiency rather than of contamination.

TABLE I

Summary of results of culture tests using different strains of Artemia *for different predators: only those culture tests are considered in which the source of* Artemia *is relevant; +, ± and − refer to relative values, i.e. good, medium and inferior results, respectively, adjudged by interpretation of the given data; these annotations compare values within one experiment, so that no absolute comparison may be made between different experiments even when the same test organism was used; *see Sorgeloos (1980b); n.s., source not specified*

Artemia source	Species tested	Survival	Growth	Development and/or metamorphosis	Remarks	Reference
AUSTRALIA						
Shark Bay (No. 113)	*Menidia menidia*	+	+			Beck *et al.*, 1980
Shark Bay (No. 113)	*Cancer irroratus*	+	+			Johns *et al.*, 1980
Shark Bay (No. 113)	*Rhithropanopeus harrisii*	+	+			Johns *et al.*, 1980
Shark Bay (No. 113)	*Pseudopleuronectes americanus*	+	+			Klein-MacPhee *et al.*, 1980
Shark Bay (No. 113)	*Mysidopsis bahia*	+	+			Johns *et al.*, 1981a
Shark Bay (No. 113)	*Cyprinus carpio*	+	+			Vanhaecke & Sorgeloos, 1983b
BRAZIL						
Macau (No. 871172, 1978)	*Menidia menidia*	+	±			Beck *et al., 1980*
Macau (No. 871172, 1978)	*Cancer irroratus*	+	+			Johns *et al.*, 1980
Macau (No. 871172, 1978)	*Rhithropanopeus harrisii*	+	+			Johns *et al., 1980*
Macau (No. 871172, 1978)	*Pseudopleuronectes americanus*	+	±			Klein-MacPhee *et al.*, 1980
Macau (No. 871172, 1978)	*Mysidopsis bahia*	+	+			Johns *et al.*, 1981a;
Macau (No. 871172, 1978)	*Cyprinodon variegatus*	+				Usher & Bengtson, 1981
Macau (No. 871172, 1978)	*Cyprinus carpio*	+	±			Vanhaecke & Sorgeloos, 1983b
Macau	*Thalamita crenata*	+	+	+		Krishnan, unpubl.
n.s.	*Libinia emarginata*	+				Goy & Costlow, 1980

	Menippe mercenaria	+					Goy & Costlow, 1980
	Palaemonetes pugio	+			+		Goy & Costlow, 1980
	Rhithropanopeus harrisii	+					Goy & Costlow, 1980
n.s.	*Scophthalmus maximus*	+	+			Better growth, survival and acceptance of dry food as compared with San Francisco Bay nauplii.	Howell *et al.*, 1981
n.s.	*Gadus* sp.	+	+			Better nutritional value than San Francisco Bay *Artemia* nauplii, because they contain certain polyunsaturated fatty acids.	Anonymous, 1982
	Scophthalmus sp.	+	+			Better nutritional value than San Francisco Bay *Artemia* nauplii, because they contain certain polyunsaturated fatty acids.	Anonymous, 1982
n.s.	*Morone saxatilis*	+					Westin *et al.*, 1983, 1985
BULGARIA							
Burgas-Pomorije	"freshwater fish"	+	+				Ljudskanova & Joshev, 1972
CANADA							
Chaplin Lake (No. 5002)	*Panulirus interruptus*	+	+			Considerable variation with regard to source of *Artemia*; Chaplin Lake most successful (other sources not defined).	Dexter, 1972
Chaplin Lake	*Palaemonetes pugio*	+			+	At least equal in quality to San Francisco Bay *Artemia* nauplii.	Provenzano & Goy, 1976
Chaplin Lake (1979)	*Menidia menidia*	±	−				Beck & Bengtson, 1982
Chaplin Lake (1979)	*Pseudopleuronectes americanus*	+	±				Klein-MacPhee *et al.*, 1982
Chaplin Lake (1979)	*Rhithropanopeus harrisii*	±	±				Seidel *et al.*, 1982
Chaplin Lake (1979)	*Cyprinus carpio*	+	−				Vanhaecke & Sorgeloos, 1983b
Chaplin Lake (1979)	*Mysidopsis bahia*	±	±				Léger & Sorgeloos, 1984
Lake Saskatchewan	*Pagrus major*	+	±			Good survival but slightly slower growth may be observed.	Fujita *et al.*, 1980 Watanabe *et al.*, 1980

TABLE I—*continued*

Artemia source	Species tested	Survival	Growth	Development and/or metamorphosis	Remarks	Reference
CHINA P.R.						
Tientsin (1979)	*Menidia menidia*	+	+			Beck & Bengtson, 1982
Tientsin (1979)	*Pseudopleuronectes americanus*	+	+			Klein-MacPhee *et al.*, 1982
Tientsin (1979)	*Rhithropanopeus harrisii*	+	+			Seidel *et al.*, 1982
Tientsin (1979)	*Cyprinus carpio*	+	+			Vanhaecke & Sorgeloos, 1983b
Tientsin (1979)	*Mysidopsis bahia*	+	+			Léger & Sorgeloos, 1984
n.s.	*Macrobrachium rosenbergii*	–			Larvae died within a few days, probably due to high levels of BHCs and DDT.	Matsuoka, 1975
n.s.	*Libinia emarginata*	+		+		Goy & Costlow, 1980
	Menippe mercenaria	+		+		Goy & Costlow, 1980
	Rhithropanopeus harrisii	+		+		Goy & Costlow, 1980
n.s.	'Sobaity'	+	+		Better survival and growth than with Great Salt Lake nauplii.	James *et al., 1982*
COLOMBIA						
Galera Zamba	*Mysidopsis bahia*	+	+			Léger, unpubl.
Manaure	*Mysidopsis bahia*	–	–		Good source after HUFA-enrichment.	Léger, unpubl.
Manaure	*Thalamita crenata*	–		–		Krishnan, pers. comm.
CYPRUS						
	Dicentrarchus labrax	+	+		Comparable with San Francisco Bay *Artemia* nauplii.	Person-Le Ruyet & Salaun, 1977
	Solea solea	+	+		Comparable with San Francisco Bay *Artemia* nauplii.	Person-Le Ruyet & Salaun, 1977

FRANCE						
Salins du Midi	*Palaemon serratus*	–	–	–	Abnormal development and lack of coordination.	Campillo, 1975
Salins du Midi	*Solea solea*	+	+		Comparable with San Francisco Bay nauplii and even some better growth noted.	Fuchs, 1976
Salins du Midi	*Dicentrarchus labrax*	+	+		Comparable with San Francisco Bay nauplii and even some better growth noted.	Fuchs & Person-Le Ruyet, 1976
	Scophthalmus maximus	+	+		Comparable with San Francisco Bay nauplii and even some better growth noted.	Fuchs & Person-Le Ruyet, 1976
Salins du Midi	*Dicentrarchus labrax*	+	+		Not different from San Francisco Bay nauplii.	Godeluck, 1981
Salins du Midi, Lavalduc (1979)	*Menidia menidia*	±	+			Beck & Bengtson, 1982
Salins du Midi, Lavalduc (1979)	*Pseudopleuronectes americanus*	±	+			Klein-MacPhee *et al.*, 1982
Salins du Midi, Lavalduc (1979)	*Rhithropanopeus harrisii*	+	+			Seidel *et al.*, 1982
Salins du Midi, Lavalduc (1979)	*Cyprinus carpio*	+	+			Vanhaecke & Sorgeloos, 1983b
Salins du Midi, Lavalduc (1979)	*Mysidopsis bahia*	+	+			Léger & Sorgeloos, 1984
n.s.	*Libinia emarginata*	+				Goy & Costlow, 1980
	Menippe mercenaria	+				Goy & Costlow, 1980
	Rhithropanopeus harrisii	+				Goy & Costlow, 1980
INDIA						
Kutch-Mundra (1979)	*Mysidopsis bahia*	+	+		Cysts originating from SFB2596 inoculation.	Vos *et al., 1984*
IRAN						
Ormia Lake	*Acipenser* sp.	+	+			Azari Takami, 1976, 1985
ITALY						
Margherita di Savoia (1977)	*Menidia menidia*	–	+			Beck *et al.*, 1980
Margherita di Savoia (1977)	*Cancer irroratus*	+	+			Johns *et al.*, 1980
Margherita di Savoia (1977)	*Rhithropanopeus harrisii*	+	+			Johns *et al., 1980*
Margherita di Savoia (1977)	*Pseudopleuronectes americanus*	+	+			Klein-MacPhee *et al.*, 1980

TABLE I—*continued*

Artemia source	Species tested	Survival	Growth	Development and/or metamorphosis	Remarks	Reference
Margherita di Savoia (1977)	*Mysidopsis bahia*	+	+			Johns *et al.*, *1981a*
Margherita di Savoia (1977)	*Cyprinus carpio*	+	+			Vanhaecke & Sorgeloos, 1983b
Margherita di Savoia (1977)	*Morone saxatilis*	+				Westin *et al.*, 1983, 1985
n.s.	*Libinia emarginata*	–		–	Abnormal megalopae.	Goy & Costlow, 1980
	Menippe mercenaria	–		–	Abnormal megalopae.	Goy & Costlow, 1980
	Palaemonetes pugio	–		–		Goy & Costlow, 1980
	Rhithropanopeus harrisii	–		–	Abnormal megalopae.	Goy & Costlow, 1980
PERU						
Piura	*Mysidopsis bahia*	+	+			Léger, unpubl.
PHILIPPINES						
Barotac Nuevo (1978)	*Mysidopsis bahia*	+	+		Cyst originating from SFB2596 inoculation.	Vos *et al.*, 1984
Jaro	*Mysidopsis bahia*	±	–		Cysts originating from Barotac Nuevo *inoculation (deficient in 20:5ω3)*.	Vos *et al.*, 1984
REFERENCE ARTEMIA CYSTS*						
	Menidia menidia	+	+			Beck & Bengtson, 1982
	Pseudopleuronectes americanus	+	+			Klein-MacPhee *et al.*, *1982*
	Rhithropanopeus harrisii	+	+			Seidel *et al.*, 1982
	Mysidopsis bahia	+	+			Léger & Sorgeloos, 1984
SPAIN						
Cadiz	*Penaeus kerathurus*	+	+	+		Rodriguez, 1975
Cadiz	*Penaeus kerathurus*	+	+	+		Yufera *et al.*, 1984

THAILAND						
Bangpakong (1979)	*Mysidopsis bahia*	+	+		Cysts originating from SFB1728 inoculation.	Vos *et al.*, 1984
TUNISIA						
Mégrine	*Mysidopsis bahia*	+	+			Van Ballaer *et al.*, 1985
Sfax	*Mysidopsis bahia*	+	+			Van Ballaer *et al.*, 1985
TURKEY						
Izmir	*Dicentrarchus labrax*		+			Uçal, 1979
U.S.A.						
Great Salt Lake	*Palaemon serratus*	–	–		Larvae did well until metamorphosis when heavy mortalities occurred; the value of Great Salt Lake *Artemia* could be improved by adding San Francisco Bay nauplii or by pre-feeding them on *Isochrysis*.	Forster & Wickins, 1967
Great Salt Lake	*Pleuronectes platessa*	–	–		Inability to support good growth and survival in flatfish larvae.	Shelbourne, 1968
	Solea solea	–	–		Inability to support good growth and survival in flatfish larvae.	Shelbourne, 1968
Great Salt Lake	*Palaemon macro-dactylus*	–	–		Inferior to San Francisco Bay *Artemia*.	Little, 1969
Great Salt Lake	*Cancer magister*	–	–			Reed, 1969
Great Salt Lake	*Palaemon serratus*	–	–		Larvae did well until metmorphosis and then became lethargic and died during metamorphosis.	Reeve, 1969a,b
Great Salt Lake	*Callinectes sapidus*	+		+	Only 1 abnormal crab stage, survival comparable with San Francisco Bay *Artemia* nauplii.	Bookhout & Costlow, 1970
	Hexapanopeus angustifrons	–		–	Abnormal megalopae, none reached 1st crab stage.	Bookhout & Costlow, 1970
	Labinia emarginata	–		–	Only 5% reached first crab.	Bookhout & Costlow, 1970
	Rhithropanopeus harrisii	–		–	Abnormal megalopae, none reached 1st crab stage.	Bookhout & Costlow, 1970

TABLE I—*continued*

Artemia source	Species tested	Survival	Growth	Development and/or metamorphosis	Remarks	Reference
Great Salt Lake	*Palaemon elegans*	–		–	Only a few unhealthy postlarvae passed through metamorphosis compared with those which were fed San Francisco Bay nauplii.	Wickins, 1972
	Palaemon serratus	–		–	*Abnormalities, poor metamorphosis success* and low survival could be significantly improved by adding *Isochrysis* to the culture tank, or by feeding *Artemia* on this alga, or by replacing Utah nauplii by San Francisco Bay nauplii after 12 days; starving the nauplii did not improve their quality.	Wickins, 1972
	Poecilia reticulata	+	+		Good survival but significantly smaller larvae than when fed San Francisco Bay *Artemia*.	Wickins, 1972
Great Salt Lake	*Palaemonetes pugio*	–		±	Appearance of supernumerary stages but not with San Francisco Bay and Chaplin Lake *Artemia*.	Provenzano & Goy, 1976
Great Salt Lake	*Lepomis macrochirus*	–			Early mortality, presumably from starvation due to nauplius size.	Smith, 1976
Great Salt Lake (Lot 185, 1977)	*Menidia menidia*	+	+			Beck *et al.*, 1980
Great Salt Lake	*Libinia emarginata*	+		+		Goy & Costlow, 1980
	Menippe mercenaria	+		+		Goy & Costlow, 1980
	Palaemonetes pugio	+		+		Goy & Costlow, 1980
	Rhithropanopeus harrisii	+		+		Goy & Costlow, 1980
Great Salt Lake	*Cancer irroratus*	–	–		Total mortality.	Johns *et al.*, 1980
Great Salt Lake (Lot, 185, 1977)	*Rhithropanopeus harrisii*	–	–		Total mortality	Johns *et al.*, 1980
Great Salt Lake	*Pseudopleuronectes americanus*	–	±			Klein-MacPhee *et al.*, 1980
	Mysidopsis bahia	+	+			Johns *et al.*, 1981a

Great Salt Lake	'Sobaity'	–	±		Results were inferior compared with Chinese nauplii.	James *et al.*, 1982
Great Salt Lake	*Cyprinus carpio*	+	+			Vanhaecke & Sorgeloos, 1983b
Great Salt Lake (North arm)	*Dicentrarchus labrax*	–	–		Good source after HUFA-enrichment.	Van Ballaer *et al.*, 1985
	Mysidopsis bahia	–	–		Good source after HUFA-enrichment.	Léger, unpubl.
Great Salt Lake (North and South arm)	*Penaeus stylirostris*	–	–	–	Good source after HUFA-enrichment.	Léger, unpubl.
	Penaeus vannamei	–	–	–	Good source after HUFA-enrichment.	Léger, unpubl.
San Francisco Bay	*Palaemon serratus*	+	+	+	Good survival and metamorphosis compared with Great Salt Lake *Artemia*.	Forster & Wickins, 1967
San Francisco Bay	*Pleuronectes platessa*	+	+	+		Shelbourne, 1968
	Solea solea	+	+	+		Shelbourne, 1968
San Francisco Bay	*Palaemon serratus*	+	+	+	Good survival and metamorphosis compared with Great Salt Lake *Artemia*.	Reeve, 1969a,b
San Francisco Bay	*Callinectes sapidus*	+		+		Bookhout & Costlow, 1970
	Hexapanopeus angustifrons	+		+		Bookhout & Costlow, 1970
	Libinia emarginata	+		+		Bookhout & Costlow, 1970
	Rhithropanopeus harrisii	+		+		Bookhout & Costlow, 1970
San Francisco Bay	*Palaemon elegans*	+		+	Good survival and metamorphosis compared with Great Salt Lake *Artemia*.	Wickins, 1972
	Palaemon serratus	+	+	+	Good survival and metamorphosis compared with Great Salt Lake *Artemia*.	Wickins, 1972
	Poecilia reticulata	+	+		Good survival and better growth than when fed Great Salt Lake *Artemia*.	Wickins, 1972
San Francisco Bay	*Palaemon serratus*	+	+	+	Satisfactory source.	Campillo, 1975
San Francisco Bay	*Uca puligator*	+		+		Christiansen & Yang, 1976
San Francisco Bay	*Dicentrarchus labrax*	+	+		Comparable with French nauplii.	Fuchs & Person-Le Ruyet, 1976
	Scophthalmus maximus	+	+		Comparable with French nauplii.	Fuchs & Person-Le Ruyet, 1976
	Solea solea	+	+		Comparable with French nauplii.	Fuchs & Person-Le Ruyet, 1976
San Francisco Bay	*Palaemonetes pugio*	+		+		Provenzano & Goy, 1976
San Francisco Bay	*Lepomis macrochirus*	+	+		Older nauplii too large as first food.	Smith, 1976

TABLE I—*continued*

Artemia source	Species tested	Survival	Growth	Development and/or metamorphosis	Remarks	Reference
San Francisco Bay	*Dicentrarchus labrax*	+	+		Comparable with Cyprus-*Artemia*.	Person-Le Ruyet & Salaun, 1977
	Solea solea	+	+		Comparable with Cyprus-*Artemia*.	Person-Le Ruyet & Salaun, 1977
San Francisco Bay	*Callinectes sapidus*	+		+		Bigford, 1978
	Libinia emarginata	+		+		Bigford, 1978
San Francisco Bay (marine type)	*Pagrus major*	+	+			Watanabe *et al.*, 1978a, 1980, 1982
San Francisco Bay (freshwater type)		–	–		'Freshwater type' nauplii exhibit a high mortality at the 6th day and a shock syndrome during the activity test; their nutritional value is enhanced after feeding on marine *Chlorella*, ω-yeast, or emulsified cuttlefish liver oil; essential fatty acids are the principal factor for the food value of *Artemia* nauplii.	Watanabe *et al.*, 1978a, 1980, 1982
San Francisco Bay	*Gadus morhua*	–			Rearing through metamorphosis is enhanced when nauplii are pre-fed on *Isochrysis* (2 days), plus addition of the same algae and *Pavlova* to the tanks.	Howell, 1979b
San Francisco Bay (No. 198)	*Cyprinus carpio*	+	+			Bryant & Matty, 1980
San Francisco Bay	*Libinia emarginata*	–		+		Goy & Costlow, 1980
	Menippe mercenaria	–		+		Goy & Costlow, 1980
	Palaemonetes pugio	+	+	+		Goy & Costlow, 1980
	Rhithropanopeus harrisii	–		+		Goy & Costlow, 1980
San Francisco Bay (1977)	*Libinia emarginata*	–		–	Abnormal megalopae.	Goy & Costlow, 1980

	Menippe mercenaria	–		–	Higher incidence of supernumerary stages.	Goy & Costlow, 1980
	Palaemonetes pugio	–		–	Higher incidence of supernumerary stages, abnormal megalopae.	Goy & Costlow, 1980
	Rhithropanopeus harrisii	–		–	Abnormal megalopae.	Goy & Costlow, 1980
San Francisco Bay	*Scophthalmus maximus*	±	±		Survival and growth of larvae was inferior to the ones fed with Brazilian *Artemia*; subsequent acceptance of dry food was also inferior; some improvement when nauplii were fed (4h) on *Isochrysis*.	Howell *et al.*, 1981
San Francisco Bay	*Mylio macrocephalus*	–			No larvae survived beyond 18th day; high 18:3ω3 content may be causal.	Lee *et al.*, 1981
San Francisco Bay	*Gadus* sp.	–	–		Inferior to Brazilian *Artemia*; inadequate to rear cod; improved when fed *Isochrysis*.	Anonymous, 1982
	Scophthalmus sp.	–	–		Inferior to Brazilian *Artemia*; inadequate to rear turbot; improved when fed *Isochrysis*.	Anonymous, 1982
San Francisco Bay (No. 236–2016)	*Penaeus stylirostris*	+	+	+		Léger *et al.*, *1985a*
San Francisco Bay No. 2596)	*Mysidopsis bahia*	+	+			Vos *et al.*, *1984*
San Francisco Bay (No. 728)	*Mysidopsis bahia*	+	+			Vos *et al.*, 1984
San Francisco Bay (14 different lots)	*Mysidopsis bahia*				Varying results according to 20:5ω3 level in nauplii.	Léger *et al.*, 1985c
San Pablo Bay (No. 1628, 1978)	*Menidia menidia*	–	±			Beck *et al.*, 1980
San Pablo Bay	*Libinia emarginata*	–		–	Abnormal megalopae.	Goy & Costlow, 1980
	Menippe mercenaria	–		–	Higher incidence of supernumerary stages, abnormal megalopae.	Goy & Costlow, 1980
	Palaemonetes pugio	–		–	Higher incidence of supernumerary stages.	Goy & Costlow, 1980
	Rhithropanopeus harrisii	–		–	Abnormal megalopae.	Goy & Costlow, 1980
San Pablo Bay No. 1628, 1978)	*Cancer irroratus*	–	–	–	Total mortality.	Johns *et al.*, 1980
	Rhithropanopeus harrisii	–	–	–	Total mortality	Johns *et al.*, 1980

TABLE I—*continued*

Artemia source	Species tested	Survival	Growth	Development and/or metamorphosis	Remarks	Reference
San Pablo Bay (No. 1628, 1978)	*Pseudopleuronectes americanus*	–	–			Klein-MacPhee *et al.*, 1980, 1982
San Pablo Bay (No. 1628, 1978)	*Mysidopsis bahia*	±	–			Johns *et al.*, 1981a
San Pablo Bay (No. 1628, 1978)	*Cyprinodon variegatus*	+	+			Usher & Bengtson, 1981
San Pablo Bay (No. 1628, 1978)	*Cyprinus carpio*	+	±			Vanhaecke & Sorgeloos, 1983b
San Pablo Bay	*Morone saxatilis*	+			Survival as good as with Brazilian *Artemia*.	Westin *et al.*, 1983, 1985
San Pablo Bay (No. 1628, 1978)	*Penaeus stylirostris*	–	–	+	Survival and growth were significantly improved by HUFA-enrichment, and were also determined by the pre-*Artemia* diet quality.	Léger *et al.*, 1985a
San Pablo Bay (No. 1628, 1978)	*Mysidopsis bahia*	–	–			Léger *et al.*, 1985c
San Pablo Bay (No. 1628, 1978)	*Dicentrarchus labrax*	–	–			Van Ballaer *et al.*, 1985
U.S.S.R.						
n.s.	*Acipenser* sp.	+	+		Live nauplii are better than stored nauplii.	Gun'ko & Pleskachevskaya, 1962

Subsequently, Dexter (1972) noted that growth and survival of *Panulirus interruptus* varied with source of *Artemia* but stated, without mentioning other sources, that the best results were obtained with Chaplin Lake (Canada) *Artemia* nauplii. Provenzano & Goy (1976) found Chaplin Lake *Artemia* nauplii at least equal in quality to San Francisco Bay *Artemia* nauplii. *Palaemon serratus* larvae fed nauplii from France (Salins du Midi) exhibited slower development and less successful metamorphosis to post-larvae than when fed nauplii from California (Campillo, 1975). Metamorphosis was not only retarded, but post-metamorphosis survival was also much lower. Campillo reported several other developmental abnormalities with a diet of French *Artemia*, *e.g.* perturbation of moulting sychronism, abnormal appendices, and rostrum, incomplete pigmentation, lack of co-ordination. None the less, several other authors reported good culture performance with French *Artemia*, *e.g.* Fuchs & Person-Le Ruyet (1976) for seabass (*Dicentrarchus labrax*), sole (*Solea solea*), and turbot (*Scophthalmus maximus*), and Godeluck (1981) also for seabass.

Brazilian *Artemia* have so far not been reported to be nutritionally questionable. Some authors find Brazilian *Artemia* to be even superior to San Francisco Bay *Artemia* (Howell, Bromley & Adkins, 1981; Anonymous, 1982). As to Chinese *Artemia*, Matsuoka (1975) observed that *Macrobrachium rosenbergii* larvae died within a few days when fed *Artemia* from this source, probably due to high levels of BHCs and DDT. James, Bou-Abbas & Dias (1982), on the contrary, observed equal growth and survival in larvae when fed Chinese or Great Salt Lake *Artemia* nauplii.

Investigations of the nutritional adequacy, in terms of essential fatty acids (EFA) in *Artemia* nauplii from San Francisco Bay, South America, and Canada indicated that brine shrimp nauplii can be classified into two categories, *i.e.* high in 18:3ω3, the EFA for freshwater fish, or those high in 20:5ω3, the EFA for marine fish (Watanabe *et al.*, 1978b,c). When the Canada strain (5·2% 20:5ω3) was fed to red seabream, *Pagrus major*, 68% of the fish survived, but when the San Francisco Bay strain (1·6% 20:5ω3) was fed, only 43% survived (Watanabe, Oowa, Kitajima & Fujita, 1980). When the San Francisco Bay nauplii were reared on *Chlorella* or ω-yeast for 24 h, the survival of fish to which they were fed increased to 67% and 86%, respectively. Watanabe, Ohta, Kitajima & Fujita (1982) later confirmed that larval survival in flounder (*Paralichthys olivaceus*) and rock seabream (*Oplegnathus fasciatus*) was also low when fed with low-20:5ω3 San Francisco Bay nauplii but could be improved by feeding the nauplii ω-yeast or cuttlefish liver oil (both rich in 20:5ω3) before presentation to the fish.

A systematic survey of geographical strains by the International Study on *Artemia* (ISA) has provided the bulk of the information on variation in nutritional quality of nauplii. In the ISA survey, a total of eight geographical strains were fed to several fish and crustacean species. The strains tested were from Australia (Shark Bay, lot 114), Brazil (Macau, lot 871172), Canada (Chaplin Lake, 1979 harvest), China (Tientsin, 1979 harvest), France (Lavalduc, 1979 harvest), Italy (Margherita di Savoia, 1977 harvest), and the United States (Great Salt Lake, lot 185, and San Pablo Bay, lot 1628). In addition, an ISA standard reference sample (Reference *Artemia* Cysts RAC, of undisclosed location, Sorgeloos, 1980b) was also tested. All eight strains were fed to three fish species (Atlantic silverside, *Menidia menidia*; winter flounder, *Pseudopleuronectes americanus*; and carp,

Cyprinus carpio) and two crustacean species (mud crab, *Rhithropanopeus harrisii*, and mysid, *Mysidopsis bahia*). Some of the strains were also fed to another fish (sheepshead minnow, *Cyprinodon variegatus*) and another crustacean (rock crab, *Cancer irroratus*). The survival data for the fish and crustacean larvae fed on the various ISA-strains are summarized in Table II. Patterns can be distinguished by reading rows and columns of data. For example, certain species (*Cyprinus carpio*, *Cyprinodon variegatus*) survived well regardless of *Artemia* strain, whereas other species (*Rhithropanopeus harrisii*, *Cancer irroratus*) were profoundly affected by the strains they were fed. Certain strains, *e.g.* Brazil and RAC seemed to be a good food for all the species tested, whereas some strains (*e.g.* Great Salt Lake and San Pablo Bay) were poor for several species; one strain (Italy) was poor for only one species, and one strain (Canada) was mediocre for most species. More information could also be obtained from the time course of mortality for each species. Species that undergo a pronounced metamorphosis (*Pseudopleuronectes americanus*, *Rhithropanopeus harrisii*, and *Cancer irroratus*) suffered almost all the mortality at the time of metamorphosis when fed a poor-quality strain. This phenomenon had been noticed previously for other species (Forster & Wickins, 1967; Shelbourne, 1968; Reeve, 1969a; Bookhout & Costlow, 1970; Wickins, 1972; Campillo, 1975). In most of those cases, survival was excellent up to the time of metamorphosis, when nearly 100% mortality occurred within a very few days. On the other hand, most mortality in culture tests with fish that do not undergo metamorphosis (*e.g.* *Menidia menidia*) occurred early in the experiment (Beck *et al.*, 1980) indicating that the causes of mortality in the different species may have been diverse.

Johns, Berry & McLean (1981b) designed an experiment to determine whether the nutritional factors in Great Salt Lake and San Pablo Bay *Artemia* causing deleterious effects in *Rhithropanopeus harrisii* larvae were acquired cumulatively or only during certain critical periods of development. They divided the larval development period into three parts: hatching to Day 5, Day 5 to Day 9, Day 9 to Day 11 (metamorphosis). The food source used during each part (Brazil, Great Salt Lake or San Pablo Bay) was varied to produce a total of 11 different feeding combinations, although each combination consisted of a maximum of two sources (*e.g.*, a three-part combination might be Brazil–Brazil–Great Salt Lake or San Pablo–San Pablo–Brazil). They found that total mortality of larvae at metamorphosis occurred only if the larvae received Great Salt Lake or San Pablo Bay for the first 9 days of the development. The type of food being given at the time of metamorphosis was irrelevant to the survival rate compared with what had been given during the first 9 days. This allowed Johns *et al.* (1981b) to conclude that the factor causing mortality was either cumulatively acquired with the diet or was cumulatively deficient in the diet.

In addition to the survival data, the ISA studies also provide results for several fish and crustacean species on growth, rate of development (time to metamorphosis), and reproduction. An examination of growth data for animals raised on the strains that gave poor (Great Salt Lake, San Pablo Bay) or mediocre (Canada) survival results provides a few clear-cut patterns, *i.e.* growth in *Pseudopleuronectes americanus*, *Mysidopsis bahia*, and *Cyprinus carpio* was significantly less when fed San Pablo Bay strain

TABLE II

Per cent survival of seven species of fish and crustacean larvae reared on Artemia *nauplii from eight geographical strains of* Artemia: *(1) Johns* et al., *1980; Seidel* et al., *1982; (2) Johns* et al., *1980; (3) Johns* et al., *1981a; Léger & Sorgeloos, 1984; (4) Beck* et al.,*1980; Beck & Bengtson, 1982; (5) Klein-MacPhee* et al., *1980, 1982; (6) Vanhaecke & Sorgeloos, 1983b; (7) Usher & Bengtson, 1981; (8) Sorgeloos, 1980b*

Artemia source	*Rhithropanopeus harrisii* (1)	*Cancer irroratus* (2)	*Mysidopsis bahia* (3)	*Menidia menidia* (4)	*Pseudopleuronectes americanus* (5)	*Cyprinus carpio* (6)	*Cyprinodon variegatus* (7)
Australia, Shark Bay	78	92	98	60	94	96	—
Brazil, Macau	80	95	95	89	89	96	100
Canada, Chaplin Lake	72	—	74	62	78	95	—
China P.R., Tientsin	84	—	90	71	72	97	—
France, Lavalduc	89	—	94	62	61	95	—
Italy, Margherita di Savoia	92	90	98	44	88	94	—
U.S.A., Great Salt Lake	0	0	98	72	46	93	—
U.S.A., San Pablo Bay	0	0	82	42	39	93	100
Reference *Artemia*[8]	89	—	92	82	86	—	—

(Klein-MacPhee, Howell & Beck, 1980, 1982; Johns, Berry & Walton, 1981a; Vanhaecke & Sorgeloos, 1983b), whereas no signicant differences were obtained in *Menidia menidia* and *Cyprinodon variegatus* (Beck *et al.*, 1980; Usher & Bengtson, 1981). The Great Salt Lake strain (which caused mass mortality in some species) yielded the best growth for *Mysidopsis bahia* (Johns *et al.*, 1981a) and *Menidia menidia* (Beck *et al.*, 1980) and the best reproduction for *Mysidopsis bahia* (Johns *et al.*, 1981a), but resulted in significantly less growth than obtained with the best *Artemia* strains in *Cyprinus carpio* (Vanhaecke & Sorgeloos, 1983b) and *Pseudopleuronectes americanus* (Klein-MacPhee *et al.*, 1980). Growth in Canadian-fed *P. americanus* (Klein-MacPhee, Howell & Beck, 1982), *Rhithropanopeus harrisii* (Seidel, Johns, Schauer & Olney, 1982), and *Cyprinus carpio* (Vanhaecke & Sorgeloos, 1983b) was significantly worse than when the other strains were fed. Although survival of most species was best when they were offered Brazilian *Artemia*, growth of *Menidia menidia* (Beck *et al.*, 1980), *Pseudopleuronectes americanus* (Klein-MacPhee *et al.*, 1980), and *Cyprinus carpio* (Vanhaecke & Sorgeloos, 1983b) on that strain was significantly less than optimal. In summary, concordance of survival and growth data is not necessarily apparent.

Although technically not part of the ISA studies, experiments with the same ISA strains were performed by Westin, Olney & Rogers (1983, 1985) using striped bass larvae, *Morone saxatilis,* and by Goy & Costlow (1980) using three crabs, *Rhithropanopeus harrisii*, *Menippe mercenaria*, and *Libinia emarginata*, and a shrimp, *Palaemonetes pugio*. In general, their results tended to corroborate the ISA results, except that Goy & Costlow observed good survival in organisms fed the Great Salt Lake strain[1] and poor survival in those fed the Italian strain. Westin *et al.*'s (1983) finding that survival of *Morone saxatilis* was equally good with the Brazilian and San Pablo Bay strains agrees with Usher & Bengtson (1981) and Vanhaecke & Sorgeloos (1983b) that the San Pablo Bay strain was an adequate food for organisms that can live in fresh water.

Reasons for the difference between a poor-quality and a good-quality *Artemia* strain are undoubtedly complex, because they must explain different patterns of mass mortality (at metamorphosis compared with during the first few days post-hatch) as well as account for the lack of congruence between growth and survival data. Attempts at explanation are further hampered by the lack of knowledge of the nutritional requirements for the species used in the ISA studies. Nevertheless, an attempt was made to relate the ISA biological data on growth and survival with biochemical data (*e.g.* fatty acids by Schauer, Johns, Olney & Simpson, 1980 and Seidel *et al.*, 1982; amino acids by Seidel, Kryznowek & Simpson, 1980b) and biometrical data (Vanhaecke & Sorgeloos, 1980) in the hope that hypotheses could be developed to explain differences in the food value of the strains.

The most immediately apparent connection that could explain mortality was between the size of the *Artemia* nauplii and mortality of *Menidia menidia* in the first 5 days after hatching (Beck & Bengtson, 1982) (see above). The length of nauplii from eight strains ranged from about 440 to 520 μm and it was calculated that when newly-hatched nauplii >480 μm

[1]It was later found that they were using a different batch of Great Salt Lake cysts.

were fed >20% mortality of *M. menidia* larvae could be expected. Thus, a good part of the mortality when this species was raised on the large, parthenogenetic strains from France, China, and especially Italy was due to the simple fact that many of the fish larvae could not ingest the food. The same phenomenon may account for some of the mortality in *Pseudopleuronectes americanus* reared on the French strain (Klein-MacPhee, Howell & Beck, 1982) and in *Morone saxatilis* reared on the Italian strain (Westin *et al.*, 1985). Because of the hypothesis of Bookhout & Costlow (1970) the ISA group originally suspected that chlorinated hydrocarbons (CHCs) such as DDT might be a cause of mortality. If organisms such as crab larvae (*Rhithropanopeus harrisii*) do accumulate CHCs from their *Artemia* diet, the toxic effect might be expressed as a mass mortality at the time of the major morphological restructuring, *i.e.* at metamorphosis. Olney *et al.* (1980), Johns, Peters & Beck (1980), and Seidel *et al.* (1982) concluded, however, that DDT was unlikely to be the causative agent, because the two strains with the highest DDT concentrations (Italy, 422 $\mu g \cdot g^{-1}$; China, 172 $\mu g \cdot g^{-1}$) yielded excellent survival of *Rhithropanopeus harrisii* larvae, whereas the strains that caused mass mortality of *R. harrisii* at metamorphosis had much lower DDT concentrations (San Pablo Bay, 42 $\mu g \cdot g^{-1}$; Great Salt Lake, 7·3 $\mu g \cdot g^{-1}$). On the other hand, bioaccumulation data for *Menidia menidia* fed on the various strains (Olney *et al.*, 1980) suggested that chlordane or dieldrin, the former found at its highest concentration in the San Pablo Bay strain, might be a causative factor for the observed mortalities. In two follow-up studies (Johns *et al.*, 1981b, McLean, Olney, Klein-MacPhee & Simpson, 1985), *Rhithropanopeus harrisii* larvae and newly-metamorphosed *Pseudopleuronectes americanus* were fed *Artemia* nauplii that had been contaminated on purpose with chlordane and dieldrin. *Rhithropanopeus harrisii* larvae did not die at metamorphosis even when the chlordane and dieldrin levels in the nauplii were one to two orders of magnitude higher than the maximum measured in the eight ISA strains. *Pseudopleuronectes americanus* showed no mortality after having been raised for 30 days on the contaminated *Artemia*, but it should be emphasized that the experiment was started with metamorphosed fish. In summary, it is likely that chlordane and dieldrin, like DDT, were not causative factors for the poor culture performances observed with some ISA strains. Westin *et al.* (1985) fed three strains of *Artemia* (Brazil, Italy, San Pablo Bay) containing different concentrations of four CHCs to *Morone saxatilis* larvae and found that they caused no significant differences in larval survival; what was observed was a parental effect, *i.e.* concentrations of those four CHCs in the eggs from which the fish larvae hatched affected their survival.

Another relationship that merits examination (based on the work of Watanabe *et al.*, 1978c, 1980), is that of the levels of the essential fatty acids, 20:5ω3 and 18:3ω3, with growth and survival of the various species. The strain that had the lowest level of 20:5ω3 (San Pablo Bay, Schauer *et al.*, 1980), an essential fatty acid for marine organisms, normally yielded the lowest survival reates for the marine species tested. Only the species that can live in fresh water (*Cyprinus carpio*, *Cyprinodon variegatus*) exhibited survival rates >90% when fed the San Pablo Bay strain. The strain with the second lowest percentage of 20:5ω3 (Great Salt Lake, Schauer *et al.*, 1980) was similarly very poor at promoting survival in marine species. Low

20:5ω3 levels, however, cannot always be referred to as the sole argument. Indeed, marine species fared somewhat poorly with regard to survival and very poorly with regard to growth when they were fed the Canadian strain, which contains more 20:5ω3 (Seidel *et al.*, 1982) than even the best strains from Brazil and RAC. The culture results with Canadian *Artemia* have to be considered separately since recent experiments (Léger, Sorgeloos, Millamena & Simpson, 1985c) have demonstrated a very good correlation between 20:5ω3 levels in several batches of San Francisco Bay *Artemia* and biomass production in *Mysidopsis bahia* reared on those batches. A similar correlation can be seen in the data of Vos, Léger, Vanhaecke & Sorgeloos (1984), who fed *M. bahia* on *Artemia* nauplii from production ponds in several Asian countries. Thus, the fatty acid 20:5ω3 does seem to be a major factor in the determination of *Artemia* quality, especially as a food for crustaceans, but also as a food for fish.

Further evidence for the importance of 20:5ω3 and 22:6ω3 (another essential fatty acid) to mud crab larvae was obtained from the experiments of Levine & Sulkin (1984). They fed several diets to *Eurypanopeus depressus* larvae and found that best survival to the megalopa stage was attained on a diet of *Artemia* nauplii or a diet of rotifers plus capsules containing *Artemia* lipids. Survival was significantly worse on a diet of rotifers alone or a diet of rotifers plus lipid-free *Artemia*. In a second experiment, they found that the survival achieved on *Artemia* nauplii or rotifers plus capsules containing 22:6ω3 was significantly better than that on rotifers alone. In none of their experiments, however, did they observed the catastrophic mortality at metamorphosis that Johns *et al.* (1980) and Seidel *et al.* (1982) reported.

Schauer *et al.* (1980) remarked that synergistic interaction effects between essential fatty acid and CHC levels may have been operated in the ISA strain studies. Thus, low levels of 20:5ω3 may have combined with high levels of total CHCs in the Great Salt Lake and San Pablo Bay strains to cause mortalities of mud crab larvae. Their argument was supported by the evidence that the Great Salt Lake strain, which had only a slightly lower level of 20:5ω3 than a sample of San Francisco Bay *Artemia* collected in 1975, but also a slightly lower CHC concentration, produced *Rhithropanopeus harrisii* mortalities (Johns *et al.*, 1980) whereas the San Francisco Bay strain did not (Johns, Peters & Beck, 1978). More recent and extensive experiments already mentioned above (Léger *et al.*, 1985c) indicated that the correlation between total CHC concentration and *Mysidopsis bahia* biomass production is very poor and that no interaction effects exist between the 20:5ω3 level and total CHCs with regard to *M. bahia*.

The analysis done on the ISA *Artemia* strains for amino acids (Seidel *et al.*, 1980b), heavy metals (Olney *et al.*, 1980), caloric content (Schauer *et al.*, 1980), and carotenoids (Soejima, Katayama & Simpson, 1980) yielded no data that could be related in any way with the biological data on test species' growth and survival. Thus, the single most important factor so far identified in defining nutritional quality of *Artemia* nauplii for marine fish and crustaceans is the content of essential fatty acids such as 20:5ω3. If one examines all the ISA studies together, a good-quality batch of *Artemia* can be considered to have a fatty-acid profile with a 20:5ω3 content of higher than 4% of the total fatty acid methyl esters. Batches with a 20:5ω3 content between 3 and 4% may or may not be good depending on other unknown factors. Batches with less than 3% 20:5ω3 consistently yield poor growth and survival of marine organisms. The exception of this rule, however, is the Canadian strain, which was the only sulphate-lake strain tested. As Léger & Sorgeloos (1984) pointed out, more research needs to be done on the sulphate-lake strains to determine what governs their quality as a food for marine organisms. It is important to reiterate here that considerable temporal variation in 20:5ω3 content can exist within a given geographical strain (see later). Watanabe *et al.* (1980, 1982) reported large fluctuations in the quantity of 20:5ω3 during a year or between years for *Artemia* from San Francisco Bay, Brazil, and China. Léger *et al.* (1985c) reported similar variability for batches collected over several years from San Francisco Bay.

THE ENHANCEMENT OF THE DIETARY VALUE OF *ARTEMIA* NAUPLII

The enrichment of Artemia *nauplii as a solution for their nutritional deficiencies*

In the previous section we demonstrated that several factors determine directly or indirectly the food value of *Artemia* nauplii. Indirect factors may be called those that are not immediately related to the nature of the nauplius, *e.g.* presence in the culture tank of unhatched cysts, shells, and other contaminants as a result of insufficient separation and washing of the nauplii. Feeding regime and its attendant use of older instar-stages may also be considered as indirect factors affecting the dietary value of *Artemia*. Reduction of the suitability and dietary value of *Artemia* due to indirect factors may be quite easily corrected as shown above. Direct factors, however, such as size of the instar I nauplii and their nutritional composition may in practice be more problematic. When size of nauplius is critical one should select a strain that produces small nauplii; indeed there is a considerable variation between different strains and cyst size, which is correlated with length of nauplius, and is principally genetically determined.

Small *Artemia* are mainly found on the American continent (Vanhaecke & Sorgeloos, 1980; Vanhaecke, 1983). It is interesting to know that also on other continents one can produce small cysts through inoculation of a properly selected natural strain (Vos *et al., 1984). Thanks to fast developing* progress in the field of genetic selection or manipulation, artificial production of "mini-cysts" and subsequent large scale inoculation and production in suitable environments may offer unique opportunities for the near future.

Nutritional variability between different *Artemia* strains and even between harvests from the same strain may look the most insuperable drawback with regard to the use of *Artemia* in the culture of larvae. Nevertheless, the recent progress in the characterization of *Artemia* and the better understanding of at least some larval nutritional requirements, has resulted in a major breakthrough in the enhancement of the nutritional value of *Artemia*.

Not considering Hauenschild's finding that the nutritional value of *Artemia* nauplii for Hydrozoa was improved by naupliar starvation (Hauenschild, 1954, 1956), the first application of the technique of nutritional enhancement of *Artemia* nauplii was suggested by Morris (1956). As pointed out earlier, he found that marine fish larvae did not prosper in his rearing trials when fed only *Artemia* metanauplii which had consumed their yolk reserves. He noticed however, that the loss in food value in *Artemia* mentanauplii could be restored by allowing them to feed on so-called "secondary foods". These include items which are too small to be directly fed upon by fish larvae, but may be incidentally ingested or delivered by "primary foods", such as *Artemia*. Morris (1956) indeed observed that when an *Artemia* nauplius was ingested by a larva the *Artemia* squirmed violently for some minutes prior to death. These vigorous movements cause the *Artemia* to void much of its gut contents into the alimentary tract of the fish larva. Morris (1956) added algae, *e.g. Stichococcus* and *Dunaliella*, suspensions of Fleishmann yeast or boiled egg yolk to the rearing tank along with the *Artemia* nauplii and observed that these products were readily ingested; as a result the nutritional quality of the *Artemia* was more adequate. One decade later Forster & Wickins (1967) demonstrated that the food value of *Artemia* nauplii of Great Salt Lake origin could be improved for *Palaemon serratus* larvae. Several methods resulted in successful metamorphosis compared with total mortality in the controls fed Great Salt Lake nauplii only, *e.g.* substitution by at least 50% San Francisco Bay *Artemia* nauplii, addition of *Isochrysis* to the culture tank, or 4 days prefeeding of the *Artemia* with *Isochrysis*. The experiments of Forster & Wickins (1967) further indicated that improved metamorphosis success was achieved by the enrichment of *Artemia* and not through direct ingestion of algae by the shrimp larvae. Wickins (1972) obtained similar improvements in metamorphosis success by 24 h pre-feeding Great Salt Lake nauplii at a density of $10\,000 \cdot l^{-1}$ in an algal suspension of 300 cells$\cdot \mu l^{-1}$. In order to avoid wastage of expensive algae and to prevent the risk that *Artemia* would grow to an unacceptable size, he determined the time at which newly hatched nauplii started to feed and their feeding rate. He noticed that the number of cells ingested increased continuously in the 48 to 60 hours after cyst incubation at 20 °C. During this period algal consumption increased from less than 500 to over 7000 cells$\cdot$ nauplius$^{-1} \cdot h^{-1}$; as a result each nauplius could ingest more than 30 000 cells within 24 h. Higher cell densities were not recommended because of the risks of producing too large metanauplii.

The same technique was successfully applied for *Macrobrachium* larvae (Monaco, 1974; Wickins, 1976). On the contrary, Maddox & Manzi (1976) demonstrated that freshly hatched nauplii were a more superior food for *Macrobrachium* than older metanauplii whether they were fed algae or not. The idea of pre-feeding *Artemia* for the purpose of quality enhancement was tested for *Pleuronectes platessa* and *Gadus morhua* by Nordeng & Bratland (1971). Analysing the guts of wild fish larvae, they assumed that phytoplankton could be an essential source of nourishment of which laboratory larvae were deprived when fed *Artemia* nauplii alone. In their culture tests fish larvae were offered additional nutrients by means of *Artemia* which had been pre-fed for 24 h. For this they used marine *Chlamydomonas* sp., ω-yeast (*Saccharomyces cerevisiae*), and ground trout food. All three groups of pre-fed metanauplii were given alternately in order to ensure that the larvae received a varied diet. With plaice, metamorphosis, pigmentation, and general condition of the larvae were optimal. Although Nordeng & Bratland (1971) failed with cod, Howell (1979b) obtained a good survival in cod (*Gadus morhua*) larvae when they were given *Artemia* nauplii that were pre-fed for 2 days on *Isochrysis galbana*, while simultaneously adding the same alga plus *Pavlova lutheri* in the tanks. *Artemia* were inadequate when not pre-fed. When Howell *et al.* (1981) pre-fed *Artemia* with *Isochrysis* for only 4 h, *i.e.* a period sufficiently long to fill up their gut, the food value of these *Artemia* for turbot (*Scophthalmus maximus*) larvae improved only appreciably when this alga was also added to the larval rearing tank. Since no evidence was found for direct utilization of the algae by the larval turbot, Howell *et al.* (1981) suggested that the *Artemia*, in order to become an effective diet, had to digest the algae first. This reminds us of the earlier observations of Morris (1956).

Kelly *et al.* (1977) also obtained a better growth in *Pandalus platyceros* by adding *Phaeodactylum tricornutum* to the culture tank along with the *Artemia* nauplii. Bromley (1978) was more successful in weaning *Scophthalmus maximus* when *Pseudoisochrysis paradoxa* was supplemented in the culture tanks as food for the rotifers and *Artemia* nauplii. The beneficial effect of adding algae along with, or 'encapsulated' in *Artemia* was recognized by many authors, but an explanation for the observed nutritional enhancement of the nauplii was not given. In 1979, however, Howell (1979a) pointed out that the choice of algae used was important; *i.e.* much better results were obtained with *Scophthalmus maximus* when using *Isochrysis galbana* instead of *Dunaliella tertiolecta*. This made him suggest that the effect of adding algae was probably more related to nutrition than to their stabilizing action on water quality with which they are often credited (*cf.* green water technique in *Macrobrachium* culturing). The use of algae of inferior 'nutritional-enhancement-quality' may explain some previous reports that no improvement was noticed after pre-feeding the *Artemia* nauplii and/or adding algae. In the same year Scott & Middleton (1979) and Scott & Baynes (1979) confirmed Howell's observation, *i.e.* addition of *Dunaliella tertiolecta* during the live food phase in the culture of *Scophthalmus maximus* larvae resulted in stunted growth and high mortality. It appeared that this effect was not an expression of toxicity but of poor nutrition, probably due to a deficiency of long chain polyunsaturated fatty acids as confirmed by the fatty acid profile of this alga. Several studies in the 1970s have indeed revealed that long chain polyunsaturated fatty acids are essential for a variety of marine animals. More particularly the ω3-highly unsaturated fatty acids (ω3-HUFA) 20:5ω3 and 22:6ω3 seem to be required by marine fish and crustaceans (Owen, Adron, Sargent & Cowey,1972; Owen, Adron, Middleton & Cowey, 1975; Sick & Andrews, 1973; Yone & Fujii, 1975; Castell & Covey, 1976; Cowey, Owen, Adron & Middleton, 1976; Guary, Kayama, Murakami & Ceccaldi, 1976; Sandifer & Joseph, 1976; Gatesoupe *et al.*, 1977; Kanazawa, Teshima & Tokiwa, 1977; Kanazawa, Teshima & Ono, 1979; Yone, 1978; Castell & Boghen, 1979, Léger *et al.*, 1979).

Analyses of the fatty acid profile of different sources of *Artemia* and different lots from the same source by Watanabe and co-workers revealed striking differences in ω3-HUFA content (see Table III). Based on the relationship between the dietary value of *Artemia* and their ω3-HUFA content Watanabe *et al.* (1978c) proposed the following classification: 20:5ω3-rich *Artemia* sources (so-called "marine type" *Artemia*) which are a good food source for red seabream juveniles and 20:5ω3-poor sources (so-called "freshwater type" *Artemia*) which yield poor culture success in red seabream larvae. It was also demonstrated that the ω3-HUFA content in the *Artemia* could be substantially increased by feeding them for 24 to 72 h with ω3-HUFA-rich food sources, such as marine *Chorella minutissima* and ω-yeast (Imada *et al.*, 1979).

As can be seen from Table III, ω3-HUFA-enriched *Artemia* were converted into an excellent food source for red seabream juveniles. On the other hand, the ω3-HUFA content of nauplii fed diets lacking ω3-HUFA, such as baker's yeast, did not differ from starved nauplii, and no improvement in food value was noted for red seabream juveniles. The most pronounced differences between the fish fed marine type or ω3-HUFA-enriched *Artemia* and freshwater type *Artemia* were revealed in the activity test as applied by Watanabe *et al.* (1980), *i.e.* survival is determined in fish larvae 24 h after being scooped out for 5 seconds from the culture vessel and transferred into another tank; (physiologically) weak fish show a shock syndrome and die. Watanabe and colleagues concluded that not protein quality, including amino-acid profile, nor mineral composition, but the presence of essential fatty acids was the principal factor which determined the food value of *Artemia* nauplii for fish larvae. Léger (unpubl.) confirmed those findings for marine crustacean larvae by pre-feeding freshly hatched San Pablo Bay (No. 1628) *Artemia* nauplii for 24 h on micronized and defatted ricebran which was coated (GLC-stationary phase coating technique) with either cod liver oil (CLO) or rice oil (RO). When CLO-rice bran was used for enrichment, the levels of ω3-HUFA in *Artemia*

TABLE III

ω3-HUFA content of "marine type" Artemia (*Canadian and enriched San Francisco Bay* Artemia) *and "freshwater"* Artemia (*San Francisco Bay*) *and their effect on survival and growth of red seabream juveniles* (*data from Watanabe* et al., *1980*): **20:3*<ω3 fatty acids

	Artemia treatment			
	Canada	San Francisco Bay		
	Newly hatched	Newly hatched	Fed *Chlorella* for 24 h	Fed ω-yeast for 24 h
ω3-HUFA content				
20:5ω3	5·2	1·6	3·2	3·4
22:6ω3	—	—	—	1·1
Σ ω3-HUFA*	5·8	2·4	4·1	5·1
Red seabream culture test				
Survival (%)	68·4	43·4	66·8	86·4
Survival after activity test (%)	37·5	24·1	46·1	50·0
Final length (mm)	9·57	10·13	11·13	11·67

markedly increased during pre-feeding and these *Artemia* had a high nutritional value for *Mysidopsis bahia* juveniles; on the other hand, no effect was noticed when rice oil coated ricebran was used for enrichment (see Table IV). Léger, Bieber & Sorgeloos (1985a) confirmed the beneficial effect of using ω3-HUFA-enriched San Pablo Bay (No. 1628) *Artemia* for a commercial crustacean *Penaeus stylirostris* (see Table IV). Furthermore, they observed that the pre-*Artemia* food phase (during protozoea stages) greatly affected post-larval metamorphosis success, *i.e.* the dietary quality differences between ω3-HUFA-rich and -poor *Artemia* nauplii were accentuated or attenuated, respectively, when protozeal food lacked sufficient levels of ω3-HUFA.

It is important to add that both in *Pagrus major* and *Penaeus stylirostris* the best culture results were obtained when enriched *Artemia* contained besides 20:5ω3 also substantial levels of 22:6ω3 (*e.g.* pre-fed with ω-yeast, CLO, AA18 and SEC, see Tables III and IV). In this regard, the better performance with *Acartia clausii* than with marine type *Artemia* nauplii as a food source for red seabream (Watanabe *et al.*, 1980) may thus be related not only to the higher levels of 20:5ω3 but especially to the higher content of 22:6ω3 in this marine copepod. The high amounts of both 20:5ω3 and 22:6ω3 in *Isochrysis galbana* (Watanabe & Ackman, 1974) may indeed explain the nutritional enhancements reported earlier in larval fish culture when this alga was supplemented, either directly or indirectly *via Artemia*. This further explains the improved fish culture success when, besides *Artemia*, *Tigriopus* and *Acartia*, both rich in 20:5ω3 and 22:6ω3 (Watanabe *et*

TABLE IV

ω3-HUFA content of San Francisco Bay and San Pablo Bay Artemia *nauplii, freshly hatched or pre-fed, and their nutritional value for* Mysidopsis bahia *juveniles and* Penaeus stylirostris *larvae (data from Léger* et al., *1985a,b; Léger, unpubl.): RO, rice oil coated rice bran; CLO, cod liver oil coated rice bran; AA18 and SEC, commercial enrichment diets (Artemia Systems S.A.)*

	San Francisco Bay *Artemia* (236–2016)	San Pablo Bay *Artemia* (1628)				
	Newly hatched	Newly hatched	24 h pre-fed RO	24 h pre-fed CLO	24 h pre-fed AA18	24 h pre-fed SEC
ω3-HUFA content (area %)						
20:5ω3	9·3	0·2	0·9	6·3	8·2	9·9
22:6ω3	0·2	—	—	1·5	1·5	5·9
Σω3-HUFA	11·4	0·7	1·9	8·9	10·6	17·8
Culture results with *Mysidopsis bahia*						
Survival (%)	93·3	62·0	60·0	75·0	92·5	95·8
Ind. length (μm)	5532	4587	4285	5029	5375	5254
Ind. dry weight (μg)	354	198	188	259	259	323
Biomass (mg·%)	33·0	12·3	11·3	19·4	24·0	30·9
Culture results with *Peneaus stylirostris*						
Survival (%)	47·5	34·0			45·7	63·9
Ind. wet weight (mg)	1·8	1·7			2·0	2·7
Biomass (mg·%)	85·5	57·8			91·4	172·5

al., 1978b) were also added (Fukusho, 1974). This agrees with Kuhlmann, Quantz & Witt (1981b) who found better results for turbot (*Scophthalmus maximus*) larvae when using *Eurytemora affinis* instead of *Artemia* nauplii. More evidence for the essential requirement of 22:6ω3 has recently been reported for several marine species by Holland & Jones (1981); Léger & Frémont (1981); Léger *et al.* (1985a); Bell, Henderson, Pirie & Sargent (1985); and Jones, Holland & Jaborie (in press). Because *Artemia* nauplii generally contain at most only marginal levels of 22:6ω3, ω3-HUFA-enrichment should be generally recommended for all *Artemia* sources.

The varying and low levels of ω3-HUFAs in *Artemia* are probably related to the exceptional tropical conditions under which the *Artemia* are found in nature, *i.e.* very high and changing salinity levels which favour various species of blue-greens and flagellates; contrary to the diatoms and flagellates usually found in natural sea water the blue-greens are low in ω3-HUFAs (Scott & Middleton, 1979). Indeed several authors have reported that *Artemia* and other zooplankton mainly reflect the fatty acid profile of their food (Kayama, Tsuchiya & Mead, 1963; Jezyck & Penicnak, 1966; Malins & Wekell, 1969; Ackman *et al.*, 1970; Culkin & Morris, 1969; Hinchcliffe & Riley, 1972; Bottino, 1974; Watanabe & Ackman, 1974; Sick, 1976; Claus *et al.*, 1979; Bottino *et al.*, 1980). Using a culture system for the controlled production of *Artemia* offspring (Lavens & Sorgeloos 1984, 1985) it has been demonstrated that the fatty acid profile of *Artemia* cysts and/or ovoviviparous nauplii reflects the profile in the food of the parental population. Moreover the ω3-HUFA content in the cysts and nauplii could be increased by feeding ω3-HUFA-fortified diets to the parental stock (see Table V, Lavens *et al.*, unpubl.).

Vos *et al.* (1984) studied the quality of *Artemia* produced in Southeast Asian saltponds and found that cysts produced in ponds fertilized with inorganic fertilizer had low levels of 20:5ω3 whereas those produced in ponds with water intake from mangrove waters (*i.e.* high food diversity) showed considerable levels of 20:5ω3 and sometimes traces of 22:6ω3; a similar observation was made when organic fertilizers such as poultry manure were applied (Léger, unpubl.). Watanabe *et al.* (1978b) analysed high levels of ω3-HUFA in *Moina* cultured on poultry manure. Similarly *Artemia* might accumulate ω3-HUFA directly from the manure or indirectly from algal blooms induced by this fertilizer; in this regard Jumalon & Ogburn (1985) and Jumalon, Estenor & Ogburn (1985) noticed that *Artemia* production ponds fertilized with poultry manure consistently showed blooms of *Tetraselmis* which is usually rich in ω3-HUFA (Millamena, Bombeo, Jumalon & Simpson, 1985). Fertilizer control of algal composition might be feasible in small production ponds (*e.g.* solar salt operations in Southeast Asia, Central America, *etc.*). This practice is, however, not conceivable in large solar salt operations (*e.g.* Mexico, Brazil, Australia, *etc.*) nor in the hugh lakes found all over the world. In the lakes the available algae may be suitable, unsuitable or subject to a considerable variation in quality. For years the dominant species in the Great Salt Lake (Utah, U.S.A.) has been *Dunaliella* (Stephens & Gillespie, 1976; Post, 1977), which is poor in ω3-HUFA (Scott & Middleton, 1979; Millamena *et al.*, 1985). As opposed to other strains the 20:5ω3 content in Great Salt Lake *Artemia* is remarkably constant, *e.g.* 1·8–3·6% in cysts collected from the Southern arm and 0·2–0·3% in Northern arm *Artemia* cysts (see Table XII, p. 597).

TABLE V

ω3-HUFA content in parental cysts and 1st generation offspring (cysts or nauplii) of Artemia *from two strains cultured on rice bran, either untreated (diet RBO) or HUFA-enriched diet (RBA) (data from Lavens & Léger, unpubl.): a, cysts were analysed after decapsulation; b, sum of ω3-highly unsaturated fatty acids (20:3<ω3 fatty acids); c, per cent fatty acid methyl ester of total fatty acid methyl esters; d, mg fatty acid methyl ester per g dry wt; e, cysts used for the production of the parental population*

Artemia source	Type of material[a]	Diet	ω3-HUFA content 20:5ω3 %[c]	20:5ω3 $mg \cdot g^{-1,d}$	22:6ω3 %	22:6ω3 $mg \cdot g^{-1}$	Σω3-HUFA[b] %	Σω3-HUFA[b] $mg \cdot g^{-1}$
France, Lavalduc								
	Parental cysts[e]	—	4·8	5·4	—	—	5·7	6·4
	1st generation cysts	RBO	0·2	0·3	—	—	0·5	0·6
	1st generation cysts	RBA	7·5	8·6	0·4	0·4	8·8	10·0
U.S.A., Great Salt Lake								
	Parental cysts	—	2·1	3·9	—	—	3·1	4·6
	1st generation cysts	RBO	0·3	0·3	—	—	0·4	0·4
	1st generation nauplii	RBO	0·5	0·6	—	—	1·3	1·7

The variability in ω3-HUFA content in the other strains may be explained by seasonal changes in algal species composition (*cf.* species diversity in San Francisco Bay and Saskatchewan Lakes, Carpelan, 1957; Haynes & Hammer, 1978) or variability in ω3-HUFA content within the same algal species (*cf.* Scott & Middleton, 1979). It has indeed been demonstrated that the nutritional composition of algae may change according to varying abiotic conditions (D'Agostino & Provasoli, 1968, 1970; Dickson, Galloway & Patterson, 1969; Provasoli, Conklin & D'Agostino, 1970; Moal, Samain & Le Goz, 1978; Scott, 1980; Enright, 1984). As a result man will always be dependent on the caprices of nature, providing ecologists and aquaculturists at one time with a present of excellent quality cysts and at other times with an inferior quality of their preferred live food source. Again, the enrichment of *Artemia* nauplii eliminates the effects of such caprices.

Enrichment techniques

Table VI summarizes the results of enrichment and culture experiments as described in the references cited. Over the past decades several techniques have been elaborated for *Artemia* nauplii enrichment. They may be classified in four groups, *i.e.* the British technique, with algae; the Japanese technique, with ω-yeast or emulsions; the French technique, with compound diets; and the Belgian technique with coated micro-particles or self-emulsifying concentrates.

The British technique. This technique has been pioneered by Forster & Wickins (1967), and Wickins (1972); *Artemia* nauplii are cultured for 24 h (Wickins, 1972) or 4 days (Forster & Wickins, 1967) on an algal suspension, mostly *Isochrysis galbana* at up to 1000 cells$\cdot \mu l^{-1}$. The same alga was in many cases also added to the larval culture tank. A density of 10 000 nauplii $\cdot l^{-1}$ in an algal suspension of 300 cells$\cdot \mu l^{-1}$ for an enrichment period of 24 h appeared to be a suitable regime to make the nauplii an adequate food for prawn larvae (Wickins, 1972). This technique may well be suited when algae have to be cultured as a food source for first-feeding larvae. Setting up an algal culture only for live food enrichment looks, however, hardly justified, especially as algal quality is variable and alternatives are available (see later).

The Japanese technique. The so-called "indirect method" developed by Watanabe *et al.* (1978c, 1980, 1982, 1983a) at first resembled the British technique. Indeed, marine algae (*Chlorella minutissima*) were used to pre-fed freshly hatched (up to 48 h hatching incubation) *Artemia* nauplii for 24 h (up to 72 h). Algal densities ranged between 14×10^6 to 18×10^6 cells$\cdot ml^{-1}$. Details on densities of nauplii, however, were not given. A similar procedure was adopted using so-called ω-yeast (0·38 mg$\cdot ml^{-1}$ or 9×10^6 cells$\cdot ml^{-1}$) as a substitute for the algae. This special yeast preparation is produced by adding cuttle fish liver oil at a 15% level to the culture medium of baker's yeast (*Saccharomyces cerevisiae*) (Imada *et al.*, 1979). Similarly to the application with algae, ω-yeast is pre-fed in newly hatched *Artemia* nauplii for 24 h.

TABLE VI

Summary of enrichment procedures for Artemia *nauplii, ω-HUFA content in* Artemia, *and results of comparative culture tests with enriched* Artemia: *this table does not include enrichment trials where the enrichment diet (mostly algae in these cases) were only added directly to the tank; experiments without a control treatment (not enriched) were not considered; the quotation +,○,±,−, refer to relative values i.e. very good, good, average, and poor culture results, respectively adjudged by interpretation of the given data; these annotations compare values within one experiment, so that no absolute comparison may be made between different experiments even when the same test organism was used; open spaces refer to same conditions (t, T) as first treatment within an experiment, or to lack of data (t, T, ω3-HUFA content, culture test): SFB, San Francisco Bay, U.S.A.; BRA, Brazil; CAN, Canada; SAM, South America; GSL, Great Salt Lake, U.S.A. (south arm); Na, North arm; SPB, San Pablo Bay, U.S.A.; t, time period in h; T, temperature in °C; %, per cent fatty acid methyl ester of total fatty acid methyl esters (area%); mg·g^{-1}, mg fatty acid methyl ester per g dry wt; Σ ω3-HUFA, sum of ω3 highly unsaturated fatty acids (note: this generally refers to the sum of ω3 fatty acids with 20 or 22 carbons and 3 or more double bounds); na, no pre-enrichment or enrichment diet was applied; tr, trace; nd, not detected; *, dry or wet basis not specified; diets A to J, see Table VII*

Artemia	Hatching		Pre-enrichment			Enrichment			ω3-HUFA-content						Culture test		Reference
source	t	T	Diet	t	T	Diet	t	T	20:5ω3		22:6ω3		Σω3-HUFA		Animal	Performance	
									%	mg·g^{-1}	%	mg·g^{-1}	%	mg·g^{-1}			
GSL			na			na									*Palaemon*	−	Forster & Wickins, 1967
			na			*Isochrysis*	96									+	
GSL			na			na									*Palaemon*	−	Wickins, 1972
			na			Starved	24									+	
			na			*Isochrysis* 300 cells·μl^{-1} or 500 cells·μl^{-1}										+	
SFB	48		na			na			2·0		tr				*Pagrus*	−	Watanabe *et al.*, 1978c
(1976)			na			Starved	24		3·0		0·1						
							48		5·1		0·4						
							72		7·0		0·7						
			na			Marine *Chlorella* (18 × 10 cells·ml^{-1})	24		3·5		0·2					+	

TABLE VI—*continued*

Artemia	Hatching		Pre-enrichment			Enrichment			ω3-HUFA-content						Culture test		Reference
									20:5ω3		22:6ω3		Σω3-HUFA				
source	t	T	Diet	t	T	Diet	t	T	%	mg·g−1	%	mg·g−1	%	mg·g−1	Animal	Performance	
							48		7·3		0·1						
							72		10·9		0·3						
			na			Baker's yeast (9 × 10^6	24		3·0		0·1					–	
						cells·ml−1)	72		4·3		0·4						
						ω-yeast (9 × 10^6	24		4·5		0·4					+	
						cells·ml−1)	48		7·3		1·1						
							72		8·8		1·5						
SFB			na			na			9·5		nd		9·9				
(1977)			na			Starved	24		10·7		nd		11·0				
						Marine *Chlorella*			15·0		0·3		15·5				
						Baker's yeast			13·0		0·2		13·3				
						ω-yeast			10·8		2·3		13·8				
SAM			na			na			0·3		nd		1·0				
			na			Starved	24		0·8		nd		1·6				
			na			Marine *Chlorella*			1·7		nd		2·5				
			na			Baker's yeast			0·9		nd		1·6				
			na			ω-yeast			6·6		1·7		8·9				
CAN			na			na			12·1		nd		12·3				
						Starved	24		12·8				12·8				
			na			Marine *Chlorella*			12·0		nd		12·0				
			na			Baker's yeast			14·3		nd		14·3				
			na			ω-yeast			10·5		1·9		13·1				
SFB-1	30-48	23-28	na			na			1·6		nd		2·4		*Pagrus*	–	Watanabe *et al.*, 1980
(78 C)			na			Marine *Chlorella* 14 × 10^6 cells·ml−1	24	20	3·2		nd		4·1			±	
			na			ω-yeast. 0·38 mg·ml−1			3·4		1·1		5·1			+	
			na			Starved			2·4		nd		3·2				
CAN			na			na			5·2		nd		5·8			±	
SFB-2			na			na			7·0		nd		7·5			o	
(78B)			na			*Spirulina* 0·5 mg·ml−1	24	20	5·9		nd		5·9			o	
			na			ω-yeast			7·3		0·9		8·4			+	
			na			Starved			7·6		nd		8·4				

SFB	24	28	na			Starved	4	16-18		*Scophthalmus*	–	Howell *et al.*, 1981
			na			*Isochrysis* (excess)		28			–	
BRA			na			Starved		16-18			+	
GSL (1977)	25-29	25	na			na				*Rhithropanopeus*	–	Johns *et al.*, 1981a
			na			Starved	24				–	
			na			*Isochrysis, cf.* Wickins, 1972					–	
BRA			na			na					+	
SFB			na			A	48			*Dicentrarchus*	–	Robin *et al.*, 1981
			na			B or C					+	
BRA			na			B					o	
			na			A					o	
			F	48		I (2·5 or 5 g·10^6 metanauplii)	0·5			*Scophthalmus*	+	Gatesoupe, 1982
SFB	48	24	D	48	24	na				*Dicentrarchus*	–	Robin, 1982
			D			H	0·5				±	
			E			H					+	
			E			na					+	
SFB	48	24-26	na			na			0·5*	*Paralichthys*	–	Watanabe *et al.*, 1982
			na			Cuttle fish liver oil emulsion = 1·5 g cuttle fish liver oil, 0·3 g egg yolk, 20 ml water, Baker's yeast (same weight as nauplii) per 30 l	15-19	24-26	4·0*	(Exp. I)	+	
			na			*Idem,* but corn oil instead of cuttle fish liver oil (= corn oil emulsion)			0·5*		–	
	48	24-26	na			na			1·0*	*Oplegnathus*	–	
			na			Cuttle fish liver oil emulsion *(cf. supra)*	15-19	24-26	3·1*	(Exp. II)	+	
			na			ω-yeast alone			3·0*		+	
			na			Baker's yeast alone			0·8*		–	
			na			Baker's yeast	15-19	24-26	1·2*	*Pagrus*	–	
			na			Corn oil emulsion			0·3*	(Exp. III)	–	
			na			Pollock liver oil emulsion			2·1*		+	
			na			Cuttle fish liver oil emulsion			7·7*		+	
			na			Methyl ω3-HUFA emulsion			7·1*		+	
	48	24-26	na			Baker's yeast	15-19	24-26	0·9*	*Pagrus*	–	
			na			Corn oil emulsion			0·7*	(Exp. IV)	–	
			na			Pollock liver oil emulsion			1·5*		±	
			na			Pollock & cuttle fish liver oil emulsion			3·0*		+	

TABLE VI—*continued*

Artemia source	Hatching		Pre-enrichment			Enrichment			ω3-HUFA-content						Culture test		Reference
	t	T	Diet	t	T	Diet	t	T	20:5ω3		22:6ω3		Σω3-HUFA		Animal	Performance	
									%	mg·g^{-1}	%	mg·g^{-1}	%	mg·g^{-1}			
			na			Cuttle fish liver oil emulsion								3·3*		+	
			na			Methyl ω3-HUFA emulsion								10·1*		+	
SFB	48	23-24	na			5 ml lipid (pollock liver oil, cuttle fish liver oil or HUFA-methylester mixture), 1 g raw egg yolk or other emulsifier, 100 ml sea water, 12 g baker's yeast per 60 l	24	24-26									Watanabe *et al.*, 1983b
SFB (No 1628)	24	28	na			na			0·5	0·5	nd	nd	2·7	2·8	*Penaeus*	–	Léger *et al.*, 1985a
			na			AA 18 (Artemia Systems S.A.); micronized powder containing 8·5% 20:5ω3 and 9·9% 22:6ω3 0·6 g·3 × 10^5 nauplii (assuming a hatching efficiency of 150 000 n·g^{-1})	24	30	5·9	7·9	3·4	4·6	11·2	15·1		+	
SFB No. 236-2016)			na			na			8·8	11·2	0·6	0·8	9·9	13·1		+	
SFB (No. 2257)	48	24	na			na							5·7	9			Robin *et al.*, 1984
			G 100 g (Day 0) + 120 g (Day 1)·m^{-3}; 15 nauplii·ml^{-1}	48	24	na							12·1	8			
			G			J(5 g·10^{-6} metanauplii)	0·5						14·9	16			
SFB			na			na									*Pomatoschistus*	o	Jones *et al.*, in press
			na			Microcapsules (10-30 μm, 300-500 ml^{-1}) containing cod oil	24									±	
			na			*Idem*, containing pollack oil										+	
GSL (Na)	24	25	na			na			0·3	0·5	nd	nd	1·2	1·9	*Mysidopsis*	–	Léger *et al.*, 1985b
	36	28-30	na			Self-emulsifying ω3-HUFA concentrate (SUPAR, Artemia Systems S.A.) added to			4·6	8·0	4·0	6·8	9·7	16·7		+	

				hatching medium from start of incubation											
	24	28-30	na	Self-emulsifying ω3-HUFA concentrate (SELCO Artemia Systems S.A.), 0·6 g·l^{-1} added to hatching medium after 24 h incubation of 1·5 g cysts·l^{-1}	12	28	4·5	6·4	2·4	3·3	8·2	11·2			
					24	28	7·0	12·1	4·4	7·5	13·3	22·1			
					48	28	12·0	22·3	6·4	11·9	21·0	38·3			
	24	28-30	na	self-emulsifying ω3-HUFA concentrate (SELCO), 0·6 g·l^{-1}, added after separation of nauplii	12	28	5·2	7·9	2·9	4·4	9·8	14·4			
							9·9	21·3	5·9	12·7	17·8	37·4		+	
				(3 × 10^5 n·l^{-1})	24	28	13·5	35·2	7·0	18·1	23·0	58·6			
SPB (No. 1628)	24	25	na	na	48	28	0·5		nd		2·7		*Mysidopsis*	–	Léger unpubl.
			na	Starved	24	24	1·4		0·6		3·5			–	
			na	Cod liver oil coated rice bran (5 cm Secchi-disk reading)			6·3		1·5		8·9			+	
			na	Rice oil coated rice bran (5 cm Secchi)			0·9		nd		1·9			–	
SFB (No. 236-2016)	24	25	na	na			9·3		0·2		11·4			+	

The advantage of using ω-yeast is mainly that one has a better control of the ω3-HUFA content since fish oils are generally rich in both 20:5ω3 and 22:6ω3. The disadvantage of this technique, however, is that as ω-yeast is required to be always in a living condition this technique can only be applied at places close to a production centre (Watanabe, pers. comm.).

Watanabe *et al.* (1982, 1983b) have also developed a "direct method" in which emulsified fish oils in combination with baker's yeast are pre-fed in *Artemia* nauplii. Indeed, *Artemia* nauplii are able to pick up emulsified lipids very easily from their culture medium. After 6 to 12 h enrichment a maximal ω3-HUFA incorporation was demonstrated. The emulsion is made up by blending 1·5 g lipid (*e.g.* cuttle fish liver oil) with 0·3 g raw egg yolk and 20 ml sea water for 3 min for use in a 30-l tank. Baker's yeast is added in an equivalent weight to the nauplii in the tank (Watanabe *et al.*, 1982). In later experiments Watanabe *et al.* (1983b) outlined a similar enrichment technique: 5 ml lipid are emulsified (lipid: egg yolk: water = 5:1:95) with a blender for 1 min and added to a 60-l enrichment tank together with 12 g baker's yeast and *Artemia* nauplii harvested from the hatching tank (48 h incubation); enrichment lasts for 24 h at 24-26 °C. Comparing raw egg yolk, soybean lecithin, and casein-Na as emulsifiers, no significant differences were noted in ω3-HUFA accumulation in the *Artemia* nauplii.

The incorporation of ω3-HUFA in *Artemia* appeared to be much lower than in rotifers: *i.e.* using an emulsified methyl ester mixture containing 85% ω3-HUFA, the incorporation rate in rotifers could yield 60% of total fatty acids within 3 h whereas in *Artemia* nauplii a minimum of 12 h were required to reach the 20% level. When using emulsified cuttle fish liver oil Watanabe *et al.* (1982) report ω3-HUFA levels from 0·31 to 0·77% (dry or wet weight basis not specified), with pollock liver oil 0·15 to 0·21%, and with ω3-HUFA methyl ester mixture 0·75 to 1·01%. They attributed these ranges in incorporation rate to varying culture conditions (*e.g.* water temperature) and density and activity of the nauplii used. It was also observed that the survival rate of the *Artemia* nauplii during enrichment fluctuated, *e.g.* 69·3% with pollock liver oil, 56·2% with cuttle fish liver oil, and 84·0% with ω3-HUFA mixture emulsion.

From their experiments Watanabe and colleagues concluded that *Artemia* containing at least 0·3% ω3HUFA (dry or wet weight basis not stated) may be a satisfactory single feed for marine fish. They added, however, that *Artemia* enrichment should always be applied since lipid contents in *Artemia* gradually decrease after hatching.

The French technique. Robin, Gatesoupe & Ricardez (1981) succeeded in improving the dietary value of San Francisco Bay *Artemia* for seabass (*Dicentrarchus labrax*) larvae by pre-feeding them for 2 days on a compound diet composed of *Spirulina* powder, I.F.P. yeast (a methanol yeast, used to reduce the quantity of the expensive *Spirulina*), DL-menthionine, choline chloride, D-glucosamine HCL, cholesterol, cod liver oil, and a vitamin premix (see Table VII, Diets B and C). No further improvement was achieved when enriching good quality Brazilian *Artemia* with the same diet. In another experiment Robin (1982) and Robin *et al.* (1984) designed a 2-step enrichment technique which consists in pre-feeding newly hatched nauplii (48 h cyst incubation) for 48 h on a compound diet

(Table VII, Diet E or G) after which the nauplii are transferred into another container for a 30-min enrichment with another compound diet, consisting mainly of fish autolysate, cod liver oil, vitamins, and minerals (Table VII, Diet H). The 2 days pre-feeding idea originates from the observation of Anderson (1967) that feeding of *Artemia* is impossible before their second moult which takes place 30 h after hatching at 20 °C. When fish larvae were fed *Artemia* nauplii which were 48-h pre-fed on brewer's yeast (Diet D), survival and growth was inferior to any other case where a compound diet was pre-fed (Diet E) followed or not by a subsequent enrichment batch (30 min, diet H). Application of an enrichment bath (Diet H) after 48 h pre-feeding on brewer's yeast did significantly improve the nutritional value of *Artemia* nauplii but larval growth was superior in those cases where the compound diet was pre-fed. An extra enrichment batch (30 min Diet H) in the latter treatment did not further improve its quality. After 48 h pre-feeding San Francisco Bay nauplii on Diet G, the ω3-HUFA content increased from 5·7% ($9 mg \cdot g^{-1}$) to 12·1% ($8 mg \cdot g^{-1}$); after subsequent enrichment for 30 min with Diet J ω3-HUFA levels reached 14·9% ($16 mg \cdot g^{-1}$; all data expressed on a dry weight basis).

Gatesoupe (1982) demonstrated that for larval turbot (*Scophthalmus maximus*) post-weaning survival and growth are largely improved when live food organisms (*Brachionus* and *Artemia*) are enriched. *Artemia* were first pre-fed for 48 h on a compound *Artemia* diet (Diet F screened through a 48-μm mesh screen) followed by a 30-min enrichment batch (Diet I). The enriched nauplii are offered to the turbot larvae along with the enrichment diet using a drip supply. The feeding of enriched rotifers and *Artemia* is particularly important in stress situations—both occasional stress (*e.g.* an infection) or the inevitable stress of weaning. Incorporation of antibacterial drugs in rotifers as applied by Gatesoupe (1982) using the same enrichment procedures might be equally well applicable to *Artemia*.

The Belgian technique. The Belgian enrichment technique consisted at first in pre-feeding newly hatched *Artemia* nauplii with ω3-HUFA coated micro-particles (5 cm Secchi-transparency or $0 \cdot 6 g \cdot l^{-1}$ for 3×10^5 nauplii$\cdot l^{-1}$; Léger, unpubl.). These micro-particles, *e.g.* micronized rice bran, were coated with various fish oils using a similar technique as used in preparing stationary phases for packed column gas-liquid-chromatography. Later, a compound analogue was formulated for larger scale testing in shrimp and fish hatcheries (Léger *et al.*, 1985a; Van Ballaer *et al.*, 1985). Using this compound analogue diet maximal ω-HUFA build-ups in *Artemia* within 24 h after hatching were at least as good as what had been reported in literature (see Table VI). The preparation of coated micro-particles is, however, complex and expensive. Therefore, another even more effective enrichment diet was developed in the form of a self-emulsifying enrichment concentrate (Léger *et al.*, 1985b). This diet is a self-dispersing complex mixture of mainly ω3-HUFA sources, vitamins, carotenoids, phospholipids, steroids, and emulsifiers. After simple dilution in water aerated by an airstone it produces finely dispersed globules which are readily available for ingestion by the nauplii. The advantages of this formulation are its ease in use and its effectiveness, *i.e.* ω3-HUFA accumulation rates in *Artemia* nauplii, especially the levels of 22:6ω3, largely surpass the figures reported in literature (see Table VI).

TABLE VII

Composition (%) of pre-enrichment and enrichment diets used in the French technique for Artemia *enrichment: A, B, C, after Robin* et al., *1981; D, E. H, after Robin, 1982; F, I, after Gatesoupe, 1982; G, J, after Robin* et al., *1984; diets A and D served as control diets*

	Pre-enrichment diets (48 h)							Enrichment bath (0·5 h)		
	A	B	C	D	E	F	G	H	I	J
Spirulina (dry)	100	87·9	40			40				
I.F.P.-yeast			40			40				
Brewer's yeast				100	89·4		89·4			
Fish autolysate								73	73	73
Cod liver oil		4	4		4	4	4	10	10	10
D-glucosamine HCl		0·5	0·5			0·5				
Cholesterol		1	1			1				
Choline chloride		2	2		2	2	2	4	4	4
DL-methionine		1	1		1	1	1	2	2	2
Vitamin premix		3·6	3·6		3	3·2		10	9·6	
Vitamin mineral premix							3·6			11
Corn starch						7·9				
$CaHPO_4$					0·5	0·3		0·8	1	
$FeSO_4 \cdot 7H_2O$					0·1	0·1			0·4	
$FeCl_2 \cdot 7H_2O$								0·2		

Different application procedures have been proposed, *i.e.* enrichment can be done after separation or without separation of the nauplii from the hatching debris. The latter technique indeed simplifies enrichment procedures for large scale applications, for which after all they ought to be developed. A first technique consists in incubating cysts, pretreated with a self-emulsifying concentrate, for 36 h at 28–30 °C. After this, the enriched *Artemia* nauplii are harvested and ready to be fed to the predator. Applying this technique, hatching and enrichment occur in the same tank without extra manipulations. Enrichment levels are high ($\Sigma\,\omega$3-HUFA = 16·7 $mg \cdot g^{-1}$ dry wt) for a total incubation time which is considerably shorter than the time periods (hatching + enrichment) claimed for the previously described techniques.

A second technique implies the addition of a self-emulsifying concentrate into the hatching tank after 24 h hatching incubation at 28–30 °C. Separation of the enriched nauplii is done after a 36 h total incubation period. After this period enrichment levels ($\Sigma\,\omega$3-HUFA = 11·2 $mg \cdot g^{-1}$) will further increase but separation of the nauplii from the hatching debris becomes difficult.

A third technique resembles French and Japanese techniques, *i.e.* after hatching and separation nauplii are incubated in a separate enrichment tank. Nauplius density, however, is higher (up to $3 \times 10^5 \cdot l^{-1}$) and mortality after 24 h enrichment is minimal. Enriched metanauplii are harvested after 12 h enrichment ($\Sigma\,\omega$3-HUFA = 14·4 $mg \cdot g^{-1}$), 24 h enrichment ($\Sigma\,\omega$3-HUFA = 37·4 $mg \cdot g^{-1}$) or 48 h enrichment ($\Sigma\,\omega$3-HUFA = 58·6 $mg \cdot g^{-1}$). For the last case lower naupliar densities are recommended. These high ω3-HUFA accumulation rates, which however may vary according to the ω3-HUFA-source used and to the enrichment conditions (*e.g.* temperature, aeration, naupliar density) are the result not only of optimal diet composition and presentation, but also of proper enrichment procedures. The first difference with other techniques is indeed the shorter hatching incubation period (24 h instead of mostly 48 h). Hatching conditions are optimized and controlled to such an extent that a maximal hatch is achieved within a minimal time. The advantage of this is that the energy decrease in the nauplii will never drop beyond a minimal loss, which inevitably occurs during yolk absorption. Indeed, attention is necessary so that the enrichment diet is available in the hatching medium at the moment of first feeding (instar II stage). Moreover poor hatching synchrony in *Artemia* cysts (*e.g.* time lapse between appearance of first and last hatching nauplius can vary from 5 h to 17 h at 25 °C, Vanhaecke, 1983) implies that first feeding time of nauplii will also be spread. In this regard nauplii should be transferred as soon as possible, before first feeding, into the enrichment medium. Application of these enrichment procedures will result not only in high ω3-HUFA accumulation rates, but also in minimal size increases of enriched nauplii, *e.g. Artemia* enriched according to Japanese and French techniques reach >900 μm, whereas Belgian procedures result in similar and higher enrichment levels in nauplii measuring 660 μm (12 h enrichment) to 790 μm (48 h enrichment).

Conclusions

The application of pre-feeding *Artemia* nauplii on ω3-HUFA enrichment diets has been shown to be effective in enhancing the dietary value of several strains and lots of *Artemia*. Enriched nauplii have an improved nutritional composition since they have a higher energy content and contain all essential fatty acids especially 22:6ω3 which is mostly absent in nauplii from whatever strain. The same enrichment techniques can also be used to transfer other nutrients, prophylactics and therapeutics into the predator larvae *via* the *Artemia*.

The use of enriched *Artemia* in larval culture is reflected in improved performances in terms of both survival and growth. Consequently, culture performance in later stages will also be improved. Fish and shrimp larvae fed enriched *Artemia* are indeed healthier and more resistent to stress conditions, *e.g.* infections, weaning, and transfer from indoor fully controlled hatchery tanks to the wild environment in nursery ponds. The effect of *Artemia* quality on culture performance in later stages has indeed been reported by several authors (New, 1976; Meyers, in Hanson & Goodwin, 1977; Ablett & Richards, 1980; Howell *et al.*, 1981; Gatesoupe, 1982; Bromley & Howell, 1983; Conklin, D'Abramo & Norman-Boudreau, 1983; Wilkenfeld *et al.*, 1984; Geiger & Parker, 1985). The only disadvantage of using enriched *Artemia* is their larger size which may limit their use in the early larval stages. In this cause freshly hatched high quality nauplii should be fed for the first days before gradually switching to enriched metanauplii. Optimized enrichment procedures may, however, reduce the disadvantage of size.

THE SEARCH FOR SUBSTITUTES AND REDUCED DEPENDENCE ON *ARTEMIA* CYSTS

The availability of sufficient quantities of food organisms is a prerequisite for any successful rearing attempt (May, 1970; Barnabé, 1976; Girin & Person-Le Ruyet, 1977; Paulsen, 1980). In this regard, the availability of *Artemia* under the form of storable dry cysts as an off-the-shelf live food has to a great extent accounted for its success in larval rearing. World cyst demand was estimated to be 60 metric tons (MT) in 1981 (Sorgeloos, 1981), 80–90 MT in 1985 and 150–170 (MT) in 1990 (Lai & Lavens, 1985). Current cyst supplies (different quality products) reach over 200 MT (Lai & Lavens, 1985) and thus exceed by far actual demands. In the 1970s the use of *Artemia* in aquaculture was, however, questioned because of an unreliable availability and high price (Bardach *et al.*, 1972; Roberts, 1974; Person-Le Ruyet, 1976; Wickins, 1976; ASEAN, 1977; Gatesoupe *et al.*, 1977; Goodwin & Hanson, 1977; Bigford, 1978; Glude, 1978a,b; Murai & Andrews, 1978; Smith *et al.*, 1978; Girin, 1979; Meyers, 1979; Manzi & Maddox, 1980; Sorgeloos, 1980c). This situation has generated efforts to substitute *Artemia* by other live food organisms and by artificial diets. Furthermore, research has and is being conducted to reduce the dependence on *Artemia* cysts by optimization of feeding levels and techniques, selecting the most bioeconomical strains, using supplemental diets, applying early weaning techniques and using decapsulated cysts and on-grown *Artemia*. A review of the results of these efforts is beyond the purpose of this review. A brief summary will, however, accentuate once more the versatility in use and nutritional quality of *Artemia* nauplii.

The substitution of Artemia

In summary we may state that for most fish and crustacean species studied complete substitution of *Artemia* nauplii by other food organisms or artificial diets has not been yet achieved.

The collection of wild plankton and other organisms may in some cases indeed provide a welcome supplement to high quality live food, but this method is hardly dependable beyond a laboratory scale (Fabre-Domergue & Bietrix, 1905; Dexter, 1972; Rakowicz, 1972; Houde, 1973; Girin & Person-Le Ruyet, 1977; Nellen *et al.*, 1981). Similarly, the intensive culture of wild food organisms still has to prove its year-round reliability on an industrial scale. None the less, interesting results have been obtained on a small scale with copepods (Kahan, 1980; Watanabe *et al.*, 1980; Kuhlmann *et al.*, 1981a,b, 1982; Kahan, Uhlig, Schwenzer & Horowitz, 1981/1982; Lee, Hu & Hirano, 1981; Kuronuma & Fukusho, 1984; Nellen *et al.*, 1981; Witt, Quantz & Kuhlmann, 1984), amphipods (Good, Bayer, Gallagher & Rittenburg, 1982), mysids (Ogle & Price, 1976; Kuhlmann *et al.*, 1981b), rotifers (Berrigan, Willis & Halscott, 1978; Yamasaki & Hirata, 1982), and nematodes (Kahan, 1979; Wilkenfeld *et al.*, 1984).

Not all trials using other live food as a substitute for *Artemia* nauplii were equally promising or successful for fish and crustacean larvae (Kurata, 1959; Gun'ko & Pleskachevskaya, 1962; May, 1970; Campillo, 1975; Fukusho, 1979; Beck, 1979; Flüchter, 1980; Hogendoorn, 1980; Dejarme, 1981; Anonymous, 1984; Emmerson, 1984). Kanazawa (1984) further stated that the mass culture of other live food organisms not only requires much labour and expensive equipment but its success also fluctuates with climatic conditions. Besides, the nutritional value of planktonic organisms is occasionally variable which restricts their possible utilization on a large scale. Following Kanazawa (1984) the development of artificial diets is one of the most important research areas for intensive larval culture. Along with this author all people involved with larval rearing will agree on the need of developing suitable artificial diets for substituting live food organisms.

Several types of artificial diets have been formulated ranging from natural products, compound diets to micro-encapsulated diets. Artificial diets are indeed appealing because of year-round availability, ease of handling and storage, uniform and constant nutritional quality, optimal size, possible germ-free formulation, no need to wean larvae, *etc.* On the other hand, some inherent problems still have to be solved: *e.g.* optimal nutritional composition (since larval requirements are as yet far from known), buoyancy, nutrient leaching, water quality problems, digestibility, production complexity and cost. Using formulated diets as a substitute for *Artemia*, promising and some successful results have been obtained (Adron, Blair & Cowey, 1974, 1977; L'Herroux *et al.*, 1977; Dabrowski *et al.*, 1978, 1984; Villegas & Kanazawa, 1978; Jones, Kanazawa & Rahman, 1979, unpubl.; Teshima, Kanazawa & Sakamoto, 1982; Levine, Sulkin & Van Heukelem, 1983). More numerous, however, are the less successful trials and failures (Broad, 1957; Regnault, 1969; San Feliu, 1973; Campillo, 1975; Barnabé, 1976; Gatesoupe *et al.*, 1977; Berrigan *et al.*, 1978; Murai & Andrews, 1978; Hogendoorn, 1980; Beck, 1979; Günkel, 1979; Schauer *et al.*, 1979; Manzi *et al.*, in Manzi & Maddox, 1980; Reddy & Shakuntala, 1980; Sandifer & Williams, 1980; Tacon & Cowey, 1982; D'Abramo, Baum, Bordner & Conklin, 1983; Bengtson *et al.*, 1978; Conklin, Devers & Shleser, 1975; Conklin, Goldblatt & Bordner, 1978; Dabrowski & Kaushik, 1984).

Total replacement of live food, has indeed met with limited success, *i.e.* despite the best efforts of scientists throughout the world, no artificial diet has yet been produced that supports long-term growth and survival comparable with that of live food organisms Bengtson *et al.*, 1978; (Beck, 1979; Cowey & Tacon, 1982; Bromley & Howell, 1983). Even the most advanced artificial diets such as micro-encapsulated diets have achieved only limited success in replacing live food, eventually caused by lack of acceptability due to insufficient gustatory stimulation invoking ingestion (Jones *et al.*, in press). On the other hand, the indirect use of those diets to improve the nutritional value of conventional live food such as *Artemia* and rotifers is proving much more successful. (See also Sakamoto, Holland & Jones, 1982; Jones *et al.*, in press.)

The reduced dependence on Artemia *cysts*

Although substitution of *Artemia* is not realistic yet, a reduced dependence on *Artemia* can be pursued in various ways. Optimizing feeding levels and feeding techniques constitutes the first opportunity for improvements. Indeed, in many cases *Artemia* is fed in excess, often only once a day. The consequences of this wasteful practice have been described earlier. Barahona-Fernandes & Girin (1977), therefore, rightly advise restriction in the daily amounts of *Artemia* nauplii to the intake capacity of the larvae. Bryant & Matty (1980) agree that considerable savings may be achieved by adjusting *Artemia* levels according to changing requirements with larval age.

Besides optimal feeding levels and techniques, Vanhaecke & Sorgeloos (1983b) claim that in the rearing of larval carp 10 to 75% of *Artemia* costs can be saved by selecting the best bioeconomical strain of *Artemia*. Their selection is based on the quantity of cysts needed per gram carp-biomass produced. This quantity is mainly determined by the hatching characteristics of the source of cyst used. For this, besides cyst price, hatching quality may be used as a selection criterion. When price and hatching quality are comparable, they recommend the use of *Artemia* strains producing large nauplii since these guarantee best growth in carp larvae.

As discussed earlier the nutritional quality of *Artemia* does not affect culture results as much in freshwater species as in marine species. For the latter, selection of the most bioeconomical *Artemia* strains should, therefore, also take into account differences in size and nutritional value.

A reduced dependence on *Artemia* cysts, without affecting culture performance, may also be achieved by supplementing a reduced *Artemia* ration with other foods such as artificial diets and other live, freshly killed or conserved food organisms (Meske, 1973; Sick & Beaty, 1974; De Figueiredo, 1975; Christiansen & Yang, 1976; Goodwin & Hanson, 1977; Berrigan *et al.*, 1978; Murai & Andrews, 1978; Al Attar & Ikenoue, 1979; Bengtson *et al.*, 1978; Günkel, 1979; Meyers, 1979; Conklin, D'Abramo, Bordner & Baum, 1980; Hogendoorn, 1980; Manzi & Maddox, 1980; Seidel *et al.*, 1980a; Spitchak, 1980; Soebiantoro, 1981; New & Singholka, 1982; Wilkenfeld *et al.*, 1984; Bombeo, 1985).

Of significant importance in saving on *Artemia* cysts are recent developments in the elaboration of early weaning techniques for fish larvae. These techniques aim to switch from *Artemia* nauplii to inanimate diets (*e.g.* artificial diets, freshly killed or conserved organisms) as early as possible in the development of larvae. Larval development of fish may indeed last from 45 to 90 days compared with a few weeks in shrimp. Several authors report successful trials in this regard (Bromley, 1978; Person-Le Ruyet, Alexandre, Le Roux & Nedelec, 1978; Girin, 1979; Metailler *et al.*, 1981; Cadena Roa *et al.*, 1982a,b; Gatesoupe & Luquet, 1981/1982; Bromley & Howell, 1983; Gatesoupe, 1983; Duray & Bagarinao, 1984). It is noteworthy that weaning success is to a large extent determined by the quantity and quality of *Artemia* fed during earlier development before weaning (Forster & Wickins, 1967; Bromley, 1978; Bromley & Howell, 1983).

Finally, the use of decapsulated cysts and on-grown *Artemia* (see later) may provide extra means of reducing the quantity of *Artemia* cysts needed.

THE USE OF DECAPSULATED *ARTEMIA* CYSTS

Decapsulated cysts are *Artemia* embryos surrounded only by the embryonic cuticle and the protecting outer cuticular membrane (see Fig. 8). Decapsulation is achieved by dissolving the chorion of the cysts in an alkaline hypochlorite solution. When properly carried out, the viability of the embryo is not affected.

The pioneering procedure was described in 1962 by Nakanishi, Iwasaki, Okigaki & Kato for the sterilization of *Artemia* cysts, *i.e.* they used a chilled diluted antiformin solution which was later also used by Lenhoff & Brown (1970, see above). Since then several authors have applied similar techniques; some of them noticed that at higher hypochlorite concentrations the cyst shell dissolved completely (Broch, 1965; Katsutani, 1965; Morris & Afzelius, 1967; Clegg & Golub, 1969; Slobin & Moller, 1976). A routine decapsulation technique for large-scale application was first described by Sorgeloos *et al.* (1977) and improved by Bruggeman *et al.* (1979, 1980) and Sorgeloos *et al.* (1983). This technique involves the following consecutive steps: hydration of the cysts because only fully spherical cysts can be completely decapsulated, treatment with alkaline hypochlorite to remove the chorion, washing and deactivation of the residual active chlorine, followed by direct use or dehydration for storage. The advantages of using decapsulated cysts are numerous.

(1) Decapsulated cysts are sterile thus eliminating the potential risk of introducing germs *via* hatched nauplii into the culture water of the predator. Furthermore, bacterial development during hatching incubation is significantly reduced.

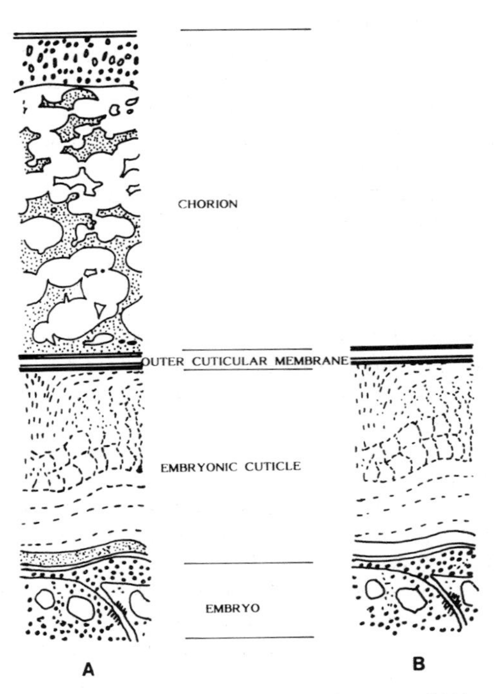

Fig. 8.—Schematic diagram of outer membranes of untreated (A) and decapsulated (B) *Artemia* cyst shell (modified from Morris & Afzelius, 1967).

(2) Because the chorion is removed separation of the nauplii from the hatching debris becomes superfluous. The only membrane discarded by the nauplius at hatching is the thin transparent embryonic cuticle which has proved to be unharmful for crabs and shrimps (Sorgeloos, 1979). As a result, after hatching of decapsulated cysts, the only procedure needed is to rinse the nauplii before feeding them to the predator.

(3) In some strains hatchability of *Artemia* cysts is significantly improved after decapsulation, *e.g.* hatching percentages increase by 1·8 to 230·3%, and because naupliar dry weights are also higher after decapsulation, hatching outputs improve by 2 to 144% (Bruggeman *et al.*, 1980; Vanhaecke & Sorgeloos, 1983a).

(4) Decapsulated cysts may be used as a direct food source for fish and crustacean larvae eliminating the need for hatching of the cysts. Several authors have indeed demonstrated the potential of using decapsulated cysts as a direct food source for decapod and fish larvae, *e.g. Scylla serrata* (Lavina in Sorgeloos, 1979), *Penaeus monodon* (Mock *et al.*, 1980a,b; Lavina & Figueroa, 1978), *P. indicus*, *Metapenaeus ensis*, *M. endeavoori*, *Macrobrachium rosenbergii* (Lavina & Figueroa, 1978), *Metapenaeus monoceros* (Royan, 1980), *Penaeus kerathurus* (Rodriguez, Martin & Rodriguez, 1980, in Sorgeloos *et al.*, 1983), *Penaeus setiferus* (Wilkenfeld *et al.*, 1984), *Chanos chanos* (De los Santos, Sorgeloos, Lavina & Bernardino, 1980; Nanayakkara, Sunderam & Royan, 1985),

Cyprinus carpio (Devrieze, 1984), *Poecilia reticulata* (Sorgeloos *et al.*, 1977), *Oreochromis niloticus, Etraplus suratensis* (Nanayakkara *et al.*, 1985), and many ornamental fish species like black mollies, red sword tails, gouramies, angles, tetras, barbs, and gold fish (Sumitra-Vijayaraghavan *et al.*, 1985). Not all larval species, however, digest decapsulated *Artemia* cysts equally well; larvae of *Solea solea* survive well on a diet of decapsulated cysts but their digestion takes 12 h and as a result growth is retarded (Dobbeleir, 1978, in Sorgeloos, 1979).

The use of decapsulated cysts as a direct food source implies several advantages.

(1) Because their diameter and volume are smaller (30 to 40%) than in freshly hatched nauplii (Vanhaecke & Sorgeloos, 1980; Vanhaecke, Steyaert & Sorgeloos, 1980; Vanhaecke, 1983) they can be fed to earlier larval stages.
(2) The energy content of decapsulated cysts is 30 to 57% higher than in freshly hatched nauplii (Vanhaecke, 1983; Vanhaecke *et al.*, 1983). This means that for an equal hunting effort a high energy intake will be achieved resulting in better growth and considerable savings in *Artemia* cysts (Anonymous, 1980; Devrieze, 1984; Nanayakkara *et al.*, 1985). Devrieze (1984) indeed demonstrated that for the production of the same carp biomass 10 to 23% *Artemia* cysts could be saved during the first week and 32 to 36% during the second week by using decapsulated cysts instead of freshly hatched nauplii.
(3) Cysts that have lost the capacity to hatch may be valuated. About 50% of present cyst stocks have a low commercial value because of their low hatchability (*e.g.* below 50%; Lai & Lavens, 1985) thus their valuation as decapsulated cysts might be more attractive.

The main problem when using decapsulated cysts as a direct food source is their fast sedimentation in sea water which makes them unavailable for planktonic larvae, unless they hatch. Their availability in the water column may be improved, at least in small scale cultures, by using conical tanks equipped with air-water-lifts. The use of dried decapsulated cysts which float and upon hydration sink only slowly may be a better solution, *e.g.* growth in carp larvae was significantly better when using dried instead of freshly decapsulated cysts (Devrieze, 1984). The same author also showed that the addition of dried decapsulated cysts at a ration of 25% of the diet significantly improved weaning success in carp larvae. In conclusion, the application of dried decapsulated cysts provides very interesting opportunities for application in intensive culture systems. A simplification of the decapsulation technique is, however, recommended if application at a larger scale is to be successful.

THE USE OF ON-GROWN AND ADULT *ARTEMIA*

In contrast to the very extensive documentation dealing with the use of *Artemia* nauplii as a food source, similar literature on the application of on-grown and adult *Artemia* (Fig. 9) is very limited. Evident reasons for this are the worldwide availability of storable *Artemia* cysts and the ease with which nauplii are obtained, whereas commercial availability of adult *Artemia* is very restricted and its cost very high; furthermore, it is only during recent years that reliable techniques have been developed for mass production of pre-adult and adult *Artemia*. *Nevertheless, several arguments support the use of on-grown and adult Artemia* as a food source.

Fig. 9.—*Artemia* pair in precopulation.

NUTRITIONAL QUALITY OF ON-GROWN AND ADULT *ARTEMIA*

As compared with freshly hatched nauplii the nutritional value of on-grown and adult *Artemia* is superior, *i.e.* protein content increases from an average of 47% in nauplii to 60% on a dry weight basis in adults; furthermore, protein quality improves as adults are rich in all essential amino acids (see later). In contrast to other food organisms, the exoskeleton of adult *Artemia* is extremely thin which facilitates digestion of the whole animal by the predator.

Prey size, however, has been the first rationale to switch from nauplius to juvenile and/or adult *Artemia*, e.g. Sick & Beaty (1975) showed that

Macrobrachium rosenbergii stage VIII could not ingest *Artemia* nauplii in sufficient amounts to give a positive energy balance. Better results in terms of relative rates of energy intake and, as a consequence, of prawn growth, developmental rate and survival were obtained with 5·5-mm juvenile *Artemia* as a food source. Purdom & Preston (1977) came to the same conclusion for turbot larvae and several other authors have applied the technique of feeding progressively larger *Artemia* to fish and crustacean larvae, *e.g.* San Feliu (1973), Dugan *et al.* (1975), Smith (1976), Cadena-Roa *et al.* (1982a,b), Ebert, Haseltine, Houk & Kelly (1983). In the case of Person-Le Ruyet *et al.* (1978), *Artemia* metanauplii cultured on dried algae or compound diets (see later) were used to weaning of fish larvae.

All lobster farming relies on adult *Artemia* as food for at least the first four larval stages, e.g. Hughes, Shleser & Tchobanoglous (1975); Van Olst, Ford, Carlberg & Dorband (1975); Carlberg & Van Olst (1976); Stewart & Castell (1976); Rosemark (1978); Conklin *et al.* (1975, 1978); Happe & Hollande (1982); Chang & Conklin (1983); Eagles, Aiken & Waddy (1984). As early as 1907 Williams noticed a better growth in *Homarus americanus* larvae when offered adult *Artemia* instead of a diet of minced clam and naturally available copepods.

Although frozen *Artemia* can be used, best results are obtained with live adults which assure better availability in the water column and do not provoke deterioration of water quality (Schuur *et al.*, 1976). The superiority of live adult *Artemia* to frozen and freeze-dried adults and artificial diets has been demonstrated repeatedly, *e.g.* Botsford, Rauch & Shleser (1974); Serflin, Van Olst & Ford (1974); Hughes *et al.* (1975); Shleser (1976); Schuur *et al.* (1976); Conklin *et al.* (1975, 1978); Happe & Hollande (1982). According to Conklin *et al.* (1978), an essential but water-soluble substance is present in live adult *Artemia* which leaches from frozen or dried brine shrimp.

Live amphipods might be used as a better alternative for adult *Artemia*; *i.e. D'Agostino (1980) reported better growth and pigmentation in lobster* juveniles when using *Calliopius leaviusculus* instead of live *Artemia*, and Good *et al.* (1982) also observed better pigmentation when *Gammarus oceanicus* was fed instead of frozen *Artemia*. Eagles *et al.* (1984), however, caution for quality control of frozen *Artemia*, *i.e.* unpigmented, fragmented and leached frozen adult *Artemia* gave less satisfactory growth and development in lobster larvae. According to Rosemark (1978) culture success in lobster can be enhanced by supplementing the *Artemia* diet with frozen natural products. Nevertheless, Happe & Hollande (1982) claim that a sole diet of *Artemia* guarantees best production results in *Homarus americanus*, *i.e.* market size can be reached in 2 years only as compared with 3 years when *Artemia* is supplemented with red crab flesh. Using only *Artemia* as food, however, makes the production cost of the lobster too high.

Besides lobster, several other species have been offered on-grown and adult *Artemia* with good results, *e.g.* the freshwater prawn *Macrobrachium rosenbergii* (Dugan *et al.*, 1975; Sick & Beaty, 1975; Perrot, 1976; Sick, 1976; Aquacop, 1977; Goodwin & Hanson, 1977; Corbin *et al.*, 1983), marine shrimp such as *Penaeus monodon* (Millamena *et al.*, 1985; Bombeo, 1985; Yashiro, 1985), *P. kerathurus* (San Feliu, 1973; Rodriguez, 1976; San

Feliu *et al.*, 1976), *P. japonicus* (Palmegiano & Trotta, 1981; Camara & De Medeiros Rocha, 1985; Guimares & De Haas, 1985; Trotta, Villani & Palmegiano, 1985), *P. aztecus* (Flores, 1985), *Palaemon serratus* (Wickins, 1972), the crab *Cancer magister* (Ebert *et al.*, 1983), several fish species such as *Pleuronectes platessa* and *Solea solea* (Shelbourne, 1968), *Solea vulgaris* (Cadena Roa *et al.*, 1982a,b), *Scophthalmus maximus* (Aronovick & Spektorova, 1971; Anonymous, 1973, 1978c; Person-Le Ruyet *et al.*, 1978), *Sparus auratus* (Alessio, 1974; San Feliu *et al.*, 1976), *Dicentrarchus labrax* (Allesio, Gandolfi & Schreiber, 1976; Barahona-Fernandes & Girin, 1977; Girin, 1976; Anonymous, 1977, 1978b; Barnabé, 1980; Trotta *et al.*, 1985), *Diplodus sargus* (Divanach, Kentouri & Paris, 1983), *Chanos chanos* (De los Santos *et al.*, 1980; Bombeo, 1985), *Acipenser* sp. (Azari Takami, 1976, 1985; Binkowski & Czeskleba, 1980), *Lepomis* sp. (Smith, 1975, 1976), and ornamental fish (Rakowicz, 1972).

The use of on-grown and adult *Artemia* has mostly been restricted to relatively small scale culture trails. During recent years, however, commercial scale use of *Artemia* biomass harvested from local salt-works (Camara & De Medeiros Rocha, 1985) or produced in manured salt-ponds (De los Santos *et al.*, 1980; Flores, 1985; Jumalon *et al.*, 1985; Tarnchalanukit & Wongrat, 1985) is gaining more and more interest especially in fish weaning and shrimp nursing. The recent finding that a diet of adult *Artemia* may induce maturation in shrimp without application of eyestalk ablation (Camara & De Medeiros Rocha, 1985; Flores, 1985) may also be of major importance in future shrimp farming.

THE USE OF INTENSIVELY PRODUCED *ARTEMIA* BIOMASS

Although the cheapest source of *Artemia* biomass is from natural and man-controlled salt-pond systems, *Artemia* produced in intensive culture systems may become more attractive especially in climates that are unsuitable for outdoor production and when quality control is critical (Sorgeloos *et al.*, 1983; Lavens *et al.*, in press). Recently much progress has been made in the development of new techniques for the high density culturing of *Artemia* using cheap agricultural by-products instead of algae as food (Bossuyt & Sorgeloos, 1980; Brisset, 1981; Brisset *et al.*, 1982; Sorgeloos *et al.*, 1983; De Meulemeester *et al.*, *1985; Lavens & Sorgeloos, 1985; Platon & Zahradnik*, 1985). Other feeds used are the marine yeast *Candida* (James, Abu-Rezeq & Dias, 1985), organic wastes (Basil & Marian, 1985), clam-meat suspension (Vishnu Bhat & Ganapathy, 1985), and dried algae (Person-Le Ruyet *et al.*, 1978).

Artemia produced in intensive culture systems appeared to be an acceptable food for the larvae of various species of fish and crustaceans (Shelbourne, 1968; Dugan *et al.*, 1975; Smith, 1976; Person-Le Ruyet *et al.*, 1978; Dobbeleir, 1979 in Sorgeloos *et al.*, 1983; Cadena Roa *et al.*, 1982a,b; Chang & Conklin, 1983; Yashiro, 1985; Trotta *et al.*, 1985; Millamena *et al.*, 1985). Contrary to what is found in wild adults, the fatty-acid profile of brine shrimp cultured on feeds of terrestrial origin (*e.g.* agricultural waste products) does not show significant levels of the essential fatty acids 20:5ω3 and 22:6ω3 (see Table XIV, p. 603).

This deficiency can, however, be remedied by application of enrichment techniques using similar diets as described earlier for the nauplii (Sakamoto *et al.*, 1982; Léger *et al.*, 1985b). In fact this technique of encapsulation provides interesting opportunities to use *Artemia* biomass not only as an attractive food but at the same time as carrier to administer various products, *e.g.* essential nutrients, pigments, prophylactics, therapeutics, hormones, *etc.* to the predator larvae (Léger *et al.*, 1985b). For various reasons *Artemia* produced in intensive cultures may be preferred over wild brine shrimp biomass; *e.g.* being produced at high salinities the latter may not survive equally long when transferred into natural sea water (Sorgeloos, (1979); moreover, wild *Artemia* can be the carriers of infectious organisms such as Cestoda (Heldt, 1926; Young, 1952; Maksimova, 1973), Spirochaeta (Tyson, 1970), Fungi (Kamienski, 1899; Lachance, Miranda, Miller & Phaff, 1976) and intracellular Procaryota (Post & Youssef, 1977). On the contrary, *Artemia* cultured on various agricultural waste products in batch systems have been shown to be relatively clean in terms of microbial contamination (Dobbeni, 1983). Another advantage of using cultured *Artemia* is that any size from 0·5 to >10 mm may be harvested and fed to the predator according to its growth.

OTHER APPLICATIONS OF ON-GROWN AND ADULT *ARTEMIA* AS FOOD SOURCE

Artemia biomass can also be applied as a dietary ingredient or gustatory attractant in artificial diets for fish and crustacean larvae (Sick & Andrews, 1973; Sick, Andrews & Baptist, 1973; Sick & Beaty, 1974, 1975; Sick, 1975, 1976; Barahona-Fernandes, Girin & Metailler, 1977; Girin, Metailler & Nedelec, 1977; Goodwin & Hanson, 1977; Metailler, Mery, Depois & Nedelec, 1977; Cadena Roa *et al.*, 1982a,b; Gatesoupe & Luquet, 1981/1982; Levine *et al.*, 1983). A most interesting application is the complete substitution of freshly hatched nauplii by freeze-dried and micronized *Artemia* biomass in the hatchery production of *Penaeus japonicus* (Guimares & De Haas, 1985), *i.e.* 1 million post-larvae could be produced with 1·8 kg *Artemia* meal.

In the future, *Artemia* biomass may also be considered as a complementary source of animal protein for terrestrial animals and even man (Helfrich, 1973; Stults, 1974; Anonymous, 1978a; Amat, 1980; Webber & Sorgeloos, 1980; Janata & Bell, 1985). A practical example was evaluated by Corazza & Sailor (1982) who tested lyophilized brine shrimp as a promising source of animal protein for broiler diets.

Dobbeni (1983), agreed that adult *Artemia* may have perspectives for human consumption and especially for intravenous feeding since its proteins have an ultra fine texture. Human consumption of brine shrimp may appear futuristic. None the less sun-dried *Artemia* was consumed centuries ago by Indian (Jensen, 1918) and African tribes (Oudney & Clapperton, 1812, in Bovill, 1968; May, 1967; Ghannudi & Tufail, 1978) and still today "pains d'*Artemia*" is on the menu of the Dawada tribe in Libya (Delga, Meunier, Pallaget & Carious, 1960; Monod, 1969; Dumont, 1979).

The idea of using *Artemia* as a food source for man is of particular interest for developing countries where animal protein is scarce and potential *Artemia* production sites abundant. Moreover, because *Artemia* occupies a lower trophic level than most farmed fish, the use of *Artemia as* a direct food source for man constitutes an economical use of live energy, which in these parts of the world is of critical importance.

THE BIOMETRICS OF *ARTEMIA*

A major advantage when using *Artemia* as food for fish and crustacean larvae is the relatively wide range of sizes from which one can chose. Indeed, in its smallest form, the decapsulated cyst, sizes ranges from around 208 to 266 μm, depending on geographical origin (Vanhaecke & Sorgeloos, 1980), freshly hatched nauplii measure from 428 to 517 μm (Vanhaecke & Sorgeloos, 1980), and when used in its adult form maximum lengths of 10 to 15 mm can be reached.

CYST DIAMETER

Vanhaecke & Sorgeloos (1980) made a detailed comparative study of the cyst biometrics in different batches of cysts from 17 geographical strains of *Artemia*. Data for the same and other strains can be found in D'Agostino (1965), Wickins (1972); Claus *et al.* (1977), Uçal (1979), Amat (1980), Vos *et al.* (1984), Nanayakkara *et al.* (1985), Van Ballaer *et al.* (1985). A compilation of cyst biometrics is provided in Table VIII. Cyst diameters differ widely, *i.e.* from 224·7 to 284·9 μm in hydrated untreated cysts and from 207·3 to 266·3 μm in hydrated decapsulated cysts. Differences between untreated and decapsulated cysts are not consistent revealing a variation in chorion thickness from 3 to 13·35 μm (Vanhaecke, 1983), which is not correlated with cyst diameter. Considering cyst diameter, American *Artemia* are relatively small when compared with the *Artemia* sources from

TABLE VIII

Biometrical data of hydrated untreated and decapsulated cysts and Instar I nauplii of different sources of Artemia *(data from Vanhaecke, 1983; Vanhaecke & Sorgeloos, 1980; Tackaert, unpubl.)*

Artemia source	Cyst diameter (μm) Untreated	Cyst diameter (μm) Decapsulated	Instar I nauplii Length (μm)	Instar I nauplii Volume ($10^{-3}\mu m^3$)
Argentina, Buenos Aires	238·2	217·4	431	7734
Australia, Adelaide	225·8	209·8		
Rockhampton	231·0			
Shark Bay	260·4	242·2	458	10249
Bahamas, Great Inagua	229·1	210·0		
Brazil, Cabo Frio	233·5	216·1		
Macau	228·7	213·8	447	8314
Bulgaria, Burgas Pomorije	281·0	263·5		
Burma	278·4			
Canada, Chaplin Lake	245·4	234·0	475	8930
China-P.R., Tientsin	274·4	257·8	515	13 097
Tsingtao	270·0	249·2		

Colombia, Galera Zamba	249·9	232·7	480	10 578
Manaure	237·0	220·8	456	8062
Cyprus, Larnaca	261·3	235·6		
Ecuador, Pacoa	226·2			
Salinas	242·3			
France, Aigues Mortes	259·6	240·8		
Lavalduc	276·3	261·5	509	12 724
Salins de Giraud	264·4			
Salins de Hyères	257·8			
Villeroy	261·2			
India, Bhayander, Bombay	258·0			
Kutch, Mundra	254·4	232·4		
Mithapur	267·7	248·0		
Tuticorin	282·9	262·7	509	
Iran, Ormia Lake	258·1	245·7		
Israel, Eilat	274·3	258·4	506	
Italy, Cervia	282·5			
Margherita di Savoia	284·9	266·3	517	13 604
Yugoslavia, Portoroz	291·7			
Kenya, Malindi	228·4			
Mexico, Bahia de Queta	224·9	207·3		
Yavaros Sonora	228·9	213·1		
Netherlands Antilles, Bonaire	236·9	219·0		
New Zealand, Lake Grassmere	231·6	216·7		
Peru, Chilca	246·9	226·7		
Virrila	227·1	208·5		
Philippines, Barotac Nuevo	228·0		429	7991
Jaro	225·2			
Pangasinan	229·7			
Portugal, Alcochete	248·4	233·6		
Puerto Rico, Bahia Salina	253·7	233·4	452	9090
Spain, Barbanera	257·3	230·6		
Delta del Ebro	277·8	258·8		
San Lucar	253·6	237·1		
Santa Pola	248·6			
Sri Lanka, Puttalam	269·8			
Tunisia, Bekalta	251·6		482·3	
Chott Ariana	268·9	245·3		
Mégrine	258·8	234·1	467·7	
Moknine	252·6			
Sfax	235·4	215·1	422·2	
Turkey, Izmir	270·4	252·9		
U.S.A., Great Salt Lake	244·2	234·8	482	9091
Jesse Lake	234·8			
Mono Lake	249·4	243·4		
Playa Tahoka	244·7	225·8		
Quemado	239·7	224·7		
Raymondville	253·9			
San Francisco Bay	224·7	210·0	428	7638
San Pablo Bay	235·6	220·4	433	8144
U.S.S.R., Azov Sea	270·2	258·9		
Bolshoe Jarovoe Lake	273·7	258·⅜		
Kujalnic Lagoon	273·5	255·9		
Mangyshlak peninsula	248·4	229·1		
Odessa	259·7	242·7		
Sivash	251·4	229·6		
Tinaki Lake	280·3	260·9		
Venezuela, Port Araya	249·0	222·6	474	9548
Tucacas	244·3	222·6		
Vietnam, Cam Ranh Bay	242·9			

the Old World. Within the American sources, considerable differences are noticed even between closely located sources, *e.g.* Chilca and Virrila in Peru. On the contrary, several American sources closely reflect the diameter of San Francisco Bay cysts (*e.g.* Great Inagua, Macau, Pacoa, Panama, Bahia de Cueta, Yavaros Sonora, and Virrila) and Great Salt Lake cysts (*e.g.* Galera Zamba, Chilca, Bahia Salinas, and Port Araya), *i.e.* the two oldest commercial strains which may have been used for (non) intentional introductions, *e.g.* San Francisco Bay *Artemia* in Macau, Brazil (Persoone & Sorgeloos, 1980). Cyst size appears to be genetically determined, *e.g.* no appreciable size differences were found between cysts from different harvests from the same source (Vanhaecke & Sorgeloos, 1980) and between cysts produced from the same inoculum in different countries (Vos *et al.*, 1984) or in laboratory-controlled systems (Lavens, unpubl.).

NAUPLIUS DIMENSIONS

Most information on nauplius lengths and volumes results again from the comparative studies of Vanhaecke & Sorgeloos (1980) and Vanhaecke (1983) (see Table XIII, p. 600). Further data can be found in D'Agostino (1965); Sorgeloos (1975); Smith (1976); Claus *et al.* (1979); Amat (1980); and Nanayakkara *et al.* (1985). According to strain origin the size of freshly hatched instar I nauplii ranges from 428 to 517 μm. The largest nauplii are produced in parthenogenetic strains with a high degree of ploidy (Vanhaecke, 1983). Vanhaecke & Sorgeloos (1980) found high degrees of positive correlation between the diameter of decapsulated cysts and nauplius length ($r = 0{\cdot}906$), and between volume of decapsulated cysts and nauplius volume. Cyst size may be an easier criterion for the selection of a proper sized *Artemia* strain either for use as food source (see above) or for *Artemia* inoculation (Vos *et al.*, 1984).

In view of the high heritability and the large variation in cyst biometrics selective breeding techniques may in the future be successful in the development of strains that produce mini-*Artemia* cysts, which would be a most welcome addition for use in early larval feeding of marine fishes and shrimps.

BIOCHEMICAL AND CHEMICAL COMPOSITION

A review of the literature on the composition of *Artemia* reveals considerable variation in amounts of the various compounds. The causes of the variation are undoubtedly several, *e.g.* different methods of extraction and analysis, different live stages of the *Artemia* studied, and different geographical populations. Although the information presented here could be averaged to portray a generalized *Artemia* composition, the most important message is that the inherent variation makes each commercially obtained batch of *Artemia* different. Scientists or aquaculturists, therefore, have the responsibility to assure that their *Artemia* provide adequate nutrition for the organisms to which they are fed.

INDIVIDUAL DRY WEIGHT AND ENERGY CONTENT

Data on the individual dry weight and energy content of newly hatched *Artemia* nauplii of different geographical origin are summarized in Table IX. The energetic content on an ash-free dry weight basis appears to be very similar for most geographical collections studied. On the contrary, individual energetic content and individual dry weight differ greatly. Not considering variability of a purely analytical origin, differences may be explained by varying hatching conditions. Von Hentig (1971) indeed demonstrated that *Artemia* hatched at a lower salinity and higher temperature contained more energy. When comparing data obtained for different *Artemia* sources hatched under the same conditions, Vanhaecke (1983) and Vanhaecke *et al.* (1983), however, still noticed considerable differences of up to 100% and more. Nevertheless, no significant differences were detected among batches from the same strain nor between cysts originating from the same parental material but produced at different localities, *e.g.* Macau (Brazil), Barotac Nuevo (Philippines) and San Francisco Bay (U.S.A.). This allowed Vanhaecke *et al.* (1983) to conclude that in *Artemia* individual dry weight and energy content are mainly genetically determined and thus strain specific. As a result nauplius dry weight and energy content are important criteria for strain selection; indeed, when size and nutritional composition are acceptable for a predator, *Artemia* with a high energy content will guarantee better predator growth, since less energy will be spent in hunting and food uptake (Vanhaecke & Sorgeloos, 1983b; Nanayakkara *et al.*, 1985).

Variability in results between authors analysing the same *Artemia* strains is most probably related to differences in hatching incubation time. Indeed, *Artemia* starts utilizing its energy reserves shortly after cyst hydration when the embryonic metabolism restarts (Urbani, 1959; Von Hentig, 1971); food-uptake only takes place after the animal has moulted into the second instar stage (Benesch, 1969). As a result significant drops in individual dry weight and energy contents have been reported in older *Artemia* metanauplii as compared with decapsulated cysts and even instar I nauplii (Paffenhöfer, 1967; Benijts *et al., 1976; Royan, 1980; Vanhaecke et al.*, 1983). According to Vanhaecke *et al.* decapsulated cysts contain 30 to 57% more energy than instar I nauplii which in their turn contain 22 to 37% more energy than instar II-III metanauplii. Metanauplius development and energy loss can be reduced to 2·5% over a period of 24 h when storing the freshly hatched nauplii at 2–4 °C (Léger *et al.*, 1983).

Data on energy content of on-grown and adult *Artemia are scarce, e.g. 7-day old Artemia* reared on *Dunaliella* contain 5854 $cal \cdot g^{-1}$(=24 499 $J \cdot g^{-1}$) (Paffenhöfer, 1967) whereas only 5100 $cal \cdot g^{-1}$(=21 344 $J \cdot g^{-1}$) was reported for frozen *Artemia* biomass (Gabaudan, Piggott & Halver, 1980). The latter result is within the same range as reported for newly hatched nauplii (Table IX). Evidently, individual energy content is much higher in adults than in nauplii, for which reason better predator growth is to be expected when on-grown *Artemia* are being fed (Sick & Beaty, 1974, 1975). Individual dry weights of 0·88 and 1·0 mg have been reported by Reeve (1963) and Tobias, Sorgeloos, Roels & Sharfstein (1980), respectively, for sexually mature animals of different origin reared on algae.

TABLE IX

Data on individual dry weight and energy content of newly hatched Artemia *nauplii from different geographical origin: c, calculated; t, transformed to SI-units*

Artemia source	Hatching conditions T(°C)	S(‰)	Individual dry weight (μg)	Energetic content ($J \cdot g^{-1}$ ash-free dry wt)	Individual energy content (J)	Reference
Argentina, Buenos Aires	25	35	1·72	23 506[c]	0·0379	Vanhaecke *et al.*, 1983
Australia, World Ocean (No. 113)	25	30		25 000		Schauer *et al.*, 1980
(No. 114)	25	35	2·47	23 575[c]	0·0576	Vanhaecke *et al.*, 1983
Brazil, Macau (1978)	25	30		23 500		Schauer *et al.*, 1980
(1978)	25	35	1·68	24 116[c]	0·0381	Vanhaecke *et al.*, 1983
(No. 871172)	25	35	1·74	23 927[c]	0·0392	Vanhaecke *et al.*, 1983
(No. 971051)	25	35	1·75			Vanhaecke, 1983
Canada, Chaplin Lake (1978)	25	35	2·04	23 488[c]	0·0446	Vanhaecke *et al.*, 1983
(1978)	25	35	1·97			Vanhaecke, 1983
(1979)	25	35	2·04			Vanhaecke, 1983
China P. R., Tientsin	25	35	3·09	23 616[c]	0·0681	Vanhaecke *et al.*, 1983
Colombia, Manaure	25	35	1·78			Vanhaecke, 1983
Galera Zamba	25	35	2·27			Vanhaecke, 1983
Cyprus	26		2·1			Person-Le Ruyet & Salaun, 1977
France, Salins du Midi	26		2·7			Fuchs & Person-Le Ruyet, 1976
Salins du Midi (Lavalduc)	25	35	3·08	23 156[c]	0·0670	Vanhaecke *et al.*, 1983
India, Tuticorin	30	35	2·80	21 934		Royan, 1980
	25	35	3·17			Vanhaecke, 1983
Israel, Eilat	25	35	3·00			Vanhaecke, 1983
Italy, Margherita di Savoia	25	30		22 400		Schauer *et al.*, 1980
	25	35	3·33	23 191[c]	0·0725	Vanhaecke *et al.*, 1983
Philippines, Barotac Nuevo	25	35	1·68	24 210[c]	0·0382	Vanhaecke *et al.*, 1983
Puerto Rico, Bahia Salinas	25	35	2·10	23 696[c]	0·0470	Vanhaecke *et al.*, 1983
Unknown				28 194[ct]		Slobodkin & Richman, 1961

Sri Lanka, Hambantota	29	35	3·29			Nanayakkara *et al.*, 1985
Tunisia, Bekalta	25	35	2·40			Van Ballaer *et al.*, 1985
Mégrine	25	35	2·61			Van Ballaer *et al.*, 1985
Sfax	25	35	1·97			Van Ballaer *et al.*, 1985
U.S.A., Great Salt Lake	20		1·65	24 913[t]		Paffenhöfer, 1967
	30	32	1·92	24 662[t]		Von Hentig, 1971
(1977)	25	30		22 400		Schauer *et al.*, 1980
(1966)	25	35	2·70	24 549[c]	0·0625	Vanhaecke, 1983
(1977)	25	35	2·42	23 698[c]	0·0541	Vanhaecke *et al.*, 1983
U.S.A., San Francisco Bay	18	30	1·5			Urbani, 1959
	26	33	2·87	27 621[t]		Dutrieu, 1960
	25	20	1·93			Clegg, 1962
	20	33	1·64			May, 1971
	28	35	1·85	23 256		Benijts *et al.*, 1976
	26		1·4			Fuchs & Person-Le Ruyet, 1976
	26		1·45			Person-Le Ruyet & Salaun, 1977
(No. 288–2606)	25	35	1·61	23 852[c]	0·0360	Vanhaecke, 1983
(No. 288–2596)	25	35	1·63	23 999[c]	0·0366	Vanhaecke *et al.*, 1983
U.S.A., San Pablo Bay (No. 1268)	25	30		23 500		Schauer *et al.*, 1980
(No. 1628)	25	35	1·92	23 660[c]	0·0429	Vanhaecke *et al.*, 1983
Venezuela, Port Araya	25	35	2·07			Vanhaecke, 1983

APPROXIMATE COMPOSITION

A summary of available information on the approximate composition of *Artemia* nauplii, pre-adults and adults again reveals considerable variation (see Table X). Protein content in nauplii ranges from 37·4 to 71·4% with an average (excluding extremes) of about 50%. Average protein content in pre-adult and adult *Artemia* is about 56%. Lipid content in nauplii also varies considerably *i.e.* from 11·6 to 30%. Sources of variation are strain differences (Schauer *et al.*, 1980) and nauplius age at analysis (Benijts *et al.*, 1976); the last authors measured a decrease in lipid content from 19·3% in the first instar stage to 13·7% in the instar II-III stage, representing a 26% loss. According to Hines, Middleditch & Lawrence (1980) instar I nauplii contain 33–38% protein, 16–22% lipid, and 8–18% carbohydrate; during 48 h post-hatch development at 18 °C all levels remained relatively constant, but after 24 h at 28 °C levels of lipids and carbohydrates had decreased.

Literature data on carbohydrate and ash content range from 10·54 to 22·7% and 4·2 to 21·4%, respectively in nauplii and from 9·25 to 17·2% and 8·89 to 29·2%, respectively in pre-adult and adult *Artemia*. Variation in ash content is particularly high in nauplii. This may be explained by the large increase in ash content as animals moult from instar I to instar II and III (*e.g.* 88%, Benijts *et al.*, 1976). Ash contents are substantially higher in adults than in nauplii.

MINERALS

The mineral content of adult brine shrimp was reported by Gallagher & Brown (1975), that of cysts was determined by Stults (1974), and that of nauplii was given by Watanabe *et al.* (1978a), Grabner *et al.* (1981/1982) and Bengtson, Beck & Simpson (in press). The studies of Watanabe *et al.* (1978a) indicate that geographic variation in mineral content is apparent, but not particularly large nor significant. Variation in the reported data seems to be due more to the investigator or method differences than to geographic variation. The range of mineral content that has appeared in the literature are: sodium (2·1–51·1 $mg \cdot g^{-1}$), phosphorus (1·1–17·5 $mg \cdot g^{-1}$), potassium (0·73–12·7 $mg \cdot g^{-1}$), magnesium (1·05–6·8 $mg \cdot g^{-1}$), calcium (0·2–4·8 $mg \cdot g^{-1}$), iron (269–2946 $\mu g \cdot g^{-1}$), zinc (75–241 $\mu g \cdot g^{-1}$), manganese (2–139 $\mu g \cdot g^{-1}$), copper (2–32 $\mu g \cdot g^{-1}$), selenium (0·83–1·4 $\mu g \cdot g^{-1}$); values compare well with the mineral content of other natural or cultured zooplankton (Watanabe *et al.*, 1978a; Grabner *et al.*, 1981/1982). At any rate, the nutritional requirements of marine fish and crustacean larvae for minerals are very poorly known and may be partially supplied by the sea water that marine fish drink (Cowey & Sargent, 1979).

AMINO ACIDS

Amino-acid profiles have been reported for *Artemia* by several authors (Gallagher & Brown, 1975; Watanabe *et al.*, 1978b; Claus *et al.*, 1979; Schauer *et al.*, 1979; Seidel *et al.*, 1980a,b; Grabner *et al.*, 1981/1982; Dabrowski & Rusiecki, 1983), but different methods of analysis and

TABLE X

Overview of published data on approximate analysis (% on dry wt basis) of Artemia *nauplii, juveniles and adults: c, recalculated from wet wt basis; n.s. source not specified.*

Artemia source	Protein	Lipid	Carbohydrate	Ash	Reference
Nauplii					
Australia, Shark Bay		18·5			Schauer *et al.*, 1980
Brazil, Macau		20·2			Schauer *et al.*, 1980
Canada	57·6[c]	17·8[c]		12·7[c]	Watanabe *et al.*, 1983a
China P.R.	47·3	12·0		21·4	Duray & Baraginao, 1984
France, Salins du Midi	55·7	12·4		15·4	Fuchs & Person-Le Ruyet, 1976
India, Tuticorin	58·0	23·3	12·8	5·7	Royan, 1980
Italy, Margherita di Savoia		15·6			Schauer *et al.*, 1980
Russia	42·5	23·2			Dutrieu, 1960
South America	71·4[c]	11·6[c]		10·9[c]	Watanabe *et al.*, 1983a
Sri Lanka, Hambantota	66·8	14·1	12·7	6·4	Nanayakkara *et al.*, 1985
U.S.A., Great Salt Lake	41·6	23·1	22·7	6·56	Von Hentig, 1971
	47·24	20·84	10·54	9·52	Claus *et al.*, 1979
		22·4			Schauer *et al.*, 1980
U.S.A., San Francisco Bay	50·3	15·9			Brick, in Helfrich, 1973
	50·0	27·2			Coehn, in Helfrich, 1973
	54·5	17·25		13·78	Fuchs & Person-Le Ruyet, 1976
		19·3		6·03	Benijts *et al.*, 1976
	59·2[c]	19·4[c]		11·7[c]	Watanabe *et al.*, 1983a
	47·26	23·53	11·24	8·17	Claus *et al.*, 1979
		17·4			Schauer *et al.*, 1980
		15·9			Schauer *et al.*, 1980
U.S.A., San Pablo Bay		16·0			Schauer *et al.*, 1980
n.s.	53·6	17·6	18·6	4·2	Coles, 1969
n.s.		30			Sulkin, 1975
n.s.	37·4	17·1		7·4	Grabner *et al.*, 1981/1982
n.s.	47·0	20·8		6·1	Bengtson *et al.*, in press

TABLE X—*continued*

Artemia source	Protein	Lipid	Carbohydrate	Ash	Reference
Adults					
Wild Adults					
U.S.A., Mono Lake	58·5	10·6		20·6	Enzler *et al.*, 1974
U.S.A., San Diego	64·0	12·0		15·4	Millikin *et al.*, 1980
San Francisco Bay	58·0	19·3		20·6	Gallagher & Brown, 1975
	57·9	12·5		12·4	Millikin *et al.*, 1980
	50·2	2·4	17·2	29·2	Good *et al.*, 1982
n.s.	51·00	8·25	9·98	17·40	Capuzzo & Lancaster, 1979
n.s.	69·02	12·84	9·25	8·89	Gabaudan *et al.*, 1980
Cultured juveniles and adults					
Brazil, Macau (at sexual maturity on *Chaetoceros*)	52·77				Tobias *et al.*, 1980
Cyprus, Larnaca Salt Lake (at sexual maturity on *Chaetoceros*)	58·07				Tobias *et al.*, 1980
France, 7 days on *Spirulina*	53·7	9·4		21·6	Fuchs & Person-Le Ruyet, 1976
U.S.A., Great Salt Lake (14 days on defatted rice bran)	56·5	19·5		9·0	Dobbeni, 1983
India, Tuticorin (at sexual maturity on *Chaetoceros*)	51·47				Tobias *et al.*, 1980
Italy, Margherita di Savoia (at sexual maturity on *Chaetoceros*)	52·03				Tobias *et al.*, 1980
U.S.A., San Francisco Bay (7 days on *Spirulina*)	62·5	10·8		19·1	Fuchs & Person-Le Ruyet, 1976
Spain, Santa Pola (at sexual maturity on *Chaetoceros*)	49·73				Tobias *et al.*, 1980

reporting of the data by different authors preclude any comparison of their results. For example, the method used by Claus *et al.* (1979) was not suitable for the detection of proline, cystine, arginine, and tryptophan, which together account for about 25% of the total amino acids reported by other authors. European authors (Claus *et al.*, 1979; Grabner *et al.*, 1981/1982; Dabrowski & Rusiecki, 1983) tend to report the content of each amino acid as a percentage of the total amino acids, whereas Japanese and American authors (Gallagher & Brown, 1975; Watanabe *et al.*, 1978b; Seidel *et al.*, 1980a,b) report it as g of each amino acid per 100 g of protein. The two methods of reporting can be approximately equivalent, but are not necessarily so, depending, for example, on whether all the amino acids can be detected and whether one is working with wet or freeze-dried material. It is appropriate here to plead for standard methods of analysis and reporting of amino-acid data.

The geographical variation in amino-acid content of *Artemia* is not large. Seidel *et al.* (1980b) found that newly-hatched nauplii from five geographical strains were relatively similar in amino-acid composition (Table XI) and that the 10 amino acids considered essential for fish (Anonymous, 1981) were generally present in sufficient quantity in the nauplii. Methionine, however, like other sulphur amino acids (Dabrowski & Rusiecki, 1983), is the first-limiting amino acid. Amino-acid composition is probably genetically controlled, not subject to much environmental variation and not a major problem in the nutritional value of *Artemia*. Dabrowski & Rusiecki (1983) demonstrated, however, that upon starvation the free amino-acid content in *Artemia* nauplii decreases. This may reduce to some extent their digestibility especially for stomachless fish larvae. Digestibility of *Artemia* protein was determined by Watanabe *et al.* (1978a) who found it to be 83% for carp and 89% for rainbow trout. Watanabe *et al.* also found high values for net protein utilization (NPU) and the protein efficiency ratio (PER).

FATTY ACIDS

Newly hatched nauplii and cysts

Although investigators routinely report on levels of 15 or more fatty acids in their profiles of *Artemia*, six of those fatty acids (16:0, 16:1ω7, 18:1ω9, 18:2ω6, 18:3ω3, and 20:5ω3) actually comprise about 80% of the total fatty acids in an *Artemia* sample. Published values (% composition as fatty acid methyl esters or FAMEs) for those six fatty acids are give in Table XII. Most of the analyses have been done on the San Francisco Bay strain, but several other strains have also been studied.

Levels of 16:0 (palmitic acid) range from 5·74 to 26·6% of total FAMEs, although most values for 16:0 approximate the mean value of 13·4%. Thus, levels of this fatty acid in *Artemia* are fairly predictable and constant (overall coefficient of variation of 24·6%, see Table XIII) compared with others that we shall examine. More variable (overall coefficient of variation of 50·4%) are the levels of 16:1ω7 (palmitoleic acid), which range from 3·12 to 30·6% of total FAMEs (overall mean of 11·7%). 44% of the values

TABLE XI

Selected data on amino-acid composition of Artemia *nauplii and adults (g amino acid per 100 g protein): a, recalculated values; b, Cys + Met; c, destroyed by HCL; d, Phe + Tyr; Ala, alanine; Arg, arginine; Asp, asparagine; Cys, cysteine; Glu, glutamine; Gly, glycine; His, histidine; Ile, isoleucine; Leu, leucine; Lys, lysine; Met, methionine; Pe, phenylalanine; Pro, proline; Ser, serine; Thr, threonine; Tyr, tyrosine; Val, valine; Tryp, tryptophan*

Artemia source	Ala	Arg	Asp	Cys	Glu	Gly	His	Ile	Leu	Lys	Met	Phe	Pro	Ser	Thr	Tyr	Val	Tryp	Reference
Nauplii([a])																			
Australia, Shark Bay	4·6	9·2	9·1	b	13·8	4·8	3·2	4·1	6·7	9·0	2·4[b]	6·5	4·6	5·0	4·6	6·2	4·6	c	Seidel *et al.*, 1980b
Brazil, Macau	3·9	9·7	9·3	b	11·1	5·1	4·1	4·7	7·5	9·9	1·9[b]	4·3	4·8	3·8	4·4	8·9	4·5	c	Seidel *et al.*, 1980b
U.S.A., San Pablo Bay	3·6	8·3	11·9	b	8·6	6·3	3·0	4·6	7·1	7·4	2·2[b]	8·8	4·1	6·5	5·1	6·5	4·7	c	Seidel *et al.*, 1980b
U.S.A., Great Salt Lake	4·1	8·2	9·5	b	11·4	5·1	2·3	5·7	8·4	7·8	3·1[b]	7·2	5·0	4·5	4·0	5·6	4·4	c	Seidel *et al.*, 1980b
Italy, Margherita di Savoia	4·1	8·3	9·5	b	12·2	6·1	3·2	5·4	8·5	9·0	3·1[b]	7·2	5·0	4·3	4·6	4·6	2·6	c	Seidel *et al.*, 1980b
Adults																			
U.S.A., San Francisco Bay (wild)	6·9	6·5	9·2	2·2	14·2	5·3	1·8	5·3	8·0	7·6	2·7	4·7	5·2	4·8	4·6	4·5	5·4	1·0	Gallagher & Brown, 1975
U.S.A., Great Salt Lake (14 days cultured on defatted rice bran)[a]	5·8	4·4	9·6		13·1	4·8	2·1	4·6	7·4	7·8	2·1	4·0	4·4	4·4	4·4	2·7	5·0		Dobbeni, 1983
Required levels for Chinook salmon	—	6·0	—	—	—	—	1·8	2·2	3·9	5·0	4·0	5·1	—	—	2·2	d	3·2	0·5	Anonymous 1981

TABLE XII

Data on per cent composition of the six major fatty acids (% fatty acid methyl ester of total fatty acid methyl esters) of Artemia *cysts and newly hatched nauplii: *, analysis performed on* Artemia *cysts; **, may include other monoënes; ***, only polar lipid fraction given; n.s., source not specified*

Artemia source		Fatty acid						Reference
		16:0	16:1ω7**	18:1ω9**	18:2ω6	18:3ω3	20:5ω3	
Australia, Shark Bay	(No. 113, 1979)	13·45	9·97	28·23	5·78	14·77	10·50	Schauer *et al.*, 1980
n.s.	(No. 1980)*	13·9	9·9	33·3	5·2	10·1	8·6	Watanabe *et al.*, 1982
Bahamas, Great Inagua		14·5	17·0	31·7	15·6	2·0	1·3	Léger, unpubl.
Brazil, Macau	(1978)	15·42	10·79	35·86	9·59	4·87	9·98	Schauer *et al.*, 1980
n.s.	(1980A)*	16·0	18·6	21·8	7·2	3·3	3·9	Watanabe *et al.*, 1982
n.s.	(1980B)*	18·2	14·4	23·7	6·4	1·1	3·5	Watanabe *et al.*, 1982
n.s.	(1980C)*	13·7	13·8	28·9	8·5	3·2	5·9	Watanabe *et al.*, 1982
n.s.	(1980D)*	18·0	14·6	16·2	3·1	0·9	4·6	Watanabe *et al.*, 1982
n.s.	(1980E)*	14·7	14·7	26·6	7·7	3·6	5·8	Watanabe *et al.*, 1982
n.s.	(1980E)	12·2	12·8	30·7	9·3	3·3	6·5	Watanabe *et al.*, 1982
n.s.	(1980F)*	13·7	14·1	28·3	11·8	2·7	5·8	Watanabe *et al.*, 1982
Guanabara	(1985)	16·4	13·1	30·5	9·2	2·7	3·3	Léger, unpubl.
Canada, n.s.		8·4	7·3	30·0	6·0	13·5	12·1	Watanabe *et al.*, 1978c
n.s.	(1978A)	9·9	10·1	32·3	5·1	14·1	5·2	Watanabe *et al.*, 1980
n.s.	(1978A)*	13·0	10·0	23·6	6·1	19·8	7·3	Watanabe *et al.*, 1980
n.s.	(1978B)*	13·5	12·8	25·4	6·4	16·0	6·7	Watanabe *et al.*, 1980
Chaplin Lake	(1979)	9·99	9·03	28·24	7·95	19·87	9·52	Seidel *et al.*, 1982
China P. R., n.s.	(1978)	13·9	23·5	23·4	3·7	7·5	7·7	Watanabe *et al., 1980*
Tientsin	(1979)	11·4	19·06	26·81	4·68	7·38	15·35	Seidel *et al.*, 1982
n.s.	(1979A)*	12·1	22·6	26·1	4·1	5·5	9·2	Watanabe *et al.*, 1982
n.s.	(1979A)*	9·7	13·6	33·5	4·4	5·3	13·0	Watanabe *et al.*, 1982
n.s.	(1979B)*	12·7	24·0	20·2	3·8	6·0	10·2	Watanabe *et al.*, 1982
n.s.	(1979)*	9·3	13·4	33·8	4·4	5·1	13·2	Watanabe *et al.*, 1982
n.s.	(1979C)*	12·7	22·4	28·3	4·3	5·1	11·3	Watanabe *et al.*, 1982
n.s.	(1980A)*	23·0	24·7	22·1	1·6	0·4	1·9	Watanabe *et al.*, 1982
n.s.	(1980B)*	21·2	22·8	17·4	2·2	0·6	1·3	Watanabe *et al.*, 1982

TABLE XII–*continued*

Artemia source		Fatty acid						Reference
		16:0	16:1ω7**	18:1ω9**	18:2ω6	18:3ω3	20:5ω3	
n.s.	(1980C)*	12·5	20·1	24·9	4·2	6·4	10·9	Watanabe *et al.*, 1982
n.s.	(1981A)*	13·1	19·1	25·3	5·0	6·6	9·3	Watanabe *et al.*, 1982
Tientsin I	(1984)	13·9	19·3	27·8	4·9	2·9	11·4	Léger, unpubl.
Tientsin II	(1985)	15·2	24·9	30·8	5·6	4·5	13·3	Léger. unpubl.
Colombia, Galera Zamba	(1983)	15·1	7·9	23·2	13·5	13·4	4·7	Léger, unpubl.
Manaure	(1983A)	13·5	9·5	29·2	13·7	1·1	1·2	Léger, unpubl.
	(1983B)	13·3	9·7	29·2	14·3	1·0	1·4	Léger, unpubl.
France, Lavalduc	(1979)	11·90	11·34	24·73	6·14	20·9	8·01	Seidel *et al.*, 1982
Lavalduc	(1981)	14·5	8·6	24·7	6·4	20·0	5·4	Léger, unpubl.
India, Mundra	(1979)	12·7	8·9	27·9	12·0	14·6	5·3	Vos *et al.*, 1984
Mithapur	(1985)	14·4	10·5	26·0	8·4	12·6	8·0	Léger, unpubl.
Tuticorin	(1985)	16·3	16·2	29·3	4·8	3·1	12·3	Léger, unpubl.
Italy, Margherita di Savoia	(1977)	15·23	10·38	29·05	6·79	6·35	13·63	Schauer *et al.*, 1980
Kenya, Malindi	(1985A)	14·1	18·3	32·3	5·5	1·6	6·9	Léger, unpubl.
	(1985B)	12·9	12·5	25·1	6·8	5·0	4·6	Léger, unpubl.
	(1985C)	14·8	17·7	27·8	8·1	1·5	6·8	Léger, unpubl.
Panama, Aguadulce I	(1984A)	14·3	15·4	27·6	8·7	2·5	7·8	Léger, unpubl.
	(1984B)	14·4	16·1	24·6	4·5	1·9	12·0	Léger, unpubl.
Aguadulce II	(1985)	14·5	16·9	27·4	6·0	3·7	9·8	Léger, unpubl.
Peru, Hierba Blanca	(1984)	14·5	10·6	27·2	5·6	11·3	6·4	Léger, unpubl.
Philippines, Barotac Nuevo	(1978)	14·4	15·9	29·6	9·1	4·2	8·6	Vos *et al.*, 1984
Jaro	(1981)	11·4	13·7	27·0	15·0	12·9	1·9	Vos *et al.*, 1984
Puerto Rico, Cabo Rojo		15·1	12·3	31·2	14·0	1·4	1·4	Léger, unpubl.
Reference *Artemia* Cysts	(1980)	12·70	16·78	30·37	9·62	2·55	8·45	Seidel *et al.*, 1982
South America, n.s.		7·9	5·8	26·3	5·2	21·0	0·3	Watanabe *et al.*, 1978c
Thailand, Bangpakong	(1979)	10·1	10·3	31·4	5·5	23·3	5·3	Vos *et al.*, 1984
Chachoengsao	(1983)	14·5	18·5	28·6	4·9	3·2	10·7	Léger, unpubl.
Fam Farm	(1985)	15·5	16·6	29·4	4·9	5·9	10·5	Léger, unpubl.

Tunisia, Sfax	(6, 1984)	15·3	7·5	24·1	8·5	20·0	2·4	Léger, unpubl.
	(502, 1984)	15·6	9·4	26·4	8·6	12·8	4·8	Van Ballaer *et al.*, 1985
Mégrine	(1984)	16·6	15·0	24·0	3·8	7·3	10·2	Van Ballaer *et al.*, 1985
U.S.A., Great Salt Lake		11·75	4·5	23·32	8·81	25·23	?	Wickins, 1972
		11·22	3·52	21·87	3·59	11·22	?	Claus *et al.*, 1979
	(S-arm 1977)	11·78	5·64	28·25	4·60	31·46	3·55	Schauer *et al.*, 1980
		15·06	5·99	30·25	6·69	28·27	1·77	Millamena *et al.*, 1985
	(S-arm 1977–18)	13·1	6·1	25·0	6·6	28·2	2·8	Léger unpubl.
	(S-arm 1977–217)	12·5	5·6	25·8	6·5	28·4	3·1	Léger, unpubl.
	(S-arm 1979–BI–1)	12·5	6·7	26·6	6·4	27·7	2·7	Léger, unpubl.
	(S-arm 1979-WC-4)	13·0	6·0	25·9	6·3	28·2	2·7	Léger, unpubl.
	(S-arm 1979–294)	13·2	5·9	26·0	6·5	28·2	2·7	Léger, unpubl.
	(S-arm 1979–185)	12·3	4·8	27·3	7·6	29·1	2·6	Léger, unpubl.
	(N-arm, 1984A)	10·9	4·1	26·5	8·4	25·5	0·3	Léger, unpubl.
	(N-arm, 1984B)	12·0	5·1	25·8	8·3	24·8	0·2	Léger, unpubl.
	(N-arm, 1984C)	11·8	5·0	23·9	7·7	22·7	0·2	Léger, unpubl.
	(N-arm, 1985A)	11·9	6·3	23·1	7·4	26·3	0·3	Léger, unpubl.
	(N-arm, 1985B)	12·3	6·3	22·9	7·5	26·1	0·3	Léger, unpubl.
U.S.A., San Francisco Bay		13·56	8·20	29·18	6·05	22·27	3·86	Wickins, 1972
		9·5	4·7	25·3	7·8	33·6	1·2	Weaver, 1974
		11·46	9·11	37·5	5·84	15·00	7·22	Schauer & Simpson, 1978
	(1975)	11·2	4·3	25·1	6·1	28·4	3·1	Watanabe *et al.*, 1978c
	(1976)	12·3	3·7	27·4	6·6	27·9	2·0	Watanabe *et al.*, 1978c
	(1977)	9·5	12·0	36·1	3·4	10·3	9·5	Watanabe *et al.*, 1978c
		5·74	3·12	14·00	2·86	7·30	?	Claus *et al.*, 1979
		11·45	16·49	34·34	4·78	4·67	13·31	Schauer *et al.*, 1979
	(No. 313/3006)	10·33	13·27	26·97	9·35	17·33	4·06	Schauer *et al.*, 1980
	(No. 321995)	12·13	19·52	31·20	3·69	5·16	12·44	Schauer *et al.*, 1980
	(1975)*	13·2	4·5	27·8	6·2	27·7	1·8	Watanabe *et al.*, 1980
	(1976)*	12·3	3·7	27·4	6·6	27·9	2·0	Watanabe *et al.*, 1980
	(1977)*	12·0	18·4	31·5	4·0	9·0	7·1	Watanabe *et al.*, 1980
	(1978A)*	19·7	30·6	14·6	5·3	2·6	6·1	Watanabe *et al.*, 1980
	(1978B)*	20·4	20·3	20·1	3·6	7·9	2·0	Watanabe *et al.*, 1980
	(1978C)*	18·9	15·3	29·2	7·8	3·8	5·4	Watanabe *et al.*, 1980
	(1978C)	14·1	13·5	33·3	9·0	3·9	7·0	Watanabe *et al.*, 1980
	(1978D)*	13·3	14·2	18·0	4·4	23·8	1·8	Watanabe *et al.*, 1980
	(1978D)	10·1	7·0	32·7	6·3	24·4	1·6	Watanabe *et al.*, 1980
	(1978E)*	13·3	11·7	27·7	5·4	21·6	1·9	Watanabe *et al.*, 1980
	***	12·6	3·6	33·6	6·9	20·0	1·0	Sakamoto *et al.*, 1982
	(1979)*	13·3	16·4	28·2	8·3	2·3	7·5	Watanabe *et al.*, 1982

TABLE XII-*continued*

Artemia source	Fatty acid						Reference
	16:0	16:1ω7**	18:1ω9**	18:2ω6	18:3ω3	20:5ω3	
(1980A)*	26·6	16·3	25·8	2·6	3·3	3·9	Watanabe *et al.*, 1982
(1980B)*	25·3	15·7	27·6	2·9	4·2	1·7	Watanabe *et al.*, 1982
(1980C)*	25·9	12·9	19·8	2·5	4·8	0·9	Watanabe *et al.*, 1982
(1980D)*	14·9	5·5	28·0	6·3	22·4	2·7	Watanabe *et al.*, 1982
(1980E)*	23·7	7·4	23·7	5·4	14·7	0·6	Watanabe *et al.*, 1982
(1980F)	9·2	14·8	19·1	8·3	5·4	6·8	Watanabe *et al.*, 1982
(1980G)	11·0	3·8	26·7	8·9	27·6	0·3	Watanabe *et al.*, 1982
(1980H)	12·2	10·4	34·9	6·6	17·2	3·5	Watanabe *et al.*, 1982
(1981A)*	15·2	10·5	28·4	7·1	17·2	3·6	Watanabe *et al.*, 1982
(1981B)*	13·6	4·3	27·1	6·1	28·1	2·4	Watanabe *et al.*, 1982
(1981C)*	10·6	5·4	26·3	7·6	27·0	2·1	Watanabe *et al.*, 1982
(1976–No. 236/2016)	12·5	20·85	34·9	3·0	7·0	8·8	Léger *et al.*, 1985c
(1976–2596)	13·0	21·9	34·1	4·7	7·8	7·9	Vos *et al.*, 1984
(1978–1728)	14·4	16·3	28·0	4·5	9·2	13·8	Vos *et al.*, 1984
	11·5	12·6	29·7	7·1	14·8	6·8	Witt *et al.*, 1984
(1976–I)	12·5	20·9	34·9	3·0	5·9	8·8	Léger *et al.*, 1985c
(1976–II)	13·0	20·0	34·7	4·7	7·5	8·2	Léger *et al.*, 1985c
(1978–V)	9·5	4·9	28·5	8·7	27·2	1·5	Léger *et al.*, 1985c
(1978–VI)	10·0	5·0	32·7	9·2	26·3	1·6	Léger *et al.*, 1985c
(1978–VIII)	9·0	4·6	28·3	9·1	27·6	1·2	Léger *et al.*, 1985c
(1978–XII)	11·5	7·7	28·2	8·2	20·9	3·6	Léger *et al.*, 1985c
(1978–XIV)	5·9	7·3	32·8	8·5	25·6	2·8	Léger *et al.*, 1985c
(1979–III)	11·3	3·1	27·6	7·9	23·6	1·8	Léger *et al.*, 1985c
(1979–IX)	10·2	4·3	28·2	9·7	26·3	0·7	Léger *et al.*, 1985c
(1979–X)	10·8	4·2	27·9	8·1	27·7	0·6	Léger *et al.*, 1985c
(1979–XI)	11·6	7·3	28·5	6·9	18·7	4·7	Léger *et al.*, 1985c
(1980–IV)	11·3	6·0	26·8	8·3	23·3	1·7	Léger *et al.*, 1985c
(1980–VII)	11·7	4·4	27·3	10·0	28·0	1·4	Léger *et al.*, 1985c
(1980–XIII)	12·2	7·1	30·8	7·5	22·2	3·4	Léger *et al.*, 1985c
(1983)	12·4	17·4	27·9	5·7	3·6	11·7	Léger, unpubl.
(1984A)	12·9	15·5	28·3	5·7	8·2	9·4	Léger, unpubl.
(1984B)	13·1	9·9	29·1	6·6	16·7	5·8	Léger, unpubl.
(1984C)	14·0	6·2	28·3	8·5	21·2	2·2	Léger, unpubl.

	(1984D)	13·2	16·6	30·7	5·0	4·5	10·0	Léger, unpubl
	(1984E)	12·6	8·7	28·0	6·8	18·7	5·0	Léger, unpubl.
	(1984F)	14·0	7·2	28·3	8·4	21·5	2·0	Léger, unpubl.
	(1985G)	13·6	6·7	28·1	9·1	22·6	1·9	Léger, unpubl.
	(1985H)	14·5	14·6	28·1	6·7	13·2	5·6	Léger, unpubl.
	(1985I)	14·9	8·9	27·5	7·9	19·1	3·8	Léger, unpubl.
	(1985J)	13·0	10·4	27·6	6·4	17·7	5·3	Léger, unpubl.
	(1985K)	13·4	7·3	29·0	8·5	22·2	2·5	Léger, unpubl.
	(1985L)	13·9	6·9	27·1	9·0	20·3	2·2	Léger, unpubl.
	(1985M)	13·9	6·3	28·6	9·3	21·9	1·8	Léger, unpubl.
U.S.A., San Pablo Bay	(1978–1628)	7·79	5·24	29·15	4·6	33·6	1·68	Schauer *et al.*, 1980
	(1978–1628)	9·3	4·8	27·0	9·9	31·0	0·2	Léger *et al.*, 1985c
Vietnam, Cam Ranh Bay	(CR, 1984)	15·4	14·2	34·8	3·6	1·6	11·0	Léger, unpubl.
	(CR, 1985)	14·8	14·4	30·7	4·9	3·9	10·5	Léger, unpubl.
	(CRHT, 1985)	15·6	14·7	31·4	5·1	3·1	10·1	Léger, unpubl.
	(CRVT, 1985)	15·8	15·5	31·1	3·8	1·8	13·7	Léger, unpubl.

TABLE XIII

Coefficient of variation of contents of particular fatty acids in Artemia *nauplii from commercial sources listed in Table XII: Sa, South arm; Na, North arm; *, may include other monoeënes; **, value of all* Artemia *strains and samples reported in Table XII*

	16:0	16:1ω7*	18:1ω9*	18:2ω6	18:3ω3	20:5ω3
Coefficient of Variation (%)						
U.S.A., San Francisco Bay	30·2	57·1	16·1	30·9	53·0	78·6
U.S.A., Great Salt Lake (Sa)	8·5	17·9	9·1	22·7	21·1	11·8
U.S.A., Great Salt Lake (Na)	4·5	17·5	6·6	5·9	5·8	21·2
Canada	19·9	20·2	12·5	16·5	18·3	18·3
Brazil	13·1	14·8	21·3	30·2	43·2	43·2
China	28·8	18·4	18·4	26·8	50·5	50·5
Overall value**	24·6	50·4	14·8	57·3	71·7	71·7

for 16:1ω7 fall between 3·0 and 9·9%, whereas another 44% of the values fall between 10·0 and 19·9%. Very often, the most abundant fatty acid in *Artemia* is 18:1ω9 (oleic acid), for which values range from 14·0–37·5% of total FAMEs (overall mean of 27·8%). Of the values listed for 18:1ω9, 96·5% are higher than 20.0%. Over all variance for this fatty acid is the lowest when compared with the other main fatty acids.

To summarize, we find that the major saturated and monoene FAMEs (16:0, 16:1ω7, and 18:1ω9) generally comprise about 40 to 60% of total FAMEs in a sample of *Artemia*. In addition, the major diene, 18:2ω6 (linoleic acid), usually contributes something <10% (range: 1·6–11·8%; overall mean of 7%) to the FAME total.

The major fatty acids of the linolenic series, 18:3ω3 (linolenic acid) and 20:5ω3 (eicosapentaenoic acid), must be considered together because of their importance as essential fatty acids (EFA) and because their levels are mostly interrelated. 18:3ω3 is considered the EFA for freshwater fish and 20:5ω3 an EFA for marine fish (see p. 560). Kanazawa *et al.* (1979) and Schauer & Simpson (1985) demonstrated that 18:3ω3 is readily converted to 20:5ω3 in freshwater fish, but the conversion by marine fish is very slight. It is, therefore, necessary to have adequate amounts of 20:5ω3 in the diet of larval marine fish and crustaceans. Although the range of values for 18:3ω3 is 0·4 to 33·6% of total FAMEs, the distribution of the values is actually bimodal. 36% of the values of 18:3ω3 are 20.0% or greater (of total FAMEs) and 43% of the values are 10.0% or less (of total FAMEs). Thus, 18:3ω3 is usually either very abundant or very scarce. This is reflected in a high overall variance (coefficient of variation of 71·7%) which is mainly due to a high variability in San Francisco Bay, Brazilian, and Chinese *Artemia*. The level of 20:5ω3 is inversely related to the level of 18:3ω3. If one examines the data for all the samples in Table XII in which the level of 18:3ω3 exceeded 20% of total FAMEs, one finds that the values of 20:5ω3 in those samples were consistently low (mean and SD of 20:5ω3 in those samples is 2·1±1·5%). By contrast, in those samples in which the level of *18:3*ω3 was <10%, the values for 20:5ω3 were substantially higher (7·9±3·9). Standard deviations are relatively high because of a few exceptions to this rule, *e.g. Artemia* from Great Inagua (Bahamas), Cabo Roya (Puerto Rico), Manaure (Colombia), and some samples from San Francisco Bay, Brazil and China have low levels of both 18:3ω3 and 20:5ω3; some *Artemia*, on the other hand, contain relatively high levels of both 18:3ω3 and 20:5ω3, *e.g.* Bangpakong (Thailand), Australia, Lavalduc (France), and Canada. In general, however, Watanabe *et al.* (1978c) were right in dividing *Artemia* samples into two categories: *i.e.* those good for freshwater organisms (high 18:3ω3, low 20:5ω3) and those good for marine organisms (low 18:3ω3, high 20:5ω3).

An examination of the 18:3ω3 and 20:5ω3 data in Tables XII and XIII from the point of view of variability between and within geographical strains is disconcerting. While there is clearly variability among strains (Schauer *et al.*, 1980; Seidel *et al.*, 1982; Léger, unpubl.), there is at least as much variability within the strain, both between years and during one year (Watanabe *et al.*, 1978c, 1980, 1982; Léger *et al.*, 1985c; Léger, unpubl.). Strains from San Francisco Bay, China, and Brazil are particularly variable in levels of 20:5ω3 (see Table XII). On the other hand, 20:5ω3 levels in Utah (Southern Arm and Northern Arm) are remarkably constant.

On-grown and adult Artemia

It is not clear whether adult *Artemia* simply reflect their diet or convert fatty acids irrespective of diet. Both indirect and direct evidence exists to show that *Artemia* can elongate 18:3ω3 to 20:5ω3. Kayama *et al.* (1963) fed phytoplankton (*Chaetoceros simplex*) lacking 20:5ω3 to *Artemia*, but the subsequent fatty-acid profile of *Artemia* included high levels of 20:5ω3. Jezyk & Penicnak (1966) obtained similar results when they reared *Artemia* on an unknown species of green algae that lacked 20:5ω3. Hinchcliffe & Riley (1972) fed *Artemia* on four separate algal species, only one of which (*Chlamydomonas* sp.) lacked 20:5ω3; nevertheless, the *Artemia* fed on *Chlamydomonas* contained 20:5ω3, although at a lower level than when fed the other algal species. The fact that, in most cases, *Artemia* did not resemble very well their diet led Hinchcliffe & Riley to conclude that the metabolic needs and conversion abilities of *Artemia* determine their fatty-acid profile. Schauer & Simpson (in press) have obtained clear evidence *via* radioactive labelling of rice-bran diets that Australian *Artemia* can elongate 18:3ω3 to 20:5ω3; however, recent evidence (Millamena & Simpson, 1985) indicates that the Utah strain may be different. Fatty-acid analyses of Utah *Artemia* grown in ponds in the Philippines show that the *Artemia* very closely resembled their live algal diets, *Chaetoceros* sp. (high 20:5ω3, low 18:3ω3) and *Dunaliella* sp. (low 20:5ω3, high 18:3ω3). These various findings are not necessarily contradictory. *Artemia* is certainly able to convert 18:3ω3 to 20:5ω3 to meet its metabolic needs, but the percentage of 20:5ω3 required to meet those needs may be much less than the levels found in some algae. From culture experiments with *Artemia* fed different diets (*e.g.* Sakamoto *et al.*, 1982; Yashiro, 1982, 1985; Millamena *et al.*, 1985; Léger, unpubl.) it is clear that 20:5ω3 levels in *Artemia* are greatly determined by the food ingested. Indeed, high 20:5ω3 levels in the diet (e.g. *Chaetoceros* sp. and fish oil based diets) are reflected in elevated levels in *Artemia*, while low dietary levels (*e.g. Dunaliella*) result in reduced concentrations in *Artemia*. Nevertheless, when 20:5ω3 lacking diets are fed (*e.g.* rice bran and other agricultural products) still a minimal 20:5ω3 level will appear in *Artemia*. This is another indication that *Artemia* is able to biosynthesize a minimal amount of 20:5ω3 to meet its metabolic requirements. Biosynthesis in *Artemia* is also noticed for 16:1 and 18:1 while 16:0, 18:2ω6, and 18:3ω3 more closely reflect dietary levels. An interesting experiment in this regard was performed by Léger (unpubl. data, see Table XIV) who cultured three *Artemia* strains (Great Salt Lake—Southern Arm, San Francisco Bay, and San Pablo Bay) that have a very different fatty-acid profile (see Table XII) on rice bran which is deficient in 18:3ω3 and 20:5ω3; after 1 week culturing the three groups of pre-adult brine shrimp ended up with a very similar fatty-acid profile. The same experiment also showed that a 20:5ω3-rich *Artemia* (SFB 236–2016) will consume its 20:5ω3 reserves up to a minimal level when fed a 20:5ω3 -lacking diet (rice bran). Similarly, 18:3ω3-rich strains (San Pablo Bay and Great Salt Lake) consume most of their 18:3ω3 reserve when fed a 18:3ω3-poor diet, even in the presence of high dietary 20:5ω3 levels (cod liver oil).

DIGESTIVE ENZYMES

Among the many explanations suggested for the superior value of live food (compared with artificial diets) for fish and crustacean larvae, one of the most intriguing is that exogenous enzymes may contribute to the digestive process. If the larval digestive tract is incompletely developed, living food eaten by the larvae may contain not only the required nutrients, but also some of the enzymes needed to digest them. The question of exogenous enzymes has been studied for both freshwater fishes (Dabrowski & Glogowski, 1977a,b) and marine shrimps (Maugle, Deshimaru, Katayama & Simpson, 1982).

Artemia nauplii possess some carbohydrase activity (Telford, 1970) with particularly strong activities on the substrates amylopectin, glycogen, maltose, and trehalose. Dabrowski & Glogowski (1977a) found relatively high proteolytic activity in *Artemia* nauplii homogenates at both acid and alkaline pH levels. The activities of amylase and trypsin in various life stages of *Artemia* have been extensively studied by Samain, Boucher & Buestel (1975) and Samain *et al.* (1980, 1985). Osuna *et al.* (1977) showed that the activity of four proteolytic enzymes in *Artemia* nauplii increased sharply after hatching and Olalla *et al.* (1978), Sillero *et al.* (1980), and Burillo, Sillero & Sillero (1982) subsequently characterized the four as alkaline proteases. An acid protease has also been discovered (Nagainis & Warner, 1979) and characterized (Warner & Shridhar, 1980) in dormant *Artemia* cysts. Burillo *et al.* (1982) pointed out that the four alkaline proteases could lyse *Artemia* yolk platelets and calculated that their activity was sufficient to account for the rate of yolk platelet degradation observed in live nauplii. Several recent publications deal with various aspects of digestive enzymes in *Artemia* (Ezquieta & Vallejo, 1985; Munuswamy, 1985; Perona & Vallejo, 1985; Samain *et al.*, 1985). Whether these enzymes operate in the digestive tracts of predators that are fed *Artemia* nauplii is unknown and is a potentially fruitful area for research.

TABLE XIV

Data on per cent composition of the six major fatty acids (as fatty acid methyl esters) of ongrown and adult Artemia: *data are expressed as percentage of total fatty acid methyl esters for each sample: *, may include other monoënes; **, only polar lipid fraction given; SFB, San Francisco Bay, U.S.A.; SPB, San Pablo Bay, U.S.A.; GSL, Great Salt Lake, U.S.A.; Sa, South arm; ML, Mono Lake, U.S.A.; n.s., source not specified*

Artemia	Source	Food	Fatty acid						Reference
			16:0	16:1ω7*	18:1ω9*	18:2ω6	18:3ω3	20:5ω3	
Wild	ML		17·0	14·5	38·8	3·9	7·4	5·9	Enzler *et al.*, 1974
	SFB		13·5	13·8	35·6	6·2	—	12·0	Gallagher & Brown, 1975
Cultured	SFB	*Chaetoceros***	15·5	19·4	30·6	2·8	3·9	12·7	Sakamoto *et al.*, 1982
		Microencapsulated diets, lipid free**	13·6	6·8	43·2	8·2	7·0	1·6	Sakamoto *et al.*, 1982
		cod liver oil**	9·4	7·2	43·7	7·8	6·9	9·2	Sakamoto *et al.*, 1982
		Tapes oil**	9·4	5·6	40·1	5·5	6·3	8·0	Sakamoto *et al.*, 1982
		soybean oil**	12·4	2·9	35·1	20·7	7·5	3·4	Sakamoto *et al.*, 1982
	SFB	Wheat flour extract	9·61	6·92	28·90	22·80	7·94	2·34	Yashiro, 1982
		Rice bran extract	12·44	4·93	34·36	26·14	4·48	2·18	Yashiro, 1982
		Milled rice extract	12·84	4·03	23·40	10·09	11·16	7·67	Yashiro, 1982
	SFB	Rice bran	15·2	10·9	33·6	21·6	1·7	0·8	Léger, unpubl.
	(No. 236–2016)	Rice bran + cod liver oil	12·2	14·4	36·4	9·1	1·2	9·2	Léger, unpubl.
	SPB	Rice bran	14·4	9·0	30·2	16·5	4·8	1·6	Léger, unpubl.
	(No. 1628)	Rice bran + cod liver oil	11·0	10·7	32·8	6·2	4·1	8·8	Léger, unpubl.
	GSL	Corn	10·62	5·87	39·54	32·03	1·63	2·18	Millamena *et al.*, 1985
		Copra	14·10	11·38	32·93	8·02	0·93	1·34	Millamena *et al.*, 1985
		Rice bran	11·91	6·76	39·17	29·08	1·90	1·19	Millamena *et al.*, 1985
		Soybean	8·97	4·29	37·30	33·12	3·47	0·98	Millamena *et al.*, 1985
		Chaetoceros	11·70	22·51	17·25	5·04	0·94	18·64	Millamena *et al.*, 1985
		Dunaliella	14·76	2·46	27·32	13·43	20·16	4·72	Millamena *et al.*, 1985

TABLE XIV—*continued*

Artemia	Source	Food	Fatty acid						Reference
			16:0	16:1ω7*	18:1ω9*	18:2ω6	18:3ω3	20:5ω3	
	GSL-Sa	Corn byproduct A	12·0	6·1	33·1	35·8	1·5	0·5	Léger, unpubl.
		Corn byproduct B	12·0	9·6	31·2	27·4	2·1	1·1	Léger, unpubl.
		Defatted rice bran	13·3	9·1	36·1	23·5	1·8	0·9	Léger, unpubl.
	n.s.	*Chaetoceros*	11·6	44·9	18·4	0·7	0·5	12·0	Kayama *et al.*, 1963
	n.s.	*Chlamydomonas*	12·0	4·4	14·0	7·7	11·9	4·6	Hinchcliffe & Riley, 1972
	n.s.	*Monochrysis*	12·9	13·4	17·8	6·5	4·4	17·3	Hinchcliffe & Riley, 1972
	n.s.	*Phaeodactylum*	9·8	9·2	21·6	10·0	9·0	11·0	Hinchcliffe & Riley, 1972
	n.s.	*Platymonas*	12·0	5·0	14·7	6·5	13·9	9·2	Hinchcliffe & Riley, 1972

CAROTENOIDS

The carotenoid composition of *Artemia* has been the subject of some controversy. Gilchrist & Green (1960) concluded that astaxanthin was the only carotenoid pigment in *Artemia*, although Gilchrist (1968) admitted that this was probably a misdiagnosis. Krinsky (1965) reported that canthaxanthin and echinenone were the major pigments present and postulated that *Artemia* converts dietary β-carotene to echinenone and thence to canthaxanthin. Subsequently, Davies, Hsu & Chichester (1965), Czygan (1966), Gilchrist (1968), Hata & Hata (1969), and Wickins (1972) all showed that the main carotenoids in *Artemia* were echinenone and canthaxanthin. Hsu, Chichester & Davies (1970) and Davies, Hsu & Chichester (1970) finally demonstrated conclusively that canthaxanthin and echinenone were the conversion products when *Artemia* were fed β-carotene and that the scheme proposed by Krinsky was most probably correct.

In all the studies mentioned in the preceding paragraph, the investigators used California *Artemia*. The controversy arose when Czygan (1968) suggested that a Canadian *Artemia* strain is able to form astacene and Czeczuga (1971) reported that cysts he had obtained from scientists in France contained mostly β-carotene (53·3%), much astaxanthin (26·8%) and almost no canthaxanthin (1·2%). Czeczuga (1971, 1980) postulated that the qualitatively different results obtained by different authors is due to differences in the food eaten by the *Artemia* and that carotenoid content of *Artemia* "eggs" depends on the carotenoid content of the adult food. Although his contention seems to be invalidated by the experiments of Hsu *et al.* (1970) and Davies *et al.* (1970), the possibility exists that the Canadian strain studied by Czygan (1968) and the (presumably) French strain studied by Czeczuga (1971) are different from the other strains. Unfortunately, Soejima *et al.* (1980) did not examine the French and Canadian strains along with the eight geographical strains that contained only echinenone and canthaxanthin. They did show, however, that astaxanthin in the diet could be absorbed and accumulated by *Artemia*. Subsequently, they also found that *Artemia* could bioaccumulate astacene from the diet (Soejima, Simpson & Katayama 1983). Recently, Nelis *et al.* (1985) analysed 19 different strains of *Artemia* and confirmed that for all strains tested canthaxanthin was the most abundant carotenoid. Some differences between strains were found in amount of total canthaxanthin, which is probably determined by environmental factors. Another difference they noticed was the relative amount of cis- and trans-canthaxanthin. Cis-canthaxanthin, which has not been isolated yet from other animals, was recently discovered by Nelis *et al.* (1984) in *Artemia* cysts and in the reproductive system of female brine shrimps.

STEROLS

Artemia are unable to synthesize sterols from acetate, but can convert several sterols to cholesterol, the only sterol found in the brine shrimp (Teshima & Kanazawa, 1971a). The dietary sterols that have been shown to be bioconverted to cholesterol by *Artemia* are ergosterol (Teshima & Kanazawa, 1971b), brassicasterol (Teshima & Kanazawa, 1972), β-sitosterol and 24-methylcholesterol (Teshima, 1971).

VITAMINS

Stults (1974) analysed *Artemia* cysts (San Francisco Bay) and found high levels of thiamin ($7 \cdot 13$ $\mu g \cdot g^{-1}$), niacin ($108 \cdot 68$ $\mu g \cdot g^{-1}$), riboflavin ($23 \cdot 15$ $\mu g \cdot g^{-1}$), pantothenic acid ($72 \cdot 56$ $\mu g \cdot g^{-1}$) and retinol ($10 \cdot 48$ $\mu g \cdot g^{-1}$ or 35 IU). These levels are higher for riboflavin and panthotenic acid and almost as high for niacin as those reported by Sparre (1962 in Stults, 1974) for whole fish meal. Stults also mentioned that vitamin losses occurring during storage of fishmeal should be zero in *Artemia* cysts as long as they remain whole and viable.

A stable form of vitamin C (L-ascorbic acid 2- sulphate) was discovered in dormant *Artemia* cysts (Mead & Finamore, 1969); Golub & Finamore (1972), however, found that during embryonic development and hatching the stable form disappears and is replaced by L-ascorbic acid.

A vitamin analysis has also been reported for adult brine shrimp (Gallagher & Brown, 1975; published in corrected form by Simpson, Klein-MacPhee & Beck, 1983). The composition compares very favourably with the minimum dietary requirement for salmonids (Ketola, 1976), but is slightly less than the recommended dietary levels for cold-water fishes (Anonymous, 1981) in niacin, pyridoxine, and riboflavin.

POLLUTANTS

Because *Artemia* grow in many areas of the world close to human populations, anthropogenic inputs to their environment such as chlorinated hydrocarbons (CHCs) and heavy metals are often found in cysts and nauplii. Bookhout & Costlow (1970) measured DDT concentrations of $2 \cdot 30$ $\mu g \cdot g^{-1}$ and $7 \cdot 05$ $\mu g \cdot g^{-1}$ in *Artemia* nauplii from California and Utah, respectively, whereas Wickins (1972) reported DDT levels of $0 \cdot 0004$–$0 \cdot 02$ $\mu g \cdot g^{-1}$ and PCB levels of $0 \cdot 04$–$0 \cdot 08$ $\mu g \cdot g^{-1}$ for nauplii from those regions. CHC concentrations in nauplii from eight geographical sources and two Reference strains (Olney *et al.*, 1980; Seidel *et al.*, 1982; Bengtson *et al.*, 1985) ranged over about two orders of magnitude (2–422 $ng \cdot g^{-1}$) for total DDTs and more than one order of magnitude (1–66 $ng \cdot g^{-}$) for total PCBs. Nauplii from Italy and China generally had the highest CHC levels and those from Brazil, Australia, and the Reference strains the lowest.

Olney *et al.* (1980) provided the only published data on heavy metal content (12 metals) in *Artemia* cysts and nauplii. They concluded that differences among geographical strains were small and that the levels observed were not particularly high. According to Blust (pers. comm.) and our own unpublished data levels of copper in Great Salt Lake *Artemia* cysts are low in the Northern Arm cysts (around 10 $\mu g \cdot g^{-1}$ on a dry weight basis) and high in commercial batches of Southern Arm cysts (80 $\mu g \cdot g^{-1}$ and more). Cyst samples collected at different sites, 40 to 60 km north of the commercial harvesting area (a major dumping site of copper ore wastes, Sanders Brine Shrimp Cy, pers. comm.) have significantly lower Cu-contents (16 to 20 $\mu g \cdot g^{-1}$); contrary to commercial batches of Great Salt Lake South Arm cysts, the latter samples appear to be an acceptable source of live food for different crab species (Goy, pers. comm.; see also p. 554).

CONCLUSIONS AND PERSPECTIVES

Although *Artemia* nauplii have already been used for a few decades as live food for culturing larvae of various fish and shrimp species, it is only during recent years that the nutritional properties of freshly hatched *Artemia* nauplii have been better understood. It had been known for some time that *Artemia* could not be considered as a 'standard' food. It was, however, only in the late 1970s when several new geographical sources of *Artemia* became available that detailed characterization work in Japan and through the International Study on *Artemia* could compare the suitability of particular sources or batches of *Artemia* cysts as a larval food source with specific *Artemia* characteristics, *e.g.* nauplius dimension, fatty-acid content, contamination level. Probably the most critical factor determining the dietary value of *Artemia*, as a food-source for marine predators, is the presence and concentration of essential fatty acids; *i.e.* the natural prey of marine fish and crustacean larvae mostly contain substantial levels of the highly unsaturated fatty acids 20:5ω3 and 22:6ω3, whereas in *Artemia* their concentration is inconsistent and minimal if present at all. This is due to the extreme as well as highly fluctuating natural environment in which *Artemia* and especially its particular diet are developed. In this regard it is very fortunate that the early pioneers in fish culturing were using a nutritionally adequate *Artemia* product from the San Francisco Bay strain; *Artemia* might never have become a widely recognized 'suitable' diet for marine organisms if Great Salt Lake *Artemia*, deficient in essential fatty acids had been the only source of *Artemia* available at that time.

It is obvious now, more than ever before, that the special value of *Artemia* as a food source is due not so much to its nutritional composition but is related to a large extent to its convenient production, its optimal physical availability as a moving prey of suitable size, and to the opportunities it provides for bioencapsulation of vital components, *i.e.* to convert it from a deficient food into a supra-natural diet. It is clear that as dietary requirements of marine fish and shrimp larvae become better known, the *Artemia* enrichment technique involving bioencapsulation of vital components will be most useful in enhancing larval nutrition. A very recent example being an improved pigmentation in flatfish larvae (Pricket, pers. comm.; Danish Aquaculture Institute, pers. comm.) through HUFA-enrichment of the live foods.

The causal relationship between high contamination levels and low nutritional quality of *Artemia* nauplii was over-estimated in the earliest publications. It is not yet clear, however, to what extent the presence of pesticides, heavy metals or other contamination products may affect the biological effectiveness of *Artemia* as a food source, especially when considering potentially delayed effects expressed in post-*Artemia* feeding stages; *e.g.* toxicity effects in larval fish during weaning when lipids in which pesticides have been accumulating are metabolized. As more and more *Artemia* production is initiated in areas where intake waters may be contaminated with industrial wastes or with the run-off waters from agricultural fields, the risks of contamination of *Artemia* cysts with persistent herbicides, pesticides, *etc.* are increasing. Because of their high tolerance for various contamination products the *Artemia* population may not be affected but bio-accumulation in the cysts will be the consequence.

The great variability in *Artemia* strains as well as batch characteristics are the origin of much confusion when trying to compare data obtained by different authors using different strains and/or batches of the same strain of *Artemia* for their culture tests. This is particularly critical in ecotoxicological testing where the bioassay results may vary as a function of the type of *Artemia* used as food for the test-animals (Bengtson *et al.*, 1984). In this regard the recommendation of the International Study on *Artemia* to use Reference *Artemia* Cysts (Sorgeloos, 1980b) as inter-calibration material should gain more interest. Reference *Artemia* Cysts are only a temporary solution as their limited stocks (from the wild) are never identical when replaced. It is hoped that the laboratory technique for controlled cyst production of Lavens & Sorgeloos (1984) can soon be scaled up to produce so-called "Standard *Artemia* Cysts" of reproducibly high nutritional quality as the inter-calibration material for future research and applications with brine shrimp.

In view of the large variation in nutritional quality of *Artemia*, not only among strains but even between batches of cysts from the same geographical origin, cyst distributors would do a great favour to their customers by providing more detailed product specifications, *i.e.* not only hatching quality characteristics but also strain origin, biometrical data, fatty-acid profiles and eventually contamination levels. In this regard it is obvious that in the future price differences for cysts will also be determined by the variation in nutritional quality.

Although cysts and nauplii still draw most attention in research on applications of *Artemia*, the potential with brine shrimp biomass is at present *under-estimated, e.g.* in nursery and maturation feeding, eventually after application of bioencapsulation enrichment, and as an animal protein source. Again in this field of research and developments, inter-calibration through product characterization (such as biochemical composition) and product processing (such as freezing technique) will be very important.

Finally, much theoretical information exists on how fish and shrimp *production can be improved, e.g.* strain selection, use of decapsulated cysts, cold stored nauplii, on-grown juveniles, *etc.* A better interaction between the academic world and the aquaculture industry is, however, essential to translate better the research findings into commercial profits. It is our conviction that this will improve as competition in this new bio-industry increases.

ACKNOWLEDGEMENTS

Our research contribution for this review has been made possible through the Belgian National Science Foundation (NFWO) grant FKFO 32.0012.82, the Institute for the Promotion of Industry and Agriculture (IWONL), the NV Artemia Systems, the Belgian Administration for Development Cooperation (ABOS), the Environmental Protection Agency (EPA) grant CR 811042–02–0 and the United States Agency for International Development (USAID)-Title XII strengthening grant AID/DSAN-XII-G-0116. P.S. is a senior scientist with the Belgian National Science Foundation.

REFERENCES

Ablett, R. F. & Richards, R. H., 1980. *Aquaculture,* **19,** 371–377.
Ackman, R. G., Eaton, C. A., Sipos, J. C., Hooper, S. N. & Castell, J. D., 1970. *J. Fish. Res. Bd Can.,* **27,** 513–533.
Adron, J. W., Blair, A. & Cowey, C. B., 1974. *Fishery Bull. NOAA.,* **72,** 353–357.
Adron, J. W., Blair, A. & Cowey, C. B., 1977. *Actes Colloq. C.N.E.X.O.,* No. 4, 67 only (Abstract).
Al Attar, M. H. & Ikenoue, H., 1979. *Kuwait Bull. Mar. Sci.,* No. 1, 32 only.
Alessio, G., 1974. *Boll. Pesca Piscic. Idrobiol.,* **29,** 133–147.
Alessio, G., Gandolfi, G. & Schreiber, B., 1976. *Etud. Rev. gen. Fish. Counc. Mediterr.,* **55,** 143–157.
Amat F., 1980. *Inf. Téc. Inst. Invest. Pesq.* No. 75, 3–24.
Anderson, D. T., 1967. *Aust. J. Zool.,* **15,** 47–91.
Anonymous, 1973. *Report of the Director of Fisheries Research: 1972–1973.* Fish. Lab., Lowestoft, U.K., 72 pp.
Anonymous, 1977. *Bull. Inf. C.N.E.X.O., Fiche Tech. Aquaculture,* No. 114, 16 pp.
Anonymous, 1978a. *Aquaculture Planning Program,* Department of Planning and Economic Development, State of Hawaii, 222 pp.
Anonymous, 1978b. *Supplément au Bulletin C.N.E.X.O.,* No. 114, 16 pp.
Anonymous, 1978c. *Report of the Director of Fisheries Research: 1974–1977.* Fish. Lab., Lowestoft, U.K., 80 pp.
Anonymous, 1980. *Fish Farmer,* No. 3, 48–49.
Anonymous, 1981. *Nutrient Requirements of Coldwater Fishes,* Natl Res. Counc., Natl Acad. Press, Washington.
Anonymous, 1982. *Report of the Director of Fisheries Research: 1977–1980.* Lowestoft, U.K., 90 pp.
Anonymous, 1984. *Boln Inst. esp. Oceanogr.,* **4,** 13–22.
Aquacop, 1977. *Actes Colloq. C.N.E.X.O.,* No. 4, 213–232.
Aronovich, T. M. & Spektorova, L. V., 1971. *Proc. All-Union Res. Inst. Mar. Fish. Ocean.,* **81,** *190*–204. *Fish. Res. Bd Can.,* Transl. Ser. No. 2385, 18 pp.
Artemia Systems, 1985. *The Brine Shrimp* Artemia, *A User's Guide.* Artemia Systems, Ghent, Belgium, 10 pp.
ASEAN, 1977, *First ASEAN Meeting of Experts on Aquaculture.* Tech. Rep., ASEAN, Semarang (Indonesia), 31 Jan.–6 Feb., 234 pp.
Austin, B. & Allen, D. A., 1981/1982. *Aquaculture,* **26,** 369–383.
Azari Takami, G., 1976. *J. Iran vet. med. Ass.,* **1,** 10–16.
Azari Takami, G., 1985. In, *Book of Abstracts, 2nd int. Symp. on the Brine Shrimp* Artemia, Antwerp, Belgium, 1–5 Sept., p. 10 only.
Barahona-Fernandes, M. H. & Girin, M., 1977. *Actes Colloq. C.N.E.X.O.,* No. 4, 69–84.
Barahona-Fernandes, M. H., Girin, M. & Metailler, R., 1977. *Aquaculture,* **10,** 53–63.
Bardach, J. E., Ryther, J. H. & McLarney, W. O., 1972. Aquaculture: the Farming and Husbandry of Freshwater and Marine Organisms. Wiley-Interscience, New York, U.S.A., 868 pp.
Barnabé, G., 1976. *Aquaculture,* **9,** 237–252.
Barnabé, G., 1980. *Synop. FAO Pêches,* No. 126, 70 pp.
Basil, J. A. & Marian, M.P., 1985. In, *Book of Abstracts, 2nd Int. Symp. on the Brine Shrimp Artemia, Antwerp, Belgium, 1*–5 Sept., p. 14 only.
Beck, A. D., 1979. In, *Cultivation of Fish Fry and its Live Food,* Europ. Maricult. Soc. Spec. Publ. No. 4, edited by E. Styczynska-Jurewicz *et al., Inst. Mar.* Scient. Res., Bredene, Belgium, pp. 63–86.

Beck, A. D. & Bengtson, D. A., 1982. In, *Aquatic Toxicology and Hazard Assessment: Fifth Conference,* edited by J. G. Pearson *et al., Amer. Soc. for Testing* and Materials, Philadelphia, U.S.A., pp. 161–169.

Beck, A. D., Bengtson, D. A. & Howell, W. H., 1980. In, *The Brine Shrimp* Artemia, *Vol. 3,* edited by G. Persoone *et al.*, Universa Press, Wetteren, Belgium, pp. 249–259.

Bell, M. V., Henderson, R. J., Pirie, B. J. S. & Sargent, J. R., 1985. *J. Fish Biol.* **26,** 181–191.

Benesch, R., 1969. *Zool. Jb (Anat.),* **86,** 307–458.

Bengtson, D. A., Beck, A. D., Lussier, S. M., Migneault, D. & Olney, C. E., 1984. In, *Ecotoxicological Testing for the Marine Environment, Vol. 2,* edited by G. Persoone *et al.*, State University of Ghent and Inst. Mar. Scient. Res., Bredene, Belgium, pp. 399–416.

Bengtson, D. A., Beck, A. D. & Poston, H. A., 1978. *Proc. 9th Ann. Meeting Maricul. Soc.,* pp. 159–174.

Bengtson, D. A., Beck, A. D. & Simpson, K. L., 1985. In, *Nutrition and Feeding in Fish,* edited by C. B. Cowey *et al.*, Academic Press, London, pp. 431–446.

Benijts, F., Vanvoorden, E. & Sorgeloos, P., 1976. In, *Proc. 10th Eur. Symp. Mar. Biol., Vol. 1,* edited by G. Persoone & E. Jaspers, Universa Press, Wetteren, Belgium, pp. 1–9.

Berrigan, M. E., Willis, S. A. & Halscott, K. R., 1978. *Completion Report, U.S. Dept of Commerce,* NOAA, NMFS, PL 88–309, No. 2–298–R–1, Job 1, unpubl.

Bigford, T. E., 1978. *Fish. Bull. NOAA,* **76,** 59–64.

Binkowski, F. P. & Czeskleba, D. G., 1980. *Paper presented at the 11th Ann. Meeting World Maricul. Soc.,* New Orleans, U.S.A., 5–8 March.

Bombeo, R. F., 1985. In, *Book of Abstracts, 2nd Int. Symp. on the Brine Shrimp* Artemia, Antwerp, Belgium, 1–5 Sept., p. 26 only.

Bookhout, C. G. & Costlow Jr, J. D., 1970. *Helgoländer wiss. Meeresunters.,* **20,** 435–442.

Bossuyt, E. & Sorgeloos, P., 1980. In, *The Brine Shrimp* Artemia, *Vol. 3,* edited by G. Persoone *et al.*, Universa Press, Wetteren, Belgium, pp. 133–152.

Botsford, L. W., Rauch, H. E. & Shleser, R. A., 1974. In, *Proc. 5th Ann. Wkshop World Maricul. Soc.,* pp. 387–401.

Bottino, R., 1974. *Mar. Biol.,* **27,** 197–204.

Bottino, N. R., Gennity, J., Lilly, M. L., Simmons, E. & Finne, G., 1980. *Aquaculture,* **19,** 139–148.

Bovill, E. W., 1968. *The Niger Explored.* Oxford University Press, London, 263 pp.

Boyd, J., 1974. *Progve Fish Cult.,* **36,** 57 only.

Brisset, P. J., 1981. Thesis, University of Lille, France, 85 pp.

Brisset, P. P., Versichele, D., Bossuyt, E., De Ruyck, L. & Sorgeloos, P., 1982. *Aquacultural Eng.,* **1,** 115–119.

Broad, A. C., 1957. *Biol. Bull. mar. biol. Lab.,* Woods Hole, **112,** 162–170.

Broch, E. S., 1965. *Cornell Univ. Agric. Expt Stn Mem.* No. 392, 48 pp.

Bromley, P. J., 1977. *Aquaculture,* **12,** 337–347.

Bromley, P. J., 1978. *Aquaculture,* **13,** 339–345.

Bromley, P. J. & Howell, B. R., 1983. *Aquaculture,* **31,** 31–40.

Bruggeman, E., Baeza-Mesa, M., Bossuyt, E. & Sorgeloos, P., 1979. In, *Cultivation of Fish Fry and its Live Food,* Europ. Maricul. Soc. Spec. Publ. No. 4, edited by E. Styczynska-Jurewicz *et al.*, Inst. Mar. Scient. Res., Bredene, Belgium, pp. 309–315.

Bruggeman, E., Sorgeloos, P. & Vanhaecke, P., 1980. In, *The Brine Shrimp* Artemia, *Vol. 3,* edited by G. Persoone *et al.*, Universa press, Wetteren, Belgium, pp. 261–269.

Bryan, P. G. & Madraisau, B. B., 1977. *Aquaculture,* **10,** 243–252.
Bryant, P. L. & Matty, A. J., 1980. *Aquaculture,* **21,** 203–212.
Burillo, S. L., Sillero, A. & Sillero, M.A.G., 1982. *Comp. Biochem. Physiol.,* **71B,** 89–93.
Cadena Roa, M., Huelvan, C., Le Borgne, Y. & Metailler, R., 1982a. *J. World Maricul. Soc.,* **13,** 246–253.
Cadena Roa, M., Menu, B., Metailler, R. & Person-Le Ruyet, J., 1982b. *I.C.E.S. Maricult. Committee* F:9, 10 pp.
Camara, M. R. & De Medeiros Rocha, R., 1985. In, *Book of Abstracts, Second International Symposium on the Brine Shrimp Artemia,* Antwerp, Belgium, 1–5 Sept., p. 30 only.
Campillo, A., 1975. *Revue Trav. Inst. Pêch. marit.,* **39,** 395–405.
Capuzzo, J. M. & Lancaster, B. A., 1979. In, *Proc. 10th Ann. Meeting World Maricult. Soc.,* pp. 689–700.
Carlberg, J. M. & Van Olst, J. C., 1976. In, *Proc. 7th. Ann. Meeting World Maricult. Soc.,* pp. 379–389.
Carpelan, L. H., 1957. *Ecology,* **38,** 375–390.
Castell, J. D., 1977. *Actes Colloq. C.N.E.X.O.,* No. 4, 277–281.
Castell, J. D. & Boghen, A. D., 1979. In, *Proc. 10th Ann. Meeting World Maricult. Soc.,* pp. 720–727.
Castell, J. D. & Covey, J. F., 1976. *J. Nutr.,* **106,** 1159–1165.
Chang, E. S. & Conklin, D. E., 1983. In, *Handbook of Mariculture, Vol. 1,* edited by J. P. McVey, CRC Press, Boca Raton, Florida, U.S.A., pp. 271–275.
Christiansen, M. E. & Yang, W. T., 1976. *Aquaculture,* **8,** 91–98.
Claus, C., Benijts, F. & Sorgeloos, P., 1977. In, *Fundamental and Applied Research on the Brine Shrimp* Artemia salina *(L.) in Belgium.* Europ. Maricult. Soc. Spec. Publ. No. 2, edited by E. Jaspers & G. Persoone, Inst. Mar. Scient. Res., Bredene, Belgium, pp. 91–105.
Claus, C., Benijts, F., Vandeputte, G. & Gardner, W., 1979. *J. exp. mar. Biol. Ecol.,* **36,** 171–183.
Clegg, J. S., 1962. *Biol. Bull. mar. biol. Lab., Woods Hole,* **123,** 295–301.
Clegg, J. S., 1964. *J. exp. Biol.,* **41,** 879–892.
Clegg, J. S. & Golub, A. L., 1969. *Devl Biol.,* **19,** 178–200.
Coleman, D. E., Nakagawa, L. K., Nakamura, R. M. & Chang, E., 1980. In, *The Brine Shrimp* Artemia, *Vol. 3,* edited by G. Persoone *et al.*, Universa Press, Wetteren, Belgium, pp. 153–157.
Coles, S. L., 1969. *Limnol. Oceanogr.,* **14,** 949–953.
Conklin, D. E., D'Abramo, L. R., Bordner, C. E. & Baum, N. A., 1980. *Aquaculture,* **21,** 243–249.
Conklin, D. E., D'Abramo, L. R. & Norman-Boudreau, K., 1983. In, *Handbook of Mariculture, Vol. 1,* edited by J. P. McVey, CRC Press, Boca Raton, Florida, U.S.A., pp. 413–423.
Conklin, D. E., Devers, K. & Shleser, R. A., 1975. In, *Proc. 6th Ann. Wkshop World Maricult. Soc.,* pp. 237–244.
Conklin, D. E., Goldblatt, M. J. & Bordner, C. E. 1978. In, *Proc. 9th Ann. Meeting World Maricult. Soc.,* pp. 243–250.
Cook, H. L. & Murphy, M. A., 1969. *Trans. Am. Fish. Soc.,* **98,** 751–754.
Corazza, L. & Sailor, W. W., 1982. *Poult. Sci.,* **62,** 846–852.
Corbin, J. S., Fujimoto, M. M. & Iwai Jr, T. Y., 1983. In, *Handbook of Mariculture, Vol. 1,* edited by J. P. McVey, CRC Press, Boca Raton, Florida, U.S.A., pp. 391–412.
Cowey, C. B., Owen, J. M., Adron, J. W. & Middleton, C., 1976. *Br. J. Nutr.,* **36,** 479–486.
Cowey, C. B. & Sargent, J. R., 1979. In, *Fish Physiology, Vol. 8,* edited by W. S. Hoar *et al.*, Academic Press, New York, pp. 1–69.

Cowey, C. B. & Tacon, A. G. J., 1982. In, *Proc. 2nd int. Conference on Aquaculture Nutrition*, World Maricult. Soc. Spec. Publ. No. 2, edited by G. D. Pruder *et al.*, pp. 13–30.
Culkin, F. & Morris, R. J., 1969. *Deep-Sea Res.*, **16**, 109–116.
Czeczuga, B., 1971. *Comp. Biochem. Physiol.*, **40B**, 47–52.
Czeczuga, B., 1980. In, *The Brine Shrimp* Artemia, *Vol. 2,* edited by G. Persoone *et al.*, Universa Press, Wetteren, Belgium, pp. 607–609 (abstract only).
Czygan, F. C., 1966. *Z. Naturf.*, **21**, 801–805.
Czygan, F. C., 1968. *Z. Naturf.*, **B23**, 1367–1368.
D'Abramo, L. R., Baum, N. A., Bordner, C. E. & Conklin, D. E., 1983. *Can. J. Fish. Aquat. Sci.*, **40**, 699–704.
Dabrowski, K., 1982. *Riv. ital. Piscic. Ittiopatol.*, **27**, 11–29.
Dabrowski, K., Charlon, N., Bergot, P. & Kaushik, S., 1984. *Aquaculture,* **41**, 11–20.
Dabrowski, K., Dabrowska, H. & Grudniewski, C., 1978. *Aquaculture,* **13**, 257–264.
Dabrowski, K. & Glogowski, J., 1977a. *Hydrobiologia,* **52**, 171–174.
Dabrowski, K. & Glogowski, J., 1977b. *Hydrobiologia,* **54**, 129–134.
Dabrowski, K. & Kaushik, S. J., 1984. *Aquaculture,* **41**, 333–344.
Dabrowski, K. & Rusiecki, M., 1983. *Aquaculture,* **30**, 31–42.
D'Agostino, A. S., 1965. Thesis, New York University, U.S.A., 83 pp.
D'Agostino, A. S., 1980. In, *The Brine Shrimp* Artemia, *Vol. 2,* edited by G. Persoone *et al.*, Universa Press, Wetteren, Belgium, pp. 55–82.
D'Agostino, A. S. & Provasoli, L., 1968. *Biol. Bull. mar biol. Lab., Woods Hole,* **134**, 1–14.
D'Agostino, A. S. & Provasoli, L., 1970. *Biol. Bull. mar. biol. Lab., Woods Hole,* **139**, 485–494.
Davies, B. H., Hsu, W.-J. & Chichester, C. O., 1965. *Biochem. J.*, **94**, 26P only.
Davies, B. H., Hsu, W.-J. & Chichester, C. O., 1970. *Comp. Biochem. Physiol.*, **33**, 601–615.
De Figueiredo, J. J., 1975. *Notas Estud. Inst. Biol. marit. Lisb.*, No. 42, 6 pp.
Dejarme, H. E., 1981. Thesis, Mindanao State University, Philippines, 46 pp.
Delga, J., Meunier, J. L., Pallaget, C. & Carious, J., 1960. *Ann Falsif. Expert. Chim,* p. 617 only.
De los Santos Jr, C., Sorgeloos, P., Lavina, E. & Bernardino, A., 1980. In, *The Brine Shrimp* Artemia, *Vol. 3*, edited by G. Persoone *et al.*, Universa Press, Wetteren, Belgium, pp. 159–163.
De Meulemeester, A., Lavens, P., De Ruyck, L. & Sorgeloos, P., 1985. In, *Book of Abstracts, 2nd int. Symp. on the Brine Shrimp* Artemia, Antwerp, Belgium, 1–5 Sept., p. 37 only.
Dendrinos, P., Dewan, S. & Thorpe, J. P., 1984. *Aquaculture,* **38**, 137–144.
Devrieze, L., 1984. Thesis, State University of Ghent, Belgium, 105 pp.
Dexter, D. M., 1972. *Calif. Fish Game,* **58**, 107–115.
Dickson, L. G., Galloway, R. A. & Patterson, G. W., 1969. *Plant Physiol.*, **44**, 1413–1416.
Divanach, P., Kentouri, M. & Paris, J., 1983. *C.r. Séanc. Acad. Sci. Ser. III,* **296**, 29–33.
Dobbeni, A., 1983. *Report PHITS COOVI,* Anderlecht, Belgium, 9 pp.
Dugan, C. C., Hagood, R. W. & Frakes, T. A., 1975. *Fla Mar. Res. Lab.*, Fla Dept Nat. Res., Publ. No. 12, 28 pp.
Dumont, H. J., 1979. Thesis, State University of Ghent, Belgium, 557 pp.
Duray, M. & Bagarinao, T., 1984. *Aquaculture,* **41**, 325–332.
Dutrieu, J., 1960. *Archs Zool. exp. gén.*, **99**, 1–134.
Dye, J. E. 1980. In, *The Brine Shrimp* Artemia, *Vol. 3,* edited by G. Persoone *et al.*, Universa Press, Wetteren, Belgium, pp. 271–276.

Eagles, M. D., Aiken, D. E. & Waddy, S. L., 1984. *J. World Maricult. Soc.*, **15**, 142–143.

Ebert, E. E., Haseltine, A. W., Houk, J. L. & Kelly, R. O., 1983. *Fish Bull. Calif.*, **172**, 259–309.

Emmerson, W. D., 1977. M. Sc. thesis, University of Port Elizabeth, Rep. South Africa, 116 pp.

Emmerson, W. D., 1980. *Mar. Biol.*, **58**, 65–73.

Emmerson, W. D., 1984. *Aquaculture*, **38**, 201–209.

Enright, C. T., 1984. Paper presented at 15th Ann. Meeting World Maricult. Soc., Vancouver B. C., Canada, 18–22 March.

Enzler, L., Smith, V., Lin, J. S. & Olcott, H. S., 1974. *J. agric. Fd Chem.*, **22**, 330–331.

Ezquieta, B. & Vallejo, C. G., 1985. In, *Book of Abstracts, 2nd int. Symp. on the Brine Shrimp* Artemia, Antwerp, Belgium, 1–5 Sept., p. 42 only.

Fabre-Domergue, P. & Bietrix, E., 1905. *Travail du Laboratoire de Zoologie Maritime de Concarneau, Vuibert et Nony*, Paris, 243 pp.

Flores, T. A., 1985. In, *Book of Abstracts, 2nd int. Symp. on the Brine Shrimp* Artemia, Antwerp, Belgium, 1–5 Sept., p. 44 only.

Flüchter, J., 1965. *Helgoländer wiss. Meeresunters*, **12**, 395–403.

Flüchter, J., 1980. *Aquaculture*, **19**, 191–208.

Flüchter, J., 1982. *Aquaculture*, **27**, 83–85.

Forster, J. R. M. & Wickins, J. F., 1967. *I.C.E.S. Maricult. Committee* E:13, 9 pp.

Forster, J. R. M. & Wickins, J. F., 1972. *Min. Agric. Fish. Food, Laboratory Leaflet*, No. 27, 32 pp.

Fuchs, J., 1976. *Rapport de Stage Optionnel.* Centre Océanologique de Bretagne, Brest, France, 20 March–15 May, 71 pp.

Fuchs, J., 1981/1982. *Aquaculture*, **26**, 321–337.

Fuchs, J. & Person-Le Ruyet, J., 1976. *I.C.E.S. Comité de L'Amélioration des Pêches* E: 24, 9 pp.

Fujimura, T. & Okamoto, H., 1970. In, *Proc. 14th Session, FAO, Indo-Pacific Fisheries Council*, 17 pp.

Fujita, S., Watanabe, T. & Kitajima, C., 1980. In, *The Brine Shrimp* Artemia, *Vol. 3*, edited by G. Persoone *et al.*, Universa Press, Wetteren, Belgium, pp. 277–290.

Fukusho, K., 1974. *Aquaculture*, **21**, 71–75.

Fukusho, K., 1979. *Nagasaki Pref. Inst. of Fisheries*, No. 6, 173 pp.

Gabaudan, J., Piggott, G. M. & Halver, J. E., 1980. *Proc. World Maricult. Soc.*, **11**, 424–432.

Gallagher, M. L. & Brown, W. D., 1975. *J. agric. Fd Chem.*, **23**, 630–632.

Gatesoupe, J., 1982. *Annls Zootech.* (*Paris*), **31**, *353*–368.

Gatesoupe, J., 1983. *Aquaculture*, **32**, 401–404.

Gatesoupe, J., Girin, M. & Luquet, P., 1977. *Actes Colloq. C.N.E.X.O.*, No. 4, 59–66.

Gatesoupe, J. & Luquet, P., 1981/1982. *Aquaculture*, **26**, 256–368.

Geiger, J. G. & Parker, N. C., 1985. *Progve Fish Cult.*, **47**, 1–13.

Ghannudi, S. A. & Tufail, M., 1978. *Lybian J. Sci.*, **8A**, 69–74.

Gilchrist, B., 1968. *Comp. Biochem. Physiol.*, **24**, 123–147.

Gilchrist, B. & Green, J., 1960. *Proc. R. Soc.*, **152**, 118–136.

Gilmour, A., McCallum, M. F. & Allan, M. C., 1975. *Aquaculture*, **6**, 221–231.

Girin, M., 1974a. *Actes Colloq. C.N.E.X.O.*, No. 1, 175–185.

Girin, M., 1974b. *Actes Colloq C.N.E.X.O.*, No. 1, 187–203.

Girin, M., 1976. *Stud. Rev. G.F.C.M.*, **55**, 133–142.

Girin, M., 1978. Thesis, Université Pierre et Marie Curie, Paris, France, 202 pp.

Girin, M., 1979. In, *Cultivation of Fish Fry and its Live Food*, Europ. Maricult. Soc. Spec. Publ. No. 4, edited by E. Styczynska-Jurewicz *et al.*, Inst. Mar. Scient. Res., Bredene, Belgium, pp. 199–209.

Girin, M., Barahona-Fernandes, M. H. & Le Roux, A. 1975. *I.C.E.S. Maricult. Committee* G:14, 8 pp.
Girin, M., Metailler, R. & Nedelec, J. 1977. *Actes Colloq. C.N.E.X.O.,* No. 4, 35–50.
Girin, M. & Person-Le Ruyet, J., 1977. *Bull. Fr. Piscic.,* **264,** 88–101.
Glude, J. B., 1978a. *The Freshwater Prawn* Macrobrachium rosenbergii. J. B. Glude, Aquaculture Consultant, Seattle, U.S.A., 59 pp.
Glude, J. B. 1978b. *The Marine Shrimp* Penaeus *spp.* J. B. Glude, Aquaculture Consultant, Seattle, U.S.A, 45 pp.
Godeluck, B., 1981. Thesis, Université Pierre et Marie Curie, Paris, France, 40 pp.
Golub, A. L. & Finamore, F. J., 1972. *Fedn Proc. Fedn Am. Soc. exp. Biol.,* **31,** 706 (Abstract).
Good, L. K., Bayer, R. C., Gallagher, M. L. & Rittenburg, J. H., 1982. *J. Shellfish Res.,* **2,** 183–187.
Goodwin, H. L. & Hanson, J. A., 1977. In, *Shrimp and Prawn Farming in the Western Hemisphere*, edited by J. A. Hanson & H. L. Goodwin, Dowden, Hutchingson & Ross, Inc., Stroudsburg, U.S.A., pp. 193–291.
Gopalakrishnan, K., 1976. *Aquaculture,* **9,** 145–154.
Goy, J. W. & Costlow, J. D., 1980. *Am. Zool.* **20,** 888 only.
Grabner, M., Wieser, W. & Lakner, R., 1981/1982. *Aquaculture,* **26,** 85–94.
Gross. F., 1937. *J. mar. biol. Ass. U.K.,* **21,** 753–768.
Guary, J. C., Kayama, M., Murakami, Y. & Ceccaldi, H. J., 1976. *Aquaculture,* **7,** 145–254.
Guimares, J. I. & De Haas, M. A. F., 1985. In, *Book of Abstracts, 2nd int. Symp. on the Brine Shrimp* Artemia, Antwerp, Belgium, 1–5 Sept., p. 50 only.
Günkel, G., 1979. In, *Cultivation of Fish Fry and its Live Food,* Europ. Maricult. Soc. Spec. Publ. No. 4, edited by E. Styczynska-Jurewicz *et al.*, Inst. Mar. Scient. Res., Bredene, Belgium, pp. 211–242.
Gun'ko, A. F., 1962. *Tr. Azovsk. Nauchn. Issled Inst. Rybn. Khoz.,* **5,** 73–96.
Gun'ko, A. F. & Pleskachevskaya, T. G., 1962. *Vaprossy ichthiologii,* **2,** 371–374.
Hanson, J. A. & Goodwin, H. L., 1977. In, *Shrimp and Prawn Farming in the Western Hemisphere*, edited by J. A. Hanson & H. L. Goodwin, Dowden, Hutchingson & Ross, Inc., Stroudsburg, U.S.A., pp. 1–192.
Happe, A. & Hollande, M., 1982. Thesis, Institut Supérieure d'Agriculture, Lille, France, 27 pp.
Hata, M. & Hata, M., 1969. *Comp. Biochem. Physiol.,* **29,** 985–994.
Hauenschild, C., 1954. *Wilhelm Roux Arch. EntwMech. Org.,* **147,** 1–41.
Hauenschild, C., 1956. *Z. Naturf.,* **11B,** 132–138.
Haynes, R. C. & Hammer, U. T., 1978. *Int. Revue ges. Hydrobiol.,* **63,** 337–351.
Heinen, J. M., 1976. In, *Proc. 7th. Ann. Meeting World Maricult. Soc.,* pp. 333–344.
Heldt, H., 1926. *Stn Océanogr. Salammbo,* Notes, **5,** 3–8.
Helfrich, P., 1973. *Seagrant Tech. Rep.,* UNIHI–SEAGRANT–TR–73–02, 173 pp.
Hentschel, E., 1968. *Zool. Anz.,* **180,** 372–384.
Herald, E. S. & Rakowicz, M., 1951. *Aquarium J.,* **22,** 234–242.
Hinchcliffe, P. R. & Riley, J. P., 1972. *J. mar. biol. Ass. U.K.,* **52,** 203–211.
Hines, H. B., Middleditch, B. S. & Lawrence, A. L., 1980. In, *The Brine Shrimp* Artemia, *Vol. 2,* edited by G. Persoone *et al.*, Universa Press, Wetteren, Belgium, pp. 169–184.
Hirano, R. & Oshima, Y., 1963. *Bull. Jap. Soc. scient. Fish.,* **29,** 282–297.
Hogendoorn, H., 1980. *Aquaculture,* **21,** 233–241.
Holland, D. L. & Jones, D. A., 1981. *Fish Farming Int.,* Dec. 1981, 17 only.
Houde, E. D., 1973. *Proc. World Maricult. Soc.,* **3,** 83–112.
Houde, E. D., 1975. *J. Fish Biol.,* **7,** 115–127.
Howell, B. R., 1971. *I.C.E.S., Fisheries Improvements Committee,* E:26, 6 pp.

Howell, B. R., 1979a. *Aquaculture,* **18,** 215–225.
Howell, B. R., 1979b. *I.C.E.S. Maricult. Committee* F:17, 4 pp.
Howell, B. R., Bromley, P. J. & Adkins, T. C., 1981. *I.C.E.S. Maricult. Committee* F:10, 4 pp.
Hsu, W. J., Chichester, C. O. & Davies, B. H., 1970. *Comp. Biochem. Physiol.,* **32,** 69–79.
Hughes, J. T., Shleser, R. A. & Tchobanoglous, G., 1975. *Progve Fish Cult.,* **39,** 129–132.
Huisman, E. A., 1974. Thesis, University of Wageningen, The Netherlands, 95 pp.
Imada, O., Kageyama, Y., Watanabe, T., Kitajima, C., Fujita, S. & Yone, Y., 1979. *Bull. Jap. Soc. scient. Fish.,* **45,** 955–959.
James, C. M., Abu-Rezeq, T. S. & Dias, P. 1985. In *Book of Abstracts, 2nd int. Symp. on the Brine Shrimp* Artemia, Antwerp, Belgium, 1–5 Sept., p. 52 only.
James, C. M., Bou-Abbas, M. & Dias, P., 1982. *Ann. Res. Rep.,* Kuwait Inst. Scient. Res., 1982, 113–115.
Janata, W. R. & Bell, D. J., 1985. In, *Book of Abstracts, 2nd int. Symp. on the Brine Shrimp* Artemia, Antwerp, Belgium, 1–5 Sept., p. 53 only.
Jensen, A. C., 1918. *Biol. Bull. mar. biol. Lab., Woods Hole,* **34,** 18–28.
Jezyck, P. F. & Penicnak, A. J., 1966. *Lipids,* **1,** 427–429.
Johns, D. M., Berry, W. J. & McLean, S., 1981b. *J. World Maricult. Soc.,* **12,** 303–314.
Johns, D. M., Berry, W. J. & Walton, W, 1981a. *J. exp. mar. Biol. Ecol.,* **53,** 209–219.
Johns, D. M., Peters, M. E. & Beck, A. D., 1978. *Am. Zool.,* **18,** 585 (Abstract).
Johns, D. M., Peters, M. E. & Beck, A. D., 1980. In, *The Brine Shrimp* Artemia, *Vol. 3,* edited by G. Persoone *et al.,* Universa Press, Wetteren, Belgium, pp. 291–304.
Jones, A. J., 1972. *J. Cons. perm. int. Explor. Mer,* **34,** 351–356.
Jones, D. A., Holland, D. L. & Jaborie, S. S., in press. *J. appl. Biochem. Biotechn.*
Jones, D. A., Kanazawa, A. & Rahman, S. A., 1979. *Aquaculture,* **17,** 33–43.
Juario, J. V. & Duray, M. N., 1981. *ISSN-0115-4710. Tech. Rep.,* No. 10, 27 pp.
Juario, J. V., Duray, M. N., Duray, V. M., Nacario, J. F. & Almendras, J. M. E., 1985. *Aquaculture,* **44,** 91–101.
Jumalon, N. A., Estenor, D. G. & Ogburn, D. M., 1985. In, *Book of Abstracts, 2nd int. Symp. on the Brine Shrimp* Artemia, Antwerp, Belgium, 1–5 Sept., p. 54 only.
Jumalon, N. A. & Ogburn, D. M., 1985. In, *Book of Abstracts, 2nd int. Symp. on the Brine Shrimp* Artemia, Antwerp, Belgium, 1–5 Sept., p. 55 only.
Kahan, D., 1979. In, *EIFAC Workshop on Mass Rearing of Fry and Fingerlings of Fresh Water Fishes,* edited by E. A. Huisman & H. Hogendoorn, EIFAC Tech. Paper No. 35, Suppl. 1, pp. 189–202.
Kahan, D., 1980. In, *Book of Abstracts, Symposium on Coastal Aquaculture,* Cochin, India, 12–18 Jan., p. 117 only.
Kahan, D., Uhlig, G., Schwenzer, D. & Horowitz, L., 1981/1982. *Aquaculture,* **26,** 303–310.
Kamienski, T. 1899. *Trav. Soc. Imp. Natural St. Petersb.,* **30,** 363–364.
Kanazawa, A. 1984. In, *Book of Abstracts, 1st int. Conf. on the Culture of Penaeid Prawns/Shrimps,* Iloilo City, Philippines, 4–8 Dec., p. 52 only.
Kanazawa, A., Teshima, S. & Ono, K., 1979. *Comp. Biochem Physiol.,* **63B,** 295–298.
Kanazawa, A., Teshima, S. & Tokiwa, S., 1977. *Bull. Jap. Soc. scient Fish.,* **43,** 849–856.
Katsutani, K., 1965. *Okayama-ken Pref. Fish. Exp Stn,* Intermediary Report, 5 pp.
Kayama, M., Tsuchiya, Y. & Mead, J. F., 1963. *Bull. Jap. Soc. scient. Fish.,* **29,** 452–458.
Kelly, R. O., Haseltine, A. W. & Ebert, E. E., 1977. *Aquaculture,* **10,** 1–16.
Kentouri, M., 1980. *Aquaculture,* **21,** 171–180.

Ketola, H. G., 1976. *Feedstuffs,* **48**.
Kingwell, S. J., Duggan, M. C. & Dye, J. E., 1977. *Actes Colloq. C.N.E.X.O.,* No. 4, 27–34.
Kinne, O., 1977. Editor, *Marine Ecology, Vol. III, part 2.* John Wiley & Sons, New York, U.S.A., pp. 579–1293.
Kittaka, J. 1977. *Actes Colloq. C.N.E.X.O.,* No. 4, 111–117.
Klein-MacPhee, G., Howell, W. H. & Beck, A. D., 1980. In, *The Brine Shrimp* Artemia, *Vol. 3,* edited by G. Persoone *et al.*, Universa Press, Wetteren, pp. 305–312.
Klein-MacPhee, G., Howell, W. H. & Beck, A. D., 1982. *Aquaculture,* **29**, 279–288.
Krinsky, N. I., 1965. *Comp. Biochem. Physiol.,* **16**, 181–187.
Kuenen, D. J., 1939. *Archs néerl. Zool.,* **3**, 365–449.
Kuenen, D. J. & Baas-Becking, L. G. M., 1938. *Zool. Meded, Leiden,* **20**, 222–230.
Kuhlman, D., Quantz, G. & Witt, U., 1981a. *Paper presented at the World Conference on Aquaculture,* Venice, Italy, 21–25 Sept.
Kuhlman, D., Quantz, G. & Witt, U., 1981b. *Aquaculture,* **23**, 183–196.
Kuhlman, D., Quantz, G., Witt, U. & Kattner, G., 1982. *I.C.E.S. Maricult. Committee* F:6, 9 pp.
Kurata, H., 1959. *Bull. Hokkaido reg. Fish. Res. Lab.,* **20**, 117–138.
Kuronuma, K. & Fukusho, K., 1984. *IDRC–TS47c,* 109 pp.
Lachance, M. A., Miranda, M., Miller, M. W. & Phaff, H. J., 1976. *Can. J. Microbiol.,* **22**, 1756–1761.
Lai, L. & Lavens, P., 1985. *Workshop*: Artemia *as a Business Perspective, 2nd int. Symp. on the Brine Shrimp Artemia,* Antwerp, Belgium, 1–5 Sept.
Lavens, P., Baert, P., De Meulemeester, A., Van Ballaer, E., Sorgeloos, P. & Smets, J., in press. *J. World Maricult. Soc.*
Lavens, P. & Sorgeloos, P., 1984. *Aquacult. Engng,* **3**, 221–235.
Lavens, P. & Sorgeloos, P., 1985. In, *Book of Abstracts, 2nd int. Symp. on the Brine Shrimp* Artemia, 1–5 Sept. Antwerp, Belgium, 1–5 Sept., p. 58 only.
Lavina, E. M. & Figueroa, R. F., 1978. *SEAFDEC Quarterly Res. Rep.,* No. 3, 11–14.
Lee, C., Hu, F. & Hirano, R., 1981. *Progve Fish Cult.,* **43**, 121–124.
Léger, C. & Frémont, L., 1981. In, *Nutrition des Poissons,* edited by C.N.R.S, Paris, pp. 215–246.
Léger, C., Gatesoupe, F. J., Metailler, R., Luquet, P. & Frémont, L., 1979. *Comp. Biochem. Physiol.,* **64B**, 345–350.
Léger, P., Bieber, G. F. & Sorgeloos, P., 1985a. *J. World Maricult Soc.,* in press.
Léger, P., Naessens-Foucqaert, E. & Sorgeloos, P., 1985b. In, *Book of Abstracts, 2nd int. Symp. on the Brine Shrimp* Artemia, Antwerp, Belgium, 1–5 Sept., p. 61 only.
Léger, P. & Sorgeloos, P., 1982. *Aquacultural Engng,* **1**, 45–53.
Léger, P. & Sorgeloos, P., 1984. *Mar. Ecol. Prog. Ser.,* **15**, 307–309.
Léger, P., Sorgeloos, P., Millamena, O. M. & Simpson, K. L. 1985c. *J. exp. mar. Biol. Ecol.,* **93**, 71–82.
Léger, P., Vanhaecke, P. & Sorgeloos, P., 1983. *Aquacultural Engng,* **2**, 69–78.
Lenhoff, H. M. & Brown, R. D., 1970. *Lab. Anim.,* **4**, 139–154.
Levine, D. M. & Sulkin, S. D., 1984. *J. exp. mar. Biol. Ecol.,* **81**, 211–223.
Levine, D. M., Sulkin, S. D. & Van Heukelem, L., 1983. In, *Culture of Marine Invertebrates,* edited by C. J. Berg Jr, Hutchingson Ross Publishing Company, Stroudsburg, Pennsylvania, U.S.A., pp. 193–203.
L'Herroux, M., Metailler, R. & Pilvin, L., 1977. *Actes Colloq. C.N.E.X.O.,* No. 4, 147–155.
Liao, I. C., Lu, Y. J., Huang, T. L. & Lin, M. C., 1971. *Fishery Ser. Chin-Am jt Comm. Rur. Reconstr.,* **11**, 1–29.
Liao, I. C., Su, H. M. & Lin, J. H., 1983. In, *Handbook of Mariculture, Vol. 1,* edited by J. P. McVey, CRC Press, Boca Raton, Florida, U.S.A., pp. 43–70.
Little, G., 1969. *Crustaceana,* **17**, 69–87.

Ljudskanova, J. & Joshev, L., 1972. *Z. Binnenfisch. D.D.R.,* **19,** 177–181.
MacDonald, G. H., 1980. In, *The Brine Shrimp* Artemia, *Vol. 3,* edited by G. Persoone *et al.*, Universa Press, Wetteren, Belgium, pp. 97–104.
MacFarlane, I. S., 1969. Thesis, University of London, London, 79 pp.
Maddox, M. B. & Manzi, J. J., 1976. In, *Proc. 7th Ann. Meeting World Maricult. Soc.,* pp. 677–698.
Maksimova, A. P., 1973. *Parazitologiya,* **7,** 347–352.
Malins, D. C. & Wekell, J. C., 1969. *Prog. Chem. Fats,* **10,** 475–486.
Manzi, J. J. & Maddox, M. B., 1980. In, *The Brine Shrimp* Artemia, *Vol. 3,* edited by G. Persoone *et al.*, Universa Press, Wetteren, Belgium, pp. 313–329.
Matsuoka, T., 1975. *Yoshoku,* **12,** 48–52.
Maugle, P. D., Deshimaru, O., Katayama, T. & Simpson, K. L., 1982. *Bull. Jap. Soc. scient. Fish.,* **48,** 1759–1764.
May, J. M., 1967. In, *Studies on Medical Geography, Vol. 1,* Hafner Publications Company, New York, p. 30 only.
May, R. C., 1970. *Calif. Mar. Res. Comm., CalCOFI Rep.,* No. 14, 76–83.
May, R. C., 1971. *NOAA Techn. Rep.,* NMFS SSRF-632, 24 pp.
McLean, S., Olney, C. E., Klein-MacPhee, G. & Simpson, K. L., 1985. In, *Book of Abstracts, 2nd int. Symp. on the Brine Shrimp* Artemia, Antwerp, Belgium, 1–5 Sept., p. 72 only.
Mead, C. G. & Finamore, F. J., 1969. *Biochemistry,* **8,** 2652–2655.
Meske, C., 1973. *Aquakultur von Warmwasser-Nutzfischen,* Verlag Eugen Ulmer, Stuttgart, Germany, 163 pp.
Metailler, R., Menu, B. & Morinière, P., 1981. *J. World Maricult. Soc.,* **12,** 111–116.
Metailler, R., Mery, C., Depois, M. & Nedelec, J., 1977. *Actes Colloq. C.N.E.X.O.,* No. 4, 93–109.
Meyers, S. P., 1979. In, *Proc. World Symp. on Finfish Nutrition and Fishfed Technology, Vol. II,* Hamburg, Berlin, 20–23 June 1978, pp. 13–20.
Millamena, O. M., Bombeo, R. F., Jumalon, N. A. & Simpson, K. L., 1985. *J. World Maricult. Soc.,* **16,** in press.
Millamena, O. M. & Simpson, K. L., 1985. *J. World Maricult. Soc.,* **16,** in press.
Miller, D. C., Lang, W. H., Marey, M., Clem, P. & Pechenik, J., 1979. In, *Book of Abstracts, Int. Symp. on the Brine Shrimp* Artemia salina, Corpus Christi, Texas, U.S.A., 20–23 Aug., p. 91 only.
Millikin, M. R., Biddle, G. N., Siewicki, T. C. & Fortner, A. R., 1980. *Aquaculture,* **19,** 149–161.
Moal, J., Samain, J. F. & Le Goz, J. R., 1978. In, *Physiology and Behaviour of Marine Organisms, Proc. of the 12th Europ. Mar. Biol. Symp.,* edited by D. S. McLusky & A. J. Berry, pp. 141–148.
Mock, C. R., Fontaine, C. T. & Revera, D. B., 1980a. In, *The Brine Shrimp* Artemia, *Vol. 3,* edited by G. Persoone *et al.*, Universa Press, Wetteren, Belgium, pp. 331–342.
Mock, C. R., Revera, D. B. & Fontaine, C. T., 1980b. *Proc. World Maricult. Soc.,* **11,** 102–117.
Monaco, G., 1974. *Aquaculture,* **4,** 309 only.
Monod, T., 1969. *Bull. Inst. fondam. Afr. noire, Ser. A,* **31,** 25–41.
Mootz, C. A., 1973. M.Sc. Thesis, University of Delaware, U.S.A.
Mootz, C. A. & Epifanio, C. E., 1974. *Biol. Bull. mar. biol. Lab., Woods Hole,* **146,** 44–55.
Morris, J. E. & Afzelius, B. A., 1967. *J. Utrastruct. Res.,* **20,** 244–259.
Morris, R. W., 1956. *Bull. Mus. océanogr.,* No. 1082, 62 pp.
Munuswamy, N., 1985. In, *Book of Abstracts, 2nd int. Symp. on the Brine Shrimp* Artemia, Antwerp, Belgium, 1–5 Sept., p. 78 only.

Murai, T. & Andrews, J. W., 1978. In, *Proc 9th Ann. Meeting World Maricult. Soc.*, pp. 189–193.
Nagainis, P. A. & Warner, A. H., 1979. *Devl Biol.*, **68**, 259–270.
Nakanishi, Y. H., Iwasaki, T., Okigaki, T. & Kato, H., 1962. *Annotnes zool. jap.*, **35**, 223–228.
Nanayakkara, M., Sunderam, R.I.M. & Royan, J. P., 1985. In, *Aquaculture and related papers*, NARA/OCC/85/1, pp. 29–43.
Nash, C. E., 1973. *Aquaculture*, **2**, 289–298.
Nash, C. E., Kuo, C. M. & McConnel, S. C., 1974. *Aquaculture*, **3**, 15–24.
Nelis, H. J. C. F., Lavens, P., Moens, L., Sorgeloos, P., Jonckheere, J. A., Criel, G. R. & De Leenheer, A. P., 1984, *J. biol. Chem.*, **259**, 6063–6066.
Nelis, H. J. C. F., Lavens, P., Sorgeloos, P., Van Steenberghe M. & De Leenheer, A. P., 1985. In, *Book of Abstracts, 2nd int. Symp. on the Brine Shrimp* Artemia, Antwerp, Belgium, 1–5 Sept., p. 81 only.
Nellen, W., Quantz, G., Witt, U., Kuhlmann, D. & Koske, H.P., 1981. *Europ. Maricult. Soc. Spec. Publ.* No. 6, 133–147.
New, M. B., 1976. *Aquaculture*, **7**, 101–144.
New, M. B. & Singholka, S., 1982. *F.A.O. Fisheries Tech. Paper* No. 225, 116 pp.
Nordeng, H. & Bratland, P., 1971. *J. Cons. perm. Int. Explor. Mer*, **34**, 51–57.
Ogle, J. & Price, W., 1976. *Gulf Res. Rep.*, **5**, 46–47.
Olalla, A., Osuna, C., Sebastian, J., Sillero, A. & Sillero, M. A. G., 1978. *Biochim. Biophys. Acta*, **523**, 181–190.
Oleinikova, F. A. & Pleskachevskaya, T. G., 1979. *Proc. 7th Japan-Soviet Joint Symp. on Aquaculture*, Tokyo, Japan, Sept. 1978, pp. 35–38.
Olesen, J. O. & Minck, F., 1983. *Aquacultural Engng*, **2**, 1–12.
Olney, C. E., Schauer, P. S., McLean, S., Lee, Y. & Simpson K. L., 1980. In, *The Brine Shrimp* Artemia, *Vol. 3*, edited by G. Persoone *et al.*, Universa Press, Wetteren, Belgium, pp. 341–352.
Omori, M., 1971. *Mar. Biol.* **9**, 228–234.
Oppenheimer, C. H. & Moreira, G. S., 1980. In, *The Brine Shrimp* Artemia, *Vol. 2*, edited by G. Persoone *et al.*, Universa Press, Wetteren, Belgium, pp. 609–613.
Osuna, C., Olalla, A., Sillero, A., Sillero, M. A. G. & Sebastian, J., 1977. *Devl Biol.*, **61**, 94–103.
Owen, J. M., Adron, J. W., Middleton, C. & Cowey, C. B., 1975. *Lipids*, **10**, 528–531.
Owen, J. M., Adron, J. W., Sargent, J. R. & Cowey, C. B., 1972. *Mar. Biol.*, **13**, 160–166.
Paffenhöfer, G. A., 1967. *Helgoländer wiss. Meeresunters.*, **16**, 130–135.
Palmegiano, G. B. & Trotta, P., 1981. *Contributed papers, World Conference on Aquaculture*, Venice, Italy, 21–25 Sept. Poster No. 121.
Paulsen, C. L., 1980. Paper presented at the 10th Ann. Meeting of the World Maricult. Soc., New Orleans, Louisiana, U.S.A, 5–8 March.
Perona, R. & Vallejo, C. G., 1985. In, *Book of Abstracts, 2nd int. Symp. on the Brine Shrimp* Artemia, Antwerp, Belgium, 1–5 Sept., p. 84 only.
Perrot, J., 1976. FIR: AQ/Conf/76/R–12, 20 pp.
Person-Le Ruyet, J., 1976. *Aquaculture*, **8**, 157–167.
Person-Le Ruyet, J., Alexandre, J. C., Le Roux, A. & Nedelec, G., 1978. *I.C.E.S.* Comité des Poissons de Fond et de Mariculture G: 55, 29 pp.
Person-Le Ruyet, J. & Salaun, A., 1977. *I.C.E.S.* Comité de l'Amélioration des Pêches E: 32, 13 pp.
Person-Le Ruyet, J. & Verillaud, P., 1980. *Aquaculture*, **20**, 351–370.
Persoone, G. & Sorgeloos, P., 1972. *Helgoländer wiss Meeresunters.*, **23**, 243–247.
Persoone, G. & Sorgeloos, P., 1980. In, *The Brine Shrimp* Artemia, *Vol. 3*, edited by G. Persoone *et al.*, Universa Press, Wetteren, Belgium, pp. 3–24.

Persoone, G., Sorgeloos, P., Roels O. & Jaspers E., 1980. In, *The Brine Shrimp Artemia, Vol. 1, Vol. 2, Vol. 3,* edited by G. Persoone *et al.*, Universa Press, Wetteren, Belgium, p. xvii only.
Platon, R. R. & Zahradnik, J. W., 1985. In, *Book of Abstracts. 2nd int. Symp. on the Brine Shrimp* Artemia, Antwerp, Belgium, 1–5 Sept., p. 88 only.
Post, F. J., 1977. *Microb. Ecol.,* 143–165.
Post, F. J. & Youssef, N. N., 1977. *Can. J. Microbiol.,* **23**, 1232–1236.
Provasoli, L., Conklin, D. E. & D'Agostino, A., 1970. *Helgoländer wiss. Meeresunters.,* **20**, 443–454.
Provasoli, L. & Shiraishi, K., 1959. *Biol. Bull. mar. biol. Lab., Woods Hole,* **117**, 347–355.
Provenzano Jr, A. J., 1967. In, *Proc. Symposium on Crustacea, Part 2,* Ernakulam, 1965, Bangalore Press, Bangalore, India, pp. 940–945.
Provenzano, A. J. & Goy, J. W., 1976. *Aquaculture,* **9**, 343–350.
Purdom, C. E. & Preston, A., 1977. *Nature, Lond.,* **266**, 396–397.
Radhakrishnan, E. V. & Vijayakumaran, M., 1980. In, *Book of Abstracts, Symposium on Coastal Aquaculture,* Cochin, India, 12–18 Jan., pp. 132–133.
Rakowicz, M., 1972. *Aquar. Dig. int.,* **1**, 16–18.
Reddy, S. R. & Shakuntala, K., 1980. In, *Book of Abstracts, Symposium on Coastal Aquaculture,* Cochin, India, 12–18 Jan., p. 131 only.
Reed, P. H., 1969. *Proc. Natl Shellfish Ass.,* **59**, 12 only.
Reeve, M. R., 1963. *Biol. Bull. mar. biol. Lab., Woods Hole,* **125**, 133–145.
Reeve, M. R., 1969a. *Fishery Invest., Lond.,* Ser. II, **26**, No. 1, 38 pp.
Reeve, M. R., 1969b. *J. mar. biol. Ass. U.K.,* **49**, 77–96.
Regnault, M., 1969. *Int. Revue ges. Hydrobiol.,* **54**, 749–764.
Rice, A. L. & Williamson, O. I., 1970. *Helgoländer wiss Meeresunters.,* **20**, 417–434.
Riley, J. D., 1966. *J. Cons. perm. int. Explor. Mer,* **30**, 204–221.
Roberts Jr, M. H., 1972. In, *Culture of Marine invertebrate Animals,* edited by W. L. Smith & M. H. Chanley, Plenum Press, New York, pp. 209–220.
Roberts Jr, M. H., 1974. *Biol. Bull. mar. biol. Lab., Woods Hole,* **146**, 67–77.
Robin, J. H., 1982. *I.C.E.S. Maricult. Committee* F: 13, 11 pp.
Robin, J. H., Gatesoupe, F. J. & Ricardez, R., 1981. *J. World Maricult. Soc.,* **12**, 119–120.
Robin, J. H., Gatesoupe, F. J., Stephan, G., Le Delliou, H. & Salaun, G., 1984. *Journées Aquariologiques de l'Institut océanographique, Oceanis,* **10**, 497–504.
Rodriguez, A. M., 1975. *Publ. Téc. Junta Est. Pesca,* **11**, 367–386.
Rodriguez, A. M., 1976. *Etud. Rev. gen. Fish. Counc. Mediterr.,* **55**, 49–62.
Rollefsen, G., 1939. *Rapp. P.-V. Réun. Cons. perm. int. Explor. Mer,* **109**, 3eme Partie, 133 only.
Rosemark, R., 1978. In, *Proc. 9th Ann. Meeting World Maricult. Soc.,* pp. 251–258.
Rosenthal, H., 1969. *Mar. Biol.,* **3**, 208–221.
Royan, J. P., 1980. In, *Book of Abstracts, Symp. on Coastal Aquaculture,* Cochin, India, 12–18 Jan., p. 133 only.
Sakamoto, M., Holland, D. L. & Jones, D. A., 1982. *Aquaculture,* **28**, 311–320.
Salser, B. R. & Mock, C. R., 1974. Paper presented at V Congreso Nacional de Oceanografia, Mexico, 15 pp.
Samain, J. F., Boucher, J. & Buestel, D., 1975. In, *10th Europ Mar. Biol. Symp. Vol. 1,* edited by G. Persoone & E. Jaspers, Universa Press, Wetteren, Belgium, pp. 391–417.
Samain, J. F., Hernandorena, A., Moal, J., Daniel, J. Y. & Le Coz, J. R., 1985. *J. exp. mar. Biol. Ecol.,* **86**, 255–270.
Samain, J. F., Moal, J., Daniel, J. Y., Le Coz, J. R. & Jezequel, M., 1980. In, *The Brine Shrimp* Artemia, *Vol. 2,* edited by G. Persoone *et al.,* Universa Press, Wetteren, Belgium, pp. 239–258.

Sandifer, P. A. & Joseph, J. D., 1976. *Aquaculture,* **8,** 129–138.

Sandifer, P. A. & Williams, J. D., 1980. In, *The Brine Shrimp* Artemia, *Vol. 3,* edited by G. Persoone *et al.,* Universa Press, Wetteren, Belgium, pp. 353–364.

San Feliu, J. M., 1973. *Inf. Tecn. Inst. Invest. Pesq.,* **14,** 87–98.

San Feliu, J. M., Munor, F., Amat, F., Ramos, J., Pena, J. & Sanz, A., 1976. *Inf. Tecn. Inst. Invest. Pesq.,* **36,** 3–47.

Schauer, P. S., Johns, D. M., Olney, C. E. & Simpson, K. L. 1980. In, *The Brine Shrimp* Artemia, *Vol. 3,* edited by G. Persoone, *et al.,* Universa Press, Wetteren, Belgium, pp. 365–373.

Schauer, P. S., Richardson, L. M. & Simpson, K. L. 1979. In, *Cultivation of Fish Fry and its Live Food,* Europ. Maricult., Soc. Spec. Publ. No. 4, edited by E. Styczynska-Jurewicz *et al.*, Inst. Mar. scient. Res., Bredene, Belgium, pp. 159–176.

Schauer, P. S. & Simpson, K. L. 1978. *Proc. World Maricult. Soc.,* **9,** 175–187.

Schauer, P. S. & Simpson, K. L., 1979. In, *Finfish Nutrition and Fishfeed Technology, Vol. 1,* edited by J. E. Halver & K. L. Tiews, Heenemann Verlagsgesellschaft mbH, Berlin, F.D.R., pp. 565–590.

Schauer, P. S. & Simpson, K. L., 1985. *Can. J. Fish. Aquat. Sci.,* **42,** 1430–1438.

Schuur, A., Fisher, W. S., Van Olst, J. C., Carlberg, J., Hughes, J. T., Shleser, R. A. & Ford, R. F., 1976. I. M. R. Ref. 76–6 Seagrant Publ. No. 48, 21 pp.

Scott, J. M., 1980. *J. mar. biol. Ass. U.K.,* **60,** 681–702.

Scott, J. M. & Baynes, S. M., 1979. In, *Finfish Nutrition and Fishfeed Technology, Vol. 1,* edited by J. E. Halver & K. Tiews, Heenemann Verlagsgesellschaft mbH, Berlin, F.D.R., pp. 423–433.

Scott, J. M. & Middleton, C., 1979. *Aquaculture,* **18,** 227–240.

Seale, A., 1933. *Trans. Am. Fish. Soc.,* **63,** 129–130.

Seidel, C. R., Johns, D. M., Schauer, P. S. & Olney, C. E., 1982. *Mar. Ecol. Prog. Ser.,* **8,** 309–312.

Seidel, C. R., Kryznowek, J. & Simpson, K. L., 1980b. In, *The Brine Shrimp* Artemia, *Vol. 3,* edited by G. Persoone *et al.*, Universa Press, Wetteren, Belgium, pp. 375–382.

Seidel, C. R., Schauer, P. S., Katayama, T. & Simpson, K. L., 1980a. *Bull. Jap. Soc. scient. Fish.,* **46,** 237–245.

Serflin, S. A., Van Olst, J. C. & Ford, R. F., 1974. *Aquaculture,* **3,** 311–314.

Shelbourne, J. E., 1968. Thesis, University of London, London, 143 pp.

Shelbourne, J. E., Riley, J. D. & Thacker, G. T., 1963. *J. Cons. perm. int. Explor. Mer,* **28,** 50–69.

Shigueno, K., 1975. In, *Shrimp Culture in Japan,* edited by *Association for International Technical Promotion,* Tokyo, Japan, 153 pp.

Shleser, R. A., 1976. In, *Proc. 10th Europ. Mar. Biol. Symp. Vol. 1,* edited by G. Persoone & E. Jaspers, Universa Press, Wetteren, Belgium, pp. 455–471.

Sick, L. V., 1975. In, *Proc. 1st int. Conf. Aquaculture Nutrition,* Lewes, Rehoboth, Delaware, U.S.A., Oct., 1975, pp. 215–228.

Sick, L. V., 1976. *Mar. Biol.,* **35,** 69–78.

Sick, L. V. & Andrews, J. W., 1973. In, *Proc. 4th Ann. Wkshop World Maricult. Soc.,* pp. 263–276.

Sick, L. V. Andrews, J. W. & Baptist, G., 1973. *Progve Fish Cult.,* **35,** 22–26.

Sick, L. V. & Beaty, H., 1974. *Ga Mar. Sci. Cent. Tech. Rep. Ser.,* No. 74, 30 pp.

Sick, L. V. & Beaty, H., 1975. In, *Proc. 6th Ann. Workshop World Maricult. Soc.,* pp. 89–101.

Sillero, M. A. G., Burillo, S. L., Dominguez, E., Olalla, A., Osuna, C., Renart, J., Sebastian, L. & Sillero, A., 1980. In, *The Brine Shrimp* Artemia, *Vol. 2,* edited by G. Persoone *et al.*, Universa Press, Wetteren, Belgium, pp. 345–354.

Simpson, K. L., Klein-MacPhee, G. & Beck, A. D., 1983. In, *Proc. 2nd int. Conf. on Aquaculture Nutrition,* World Maricult. Soc. Spec. Publ., No. 2, edited by G. D. Pruder *et al.*, Louisiana State University Press, Baton Rouge, U.S.A., pp. 180–201.

Sleet, R. B. & Brendel, K., 1983. *J. Aquaricult. Aquat. Sci.,* **3**, 76–83.
Slobin, L. I. & Moller, W., 1976. *Europ. J. Biochem.,* **69**, 351–366.
Slobodkin, L. B., 1968. *Biol. Sci. Tokyo,* **18**, 16–23.
Slobodkin, L. B. & Richman, S., 1961. *Nature, Lond.,* **191**, 299 only.
Smith, T. I. J., Hopkins, J. S. & Sandifer, P. A., 1978. In, *Proc. 9th Ann. Meeting World Maricult. Soc.,* pp. 701–714.
Smith, W. E., 1975. *Progve Fish Cult.,* **37**, 227–229.
Smith, W. E., 1976. *Progve Fish Cult.,* **38**, 95–97.
Soebiantoro, B., 1981. Ph.D. Thesis, Auburn University, Alabama, U.S.A., 90 pp.
Soejima, T., Katayama, T. & Simpson, K. L., 1980. In, *The Brine Shrimp* Artemia, *Vol. 2,* edited by G. Persoone *et al.*, Universa Press, Wetteren, Belgium, pp. 613–622.
Soejima, T., Simpson, K. L. & Katayama, T., 1983. *Bull. Jap. Soc. scient. Fish.,* **49**, 137–139.
Sorgeloos, P., 1975. Ph.D. thesis, State University of Ghent, Belgium, 235 pp.
Sorgeloos, P., 1979. Thesis, State University of Ghent, Belgium, 319 pp.
Sorgeloos, P., 1980a. In, *The Brine Shrimp* Artemia, *Vol. 1, Vol. 2, Vol. 3,* edited by G. Persoone *et al.*, Universa Press, Wetteren, Belgium, pp. xix–xxiii.
Sorgeloos, P., 1980b. *Mar. Ecol. Prog. Ser.,* **3**, 363–364.
Sorgeloos, P., 1980c. In, *The Brine Shrimp* Artemia, *Vol. 3,* edited by G. Persoone *et al.*, Universa Press, Wetteren, Belgium, pp. 25–46.
Sorgeloos, P., 1981. Paper presented at World Conference on Aquaculture, Venice, Italy, 21–25 Sept.
Sorgeloos, P., Bossuyt, E., Lavens, P., Léger, P., Vanhaecke, P. & Versichele, D., 1983. In, *Handbook of Mariculture, Vol. 1,* edited by J. P. McVey, CRC Press, Boca Raton, Florida U.S.A., pp. 71–96.
Sorgeloos, P., Bossuyt, E., Lavina, E., Balza-Mesa, M. & Persoone, G., 1977. *Aquaculture,* **12**, 311–315.
Spectorova, L. V. & Doroshev, S. I., 1976. *Aquaculture,* **9**, 275–286.
Spitchak, M. K., 1980. In, *The Brine Shrimp* Artemia, *Vol. 3,* edited by G. Persoone *et al.*, Universa Press, Wetteren, Belgium, pp. 127–128 (abstract).
Stephens, D. W. & Gillespie, D. M., 1976. *Limnol. Oceanogr.,* **21**, 74–87.
Stewart, J. E. & Castell, J. D., 1976. Paper presented at *FAO Techn. Conf. on Aquaculture,* Kyoto, Japan, 26 May–2 June.
Stroband, H. W. J. & Dabrowski, K., 1981. In, *La Nutrition des Poissons,* CNERNA, Paris, pp. 353–376.
Stults, V. J., 1974. Thesis, Michigan State University, East Lansing, U.S.A, 110 pp.
Sulkin, S. D., 1975. *J. exp. mar. Biol. Ecol.,* **20**, 119–135.
Sulkin, S. D., 1978. *J. exp. mar. Biol. Ecol.,* **34**, 29–41.
Sulkin, S. D. & Epifanio, C. E., 1975. *Estuar. cstl. mar. Sci.,* **3**, 109–113.
Sumitra-Vijayaraghavan, Kuruppu, M. M., Grero, J. J. & Asoka Perera, 1985. *NARA/OCC/85/1,* pp. 58–77.
Tabb, D. C., Yang, W. T., Hirono, Y. & Heinen, J., 1972. *NOAA Seagrant N 2 - 35147 - Seagrant Spec. Bull.* No. 7, 59 pp.
Tacon, A. G. J. & Cowey, C. B., 1982. In, *Proc. 2nd int. Conf. Aquaculture Nutrition,* World Maricult. Soc. Spec. Publ. No. 2, edited by E. D. Pruder *et al.*, Louisiana State University Press, Baton Rouge, U.S.A., pp. 13–30.
Tarnchalanukit, W. & Wongrat, L., 1985. In, *Book of Abstracts, 2nd int. Symp. on the Brine Shrimp* Artemia, 1–5 Sept., Antwerp, Belgium, p. 117 only.
Telford, M., 1970. *Comp. Biochem, Physiol.,* **34**, 81–90.
Teshima, S., 1971. *Comp. Biochem. Physiol.,* **39B**, 815–822.
Teshima, S. & Kanazawa, A., 1971a. *Bull. Jap. Soc. scient. Fish.,* **37**, 720–723.
Teshima, S. & Kanazawa, A., 1971b. *Comp. Biochem. Physiol.,* **38B**, 603–607.
Teshima, S. & Kanazawa, A., 1972. *Bull. Jap. Soc. scient. Fish.,* **38**, 1305–1310.
Teshima, S. & Kanazawa, A. & Sakamoto, M., 1982. *Min. Rev. Data File Fish. Res.,* **2**, 67–86.

Tholasilingam, T. & Rangarajan, K., 1980. In, *Book of Abstracts, Symposium on Coastal Aquaculture,* 12–18 Jan., Cochin, India, p. 95 only.
Tobias, W. J., Sorgeloos, P., Roels, O. A. & Sharfstein, B. H., 1980. In, *The Brine Shrimp* Artemia, *Vol. 3,* edited by G. Persoone *et al,* Universa Press, Wetteren, Belgium, pp. 383–392.
Trotta, P., Villani, P. & Palmegiano, G. B., 1985. In, *Book of Abstracts, 2nd int. Symp. on the Brine Shrimp* Artemia, Antwerp, Belgium, 1–5 Sept., p. 124 only.
Tyson, G. E., 1970. *J. invert. Pathol.,* **15,** 145–147.
Uçal, O., 1979. *Rapp. Commn int. Mer Médit.,* No. 25/26, 127–128.
Uno, Y., 1971. *La Mer (Bull. Soc. Franco-Japonaise L'Océanogr.),* **9,** 123–128.
Urbani, E., 1959. *Acta Embryol. Morph. exp.,* **2,** 171–194.
Usher, R. R. & Bengtson, D. A., 1981. *Progve Fish Cult.,* **43,** 102–105.
Van Ballaer, E., Amat, F., Hontoria, F., Léger, P. & Sorgeloos, P., 1985. *Aquaculture,* **49,** 223–229.
Van Ballaer, E., Versichele, D., Vanhaecke, P., Léger, P., Ben Abdelkader, N., Turki, S. & Sorgeloos, P., 1985. In, *Book of Abstracts, 2nd int. Symp. on the Brine Shrimp* Artemia, Antwerp, Belgium, 1–5 Sept., p. 126 only.
Vanhaecke, P., 1983. Ph.D. thesis, State University of Ghent, Belgium, 420 pp.
Vanhaecke, P., Lavens, P. & Sorgeloos, P., 1983. *Annls Soc. r. zool. Belg.,* **113,** 155–164.
Vanhaecke, P. & Sorgeloos, P. 1980. In, *The Brine Shrimp* Artemia, *Vol. 3,* edited by G. Persoone *et al.*, Universa Press, Wetteren, Belgium, pp. 393–405.
Vanhaecke, P. & Sorgeloos, P., 1983a. *Aquaculture,* **30,** 43–52.
Vanhaecke, P. & Sorgeloos, P., 1983b. *Aquaculture,* **32,** 285–293.
Vanhaecke, P., Steyaert, H. & Sorgeloos, P., 1980. In, *The Brine Shrimp* Artemia, *Vol. 1,* edited by G. Persoone *et al.*, Universa Press, Wetteren, Belgium, pp. 107–115.
Vanhaecke, P., Tackaert, W. & Sorgeloos, P., 1985. In, *Book of Abstracts, 2nd int. Symp. on the Brine Shrimp* Artemia, Antwerp, Belgium, 1–5 Sept., p. 133 only.
Van Olst, J. C., Ford, R. F., Carlberg, J. M. & Dorband, W. R., 1975. In, *Power Plant Waste Heat Utilization in Aquaculture*—Workshop 1, PSE & G, Newark, New York, U.S.A., pp. 71–97.
Vijayakumaran, M. & Radhakrishnan, E. V., 1980. In, *Book of Abstracts, Symposium on Coastal Aquaculture,* Cochin, India, 12–18 Jan., p. 132 only.
Villegas, C. T. & Kanazawa, A., 1978. *SEAFDEC Quarterly Res. Rep.* No. 11, 24–29.
Vishnu Bhat, B. & Ganapathy, R., 1985. In, *Book of Abstracts, 2nd int. Symp. on the Brine Shrimp* Artemia, Antwerp, Belgium, 1–5 Sept., p. 137 only.
Von Hentig, R., 1971. *Mar. Biol.,* **9,** 145–182.
Vos, J., Léger, P., Vanhaecke, P. & Sorgeloos, P., 1984. *Hydrobiologia,* **108,** 17–23.
Ward, W. W., 1974. *Chesapeake Sci.,* **15,** 116–118.
Warner, A. H. & Shridhar, V., 1980. In, *The Brine Shrimp* Artemia, *Vol. 2,* edited by G. Persoone *et al.*, Universa Press, Wetteren, Belgium, pp. 355–364.
Watanabe, T. & Ackman, R. G., 1974. *J. Fish Res. Bd Can.,* **31,** 403–409.
Watanabe, T., Arakawa, T., Kitajima, C. & Fujita, S., 1978a. *Bull. Jap. Soc. scient. Fish.,* **44,** 985–988.
Watanabe, T., Arakawa, T., Kitajima, C., Fukusho, K. & Fujita, S., 1978b. *Bull. Jap. Soc. scient. Fish.,* **44,** 1223–1227.
Watanabe, T., Kitajima, C. & Fujita, S., 1983a. *Aquaculture,* **34,** 115–143.
Watanabe, T., Ohta, M., Kitajima, C. & Fujita, S., 1982. *Bull. Jap. Soc. scient. Fish.,* **48,** 1775–1782.
Watanabe, T., Oowa, F., Kitajima, C. & Fujita, S., 1978c. *Bull. Jap. Soc. scient. Fish.,* **44,** 1115–1121.
Watanabe, T., Oowa, F., Kitajima, C. & Fujita, S., 1980. *Bull. Jap. Soc. scient. Fish.,* **46,** 35–41.

Watanabe, T., Tamiya, T., Oka, A., Hirata, M., Kitajima, C. & Fujita, S., 1983b. *Bull. Jap. Soc. scient. Fish.*, **49**, 471–479.
Weaver, J. E., 1974. *Trans. Am. Fish. Soc.*, **2**, 382–386.
Webber, H. H. & Sorgeloos, P. 1980. In, *The Brine Shrimp Artemia, Vol. 3,* edited by G. Persoone *et al.*, Universa Press, Wetteren, Belgium, p. 413 only.
Welch, J. & Sulkin, S. D., 1975. *J. Elisha Mitchell Sci. Soc.*, **90**, 69–72.
Werner, B., 1968. *Helgoländer wiss Meeresunters,* **18**, 136–168.
Westin, D. T., Olney, C. E. & Rogers, B. A., 1983. *Bull. environ. Contam. Toxicol.*, **30**, 50–57.
Westin, D. T., Olney, C. E. & Rogers, B. A., 1985. *Trans. Am. Fish. Soc.*, **114**, 125–136.
Wheeler, R., Yudin, A. I. & Clark Jr, W. H., 1979. *Aquaculture,* **18**, 59–67.
White, D. B. & Stickney, R. R., 1973., *Ga Mar. Sci. Cent., Tech. Rep. Ser.* 73–7, (unpubl. rep.).
Wickins, J. F., 1972. *J. exp. mar. Biol. Ecol.*, **10**, 151–170.
Wickins, J. F., 1976. *Oceanogr. Mar. Biol. Ann. Rev.*, **14**, 435–507.
Wilkenfeld, J. S., Lawrence, A. L. & Kuban, F. D., 1984. *J. World Maricult. Soc.*, **15**, 31–49.
Williams, K. W., 1907. In, *The 37th Annual Report of the Commissioners of Inland Fisheries,* Providence, Rhode Island, U.S.A., pp. 20–178.
Witt, U., Quantz, G. & Kuhlmann, D., 1984. *Aquacultural Engng.*, **3**, 177–190.
Yamasaki, S. & Hirata, H., 1982. *Min. Rev. Data File Fish. Res.*, No. 2, 87–89.
Yashiro, R., 1982. M. Sc. thesis. College of Fisheries, University of the Philippines in the Visayas, Philippines, 48 pp.
Yashiro, R. 1985. In, *Book of Abstracts, 2nd int. Symp. on the Brine Shrimp* Artemia, Antwerp, Belgium, 1–5 Sept., p. 149 only.
Yone, Y., 1978. In, *Dietary Lipids in Aquaculture,* edited by Japan Soc. scient. Fish., Koseisha-Koseikaku, Japan, pp. 43–59.
Yone, Y. & Fujii, M., 1975. *Bull. Jap. Soc. scient. Fish.*, **41**, 73–77.
Young, R. T., 1952. *J. Wash. Acad. Sci.*, **42**, 385–388.
Yufera, M., Rodriguez, A. & Lúbian, C. M., 1984. *Aquaculture,* **42**, 217–224.
Zimmerman, S. T., 1973. *Pacif. Sci.*, **27**, 247–259.

AUTHOR INDEX

Reference to complete articles are given in heavy type; reference to pages are given in normal type; references to bibliographical lists are given in italics.

SYSTEMATIC INDEX

References to complete articles are given in heavy type; references to pages are given in normal type.

SUBJECT INDEX

References to complete articles are given in heavy type; references to sections of articles are given in italics; references to pages are given in normal type.